CW00642792

paraconsistency

PURE AND APPLIED MATHEMATICS

A Program of Monographs, Textbooks, and Lecture Notes

LECTURE NOTES IN PURE AND APPLIED MATHEMATICS

1. *N. Jacobson*, Exceptional Lie Algebras
2. *L.-Å. Lindahl and F. Poulsen*, Thin Sets in Harmonic Analysis
3. *I. Satake*, Classification Theory of Semi-Simple Algebraic Groups
4. *F. Hirzebruch et al.*, *Differentiable Manifolds and Quadratic Forms*
5. *I. Chavel*, Riemannian Symmetric Spaces of Rank One
6. *R. B. Burckel*, Characterization of C(X) Among Its Subalgebras
7. *B. R. McDonald et al.*, Ring Theory
8. *Y.-T. Siu*, Techniques of Extension on Analytic Objects
9. *S. R. Caradus et al.*, Calkin Algebras and Algebras of Operators on Banach Spaces
10. *E. O. Roxin et al.*, Differential Games and Control Theory
11. *M. Orzech and C. Small*, The Brauer Group of Commutative Rings
12. *S. Thomier*, Topology and Its Applications
13. *J. M. Lopez and K. A. Ross*, Sidon Sets
14. *W. W. Comfort and S. Negrepontis*, Continuous Pseudometrics
15. *K. McKennon and J. M. Robertson*, Locally Convex Spaces
16. *M. Carmeli and S. Malin*, Representations of the Rotation and Lorentz Groups
17. *G. B. Seligman*, Rational Methods in Lie Algebras
18. *D. G. de Figueiredo*, Functional Analysis
19. *L. Cesari et al.*, Nonlinear Functional Analysis and Differential Equations
20. *J. J. Schäffer*, Geometry of Spheres in Normed Spaces
21. *K. Yano and M. Kon*, Anti-Invariant Submanifolds
22. *W. V. Vasconcelos*, The Rings of Dimension Two
23. *R. E. Chandler*, Hausdorff Compactifications
24. *S. P. Franklin and B. V. S. Thomas*, Topology
25. *S. K. Jain*, Ring Theory
26. *B. R. McDonald and R. A. Morris*, Ring Theory II
27. *R. B. Mura and A. Rhemtulla*, Orderable Groups
28. *J. R. Graef*, Stability of Dynamical Systems
29. *H.-C. Wang*, Homogeneous Branch Algebras
30. *E. O. Roxin et al.*, Differential Games and Control Theory II
31. *R. D. Porter*, Introduction to Fibre Bundles
32. *M. Altman*, Contractors and Contractor Directions Theory and Applications
33. *J. S. Golan*, Decomposition and Dimension in Module Categories
34. *G. Fairweather*, Finite Element Galerkin Methods for Differential Equations
35. *J. D. Sally*, Numbers of Generators of Ideals in Local Rings
36. *S. S. Miller*, Complex Analysis
37. *R. Gordon*, Representation Theory of Algebras
38. *M. Goto and F. D. Grosshans*, Semisimple Lie Algebras
39. *A. I. Arruda et al.*, Mathematical Logic
40. *F. Van Oystaeyen*, Ring Theory
41. *F. Van Oystaeyen and A. Verschoren*, Reflectors and Localization
42. *M. Satyanarayana*, Positively Ordered Semigroups
43. *D. L Russell*, Mathematics of Finite-Dimensional Control Systems
44. *P.-T. Liu and E. Roxin*, Differential Games and Control Theory III
45. *A. Geramita and J. Seberry*, Orthogonal Designs
46. *J. Cigler, V. Losert, and P. Michor*, Banach Modules and Functors on Categories of Banach Spaces
47. *P.-T. Liu and J. G. Sutinen*, Control Theory in Mathematical Economics
48. *C. Byrnes*, Partial Differential Equations and Geometry
49. *G. Klambauer*, Problems and Propositions in Analysis
50. *J. Knopfmacher*, Analytic Arithmetic of Algebraic Function Fields
51. *F. Van Oystaeyen*, Ring Theory
52. *B. Kadem*, Binary Time Series
53. *J. Barros-Neto and R. A. Artino*, Hypoelliptic Boundary-Value Problems
54. *R. L. Sternberg et al.*, Nonlinear Partial Differential Equations in Engineering and Applied Science
55. *B. R. McDonald*, Ring Theory and Algebra III
56. *J. S. Golan*, Structure Sheaves Over a Noncommutative Ring
57. *T. V. Narayana et al.*, Combinatorics, Representation Theory and Statistical Methods in Groups
58. *T. A. Burton*, Modeling and Differential Equations in Biology
59. *K. H. Kim and F. W. Roush*, Introduction to Mathematical Consensus Theory

60. *J. Banas and K. Goebel,* Measures of Noncompactness in Banach Spaces
61. *O. A. Nielson,* Direct Integral Theory
62. *J. E. Smith et al.,* Ordered Groups
63. *J. Cronin,* Mathematics of Cell Electrophysiology
64. *J. W. Brewer,* Power Series Over Commutative Rings
65. *P. K. Kamthan and M. Gupta,* Sequence Spaces and Series
66. *T. G. McLaughlin,* Regressive Sets and the Theory of Isols
67. *T. L. Herdman et al.,* Integral and Functional Differential Equations
68. *R. Draper,* Commutative Algebra
69. *W. G. McKay and J. Patera,* Tables of Dimensions, Indices, and Branching Rules for Representations of Simple Lie Algebras
70. *R. L. Devaney and Z. H. Nitecki,* Classical Mechanics and Dynamical Systems
71. *J. Van Geel,* Places and Valuations in Noncommutative Ring Theory
72. *C. Faith,* Injective Modules and Injective Quotient Rings
73. *A. Fiacco,* Mathematical Programming with Data Perturbations I
74. *P. Schultz et al.,* Algebraic Structures and Applications
75. *L Bican et al.,* Rings, Modules, and Preradicals
76. *D. C. Kay and M. Breen,* Convexity and Related Combinatorial Geometry
77. *P. Fletcher and W. F. Lindgren,* Quasi-Uniform Spaces
78. *C.-C. Yang,* Factorization Theory of Meromorphic Functions
79. *O. Taussky,* Ternary Quadratic Forms and Norms
80. *S. P. Singh and J. H. Burry,* Nonlinear Analysis and Applications
81. *K. B. Hannsgen et al.,* Volterra and Functional Differential Equations
82. *N. L. Johnson et al.,* Finite Geometries
83. *G. I. Zapata,* Functional Analysis, Holomorphy, and Approximation Theory
84. *S. Greco and G. Valla,* Commutative Algebra
85. *A. V. Fiacco,* Mathematical Programming with Data Perturbations II
86. *J.-B. Hiriart-Urruty et al.,* Optimization
87. *A. Figa Talamanca and M. A. Picardello,* Harmonic Analysis on Free Groups
88. *M. Harada,* Factor Categories with Applications to Direct Decomposition of Modules
89. *V. I. Istrătescu,* Strict Convexity and Complex Strict Convexity
90. *V. Lakshmikantham,* Trends in Theory and Practice of Nonlinear Differential Equations
91. *H. L. Manocha and J. B. Srivastava,* Algebra and Its Applications
92. *D. V. Chudnovsky and G. V. Chudnovsky,* Classical and Quantum Models and Arithmetic Problems
93. *J. W. Longley,* Least Squares Computations Using Orthogonalization Methods
94. *L. P. de Alcantara,* Mathematical Logic and Formal Systems
95. *C. E. Aull,* Rings of Continuous Functions
96. *R. Chuaqui,* Analysis, Geometry, and Probability
97. *L. Fuchs and L. Salce,* Modules Over Valuation Domains
98. *P. Fischer and W. R. Smith,* Chaos, Fractals, and Dynamics
99. *W. B. Powell and C. Tsinakis,* Ordered Algebraic Structures
100. *G. M. Rassias and T. M. Rassias,* Differential Geometry, Calculus of Variations, and Their Applications
101. *R.-E. Hoffmann and K. H. Hofmann,* Continuous Lattices and Their Applications
102. *J. H. Lightboume III and S. M. Rankin III,* Physical Mathematics and Nonlinear Partial Differential Equations
103. *C. A. Baker and L. M. Batten,* Finite Geometrics
104. *J. W. Brewer et al.,* Linear Systems Over Commutative Rings
105. *C. McCrory and T. Shifrin,* Geometry and Topology
106. *D. W. Kueke et al.,* Mathematical Logic and Theoretical Computer Science
107. *B.-L. Lin and S. Simons,* Nonlinear and Convex Analysis
108. *S. J. Lee,* Operator Methods for Optimal Control Problems
109. *V. Lakshmikantham,* Nonlinear Analysis and Applications
110. *S. F. McCormick,* Multigrid Methods
111. *M. C. Tangora,* Computers in Algebra
112. *D. V. Chudnovsky and G. V. Chudnovsky,* Search Theory
113. *D. V. Chudnovsky and R. D. Jenks,* Computer Algebra
114. *M. C. Tangora,* Computers in Geometry and Topology
115. *P. Nelson et al.,* Transport Theory, Invariant Imbedding, and Integral Equations
116. *P. Clément et al.,* Semigroup Theory and Applications
117. *J. Vinuesa,* Orthogonal Polynomials and Their Applications
118. *C. M. Dafermos et al.,* Differential Equations
119. *E. O. Roxin,* Modern Optimal Control
120. *J. C. Díaz,* Mathematics for Large Scale Computing

121. *P. S. Milojević* Nonlinear Functional Analysis
122. *C. Sadosky,* Analysis and Partial Differential Equations
123. *R. M. Shortt,* General Topology and Applications
124. *R. Wong,* Asymptotic and Computational Analysis
125. *D. V. Chudnovsky and R. D. Jenks,* Computers in Mathematics
126. *W. D. Wallis et al.,* Combinatorial Designs and Applications
127. *S. Elaydi,* Differential Equations
128. *G. Chen et al.,* Distributed Parameter Control Systems
129. *W. N. Everitt,* Inequalities
130. *H. G. Kaper and M. Garbey,* Asymptotic Analysis and the Numerical Solution of Partial Differential Equations
131. *O. Arino et al.,* Mathematical Population Dynamics
132. *S. Coen,* Geometry and Complex Variables
133. *J. A. Goldstein et al.,* Differential Equations with Applications in Biology, Physics, and Engineering
134. *S. J. Andima et al.,* General Topology and Applications
135. *P Clément et al.,* Semigroup Theory and Evolution Equations
136. *K. Jarosz,* Function Spaces
137. *J. M. Bayod et al., p*-adic Functional Analysis
138. *G. A. Anastassiou,* Approximation Theory
139. *R. S. Rees,* Graphs, Matrices, and Designs
140. *G. Abrams et al.,* Methods in Module Theory
141. *G. L. Mullen and P. J.-S. Shiue,* Finite Fields, Coding Theory, and Advances in Communications and Computing
142. *M. C. Joshi and A. V. Balakrishnan,* Mathematical Theory of Control
143. *G. Komatsu and Y. Sakane,* Complex Geometry
144. *I. J. Bakelman,* Geometric Analysis and Nonlinear Partial Differential Equations
145. *T. Mabuchi and S. Mukai,* Einstein Metrics and Yang–Mills Connections
146. *L. Fuchs and R. Göbel,* Abelian Groups
147. *A. D. Pollington and W. Moran,* Number Theory with an Emphasis on the Markoff Spectrum
148. *G. Dore et al.,* Differential Equations in Banach Spaces
149. *T. West,* Continuum Theory and Dynamical Systems
150. *K. D. Bierstedt et al.,* Functional Analysis
151. *K. G. Fischer et al.,* Computational Algebra
152. *K. D. Elworthy et al.,* Differential Equations, Dynamical Systems, and Control Science
153. *P.-J. Cahen, et al.,* Commutative Ring Theory
154. *S. C. Cooper and W. J. Thron,* Continued Fractions and Orthogonal Functions
155. *P. Clément and G. Lumer,* Evolution Equations, Control Theory, and Biomathematics
156. *M. Gyllenberg and L. Persson,* Analysis, Algebra, and Computers in Mathematical Research
157. *W. O. Bray et al.,* Fourier Analysis
158. *J. Bergen and S. Montgomery,* Advances in Hopf Algebras
159. *A. R. Magid,* Rings, Extensions, and Cohomology
160. *N. H. Pavel,* Optimal Control of Differential Equations
161. *M. Ikawa,* Spectral and Scattering Theory
162. *X. Liu and D. Siegel,* Comparison Methods and Stability Theory
163. *J.-P. Zolésio,* Boundary Control and Variation
164. *M. Křížek et al.,* Finite Element Methods
165. *G. Da Prato and L. Tubaro,* Control of Partial Differential Equations
166. *E. Ballico,* Projective Geometry with Applications
167. *M. Costabel et al.,* Boundary Value Problems and Integral Equations in Nonsmooth Domains
168. *G. Ferreyra, G. R. Goldstein, and F. Neubrander,* Evolution Equations
169. *S. Huggett,* Twistor Theory
170. *H. Cook et al.,* Continua
171. *D. F. Anderson and D. E. Dobbs,* Zero-Dimensional Commutative Rings
172. *K. Jarosz,* Function Spaces
173. *V. Ancona et al.,* Complex Analysis and Geometry
174. *E. Casas,* Control of Partial Differential Equations and Applications
175. *N. Kalton et al.,* Interaction Between Functional Analysis, Harmonic Analysis, and Probability
176. *Z. Deng et al.,* Differential Equations and Control Theory
177. *P. Marcellini et al.* Partial Differential Equations and Applications
178. *A. Kartsatos,* Theory and Applications of Nonlinear Operators of Accretive and Monotone Type
179. *M. Maruyama,* Moduli of Vector Bundles
180. *A. Ursini and P. Aglianò,* Logic and Algebra
181. *X. H. Cao et al.,* Rings, Groups, and Algebras
182. *D. Arnold and R. M. Rangaswamy,* Abelian Groups and Modules
183. *S. R. Chakravarthy and A. S. Alfa,* Matrix-Analytic Methods in Stochastic Models

184. *J. E. Andersen et al.*, Geometry and Physics
185. *P.-J. Cahen et al.*, Commutative Ring Theory
186. *J. A. Goldstein et al.*, Stochastic Processes and Functional Analysis
187. *A. Sorbi*, Complexity, Logic, and Recursion Theory
188. *G. Da Prato and J.-P. Zolésio*, Partial Differential Equation Methods in Control and Shape Analysis
189. *D. D. Anderson*, Factorization in Integral Domains
190. *N. L. Johnson*, Mostly Finite Geometries
191. *D. Hinton and P. W. Schaefer*, Spectral Theory and Computational Methods of Sturm–Liouville Problems
192. *W. H. Schikhof et al.*, *p*-adic Functional Analysis
193. *S. Sertöz*, Algebraic Geometry
194. *G. Caristi and E. Mitidieri*, Reaction Diffusion Systems
195. *A. V. Fiacco*, Mathematical Programming with Data Perturbations
196. *M. Křížek et al.*, Finite Element Methods: Superconvergence, Post-Processing, and A Posteriori Estimates
197. *S. Caenepeel and A. Verschoren*, Rings, Hopf Algebras, and Brauer Groups
198. *V. Drensky et al.*, Methods in Ring Theory
199. *W. B. Jones and A. Sri Ranga*, Orthogonal Functions, Moment Theory, and Continued Fractions
200. *P. E. Newstead*, Algebraic Geometry
201. *D. Dikranjan and L. Salce*, Abelian Groups, Module Theory, and Topology
202. *Z. Chen et al.*, Advances in Computational Mathematics
203. *X. Caicedo and C. H. Montenegro*, Models, Algebras, and Proofs
204. *C. Y. Yıldırım and S. A. Stepanov*, Number Theory and Its Applications
205. *D. E. Dobbs et al.*, Advances in Commutative Ring Theory
206. *F. Van Oystaeyen*, Commutative Algebra and Algebraic Geometry
207. *J. Kakol et al.*, *p*-adic Functional Analysis
208. *M. Boulagouaz and J.-P. Tignol*, Algebra and Number Theory
209. *S. Caenepeel and F. Van Oystaeyen*, Hopf Algebras and Quantum Groups
210. *F. Van Oystaeyen and M. Saorin*, Interactions Between Ring Theory and Representations of Algebras
211. *R. Costa et al.*, Nonassociative Algebra and Its Applications
212. *T.-X. He*, Wavelet Analysis and Multiresolution Methods
213. *H. Hudzik and L. Skrzypczak*, Function Spaces: The Fifth Conference
214. *J. Kajiwara et al.*, Finite or Infinite Dimensional Complex Analysis
215. *G. Lumer and L. Weis*, Evolution Equations and Their Applications in Physical and Life Sciences
216. *J. Cagnol et al.*, Shape Optimization and Optimal Design
217. *J. Herzog and G. Restuccia*, Geometric and Combinatorial Aspects of Commutative Algebra
218. *G. Chen et al.*, Control of Nonlinear Distributed Parameter Systems
219. *F. Ali Mehmeti et al.*, Partial Differential Equations on Multistructures
220. *D. D. Anderson and I. J. Papick*, Ideal Theoretic Methods in Commutative Algebra
221. *Á. Granja et al.*, Ring Theory and Algebraic Geometry
222. *A. K. Katsaras et al.*, *p*-adic Functional Analysis
223. *R. Salvi*, The Navier-Stokes Equations
224. *F. U. Coelho and H. A. Merklen*, Representations of Algebras
225. *S. Aizicovici and N. H. Pavel*, Differential Equations and Control Theory
226. *G. Lyubeznik*, Local Cohomology and Its Applications
227. *G. Da Prato and L. Tubaro*, Stochastic Partial Differential Equations and Applications
228. *W. A. Carnielli et al.*, Paraconsistency

Additional Volumes in Preparation

paraconsistency

the logical way to the inconsistent

proceedings of the world congress held in São Paulo

edited by

Walter A. Carnielli
Marcelo E. Coniglio
Itala M. Loffredo D'Ottaviano
State University of Campinas
Campinas, S.P., Brazil

LONDON AND NEW YORK

First published 2002 by Marcel Dekker, Inc .

Published 2018 by Routledge
2 Park Square, Milton Park, Abingdon, Oxon, OX14 4RN
52 Vanderbilt Avenue, New York, NY 10017

First issued in hardback 2018

Routledge is an imprint of the Taylor & Francis Group, an informa business

ISBN 13: 978-1-138-46690-6 (hbk)
ISBN 13: 978-0-8247-0805-4 (pbk)

Newton C. A. da Costa

Preface

1. The Brazil Paraconsistency Conference in Honour of Newton da Costa

The "I World Congress on Paraconsistency" was held in Ghent, Belgium, in 1997. As a sequel, in 1998, the "Stanisław Jaśkowski Memorial Symposium" was held in Toruń, Poland.

The "II World Congress on Paraconsistency" (WCP'2000) was held in Juquehy - São Sebastião, São Paulo State, in Brazil, from May 12-19, 2000, and was attended by more than 80 people, from about 20 countries.

The conference was dedicated to Newton Carneiro Affonso da Costa, in honour of his 70th birthday, and was organized by the Brazilian Logic Society and the Centre for Logic, Epistemology and the History of Science of the State University of Campinas, under the sponsorship of the Association for Symbolic Logic. Both the conference and this volume are intended to recognize the value of da Costa's scientific work, his logical creativity, and the importance of his rôle in the development of logic in Brazil and Latin America.

Newton Carneiro Affonso da Costa was born in Curitiba, State of Paraná, Brazil, in 1929. He received his B.S. degree in Civil Engineering and Mathematics at the Federal University of Paraná. He received the Doctoral degree in Mathematics in 1961 and became Full Professor of Mathematical Analysis and Superior Analysis in 1965 at the same university.

In 1968 he moved to the Mathematics and Statistics Institute of the University of São Paulo (IME,USP) and to the Institute of Mathematics, Statistics and Computer Science of the State University of Campinas (IMECC,UNICAMP).

In 1985 he began to work at the Department of Philosophy of the University of São Paulo, where he became Full Professor of Philosophy in 1991.

In 1985 he retired from UNICAMP, and in 1999, at age 70, he retired from USP, where he continues to supervise graduate students and coordinate a logic research group.

Da Costa published more than 200 papers and books, some of them translated into several languages, and he has been visiting professor, researcher and lecturer in many Latin American universities and other universities around the world.

His work covers several branches of knowledge, from model theory to the foundations of set theory and the foundations of mathematics, from the philosophy of science to the foundations of physics and the foundations of dynamical systems, as well as the theory of computation.

He was unanimously elected a member of the Institut International de Philosophie de Paris; in 1993 he received the "Prêmio Moinho Santista" (Moinho Santista Prize), one of the most important awards given to scientists and artists in Brazil. In 1998, during the Stanisław Jaśkowski Memorial Symposium, he was awarded the "Nicolas Copernicus Medal", one of the most important honours of Polish universities.

As the patron of logic in Brazil, his effort and work have been essential for the creation of a School of Logic in Brazil and in Latin America. He was the founder of the Brazilian Logic Society and its first President, and for several years he was also the President of the Latin American Committee on Logic of the Association for Symbolic Logic. With Ayda I. Arruda and Rolando Chuaqui, dear friends and co-workers, he was formerly responsible for the Latin American Symposia on Mathematical Logic.

As for his research, da Costa developed, with collaborators, a broad theory of valuations that can be seen as a general theory of logical systems. The concept of pragmatic truth presented by him and co-workers, has been applied to a variety of situations, once it captures the notion that there can be cases in which we don't have sufficient information about the validity (or non-validity) of certain relations between the objects of a given domain, in contrast with usual mathematics, where every relation is total.

His theory of quasi-truth was applied to the logic of induction, to the question of the acceptance of scientific theories and to the admissibility of incompatible theories.

He has extended Gödel's Incompleteness Theorems to physical theories, showing that notions as undecidability and incompleteness are more general than one might suppose, pervading practically every organized and consistent discipline that involves at least the elementary arithmetics.

The book **Mathematics: Frontiers and Perspectives**, published in 2000 by the American Mathematical Society, under the sponsorship of the International Mathematical Union, announces the perspectives of the new century mathematics, in the style of Hilbert's 23 problems of 1900. In Steve Smale's paper *Mathematical Problems for the Next Century*, Newton da Costa is mentioned for his work on undecidability and incompleteness in classical mechanics.

It is also important to mention da Costa's recent work about the dimensions of scientific rationality.

In the field of non-classical logics, da Costa is one of the founders of paraconsistent logic, together with the Polish logician Stanisław Jaśkowski. Da Costa, his disciples and co-workers, in Brazil and several other countries, in particular the Brazilians Ayda I. Arruda and Antonio M. Sette, the above mentioned Chilean Rolando Chuaqui and the Australian Richard Routley, all these prematurely dead, have deeply contributed to the development of paraconsistent logic and set theory, model theory, proof theory, algebraic paraconsistent logic, its philosophical foundations and its application in mathematics, computer science and philosophy. Da Costa is one of those most responsible for the inclusion of the entry Paraconsistent Logic in the subjects classification of *Mathematical Reviews*, in 1990.

For the generation he helped to form and which has now the responsibility for pursuing his work, da Costa is an example.

2. Paraconsistency and Paraconsistent Logics

A logic is paraconsistent if it can be used as the underlying logic for inconsistent but nontrivial theories, called *paraconsistent theories*.

A theory is *explosive* if the addition to it of any contradiction is sufficient to make it trivial. Paraconsistent logics are not, in general, explosive.

In paraconsistent logics the scope of the principle of (non-)contradiction is, in a certain sense, restricted. We may even say, as da Costa does, that if the strength of this principle is restricted in a system of logic, then the system belongs to the class of paraconsistent logics.

Paraconsistent logic is closely related to other kinds of non-classical logics, especially to dialectical and relevant logic, many-valued logics and fuzzy logic, the general theory of vagueness, Meinong's theory of objects, as well as to the logical theses of the later Wittgenstein.

The study of paraconsistent logics, besides allowing the construction of paraconsistent theories, allows the direct study of logical and semantical paradoxes; the study of certain principles in their full strength, as for example the principle of comprehension in set theory; and permits a better understanding of the concept of negation.

The status of contradiction in logic, philosophy and mathematics has a long history. It was discussed by the presocratics, and received momentum from Aristotle in his defense of non-contradiction. It was deeply discussed by Łukasiewicz and others from Russell to Meinong. It is likely to proceed for centuries, as far as the philosophical aspects of the dispute are concerned.

The task of paraconsistency is to study the behavior of contradictory yet non-trivial theories. Paraconsistent logics, on the other hand, try to tackle another task: is it possible, and philosophically defensible, to have a contradictory theory from which reasonable inferences can be made?

The two forefathers of paraconsistency are Lukasiewicz and Vasil'ev. But the two first logicians to construct systems of paraconsistent logics are Jaśkowski and da Costa.

Jaskowśki, in his well-known paper of 1948, published in English only in 1969, proposed the problem of constructing propositional calculi with the following properties: 'when applied to contradictory systems would not always entail their over-completeness; it would be rich enough to enable practical inferences; it would have an intuitive justification'. Jaśkowski presented his own solution only at the propositional level, having introduced his discussive logic D2.

In the fifties, without knowing Jaśkowski's work, da Costa began to develop his ideas about the importance of the study of contradictory theories. In 1958 da Costa proposes his Principle of Tolerance in Mathematics: 'From the syntactical and semantical points of view, every theory is permissible, since it is not trivial'. Da Costa's ideas were worked out in 1963, when he began to publish a series of papers containing his hierarchies of logics for the study of inconsistent but non-trivial theories. In 1989 Graham Priest and Richard Routley wrote: 'with da Costa's work we arrive at something strikingly different from what had gone before, deliberately fashioned paraconsistent systems - not overtly matrix logics or translations of modal logics'.

Da Costa first constructed a hierarchy of propositional calculi $C_n, 1 \leq n \leq \omega$, satisfying the following conditions: 'the principle of contradiction, in the form $\neg(A\&\neg A)$, should not be valid in general; from two contradictory premises A and $\neg A$, we should not deduce any formula whatever; they should contain the most important schemes and rules of classical logic compatible with the first two conditions'.

The hierarchy C_n was extended to a hierarchy of first-order predicate calculi with equality $C_n^=, 1 \leq n \leq \omega$, and to a hierarchy of calculi of descriptions $D_n, 1 \leq n \leq \omega$, and all of then applied to the construction of a hierarchy of inconsistent but non-trivial set theories.

Many other paraconsistent systems have been introduced and studied by several logicians, and among the pioneer logicians we must also mention D. Nelson and Florencio Asenjo.

Contradictory theories do exist. Whether that is a consequence of the incorrect description of a contradictory world, or just a temporary state of our knowledge, or perhaps the result of a particular language that we have chosen to describe the world, conflicting observational criteria, or superpositions of world views, contradictions are apparently unavoidable in our theories.

3. Contents of this Volume

The papers presented at the WCP'2000 covered most of the aspects of paraconsistency, including several systems in paraconsistent logics, paraconsistency as related to logics, and philosophical aspects of paraconsistency. The present volume contains some of these papers.

The first part of the book, "Paraconsistency and Systems of Paraconsistent Logics", contains several approaches to paraconsistency.

In "A Taxonomy of **C**-systems" by **W. A. Carnielli** and **J. Marcos**, the authors present a study on the foundations of a large class of paraconsistent logics from the point of view of the *logics of formal inconsistency* (**LFI**s). Those are paraconsistent logics which do not obey the full Principle of Pseudo-Scotus (PPS, also called Principle of Explosion) but do follow a weaker version of PPS. This results in a novel approach to the model-theoretical notion of consistency, one of its effects being the possibility of logically separating the notions of contradictoriness and inconsistency. A subclass of **LFIs** called **C**-*systems* is singled out, characterized as the **LFIs** that are built on some given positive basis and which can express consistency and inconsistency by means of new linguistic operators. They point out that the gist of paraconsistent logic lies in the Principle of Explosion, rather than in the Principle of Non-Contradiction, also to be distinguished from the Principle of Non-Triviality. The authors show how several well-known paraconsistent logics in the literature can be recast as **C**-systems. The paper also defines a particular subclass of the **C**-systems, the **dC**-*systems*, as the ones in which the new operators of consistency and inconsistency can be dispensed - for example, the well-known hierarchy of da Costa's paraconsistent calculi is shown to be constituted by some very particular **dC**-systems. The paper points out the connections to other studies by several authors, sets some open problems and suggests a few directions of continuation.

In **B. Brown**'s "Paraconsistent Classical Logic" a paraconsistent system is presented which is *classical*, in the sense that its semantics is fully classical, and in the more significant sense that it results from a well-motivated type-raising of a fully classical sentence-sentence consequence relation to a set-set consequence relation. Some additional weakly aggregative features such as: $\{p,q\} \nvdash \{(p \wedge q)\}$ and $\{(p \vee q)\} \nvdash \{p,q\}$ have interesting connections to graph and hyper-graph colourings.

"The Logic of Opposition", by **F. Asenjo**, proposes a logical system intended to be an instrument with which to place the study of antinomies in its most general setting. The main argument of Asenjo is that many antinomies have nothing to do with negation but rather with some basic, more general notion of opposition relative to which negation is only a particular case. Several arguments are given about the foundations of mathematics which go beyond the issue of antinomies and which reflect the future changes that are and will be taking place in mathematics as a whole, taking in consideration the impact that computers are having on the way mathematics is practiced.

The paper "Categorical Consequence for Paraconsistent Logic", by **F. Johnson** and **P. Woodruff**, extends the work of Rumfitt, Smiley and others on consequence relations employing sentences with force operators. In this paper, the authors investigate multiple-conclusion consequence relations generated by sets of valuations whose members include those that admit truth value gluts (there are sentences that are both true and false on some valuations), as well as gaps (there are sentences that are neither true nor false on some valuations). The question of categoricity for these consequence relations is answered positively.

In "Ontological Causes of Inconsistency and a Change-Adaptive, Logical Solution" **G. Vanackere** argues that an implicit ontological assumption that is commonly made results in blurring of the distinction between 'the object a' and 'the object a at a given moment'. This causes many inconsistencies, and the use of the non-monotonic, paraconsistent, change-adaptive logic **CAL2** is proposed here to circumvent this problem, by the introduction of names for *objects at a moment*.

On the other hand, adaptative logics **APV** and **APT** are proposed by **J. Meheus** in her essay "An Adaptive Logic for Pragmatic Truth", as substitutes for the logics (of da Costa, Bueno and Béziau) **PV** and **PT** of *pragmatic validity* and *pragmatic truth*. The idea behind these systems is to interpret a (possibly inconsistent) theory Γ 'as pragmatically as possible', by localizing the 'consistent core' of Γ, and delivering all sentences that are compatible with this core. The semantics and a dynamic proof theory of the systems is presented.

A possible-worlds semantics for theories that are represented in R. Reiter's default logic style is investigated in the article "A Multiple Worlds Semantics for a Paraconsistent Nonmonotonic Logic" by **A. T. Martins**, **M. Pequeno** and **T. Pequeno**. The system of logics studied, called IDL & LEI, is well-suited to formalize reasoning under incomplete knowledge, and combines nonmonotonicity to model inferences on the basis of partial evidence, with paraconsistency to deal with the inconsistencies introduced by extended inferences. The semantics proposed by the authors is shown to be sound and complete.

N. C. A. da Costa and **D. Krause** adopt in "An Inductive Annotated Logic" an alternative point of view to non-monotonic reasoning. They criticize those kind of logics that deal only with non-doxastic states of inputs and outputs, i.e., there are no references to degrees of belief, or confidence, about the states of the data. (However, the system IDL & LEI mentioned above goes in a somewhat related direction.) The authors propose a certain kind of paraconsistent logic, called *annotated logic*, for dealing with propositions that are vague in a sense but that, despite their vagueness, can be "believed" with a certain degree of confidence.

The first part of the book ends with "On NCG_ω: a paraconsistent sequent calculus" by **J.E. de A. Moura** and **I.M.L. D'Ottaviano**. The system NCG_ω, a sequent calculus formulation equivalent to da Costa's system C_ω, is introduced. This new logic is suitable for the application of Gentzen techniques and allows the proof of cut elimination, decidability and consistency theorems for C_ω.

The second part of the book, "Paraconsistency as Related to Other Logics" presents several systems of non-standard logics with paraconsistent features.

"**A**, Still Adorable", by **R. K. Meyer** and **J. K. Slaney**, opens the section.

The paper expands on an earlier paper by the same authors, "Abelian Logic (From **A** to **Z**)". The main result is that the Abelian logic **A** is rejection-complete, i.e. if a formula α is not provable in **A** then it is rejectable by applying a Łukasiewicz-style rejection axiom and rules added to **A**. In addition to the presentation of an interesting normal form for **A**, the authors also show that the implicational fragment of the logic admits a single axiom axiomatization, and they prove the finite model property.

In **G. Priest**'s "Fuzzy Relevant Logic" the author addresses the problem of defining logics dealing with two notions related to paraconsistency: vagueness (fuzziness) and relevance. Two possible approaches to define fuzzy relevant logics are presented in the paper: the first one by providing fuzzy Kripke semantics for relevant logics, changing the discrete truth values to continuum-valued ones, which gives sublogics of the corresponding standard relevant logics. The second strategy consists in a reinterpretation of the algebraic semantics for relevant logic, thinking of the algebraic values as degrees of truth; thus standard relevant logics can be thought of as fuzzy logics.

The next paper is "On some Remarkable Relations between Paraconsistent Logics, Modal Logics, and Ambiguity Logics", by **D. Batens**. The paper establishes new relationships between paraconsistent logics, ambiguity logics and the modal logic **S5**, through three main directions: first, a paraconsistent logic **A** is presented that has the same expressive power as **S5**. Next, paraconsistent logics (such as Priest's **LP**) are defined from **S5** and ambiguity logic **AL**. Finally, it is shown that some paraconsistent logics and inconsistency-adaptive logics serve exactly the same purpose as some modal logics and ampliative adaptive logics based on **S5**.

An adaptative approach to paraconsistency is adopted in the paper "The Dialogical Dynamics of Adaptive Paraconsistency", by **S. Rahman** and **J.-P. Van Bendegem**, based on earlier work of the authors, where they suggested that an adaptive version of paraconsistency is the natural way to capture the inherent dynamics of dialogues. The main objective of the paper is to obtain a formulation of dialogical paraconsistent logic in the spirit of an adaptive perspective which explores the possibility of eliminating inconsistencies by means of logical preference strategies.
strategies.

Adaptative logic is also present in "An inconsistency-adaptive proof procedure for logic programming", by **T. Vermeir**, where a paraconsistent proof procedure is introduced that combines logic programming and inconsistency-adaptive logics. The method profits from the ease of computing in logic programming. On the other hand, the author maintains paraconsistency, dynamics and non-monotonicity from adaptive logics. This is done by combining the notion of *competitor* from logic programming with the *conditionallity* which characterizes all the adaptive proofs.

The aim of the paper "Referential and inferential many-valuedness", by **G. Malinowski**, is to present two approches to the problem of many-valuedness, referential and inferential. In the former, many-valuedness may be received as the result of multiplication of semantic correlates of sentences, and not logical values. In the latter, three-valuedness is the metalogical property of inference which leads from non-rejected assumptions to accepted conclusions. Two applications of the inferential framework are given: to the Ł-modal logic and to a paraconsistent version of the three-valued Łukasiewicz propositional logic.

The main subject of **M. Finger**'s paper "When is a Substructural Logic Paraconsistent? Structural conditions for paraconsistency in ternary frames" is how violations of certain semantic consistency conditions leads to paraconsistency. A remarkable point of his approach is the combination of the usage of a ternary (and not binary) accessibility relation for substructural logics and the notion of correspondence theory previously developed only for modal logics by R. Routley and R. Meyer. The author focuses on different definitions of what constitutes a consistency axiom and the respective first-order restrictions over ternary frames caused by the acceptations of those definitions.

R.E. Jennings and D. Sarenac, in "Beyond Truth(-Preservation)", present a preservationist treatment of implication, including remarks on Heyting's Intuitionist 3-valued matrix for implication $\rightarrow$ as well as a new preservationist matrix for implication that resists trivialization even with nested antecedents. The connections of their system SX with respect to paraconsistency are discussed as well.

In "Paraconsistency in Chang's Logic with Positive and Negative Truth Values", **R. Lewin** and **M. Sagastume** start from $[-1,1]$-valued logic $Ł^*$, which C. C. Chang introduced as a generalization of $[0,1]$-valued logic of Łukasiewicz, and provide an axiomatization of all sentences in $Ł^*$ zero-valued by every valuation. This is the set of *paraconsistent* sentences, in the sense that $v(\phi) = v(\neg\phi) = 0$ and $v(\phi) \rightarrow (\neg v(\phi) \rightarrow v(\psi))$ is not a tautology. They prove the soundness and the completeness of this logic, and establish the categorical equivalence between MV-algebras (the algebras of $[0,1]$-valued Łukasiewicz logic) and MV*-algebras (the algebras of $[-1,1]$-valued Chang logic).

MV-algebras also appear in **D. Mundici**'s paper "Fault-tolerance and Rota-Metropolis cubic logic", where the author discusses the relationship between the cubic algebras of Rota and Metropolis, the three-valued MV-algebras with a self-negated element, and the Ulam-Rényi game of twenty questions with one lie (at most). The latter is a chapter of fault-tolerant search, and of error-correcting codes with feedback, first considered by Dobrushin and Berlekamp. The paper relates different fields of mathematics, like finite geometry, algebra, logic and the theory of error-correcting codes, with particular reference to the issue of non-triviality vs. inconsistency-tolerance.

The next contribution, "The Annotated Logics $OP_{\mathbf{BL}}$", by **G. Ortiz Rico**, introduces the systems $OP_{\mathbf{BL}}$ and $COP_{\mathbf{BL}}$ of annotated logics which are related with systems SP_{τ} and SAL defined by Lewin-Mikenberg-Schwarze. These new systems have a simpler axiomatization and differences in the so-called *well behaved formulas* with respect to earlier approaches. They allow annotations of annotations, defined in bilattices. These systems could be a basis of a programming language for reasoning about databases that contain inconsistencies, according to the aim of annotated logics.

L. Maksimova's "Definability and Interpolation in Extensions of Johansson's Minimal Logic" studies the projective Beth property (PBP) and the Craig interpolation property (CIP) for the extensions of Johansson's minimal logic J as well as the extensions, in the positive language $(\bot, \wedge, \vee, \rightarrow)$, of what the author calls the positive fragment J^+ of J. The properties for extensions of J^+, in the positive language, are reduced to the corresponding properties for some superintuitionistic logics. The extensions of J by positive axioms and PBP are described as well as the ones among them with CIP.

This part of the book ends with "Toward a Mathematics of Impossible Pictures", by **C. Mortensen**. In the essay the author gives a brief account of the recent history of the so-called *impossible pictures*, made popular by Escher's drawings. It is argued that the perception of an impossible picture is represented by an inconsistent theory joining visual experience with incompatible geometrical expectations. It is also suggested that the inconsistency may reduce to identifying and separating simultaneously certain points of the figures or the underlying space, as in the theory of "heaps" of natural or real numbers by Meyer and others. Finally, some adjoint functors between consistent and inconsistent theories are described.

The last part of the book, "Philosophical Aspects of Paraconsistency" contains several essays analyzing the subject of paraconsistency from a philosophical perspective.

B. H. Slater, with "Ambiguity is not Enough", opens the section. As suggested by B. Brown, it is possible to reinterpret G. Priest's *Logic of Paradox* (**LP**) in a preservationist manner, showing that the **LP** approach to paraconsistency is very closely related to treating sentence letters as ambiguous. The aim of Slater's paper is to show that such forms of disambiguation will not handle all paraconsistent situations, in connection with some of the classical paradoxes of self–reference.

"Are paraconsistent negations negations?", by **Jean-Yves Béziau**, is a discussion of a pressing issue, namely whether any "paraconsistent negation" in the current literature is a genuine negation. In his view, none are. In order to argue for that Béziau first considers philosophical issues, and shows that some confusion has been made in the attempts to articulate an account of paraconsistent negation. In the second part of the paper he provides a critical look at the main paraconsistent negations as they appear in the literature.

A different point of view is adopted by **Max Urchs** in his paper "On the role of adjunction in para(in)consistent logic", where rejection of the principle of contradiction is contrasted with the rejection of the principle of *ex falso quodlibet*. He suggests that Jaskowski's construction of discussive logic is one of the most interesting and promising approaches to inconsistency-tolerant reasoning, and numerous classical ideas of Leibniz or Kant can be restated within that setting. Urchs argues that at the very core of paraconsistency lies not negation, but conjunction: at least for those approaches which are directed toward application, conjunction seems to be the most important connective.

Don Faust, in his paper "Between Consistency and Paraconsistency: Perspectives from Evidence Logic", sketches the formal basis and philosophical background of so-called *Evidence Logic* (**EL**). In this logic every atomic predicate formula $P\ (t,...)$ is made into two formulas: $P_c(t,...)$ (confirmatory evidence for P) and $P_r(t,...)$ (refutatory evidence for P) (additionally, evidence is relativised to some finitary level). Consequently, $\neg P_c(t,...)$ means absence of evidence and $P_r(t,...)$ means the evidence of absence. This illuminates the distinction in Priest's dialectic between rejecting φ and accepting $\neg\varphi$, and provides some insight into foundational aspects of paraconsistency.

"Kinds of Inconsistency", the contribution of **G. Wheeler** to this volume, discusses two approaches to inconsistency: the ontological argument, which claims that inconsistent objects or events of some kind demand a logic able to reason about them, and the epistemological argument, which denies that language or the world is infected with inconsistency but claims instead that the problem is all in a reasoning agent's understanding. This essay argues against the practice of linking epistemological concerns with weak paraconsistent logics, and it introduces an epistemological motivation for adopting a strong paraconsistent logic by considering the results of measuring physical objects and how we might go about reasoning with those results. From the author's point of view, measurement is a source of inconsistency that is neither best understood as a problem between agents nor the result of either a paradoxical property of language or the world.

The comparison between paraconsistent and Meinongian logics is studied by **J. Paśniczec** in "Paraconsistent logic vs. Meinongian logic". While paraconsistent logics deal with inconsistency mostly on the propositional level, Meinong's theory of objects explains formal features of intentional objects; in particular, they account for contradictions encapsulated in these objects. In other words, these logics deal with inconsistency within the subject-predicate structure. The main thesis of the paper is that although paraconsistent and Meinongian logics need not reduce to each other, one can trace various mutual inspirations of the philosophical and logical character.

Finally, **O. Bueno**'s contribution, "Can a paraconsistent theorist be a logical monist?", analyses the debate about logical pluralism and monism with respect to the understanding of paraconsistent logic. Reasoning along the lines articulated by da Costa, the author puts forward a defense of logical pluralism as the best perspective for the paraconsistent logician. In his opinion, paraconsistent theorists cannot make sense of their own practice in a logical monist setting.

Acknowledgments

The preparation, selection and editing of this volume would not have been possible without the careful attention, time and energy of the staff of the Centre for Logic, Epistemology and the History of Science of UNICAMP, who have assisted us on numerous occasions. The staff of Marcel Dekker, New York, particularly Maria Allegra, has been a great help in guiding us through the editorial procedures, and we are very grateful for their help. We are also very much indebted to the following colleagues, who helped to select the articles, and improve the selected ones:

Nancy Amato, Sandra de Amo, Ofer Arieli, Arnon Avron, Ronen Basri, Diderik Batens, J.C. Beall, Philippe Besnard, Jean-Yves Béziau, Ricardo Bianconi, Guilherme Bittencourt, Maria Paola Bonacina, Ross Brady, Manuel Bremer, Bryson Brown, Otávio Bueno, Martin Bunder, Xavier Caicedo, Agata Ciabattoni, Mark Colyvan, Carlos V. Damásio, Antonio Di Nola, Carlos Di Prisco, Richard L. Epstein, Christian Fermueller, Marcelo Finger, George Francis, Steven French, Lluis Godo, Siegfried Gottwald, John Grant, Marcel Guillaume, Petr Hajek, Gerhard Heinzmann, Andreas Herzig, Ramon Jansana, Michael Kaminski, Beata Konikowska, Tomasz Kowalski, Fred Kroon, Renato Lewin, E.G. K.Lopes-Escobar, Eliezer Lozinski, Grzegorz Malinowski, Jacek Malinowski, Joo Marcos, Victor W. Marek, Robert K. Meyer, David Miller, Francisco Miraglia, Franco Montagna, Larry Moss, Daniele Mundici, Sara Negri, Donald Nute, Sergey Odincow, Hiroakira Ono, Jeffrey Paris, Jacek Pasniczek, Andrzej Pelc, Nicolas Peltier, Don Perlis, John L. Pollock, Graham Priest, Shahid Rahman, Marta Sagastume, Ralf Dieter Schindler, Peter Schotch, Barry H. Slater, Timothy Smiley, Max Urchs, Alasdair Urquhart, Guido Vanackere,

Jean Paul van Bendegem, Vladimir Vasyukov, Heinrich Wansing, Gunnar Wilken, Paul Wong, Peter Woodruff and Alberto Zanardo .

It is a great pleasure to have been supported by such a formidable assembly of specialists. To all them our most sincere thanks.

Walter Alexandre Carnielli
Marcelo E. Coniglio
Itala M. Loffredo D'Ottaviano
Editors

Organizing Committee of WCP'00
Itala M. Loffredo D'Ottaviano (State University of Campinas)
Walter A. Carnielli (State University of Campinas)
Marcelo E. Coniglio (State University of Campinas)
Daniel D.P. Alves (State University of Campinas)
João Marcos (State University of Campinas)

Contents

Contributors

Florencio G. Asenjo University of Pittsburgh, Pittsburgh, Pennsylvania

Diderik Batens Ghent University, Ghent, Belgium

Jean-Yves Béziau Stanford University, Stanford, California

Bryson Brown University of Lethbridge, Alberta, British Columbia, Canada

Otávio Bueno California State University, Fresno, California

Walter A. Carnielli State University of Campinas, Campinas, SP, Brazil

Itala M. Loffredo D'Ottaviano State University of Campinas, Campinas, SP, Brazil

Newton C. A. da Costa University of São Paulo, São Paulo, SP, Brazil

Don Faust Northern Michigan University, Marquette, Michigan

Marcelo Finger Universidade de São Paulo, São Paulo, SP, Brazil

R. E. Jennings Simon Fraser University, Burnaby, British Columbia, Canada

Fred Johnson Colorado State University, Fort Collins, Colorado

Décio Krause Federal University of Santa Catarina, Florianopolis, SC, Brazil

Renato A. Lewin Pontificia Universidad Católica de Chile, Santiago, Chile

Larisa Maksimova Institute of Mathematics, Siberian Division of the Russian Academy of Sciences, Novosibirsk, Russia

Grzegorz Malinowski University of Lodz, Lodz, Poland

João Marcos State University of Campinas, Campinas, SP, Brazil

Ana T. Martins Federal University of Ceará, Fortaleza, CE, Brazil

Joke Meheus Ghent University, Ghent, Belgium

Robert K. Meyer Australian National University, Canberra, ACT, Australia

Chris Mortensen The University of Adelaide, Adelaide, Australia

José E. Moura Federal University of Rio Grande do Norte, Natal, RN, Brazil

Daniele Mundici University of Milan, Milan, Italy

Jacek Pásniczek Maria Curie-Sklodowska University, Lublin, Poland

Marcelino Pequeno Federal University of Ceará, Fortaleza, CE, Brazil

Tarcísio Pequeno Federal University of Ceará, Fortaleza, CE, Brazil

Graham Priest University of Queensland, Queensland, Australia

Shahid Rahman University of Saarlandes, Saarlandes, Germany

Guillermo Ortiz Rico Pontificia Universidad Católica de Chile, Santiago, Chile

Marta S. Sagastume Universidad Nacional de La Plata, La Plata, Argentina

Darko Sarenac Simon Fraser University, Burnaby, British Columbia, Canada

John K. Slaney Australian National University, Canberra, Australia

Barry H. Slater University of Western Australia, Crawley, Australia

Max Urchs University of Konstanz, Konstanz, Germany

Jean P. van Bendegen University of Brussels, Brussels, Belgium

Guido Vanackere University of Ghent, Ghent, Belgium

Timothy Vermeir Ghent University, Ghent, Belgium

Gregory R. Wheeler University of Rochester, Rochester, New York

Peter W. Woodruff University of California, Irvine, Irvine, California

A Taxonomy of C-systems °•

WALTER A. CARNIELLI CLE and IFCH, Unicamp, Brazil
carniell@cle.unicamp.br

JOÃO MARCOS RUG, Ghent, Belgium, and IFCH, Unicamp, Brazil
vegetal@cle.unicamp.br

Abstract
The logics of formal inconsistency (**LFI**s) are paraconsistent logics which permit us to internalize the concepts of consistency or inconsistency inside our object language, introducing new operators to talk about them, and allowing us, in principle, to logically separate the notions of contradictoriness and of inconsistency. We present the formal definitions of these logics in the context of General Abstract Logics, argue that they in fact represent the majority of all paraconsistent logics existing up to this point, if not the most exceptional ones, and we single out a subclass of them called **C**-systems, as the **LFI**s that are built over the positive basis of some given consistent logic. Given precise characterizations of some received logical principles, we point out that the gist of paraconsistent logic lies in the Principle of Explosion, rather than in the Principle of Non-Contradiction, and we also sharply distinguish these two from the Principle of Non-Triviality, considering the next various weaker formulations of explosion, and investigating their interrelations. Subsequently, we present the syntactical formulations of some of the main **C**-systems based on classical logic, showing how several well-known logics in the literature can be recast as such a kind of **C**-systems, and carefully study their properties and shortcomings, showing for instance how they can be used to faithfully reproduce all classical inferences, despite being themselves only fragments of classical logic, and venturing some comments on their algebraic counterparts. We also define a particular subclass of the **C**-systems, the **dC**-systems, as the ones in which the new operators of consistency and inconsistency can be dispensed. A survey of some general methods adequate to provide these logics with suitable interpretations, both in terms of valuation semantics and of possible-translations semantics, is to be found in a follow-up, the paper [42]. This study is intended both to fully present and characterize, from scratch, the field into which it inserts, hinting of course to the connections with other studies by several authors, as well as to set some open problems, and to point to a few directions of continuation, establishing on the way a unifying theoretical framework for further investigation for researchers involved with the foundations of paraconsistent logic.

°• Carnielli acknowledges financial support from CNPq / Brazil and from the A. von Humboldt Foundation, and thanks colleagues from the Advanced Reasoning Forum present at the Bucharest meeting in 2000 for the opportunity of discussing some aspects of this work. Those discussions gave rise to the pamphlet [40], an embryonic version of the present paper. Marcos acknowledges, first, the financial support received from CNPq / Brazil, and, later, from a Dehousse doctoral grant in Ghent, Belgium. Both authors acknowledge support also from a CAPES / DAAD grant for a ProBrAl project Campinas / Karlsruhe, and are indebted to all the colleagues present at the II World Congress on Paraconsistency, and especially to Newton da Costa, for his achievements, ideas, and his enthusiasm, both enduring and contagious. And, of course, to the excellent comments by Chris Mortensen, Dirk Batens and Jean-Yves Béziau on a beta-version of this paper, plus the last minute corrections and comments by the friends Carlos Caleiro and Marcel Guillaume, as well as to all the patient people who have waited long enough for this study to be concluded. As we already clarified above, a complete semantical study of the systems here presented is soon to be found in [42] (as an outcome of [76]). Dividing it into two papers, we have tried to keep the length and termination of this study a bit more reasonable. All comments are welcome in the meanwhile.

1 THOU SHALT NOT TRIVIALIZE!

> On account of the classical principle of [non-]contradiction, a proposition and its negation cannot be both simultaneously true; thanks to this, it is not possible that a theory which is valid under the philosophical (or logical) point of view includes internal contradictions. To suppose the contrary would seemingly constitute a philosophical error.
> —Newton C. A. da Costa, [46], p. 6–7, 1958.

In the dawn of the XXI century, debates on the statute of contradiction in logic, philosophy and mathematics are still likely to raise the most diverse and animated sentiments. And this is an old story, whose first dramatic strokes can be traced back to authors as early as Aristotle (for the defense of non-contradiction), or Heraclitus (for the contrary position). Be that as it may, the fact is that in the beginning of the last century essentially the same dispute was still taking place, this time contraposing Russell to Meinong. And so it could still proceed, for centuries, if only the philosophical aspects of the dispute were touched. Even on more technical grounds, logicians of caliber, such as Alfred Tarski, would eventually speculate about that (cf. [106]):

> I do not think that our attitude towards an inconsistent theory would change even if we decided for some reason to weaken our system of logic so as to deprive ourselves of the possibility of deriving every sentence from any two contradictory sentences. It seems to me that the real reason of our attitude is a different one: We know (if only intuitively) that an inconsistent theory must contain false sentences; and we are not inclined to regard as acceptable any theory which has been shown to contain such sentences.

Against such suspicions, the philosopher Wittgenstein, who had devoted almost half of his late work to the philosophy of mathematics and used to refer to it as his 'main contribution' (cf. the entry *Mathematics*, in [63]), would have had something to say. Indeed, he often felt puzzled about 'the superstitious fear and awe of mathematicians in face of contradiction' (cf. [109], Ap.III–17), and asked himself: 'Contradiction. Why just this *one* spectre? This is surely much suspect.' (id., IV–56). His point was that 'it is one thing to use a mathematical technique consisting in the avoidance of contradiction, and another thing to philosophize against contradiction in mathematics' (id., IV–55), and that it was necessary to remove the 'metaphysical thorn' stuck here (id., VII–12). In this respect, the philosopher described his own objective as precisely that of altering the *attitude* of mathematicians concerning contradictions (id., III–82).

The above passage from da Costa's [46] could also be directed upon criticizing a position such as the above one of Tarski. The presupposition to be challenged here, of course, is that of an inconsistent theory obligatorily containing false sentences. Thus, if models may be described of structures in which some (but not all) contradictory sentences are simultaneously true, we will have a technical point against such suspicions of impossibility or implausibility of maintaining contradictory sentences inside of some theory and still being able to perform reasonable inferences from that, *instead* of being able to derive arbitrarily other sentences. This is sure to make a point in conferring to the task of studying the behavior of contradictory yet non-trivial theories —the task of *paraconsistency*— some respectability. And it *is* indeed possible to assign models for inconsistent non-trivial theories, even if these were to be regarded by some as epistemologically puzzling, or ontologically perplexing! Obtaining models and understanding their role is certainly an extraordinarily important mathematical enterprise: Enourmous efforts from the most brilliant minds and more than twenty centuries were required until mathematicians would allow themselves to consider models in which, given a straight line S and a point P outside of it, one could draw not just one line, but infinite, or no parallel lines to S passing through P, as in the well-known case of non-Euclidean geometries. In the present case, then, the problem will not be that of *validating falsities*, but that of *extending our notion of truth* (an idea further explored, for instance, in [28]).

At that same decisive moment, in the first half of the last century, there were in fact these other people like Łukasiewicz or Vasiliev who were soon proposing relativizations of the idea of non-contradiction, offering formal interpretations to formal systems in which this idea did not hold, and in which contradictions could make sense. And in between the 40s and the 60s the world would finally be watching the birth of the first real operative systems of paraconsistent logic (cf. Jaśkowski's [67], Nelson's [86], and da Costa's [49]). But the paleontology of paraconsistent logic will not be our main subject here —for that we prefer to redirect the reader to some of the following articles [6], [59], [55], and those in section 1 of [95], plus the book [26].

1.1 Contradictory theories do exist. Be them a consequence of the only correct description of a contradictory world (as assumed in [90]), be them just a temporary state of our knowledge, or again the outcome of a particular language that we have chosen to describe the world, the result of conflicting observational criteria, superpositions of worldviews, or simply, in science, because they result from the best theories available at a given moment (cf. [14]), contradictions are presumably unavoidable in our theories. Even if contradictory theories were to appear only by mistake, or perhaps by some Janus-like crooked behavior of their proposers, it is hard to see, given for instance results such as Gödel's incompleteness theorems, how contradictions could be prevented from even being taken into consideration. So it should be clear that the point here is not about the *existence* of contradictory theories, but about *what we should do* with them! Should these theories be allowed to explode and derive anything else, as in classical logic, or rather should we try to substitute the underlying logic, in (potentially) critical situations, in order to still be able to draw (if only temporarily, if you want) reasonable conclusions from those theories?

At this point it is interesting to consider the following motto set down by Newton da Costa, one of the founders of modern paraconsistent logic (cf. [47]):

> From the syntactical-semantical standpoint, every mathematical theory is admissible, unless it is trivial.

Da Costa designated that motto 'Principle of Tolerance in Mathematics', in analogy to the 'syntactical' principle proposed before by Carnap (cf. [35], p.52). According to this, the dividing line in between systems worthy of investigation and those that do not 'make a difference' (cf. [55]), nor convey any information (cf. [14]), should be drawn around non-triviality, rather than in the vicinity of non-contradictoriness. This will give us the first key to paraconsistency: if there are no contradictions around, then everything is under control, once we are inside of a consistent environment; but if contradictions are allowed, non-triviality should be the aim —but then what we must control is the *explosive* character of our underlying logic. Indeed, inside of a consistent logic we know that contradictions are dangerous in a theory precisely because they will give sufficient reason for this theory to explode, deducing anything else!

So, given a logic whose language includes a negation symbol $\neg$, let's call *contradictory* a theory from which some formula A and its negation $\neg A$ can be derived by way of the underlying logic. Let's also call a theory *trivial* if any formula B can be derived from it by way of the underlying logic, and call a theory *explosive* if the addition to it of any contradiction A and $\neg A$ is sufficient to make it trivial. The underlying logic, in its own right, will also be called *contradictory, trivial,* or *explosive* if, respectively, all of the theories about which it can talk are contradictory, trivial, or explosive. To be sure, any trivial theory / logic will also turn to be contradictory, whenever there is a negation available (anything is derived from it, in particular all pairs of formulas of the form A and $\neg A$). Inside classical or intuitionistic logic, and, in a general way, inside any 'consistent' logic (this will be defined in what follows), the contradictory and the trivial theories simply coincide, by way of their explosive character. *Paraconsistent logics* were then proposed to be the logics to underlie those contradictory theories which were still to be kept non-trivial, and what those logics must of course effect to such an end is weakening or annulling the explosive character of these theories.[1] So, all at once, paraconsistency comes and provides a sharp distinction in between the logical notions of contradictoriness, explosiveness, and triviality.

Anyone working as a knowledge engineer, assembling and managing knowledge databases, will be perfectly aware that gathering inconsistent information is the rule rather than the exception. And again, either if you assume, by some sort of methodological requirement, inconsistent theories to be problematic (cf. [14]) or not (cf. [90]), this does not prevent you from also assuming them to be, in general, quite *informative*, and wanting to *reason* from them in a sensible way. Consider, for instance, this very simple situation (cf. [40]) in which you ask two people, in the due course of an investigation, a 'yes-no' question such as 'Does Dick live in Arizona?', so that what will result will be exacly one of the three following different possible scenarios: they might both say 'yes', they might both say 'no', or else one of them might say 'yes' while the other says 'no'. Now, it happens that in neither situation you may be sure about where Dick really lives (unless you trust some of the interviewees more than the other), but only in the last scenario, where an inconsistency appears, *are you sure* to have received wrong information from one of your sources!

Our next point is that also the logical notions of *inconsistency* and of *contradictoriness* can and should be *distinguished* in a purely abstract way. Distinctions have already been proposed, in the literature, among the notions of *paradoxical* and of *antinomical* theories (cf., for instance, Arruda's [6], p.3, or da Costa's [51], p.194), the paradoxical ones being identified with those theories in which inconsistencies could occur without necessarily leading to trivialization, and the antinomical ones identified with those in which any occurring contradiction turns out to be fatal, as in the case of Russell's antinomy in naive set theory. Let us, here, insist on this distinction for a moment and stretch it a bit further. One first difficulty to be confronted with is that of some English technical terms: It is such a pity that techniques and results such as Hilbert's witch-hunt programme in search of a *Widerspruchfreiheitbeweis* for Arith-

[1] Surprising as it may seem, this would also have been the advice given by Wittgenstein on how to proceed in the presence of contradictions: 'The contradiction does not even falsify anything. Let it lie. Do not go there.' (cf. [110], XIV, p.138) For the relations and non-relations between Wittgenstein and the paraconsistent enterprise the reader may consult, for instance, [75], [64] or [77].

metic were to be eventually translated into the search for a 'consistency proof', given that what it literally means is something much more precise, namely a 'proof of freedom from contradictions'! More often than not, German language indeed shows itself to be exceedingly precise, so that we should rather stick here to the literal meaning of *Widerspruchfreiheit* as non-contradictoriness, and associate inconsistency, if we may, with something like the term *Unbeständigkeit* (or any other synonym of *Inkonsistenz* together perhaps with some terms opposed to *Beschaffenheit* and to *Widerstandsfähigkeit*). Now, antinomies will be related to the presence of 'strong' contradictions —those with explosive behavior—, while paradoxes will be related to the presence of inconsistencies, which do not necessarily depend on negation, such as in the case of the well-known Curry's paradox (cf. [45]). Let us try to summarize this whole story in a picture (maybe you do not agree on our choice of names, but we beg you to stick to our terminology for the moment):

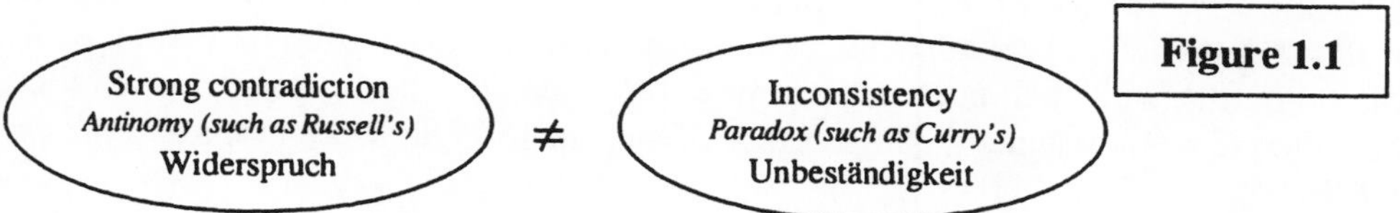

Figure 1.1

The above distinctions, of course, are more illustrative than formal (nothing prevents you, for instance, from thinking of Russell's antinomy as something not as destructive as it was, if you just change the underlying logic of its theory so as to make it only paradoxical, but the distinction between antinomies as involving the notion of strong negation, on the one hand, and inconsistencies, on the other hand, as something more general and in principle independent of negation, should be taken more seriously).

Now, whatever an inconsistency might *mean*, be it more general or not than a simple contradiction, we may certainly presuppose that a contradiction is at least an *example* of an inconsistency, be it the only possible one or not. Traditionally, as we have noted a few paragraphs above, the contradictoriness of a given theory / logic was to be identified with the fact that it derives at least some pairs of formulas of the form A and $\neg A$, while inconsistency was usually talked about as a model-theoretic property to be guaranteed so that our theories can make sense and talk about 'real existing structures'. Of course, any trivial theory / logic, thus, given our assumption above that contradictions entail inconsistencies, will also be both contradictory and inconsistent. Now, if explosiveness does not hold, as we shall see, in the scope of paraconsistent logics, there is in principle no reason to suppose that the converse would also be the case, and that a contradiction would always lead to trivialization. How to reconcile these concepts then? Da Costa's idea, when proposing his first paraconsistent calculi (cf. [49]), was that the 'consistency' and the 'classic-like behavior' (he called that 'well-behavior') of a given formula, as a sufficient requisite to guarantee its explosive character, could be represented as simply another formula of its underlying logic (he chose, for his first calculus, C_1, to represent the consistency of a formula A by the formula $\neg(A\wedge\neg A)$, and referred to this last formula —to be intuitively read as saying that 'it is not the case that both A and $\neg A$ are true'—, as a realization of the 'Principle of Non-Contradiction', conventions that we will here, in general, *not* follow —neither will we follow, necessarily, the identification of consistency with 'classic-like behavior'). In fact, our proposal here, inspired by da Costa's idea, is exactly that of introducing consistency as a *primitive notion* of our logics: the paraconsistent logics which internalize the notion of consistency so as to introduce it already at the object level will be called *logics of formal inconsistency* (**LFI**s). And, given a consistent logic **L**, the **LFI**s which extend the positive basis of **L** will be said to constitute **C**-*systems based on* **L**. Our main aim in this paper, besides making all the above definitions and their multiple shades and interrelations entirely clear, will be that of studying a large class of **C**-systems based on classical logic (of which the calculi C_n of da Costa will be but very particular examples).

On what concerns this story about regarding consistency as a primitive notion, the status of points, lines and planes in geometry may immediately be thought of, but the case of (imaginary) complex numbers seems to make an even better comparison: even if we do not know what they are, and may even suspect there is little sense in insisting on which way they can exist in the 'real' world, the most important aspect is that it is possible to calculate with them. Girolamo Cardano, who first had the idea of computing with such numbers, seems to have seen this point clearly —he failed, however, to acknowledge the importance of this; in 1545 he wrote in his *Ars Magna* (cf. [87]):

> Dismissing mental tortures, and multiplying $5+\sqrt{-15}$ by $5-\sqrt{-15}$, we obtain $25-(-15)$. Therefore the product is 40. ...and thus far does arithmetical subtlety go, of which this, the extreme, is, as we have said, so subtle that it is useless.

His discovery, that one could operate with a mathematical concept independent of what our intuition would say and that usefulness (or something else) could be a guiding criterion for accepting or rejecting experimentation with mathematical objects, definitely contributed to the proof of the Fundamental Theorem of Algebra by C. F. Gauss in 1799, before which complex numbers were not fully accepted.

To make matters clear, the basic idea behind the internalization of consistency inside our logics will be, in general, accomplished by the addition of a unary connective expressing consistency (and usually also another connective to express inconsistency), plus the following important assumption, that *consistency* is exactly what a theory might be lacking in order to deliver triviality when exposed to a contradiction.[2] Recapitulating: as we said before, triviality entails contradictoriness (if a negation is present), and contradictoriness entails inconsistency (or, to be more precise, contradictoriness entails 'non-consistency', for it may happen, as we will see, that consistency and inconsistency are not exactly dual in some of our logics, if we take both notions as primitive); now we just add to this the assumption that contradictoriness *plus* consistency implies triviality! We are in fact introducing, in this way, a novel definition of consistency, more fine-grained than the usual model-theoretic one: for a large class of logics (see FACT 2.14(ii)) it will turn out that consistency may be identified with the presence of *both* non-contradictoriness and explosive features.

---

[2] It is interesting to notice, by the way, that this assumption is remarkably compatible with Jaśkowski's intuition on the matter. As he put it, 'in some cases we have to do with a system of hypotheses which, *if subjected to a too consistent analysis*, would result in a contradiction between themselves or with a certain accepted law, but which we use in a way that is restricted so as not to yield a self-evident falsehood' (our italics, see [67], p.144). It is clear that we can give this at least one reading according to which Jaśkowski seemed already to have been worried about the effects of consistent contradictions!

Now, non-contradictoriness will be a necessary but *no more* a sufficient requirement for us to prove consistency. In the case of explosive logics, of course, the concepts of non-contradictoriness and non-triviality will coincide, so that non-contradictoriness and consistency are also to be identified. Paraconsistent logics are situated exactly in that terra incognita which lays in between non-explosive logics and trivial ones, and they comprehend exactly those logics which are both non-explosive and non-trivial (examples of such logics are provided by the whole of the literature on paraconsistent logics)! So, again, consistency divides the logical space in between consistent (and so, explosive and non-contradictory) logics, and inconsistent ones, and these last ones may, at their turn, be either paraconsistent (and so, non-explosive, and possibly even contradictory), or trivial.

1.2 Paraconsistent, but not contradictory! In fact, there is another point that we want to stress here, for it seems that much confusion has been unnecessarily raised around it. In general, paraconsistent logics do *not* validate contradictions or invalidate anything like the 'Principle of Non-Contradiction' (though there are a few that do). Most paraconsistent logics, actually, are just fragments of some other given consistent logic (such as some version of classical logic, or else some normal modal logic), so that they *cannot*, in any case, be contradictory! However, a good way of making this whole point much less ambiguous (even though still open to dispute, but now on a different level) is by considering formal definitions of those so-called (meta)logical *principles*.

Let us say that a logic respects the *Principle of Non-Contradiction*, (PNC), if it is non-contradictory, according to our previous definitions, that is, if it has non-contradictory theories, that is, theories in which no contradictory pair of formulas A and $\neg A$ may be inferred. Let us also say that a logic respects the *Principle of Non-Triviality*, (PNT), (a realization of da Costa's Principle of Tolerance inside of the logical space) if it is non-trivial, thus possessing non-trivial theories, and say that a logic respects the *Principle of Explosion*, or *Pseudo-Scotus*, (PPS), if it is explosive, that is, if all of its theories explode when in contact with a contradiction. It is clear now that all paraconsistent logics, by their very nature, must disrespect (PPS), aiming to retain (PNT), but it is also clear that they cannot disrespect (PNC) as long as they are defined as fragments of other logics that do respect (PPS)! The gist and legacy of paraconsistent logic indeed lies in showing that logics may be constructed in which the Principle of Pseudo-Scotus is controlled in its power, and this has 'in principle' *nothing to do* with the validity or not of the Principle of Non-Contradiction as we understand it. Yet a few logics exist which are not only paraconsistent, but that in fact disrespect (PNC). Such logics are usually put forward in order to formalize some dialectical principles, and are accordingly known as *dialectical logics*. Being able to infer contradictions, however, such dialectical logics cannot be fragments of any consistent logic, and in order to avoid trivialization they should also usually assume, for instance, the failure of Uniform Substitution, at least when applied to some specific formulas, such as the contradictions that those logics can infer (or else any other contradiction, and thus any other formula, would be inferable). Much weaker versions of the Principle of Non-Contradiction have nevertheless been considered in the literature, as for instance the following one, deriving from semantical approaches to the matter: a logic is said to respect the *Principle of Non-Contradiction, second form*, (PNC2), if it has non-trivial models for pairs of contradictory formulas. But then, of course, every model for the falsification of (PPS), that is, every model for a paraconsistent logic, would also satisfy (PNC2), and vice-versa, so that not only would (PNC2) be unnecessary as a new principle, but there would also be no principle dealing specifically with the existence of dialectical logics. Too bad! And, of course, there is a BIG difference in having models for some specific contradictions and having *all* models of a given logic validating some contradictory pair of formulas — this amounts, in the end, to the same difference which exists, in classical logic, in between contingent formulas, on the one hand, and (tautological or) contradictory ones, on the other hand...

The above definition of (PNC) will also prevent us from identifying this principle, inside some arbitrary given logic **L**, with the validity in **L** of some particular formula, such as $\neg(A\wedge\neg A)$ (as in da Costa's first requisite for the construction of his paraconsistent calculi —check the subsection **3.8**). But it *is* true that such a formula can, as well as many other formulas, be identified, in some situations, to the expression of consistency inside of some specific logics, such as da Costa's C_1! Let us, in general, say that a theory Γ is *gently explosive* when there is always a way of expressing the consistency of a given formula A by way of formulas which depend only on A, that is, when there is a (set of) formula(s) constructed using A as their sole variable and that cannot be added to Γ together with a contradiction A and $\neg A$, unless this leads to triviality. A gently explosive logic, then, is exactly a logic having only gently explosive theories, and we can now formulate (gPPS), a 'gentle version' of (PPS), for a given logic **L**, asserting that this logic must be gently explosive. Gently explosive paraconsistent logics, thus, are precisely those logics that we have above dubbed **LFI**s, the logics of formal inconsistency. In the logics we will be studying in this paper, we will in general assume that the consistency of each formula A can be expressed by operators already at their linguistic level, and in the simplest case this will be written as $\circ A$, where '$\circ$' is the 'consistency connective'. The **C**-systems (in this paper they will be supposed to be based on classical logic), will be particular **LFI**s illustrating some different ways in which one can go on to axiomatize the behavior of this new connective.

There are also some other forms of explosion, as the *partial* one, which does not trivialize the whole logic, but just part of it (for instance, when a contradiction does not prove every other formula, but does prove every other *negated* formula). We will let our paraconsistent logics also reject this kind of explosion. There is *ex falso*, which asserts that at least one element should exist in our logics so that everything follows from it (a kind of *falsum*, or *bottom particle*). There is *controllable* explosion, which states that, if not all, at least some of our formulas should lead to trivialization when taken together with their negations. And, finally, there is *supplementing* explosion, which states that our logics should possess, or be able to define, a *supplementing*, or *strong* negation, to the effect that strongly negated propositions (that we have above called strong contradictions) should explode. (There are also all sorts of combinations of these forms of explosions, and perhaps some other forms still to be uncovered, but these are the ones we will concentrate on, here.) All of these alternative forms of explosion can be turned into logical (meta)principles, and none of these rejects, by their own right, 'full' Pseudo-Scotus —all of them, nonetheless, can still be held even when the Pseudo-Scotus does not hold! The paraconsistent logics studied in this paper will, of course, disrespect Pseudo-Scotus, and in addition to that they will also disrespect the principle regarding partial explosion, while, in most cases, they will still respect the principles regarding gentle explosion, *ex falso*, supplementing explosion, and, often, controllable explosion as well. This will be made much clearer in section **2**, where this study will be made more precise, and the interrelations between all of those principles will be more deeply investigated.

1.3 What do you mean? Let's now briefly describe the exciting things that await the reader in the next sections (we will skip section **1** in our description —you are reading it—, but do not stop here!).

Section **2** is *General Abstract Nonsense*. No particular systems of paraconsistent logic are studied here (though some are mentioned), but most of the definitions and preprocessed material that you will need to understand the rest are to be found in this section. There is nothing for you to lose your appetite —you can actually intensify it, even if, or especially if, you do not agree with some of our positions. We first make clear what we mean by *logics*, introduced by their *consequence relations*, and what we mean by *theories* based on these logics, and on the way you will also learn what *closed theories* and *monotonicity* mean, and what it means to say that a logic is a *fragment* or an *extension* of another logic. This is just preparatory work. We then introduce the logical notions of *contradictoriness*, *explosiveness* and *triviality*, concerning theories and logics, and pinpoint some immediate connections between these notions. This already takes us to the subsection **2.1**, where the first logical (meta)-principles are introduced, namely the principles of non-contradiction, (PNC), of non-triviality, (PNT), and of explosion, (PPS) (a.k.a. Pseudo-Scotus, or *ex contradictione*). You will even learn a little bit about the (pre-)history of these principles, their interrelations, and some confusions about them which lurk around. Some of their ontological aspects are also lightly touched. The subsection **2.2** brings us to *paraconsistent logics*, formulated in two equivalent (but not necessarily so) presentations, one of them saying that they should allow for contradictory non-trivial theories, the other one saying that they must disrespect (PPS). After you learn what it means to say that two given formulas / theories are *equivalent* inside some given logic, FACT 2.8 will call your attention to the discrimination that paraconsistent logics ought to make between contradictions: they cannot be all equivalent inside such logics. *Dialectical logics*, being those logics disrespecting (PNC), are mentioned to fill the gaps in the general picture, but they will not be studied here. In the subsection **2.3** we start talking about finite trivializability, and look at some remarkable examples of this phenomenon, as for example the one of a logic having *bottom particles* —thus respecting a principle that we call *ex falso*, (ExF)—, and the one of a logic having *strong negations* —and respecting a principle we call *supplementing explosion*, (sPPS). We also consider some properties of adjunction (and so, of conjunction), and in the end we draw a map to show the relationships between (PPS), finite trivializability, (ExF) and (sPPS), noting that no two of these principles are to be necessarily identified (and, in particular, *ex falso* does not coincide with *ex contradictione*). Pay special attention to FACT 2.10(ii), in which all non-trivial logics respecting *ex falso* are shown to have strong negations. Subsection **2.4** considers what happens when one says farewell to (PPS) but still maintains some of the other special forms of explosion exposed before and hints are given as to some disadvantages presented by paraconsistent logics which disrespect all of those principles at once. Some other misunderstandings about the construction of paraconsistent logics are discussed, and the difference between contradictoriness and inconsistency is finally called into scene. Logics respecting a so-called principle of *gentle explosion*, (gPPS), are introduced as the ones in which *consistency* can be expressed, and even a *finite* version of gentle explosion, (fgPPS), is considered, as a particular case of finite trivialization. *Logics of formal inconsistency*, **LFI**s, are then defined to be exactly those respecting (gPPS) while disrespecting (PPS), and the great majority of the **LFI**s that we will be studying in the following will actually also respect (fgPPS). The new definition of consistency that we introduce is shown, for a given logic, in general to coincide simply with the sum of (PPS) and (PNT). Systems of paraconsistent logic known as *discussive* (or *discursive*) *logics* are shown to be representable as **LFI**s. In the subsection **2.5**, the principles of *partial explosion*, (pPPS), and *controllable explosion*, (cPPS), are finally introduced. A *boldly* paraconsistent logic is defined as one in which not only (PPS) but also (pPPS) is disrespected, and we try to concentrate exclusively on such logics. Classical logic respects all of the above principles, but for each of those principles, except (PNT), examples will be explicitly presented, or at least referred to, at some point or another, of logics disrespecting it. Multiple connections between those principles are exhibited not only along these lines but also in the section **3**. In this respect, pay also special attention to FACT 2.19 and the comments around it, in the subsection **2.6**, which show that the **LFI**s are ubiquitous: an enormous subclass of the already known paraconsistent logics can have its members recast as logics of formal inconsistency. **C**-systems are also introduced in this last subsection, and the map of the paraconsistent land presented in the subsection **2.4** gets richer and richer.

Section **3** brings a very careful syntactical study of a large class of **C**-systems based on classical logic. Each new axiom is justified as it is introduced, and its effects and counter-effects are exhibited and discussed. The systems presented are initially linearly ordered by extension, but soon spread out in many directions. The remarkable unifying character of our approach in terms of **LFI**s is made clear while most logics produced by the 'Brazilian school' in the last forty years or so are shown to smoothly fit the general schema and together make up a whole coherent map of **C**-land. Subsection **3.1** presents a kind of minimal paraconsistent logic (for our purposes), called C_{min} and constructed from the positive part of classical logic by the addition of (the axiom which represents the principle of) excluded middle, plus an axiom for double negation elimination. The Deduction Metatheorem holds for this logic and its extensions, and C_{min} is shown to be paraconsistent. Comparisons are drawn between C_{min} and one of its fragments, da Costa's C_ω, and the facts that no strong negation, or bottom particle, or finitely trivializable theory, or negated theorem are to be found in these logics are mentioned. You will also learn that, in these logics, no two different negated formulas are provably equivalent. Of course, as a consequence of these last facts, these logics cannot be **LFI**s, what to say **C**-systems, but the **C**-systems which will be studied in the following subsections are all extensions of them. We make some observations about versions of *proof by cases* provable in these logics, by way of excluded middle, and we adjust some of its axioms to better suit deduction. A way to turn these logics into classical logic, simply by adding back the Pseudo-Scotus to them, is also demonstrated.

In subsection **3.2**, we introduce the *basic logic of (in)consistency*, called **bC**, by adding a new axiom to C_{min}, and we show how to immediately extract from this axiom a strong negation and a bottom particle. We now have '∘', the *consistency* connective, at our disposal as a new primitive constructor in our language, realizing the finite gentle explosion. The logic **bC**, which is, in fact, a conservative extension of C_{min}, is shown already to have negated theorems and equivalent negated formulas, but on the other hand it does not have any provably consistent formulas. Sufficient and necessary conditions for a **bC**-theory to behave classically are presented. The axiom defining **bC** define a kind of restricted *Pseudo-Scotus*, as obvious. Some related restricted forms of *reductio ad absurdum* which are also present are studied, and the elimination of double negation shows its purpose in THEOREM 3.13, where you will learn that some forms of partial trivialization are avoided by all paraconsistent extensions of **bC**. Restricted forms of *reductio* deduction and inference rules are shown to be present in **bC**, and some other rules relating contradictions and consistency are exhibited. No paraconsistent extensions of **bC** will contain the formula $(A \wedge \neg A)$ as a bottom particle (but some other **LFI**s, such as Jaśkowski's **D2** —at least under some presentations— do have it as a bottom particle). A formula such as $\neg(A \wedge \neg A)$ is also not provable in **bC**, but can be proved in some of its extensions, such as the three-valued maximal paraconsistent logics **LFI1** and **LFI2**.

Subsection **3.3** is mostly composed of negative results. It starts by showing that not many rules making the interdefinition of connectives possible hold in **bC**. The reader will also learn about the obligatory failure of *disjunctive syllogism* in vast extensions of the paraconsistent land, and the failure of 'full' *contraposition* inference rules in **bC** and all of its paraconsistent extensions, though some *restricted* forms of it had already been shown to hold, in the previous subsection. The uses of disjunctive syllogism and of contraposition to derive the Pseudo-Scotus had already been pointed out a long ago, respectively, by C. I. Lewis (and, much before, by the 'Pseudo-Scotus' himself), and by Popper. Some asymmetries related to negation of equivalent formulas are pointed out, and as a result it will not be possible to prove a *replacement* theorem for **bC**, which would establish the validity of *intersubstitutivity of provable equivalents*, (IpE), and the same phenomenon will be observed in most, but not all, of **bC**'s extensions. Reasons for all these failures, and possible solutions for them, are discussed.

In subsection **3.4** the problem of adding an inconsistency connective '•' to **bC**, intended as dual to '∘', making inconsistency coincide with non-consistency, and consistency coincide with non-inconsistency, is shown to be not as easy as it may seem, as a consequence of the last negative results. Some intermediary logics obtained in the strive towards the solution of this problem are exhibited, and hints are given on how this solution should look. A first solution, adopted, in fact, in the whole literature, is to make inconsistency equivalent to contradiction, and this is exactly what the logic **Ci**, introduced in the subsection **3.5**, does. It will *not* be the case, however, that consistency in **Ci** can be identified with the negation of a conjunction of contradictory formulas, that is, the consistency of a formula A will not be equivalent to any formula such as $\neg(A \wedge \neg A)$. New forms of gentle explosion and restricted contraposition deduction rules are shown to hold in **Ci**, and provably consistent formulas in **Ci** are shown to exist. Indeed, the notable FACT 3.32 shows that provably consistent formulas in **Ci** coincide with the formulas causing controllable explosion in this logic and in all of its extensions. Some more restricted forms of contraposition inference rules introduced by **Ci** are also exhibited, and the failure of (IpE) also for this logic is pointed out. In fact, as we already know that full contraposition cannot be added to **Ci** in order to get (IpE), some weaker contraposition deduction rules which would also do the job are tested, and these are also shown to lead to collapse into classical logic (see FACT 3.36). But there still can be a chance of obtaining (IpE) in extensions of **Ci** by the addition of even weaker forms of contraposition deduction rules, as the reader is going to see at the end of subsection **3.7**, where positive results for this are shown for extensions of **bC**. The connectives '$\circ$' and '$\bullet$' have a good behavior inside of **Ci**, and we show that any formula having them as the main operator is consistently provable, and that consistency propagates through negation (and inconsistency back-propagates through negation). In addition to that, schemas such as $(A \to \neg\neg A)$, which are shown *not* to hold in the general case, are indeed shown to hold if A has the form $\circ B$, for some B. All of this comes either as an effect or a consequence of the fact that a restricted replacement is valid in **Ci**, as proved in the subsection **3.6**, to the effect that inconsistency here can be really introduced, by definition, as non-consistency, or else consistency can be introduced, by definition, as non-inconsistency.

Subsection **3.7** shows how to compare the previously introduced **C**-systems with an extended version of classical logic, **eCPL** (adding innocuous operators for consistency and inconsistency). As a result, we can show that the strong negation that we had defined for **bC** does not have all properties of classical negation, but another strong negation can be defined in **bC**, which *does* have a classical character. In **Ci** these two negations are shown to be equivalent, but the interesting output of a strong classical negation is making it possible for us to conservatively translate classical logic inside of all our **C**-systems (THEOREM 3.46 and comments after THEOREM 3.48), so that any classical inference can be faithfully reproduced, up to a translation, inside of **bC** or of any of its extensions. About **Ci**, we can now prove that it has some redundant axioms, and the remarkable FACT 3.50, showing that only consistent or inconsistent formulas can themselves be consistently provable in this logic, and so these are the only formulas that can cause controllable explosion in **Ci**. An even more remarkable THEOREM 3.51 shows several conditions which *cannot* be fulfilled by paraconsistent systems in order to render the proof of full replacement, (IpE), possible. But paraconsistent extensions of **bC** in which (IpE) holds are indeed shown to exist, the same task remaining open for extensions of **Ci**.

Subsection **3.8** presents the **dC**-*systems*, which are the **C**-systems in which the connectives '$\circ$' and '$\bullet$' can be dispensed, definable from some combination of the remaining connectives. The particular combinations chosen by da Costa in the construction of his calculi C_n are surveyed, and we start concentrating more and more on general parallels of da Costa's original requisites for the construction of paraconsistent calculi (which does not mean that we shall feel obliged to obey them *ipsis litteris*). Criticisms on the particular choices made by da Costa and some of their consequences are surveyed. Again, in a particular case, that of da Costa's C_1, the consistency of a formula A is identified, as we have said before, with the formula $\neg(A \wedge \neg A)$, and the extension of **Ci** which makes this identification is called **Cil**. This system is shown still to suffer some strange asymmetries related to the negations of equivalent formulas (for instance, $\neg(\neg A \wedge A)$ is not equivalent to $\neg(A \wedge \neg A)$), and some partial or full solutions to that are discussed at that point and below. Connections between our **C**-systems and *relevance* or *intuitionistic* logics are touched. In particular, the problem of defining **C**-systems based on intuitionistic, rather than classical, logic is touched, but no interesting solutions are presented (because they seem still not to exist, but perhaps the reader will have the pleasure of finding them in the future).

In subsection **3.9**, **dC**-systems are put aside for a moment and the addition of an axiom for the introduction of double negation is considered, together with its consequences. Arguments, both positive and negative, for the 'proliferation of inconsistencies' that such an axiom could cause are presented, and rejected. Subsection **3.10** surveys various ways in which consistency (and inconsistency) can propagate from simpler to more complex formulas and vice-versa. One of these forms, perhaps the most basic one, is illustrated by an extension of **Ci**, the logic **Cia** (or else an extension of **Cia**, the logic **Cila**, which was recorded into history under the name C_1), and this logic will be shown to make possible a new and interesting conservative translation from classical logic inside of it, or any of its extensions. If all the reader wants to know about is da Costa's original calculi C_n, this subsection is the place (together with some earlier comments in the subsection **3.7**), but in that case be warned: You may miss most of the fun! Properties of the C_n are surveyed, and the problem of finding a real *deductive limit* to this hierarchy is presented together with its solution, so that the reader can forget once and for all any ideas they may have had about the logic C_ω having its place as part of this hierarchy. In particular, the deductive limit of the C_n, the logic C_{Lim}, is shown to constitute an **LFI**, though we are not sure if its form of gentle explosion can be made finite. Again, (IpE) is shown not to hold for the calculi C_n, so that *Lindenbaum-Tarski*-like algebraizations for these logics can be forgotten, but the situation for them is actually worse, for it has been proven that they just cannot define any non-trivial congruence, putting aside also the possibility of finding *Blok-Pigozzi*-like algebraizations for them. But several extensions of the C_n can indeed fix this last problem, and so we try to concentrate on some stronger forms of propagation of consistency which will help us with this. In particular, the logics C_1^+ (later proposed by da Costa and his disciples), as well as five three-valued logics, $\mathbf{P}^1$, $\mathbf{P}^2$, $\mathbf{P}^3$, **LFI1** and **LFI2**, proposed in several studies, are also axiomatized and studied as extensions of **Ci**. An increasingly detailed and clear map of the **C**-systems based on classical logic is being drawn.

In the subsection **3.11**, the last five three-valued logics are shown to constitute part of a much larger family of 8,192 three-valued paraconsistent logics, each of them proven to be axiomatizable as extensions of **Ci** containing suitable axioms for propagation of consistency. Each of these three-valued logics can also be shown to be *distinct* from all of the others, and *maximal* relative to classical logic, **eCPL**, solving one of the main requisites set down by da Costa, to the effect that 'most rules and schemas of classical logic' should be provable in a 'good' paraconsistent logic. We also count how many of those 8,192 logics are in fact **dC**-systems, and not only **C**-systems, and show many *connections* between them and the other logics presented before, all of them fragments of some of these three-valued maximal paraconsistent logics. Interestingly, $\mathbf{P}^1$ is shown to be conservatively translatable inside of any of the other three-valued logics, and all of these are shown to be conservatively translatable inside of **LFI1**. (IpE) is proven not to hold in any of these logics, but there are some other interesting connectives which they can define, as some sort of *'highly' classical* negation, and *congruences* which will make possible, in subsection **3.12**, the definition of non-trivial (Blok-Pigozzi) algebraizations of all of these three-valued logics. Indeed, the subsection **3.12** surveys positive and negative results regarding algebraizations of the **C**-systems.

Section **4** sets some exciting open problems and directions for further research, for the reader's recreation.

1.4 Standing on the shoulders of each other. It would be very unwise of us to present this study, which includes a technical survey of its area, without trying to connect it as much as possible to the rest of the related literature. But we pledge to have done our very best to highlight, wherever opportune, some of the relevant papers which come close or very close to our points, or on which we simply base our study at some points! As in the case of the famous legendary caliph who set the books of the library of Alexandria on fire, we could say that the relevant papers which are not cited here at some point or another are, in most cases, either *blasphemous* (meaning that there is no context here for them to be mentioned), or *unnecessary* (meaning that in general you have to go no farther than the bibliography of our bibliography to learn about them). Other options would be our total *ignorance* about such and such papers at the moment of writing this (about which we thank for any enlightenment that we might receive), or else because we felt it was already well represented by another publication on the same matter, or because it integrates our list of *future research* (that was a good excuse, wasn't it?). Or perhaps it was *not* relevant at all! (You wouldn't know that, would you?) Read the text and judge for yourself. We just want to mention in this subsection a few other papers whose structural or methodological similarity (or dissimilarity) to some of our themes is most striking —so that we can better highlight our *own* originality on some topics, whenever it becomes the case.

Our study in section **2** is totally situated at the level of a general theory of consequence relations, a field sometimes referred to as that of *General Abstract Logics* (cf. [111]), or *Universal Logic* (cf. [19]). There are a few (rare) papers dealing with the definitions of the logical principles at a purely logical level. One of them is Restall's [99], where an approach to the matter quite different from ours is tackled. Also starting from the definition of logic as determined by its consequence relation (even though monotonicity is not pressuposed), and assuming from the start that an adjunctive conjunction is available, the author also requires one sort of contraposition deduction rule to be valid for all of the negations he considers (something that, in most logics herein studied, does not hold), and fixes the relevance logic R in the formulation of most of his results. Several versions of the 'law' of non-contradiction are then presented, starting from the outright identification of this principle with the principle of explosion, passing through the identification of the principle of non-contradiction with the principle of excluded middle, or with the validity of the formula $\neg(A\wedge\neg A)$, and going up to some sort of difference in degree between accepting the inference of *all* propositions at once from some given formula $(A\wedge\neg A)$ instead of accepting *each* at a turn. It is clear that the outcome of all this is completely diverse from what we propose here. Some other studies go so far as to also study some of the alternative forms of explosion that we concentrate on here. This is the case, for instance, for Batens's [10], and Urbas's [108]. We are unaware, however, of any study which has taken these alternative forms as far as we do, and have studied them in precise and detailed terms. Such a study is presumed to be essential to help clarify the foundations, the nature and the reach of paraconsistent logics. There have been, for instance, arguments to the effect that the negations of paraconsistent logics are not (or may not be) negation operators after all (cf. Slater's [104] and Béziau's [23]). Béziau's argument amounts to a request for the definition of some minimal 'positive properties' in order to characterize paraconsistent negation really as constituting a *negation* operator, instead of something else. Slater argues for the *inexistence* of paraconsistent logics, given that their negation operator is not a 'contradictory forming functor', but just a 'subcontrary forming one', recovering and extending an earlier argument from Priest & Routley in [93]. Evidently, the same argument about not being a contradictory forming functor applies as well to intuitionistic negation, or in general to any other negation which does not have a classical behavior. Regardless of whether you wish to call such an operator 'negation' or something else, the negations of paraconsistent logics had better be studied under a less biased perspective, by the investigation of general properties that they can or cannot display inside paraconsistent logics. Some good examples of that kind of critical study may be found not only in the present paper but also in Avron's [9], Béziau's [18], and especially Lenzen's [70], among others.

In section **3** we investigate **C**-systems based on classical logic. In this respect, the present study has at least one very important ancestor, namely Batens's [10], where a general investigation of logics extending the positive classical logic (not all of them being **C**-systems!) is presented. This same author has also presented, elsewhere one of the best arguments that may be used to support our approach in terms of logics of formal inconsistency, **LFI**s. Criticizing Priest's logic *LP* (cf. [90]), Batens insists that:

> There simply is no way to express, within this logic, that *B* is *not* false or that *B* behaves consistently. (Cf. [13], p.216)

Asserting that 'paraconsistent negation should not and cannot express rejection' (id., p.223), Batens wants to say that it is not because a negated sentence $\neg B$ is inferred from some non-trivial theory of a paraconsistent logic that we can conclude that B is not also to be inferred from that, i.e. $\neg B$ does not express the *rejection* of B. From that he will draw several lessons along his article, such as that: (i) the presence of a strong negation (he writes 'classical negation', but this is clearly an extrapolation —see our note 15, in the subsection **3.7**) inside of a paraconsistent logic is not only a sufficient requisite but also a necessary one to express (classical) rejection; (ii) that one needs a controllably explosive paraconsistent logic (he calls it 'non-strictly paraconsistent') to be able to 'fully describe classical logic' (id., p.225); (iii) that the existence of a bottom particle is also sufficient for the above purposes, for it may define a strong negation (this appeared in an addenda to the paper). So, all at once, this author argues for the validity of three of our alternative explosion principles: (sPPS), (cPPS), and (ExF). From these, of course, we already know (see FACT 2.19) that our principle (gPPS) will often follow, so that according to his recommendations we are finally left with **LFIs**, instead of something like Priest's logic, which, again according to Batens, and for the above reasons, 'fails to capture natural thinking' (though it was proposed to such an effect), and does not provide sufficient environment for us to do 'paraconsistent mathematics'. Priest's response to such criticisms seems to us to be somewhat of a cheat, for he proposes to introduce such a strong negation using an ill-defined bottom particle (see note 25, in the subsection **3.10**). The only point where Batens goes too far to be right seems to be on his argument about paraconsistent logics not being adequate to be used on our metalanguage, because we would be in need of strong negations to complete any consistent description of the world. But now we know that an **LFI** would be more than enough to such an end, being able to fully reproduce all the classical inferences (THEOREM 3.46). And do remember that **LFI**s are especially tailored in order to *express* the fact that B behaves consistently, attending, thus, to Batens's requisite above (and also to his praxis, given that he has already been using in his articles, since long, some symbol to express inconsistency in the object language, be it just an abbreviation or some sort of metaconnective —check the symbol '!' in [15]).

The very idea of a paraconsistent logic still has, nowadays, as strong defenders as attackers. Though, as we know, many attacks are but misunderstandings, many defenses are also poor or unsound. We hope here to contribute to this debate, in one way or another, combining as much precision and clarity as we can. At a more fundamental philosophical level, also, paraconsistency has raised diverse excited opinions about the contribution (or damage) it makes to the very notion of *rationality*. An author such as Mario Bunge will on the one hand compare the Pseudo-Scotus with some sort of *cancer* (cf. [32], p.17), and on the other hand observe that 'a refined symbolism can hide a brazen irrationalism' (id., p.23). He asserts that paraconsistent logic is non-rational by definition, because 'it does not include the principle of non-contradiction' (id., p.24). About this same point other authors will concede similar verdicts, and yet arrive at different conclusions. As Gilles-Gaston Granger put it, paraconsistent logics can be seen as a 'provisory recourse to the irrational', for maintaining an indicium of the rational (sic), namely the principle of non-trivialization, while also maintaining an indicium of the irrational, namely the possible presence of contradictions, to be 'philosophically justified' (cf. [65], p.175). Yet some other authors, such as Newton da Costa, defended that, according to some pragmatic principles of reason which 'seem to be present in all processes of systematization of rational knowledge', this same rational knowledge can be said, among other things, to be both intuitive and discursive, to result from the interaction of the spirit and its environs, and not to be identifiable with a particular system of logic. About reason, on its own turn, he maintained that it is tied to its historical evolution, has its range of application determinable only pragmatically, and is always expressible by way of some logic, which, in each case, is supposed to be uniquely determined by each given context, as being precisely the logic that is most adequate to that context. To determine the concept of adequacy, finally, da Costa recurs again to pragmatic factors, such as simplicity, convenience, facility, economy, and so on (cf. [51]). This is in fact a very thought-provoking issue, and several other authors have advanced positions on the relations of paraconsistency and rationality, such as Francisco Miró Quesada (in [81] and [80]), Nicholas Rescher (in [98]), Jean-Yves Béziau (in [23]), and Bobenrieth (in [26]). The reader is invited to read those authors directly, if this is their interest. We will not venture here any further steps in this slippery slope, for our aim is much less ambitious. After reading this comprehensive technical survey, however, we hope that the reader will feel illuminated enough to risk their own rationally-based judgements on the matter.

2 A PARACONSISTENT LOGIC IS A PARACONSISTENT LOGIC IS...

> Logic is the chosen resort of clear-headed people, severally convinced of the complete adequacy of their doctrines. It is such a pity that they cannot agree with each other.
> —A. N. Whitehead, "Harvard: The Future", Atlantic Monthly 158, p. 263.

Many a logician will agree that the fundamental notion behind logic is the notion of 'derivation', or rather should we say the notion of 'consequence'. On that account, in our common heritage it is to be found the Tarskian notion of a *consequence relation*. As usual, given a set *For* of formulas, we say that $\Vdash \subseteq \wp(For) \times For$ defines a consequence relation on *For* if the following clauses hold, for any formulas A and B, and subsets Γ and Δ of *For:* (formulas and commas at the left-hand side of $\Vdash$ denote, as usual, sets and unions of sets of formulas)

(Con1) $A \in \Gamma \Rightarrow \Gamma \Vdash A$ (reflexivity)
(Con2) $(\Delta \Vdash A \text{ and } \Delta \subseteq \Gamma) \Rightarrow \Gamma \Vdash A$ (monotonicity)
(Con3) $(\Delta \Vdash A \text{ and } \Gamma, A \Vdash B) \Rightarrow \Delta, \Gamma \Vdash B$ (transitivity)

So, a logic **L** will here be defined simply as a structure of the form $<For, \Vdash>$, containing a set of formulas and a consequence relation defined on this set. We need not suppose at this point that the set *For* should be endowed with any additional structure, like the usual algebraic one, but we will hereby suppose, for convenience, that *For* is built on a denumerable language having $\neg$ as its (primitive or defined) negation symbol, and we will also suppose the connectives to be constructing operators on the set of formulas. Any set $\Gamma \subseteq For$ is called a *theory* of **L**. A theory Γ is said to be *proper* if $\Gamma \neq For$, and a theory Γ is said to be *closed* if it contains all of its consequences, i.e. if the converse of (Con1) holds: $\Gamma \Vdash A \Rightarrow A \in \Gamma$. Whenever we have, in a given logic, that $\Gamma \Vdash A$, for a given theory Γ and some formula A, we will say that A is *inferred* from Γ (in this logic); if, for all Γ, we have that $\Gamma \Vdash A$, that is, if A is inferred from any given theory, we will say that A is a *thesis* (of this logic).

Not all known logics respect all the above clauses, or only them. For instance, those logics in which (Con2) is either dropped out or substituted by a form of 'cautious monotonicity' are called *non-monotonic*, and the logics whose consequence relations are closed under substitution are called *structural*. Unless explicitly stated to the contrary, we will from now on be working with some fixed arbitrary logic $\mathbf{L} = <For, \Vdash>$, and with some fixed arbitrary theory Γ of **L**. Properties (Con1)–(Con3) will be assumed to hold irrestrictedly, and they will be used in some proofs here and there. Some interesting and quite immediate consequences from (Con2) and (Con3) which we shall make use of are the following:

FACT 2.1 The following properties hold for any logic, any given theories Γ and Δ, and any formulas A and B:
(i) $\Gamma, \Delta \nVdash A \Rightarrow \Gamma \nVdash A$;
(ii) $(\Gamma \Vdash A \text{ and } A \Vdash B) \Rightarrow \Gamma \Vdash B$;
(iii) $(\Gamma \Vdash A \text{ and } \Gamma, A \Vdash B) \Rightarrow \Gamma \Vdash B$.

Proof: (i) follows from (Con2); (ii) and (iii), from (Con3). □

Given two logics $\mathbf{L1} = <For_1, \Vdash_1>$ and $\mathbf{L2} = <For_2, \Vdash_2>$, we will say that **L1** is a *linguistic extension* of **L2** if For_2 is a proper subset of For_1, and we will say that **L1** is a *deductive extension* of **L2** if $\Vdash_2$ is a proper subset of $\Vdash_1$. Finally, if **L1** is both a linguistic and deductive extension of **L2**, and if the restriction of **L1**'s consequence relation $\Vdash_1$ to the set For_2 will make it identical to $\Vdash_2$ (that is, if $For_2 \subset For_1$, and for any $\Gamma \cup \{A\} \subseteq For_2$ we have that $\Gamma \Vdash_1 A \Leftrightarrow \Gamma \Vdash_2 A$) then we will say that **L1** is a *conservative extension* of **L2**. In any of the above cases we can more generally say that **L1** is an *extension* of **L2**, or that **L2** is a *fragment* of **L1**. These concepts will be used mainly in the next section, where we will build and compare a number of paraconsistent logics. Just as a guiding note to the reader, however, we could remark that usually, but not obligatorily, linguistic extensions are also deductive ones, but it is quite easy to find in the realm of non-classical logics, on the other hand, deductive fragments which are not linguistic ones (like intuitionistic logic is a deductive fragment of classical logic). Most paraconsistent logics in the literature are also deductive fragments of classical logic themselves, but the ones we shall be working on here, the **C**-*systems*, are in general deductive fragments only of a conservative extension of classical logic —by the addition of (explicitly definable) connectives expressing consistency / inconsistency). A particular case of them, the **dC**-systems, will nevertheless be shown to be characterizable as deductive fragments of good old classical logic, dispensing its mentioned extension. But these assertions will be made much clearer in the near future.

Let Γ be a theory of **L**. We say that Γ is *contradictory with respect to* $\neg$, or simply *contradictory*, if it is such that, for some formula A, we have $\Gamma \Vdash A$ and $\Gamma \Vdash \neg A$. With some abuse of notation, but (hopefully) no risk of misunderstanding, we will from now on write these sort of sentences in the following way:

$$\exists A\, (\Gamma \Vdash A \text{ and } \Gamma \Vdash \neg A). \tag{D1}$$

For any such formula A we may also say that Γ is A-*contradictory*, or simply that A is *contradictory* for such a theory Γ (and such an underlying logic **L**). It follows that:

FACT 2.2 For a given theory Γ: (i) If $\{A, \neg A\} \subseteq \Gamma$ then Γ is A-contradictory. (ii) If Γ is both A-contradictory and closed, then $\{A, \neg A\} \subseteq \Gamma$.

Proof: Part (i) comes from (Con1), part (ii) from the very definition of a closed theory. □

A theory Γ is said to be *trivial* if it is such that:

$$\forall B\, (\Gamma \Vdash B). \tag{D2}$$

Hence, a trivial theory can make no difference between the formulas of a logic — all of them may be inferred from it. Of course, using (Con1) we may notice that the non-proper theory *For* is trivial. We may also immediately conclude that:

FACT 2.3 Contradictoriness is a necessary condition for triviality in a given theory. (D2) ⇒ (D1)

A theory Γ is said to be *explosive* if:

$$\forall A\, \forall B\, (\Gamma, A, \neg A \Vdash B). \tag{D3}$$

Thus, a theory is called explosive if it trivializes when exposed to a pair of contradictory formulas. Evidently:

FACT 2.4 (i) If a theory is trivial, then it is explosive. (ii) If a theory is contradictory and explosive, then it is trivial. (D2) ⇒ (D3); (D1) and (D3) ⇒ (D2)

Proof: Use (Con2) in the first part and FACT 2.1(iii) in the second. □

2.1 A question of principles. Now, remember that talking about a logic is talking about the inferential behavior of a set of theories. Accordingly, using the above definitions, we will now say that a given logic **L** is *contradictory* if all of its theories are contradictory. (D4)

In much the same spirit, we will say that **L** is *trivial*, or *explosive*, if, respectively, all of its theories are trivial, or explosive. respect. (D5), (D6)

The empty theory may be here regarded as playing an important role, revealing some intrinsic properties of a given logic, in spite of the behavior of any of its specific non-empty theories (also called 'non-logical axioms'). Indeed:

FACT 2.5 A monotonic logic **L** is contradictory / trivial / explosive if, and only if, its empty theory is contradictory / trivial / explosive.

We can now tackle a formal definition for some of the so-called *logical principles* (relativized for a given logic **L**), namely:

PRINCIPLE OF NON-CONTRADICTION (PNC)
L must be non-contradictory: $\exists\Gamma\,\forall A\,(\Gamma \nVdash A \text{ or } \Gamma \nVdash \neg A)$.

PRINCIPLE OF NON-TRIVIALITY (PNT)
L must be non-trivial: $\exists\Gamma\,\exists B\,(\Gamma \nVdash B)$.

PRINCIPLE OF EXPLOSION, or PRINCIPLE OF PSEUDO-SCOTUS[3] (PPS)
L must be explosive: $\forall\Gamma\,\forall A\,\forall B\,(\Gamma, A, \neg A \Vdash B)$.

This last principle is also often referred to as *ex contradictione sequitur quodlibet*.

The reader will immediately notice that these principles are somewhat interrelated:

FACT 2.6 (i) An explosive logic is contradictory if, and only if, it is trivial. (ii) A trivial logic is both contradictory and explosive. (iii) A logic in which the Principle of Explosion holds is a trivial one if, and only if, the Principle of Non-Contradiction fails. (D6) ⇒ [(D4) ⇔ (D5)]; (D5) ⇒ [(D4) and (D6)]; (PPS) ⇒ [not-(PNT) ⇔ not-(PNC)]

Proof: Just consider FACT 2.4, and the definitions above. □

A trivial logic, i.e. a logic in which (PNT) fails, cannot be a very interesting one, for in such a logic anything could be inferred from anything, and any intended capability of modeling 'sensible' reasoning would then collapse. Of course, (PPS) would still hold in such a logic, as well as any other universally quantified sentence dealing with the behavior of its consequence relation, but this time only because they would be unfalsifiable! It is readily comprehensible then that triviality might have been regarded as the mathematician's worst nightmare. Indeed, (PNT) constituted what Hilbert called 'consistency (or compatibility) principle', with which proof his Metamathematical enterprise was crafted to cope. Well aware of the preceding fact, and working inside the environment of an explosive logic such as classical logic, Hilbert transposed the 'problem of consistency' (that is, the problem of non-triviality) to the problem of proving that there were no contradictions among the axioms of arithmetic and their consequences (this was Hilbert's Second Problem, cf. [66]). By the way, this situation would eventually lead Hilbert to the formulation of a curious criterion according to which the non-contradictoriness of a mathematical

---

[3] Which was made visible by a reedition of a collection of commentaries on Aristotle's *Prior Analytics*, long attributed, in error, to Johannes Duns Scotus (1266-1308), in the twelve books of the *Opera Omnia*, 1639 (reprint 1968). The current most plausible conjecture about the authorship of these books will trace them back to John of Cornwall, around 1350. See [26] for more on its history.

object is a necessary and sufficient condition for its very *existence*.[4] Perhaps he would have never proposed such a criterion if he had only considered the existence of non-explosive logics, with or without (PNC), logics in which contradictory theories do not necessarily lead to trivialization! The search for such logics would give rise, much later, to the 'paraconsistent enterprise'.[5]

In classical logic, of course, all the three principles above hold, and one could naively speculate from such that they are all 'equivalent', in some sense. Indeed, they have all been now and again confused in the literature and each one of them has, in turn, been identified with the 'Principle of Non-Contradiction' (and these will not exhaust all formulations of this last principle that have been proposed here and there). The emergence of paraconsistent logic, as we shall see, will serve to show that this equivalence is far from being necessary, for an arbitrary logic **L**.

2.2 The paraconsistency predicament. Some decades ago, S. Jaśkowski ([67]) and N. C. A. da Costa ([49]), the founders of paraconsistent logic, proposed, independently, the study of logics which could accommodate contradictory yet non-trivial theories. Accordingly, a *paraconsistent logic* (a denomination which would be coined only in the seventies, by Miró Quesada) would be initially defined as a logic such that:

$$\exists\Gamma\exists A\exists B\ (\Gamma \Vdash A \text{ and } \Gamma \Vdash \neg A \text{ and } \Gamma \nVdash B). \tag{PL1}$$

Attention: This definition says *not* that (PNC) is not to hold in such a logic, for it says nothing about *all* theories of a paraconsistent logic being contradictory, but only that *some* of them should be contradictory, and yet non-trivial. As a consequence, following our definitions above, the notion of paraconsistent logic has, in principle, nothing to do with the rejection of the Principle of Non-Contradiction, as it is commonly held! On the other hand, it surely has something to do with the rejection of explosiveness. Indeed, consider the following alternative definition of a paraconsistent logic, as a logic in which (PPS) fails:

$$\exists\Gamma\exists A\exists B\ (\Gamma, A, \neg A \nVdash B). \tag{PL2}$$

Now one may easily check that:

FACT 2.7 (PL1) and (PL2) are equivalent ways of defining a paraconsistent logic, if its consequence relation is reflexive and transitive.

[(Con1) and (Con3)] $\Rightarrow$ [(PL1) $\Leftrightarrow$ (PL2)]

[4] Girolamo Saccheri (1667-1733) had already paved the way much before to set non-contradictoriness, instead of intuitiveness, as a sufficient, other than necessary, criterion for the legitimateness of a mathematical theory (cf. [1]). The so-called 'Hilbert's criterion for existence in mathematics' seems thus to constitute a further step in taking this method to its ultimate consequences.

[5] Assuming intuitively (cf. [46], p.7) that a contradiction could painlessly be admitted in a given theory if only this theory was not to be trivialized by it, even some years before the actual proposal of his first paraconsistent systems, da Costa was eventually led to trace the Metamathematical's problem about the utility of a formal system back to (PNT). At that point da Costa was even to suggest that Hilbert's criterium for existence in mathematics should be changed, and that *existence*, in mathematics, should be equated with non-triviality, rather than with non-contradictoriness (cf. [47], p.18). To be more precise about this point, da Costa has in fact recovered Quine's motto ([96], chapter I): 'to be is to be the value of a variable' —from which follows that the ontological commitment of our theories is to be measured by the domain of its variables—, and then proposed the following modification to it: 'to be is to be the value of a variable, in a given language of a given logic' (cf. [52], and the entry 'Paraconsistency' in [33]). This was meant to open space for the appearance of different ontologies based on different kinds of logic, analogously to what had happened in the XIX century with the appearance of different geometries based on different sets of axioms.

Proof: To show that (PL1) implies (PL2), use (Con3), or directly FACT 2.1; to show the converse, use (Con1). □

Say that two formulas A and B are *equivalent* if each one of them can be inferred from the other, that is:

$$(A \Vdash B) \text{ and } (B \Vdash A). \quad \text{(Eq1)}$$

In a similar manner, say that two sets of formulas Γ and Δ are equivalent if:

$$\forall A \in \Delta\,(\Gamma \Vdash A) \text{ and } \forall B \in \Gamma\,(\Delta \Vdash B). \quad \text{(Eq2)}$$

We will alternatively denote these facts by writing, respectively, $A \dashVdash B$, and $\Gamma \dashVdash \Delta$.

Now, an essential trait of a paraconsistent logic is that it does not see all contradictions at the same light —each one is a different story. Indeed:

FACT 2.8 Given any arbitrary transitive paraconsistent logic, it cannot be the case that all of its contradictions are equivalent.

Proof: If, for whatever formulas A and B, we have that $\{A, \neg A\} \dashVdash \{B, \neg B\}$, then any A-contradictory theory, would also be, by transitivity and definition (Eq2), a B-contradictory theory. But if a theory infers every pair of contradictory formulas, it infers, in particular, any given formula at all, and so it is trivial. □

Once again, the reader should note that the existence of a paraconsistent logic **L** presupposes only the existence of *some* non-explosive theories in **L**; this does not mean that *all* theories of **L** should be non-explosive —and how could they all be so? (recall FACT 2.4(i)) Moreover, once more according to our proposed definitions, the reader will soon notice that the great majority of the paraconsistent logics found in the literature, and all the paraconsistent logics studied in this paper, are non-contradictory (i.e. 'consistent', following the usual model-theoretic connotation of the word). In particular, they usually have non-contradictory empty theories, which means, from a proof-theoretical point of view, that they bring no built-in contradiction in their axioms, and that their inference rules do not generate contradictions from these axioms. Even so, because of their paraconsistent character, they can still be used as underlying logics to extract some sensible reasoning of some theories that are contradictory and are still to be kept non-trivial. This phenomenon is no miracle, and certainly no sleight of hand, as the reader will understand below, but is obtained from suitable constraints on the power of explosiveness, (PPS). So, all paraconsistent logics which we will present here are in some sense 'more conservative' than classical logic, in the sense that they will extract less consequences than classical logic would extract from some given classical theory, or at most the same set of consequences, but never more. Our paraconsistent logics then (as most paraconsistent logics in the literature) will not validate any bizarre form of reasoning, and will not extract any contradictory consequence in the cases where classically there were no such consequences. It is nonetheless possible to also build logics which disrespect both (PPS) *and* (PNC), and thus might be said to be 'highly' non-classical, in a certain sense, once they *do* have theses which are not classical theses. Such logics will constitute a particular case of paraconsistent logics that are generally dubbed *dialectical logics*, or *logics of impossible objects*, and some specimens of these may be found, for instance, in [56], [83], [88], and [100]. We will not study these kind of logics here.

We shall, from now on, make use of either one of the above definitions for paraconsistent logic, indistinctly.

2.3 The trivializing predicament. Given (PL1), we know that any paraconsistent logic must possess contradictory non-trivial theories, and from (PL2) we know that these must be non-explosive. Evidently, not all theories of a given logic can be such: we already also know that any trivial theory is both contradictory and explosive, and every logic has trivial theories (consider, for instance, the non-proper theory *For*, i.e. the whole set of formulas). It is possible, though, and in fact very interesting, to further explore this no man's land which lays in between plain non-explosiveness and outright explosiveness, if one considers some paraconsistent logics having some suitable explosive proper theories. That is what we will do in the following subsections.

A logic **L** is said to be *finitely trivializable* when it has finite trivial theories. (D7)

Evidently:

FACT 2.9 If a logic is explosive, then it is finitely trivializable. (D6) ⇒ (D7)

Proof: All theories of an explosive logic are explosive, in particular the empty one. Thus, for any A, the finite theory $\{A, \neg A\}$ is trivial. □

This same fact does not hold for non-explosive logics. In fact, we will present, in the following, a few paraconsistent logics which are *not* finitely trivializable, although these shall, in general, not concern us in this article, for reasons which will soon be made clear. Let us first state and study some few more simple definitions.

A logic **L** has a *bottom particle* if there is some formula C in **L** that can, by itself, trivialize the logic, that is:

$$\exists C \forall \Gamma \forall B\ (\Gamma, C \Vdash B). \qquad \text{(D8)}$$

We will denote any fixed such particle, when it exists, by $\bot$. Evidently, no arbitrary monotonic and transitive logic can have a bottom particle as a thesis, under pain of turning this logic into a trivial logic —in which, of course, all formulas turn to be bottom particles.

It is instructive here to remember another formulation of (PPS) which sometimes shows up in the literature:

PRINCIPLE OF 'EX FALSO SEQUITUR QUODLIBET' (ExF)

L must have a bottom particle.

Now, if we are successful in isolating logics that disrespect (PPS) while still respecting (ExF) we will show that *ex contradictione* (*sequitur quodlibet*) does not need to be identified with *ex falso* (*sequitur quodlibet*), as is quite commonly held.[6]

We say that a logic **L** has a *top particle* if there is some formula C in **L** that is a consequence of every one of its theories, no matter what, that is:

$$\exists C \forall \Gamma\ (\Gamma \Vdash C). \qquad \text{(D9)}$$

We will denote any fixed such particle, when it exists, by $\top$. Evidently, given a monotonic logic, any of its theses will constitute such a top particle (and logics with no theses, like Kleene's three-valued logic, will have no such particles). Also, given transitivity and monotonicity, it is easy to see that the addition of a top particle to a given theory is pretty innocuous, for in that case $(\Gamma, \top \Vdash B)$ if and only if $(\Gamma \Vdash B)$.

Let **L** now be some logic, let σ: *For* → *For* be a mapping (if *For* comes equipped with some additional structure, we will require σ to be an endomorphism), and let this mapping be such that $\sigma(A)$ is to denote a formula which *depends only on A*. By this we shall mean that $\sigma(A)$ is a formula constructed using but A itself and some

[6] And yet this separation between these two principles can already be found in the work of the Pseudo-Scotus (see [26], chapter V, section 2.3).

purely logical symbols (such as connectives, quantifiers, constants). In more general terms, given any sequence of formulas $A_1, A_2, \ldots, A_n$, we will let $\sigma(A_1, A_2, \ldots, A_n)$ denote a formula which depends only on the formulas of the sequence. Similarly, we will let $\Gamma(A_1, A_2, \ldots, A_n)$ denote a set of formulas each of which depends only on the sequence $A_1, A_2, \ldots, A_n$. In some situations it will help to assume this σ to be a *schema*, that is, that given any two sequences $A_1, A_2, \ldots, A_n$ and $B_1, B_2, \ldots, B_n$, we must have that $\sigma(A_1, A_2, \ldots, A_n)$ will be made identical to $\sigma(B_1, B_2, \ldots, B_n)$ if we only change each A_i for B_i, in $\sigma(A_1, A_2, \ldots, A_n)$ (this means, in some sense, that all these σ-formulas will share some built-in *logical form*). Usually, when saying that we have a formula, or set of formulas, depending only on some given sequence of formulas, we further presuppose that this dependency is schematic, but this supposition will in general be not strictly necessary to our purposes.

We say that a logic **L** has a *strong* (or *supplementing*) *negation* if there is a schema $\sigma(A)$, depending only on A, that does not consists, in general, of a bottom particle and that cannot be added to any theory inferring A without causing its trivialization, that is:

(a) $\exists A$ such that $\sigma(A)$ is not a bottom particle, and
(b) $\forall A\, \forall \Gamma\, \forall B\, [\Gamma, A, \sigma(A) \Vdash B]$. (D10)

We will denote the strong negation of a formula A, when it exists, by $\sim A$.

Parallel to the definition of contradictoriness with respect to $\neg$, we might now define a theory Γ to be *contradictory with respect to* $\sim$ if it is such that:

$$\exists A\, (\Gamma \Vdash A \text{ and } \Gamma \Vdash \sim A). \quad \text{(D11)}$$

Accordingly, a logic **L** is said to be *contradictory with respect to* $\sim$ if all of its theories are contradictory with respect to $\sim$. (D12)

Here we may of course introduce yet another version of (PPS):

SUPPLEMENTING PRINCIPLE OF EXPLOSION (sPPS)
L must have a strong negation.[7]

Some immediate consequences of the last definitions are:

FACT 2.10 (i) If a logic has either a bottom particle or a strong negation, then it is finitely trivializable. (ii) If a non-trivial logic has a bottom particle, then it admits a strong negation. (iii) If a logic is explosive and non-trivial, then it is supplementing explosive. [(D8) or (D10)] $\Rightarrow$ (D7); [not-(D5) and (D8)] $\Rightarrow$ (D10); [(PNT) and (ExF)] $\Rightarrow$ (sPPS); [(PPS) and (PNT)] $\Rightarrow$ (sPPS)

Proof: (i) is obvious. To prove (ii), define the strong negation $\sim A$ of a formula A by stipulating that, for any theories Γ and Δ, we have (a) ($\Gamma, \Delta \Vdash \sim A$) iff ($\Gamma, \Delta, A \Vdash \bot$), and (b) ($\Gamma, \Delta, \sim A \Vdash \bot$) iff ($\Gamma, \Delta \Vdash A$). By (Con1), we have that ($\Gamma, \sim A \Vdash \sim A$), and so, part (a) will give us ($\Gamma, \sim A, A \Vdash \bot$), choosing $\Delta = \{\sim A\}$. But $\bot \Vdash B$, for any formula B, once $\bot$ is a bottom particle. So, by FACT 2.1(ii), we conclude that ($\Gamma, A, \sim A \Vdash B$), for any B. Now, to check that such a strong negation, thus defined, cannot be always a bottom particle, notice that part (b) will give us $\Vdash A$ iff $\sim A \Vdash \bot$, choosing both Γ and Δ to be empty. So, if $\sim A$ were a bottom particle, $\sim A \Vdash \bot$ would be the case, and hence any A would be a thesis of this logic, which is not the case, once we have supposed it to be non-trivial.[8] To check (iii), just note that a non-trivial ex-

[7] A strong negation should *not* be confused with a 'classical' one! Take a look at THEOREM 3.42.

[8] In the presence of a convenient implication, for instance, obeying the Deduction Metatheorem (THEOREM 3.1) such an 'implicit' definition of a strong negation from a bottom particle can be internalized by the underlying logic as an 'explicit' definition (as in the case of intuitionistic logic). Check also our remarks about this matter in our discussion of Beth Definability Property, at the end of the subsection **3.12**.

plosive logic will come already equipped with a built-in strong negation, coinciding with its own primitive negation. □

FACT 2.11 Let **L** be a logic with a strong negation ~. (i) Every theory which is contradictory with respect to ~ is explosive. (ii) A logic is contradictory with respect to ~ if, and only if, it is trivial. (D11) ⇒ (D3); (D12) ⇔ (D5)

A logic **L** is said to be *left-adjunctive* if for any two formulas A and B there is a schema $\sigma(A, B)$, depending only on A and B, with the following behavior:

$$\begin{array}{l}\text{(a) } \exists A\, \exists B \text{ such that } \sigma(A, B) \text{ is not a bottom particle, and}\\ \text{(b) } \forall A\, \forall B\, \forall\Gamma\, \forall D\, [\Gamma, A, B \Vdash D \Rightarrow \Gamma, \sigma(A, B) \Vdash D].\end{array} \qquad \text{(D13)}$$

Such a formula, when it exists, will be denoted by $(A \wedge B)$, and the sign $\wedge$ will be called a *left-adjunctive conjunction* (but it will not necessarily have, of course, all properties of a classical conjunction). Similarly, a logic **L** is said to be *left-disadjunctive* if there is a schema $\sigma(A, B)$, depending only on A and B, such that (D12) is somewhat inverted, that is:

$$\begin{array}{l}\text{(a) } \exists A\, \exists B \text{ such that } \sigma(A, B) \text{ is not a top particle, and}\\ \text{(b) } \forall A\, \forall B\, \forall\Gamma\, \forall D\, [\Gamma, \sigma(A, B) \Vdash D \Rightarrow \Gamma, A, B \Vdash D].\end{array} \qquad \text{(D14)}$$

In general, whenever there is no risk of misunderstanding, we might also denote this formula, when it exists, by $A \wedge B$, and we will accordingly call $\wedge$ a *left-disadjunctive conjunction*. Now, one should be aware of the fact that, in principle, a logic can have just one of these conjunctions, or it can have both a left-adjunctive and a left-disadjunctive conjunction without the two of them coinciding.

To convince themselves of the naturalness of these definitions and the comments we made about them, we invite the reader to consider the following two more 'concrete' properties of conjunction:

$$\begin{array}{l}\text{(a) } \exists A\, \exists B \text{ such that } A \wedge B \text{ is not a bottom particle, and}\\ \text{(b) } \forall\Gamma\, \forall A\, \forall B\, (\Gamma, A \wedge B \Vdash A \text{ and } \Gamma, A \wedge B \Vdash B).\end{array} \qquad \text{(pC1)}$$

$$\begin{array}{l}\text{(a) } \exists A\, \exists B \text{ such that } A \wedge B \text{ is not a top particle, and}\\ \text{(b) } \forall\Gamma\, \forall A\, \forall B\, (\Gamma, A, B \Vdash A \wedge B).\end{array} \qquad \text{(pC2)}$$

Now, it is easy to see that:

FACT 2.12 Let **L** be a logic obeying (Con1)–(Con3). (i) A conjunction in **L** is left-adjunctive iff it respects (pC1). (ii) A conjunction in **L** is left-disadjunctive iff it respects (pC2). [(Con1)–(Con3)] ⇒ {[(D13) ⇔ (pC1)] and [(D14) ⇔ (pC2)]}

Proof: To prove that a left-adjunctive conjunction respects (pC1) and that a left-disadjunctive conjunction respects (pC2), use (Con1) and (Con2). For the converses, use (Con3). □

The reader might mind to notice that a conjunction which is both left-adjunctive and left-disadjunctive is sometimes called, in the scope of relevance logic, an *intensional* conjunction, and in the scope of linear logic such a conjunction is said to be a *multiplicative* one (also, in [9], this is what the author calls an *internal* conjunction).

We may now check that:

FACT 2.13 Let **L** be a left-adjunctive logic. (i) If **L** either is finitely trivializable or has a strong negation, than it has a bottom particle. (ii) If **L** is finitely trivializable, then it will be supplementing explosive. (iii) If **L** respects *ex contradictione*, then it will respect *ex falso*. (D13) ⇒ {[(D7) or (D10)] ⇒ (D8)}; [(D13) and (D7)] ⇒ (sPPS); (D13) ⇒ [(PPS) ⇒ (ExF)]

Proof: To prove (i), note that if **L** has a finite trivial theory Γ, one may define a bottom particle from the conjunction of all formulas in Γ; in case it has a strong negation, any formula in the form $(A\wedge{\sim}A)$, for some formula A of **L**, will suffice. Parts (ii) and (iii) are immediate.
□

Consider now the *discussive logic* proposed by Jaśkowski in [67], **D2**, which is such that $\Gamma \vDash_{D2} A$ iff $\Diamond\Gamma \vDash_{S5} \Diamond A$, where $\Diamond\Gamma = \{\Diamond B$: for all $B \in \Gamma\}$, $\Diamond$ denotes the possibility operator, and $\vDash_{S5}$ denotes the consequence relation defined by the well-known modal logic $S5$. It is easy to see that in **D2** one has that $(A, \neg A \vDash_{D2} B)$ *does not* hold in general, though $(A\wedge\neg A) \vDash_{D2} B$ *does* hold, for any formulas A and B. This phenomenon can only happen because (pC1) holds while (pC2) does not hold in **D2**, and so its conjunction is left-adjunctive but not left-disadjunctive, while $(A\wedge\neg A)$ defines a bottom particle. Hence, the fact above still holds for **D2**, and this logic indeed displays a quite immediate example of a logic respecting (ExF) but not (PPS).

To sum up with the latest definitions and their consequences, we can picture the situation as follows, for some given logic **L**:

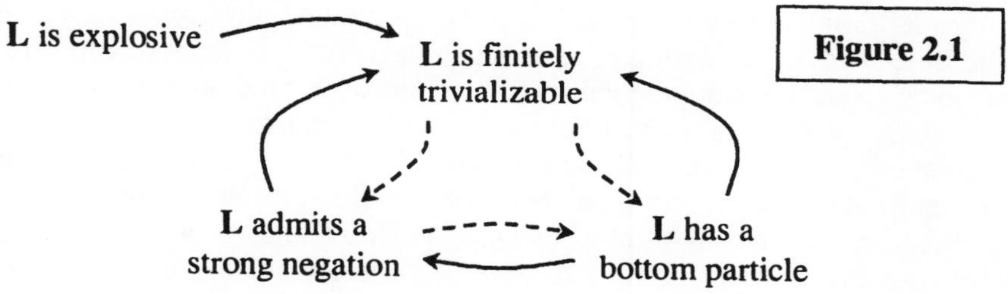

Figure 2.1

Where:

× ⟶ ✓ means that × entails ✓

× ⇢ ✓ means that × plus left-adjunctiveness entails ✓

2.4 Huge tracts of the logical space. Lo and behold! If now the reader only learns that all properties mentioned in the last subsection *are* compatible with the definition of a paraconsistent logic, they are sure to obtain a wider view of the paraconsistent landscape. Indeed, general non-explosive logics, that is, logics in which not all theories are explosive, can indeed uphold the existence either of finitely trivializable theories, strong negations, or bottom particles! (A rough map of this brave new territory may be found in **Figure 2.2**.) Logics which are paraconsistent but nevertheless have some special explosive theories, such as the ones just mentioned, will constitute the focus of our attention from now on, for, as we shall argue, they may let us explore some fields into which we would not tread in the lack of those theories. Some interesting new concepts can now be studied —this is the case of the notion of *consistency* (and its dual, the notion of *inconsistency*), as we shall argue.

Consider for instance the logic *Pac*, given by the following matrices:

$\wedge$	1	½	0
1	1	½	0
½	½	½	0
0	0	0	0

$\vee$	1	½	0
1	1	1	1
½	1	½	½
0	1	½	0

$\rightarrow$	1	½	0
1	1	½	0
½	1	½	0
0	1	1	1

	$\neg$
1	0
½	½
0	1

where both 1 and ½ are distinguished values. This is the name under which this logic appeared in Avron's [8] (section 3.2.2), though it had previously appeared, for instance, in Avron's [7], under the name $RM_3^{\supset}$, and, even before than that, in Batens's [10], under the name PI^s. It is easy to see that, in such a logic, for no formula A it can be the case that $A, \neg A \vDash_{Pac} B$, for all B. So, *Pac* is a non-explosive, thus paraconsistent, logic. Conjunction, disjunction and implication in *Pac* are fairly classical connectives: in fact, the whole positive classical logic is validated by its matrices. But the negation in *Pac* is in some sense strongly non-classical in its surrounding environment, and the immediate consequence of this is that *Pac* does not have any explosive theory as the ones mentioned above. If such a three-valued logic would define a negation having all properties of classical negation, the table at the right shows how it would look. It is very easy to see that such a negation (in fact, a strong negation with all classical properties) is *not* definable in *Pac*, for any truth-function of this logic having only ½'s as input will also have ½ as output. As a consequence, *Pac* will not respect *ex falso*, having no bottom particle (being unable, thus, as we shall argue before the end of this section, to express the consistency of its formulas), and once it is evidently a left-adjunctive logic as well, it will not even be finitely trivializable at all. One could then criticize such a logic for providing a very weak interpretation for negation, once in this logic all contradictions are admissible. This has some weird consequences and is certainly too light a way of obtaining a paraconsistent logic (this is also the central point of Batens's criticism of Priest's *LP*,[9] see [13]): if some contradictions will give you trouble just assume, then, that no contradiction at all can ever really hurt your logic! Under our present point of view, proposing a logic in which no single contradiction can ever have a harmful effect on their underlying theories is quite an extremist position, and may take us too far away from any classical form of reasoning.[10]

	~
1	0
½	0
0	1

Now, if one endows the language of *Pac* either with such a strong negation or a *falsum* constant (a bottom particle), with the canonical interpretation, what will result is a well-known conservative extension of it, called $\mathbf{J}_3$, which is still paraconsistent but has all those special explosive theories neglected by *Pac*. This logic $\mathbf{J}_3$ was first introduced by D'Ottaviano and da Costa in 1970 (cf. [60]) as a 'possible solution to the problem of Jaśkowski', and reappeared quite often in the literature after that. The first presentation of $\mathbf{J}_3$ did not bring the strong negation ~ as a primitive connective, but displayed instead a primitive 'possibility connective' ∇ (see its table to the right). In [61] it was once more presented, but this time having also a sort of 'consistency connective' ∘ as primitive (table to the right), and in [44] we have explored more deeply the expressive and inferential power of this logic, and the possibility of applying it to the study of inconsistent

	∇	∘
1	1	1
½	1	0
0	0	1

[9] By the way, Priest's logic *LP* is nothing but the implicationless fragment of *Pac* (cf. [90]).

[10] This constitutes indeed the kernel of a long controversy between H. Jeffreys and K. Popper. The first author argued in 1938 that contradictions should not be reasonably supposed to imply anything else, to which the second author replied in 1940 saying that contradictions are fatal and should be avoided at all costs, to prevent science from collapsing. Jeffreys aptly reiterated, in 1942, that he was not suggesting that *all* contradictions should be tolerated, but at least *some*. Popper responded to this successively in 1943, 1959 and 1963, saying that he himself had thought about a system in which contradictory sentences were not 'embracing', that is, did not explode, but he abandoned this system because it turned out to be too weak (lacking, for instance, *modus ponens*), and he hastily concluded from that that no useful such a system could ever be attained. See more details and references about this dispute in [26], chapter VI.

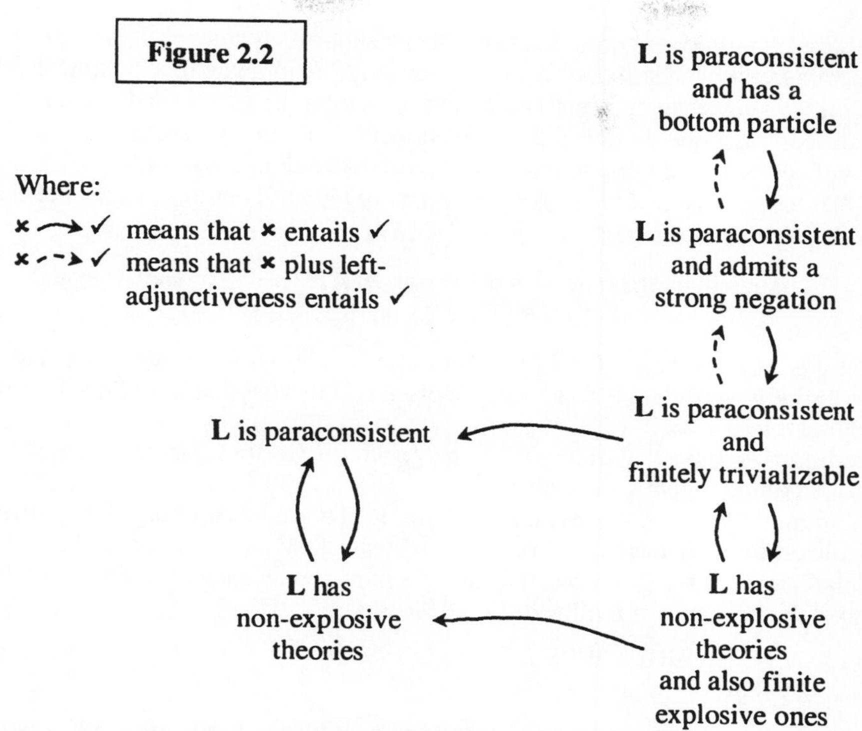

databases, abandoning ~ and ∇ but still maintaining ∘ as primitive. As a result, we have argued that this logic (now renamed **LFI1**, one of our main 'logics of formal inconsistency') has been shown to be perfectly adequate, among other options, for the task of formalizing the notion of (in)consistency in a very strong and sensible way. But we will have much more to say about this further on.

The reader could now certainly ask himself: If paraconsistency is about non-explosiveness, why are you so interested in having these special explosive theories? Because our interest lies much further than the simple control of the explosive power of contradictions —we want to be able to retain classical reasoning, if only under some suitable interpretation of a fragment of our paraconsistent logics, and we also want to use these paraconsistent logics not only to reason under conditions which do not presuppose consistency, but we want to be able to take hold of the very notion of consistency inside of our logics! From this point of view, the paraconsistent logics which shall interest us are exactly those which permit us to formalize, and get a good grip on, the intricate phenomenon of *inconsistency*, as opposed to mere cut and dried *contradictoriness*.

Whatever inconsistency might mean, by our previous analysis, we might surely suppose a trivial theory to be not only contradictory but inconsistent as well. But yet, a contradiction is certainly one of the many guises of inconsistency! So one may conjecture that *consistency* is exactly what a contradiction might be lacking to become explosive —if it was not explosive from the start. Roughly speaking, we are going to suppose that a 'consistent contradiction' is likely to explode, even if a 'regular' contradiction is not. In logics such as classical logic, consistency is well established, and indeed all theories are explosive; therefore, in any given classical theory, a contradiction turns out to be not only a necessary but also a sufficient condition for triviality.

Now, based on the above considerations, let us suppose in general that a proposition *can* be contradictory and still does not cause much harm, in general, in a paraconsistent logic, if only its consistency is not guaranteed, or cannot be established. Thus, an 'inconsistent' contradiction will be allowed to show up with no big commotion, but still a 'consistent' one should behave classically, and explode! This is how we will put it in formal terms. Let $\Delta(A)$ here denote a (possibly empty) set of schemas depending only on A. We will call a theory Γ *gently explosive* if:

(a) $\exists A$ such that $\Delta(A)\cup\{A\}$ is not trivial, $\Delta(A)\cup\{\neg A\}$ is not trivial, and
(b) $\forall A\, \forall B\, [\Gamma, \Delta(A), A, \neg A \Vdash B]$. (D15)

The gently explosive theory Γ will be said to be finitely so when $\Delta(A)$ is a finite set, so that a finitely gently explosive theory will be simply one that is finitely trivialized in a very distinctive way. (D16)

Accordingly, a logic **L** will be said to be *[finitely] gently explosive* when all of its theories are [finitely] gently explosive. [(D17)] (D18)

Thus, in any such a gently explosive logic, given a contradictory theory there is always something 'reasonable' —to wit, consistency— which one can add to it in order to guarantee that it will become trivial. We may now consider the following gentle versions of the Principle of Explosion:

[FINITE] GENTLE PRINCIPLE OF EXPLOSION [(fgPPS)] (gPPS)
L must be [finitely] gently explosive.

So, according to the interpretation proposed above, what we are implicitly assuming in the above principles is that, for any given formula A, the (finite) set $\Delta(A)$ will express, in a certain sense, the *consistency* of A relative to the logic **L**.

Based on that, we may define the consistency of a logic in the following way. **L** will be said to be *consistent* if:

(a) **L** is gently explosive, and (b) $\forall A\, \forall\Gamma\, (\forall B\in \Delta(A))(\Gamma \Vdash B)$. (D19)

It immediately follows, from these definitions and the preceding ones, that:

FACT 2.14 (i) Any non-trivial explosive theory / logic is finitely gently explosive. (ii) Any transitive logic is consistent if, and only if, it is both explosive and non-trivial. (iii) Any transitive consistent logic is finitely gently explosive. (iv) Any left-adjunctive finitely gently explosive logic is supplementing explosive.

[not-(D2) and (D3)] $\Rightarrow$ (D16); [(PNT) and (PPS)] $\Rightarrow$ (fgPPS);
(Con3) $\Rightarrow$ {(D19) $\Leftrightarrow$ [(D6) and not-(D5)]};
[(Con3) and (D19)] $\Rightarrow$ (D17);
[(D13) and (fgPPS)] $\Rightarrow$ (sPPS)

Proof: To check (i), just let $\Delta(A)$ be empty, for every formula A. This result evidently parallels FACT 2.10(iii), about supplementing explosive logics. To see, in (ii), that any given consistent logic is explosive use transitivity whenever you meet a non-empty Δ. Part (iii) follows from (i) and (ii), and part (iv) simply reflects FACT 2.13(ii). □

So, based on the above definition of a consistent logic and the subsequent fact, if we were to define a so-called *Principle of Consistency*, it would then simply coincide with the sum of (PNT) and (PPS), for logics obeying transitivity. We shall, therefore, not insist in explicitly formulating here such a principle.

We may now finally define what we will mean by a *logic of formal inconsistency* (**LFI**), which will be nothing more than a logic that allows us to 'talk about consistency' in a meaningful way. We will consider, of course, an *inconsistent* logic to be simply one that is not consistent. This assumption, together with FACT 2.14(ii), explains why paraconsistent logics were early dubbed, by da Costa, 'inconsistent formal systems', once all paraconsistent logics are certainly inconsistent in the sense of not respecting (D19), even though they are always also non-trivial and quite often they are non-contradictory as well. Those inconsistent logics which went so far as to be trivial, and thus no more paraconsistent at all, were dubbed, by Miró Quesada, *absolutely inconsistent* logics (cf. [80]). Now, an **LFI** will be any non-trivial logic in which consistency does not hold, but can still be expressed, thus being a gently explosive and yet non-explosive logic, that is, a logic in which:

(a) (PPS) does not hold, but (b) (gPPS) holds. (D20)

Classical logic, then, will not be an **LFI** just because (PPS) holds in it. *Pac* will also not be an **LFI**, even though it is paraconsistent, for *Pac* is not finitely trivializable. But D'Ottaviano & da Costa's $\mathbf{J}_3$ (and, consequently, our **LFI1**), which conservatively extends *Pac*, will *indeed* be an **LFI**, where consistency is expressed by the connective $\circ$ (see above), and inconsistency, as usual, is expressed, by the negation of this connective. Also, Jaśkowski's **D2** will constitute an **LFI**, as the reader can easily check, where the consistency of a formula A can be expressed by the formula $(\Box A \vee \Box\neg A)$, written in terms of the necessity operator $\Box$ of *S5*.[11]

'Only' **LFI**s —though these seem to comprise by far the *great majority* of all known paraconsistent logics— will interest us in this study.

2.5 DEFCON 2: one step short of trivialization. The distinction between the original formulation of explosiveness, its formulation in terms of *ex falso*, and its supplementing and gentle formulations offered above does not tell you everything you need to know about the ways of exploding. Indeed, there are more things in the realm of explosiveness, dear reader, than are dreamt of in your philosophy! Thus, for instance, a not very interesting scenario seems to unfold if contradictions are still prevented from rendering a given theory trivial but nevertheless are allowed to go half the way, causing some kind of 'partial trivialization'. So, a theory Γ will be said to be *partially trivial with respect to* a given schema $\sigma(C_1, \ldots, C_n)$, or σ*-partially trivial*, if:

(a) $\exists C_1 \ldots \exists C_n$ such that $\sigma(C_1, \ldots, C_n)$ is not a top particle, and
(b) $\forall C_1 \ldots \forall C_n\ [\Gamma \Vdash \sigma(C_1, \ldots, C_n)]$. (D21)

Following this same path, a theory Γ will be said to be *partially explosive with respect to* the schema $\sigma(C_1, \ldots, C_n)$, or σ*-partially explosive*, if:

(a) $\exists C_1 \ldots \exists C_n$ such that $\sigma(C_1, \ldots, C_n)$ is not a top particle, and
(b) $\forall C_1 \ldots \forall C_n \forall A\ [\Gamma, A, \neg A \Vdash \sigma(C_1, \ldots, C_n)]$. (D22)

Of course, a logic **L** will be said to be σ-partially trivial / σ-partially explosive if all of its theories are σ-partially trivial / σ-partially explosive. respect. (D23), (D24)

More simply, a theory, or a logic, can now be said to be *partially trivial / partially explosive* if this theory, or logic, is σ-partially trivial / σ-partially explosive, for some

[11] This needs to be qualified. Among the various formulations among which **D2** has appeared in the literature, it is not completely clear if its language has a necessity operator available so as to make this definition possible, or not. If this is not available, it may well be that **D2** is not characterizable as an **LFI** after all (even though a situation for a necessity operator would quite naturally appear, to all practical purposes, in the trivial case in which there is just one person 'discussing', or even more unlikely, a situation in which all contenders just agree with each other).

schema σ. We can now immediately formulate the following new version of the Principle of Explosion:

PRINCIPLE OF PARTIAL EXPLOSION (pPPS)
L must be partially explosive.

One may immediately conclude that:

FACT 2.15 (i) Any partially trivial theory / logic is partially explosive. (ii) Any explosive logic is partially explosive. (D21) ⇒ (D22); (D23) ⇒ (D24); (PPS) ⇒ (pPPS)

A well-known example of a logic which is not explosive but is partially explosive even so, is given by Kolmogorov & Johánsson's Minimal Intuitionistic Logic, **MIL**, which is obtained by the addition to the positive part of intuitionistic logic of some forms of *reductio ad absurdum* (cf. [68] and [69]). What happens, in this logic, is that $\forall\Gamma\,\forall A\,\forall B\,(\Gamma, A, \neg A \Vdash B)$ is *not* the case, but still it *does* hold that $\forall\Gamma\,\forall A\,\forall B\,(\Gamma, A, \neg A \Vdash \neg B)$. This means that **MIL** is paraconsistent in a broad sense, for contradictions do not explode, but still all *negated* propositions can be inferred from any given contradiction!

It is something of a consensus that an interesting paraconsistent logic should not only avoid triviality but also partial triviality. Thus, the following definition now comes in handy. A logic **L** will be said to be *boldly paraconsistent* if:

(pPPS) fails for **L**. (BPL)

Evidently:

FACT 2.16 A boldly paraconsistent logic is paraconsistent. (BPL) ⇒ (PL2)

Now, let's tackle a somewhat inverse approach. Call a theory Γ *controllably explosive in contact with* a given schema $\sigma(C_1, \ldots, C_m)$ if:

(a) $\exists C_1 \ldots \exists C_m$ such that $\sigma(C_1, \ldots, C_m)$ and $\neg\sigma(C_1, \ldots, C_m)$ are not bottom particles, and (b) $\forall C_1 \ldots \forall C_n \forall B\,[\Gamma, \sigma(C_1, \ldots, C_m), \neg\sigma(C_1, \ldots, C_m) \Vdash B]$. (D25)

Accordingly, a logic **L** will be said to be *controllably explosive in contact with* $\sigma(C_1, \ldots, C_m)$ when all of its theories are controllably explosive in contact with this schema. (D26)

Some given theory / logic can now more simply be called *controllably explosive* when this theory / logic has some schema in contact with which it is controllably explosive. An immediate new version of the Principle of Explosion that suggests itself then is:

CONTROLLABLE PRINCIPLE OF EXPLOSION (cPPS)
L must be controllably explosive.

Similarly to the case of FACT 2.14, parts (i) and (iii), it follows here that:

FACT 2.17 (i) Any non-trivial explosive theory / logic is controllably explosive. (ii) Any transitive consistent logic is controllably explosive. [not-(D2) and (D3)] ⇒ (D25)
[(PNT) and (PPS)] ⇒ (cPPS); [(Con3) and (D19)] ⇒ (D26)

By the way, we may also now emend FACT 2.9 so as to immediately conclude that:

FACT 2.18 Any finitely-gently / controllably explosive logic is finitely trivializable, and yet non-trivial. [(D17) or (D26)] ⇒ [(D7) and not-(D5)]

This fact can be used to update and complement the information conveyed in **Figure 2.1**.

Now, there seems to be no good reason to rule out controllably explosive theories, as we did in the case of partially explosive theories by way of the bold definition of paraconsistency, (BPL). In fact, it seems that most, if not all, finitely gently explosive logics *are* controllably explosive, and vice-versa! We will see, later on, many examples of paraconsistent logics —indeed, of **LFIs**— which not only are obviously gently explosive, but are also controllably explosive in contact with schemas such as $(A \wedge \neg A)$, or such as $\circ A$, where $\circ$, we recall, is a connective expressing consistency (Jaśkowski's **D2**, for instance, may already be one of these, but the logic **LFI1**, on the other hand, explodes only in contact with the second of these schemas). There are even logics which controllably explode in contact with large classes of non-atomic propositions (see [78], and ahead, for a number of them). An extreme case of these, as we shall see, is given by Sette's three-valued logic $\mathbf{P}^1$ (cf. [103]), which controllably explodes in contact with *any* complex formula, and so can be said to behave paraconsistently only at the level of its atoms. It is also not uncommon for some paraconsistent logic **L** having a strong negation $\div$ to be controllably explosive. In fact, it suffices that such a logic is transitive and infers $\neg\div A \Vdash A$, and of course it will turn out to be controllably explosive in contact with $\div A$, or at least in contact with $\div\div A$ (see, for instance, FACT 3.76, or THEOREM 3.51(i) and FACT 3.66). Many **LFIs** will moreover be controllably explosive in contact with any consistent formula (see FACT 3.32). And so on, and so forth.

A range of variations on the above versions of the Principle of Explosion can be obtained if we only mix the ones we already have. We shall nevertheless not investigate this theme here any further, but only notice that the multiple relations, hinted above, between (sPPS), (gPPS) and (cPPS), the supplementing, the gentle and the controllable forms of explosion, certainly deserve a closer and more attentive look by the 'paraconsistent community' and sympathizers.

2.6 C-systems. Given a logic $\mathbf{L}=\langle For, \Vdash\rangle$, let $For^+ \subseteq For$ denote the set of all *positive formulas* of **L**, that is, the *negationless* fragment of *For*, or, in still other words, the set of all formulas in which no negation symbol $\neg$ occurs. The logic $\mathbf{L1}=\langle For_1, \Vdash_1\rangle$ is said to be *positively preserving relative to* the logic $\mathbf{L2}=\langle For_2, \Vdash_2\rangle$ if:

$$\text{(a) } For_1^+ = For_2^+, \text{ and (b) } (\Gamma \Vdash_1 A \;\Leftrightarrow\; \Gamma \Vdash_2 A), \text{ for all } \Gamma \cup \{A\} \subseteq For_1^+. \qquad \text{(D27)}$$

So, if **L1** is positively preserving relative to **L2**, then it will in general be a conservative extension of the positive fragment of **L2**. Now, as an example of the ubiquity of **LFIs** inside the realm of paraconsistent logics, just notice that:

FACT 2.19 Any paraconsistent logic that is positively preserving relative to classical logic and has a bottom particle can be characterized as an **LFI**.

Proof: Just define $\circ A$ as $(A \to \bot) \vee (\neg A \to \bot)$, and check that, in general, $\circ A$ is not a top particle, $\{\circ A, A\}$ is not always trivial, and $\{\circ A, \neg A\}$ is not always trivial, but that, in any case, $\{\circ A, A, \neg A\}$ is indeed a trivial theory. This result actually holds for any logic having a *left-adjunctive disjunction*, that is, a binary connective $\vee$ such that $(B \vee C)$ is not a bottom particle, for some formulas B and C, and such that $\forall B \forall C \forall \Gamma \forall \Delta \forall D\ \{(\Gamma, B \Vdash D) \text{ and } (\Delta, C \Vdash D)] \Rightarrow [\Gamma, \Delta, (B \vee C) \Vdash D]\}$ (for a particular consequence of this feature, see FACT 3.7), and having *modus ponens*: $\forall \Gamma \forall A \forall B\ [\Gamma, A, (A \to B) \vdash B]$. You just have to choose $\Gamma = \{A\}$, $B = (A \to \bot)$, $\Delta = \{\neg A\}$, and $C = (\neg A \to \bot)$, and notice that, in this case, both $(\Gamma, B \Vdash \bot)$ and $(\Delta, C \Vdash \bot)$, by *modus ponens*.

□

This last result shows that any paraconsistent logic conservatively extending the positive classical logic and respecting either one of the principles of *ex falso* or of supplementing explosion will be finitely gently explosive as well, throwing some light on some hitherto unsuspected connections between (ExF), (sPPS) and (fgPPS), and consequently any such a logic can be easily recast as an **LFI** (take another look at **Figure 2.2**). Consequently, for all such logics, it amounts to be more or less the same starting either with a consistency operator, or with a strong negation, or with a bottom particle: each of these can be used to define the others. This does not mean, however, that 'only' such logics are **LFI**s (see the case of C_{Lim}, in the subsection **3.10**).

To specialize a little bit from this very broad definition of **LFI**s above we will now define the concept of a **C**-system. The logic **L**1 will be said to be a **C**-*system based on* **L**2 if:

(a) **L**1 is an **LFI** in which consistency or inconsistency
are expressed by operators (at the object language level),
(b) **L**2 is not paraconsistent, and
(c) **L**1 is positively preserving relative to **L**2. (D28)

Any logic constructed as a **C**-system based on some other logic will more generally be identified simply as a **C**-*system*. In the next section we will study various logics which are **C**-systems, and pinpoint some which are not.

Jaśkowski's **D2**, as we have already seen in the above subsections, *is* an **LFI** and *can* define an operator expressing consistency —at least under some presentations (see note 11). But, in order for it to be characterized as a **C**-system it would still have to be clarified on which logic it is based, that is, where does its peculiar positive (non-adjunctive) part come from! This same question arises with respect to all other logics that are left-adjunctive but not left-disadjunctive, as well as with respect to many relevance logics.

All **C**-systems we will be studying below are inconsistent, non-contradictory and non-trivial. Furthermore, they are boldly paraconsistent (though the proof of *this* fact will be left for [42]), and often controllably explosive as well, they have strong negations and bottom particles, and are positively preserving relative to classical propositional logic —so, that they will respect (PNC), (PNT), (ExF), (sPPS), (gPPS) and often (cPPS), but they will not respect neither (PPS) nor (pPPS). Let's now jump to them.

3 COOKING THE C-SYSTEMS ON A LOW FLAME

> Indeed, even at this stage I predict a time when there will be mathematical investigations of calculi containing contradictions, and people will actually be proud of having emancipated themselves even from consistency.
> —Wittgenstein, Philosophical Remarks, p.332.

Underlying the original approach of da Costa to the concoction of a propositional calculus capable of admitting contradictions, yet remaining sensible to performing reasonable deductions, laid the idea of maintaining the positive fragment of classical logic unaltered. This explains why his approach to paraconsistency has eventually received the inelegant label of 'positive (logic) plus approach' and, more recently, the not much descriptive (and in some cases plainly inadequate) label of 'non-truth-functional approach' (cf., respectively, [92] and [94]). Surely, competitive approaches do exist, like the one stemming from Jaśkowski's or Rescher & Brandom's investigations, which rejects left-disadjunction, and is usually referred to as a 'non-adjunctive approach' (cf. [67] and [98]), and which has more recently been tentatively dubbed, by J. Perzanowski, as 'parainconsistent logic'.[12] Another megatrend comes from the 'relevance approach' to paraconsistency, captained by the American-Australian school, whose concern is not so much with negation as with implication, giving rise to 'relevance logics' (cf. [3]). Still another very interesting proposal came from Belgium, under the appellation of 'adaptive logics', which do not worry so much about proving consistency, but assume it instead from the very start, as some kind of default (cf. [11] and [12]). Now, let us make it crystal clear that our concentration in this study on the investigation of **C**-systems, born from the first approach mentioned above, wishes not to diminish the other approaches, nor affirms that they should be held as mutually exclusive. Our intention, indeed, is but to present the **C**-systems under a more general and suggestive background, and from now on we shall draw on the other approaches only when we feel that as a really necessary or instructive step. To the reader particularly interested in them, we prefer simply to redirect them to the competent sources.

3.1 Paleontology of C-systems. All definitions and remarks made above were set forth directing an arbitrary consequence relation $\Vdash$, be it syntactical, semantical or defined in any other mind-boggling way. Once the surfacing of contradictions on a theory involves negation, and nothing but that, it is appealing to consider and explore the intuitive idea that an interesting class of paraconsistent logics is to be given by the ones which are positively preserving relative to classical logic, differing from classical logic only in the behavior of formulas involving negation. This is the idea into which we will henceforth be digging, by axiomatically proposing a series of logics characterized by their syntactical consequence relations, $\vdash$, and containing all rules and schemas which hold in the positive part of classical logic. Thus, let's initially consider $\wedge$, $\vee$, $\rightarrow$, and $\neg$ to be our primitive connectives, and consider the set of formulas *For*, as usual, to be the free algebra generated by these connectives. We will start our journey from the following set of axioms:

(Min1) $\vdash_{min} (A \rightarrow (B \rightarrow A))$;
(Min2) $\vdash_{min} ((A \rightarrow B) \rightarrow ((A \rightarrow (B \rightarrow C)) \rightarrow (A \rightarrow C)))$;
(Min3) $\vdash_{min} (A \rightarrow (B \rightarrow (A \wedge B)))$;
(Min4) $\vdash_{min} ((A \wedge B) \rightarrow A)$;
(Min5) $\vdash_{min} ((A \wedge B) \rightarrow B)$;
(Min6) $\vdash_{min} (A \rightarrow (A \vee B))$;
(Min7) $\vdash_{min} (B \rightarrow (A \vee B))$;
(Min8) $\vdash_{min} ((A \rightarrow C) \rightarrow ((B \rightarrow C) \rightarrow ((A \vee B) \rightarrow C)))$;
(Min9) $\vdash_{min} (A \vee (A \rightarrow B))$;
(Min10) $\vdash_{min} (A \vee \neg A)$;
(Min11) $\vdash_{min} (\neg\neg A \rightarrow A)$.

Here, by writing $\vdash_{min} (A \rightarrow (B \rightarrow A))$ we will be abbreviatedly denoting that:

$$\forall \Gamma \forall A \forall B\, [\Gamma \vdash_{min} (A \rightarrow (B \rightarrow A))],$$

[12] On his conference delivered at the Jaśkowski's Memorial Symposium, held in Toru , Poland, July 1998.

and so on, for the other axioms. The only inference rule, as usual, will be *modus ponens*, (MP): $\forall\Gamma\forall A\,\forall B\;[\Gamma, A, (A\rightarrow B) \vdash_{min} B]$. The logic built using such axioms, plus (MP) and the usual notion of proof from premises (we may now be calling *proofs*, *theorems* and *premises* which we have previously called, respectively, inferences, theses and theories) was called $C_{min}=<For, \vdash_{min}>$ and studied by the authors in [39].

First of all, let us observe that the so-called *Deduction Metatheorem* is here valid:

THEOREM 3.1 $[\Gamma, A \vdash_{min} B \;\Rightarrow\; \Gamma \vdash_{min} (A\rightarrow B)]$.[13]

Proof: It is a familiar and straightforward procedure to show that the Deduction Metatheorem holds for any logic containing (Min1) and (Min2) as provable schemas and having only *modus ponens* as a primitive rule. □

Evidently, by monotonicity and transitivity, *modus ponens* already gives us the converse of THEOREM 3.1. This makes it possible for us to introduce all axioms as some sort of axiomatic inference rules, and this is what we shall do from now on. Moreover, using the Deduction Metatheorem and its converse, one could now equivalently represent, in C_{min}, the fact that (PPS) (the Principle of Explosion) does not hold by the unprovability of the theorem (tPS): $(A\rightarrow(\neg A\rightarrow B))$. And indeed:

THEOREM 3.2 (tPS) is not provable by C_{min}.

Proof: Use the matrices of *Pac*, in the subsection **2.4**, to check that all axioms above are validated and that (MP) preserves validity, while (tPS) is not always validated. This shows that C_{min} is a fragment of *Pac*, and so it also cannot prove (tPS). In fact, (tPS) is more than non-provable, it is *independent* from C_{min} (and *Pac*) for its negation is not even classically provable, and *Pac* is a deductive fragment of classical logic. □

As usual, *bi-implication*, $\leftrightarrow$, will be defined by setting $(A\leftrightarrow B) \stackrel{\text{def}}{=} ((A\rightarrow B)\wedge (B\rightarrow A))$. Note that, by the above considerations, $\vdash_{min} (A\leftrightarrow B)$ if, and only if, $A \vdash_{min} B$, and $B \vdash_{min} A$, which is the same as writing $A \dashv\vdash_{min} B$. So, bi-implication holds between two formulas if, and only if, they are (provably) equivalent (see (Eq1), in the subsection **2.2**). Nevertheless, as the reader shall see below, having two equivalent formulas, in the logics we will be studying here, usually does *not* mean, as in classical logic, that these formulas can be freely intersubstituted everywhere (take a look, ahead, for instance, at results 3.22, 3.35, 3.51, 3.58, 3.65, and 3.74).

Axioms (Min1)–(Min8) are known at least since Gentzen's [62] as providing an axiomatization for the so-called 'positive logic'. Of course, they immediately tell us, among other things, that the conjunction of this logic is both left-adjunctive and left-disadjunctive (just take a look at axioms (Min3)–(Min5)). Nevertheless, (Min9): $(A\vee(A\rightarrow B))$, which *is* a positive schema, is *not* provable even if one uses (Min10) and (Min11) in addition to (Min1)–(Min8) and (MP) (i.e. the logic axiomatized as C_{min} minus the axiom (Min9))! Indeed:

THEOREM 3.3 (Min9) is not provable by $C_{min}\backslash\{(\text{Min9})\}$.

Proof: Use the following matrices (cf. [2]) to check that (Min9) is independent from $C_{min}\backslash\{(\text{Min9})\}$:

---

[13] Read this kind of sentence as a universally quantified one —in this case, for example, it would be $\forall\Gamma\,\forall A\,\forall B\,(\Gamma, A \vdash_{min} B \;\Rightarrow\; \Gamma \vdash_{min} (A\rightarrow B))$

∧	1	½	0
1	1	½	0
½	½	½	0
0	0	0	0

∨	1	½	0
1	1	1	1
½	1	½	½
0	1	½	0

→	1	½	0
1	1	½	0
½	1	1	0
0	1	1	1

	¬
1	0
½	1
0	1

where 1 is the only distinguished value. □

So, what is this thing that Gentzen (and Hilbert before him) have dubbed 'positive logic', if even a deductive extension of it is unable to prove all positive theorems of classical logic? Here is the trick: Gentzen referred of course to positive *intuitionistic* logic, and not to the classical logic! So, this logic $C_{min}\backslash\{(\text{Min9})\}$, which was proposed by da Costa (cf. [49]) and called C_ω by him, turns out to be only positively preserving relative to intuitionistic logic, and not relative to classical logic. In [39] we have proven that its deductive extension C_{min}, obtained by adjoining (Min9) to C_ω, is indeed positively preserving relative to classical logic, and moreover:

THEOREM 3.4 C_{min} does have neither a strong negation nor a bottom particle, and is not finitely trivializable.
Proof: PROPOSITION 2.5, in [39], shows that C_{min} does not have a bottom particle, and so, by left-disadjunction and FACT 2.13, it does not have neither a strong negation nor is it finitely trivializable. □

Moreover, in [39] we also proved that:

THEOREM 3.5 C_{min} does not have any negated theorem, i.e. $(\nvdash_{min} \neg A)$.

Of course, both results above are valid, *a fortiori*, for C_ω. Indeed, as shown by Urbas (cf. [107]), these logics are very weak with respect to negation, so that the following holds:

THEOREM 3.6 No two different negated formulas of C_{min} are provably equivalent.
Proof: The THEOREM 2, in [107], shows that $\neg A \dashv\vdash \neg B$ is derivable in C_ω if and only if A and B are the same formula. It is straightforward to adapt this result also to C_{min}. □

Much more about the provability (or validity) of negated theorems will be seen in the paper [42], which brings semantics to most logics here studied.

THEOREM 3.4 shows that C_{min}, or C_ω, *cannot* be **C**-systems based on classical logic, or intuitionistic logic, once they are both *compact* (all proofs are finite) and not finitely gently explosive, so that they cannot be gently explosive at all, and thus cannot formalize 'consistency', in the precise sense formulated in the subsection **2.4**. We had better then make them deductively stronger in order to get what we want.

We make a few more important remarks before closing this subsection. First, note that (Min10): $(A\vee\neg A)$ was added in order to keep C_{min} and C_ω from being *paracomplete* as well as paraconsistent (let's investigate one deviancy at a time!), and this axiom can indeed be pretty useful in providing us with a form of *proof by cases*:

FACT 3.7 $[(\Gamma, A \vdash_{min} B)$ and $(\Delta, \neg A \vdash_{min} B)] \Rightarrow (\Gamma, \Delta \vdash_{min} B)$.
Proof: From (Min8) and (Min10), by *modus ponens*, monotonicity, and the Deduction Metatheorem (from now on, we will not mention these last three every time we use them anymore). □

It will also be practical here and there to use $[(A \to B), (B \to C) \vdash_{min} (A \to C)]$ (a kind of logical version for the transitivity property) as an alternative form of the axiom (Min2). Indeed:

FACT 3.8 (Min2) can be substituted, in C_{min}, by $[(A \to B), (B \to C) \vdash_{min} (A \to C)]$.

We shall often make use of both these forms without discriminating which.

In the next subsection (see THEOREM 3.13) we will learn about the utility of (Min11): $(\neg\neg A \to A)$, which was added by da Costa as a way of rendering the negation of his calculi a bit stronger, using as an argument the intended duality with the logics arising from the formalization of intuitionistic logic, in which usually only the converse of (Min11), i.e. the formula $(A \to \neg\neg A)$, is valid.

It is quite interesting as well to notice that the addition of the 'Theorem of Pseudo-Scotus', (tPS), to C_{min} as a new axiom schema will not only prevent the resulting logic from being paraconsistent, but it will also provide a complete axiomatization for the *classical propositional logic* (hereby denoted **CPL**). In fact, it is a well-known fact that:

THEOREM 3.9 Axioms (Min1)–(Min11) plus (tPS): $(A \to (\neg A \to B))$, and (MP), provide a sound and complete axiomatization for **CPL**.

Actually, the axiom (Min11) can be discharged from the above axiomatization, being proved from the remaining ones. Axiom (Min9) also turns to be redundant (take a look at the FACT 3.45, below).

3.2 The basic logic of (in)consistency. Let's consider an extension of our language by the addition of a new unary connective, $\circ$, representing *consistency*. Let's now also add, to C_{min}, a new rule, realizing the Finite Gentle Principle of Explosion:

(bc1) $\circ A, A, \neg A \vdash_{bC} B$. 'If A is consistent and contradictory, then it explodes'

We will call this new logic, characterized by axioms (Min1)–(Min11) and (bc1), plus (MP), the *basic logic of (in)consistency*, or **bC**. Clearly, thanks to (bc1), we know that **bC** is indeed an **LFI**, i.e. a *logic of formal inconsistency*, and so it is in fact a **C**-*system* based on **CPL**. A strong negation, ~, for a formula A can now be easily defined by setting $\sim A \overset{def}{=} (\neg A \wedge \circ A)$, and evidently we will have $[A, \sim A \vdash_{bC} B]$, as expected. A *bottom particle*, of course, is given by $(A \wedge \sim A)$, for any A. For alternative ways of formulating **bC**, consider FACT 2.19 and the comments which follow it.

We can already show that THEOREMS 3.5 and 3.6 do *not* hold for **bC**:

THEOREM 3.10 **bC** does have negated theorems, and equivalent negated formulas (but, on the other hand, it has no consistent theorems, that is, theorems of the form $\circ A$).

Proof: Consider any bottom particle $\bot$ of **bC**. By definition, it must be such that $(\bot \vdash_{bC} B)$, for any formula B, and so, in particular, $(\bot \vdash_{bC} \neg\bot)$. But we also have that $(\neg\bot \vdash_{bC} \neg\bot)$, and proof by cases (FACT 3.7) tells us then that $\vdash_{bC} \neg\bot$. By the way, this result also transforms THEOREM 3.4 into a corollary of THEOREM 3.5 —if a reflexive logic has proof by cases and no negated theorems, then it cannot contain a bottom particle. Evidently, any bottom particle is equivalent to any other. To check that no formula of the form $\circ A$ is provable, one may just use the classical matrices for $\wedge$, $\vee$, $\to$ and $\neg$, and pick for $\circ$ a matrix with value constant and equal to 0. □

Now, it is easy to see that, in such logic **bC**, if the consistency of the right formulas is guaranteed, than its inferences will behave exactly like in **CPL**. Indeed:

THEOREM 3.11 $[\Gamma \vdash_{\mathbf{CPL}} A] \Leftrightarrow [\circ(\Delta), \Gamma \vdash_{\mathbf{bC}} A]$, where $\circ(\Delta) = \{\circ B : B \in \Delta\}$, and Δ is a finite set of formulas.

Proof: On the one hand, one may just reproduce line by line a **CPL** proof in **bC**, and when it comes to an application of (tPS) —see an axiomatization of **CPL** in the THEOREM 3.9— one will have to use (bc1) instead, and add as a further assumption the consistency of the formula in the antecedent. The converse is immediate. □

We know that **bC** is a both a linguistic and a deductive extension of C_{min}, once it not only introduces a new connective but has an axiomatic rule telling you what to do with it. But we know more than that:

THEOREM 3.12 **bC** is a conservative extension of C_{min}.

Proof: Indeed, if you consider the **bC**-inferences in the language of C_{min}, you can no more use (bc1) along a proof, and so you can prove nothing more than you could prove before. □

What we have then, in (bc1), is a sort of rough logical clone for the finite gentle rule of explosion. Now, da Costa, in the original presentation of his calculi, which guides us here, has never used a gentle form of explosion but used instead a gentle form of *reductio ad absurdum*:

(RA0) $\circ B,\ (A \to B),\ (A \to \neg B) \vdash \neg A$.

'If supposing A will bring us to a consistent contradiction, then $\neg A$ should be the case'

Notice, by the way, that $((A \to B) \to ((A \to \neg B) \to \neg A))$ was exactly the form of *reductio* used by Kolmogorov and Johánsson in the proposal of their Minimal Intuitionistic Logic, mentioned above as an example of a logic which is paraconsistent and yet not boldly paraconsistent. Now, the reader might be suspecting that it would really make no difference whether we used (bc1) or (RA0) in the characterization of **bC**. They are right, but this assertion could be made more precise. Indeed, consider the two following alternative versions of these rules:

(bc0) $\circ A,\ A,\ \neg A \vdash \neg B$;

'If A is consistent and contradictory, then it partially explodes with respect to negated propositions'

(RA1) $\circ B,\ (\neg A \to B),\ (\neg A \to \neg B) \vdash A$,

'If supposing $\neg A$ will bring us to a consistent contradiction, then A should be the case'

and consider the logic *PI* (that is how it was called when it appeared in [10]), characterized simply by (Min1)–(Min10) plus (MP), that is, C_{min} deprived of the schema (Min11): $(\neg\neg A \to A)$. Then we can prove that:

THEOREM 3.13 (i) It does *not* have the same effect adding either (bc1) or (RA0) to *PI*. (ii) It *does* have the same effect adding to *PI*: a) (bc0) or (RA0); b) (bc1) or (RA1). (iii) It *does* have the same effect adding to C_{min} whichever of the schemas (bc0), (bc1), (RA0) or (RA1). (iv) **bC** cannot be extended into a $\neg$-partially explosive paraconsistent logic.

Proof: To check part (i), use the classical matrices (with values 1 and 0) for $\wedge$, $\vee$ and $\to$, but let both $\neg$ and $\circ$ have matrices constant and equal to 1 —this way you will see that (bc1) is not provable by the logic obtained from the addition of (RA0) to *PI*. Part (ii) is easy: use FACT 3.8 to prove (bc0) in *PI* plus (RA0), and to prove (bc1) in *PI* plus (RA1); use (Min1) and the proof by cases to prove (RA0) in *PI* plus (bc0), and to prove (RA1) in *PI* plus (bc1). We leave part (iii) as an even easier exercise to the reader (*hint*: use (Min11)). (iv) is an immediate consequence of (iii). □

So, this last result gives one reason for us to have our study started from C_{min} rather than from *PI*: we will be avoiding that paraconsistent extensions of our initial logic might turn out to be partially explosive with respect to negated propositions in general, as what occurred with **MIL**, the Minimal Intuitionistic Logic (recall the subsection **2.5**). This feature will help in making many results below more symmetrical. But, to be sure, this does not guarantee that all such extensions will be boldly paraconsistent as well!

The reader should notice that there are, however, some restricted forms of 'reasoning by absurdum' left in **bC**. For example:

FACT 3.14 The following *reductio* deduction rules hold in **bC**:
(i) $[(\Gamma \vdash_{bC} \circ A)$ and $(\Delta, B \vdash_{bC} A)$ and $(\Lambda, B \vdash_{bC} \neg A)] \Rightarrow (\Gamma, \Delta, \Lambda \vdash_{bC} \neg B)$;
(ii) $[(\Gamma, B \vdash_{bC} \circ A)$ and $(\Delta, B \vdash_{bC} A)$ and $(\Lambda, B \vdash_{bC} \neg A)] \Rightarrow (\Gamma, \Delta, \Lambda \vdash_{bC} \neg B)$;
(iii) $[(\Gamma, \neg B \vdash_{bC} \circ A)$ and $(\Delta, \neg B \vdash_{bC} A)$ and $(\Lambda, \neg B \vdash_{bC} \neg A)] \Rightarrow (\Gamma, \Delta, \Lambda \vdash_{bC} B)$.
Proof: Part (i) comes immediately from (RA0), part (ii) comes from part (i) using reflexivity and proof by cases, part (iii) comes as a variation of (ii), if you use (Min11). □

But we still have not mentioned some of the most decisive features of **bC**! We are now ready for this. Consider, to start with, the following result:

THEOREM 3.15 (i) $(A \wedge \neg A)$ is not a bottom particle in any paraconsistent extension of **bC**. (ii) $\neg(A \wedge \neg A)$ and $\neg(\neg A \wedge A)$ are not top particles in **bC**.
Proof: For part (i), just use left-disadjunction and THEOREM 3.2 (but the reader might recall from the subsection **2.3** that this formula *is* a bottom particle in some non-left-disadjunctive paraconsistent logics such as Jaśkowski's **D2**). To check part (ii) use the following matrices to confirm that neither $\neg(A \wedge \neg A)$ nor $\neg(\neg A \wedge A)$ are provable by **bC**:

∧	**1**	**½**	**0**
1	1	1	0
½	1	1	0
0	0	0	0

∨	**1**	**½**	**0**
1	1	1	1
½	1	1	1
0	1	1	0

→	**1**	**½**	**0**
1	1	1	0
½	1	1	0
0	1	1	1

	¬	∘
1	0	1
½	1	0
0	1	1

where 1 and ½ are the distinguished values. By the way, the matrices of ∧, ∨, →, and ¬ are exactly the same matrices which originally defined the maximal three-valued logic $\mathbf{P}^1$, proposed in [103], and mentioned in the subsection **2.5** as a logic which is paraconsistent and yet controllably explosive when in contact with any non-atomic formula. □

As to the relations between contradictions and inconsistencies what we will find here are some variations on the intuitive idea that a contradiction should not be consistent (but not necessarily the other way around):

FACT 3.16 These are some special rules of **bC**, relating contradiction and consistency:
(i) $A, \neg A \vdash_{bC} \neg\circ A$;
(ii) $(A \wedge \neg A) \vdash_{bC} \neg\circ A$;
(iii) $\circ A \vdash_{bC} \neg(A \wedge \neg A)$;
(iv) $\circ A \vdash_{bC} \neg(\neg A \wedge A)$.
The converses of these rules do *not* hold in **bC**.

Proof: Use FACT 3.14 to prove (i), and left-adjunction to jump from this fact to (ii); play similarly to prove (iii) and (iv). To show that none of the converses of (ii)– (iv) are provable by **bC**, use the same matrices as in THEOREM 3.15(ii), substituting only the matrix for negation by this one to the right. □

	$\neg$
1	0
½	0
0	1

The significance of stating both (iii) and (iv) is to draw attention to the fact that, in what follows, logics will be shown in which, due to some unexpected asymmetry, only one of their converses hold. This is the case, for instance, for C_1, the first logic of the pioneering hierarchy of paraconsistent logics, C_n, $1 \leq n < \omega$, proposed by da Costa (cf. [49] or [50]). As we shall see, the converse of (iii) holds in C_1, while the converse of (iv) fails, so that $\neg(A \wedge \neg A)$ and $\neg(\neg A \wedge A)$ are *not* equivalent formulas in this logic (in this respect, see also THEOREM 3.21(iii)).

As the reader will learn in the next subsection (THEOREM 3.20), the regular forms of 'reasoning by contraposition' cannot be valid in any logic which is, as **bC** and its extensions (cf. THEOREM 3.13(iv)), both positively preserving with respect to classical logic and not partially explosive with respect to negation. But there are some restricted forms of it that hold already in **bC**:

FACT 3.17 These are some restricted forms of contraposition that hold in **bC**:

(i) $\circ B, (A \to B) \vdash_{\mathbf{bC}} (\neg B \to \neg A)$;
(ii) $\circ B, (A \to \neg B) \vdash_{\mathbf{bC}} (B \to \neg A)$;
(iii) $\circ B, (\neg A \to B) \vdash_{\mathbf{bC}} (\neg B \to A)$;
(iv) $\circ B, (\neg A \to \neg B) \vdash_{\mathbf{bC}} (B \to A)$.

Proof: To check (i), let $\Gamma = \Delta = \Lambda = \{\circ B, (A \to B), \neg B\}$ and apply FACT 3.14(ii) to $\Gamma \cup \{A\}$, so as to obtain $\Gamma \vdash_{\mathbf{bC}} \neg A$. From this it follows that $[\circ B, (A \to B) \vdash_{\mathbf{bC}} (\neg B \to \neg A)]$. Part (ii) is similar to (i). For parts (iii) and (iv) apply FACT 3.14(iii). □

Now, may the reader be aware that rules such as $[\circ A, (A \to B) \vdash_{\mathbf{bC}} (\neg B \to \neg A)]$ *do not* hold in this logic!

3.3 On what one cannot get. If 'logic is about trade-offs', as Patrick Blackburn likes to put it, let us now start counting the dead bodies to see what we have irremediably lost, up to now. The connectives $\wedge$, $\vee$ and $\to$ of **bC**, for example, show up as quite independent from one another, and cannot be interdefined as in the classical case:

THEOREM 3.18 The following rule holds in **bC**:

(i) $(\neg A \to B) \vdash_{\mathbf{bC}} (A \vee B)$,

but none of the following rules hold in **bC**:

(ii) $(A \vee B) \vdash_{\mathbf{bC}} (\neg A \to B)$;
(iii) $\neg(\neg A \to B) \vdash_{\mathbf{bC}} \neg(A \vee B)$;
(iv) $\neg(A \vee B) \vdash_{\mathbf{bC}} \neg(\neg A \to B)$;
(v) $(A \to B) \vdash_{\mathbf{bC}} \neg(A \wedge \neg B)$;
(vi) $\neg(A \wedge \neg B) \vdash_{\mathbf{bC}} (A \to B)$;
(vii) $\neg(A \to B) \vdash_{\mathbf{bC}} (A \wedge \neg B)$;
(viii) $(A \wedge \neg B) \vdash_{\mathbf{bC}} \neg(A \to B)$;
(ix) $\neg(A \wedge B) \vdash_{\mathbf{bC}} (\neg A \vee \neg B)$;
(x) $(\neg A \vee \neg B) \vdash_{\mathbf{bC}} \neg(A \wedge B)$;
(xi) $\neg(\neg A \vee \neg B) \vdash_{\mathbf{bC}} (A \wedge B)$;
(xii) $(A \wedge B) \vdash_{\mathbf{bC}} \neg(\neg A \vee \neg B)$.

Proof: This is much easier to directly check after you take a look at the semantics and decision procedure of **bC**, in the paper [42]. But it also comes as a consequence from the fact that this is already valid for C_{min}, as we have proved in [39], and that **bC** is a conservative extension of it (THEOREM 3.12). □

Notice that any uniform substitution of a component formula C for its negation $\neg C$, or vice-versa, will not alter the fact that the above rules hold or not in **bC**. That is to say, for instance, that $(A \to \neg B) \vdash_{bC} (\neg A \vee \neg B)$ does hold but $(\neg A \vee B) \vdash_{bC} (A \to B)$ does not. Of course, the failure of a rule such as $(A \vee \neg B) \vdash_{bC} \neg(A \wedge \neg B)$ was already to be expected from the fact that $(A \vee \neg A)$ is provable (it is (Min10)) but $\neg(A \wedge \neg A)$ is not (see THEOREM 3.15(ii)).

Now, it should be crystal-clear that the above fact is only about **bC**, and that it does not necessarily carry on to stronger logics. In fact, it is not hard at all to check, for instance, that the three-valued maximal logic **LFI1**, whose matrices were presented in the subsection **2.4**, both extends **bC** and validates all the rules above, except for (ii) and (vi). Once more, the non-validity of (vi) is barely circumstantial, for there are logics extending **bC** in which it holds, such as the above mentioned $\mathbf{P}^1$ (see also the results 3.68 and 3.70, below). Still and all, there *is* a very good reason for the failure of (ii)! Indeed, this is a consequence of the following fact:

THEOREM 3.19 The *disjunctive syllogism*, $[A, (\neg A \vee B) \vdash B]$, cannot hold in any paraconsistent extension of positive (classical or intuitionistic) logic.

Proof: Assume that it held. From (Min6), we would have that $[\neg A \vdash (\neg A \vee B)]$ and so, ultimately, we would conclude, by the transitivity of $\vdash$, that $[A, \neg A \vdash B]$. □

Finally, as we have already advanced above, 'full' contraposition is lost (cf. [54]):

THEOREM 3.20 The regular forms of *contraposition*, such as $[(A \to B) \vdash (\neg B \to \neg A)]$, cannot hold irrestrictedly in any paraconsistent extension of **bC**. Furthermore, they cannot hold in any extension of the positive classical logic which happens to be not $\neg$-partially explosive.

Proof: If the above rule held in a logic **L** that extends the positive classical logic, from (Min1) we would obtain $[B \vdash (A \to B)]$, and from (MP) we obtain $[(A \to B), \neg B \vdash \neg A]$. These two rules would ultimately lead to $[B, \neg B \vdash \neg A]$, and so **L** would be partially explosive with respect to negated propositions. If we assume **L** to be **bC**, then a particular case of $[B, \neg B \vdash \neg A]$ would be $[B, \neg B \vdash \neg\neg C]$, taking A as $\neg C$, and (Min11) would then give $[B, \neg B \vdash C]$, and so it would not be paraconsistent at all. Indeed, this addition of contraposition to **bC** would simply cause the collapse of the resulting logic into classical logic (by THEOREM 3.9). Still some other forms of this contraposition rule, such as $[(\neg A \to B) \vdash (\neg B \to A)]$, could be ruled out even without recurring to (Min11), or to partial explosion. □

The use of the disjunctive syllogism (THEOREM 3.19) constitutes indeed the kernel of the well-known argument laid down by C. I. Lewis for the derivation of (PPS) in classical logic (cf. [73], pp.250ff), and this was, in fact, a rediscovery of an argument used by the Pseudo-Scotus, much before.[14] The use of contraposition (THEOREM 3.20) to the same purpose was pointed out in an argument by Popper (cf. [89], pp.320ff). Of course, in a logic where both the disjunctive syllogism and contraposition are invalid derivations, these arguments do not apply as such.

[14] See Duns Scotus's *Opera Omnia*, pp.288ff. Cf. also note 3.

The failure of contraposition gives us a good reason for having doubts also about the validity of the *intersubstitutivity of provable equivalents*, which states that, given a schema $\sigma(A_1, \ldots, A_n)$:

$$\forall B_1 \ldots \forall B_n\ [(A_1 \dashv\vdash B_1) \text{ and } \ldots \text{ and } (A_n \dashv\vdash B_n)] \Rightarrow [\sigma(A_1, \ldots, A_n) \dashv\vdash \sigma(B_1, \ldots, B_n)]. \qquad \text{(IpE)}$$

Now, as a particular example, if we had (IpE), from $A \dashv\vdash B$ we would immediately derive, for instance, $\neg A \dashv\vdash \neg B$. But this is not the case here. Indeed, in what follows we exhibit some samples of that failure in **bC**:

THEOREM 3.21 In **bC**:

(i) $(A \wedge B) \dashv\vdash_{bC} (B \wedge A)$ holds, but $\neg(A \wedge B) \dashv\vdash_{bC} \neg(B \wedge A)$ does not;
(ii) $(A \vee B) \dashv\vdash_{bC} (B \vee A)$ holds, but $\neg(A \vee B) \dashv\vdash_{bC} \neg(B \vee A)$ does not;
(iii) $(A \wedge \neg A) \dashv\vdash_{bC} (\neg A \wedge A)$ holds, but $\neg(A \wedge \neg A) \dashv\vdash_{bC} \neg(\neg A \wedge A)$ does not.

Proof: The parts which hold are easy, using positive classical logic. Now, to check that none of the other parts hold, even if axioms and rules of **bC** are taken into consideration, use the same matrices and distinguished values as in THEOREM 3.15(ii), changing only the values of $(1 \wedge \frac{1}{2})$ and $(1 \vee \frac{1}{2})$ from 1 to ½ (but leaving the values of $(\frac{1}{2} \wedge 1)$ and $(\frac{1}{2} \vee 1)$ as they are, equal to 1). □

COROLLARY 3.22 (IpE) does not hold for **bC**.

The reader should keep in mind that this last result is, initially, only about **bC**, and that some deductive extensions of it may fix some or even all the counter-examples to intersubstitutivity. Now, given that (IpE) holds for classical logic, it will obviously hold for the positive (classical) fragment of **bC** as well, that is, for the set of formulas in which neither $\neg$ nor $\circ$ occur. Adding contraposition as a new inference rule, it is easy to see, by the transitivity of the consequence operator and the Deduction Metatheorem, that one could extend (IpE) from positive logic to include also the fragment of **bC** containing negation. But then (bold) paraconsistency would be lost, as we learn from THEOREM 3.20! What happens, though, is that the contraposition inference rule is much more than one needs in order to obtain intersubstitutivity for the consistencyless fragments of our logics. In fact, any of the following 'contraposition' deduction rules would of course do the job equally well (cf. [107] and [105]):

$$\forall A\, \forall B\ [(A \vdash B) \Rightarrow (\neg B \vdash \neg A)]; \qquad \text{(RC)}$$
$$\forall A\, \forall B\ [(A \dashv\vdash B) \Rightarrow (\neg B \vdash \neg A)]. \qquad \text{(EC)}$$

It is obvious that (EC) can be inferred from (RC), and Urbas has shown in [107] that the paraconsistent logic obtained by adding (EC) to C_ω is extended by the paraconsistent logic obtained by the addition of (RC) to C_ω (and both, of course, are extended by classical logic). So, it *is* possible to obtain paraconsistent extensions of C_ω (and also of C_{min}, for Urbas's proof of non-collapse into classical logic by the addition of (EC) also applies to this logic), but then these new logics can all still be shown to lack a bottom particle (as in THEOREM 3.4), constituting thus no **LFI**s! The question then would be if (IpE) could be obtained for *real* **LFI**s. The closest we will get to this here is showing, in THEOREM 3.53, that there are fragments of classical logic extending **bC** for which (IpE) holds, but then these specific fragments turn out not to be paraconsistent in our sense. At any rate, for various other classes of **LFIs** we will show that such intersubstitutivity results are just unattainable, as shown in THEOREM 3.51 (see also, for instance, FACT 3.74).

To be sure, one does not need to blame *paraconsistency* for these last few negative results. As the reader will see below, the eccentricities in THEOREM 3.21 can be fixed by some extensions of **bC**. As for THEOREM 3.20, one could always throw away some piece of the positive classical logic in an extreme effort to avoid its consequences. This is what is done, for instance, by some logics of relevance. This could, however, have the effect of throwing the baby out with the bath water —most such logics, if not all, will also dismiss the useful Deduction Metatheorem or, regrettably enough, *modus ponens*. Now, suppose that, driven by itches of relevance, one was taken to consider logics such that $(A, B \nVdash A)$. This would definitely mean, thus, that their consequence relations would be no more than 'cautiously reflexive'. If one still insisted that $(A \Vdash A)$ should hold, then the logics produced would be non-monotonic as well. This would mean, of course, that many of the results that we attained in the last section would not be immediately adaptable to such logics (and this remark also applies to adaptive logics, once they are also non-monotonic, even if for other reasons). These are not problems of actual relevance logics, nevertheless, as they are usually relevant only at the level of theoremhood (always invalidating $(A\to(B\to A))$, while in some cases still validating $(A\to A)$), but still not at the level of their consequence relations, as conjectured above (see, for instance, [3] or [9]) —and of course, in all such cases, the Deduction Metatheorem cannot hold. But yes, we had better push our exposition on, instead of scrubbing this matter here any further.

3.4 Letting bC talk about (dual) inconsistency. The reader may find it a bit awkward, indeed, that we would be calling **bC** a logic of formal *inconsistency*, since it only has a connective expressing *consistency*, but not its opposed concept. So, for *us* to be more consistent, let's now consider a further extension of our language, this time adding a new unary connective, $\bullet$, to represent inconsistency. The intended interpretation about the dual relation between consistency and inconsistency would require exactly that each of these concepts should be opposed to the other. But how do we formalize this? Consider the following additional axiomatic rule:

(bc2) $\neg\bullet A \vdash_{\mathbf{bbC}} \circ A.$ 'If A is not inconsistent, then it is consistent'

This is surely a must, but in fact it does not represent much of an addition. Indeed, consider its contrapositional variation:

(bc3) $\neg\circ A \vdash_{\mathbf{bbC}} \bullet A.$ 'If A is not consistent, then it is inconsistent'

The lack of contraposition (see THEOREM 3.20), despite the presence of some restricted forms of it (such as in FACT 3.17) can be partly blamed for the fact that **bC** plus (bc2) can still not prove (bc3). Indeed:

THEOREM 3.23 (bc3) is not provable by **bC** plus (bc2).

Proof: Just consider three-valued matrices such that: $v(A\wedge B)=0$ if $v(A)=0$ or $v(B)=0$, and $v(A\wedge B)=1$, otherwise; $v(A\vee B)=0$ if $v(A)=0$ and $v(B)=0$, and $v(A\vee B)=1$, otherwise; $v(A\to B)=0$ if $v(A)\neq 0$ and $v(B)=0$, and $v(A\to B)=1$, otherwise; $v(\neg A)=1-v(A)$; and the matrices for the non-classical connectives are the ones demonstrated on the right. 0 is the only non-distinguished value. □

	∘	•
1	0	1
½	0	1
0	½	0

So, let us now, for the sake of symmetry, define the logic **bbC** as given by the addition of both (bc2) and (bc3) to the basic logic of (in)consistency, **bC**. This is still not much... for consider now the converses of these rules:

(bc4) $\bullet A \vdash_{\mathbf{bbbC}} \neg\circ A$; 'If A is inconsistent, then it is not consistent'
(bc5) $\circ A \vdash_{\mathbf{bbbC}} \neg\bullet A$. 'If A is consistent, then it is not inconsistent'

Will these hold in **bbC**? The answer is once more in the negative:

THEOREM 3.24 Neither (bc4) nor (bc5) are provable by **bbC**.
Proof: Consider the same three-valued matrices for the binary connectives as in THEOREM 3.23, but let now negation be such that $\nu(\neg A)=0$ if $\nu(A)\neq 0$, and $\nu(\neg A)=1$, otherwise. The non-classical connectives will now be defined by the new matrices to the right. Once more, 0 is the only non-distinguished value. □

	∘	•
1	0	1
½	½	1
0	0	1

In reality, the situation is even worse than it may appear at first sight, though predictable. It happens that, once more, it is not enough to add just one of (bc4) or (bc5) to **bbC** —the other one would still not be provable. Indeed:

THEOREM 3.25 (i) (bc4) is not provable by **bbC** plus (bc5); (ii) (bc5) is not provable by **bbC** plus (bc4).
Proof: Consider now the four-valued matrices where $\wedge$, $\vee$, $\rightarrow$ and $\neg$ are once more defined as in THEOREM 3.23 (only that now they have a wider domain, with four values). For part (i), let ∘ and • be given by the matrices to the right. For part (ii), just modify ∘ so that $\circ(⅔)=⅓$ (and no more ⅔); modify also • in the contrary sense, so that $\bullet(⅔)=⅔$ (and no more ⅓). In both cases, only 0 should be taken to be a non-distinguished value. □

	∘	•
1	1	0
⅔	⅔	⅓
⅓	0	⅔
0	1	0

Taking the above results into account, we will now define the logic **bbbC** to be given by the addition of both (bc4) and (bc5) to the preceding **bbC**.

It is important to note that the last theorem above also shows that it is ineffective trying to introduce the inconsistency connective in the logic **bC** simply by setting, by definition, $\bullet A \stackrel{\text{def}}{=} \neg\circ A$. The reason is that, even though this would automatically guarantee that $\bullet A \dashv\vdash \neg\circ A$, and that $\neg\bullet A \dashv\vdash \neg\neg\circ A$, and so on, just by definition and reflexivity, this would *not* guarantee as well that, for instance, we would have $\circ A \vdash \neg\bullet A$. Indeed, to check this you may here just reconsider THEOREM 3.25(ii). So, the relation between ∘ and • cannot, in the cases of **bC** and **bbC**, be characterized by a simple definition. Despite this, one may now establish new presentations for some previous facts and theorems, just slightly different from before:

THEOREM 3.26 The results 3.11, 3.14, 3.15, 3.16, and 3.17 are all valid for **bbbC**, and are still valid as well if one substitutes any occurrence of ∘ for ¬•, and ¬∘ for •.
Proof: This is routine, just using (bc2)–(bc5). For 3.15 and 3.16 remember to add a matrix for •, just negating the matrix for ∘ presented in THEOREM 3.15(ii). □

So, could the relation between ∘ and • be characterized by a definition, now that we have **bbbC**? Another NO is the answer. For if a definition such as $\bullet A \stackrel{\text{def}}{=} \neg\circ A$ were feasible, this would mean, given (bc5): $\circ A \vdash_{\mathbf{bbbC}} \neg\bullet A$, that $\circ A \vdash_{\mathbf{bbbC}} \neg\neg\circ A$ should hold just by straightforward substitution. But, as it happens, this last rule does *not* hold in **bbbC**:

THEOREM 3.27 Neither $\circ A \to \neg\neg\circ A$ nor $\bullet A \to \neg\neg\bullet A$ are provable by **bbbC**.

Proof: Consider once more the same three-valued matrices for the binary connectives given in THEOREM 3.23, 0 as the only non-distinguished value, but now let the unary connectives be those pictured to the right. □

	$\neg$	$\circ$	$\bullet$
1	0	½	½
½	1	0	1
0	1	0	1

Evidently, the above matrices must also display the non-provability by **bbbC** of the schema $A \to \neg\neg A$, the converse of (Min11). But if the validity of $A \to \neg\neg A$ clearly implies the validity of the two schemas in THEOREM 3.27, the validity of those schemas certainly *does not* imply the validity of $A \to \neg\neg A$. Indeed:

THEOREM 3.28 $A \to \neg\neg A$ is not provable by **bbbC** plus $\circ A \to \neg\neg\circ A$ and $\bullet A \to \neg\neg\bullet A$.

Proof: Consider the same matrices and distinguished values as in THEOREM 3.27, only that now $\circ$ is constant and equal to 0, and $\bullet$ is constant and equal to 1. □

Now, if we added to **bbbC** the axioms $\circ A \vdash \neg\neg\circ A$ and $\bullet A \vdash \neg\neg\bullet A$ this would only shift our problem to proving that $\neg\bullet A \vdash \neg\neg\neg\bullet A$ and $\neg\circ A \vdash \neg\neg\neg\circ A$ hold, and so on, and so forth. Of course, these would be all guaranteed if we now defined **bbbbC** by the addition to **bbbC** of an infinite number of axiomatic rules, to the effect that $\neg^n\circ A \vdash_{\mathbf{bbbbC}} \neg^{n+2}\circ A$ and $\neg^n\bullet A \vdash_{\mathbf{bbbbC}} \neg^{n+2}\bullet A$, where $\neg^m$ denotes m occurrences of negation in a row. We could also solve all of this at once by fixing $A \vdash \neg\neg A$ as a new axiomatic rule, but we argue that it is a bit too early for this last solution —indeed, there is a gamut of interesting **C**-systems in which this axiom does *not* hold, and we would rather explore them first. So, let us study first, in what follows, some other forms of obtaining the intended duality between $\circ$ and $\bullet$ using a finite set of schemas, and without yet incorporating $A \vdash \neg\neg A$ as a rule.

3.5 The logic Ci, where contradiction and inconsistency meet. While strengthening **bC**, we have been trying to keep up with the intended duality between consistency and inconsistency. But, given the new version of the FACT 3.16 obtained in THEOREM 3.26 (which also applies to **bbbbC**), we know that in any of the logics **b(b(b(b)))C** a contradiction implies an inconsistency, but not the other way around —so, *this* situation has still not been changed. Now, the distinction between contradiction and inconsistency is a contribution of the present study, and we are unaware of any other formal attempts to do so in the same way as we do here. What will ha ppen then if we now introduce new axioms in order to finally obtain the identification of contradiction and inconsistency, getting closer this way to the other paraconsistent logics in the literature? Let's do it. Consider the two following axiomatic rules:

(ci1) $\bullet A \vdash_{\mathbf{Ci}} A$; 'If A is inconsistent, then A should be the case'

(ci2) $\bullet A \vdash_{\mathbf{Ci}} \neg A$. 'If A is inconsistent, then $\neg A$ should be the case'

Given the classical properties of conjunction, these two rules will evidently have the same effect as the following single one:

(ci) $\bullet A \vdash_{\mathbf{Ci}} (A \wedge \neg A)$. 'An inconsistency implies a contradiction'

So, let's call **Ci** the logic obtained by the addition of (ci1) and (ci2) (or, equivalently, the addition of (ci)) to **bbbC**, that is, the logic axiomatized by (Min1)–(Min11), (bc1)–(bc5), (ci), and (MP). In **Ci** we finally have that $\bullet A$ and $(A \wedge \neg A)$ are equivalent for-

(bc5), (ci), and (MP). In **Ci** we finally have that $\bullet A$ and $(A \wedge \neg A)$ are equivalent formulas, and we shall see that this will make a BIG difference on **Ci**'s deductive strength.

First, let us note that, even though we now have, in **Ci**, the converse of parts (i) and (ii) of FACT 3.16, the converses of parts (iii) and (iv) still do *not* hold. Indeed:

FACT 3.29 This rule does hold in **Ci**:
(i) $\neg\circ A \vdash_{\mathbf{Ci}} (A \wedge \neg A)$,
but the following rules do not:
(ii) $\neg(A \wedge \neg A) \vdash_{\mathbf{Ci}} \circ A$;
(iii) $\neg(\neg A \wedge A) \vdash_{\mathbf{Ci}} \circ A$.

Proof: The first part is obvious. For the following ones, consider, for instance, the three-valued matrices such that:

$v(A \wedge B) = min(v(A), v(B))$;
$v(A \vee B) = max(v(A), v(B))$;
$v(A \to B) = v(B)$, if $v(A) \neq 0$, and $v(A \to B) = 1$, otherwise;
$v(\neg A) = 1 - v(A)$;
$v(\circ A) = 0$, if $v(A) = v(\neg A)$, and $v(\circ A) = 1$, otherwise;
$v(\bullet A) = 1$, if $v(A) = v(\neg A)$, and $v(\bullet A) = 0$, otherwise,

where 0 is the only non-distinguished value. The attentive reader might have noticed that these are exactly the matrices defining the already mentioned **LFI1**, in the subsection **2.4**. □

So, this last theorem reminds us that, even though in **Ci** we *do* have an equivalent way of referring to inconsistency using just the classical language, this does not mean that we should also have an immediate **CPL**-linguistic equivalent manner of referring to *consistency* as well (but confront this with what happens in the case of the **dC**-systems, in the subsection **3.8**)! There are, however, many other things that we *do* have. For instance, in **Ci** the THEOREM 3.15 is still entirely valid. Indeed:

THEOREM 3.30 $\neg(A \wedge \neg A)$ and $\neg(\neg A \wedge A)$ are not top particles in **Ci** (also the formula $(A \to \neg\neg A)$ is still not provable).

Proof: Use again the matrices of $\mathbf{P}^1$ (in THEOREM 3.15(ii)), adding a matrix for '$\bullet$' by negating the matrix for '$\circ$'. □

We also have in **Ci** some new ways of formulating gentle explosion and the *reductio* deduction rules:

FACT 3.31 The following rules hold in **Ci**:
(i) $\circ A,\ \bullet A \vdash_{\mathbf{Ci}} B$;
(ii) $\circ A,\ \neg\circ A \vdash_{\mathbf{Ci}} B$;
(iii) $\bullet A,\ \neg\bullet A \vdash_{\mathbf{Ci}} B$;
(iv) $[(\Gamma, B \vdash_{\mathbf{Ci}} \circ A)$ and $(\Delta, B \vdash_{\mathbf{Ci}} \bullet A)] \Rightarrow (\Gamma, \Delta \vdash_{\mathbf{Ci}} \neg B)$;
(v) $[(\Gamma, B \vdash_{\mathbf{Ci}} \circ A)$ and $(\Delta, B \vdash_{\mathbf{Ci}} \neg\circ A)] \Rightarrow (\Gamma, \Delta \vdash_{\mathbf{Ci}} \neg B)$;
(vi) $[(\Gamma, B \vdash_{\mathbf{Ci}} \bullet A)$ and $(\Delta, B \vdash_{\mathbf{Ci}} \neg\bullet A)] \Rightarrow (\Gamma, \Delta \vdash_{\mathbf{Ci}} \neg B)$.

Proof: Part (i) comes from (ci) and (bc1), parts (ii) and (iii) come from part (i) if you use (bc2)–(bc5). Rules (iv), (v) and (vi) are variations on FACT 3.14(ii), using the previous rules. □

Parts (ii) and (iii) of FACT 3.31 simply show **Ci** to be controllably explosive in contact either with a consistent or with an inconsistent formula. In fact, in **Ci** one can go on to prove a much more intimate connection between consistency and controllable explosion, and this will reveal some even stronger consequences of the new axiomatic rule, (ci), that we now consider:

FACT 3.32 A particular given schema in **Ci** (or in any extension of this logic) is consistent if, and only if, **Ci** is controllably explosive in contact with this schema.

Proof: To show that $[(\Gamma \vdash_{Ci} \circ A) \Rightarrow (\Gamma, A, \neg A \vdash_{Ci} B)]$ just invoke axiom (bc1) and the transitivity of $\vdash$. For the converse, note that, from (ci) and (bc3), one may obtain $[\neg\circ A \vdash_{Ci} (A \wedge \neg A)]$, and so, from the supposition that $(\Gamma, A, \neg A \vdash_{Ci} B)$ it follows that $\neg\circ A$ is a bottom particle. One may then conclude, as in THEOREM 3.10, that $\vdash_{Ci} \neg\neg\circ A$, and, by (Min11), that $\vdash_{Ci} \circ A$. □

FACT 3.33 These are some special theses of **Ci**:

(i) $\vdash_{Ci} \circ\circ A$;
(ii) $\vdash_{Ci} \neg\bullet\circ A$;
(iii) $\vdash_{Ci} \circ\bullet A$;
(iv) $\vdash_{Ci} \neg\bullet\bullet A$.

Proof: Parts (i) and (iii) come directly from FACT 3.32 and from parts (ii) and (iii) of FACT 3.31. For (ii) and (iv), use (bc2) and the previous parts. □

This last result (check also [54]) implies that **Ci** will not have consistency or inconsistency appearing at different levels: both consistent and inconsistent formulas are consistent (in contrast to what happened in the case of **bC** —see THEOREM 3.10—, where no formula was provably consistent), and none of them is inconsistent (check also FACT 3.50 for a much stronger version of the last fact in **Ci**).

The reader will recall from the subsection **3.3** that contraposition inference rules not only did not hold in **bC** but could not even be added to any paraconsistent extension of it (THEOREM 3.20). **Ci** can be shown to count, nevertheless, with more restricted forms of contraposition than **bC** (compare the following with FACT 3.17):

FACT 3.34 These are some restricted forms of contraposition introduced by **Ci**:

(i) $(A \to \circ B) \vdash_{Ci} (\neg\circ B \to \neg A)$;
(ii) $(A \to \neg\circ B) \vdash_{Ci} (\circ B \to \neg A)$;
(iii) $(\neg A \to \circ B) \vdash_{Ci} (\neg\circ B \to A)$;
(iv) $(\neg A \to \neg\circ B) \vdash_{Ci} (\circ B \to A)$.

Proof: To check (i), let $\Gamma = \Delta = \{(A \to \circ B), \neg\circ B\}$ and apply FACT 3.31(v) to $\Gamma \cup \{A\}$, so as to obtain $\Gamma \vdash_{Ci} \neg A$. This will give the desired result. Alternatively, one could use directly FACT 3.17(i) and note that $\circ\circ B$ is a theorem of **Ci** (this is FACT 3.33(i)). The other parts are similar, and we leave them as easy exercises to the reader. □

Note that all rules in the last result continue to be valid if one substitutes any '$\circ$' for '$\neg\bullet$', and any '$\neg\circ$' for '$\bullet$'. On the other hand, rules such as $[(\circ A \to B) \vdash_{Ci} (\neg B \to \neg\circ A)]$ *do not* hold in this logic!

Now, we have learned from COROLLARY 3.22 that the intersubstitutivity of provable equivalents, (IpE), does not hold for **bC**. The same result is true for **Ci**, and still the same counter-examples mentioned before can be presented here:

THEOREM 3.35 (IpE) does not hold for **Ci**.

Proof: Add to the matrices on THEOREM 3.21 one matrix for $\bullet$ such that $v(\bullet A) = 1 - v(\circ A)$, and check that all the new axioms, defining **Ci** from **bC**, still hold. □

Now, in order to go one step further from the actual absence of contraposition in **Ci**, let us recall that in the subsection **3.3** it has been pointed out that the addition of some of the deduction 'contraposition' rules (EC) or (RC) would have been equally sufficient for obtaining (IpE) for consistencyless fragments of our paraconsistent logics. It seems, nevertheless, that obtaining (IpE) will not be an easy task, after all:

FACT 3.36 The addition of (RC): $[(A \vdash B) \Rightarrow (\neg B \vdash \neg A)]$ to **Ci** causes its collapse into classical logic.

Proof: From (ci1) and (ci2), plus (bc3), one obtains, respectively, that $\neg\circ A \vdash_{\mathbf{Ci}} A$, and $\neg\circ A \vdash_{\mathbf{Ci}} \neg A$. Applying (RC) and (Min11) one would have then $\neg A \vdash_{\mathbf{Ci}} \circ A$ and $\neg\neg A \vdash_{\mathbf{Ci}} \circ A$. But then, using the proof by cases, one would conclude that $\vdash_{\mathbf{Ci}} \circ A$, that is, all formulas would be consistent. Looking at THEOREM 3.9 and (bc1), one sees that this was exactly what was lacking in order for classical logic to be characterized. □

So, (RC) must be ruled out as an alternative in order to obtain (IpE), in the case of **Ci**. As for (EC), its possible addition to **Ci** will be discussed below, in THEOREM 3.51, FACT 3.52, and the subsequent commentaries on these results.

The new restricted forms of contraposition in FACT 3.34 are, in any case, strong enough for us to show that **Ci** has some redundant axioms as it is. Indeed:

FACT 3.37 In **Ci**: (i) (bc2) proves (bc3), and vice-versa. (ii) (bc4) proves (bc5), and vice-versa.

Other interesting consequences of (ci) are those that we shall call 'Guillaume's Theses', which regulate the propagation of consistency and the back-propagation of inconsistency through negation:

FACT 3.38 **Ci** also proves the following:

(i) $\circ A \vdash_{\mathbf{Ci}} \circ\neg A$;

(ii) $\bullet\neg A \vdash_{\mathbf{Ci}} \bullet A$.

Proof: From (ci) and (bc4), we have that $[\neg\circ\neg A \vdash_{\mathbf{Ci}} (\neg A \wedge \neg\neg A)]$, from C_{min} we have that $[(\neg A \wedge \neg\neg A) \vdash_{\mathbf{Ci}} (A \wedge \neg A)]$, and from FACT 3.16(ii) we know that $[(A \wedge \neg A) \vdash_{\mathbf{Ci}} \neg\circ A]$. So, ultimately, we have the rule $[\neg\circ\neg A \vdash_{\mathbf{Ci}} \neg\circ A]$. By (bc3) and (bc4) we prove part (ii) of our fact. Part (i) comes from this same rule, by an application of FACT 3.34(iv). □

This last result will provide us with some other forms for the theses in FACT 3.33, such as:

FACT 3.39 These are also some special theses of **Ci**:

(i) $\vdash_{\mathbf{Ci}} \circ\neg\circ A$;

(ii) $\vdash_{\mathbf{Ci}} \neg\bullet\neg\circ A$;

(iii) $\vdash_{\mathbf{Ci}} \circ\neg\bullet A$;

(iv) $\vdash_{\mathbf{Ci}} \neg\bullet\neg\bullet A$.

It will also be useful to note that here we have (contrasting with THEOREM 3.30, which informed us, among other things, that $[A \nvdash_{\mathbf{Ci}} \neg\neg A]$):

FACT 3.40 Here are some more special theses of **Ci**:

(i) $\circ A \vdash_{\mathbf{Ci}} \neg\neg\circ A$;

(ii) $\bullet A \vdash_{\mathbf{Ci}} \neg\neg\bullet A$.

Proof: These will follow directly if you apply FACT 3.34 twice. The reader might remember that we lacked these forms in **bbbC** (this was THEOREM 3.27). □

Now, do we obtain in **Ci** that intended duality between consistency and inconsistency? The answer is YES. This is the topic for our next subsection.

3.6 On a simpler presentation for Ci. The logic **Ci** provides us with a sufficient environment to prove a kind of restricted *intersubstitutivity* or *replacement* theorem. While we know from THEOREM 3.35 that full replacement for the formulas of **Ci** does not obtain, our present restricted forms of contraposition, nevertheless, will help us to show that intersubstitutivity *does* hold if only we are talking only about substituting some formula whose outmost operator is '$\circ$' by this same formula, but now having '$\neg\bullet$' in the place of that '$\circ$', or if we will substitute some formula whose outmost operator is '$\bullet$' by this same formula, but now having '$\neg\circ$' in the place of that '$\bullet$'. In simpler terms, what we are saying is that we can now take just one of the operators '$\circ$' and '$\bullet$' as primitive, and define the other in terms of the negation of that first one. So, we will now show that:

THEOREM 3.41 An equivalent axiomatization for **Ci** is obtained if we consider only axioms (Min1)–(Min11), (bc1), (ci), and (MP), and set one of these two definitions:

(i) $\bullet A \stackrel{\text{def}}{=} \neg\circ A$;

(ii) $\circ A \stackrel{\text{def}}{=} \neg\bullet A$.

Proof: Consider part (i) to be the case. This means that we can take (bc3): $\neg\circ A \vdash_{Ci} \bullet A$ and (bc4): $\bullet A \vdash_{Ci} \neg\circ A$, for granted, simply by definition. Now, (bc2): $\neg\bullet A \vdash_{Ci} \circ A$, will be the case if, and only if, given the definition of '$\bullet$', $\neg\neg\circ A \vdash_{Ci} \circ A$ is the case —and it is, because of (Min11). As to (bc5): $\circ A \vdash \neg\bullet A$, it will be the case if, and only if, $\circ A \vdash_{Ci} \neg\neg\circ A$ is the case —and it is, this time thanks to FACT 3.40(i). An alternative, and much simpler way, of checking that (bc2) and (bc5) should hold here is by taking FACT 3.37 into consideration. The axiomatic rule (bc1): $\circ A$, A, $\neg A \vdash_{Ci} B$ is already in the 'standard form' (we are here eliminating all occurrences of '$\bullet$'s and leaving only '$\circ$'s), and the rule (ci): $\bullet A \vdash_{Ci} (A \wedge \neg A)$ can be exchanged, by the definition of '$\bullet$', that is, by (bc3) and (bc4), for $\neg\circ A \vdash_{Ci} (A \wedge \neg A)$. Now we have shown that all occurrences of '$\bullet$' in the axioms of **Ci** can be substituted by an occurrence of '$\neg\circ$', and all occurrences of '$\neg\bullet$' in the axioms of **Ci** can be substituted by an occurrence of '$\circ$'. So, if you would have proven a formula in which, respectively, an inconsistency connective '$\bullet$' or its negated form '$\neg\bullet$' appears at some point, you can now rewrite the proof using the new versions of the axioms above and what will appear in the end will be, respectively, a negated consistency connective '$\neg\circ$', or simply the connective '$\circ$'. For part (ii) the procedure is entirely analogous, but now use FACT 3.40(ii), or FACT 3.37 again, when necessary. □

So, this last result provides us with a restricted form of replacement theorem for consistent formulas, and guarantees the intended duality between $\circ$ and $\bullet$, which could not be obtained in the subsection **3.4**, within **bbbbC** or its fragments. With such a result in hand we need make no big effort to verify that formulas such as $\neg(\circ A \wedge \neg\circ A)$, $\neg(\circ A \wedge \bullet A)$, $\neg(\neg\bullet A \wedge \neg\circ A)$ and $\neg(\neg\bullet A \wedge \bullet A)$ are all equivalent, which could, otherwise, be quite a non-trivial task!

The reason why we can obtain this new axiomatization, as the reader will make out after he is introduced to the semantics of **Ci**, in [42], is that, truth-functionally based on the non-classical behavior of negation, both the consistency and the inconsistency operators of this logic will work quite 'classically'.

3.7 Using LFIs to talk about classical logic. At this point, working with **Ci**, perhaps the question would arise as to how far we are from classical propositional logic, **CPL**. The answer is: a lot —and just a little bit. As we are not presupposing any kind of doublethinking, let us then reformulate a few things for the question, and its answer, really to make sense.

To start with, it is hard to compare two logics if they 'talk about different things', and are so disjoint that none of them is an extension of the other. For C_{min} was a conservative extension of positive classical logic, but it was a fragment, in the same language, of 'full' **CPL**, as we know, and **bC** was a conservative extension of C_{min}. Thus, **Ci**, which is a deductive extension of **bC**, happens to be written in a richer language than that of **CPL**, but it does not contain all classical inferences, and so these two logics are hardly comparable. Now, this is easy to fix. Let us also conservatively extend **CPL** by the addition of connectives for consistency and inconsistency, whose matrices will be such that $\circ$ takes always the distinguished value 1, and $\bullet$, on the contrary, is constant and equal to 0. We will designate this 'new' logic, obtained by such an extension of **CPL**, *extended classical logic*, or **eCPL**. Of course, **eCPL** can be easily axiomatized by the addition of an axiomatic schema such as:

(ext) $\vdash_{\mathbf{eCPL}} \circ A$ 'Every A is consistent'

to any axiomatization of **CPL**, like the one mentioned in THEOREM 3.9. The inconsistency connective, $\bullet$, can be here introduced as a definition: $\bullet A \stackrel{\text{def}}{=} \neg\circ A$, just as in THEOREM 3.41(i). In this way we obtain an extension of classical logic which looks as a logic of formal inconsistency (see (D20)), having an operator expressing consistency, and of course an axiomatic rule such as $[\circ A, A, \neg A \vdash_{\mathbf{eCPL}} B]$, expressing finite gentle explosion, will hold in **eCPL**. But, as it happens, given axiom (ext), we know that **eCPL** is not only finitely gently explosive (and so, non-trivial), but explosive as well. It is, in fact, a *consistent* logic (see (D19)), instead of an **LFI**.

Well and good, but is **Ci** now to be characterized as a deductive fragment of **eCPL**? Indeed! Just check that all axioms of **Ci** are validated by the matrices of **eCPL**, and that's it. So, **Ci** is in fact a fragment of an alternative formulation of classical logic, and this of course will guarantee that **Ci** is a non-contradictory logic (once **Ci** is not explosive, but it is a fragment of **eCPL**, and **eCPL** is still at least as explosive and non-trivial as **CPL** was, and consequently it cannot prove a contradiction). Is that all to it? No, because we will now see that we can still use **Ci** to reproduce in a very faithful way every inference of **CPL** (or of **eCPL**)!

How can this be done? Remember that **Ci** has a strong (or supplementing) negation, which can be defined, as in the case of **bC**, by setting $\sim A \stackrel{\text{def}}{=} (\neg A \wedge \circ A)$. But, in **bC**, even though this negation had the power of producing (supplementing) explosions, it could not still be said to have all properties of a *classical negation*. Indeed:

THEOREM 3.42 The strong negation $\sim$, in **b(b(b(b)))C**, is not classical.
Proof: Just consider once more the classical matrices for the classical connectives, as in the above definition of **eCPL**, but now exchange the matrices of $\circ$ and $\bullet$, letting $\circ$ be constant and equal to 0 (and not to 1, as before), and letting $\bullet$ be constant and equal to 1 (and not to 0, as before). It is easy to see that all axioms and rules of

b(b(b(b)))C are validated by such matrices, but (ci) and (ext) are not, and consequently formulas such as $(A\vee{\sim}A)$ and $(A\rightarrow{\sim}{\sim}A)$ (recall the definition of ${\sim}A$) are not validated as well, being independent from all logics we have exposed previous to **Ci**. □

This is an interesting result that shows that being explosive is not enough to make a negation classical.[15] But what would be enough? Well, given the axiomatization of classical logic in THEOREM 3.9, we know that any connective $\div$ added in an axiomatic environment where (Min1)–(Min9) hold and which is such that:

(Alt10) $\vdash_{\mathbf{Alt}} (A\vee\div A)$;
(Alt11) $\vdash_{\mathbf{Alt}} (\div\div A\rightarrow A)$;
(Alt12) $\vdash_{\mathbf{Alt}} (A\rightarrow(\div A\rightarrow B))$,

also hold, should behave as the classical negation. So, all we have to do now is to show that (Alt10)–(Alt12) hold in **Ci** if one substitutes $\div$ for the strong negation ${\sim}$. We could here make use of an auxiliary lemma:

LEMMA 3.43 These are some theorems of **Ci**:
(i) $\vdash_{\mathbf{Ci}} (A\vee\circ A)$;
(ii) $\vdash_{\mathbf{Ci}} (\neg A\vee\circ A)$.
Proof: For part (i), observe that, from (Min6), $[\circ A\vdash_{\mathbf{Ci}} (A\vee\circ A)]$, and, from (ci1), $[\neg\circ A\vdash_{\mathbf{Ci}} A]$, so, once more by (Min6), and transitivity, $[\neg\circ A\vdash_{\mathbf{Ci}} (A\vee\circ A)]$. Using the proof by cases one finally concludes that $[\vdash_{\mathbf{Ci}} (A\vee\circ A)]$. Part (ii) is similar to (i), but you should now use (ci2). □

THEOREM 3.44 The strong negation ${\sim}$, in **Ci**, is classical.
Proof: To check that (Alt10) holds for ${\sim}$, that is, that $[\vdash_{\mathbf{Ci}} (A\vee(\neg A\wedge\circ A))]$, notice that this last schema is equivalent to $[\vdash_{\mathbf{Ci}} (A\vee\neg A)\wedge(A\vee\circ A))]$, by positive classical logic, and the latter is provable from (Min10) and LEMMA 3.43(i), using (Min3). Now, (Alt12) is immediate, by the very definition of ${\sim}$, and to check (Alt11) you might just notice that by reflexivity we have $[{\sim}{\sim}A,\ A\vdash_{\mathbf{Ci}} A]$, and from (Alt12) we have $[{\sim}{\sim}A,\ {\sim}A\vdash_{\mathbf{Ci}} A]$; so, using a new form of proof by cases obtained from (Alt10) and (Min8) (as in FACT 3.7), we conclude that $[{\sim}{\sim}A\vdash_{\mathbf{Ci}} A]$. □

So, **Ci** is strong enough to endow its strong negation with all properties of a classical negation. This result has some immediate consequences. For instance, we could use it to show that (Min9) is redundant in **Ci** (and all other logics extending it). Notice, of course, that the two last results did not really need to use the whole positive *classical* logic, but that its *intuitionistic* fragment (which does not contain (Min9)) would have been enough. Confront the following fact with THEOREM 3.3:

FACT 3.45 The schema (Min9): $(A\vee(A\rightarrow B))$ is redundant in the axiomatization of **Ci**.
Proof: From reflexivity and (Alt12) we have that $[A\vdash_{\mathbf{Ci}} A]$ and $[{\sim}A\vdash_{\mathbf{Ci}} (A\rightarrow B)]$. But, of course, either A or $(A\rightarrow B)$, by (Min6) and (Min7), imply the above schema, $(A\vee(A\rightarrow B))$. So, using (Alt10) once more to provide a proof by cases, we are done. □

[15] There seems to be, at any rate, a widespread mistaken assumption in the literature to that effect (despite the example of intuitionistic negation, strong but not classical). Yet in some other studies, as for instance Batens's [13], note 11, a 'classical' negation, $+$, in a paraconsistent logic is assumed to be one which is not only strong but it should also be the case that $[+A\vdash\neg A]$ holds (as in axiom (bun), in the subsection **3.8**). This *is* the case, however, for **bC**'s strong negation ${\sim}$, but now we know that it is still *not* classical (it just has some kind of intuitionistic behavior). Ten years before, nevertheless, this same author (see [10], page 224) had put things more precisely, and required for that definition that $[\vdash (A\vee+A)]$ should also be the case.

We shall not list the properties of ~ in **Ci** at this point, but only mention the fact that 'it is a classical negation' when necessary, and then use any property that derives from this fact.

Now, this strong (classical) negation will give us a very interesting result. We already knew that the other binary connectives worked as their classical counterparts, and we were informed above that **Ci** comes also equipped with a negation which works like the classical one; so why don't we use **Ci** to 'talk about classical logic', that is, use **Ci**'s own stuff to reproduce any classical inference? One intuitive procedure to bring forth such an effect would be to pick any classical inference and just substitute any occurrence of a classical negation by an occurrence of a strong negation, and leave the rest as it is. And this indeed works:

THEOREM 3.46 The following mapping conservatively translates **CPL** inside of **Ci**:

(t1.1) $t_1(p)=p$, if p is an atomic formula;

(t1.2) $t_1(A\#B)=t_1(A)\#t_1(B)$, if # is any binary connective;

(t1.3) $t_1(\neg A)=\sim t_1(A)$.

So, it is the case that $[\Gamma \vdash_{\mathbf{CPL}} A] \Leftrightarrow [t_1[\Gamma] \vdash_{\mathbf{Ci}} t_1(A)]$.

Proof: Given THEOREM 3.44, we know that, by way of the above transformation, a counterpart to **CPL**'s axiomatization can be obtained inside of **Ci**. □

COROLLARY 3.47 We also have a conservative translation of **eCPL** inside of **Ci**. Just extend the above mapping by adding:

(t1.4) $t_1(\circ A)=\circ\circ t_1(A)$.

Proof: This comes from the above theorem, **eCPL**'s axiom (ext) and FACT 3.33(i). □

The above recursive translation just substitutes one negation for another, thus giving rise to a *grammatically faithful* (cf. [61], chapter X) way of reproducing classical inferences inside of **Ci**, and inside of any other logic deductively stronger than it, as the ones we will be studying below. Of course, other logics may provide yet some other sensible ways of translating classical logic inside of them (see, for instance, COROLLARY 3.62).

To be sure, we already had, in **bC**, a way of reproducing classical inferences (recall THEOREM 3.11), but at that point we had to introduce further premises in our theories —to wit, the premises that some of our propositions were consistent). A natural question which may arise then is whether this was really necessary, given that from THEOREM 3.42 we know that the 'canonical' strong negation of **bC** was not a classical one, and would then not allow the above translations to be performed inside of **bC**, or could it perhaps be the case that all strong negations are indeed strong, but some are stronger than others? This last option is indeed what occurs, for it can be easily shown, if we just recall FACT 2.10(ii), how one can define a classical negation inside of **bC**, despite the weakness of this logic, thus being able to talk about classical logic already inside of the most basic **C**-system we here present:

THEOREM 3.48 The logic **bC** does have a classical negation.

Proof: From (bc1), the axiom that realizes finite gentle explosiveness, and from left-adjunctiveness, we know that $(A\wedge(\neg A\wedge\circ A))$ is a bottom particle, for any formula A —let's choose any of these conjunctions and denote it by $\bot$, as usual. Inspired by FACT 2.10(ii) and using the Deduction Metatheorem we then define a new strong negation, $\div$, on **bC** as $\div A \stackrel{\text{def}}{=} (A\to\bot)$.[16] To check that *this* negation is

[16] This was indeed one of the many 'negations' set forth by Bunder in [30], though this author seems not to have completely understood their properties (see below the subsection **3.8**).

classical, we just need to prove that $(\dot{\sim}\dot{\sim}A \to A)$ is a theorem of **bC**. To such an end, first note that $[\vdash_{\mathrm{Ci}} (A \vee \dot{\sim}A)]$, given that this is $[\vdash_{\mathrm{Ci}} (A \vee (A \to \bot))]$, a form of axiom (Min9), and this gives us a new form of proof by cases, as in THEOREM 3.44. Next, notice that $[((A \to \bot) \to \bot), (A \to \bot) \vdash_{\mathrm{Ci}} \bot]$, by *modus ponens*, and $[\bot \vdash_{\mathrm{Ci}} A]$ by definition of the bottom particle, but also $[((A \to \bot) \to \bot), A \vdash_{\mathrm{Ci}} A]$. Thus, the new form of proof by cases will immediately give us $[((A \to \bot) \to \bot) \vdash_{\mathrm{Ci}} A]$. □

It is easy then to transform THEOREM 3.46 into a grammatically faithful translation of **CPL** already inside of **bC**, but it would be less easy to find a non-trivial analogue of COROLLARY 3.47, the translation of **eCPL**, given that **bC** is already known to have no consistent theorems (recall THEOREM 3.10) —that is, no theorems of the form $\circ A$. As to the status of the two different strong negations presented above inside of the stronger logic **Ci**, one can easily go on to show that:

FACT 3.49 In **Ci** the two strong negations above, $\sim$ and $\dot{\sim}$, are both classical, and are in fact equivalent, in a sense (but not all strong negations are classical in **Ci**).
Proof: That they are both classical is an obvious consequence from THEOREM 3.44 and THEOREM 3.48. To see that they are equivalent in **Ci**, remember, on the one hand, that $[A, {\sim}A \vdash_{\mathrm{Ci}} \bot]$, by definition, and so $[{\sim}A \vdash_{\mathrm{Ci}} (A \to \bot)]$, that is, $[{\sim}A \vdash_{\mathrm{Ci}} \dot{\sim}A]$, by the Deduction Metatheorem. On the other hand, we have both that $[(A \to \bot), A \vdash \bot]$, and thus $[(A \to \bot), A \vdash_{\mathrm{Ci}} {\sim}A]$, and that $[(A \to \bot), {\sim}A \vdash_{\mathrm{Ci}} {\sim}A]$, so the form of proof by cases offered by THEOREM 3.44 will allow us to conclude that $[\dot{\sim}A \vdash_{\mathrm{Ci}} {\sim}A]$. The reader will be right in thinking that all classical negations extending a positive classical basis are equivalent, but it is still the case, nonetheless, that **Ci** can define other strong negations that do not have a classical character, as for instance $(\neg\neg{\sim}A)$ or $(\neg\neg\dot{\sim}A)$. Take a look at [42], our paper on semantics, in the section on **Ci**, to check this claim. □

Now, the reader may perhaps think that classically negated propositions in **bC** and **Ci** (especially given the reconstruction of classical inferences inside of these logics that such negations support by way of the above mentioned conservative translations) would be classical enough so as to be consistent propositions themselves, that is, that $\circ{+}A$ would be a theorem, for instance, of **Ci**, for some classical negation $+$. We will now show that this can hardly be the case:

FACT 3.50 Only consistent or inconsistent formulas can themselves be provably consistent in **Ci**. Thus, $(\circ A)$ is a theorem of **Ci** if, and only if, A is of the form $\circ B$, $\bullet B$, $\neg\circ B$ or $\neg\bullet B$, for some B.
Proof: On the one hand, we already know from FACT 3.33 and FACT 3.39 that formulas such as $\circ B$ or $\bullet B$, and their variations, are all provably consistent in **Ci**. To see that the converse is also true, consider the following three-valued matrices, such that 0 is the only non-distinguished value and $v(A \wedge B) = ½$ if $v(A) \neq 0$ and $v(B) \neq 0$, and $v(A \wedge B) = 0$, otherwise; $v(A \vee B) = ½$ if $v(A) \neq 0$ or $v(B) \neq 0$, and $v(A \vee B) = 0$, otherwise; $v(A \to B) = ½$ if $v(A) = 0$ or $v(B) \neq 0$, and $v(A \to B) = 0$, otherwise; $v(\neg A) = 1 - v(A)$; $v(\circ A) = 1$ if $v(A) \neq ½$, and $v(\circ A) = 0$, otherwise. □

As a consequence of the last result, in particular, formulas of the form $\circ{\sim}A$ and $\circ\dot{\sim}A$ will not be provable in **Ci**, and, from FACT 3.32, we conclude that **Ci** is *not* controllably explosive in contact with (at least some) classically negated propositions. As we shall see, on the other hand, there are many extensions of this logic that do have this property, at least for some particular A's (see FACT 3.66, or FACT 3.76). But what would have happened if we had indeed theorems such as tho ones ruled out above? Let us here allow ourselves some counterfactual reasoning, and ask ourselves about the possible validity of (IpE) in some specific paraconsistent logics, like the extensions of **Ci**, or of some of its fragments (given that we know, from THEOREM 3.35, that (IpE) still does not hold in **Ci**, anyway):

THEOREM 3.51 (IpE) cannot hold in any paraconsistent extension of **Ci** in which:
(i) $(\circ{\div}{\div}A)$ holds, for some given classical negation $\div$; *or*
(ii) $\neg(A\wedge\neg A)$ or $\neg(\neg A\wedge A)$ hold; *or*
(iii) $[(\neg A\vee\neg B)\vdash\neg(A\wedge B)]$ hold; *or*
(iv) $[\neg(A\wedge B)\vdash(\neg A\vee\neg B)]$ hold.
(IpE) cannot hold in any paraconsistent extension of **bC** in which:
(v) $[\neg(A\rightarrow B)\vdash(A\wedge\neg B)]$ hold.
(IpE) cannot hold in any adjunctive paraconsistent extension of C_{min} in which:
(vi) both $[(A\wedge B)\vdash\neg(\neg A\vee\neg B)]$ and $[\neg(\neg A\vee\neg B)\vdash(A\wedge B)]$ hold.
(IpE) cannot hold in any adjunctive paraconsistent logic in which:
(vii) both $\neg(A\wedge\neg A)$ and $[(A\wedge\neg A)\dashv\vdash\neg\neg(A\wedge\neg A)]$ hold.

Proof: For part (i), given that $\div$ is a classical negation we can then assume $[A\dashv\vdash{\div}{\div}A]$ to hold. Now, if (IpE) were valid one could conclude, in particular, that $[\circ A\dashv\vdash\circ{\div}{\div}A]$, and given that $(\circ{\div}{\div}A)$ is a theorem of this logic extending **Ci**, by hypothesis, one would infer $\circ A$ as a theorem, but this is (ext), exactly the axiom that is lacking to make **Ci** collapse into **eCPL**. This generalizes a similar argument to be found in [107], Theorem 9. To check part (ii), recall that, in **Ci**, $[\bullet A\dashv\vdash(A\wedge\neg A)]$, and (IpE) would then give $[\circ A\dashv\vdash\neg(A\wedge\neg A)]$, and we are again left with the theorem $\circ A$, as in part (i). For part (iii), recall that $(\neg A\vee\neg\neg A)$ is a theorem already of C_{min}, and the problem reduces then to part (ii). For parts (iv) and (v), we will just show that $[(\neg{\div}A)\vdash A]$ is obtained, and so we may conclude that controllable explosion occurs in contact with $\div A$. Given that $[\neg(A\wedge B)\vdash(\neg A\vee\neg B)]$ holds, consider the strong negation $\sim A\stackrel{\text{def}}{=}(\neg A\wedge\circ A)$, for which one would immediately obtain $[\neg\sim A\vdash(\neg\neg A\vee\neg\circ A)]$, and so, from (Min11), (ci1) and (Min8), we get $[\neg\sim A\vdash A]$. Given that $[\neg(A\rightarrow B)\vdash(A\wedge\neg B)]$ holds, pick up $\dot{\div}A\stackrel{\text{def}}{=}(A\rightarrow\bot)$, and, from (Min4), we have that $[\neg\dot{\div}A\vdash A]$. For part (vi), given once more that $(\neg A\vee\neg\neg A)$ is a theorem of C_{min}, (IpE) would give us $[\neg(\neg A\vee\neg\neg A)\dashv\vdash\neg(\neg B\vee\neg\neg B)]$, and the rules that we here assume give us $[(A\wedge\neg A)\dashv\vdash(B\wedge\neg B)]$, so, by adjunction, we conclude in particular that $[A,\neg A\vdash B]$. This is the main result in Béziau's [21]. Finally, for part (vii), (IpE) would give us $[\neg\neg(A\wedge\neg A)\dashv\vdash\neg\neg(B\wedge\neg B)]$, and so $[(A\wedge\neg A)\dashv\vdash(B\wedge\neg B)]$, and we are in the same situation as in (vi). This is a stronger version of the main result in Béziau's [22], where actually $[A\dashv\vdash\neg\neg A]$ was assumed, instead of $[(A\wedge\neg A)\dashv\vdash\neg\neg(A\wedge\neg A)]$. Of course, a similar version of this last result arises if one just uniformly substitutes $(A\wedge\neg A)$ for $(\neg A\wedge A)$ in its statement. Notice that the rules mentioned in parts (iii) to (vi) had already shown up as the items (x), (ix), (vii), (xii) and (xi) of THEOREM 3.18.□

So far we have some negative results about the validity of (IpE) in some possible paraconsistent extensions of **bC** or **Ci**, but are there paraconsistent extensions of these logics in which (IpE) *does* hold? In the search for an answer, one could start by testing the compatibility of the addition, to those logics, of at least one of the following rules of deduction, (RC): $[(A\vdash B)\Rightarrow(\neg B\vdash\neg A)]$ or (EC): $[(A\dashv\vdash B)\Rightarrow(\neg B\vdash\neg A)]$

(see the subsection **3.3**, where these were argued to be enough for the consistency-less fragment of our language), and also of at least one of the following:

$$\forall A\, \forall B\ [(A \vdash B) \Rightarrow (\circ A \vdash \circ B)]; \qquad \text{(RO)}$$

$$\forall A\, \forall B\ [(A \dashv\vdash B) \Rightarrow (\circ A \vdash \circ B)]. \qquad \text{(EO)}$$

We have already shown, in FACT 3.36, that (RC) cannot be added to **Ci** without collapsing into classical logic. We can now actually show more:

FACT 3.52 In extensions of **Ci**, the validity of (EC) also guarantees (EO).
Proof: From $[A \dashv\vdash B]$ we conclude, by (EC), that $[\neg A \dashv\vdash \neg B]$. From these two sentences, by positive logic, we conclude that $[(A \wedge \neg A) \dashv\vdash (B \wedge \neg B)]$, but from FACT 3.16(ii) and FACT 3.29(i) we know that $[\neg \circ C \dashv\vdash (C \wedge \neg C)]$, and so we have that $[\neg \circ A \dashv\vdash \neg \circ B]$. Finally, from FACT 3.34(iv), we have that $[\circ A \dashv\vdash \circ B]$. □

The problem of finding paraconsistent extensions of **Ci** in which (IpE) holds reduces then to the problem of finding out if (EC) can be added to this logic without losing the paraconsistent character. We suspect this can be done, but shall leave it as an open problem at this point. As to extensions of **bC**, on the other hand, we can already present a (very partial) result:

THEOREM 3.53 There are fragments of **eCPL** extending **bC** in which (IpE) holds.
Proof: We already know, from COROLLARY 3.22, that (IpE) does not hold for **bC** as it is. It suffices now to show that the addition of the rules (EC) and (EO) to **bC** may still originate a paraconsistent fragment of (extended) classical logic, once these rules are evidently enough to ensure (IpE). To such an end, one may simply make use of the following matrices by Urbas ([107], Theorem 8):

$\wedge$	**1**	6/7	5/7	4/7	3/7	2/7	1/7	**0**
1	1	6/7	5/7	4/7	3/7	2/7	1/7	0
6/7	6/7	6/7	3/7	2/7	3/7	2/7	0	0
5/7	5/7	3/7	5/7	1/7	3/7	0	1/7	0
4/7	4/7	2/7	1/7	4/7	0	2/7	1/7	0
3/7	3/7	3/7	3/7	0	3/7	0	0	0
2/7	2/7	2/7	0	2/7	0	2/7	0	0
1/7	1/7	0	1/7	1/7	0	0	1/7	0
0	0	0	0	0	0	0	0	0

$\vee$	**1**	6/7	5/7	4/7	3/7	2/7	1/7	**0**
1	1	1	1	1	1	1	1	1
6/7	1	6/7	1	1	6/7	6/7	1	6/7
5/7	1	1	5/7	1	5/7	1	5/7	5/7
4/7	1	1	1	4/7	1	4/7	4/7	4/7
3/7	1	6/7	5/7	1	3/7	6/7	5/7	3/7
2/7	1	6/7	1	4/7	6/7	2/7	4/7	2/7
1/7	1	1	5/7	4/7	5/7	4/7	1/7	1/7
0	1	6/7	5/7	4/7	3/7	2/7	1/7	0

$\to$	**1**	6/7	5/7	4/7	3/7	2/7	1/7	**0**
1	1	6/7	5/7	4/7	3/7	2/7	1/7	0
6/7	1	1	5/7	4/7	5/7	4/7	1/7	1/7
5/7	1	6/7	1	4/7	6/7	2/7	4/7	2/7
4/7	1	6/7	5/7	1	3/7	6/7	5/7	3/7
3/7	1	1	1	4/7	1	4/7	4/7	4/7
2/7	1	1	5/7	1	5/7	1	5/7	5/7
1/7	1	6/7	1	1	6/7	6/7	1	6/7
0	1	1	1	1	1	1	1	1

	$\neg$
1	0
6/7	5/7
5/7	2/7
4/7	3/7
3/7	4/7
2/7	5/7
1/7	1
0	1

verse interpretations from several researchers (our own proposal on its interpretation will be found in the subsection **3.11**), the requisite **dC[i]** is in fact the one to blame for the confusion we were talking about. First of all, under our present perspective, to call the formula $\neg(A\wedge\neg A)$ 'principle of non-contradiction' is quite misleading, not only because this would bring us, as a side effect, to commit to very particular interpretations for the negation of a proposition, the conjunction of contradictory propositions, and the negation of this conjunction, but ascribe to us as well a very particular interpretation for the consistency connective (however, if —and only if— you are working in the context of some specific consistent logics, such as classical logic itself, or intuitionistic logic, we admit that this designation can indeed make sense). Evidently, FACT 3.54 is then just a consequence of such a contract.

Now, if, on the one hand, some authors have questioned the validity of $\neg(A\wedge\neg A)$ in the context of a paraconsistent logic,[18] on the other hand the construction of paraconsistent logics in which this formula does not hold has also been rather criticized since then, and for various reasons. Some of these criticisms unfold from or link to, by and large, still that same understandable and widespread confusion between the 'principle of non-contradiction' and the aims of paraconsistent logic, namely to avoid the 'principle of explosion' instead (see **dC[ii]**, and our subsections **2.1** and **2.2**) —but these more or less loose arguments can hardly be recast under our present formal definitions of those principles. A slightly more elaborate argumentation appears in Routley & Meyer's [100], where the authors are looking for some formalization of dialectical logic, and they claim to that effect not only that $\neg(A\wedge\neg A)$ is 'usually' a theorem of the 'entailment systems' that they have examined, but also that this does not conflict with other logical truths of those dialectical systems (in their words, this does not generate any 'intolerable tensions which destroy any prospect of a coherent logic'), even though these systems do have contradictory theorems (that are to be understood as 'synthetic a priori'), and validate adjunction. Moreover, they maintain that 'the orthodox Soviet position appears to retain $\neg(A\wedge\neg A)$ as a thesis', and they want to deal with it.[19] Now, none of the logics we study here are dialectical, in the sense of disrespecting the Principle of Non-Contradiction and actually proving contradictory formulas (recall the subsection **2.2**), and so the last critique above, in any case, simply falls idle.

---

[18] For instance, Béziau's [23], section 2.3, argues that, from a *philosophical* standpoint, it is hard to reconcile the validity of $\neg(A\wedge\neg A)$ and an intuitive interpretation for the negation symbol of a paraconsistent logic; as to the *technical* aspect of his criticism, it seems to consists basically of a consequence of THEOREM 3.51(vii) and the wish to obtain both adjunctiveness and the validity of (IpE).

[19] They also say some other things which seem a bit weird. First, that the 'non-orthodox' systems not containing $\neg(A\wedge\neg A)$ as a theorem are all *weaker* dialectical logics (as if the logics were all linearly ordered by strength!). Secondly, they insist on calling the formula $\neg(A\wedge\neg A)$ 'Aristotle's principle of non-contradiction', and after formally presenting their dialectical logics they argue that this formula is 'correct, both in syntactical and semantical formulations': *syntactically* correct because 'it is a theorem, hence valid, hence true' —despite the seemingly naive *petitio principii* brought therein; and *semantically* correct because '[one of its historical formulations] asserts that no statement is both true and false', and this feature, in the case of their logics, is supposed to be 'guaranteed by the bivalent features of the semantics' —now this is surely a mistake, for what guarantees this fact can only be the functional (rather than relational) character of their proposed interpretation, but in any case this last argument by these authors, even if they were kind enough to clear up the somewhat obscure relation of it with the first one, should hardly be accepted as a justification, given that any associated semantics provided to a consequence relation of a given logic is barely *circumstantial*, in a sense, and can often be recast in many apparently non-equivalent ways (if you're not happy with a particular semantics, you can always *look for another one*). A similar criticism of these points has been made before in Batens's [10], section 9.

where 1 is the only distinguished value, and add to these a matrix for $\circ$ that is constant and equal to 0. It is straightforward to check that the above matrices validate all axioms and rules of **bC** plus the two rules above, while formulas such as $(A\to(\neg A\to B))$ and $\neg(A\wedge\neg A)$ are still not validated by them.[17] □

3.8 Beyond Ci: The dC-systems. We have now come closer to the more orthodox approach to paraconsistent logics that the reader will find in the field, which *does* identify contradictoriness and inconsistency. All logics that we will be studying from here on, and, we argue, all logics of formal inconsistency presented in the literature so far, do not distinguish between these two notions. But this does *not* mean, the reader should be aware, that one can simply dispense with the new operators that have allowed us, so far, to talk about a formula being consistent or inconsistent, for, we remember from FACT 3.16, even though $[\bullet A \dashv\vdash_{Ci} (A\wedge\neg A)]$ holds, $[\circ A \dashv\vdash_{Ci} \neg(A\wedge\neg A)]$, for instance, does not hold (in fact, $[\nvdash_{Ci} \neg(A\wedge\neg A)]$)! Suppose then that we construct now the logic **CiI** exactly by adding to **Ci** (that is, (Min1)–(Min11), (bc1), (ci), (MP), plus the definition of $\circ$ in terms of $\bullet$) the following 'missing' axiomatic rule:

(cl) $\neg(A\wedge\neg A)\vdash\circ A$. 'If $\neg(A\wedge\neg A)$ is the case, then A is consistent'

Much confusion has been raised around this particular formula, $\neg(A\wedge\neg A)$. Recall, from THEOREM 3.15(ii) and THEOREM 3.30, that it was not a theorem of our previous logics, **bC** or **Ci**. Now, of course, we can immediately conclude even more:

FACT 3.54 No paraconsistent extension of **CiI** can have $\neg(A\wedge\neg A)$ as a theorem.
Proof: If so, axiom (cl) would give us consistency, thus ruining paraconsistency. □

Nonetheless, some other paraconsistent extensions of **Ci**, such as **LFI1** (see its matrices in the subsection **2.4** or at FACT 3.29, and see its axiomatization in THEOREM 3.69), *do* have this as theorem (and, as a consequence of THEOREM 3.51(ii), they must lack a full replacement theorem).

The attribution of a privileged status to the formula $\neg(A\wedge\neg A)$, using it to express consistency inside some paraconsistent logics, stems from the early requisites put forward by da Costa on the construction of his famed calculi C_n:

- **dC[i]** in these calculi the principle of non-contradiction [sic], in the form $\neg(A\wedge\neg A)$, should not be a valid schema;
- **dC[ii]** from two contradictory formulas, A and $\neg A$, it would not in general be possible to deduce an arbitrary formula B;
- **dC[iii]** it should be simple to extend these calculi to corresponding predicate calculi (with or without equality);
- **dC[iv]** they should contain the most part of the schemas and rules of the classical propositional calculus which do not interfere with the first conditions.

While the requisite **dC[ii]** is nothing but the very definition of a paraconsistent logic (recall the subsection **2.2**), **dC[iii]** is simply a claim for extensions of these logics to higher-order calculi (that we will *not* explore here, reiterating instead the popular and still powerful argument of paraconsistentists about the fact that most, if not all, innovations of paraconsistent logic can already be met at the propositional level), and **dC[iv]** is indeed somewhat vague, having received much attention and many di-

[17] Notice, nevertheless, that all that is proved here is that there certainly exist fragments of classical logic extending **bC** for which (IpE) holds. But the matrices above do not fulfill, of course, our requisite for defining a paraconsistent logic (namely, disrespecting (PPS)), so that the question is still left open as to whether there are *paraconsistent* such extensions of **bC**! (With thanks to Dirk Batens for calling our attention to that.)

But let us first explore some consequences of the new axiom (cl), before really questioning it any deeper, or looking for substitutes. The main and most far-reaching consequence is the following:

THEOREM 3.55 In **Cil** we can define the inconsistency operator as $\bullet A \stackrel{\text{def}}{=} (A \wedge \neg A)$ (from which the consistency operator will be defined as $\circ A \stackrel{\text{def}}{=} \neg(A \wedge \neg A)$).
Proof: It can immediately be seen, from the above definitions, that the axioms (bc1), (ci) and (cl) will still hold if one just substitutes all occurrences of the operators $\bullet$ and $\circ$ by their new definitions. □

COROLLARY 3.56 Given a theorem B of **Cil** we can substitute all occurrences of $\bullet$ and of $\circ$ in its subformulas according to the above definitions.
Proof: Recall from THEOREM 3.41 that all axioms of **Ci** can be written just with the use of $\bullet$, substituting $\circ$ for $\neg\bullet$, or just with the use of $\circ$, substituting $\bullet$ for $\neg\circ$. In the first case, where we have only '$\bullet$'s, the above theorem permits us to rewrite in **Cil** the proof of B using $(A \wedge \neg A)$ in the place of each formula $\bullet A$ that appears, and using $\neg(A \wedge \neg A)$ in the place of each formula $\neg\bullet A$ that occurs in the proof. In the second case, and for the same reason, we may rewrite the proof of B using $\neg(A \wedge \neg A)$ in the place of each formula $\circ A$ that appears, and using $(A \wedge \neg A)$ in the place of each formula $\neg\circ A$ that occurs in the proof (you may in this last part also wish to take FACT 3.40 and (Min11) once more into consideration). □

The above results are structurally similar to those of THEOREM 3.41 and its consequences, where a restricted form of replacement was obtained for the operators $\bullet$ and $\circ$, and we have seen that each one of them could be substituted in **Ci** by the negation of the other. But now we know more, we know that we can simply dispense with the operators $\bullet$ and $\circ$, substituting each formula $\bullet A$ and each formula $\neg\circ A$ for the formula $(A \wedge \neg A)$, each formula $\circ A$ and each formula $\neg\bullet A$ for the formula $\neg(A \wedge \neg A)$. This brings us to the definition of a particular subclass of the C-systems that we will call **dC**-*systems*, such as the C-systems in which $\bullet$ and $\circ$ can be defined in terms of the other connectives. **Cil** is the first example of a **dC**-system that we here consider; before presenting other examples let us point out some consequences of this last result for **Cil**. It is easy to see, for instance, that the restricted forms of contraposition presented in FACT 3.17 for **bC** and in FACT 3.34 for **Ci**, as well as the forms of *reductio* presented in FACT 3.14 for **bC** and in FACT 3.31 for **Ci**, and the forms of controllable explosion presented in this last fact, together with the fundamental fact relating controllable explosion and consistency in FACT 3.32, all have new versions in **Cil**, if we just change each ocurrence of $\circ$ and $\bullet$ for their definitions in THEOREM 3.55. We can also update THEOREM 3.11 with yet another way of reproducing classical inferences inside of **Cil** by the addition of the appropriate premises to its theories (namely the addition of a finite number of formulas of the form $\neg(A \wedge \neg A)$, the formula that in the present circumstances represents the consistency of A). Analogously to what we did in FACT 3.40, we can now prove that $[(A \wedge \neg A) \vdash_{\mathbf{Cil}} \neg\neg(A \wedge \neg A)]$, even though $[A \vdash_{\mathbf{Cil}} \neg\neg A]$ still does not hold.

Now, if FACT 3.50 has provided us with a very precise characterization of the consistent theorems in **Ci**, which turned out to be only consistent or inconsistent formulas themselves, namely the ones appearing in FACT 3.33 and FACT 3.39, we may now use those same results to conclude that formulas such as $\circ(A \wedge \neg A)$, and, by

FACT 3.38(i), also $\circ\neg(A\wedge\neg A)$, are theorems of **Ci1**. These last theorems have raised yet some other protests in the literature. For instance, Sylvan (cf. [105]) claims that the fact that such a logic validates some 'unjustifiable' intuitionistically invalid theorems, together with the validity of $\circ(A\wedge\neg A)$ 'defeats certain paraconsistent objectives' (too bad that he did not proceed to clear up which objectives were these...). This echoes, in one way or another, to a common, and entirely well-founded, criticism, which has been raised by various authors, both to the fact that **C**-systems such as those we have been studying do maintain the whole of positive classical logic and to the fact that many of them (but not all!) are in fact **dC**-systems, and come up with rather particular definitions for the consistency operator. However, both these aspects can be easily varied and experimented. We have here, by a matter of simplicity, set up an investigation of **C**-systems based on classical propositional logic, but it is clear that other approaches may be tackled, by the investigation of **C**-systems based on relevance logic, or intuitionistic logic, as soon as some paradoxes of relevance, or paradoxes raising from some non-constructive assumption, are decided to be avoided. This is clearly not, however, a problem of *paraconsistency* as we have it, but a further (interesting) problem which can be added to it.

Some **dC**-systems based on intuitionistic logic have in fact been defined and studied, for instance, in Bunder's [29]. $\mathbf{B}_1$, the stronger logic of the main hierarchy of calculi proposed by the author of that paper, is obtainable simply by dropping the axioms (Min9), (Min10) and (Min11) out of **Ci1**, while adding to the resulting logic the axiom:

(bun) $(A\rightarrow(\circ B\wedge(B\wedge\neg B)))\vdash\neg A$. 'If A implies a bottom particle, then $\neg A$ is the case'

It is more or less clear that the deletion of the above axioms from **Ci1** will give the resulting logic a kind of intuitionistic behavior, and that the addition of (bun) cannot recover any of the classical properties which were lost. Despite of this, most, if not all, other claims that the author advances about this logic seem to be mistaken. He conjectures, for instance, that the strong negation defined by the antecedent of (bun), by setting $\dot{\sim}A \stackrel{\text{def}}{=} (A\rightarrow(\circ B\wedge(B\wedge\neg B)))$, for some formula B, *is not* a classical negation, and *continues not to be* a classical negation even if one adds back to $\mathbf{B}_1$ the axioms (Min10) and (Min11). This is wrong, for we know from FACT 3.45 that in this case the axiom (Min9) turns to be provable, and so we obtain a logic at least as strong as **Ci** (plus (bun), if this would make any difference), but than we remember from FACT 3.49 that $\dot{\sim}$ is *indeed* a classical negation in **Ci** (and even in **bC**, as we saw in THEOREM 3.48). The author then claims, and purports to prove, that his $\mathbf{B}_1$ is *not* a subsystem of da Costa's logic C_1, which, as we will see in the subsection **3.10**, is simply an extension of **Ci1**. Once more he is wrong, and for the very same reason —FACT 3.49 shows us once more that (bun) is evidently provable in **Ci**, and so already **Ci** (and consequently C_1) extends $\mathbf{B}_1$ and all the other weaker calculi proposed by Bunder. From that point on, all of the remaining remarks made by this author on the comparison of his calculi with the ones proposed by da Costa falls apart. The only point remaining from those calculi, therefore, is that of constituting **dC**-systems based on intuitionistic, rather than classical, logic. But the author did not even try to study them any deeper, looking for instance for interpretations for these calculi!

In another paper, Bunder unwittingly produced an even bigger mistake. Starting from the reasonable idea of looking for other formulations for da Costa's version of *reductio* (that is, (RA0): $[\circ B, (A\rightarrow B), (A\rightarrow\neg B)\vdash\neg A]$, in the subsection **3.2**), the author simply proposes (in [31]) to change $\circ B$ for $\circ A$ in that formula, asserting that 'there seems to be no particular reason why, in (RA0), the B has a restriction, rather than the A'. In this case, however, we can use our THEOREM 3.13 again to see that this proposal would be equivalent to the addition of $[\circ B,\ A,\ \neg A \vdash C]$ to C_{min}, which is clearly absurd, for it would be trying to express the consistency of A by way of some foreign formula B! The author then claims that the 'paraconsistent' calculus D_1 (again, the 'strongest' one of a hierarchy D_n) that he obtains by adding this last formula as a new axiom to C_1 is 'strictly stronger than C_n', the calculi of da Costa's hierarchy, and purports to prove some facts about them. Once more these facts turn out to be mistaken, and this is easy to see if one remembers that $\circ\circ D$ is a theorem of **Ci** (see our FACT 3.33(i), or his Theorem 5), and so we are left with $[A,\ \neg A \vdash C]$, for any A and C, and explosiveness is back. So, the author was actually right about his calculi being extensions of the calculi C_n, but only because they all *collapse* into classical logic, after all...[20]

As advanced above (and we shall confirm this below, in the subsection **3.10**), that the identification of consistency with the formula $\neg(A\wedge\neg A)$ was exactly what was done in da Costa's calculus C_1, which in fact just adds to **Cil** some more axioms to deal with the 'propagation of consistency' from simpler to more complex formulas. Now, many authors have criticized this identification —'there is nothing sacrosanct about the original definition of this schema as $\neg(A\wedge\neg A)$', says Urbas in [107]—, or else its consequences, as we have mentioned above (as the 'anomalies' described, for instance, in Sylvan's [105]). One of the most unexpected consequences of this identification, in fact, has already been pointed out in Urbas's [107], Theorem 4, and was hinted above in our subsection **3.2**:

THEOREM 3.57 In **Cil** the consistency of the formula A can be expressed by the formula $\neg(A\wedge\neg A)$, but *not* by the formula $\neg(\neg A\wedge A)$. In fact, one can even add $\neg(\neg A\wedge A)$ to **Cil**, but not $\neg(A\wedge\neg A)$, without this logic losing its paraconsistent character.

Proof: That consistency is so expressed in **Cil** and that $\neg(A\wedge\neg A)$ cannot be added to it are simply consequences, respectively, of COROLLARY 3.56 and FACT 3.54. But while it is easy to see that $[\neg(A\wedge\neg A) \vdash_{\mathbf{Cil}} \neg(\neg A\wedge A)]$, the converse of this does *not* hold, as we see from the matrices in THEOREM 3.21 (those three-valued matrices are in fact a much simpler way of checking the same result for which Urbas has used six-valued ones). Notice, in particular, that these matrices in fact also validate the formula $\neg(\neg A\wedge A)$. □

The above phenomenon is a bit tricky, and has actually fooled people working with the calculi C_n for perhaps too long a time (see, for instance, [76], note 6, ch.2, p.49, or else [42]). Let us then consider the following alternatives to the 'levo-'axiom (cl):

(cd) $\neg(\neg A\wedge A) \vdash \circ A$;

(cb) $(\neg(A\wedge\neg A)\vee\neg(\neg A\wedge A)) \vdash \circ A$.

---

[20] By the way, given the above considerations, one of Bunder's main results about these systems (besides the supposed proof about all the calculi D_n constituting different systems), objected to show that the calculi D_n do not satisfy (IpE) (namely the Theorem 10 that closes his [31]) evidently must fail. It is easy, in fact, to find counter-examples for the validity of the formula $[\neg(A\wedge\neg A),\ (A\rightarrow B),\ (A\rightarrow\neg B) \vdash \neg A]$ in the matrices that he proposes (picking them up from Urbas's [107], who have used them correctly, in the case of the C_n systems): just choose $v(A)\in\{0,3\}$ and $v(B)=1$.

Evidently, the addition to **Ci** of the 'dextro-'axiom (cd), instead of the axiom (cl), would give us this logic **Cid** which has exactly the same qualities and defects as **Cil**, but which would singularize the formula $\neg(\neg A\wedge A)$ as much as the formula $\neg(A\wedge\neg A)$ has been previously singularized by **Cil**. The addition of (cb), instead, defining the logic **Cib**, would assure to both $\neg(A\wedge\neg A)$ and $\neg(\neg A\wedge A)$ the same status, and this would fix, for instance, the famed asymmetry in THEOREM 3.21(iii) (but not the ones in parts (i) and (ii)). Logics having (cb) instead of (cl) have already been studied (see [36] or [76]), but it should be noted that these still suffer from some anomalies related to the definition of consistency in terms of some operation over a conjunction of contradictory formulas. In fact, if the logics having (cb) as an axiom do identify the two formulas above, they do not necessarily identify these with some other formulas such as $\neg(A\wedge(A\wedge\neg A))$, or $\neg((A\wedge\neg A)\wedge A)$, for instance, even though all of the formulas $(A\wedge\neg A)$, $(\neg A\wedge A)$, $(A\wedge(A\wedge\neg A))$ and $((A\wedge\neg A)\wedge A)$ are equivalent on any **C**-system based on classical logic. All of this will have, of course, deep consequences when we go on to provide semantics to these logics, as the reader will see in [42]). Perhaps a good way of fixing all of this at once is by the addition of a new 'global' axiomatic rule to such **dC**-systems, such as:

(cg) $\quad (B\leftrightarrow(A\wedge\neg A))\vdash(\neg B\leftrightarrow\neg(A\wedge\neg A))$,

or else the weaker deduction rule:

(RG) $\quad [B \dashv\vdash (A\wedge\neg A)] \Rightarrow [\neg B \dashv\vdash \neg(A\wedge\neg A)]$.

Logics having such rules are yet to be more deeply investigated. In one way or another, it is clear that the mere addition of such rules is not enough to remedy the whole of THEOREM 3.21 (but compare the following result to the proposal by Mortensen, in the subsection **3.12**). Indeed, similarly to what had happened in COROLLARY 3.22 and in THEOREM 3.35:

THEOREM 3.58 (IpE) does not hold for **Cib** plus (cg) or (RG).
Proof: Check for instance that THEOREM 3.21(iii) still holds, that is, that formulas such as $\neg(A\vee B)$ and $\neg(B\vee A)$ are still not equivalent, once more by way of the same matrices and distinguished values as in THEOREM 3.15(ii), but now changing only the value of $(1\vee ½)$ from 1 to ½ (and leaving the value of $(½\vee 1)$ as it is, equal to 1). □

It is also noteworthy that da Costa in fact proposed not just one definition of consistency (the one above, from the calculus C_1), but considered instead the possibility of having weaker and weaker logics (see [49] or [50]), modifying the requirement for consistency in such a way as to produce an infinite number of logics at once (he acted more with an illustrative than with a practical purpose, but that manoeuvre has, in one way or another, produced some permanent impression). The idea is simple, namely that of having, for each C_n, $0\leq n<\omega$, more and more premises to be fulfilled in order to guarantee consistency. In the case of $n=1$ we already know that $\circ A$ (da Costa denoted it A°) abbreviated the formula $\neg(A\wedge\neg A)$, for $1<n<\omega$ it was taken to be $A^{(n)}$, where this abbreviation was recursively defined by first setting A^n, $0\leq n<\omega$, as $A^0 \stackrel{\text{def}}{=} A$ and $A^{n+1} \stackrel{\text{def}}{=} (A^n)^{\circ}$, and then setting $A^{(n)}$, $1\leq n<\omega$, as $A^{(1)} \stackrel{\text{def}}{=} A^1$ and $A^{(n+1)} \stackrel{\text{def}}{=} A^{(n)}\wedge A^{n+1}$. In other words, each of da Costa's **dC**-systems was defined by exactly the same axioms, changing only the definition of $\circ A$ in each case for $A^{(n)}$, for each given n.[21] It is clear in this way, if one really feels inclined to do it fosome reason or another, that for each **dC**-system one can go on to multiply it into an infinite number of (in principle, distinct) **dC**-systems, applying the same strategy above. Of course, the same asymmetries already pointed out in the case of **Cil**, **Cid** and **Cib** (THEOREM 3.57), may still apply in each case in appropriate forms, and the theorems obtained from these systems must also be modified, in each case, according to the specific definition of consistency brought therein.

3.9 The opposite of the opposite. Having been introduced to the **dC**-systems, a particular class of **C**-systems that can dispense with the use of the operators $\circ$ and $\bullet$, we shall not, nevertheless, dedicate ourselves in what follows exclusively to the study of **dC**-systems. All the logics we will present from this point on are still bound to be **C**-systems extending **Ci**, but only by chance will they turn out to be **dC**-systems as well. What we will consider in this subsection is the addition of the following axiomatic rule for 'expansion' of negations, converse to (Min11): $(\neg\neg A \to A)$:

(ce) $\quad A \vdash \neg\neg A$.

Let **Cie** be the logic obtained by the addition of (ce) to **Ci** (recall, from THEOREM 3.30, that this addition is *not* redundant). In the subsections **3.1** and **3.2** we have learned about the role played by (Min11) in **bC** and the logics which extend it (recall THEOREM 3.13), and suggestions were made as to the reasons why da Costa has introduced (Min11) in his first paraconsistent calculi as a dual substitute to (ce), present in intuitionistic logic, as much as (Min10): $(A \vee \neg A)$ was intended to be the dual substitute to (PPS), the explosiveness (or *reductio*) that is lost by all paraconsistent calculi. Despite this, qualified forms of both (PPS) and (ce) are retained by **bC**, in the form of the rule (bc1): $[\circ A,\ A,\ \neg A \vdash B]$, and the rule $[\circ A,\ A \vdash \neg\neg A]$, this last one being a rule of **bC** that comes immediately from (bc1) and FACT 3.14(iii). Now, it happens that only (PPS), but not (ce), is a problem of paraconsistent logic, as we put it, and, as far as we know, (ce) was only avoided by da Costa in his first calculi (see [49] and [50]), in spite of his manifest intention, on his requisite **dC[iv]** (see the last subsection), to maintain 'most rules and schemas of classical logic not conflicting with the other requisites', because there seemed to be some apprehension about the addition of (ce) leading us back to classical logic, **CPL**, or perhaps making us just lose the paraconsistency character of our logics, after all. It is, however, very easy to see that this is not the case. Indeed:

THEOREM 3.59 (tPS): $(A \to (\neg A \to B))$ is not provable by **Cie**.

---

[21] One should observe, however, that the definition of the schema $A^{(n)}$ proposed in da Costa's foundational work, [49], in reality does *not* coincide with the definitions to be found in other studies in the literature (such as the well-known da Costa's [50], or da Costa & Alves's [53]), and those that we adopt here. Indeed, on page 16 of [49] the reader will find the following definition, setting $A^{(1)} \stackrel{\text{def}}{=} A^\circ$ and $A^{(n+1)} \stackrel{\text{def}}{=} A^{(n)} \wedge (A^{(n)})^\circ$. If one follows this last definition, one ought to conclude, for instance, that $A^{(3)}$ is to denote the formula $A^\circ \wedge A^{\circ\circ} \wedge\ (A^\circ \wedge A^{\circ\circ})^\circ$, while the definition we have presented above would give instead $A^\circ \wedge A^{\circ\circ} \wedge A^{\circ\circ\circ}$. It is easy to see, however, if one just makes use of any of the semantics and decision procedures that have been associated to the calculi C_n that these two formulas are *not* equivalent in each C_n (see [36] for the semantics of a slightly stronger version of these calculi —in the case of $n=1$, for instance, axiom (cb) is used instead of (cl), and axiom (ce), which appears below, was also added; or else go to [76] and [74], for the original versions).

Proof: Use the matrices of **LFI1** again, as in FACT 3.29, or else the matrices of $\mathbf{P}^1$, as in THEOREM 3.30 —but in this last case you must change the matrix of negation, setting the value of $\neg\frac{1}{2}$ as $\frac{1}{2}$, instead of 1. In fact, it is to be remarked that this modification on the matrices of $\mathbf{P}^1$ in fact originates a new and interesting maximal three-valued paraconsistent logic, $\mathbf{P}^2$. These three logics, **LFI1**, $\mathbf{P}^1$ and $\mathbf{P}^2$, will be studied in their own right in the subsection **3.11**, as members of a larger family of similar logics. □

Some of the immediate and main syntactical results obtained by the logic **Cie** are:

FACT 3.60 **Cie** proves the following:

(i) $\circ\neg A \vdash_{\mathbf{Cie}} \circ A$;

(ii) $\bullet A \vdash_{\mathbf{Cie}} \bullet\neg A$.

Proof: Just turn the FACT 3.38 upside-down. □

Now, some people felt unease by the presence of FACT 3.60(ii), understanding that this would mean a 'proliferation of inconsistencies' —given that any formula A proved to be inconsistent would, by way of that rule, generate infinitely many 'other' inconsistencies (the negation of A, the negation of the negation of A, and so on). Be that as it may, it is still clear that in **Cie** there are no *new* inconsistencies added by way of this procedure, in a sense, given that the converse of FACT 3.60(ii) is also valid here (it is the FACT 3.38(ii)), and so, in fact, $\bullet A$ and $\bullet\neg A$ are *equivalent* formulas! In **Cile**, the logic obtained by the addition of (ce) to **Cil**, instead of **Ci** (to see that this logic is also paraconsistent, use again the matrices of $\mathbf{P}^2$ in THEOREM 3.59) we evidently obtain a new version of the above result, and FACT 3.60(ii) converts itself into $[(A\wedge\neg A) \vdash_{\mathbf{Cile}} (\neg A\wedge\neg\neg A)]$, leading, so it seems, into a 'proliferation of contradictions'. Once again, this would perhaps not be said to be the case if the formulas at the right and the left hand side are again remembered to be equivalent.

Routley & Meyer, in [100], on their attempt to define a 'dialectical logic', *DL*, meeting the standards of 'Soviet logic' and to recover the 'orthodox Marxist view of negation' have pondered the possibility of criticism coming from some dialecticians to the effect that their 'negative logic is excessively classical', and considered the constitution of a 'weaker dialectical logic', *DM*, having only (Min11), but not (ce), as an axiom. But, even in the case of their *DL*, they have met inferences such as the ones above, acknowledging the possibility of generation of an infinite number of 'distinct' contradictions from any given one, and still defended that this would be all right — one just has to remember that it is still not the case that *any* contradiction is derivable, indeed, just a very specific set of contradictions, 'forming a chain', are derivable, but this, of course, 'does not result in total system disorganization'. But how could it be that these contradictions that they obtain in *DL* are all *distinct*, as they asserted? This is a bit tricky. Let A_0 and $\neg A_0$ be two theorems of *DL* (it's a *dialectical* logic, after all —see the subsection **2.2**), and let A_n abbreviate the formula $(A_{n-1}\wedge\neg A_{n-1})$. We have already learned in the last subsection about Routley & Meyer's argument for the validity of the schema $\neg(A\wedge\neg A)$, and from this we may conclude that each A_n will be a theorem, starting from the mere fact that both A_0 and $\neg A_0$ hold. But if, on the one hand, it is clear that $(A_n\rightarrow A_{n-1})$ will be a theorem of *DL*, on the other hand it is equally clear that $[A_n \dashv\vdash A_{n-1}]$ holds, as above, similarly also to what had occurred in the cases of **Cie** and **Ciel** (FACT 3.60(ii) and its variations). So, again, how could it be that 'all these contradictions are distinct', as asserted by these authors? The point is that the propositional bases of both *DL* and *DM* are relevance logics, and so it may occur that A_n and A_{n-1} are equivalent formulas, but it is still the case that $(A_{n-1} \rightarrow A_n)$ is not a theorem of *DL*. So, after all, we see that *this* is the sense of 'distinctness' employed by those authors, determined exclusively by the validity or not of a bi-implication, and *not* by the sets of consequences of the formulas under examination.

3.10 Consistency may be contagious! Supposing we can really trust the consistency of some formulas in our theories, what can we say about the more complex formulas that one can build using the last ones as components: will *these* be also consistent? From FACT 3.38(i): $[\circ A \vdash \circ\neg A]$ we know that already in **Ci** the consistency 'propagates' through negation, that is, the consistency of A is sufficient information for us to be sure about the consistency of $\neg A$. This is essentially a consequence of (Min11): $(\neg\neg A \rightarrow A)$ and the identification of inconsistency and contradiction guaranteed by the axiom (ci). Now, what do we know about the propagation of consistency through other connectives besides negation? Not much, so far.

The idea behind the construction of the original calculi C_n by da Costa (see [49] and [50]) was that of requiring each component to be consistent as a sufficient reason to count on the consistency of the more complex formula. Bluntly speaking, da Costa's C_1 was built by the addition to **Cil** (see the beginning of the subsection **3.8**) of the following axiomatic rules:

(ca1) $(\circ A \wedge \circ B) \vdash \circ(A \wedge B)$;
(ca2) $(\circ A \wedge \circ B) \vdash \circ(A \vee B)$;
(ca3) $(\circ A \wedge \circ B) \vdash \circ(A \rightarrow B)$.

Let's call **Cila** the logic obtained by the addition of (ca1)–(ca3) to **Cil**. The difference from **Cila** and the original formulation of C_1 is only one: that the connective $\circ$ in C_1 was not taken as primitive, but $\circ A$ was instead denoted as A° and was taken more directly as an abbreviation of the formula $\neg(A \wedge \neg A)$ (recall the THEOREM 3.55). As for the other calculi in the hierarchy C_n, $1 \leq n < \omega$, they were built using the simple trick of letting $\circ A$ abbreviate more and more complex formulas (as we saw at the end of the subsection **3.8**).

As an immediate consequence of the above definitions, one can easily prove in **Cila** —and in each calculus C_n— the following 'translating' results (compare these with the less specific THEOREM 3.11 and with the generally applicable COROLLARY 3.47):

THEOREM 3.61 $[\Gamma \vdash_{\mathbf{CPL}} A] \Leftrightarrow [\circ(\Pi), \Gamma \vdash_{\mathbf{Cila}} A]$, where $\circ(\Pi) = \{\circ p : p$ is an atomic formula occurring as a subformula in $\Gamma \cup \{A\}\}$.[22]

Proof: Immediate, using (ca1)–(ca3). Note that the axiom (cl) plays no role here. □

COROLLARY 3.62 The following mapping conservatively translateseCPL inside of **Cia**:

(t2.1) $t_2(p) = \circ p$, if p is an atomic formula;
(t2.2) $t_2(A \# B) = t_2(A) \# t_1(B)$, if $\#$ is any binary connective;

[22] Might the reader observe that the first formulations of this result, on da Costa's [49], Theorem 9, page 16, and on da Costa's [50], Theorem 4, page 500, the general case in which an infinite number of atomic formulas occur in $\Gamma \cup \{A\}$ is not considered.

(t2.3) $t_2(\neg A) = \neg t_2(A)$;
(t2.4) $t_2(\circ A) = \circ t_2(A)$.

So, working with **Cila**, all we need to rely on in order to go on making 'classical inferences' is on the consistency of the atomic constituents of our formulas. As a particular consequence of that, one can now substitute each new axiomatic rule of **Cila** by an alternative version in terms of '$\bullet$'s instead of '$\circ$'s. Thus, the axiom (ca3), for instance, can be rewritten as $[\bullet(A\to B) \vdash (\bullet A \vee \bullet B)]$ (use FACT 3.34(i) and COROLLARY 3.62). And so on.

Proposing an infinite number of calculi, instead of one, as in the case of the C_n, only starts to make sense after we prove that we are not just repeating the same tune:

THEOREM 3.63 Each C_n deductively extends each C_{n+1}, for $1 \leq n < \omega$.[23]
Proof: We will not here give it a try by usual 'syntactical means'. For sure, this will be much easier to check if one just considers the semantics associated with these calculi, for instance in [53] (*corrected* in [74]) and in [36] (or [76]). □

Evidently, all these calculi C_n extend also the calculus C_ω —they even extend C_{min}, the stronger logic on which we based **bC**, our first **LFI** (recall the subsection **3.1** for the definition of these logics). This C_ω, we argue, was indeed a very bad choice as a kind of 'limit' to the hierarchy C_n, $1 \leq n < \omega$. Consider, for instance, the following result:

FACT 3.64 The only addition made by C_n (in fact, by **Cia**, for the axiom (cl) has no use in this result) to the rules provable by **bC** about the interdefinability of the binary connectives (see THEOREM 3.18) is the rule (ix): $[\neg(A \wedge B) \vdash_{\mathbf{Cia}} (\neg A \vee \neg B)]$, and its variants.
Proof: We just show that that rule holds already in **Cia**, and point the reader again to the semantical studies of the calculi C_n to check that the other formulas are still not provable in **Cila**. First of all, setting $\Gamma = \{\circ(A\wedge B), \neg(A\wedge B), A\}$ it can immediate be seen that $[\Gamma, B \vdash_{\mathbf{Cia}} \circ(A\wedge B)]$, $[\Gamma, B \vdash_{\mathbf{Cia}} (A\wedge B)]$ and $[\Gamma, B \vdash_{\mathbf{Cia}} \neg(A\wedge B)]$, so we apply FACT 3.14(ii) to obtain $[\Gamma \vdash_{\mathbf{Cia}} \neg B]$, and consequently $[\Gamma \vdash_{\mathbf{Cia}} (\neg A \vee \neg B)]$. But then, also $[\neg A \vdash_{\mathbf{Cia}} (\neg A \vee \neg B)]$, and so the proof by cases will give us $[\circ(A\wedge B), \neg(A\wedge B) \vdash_{\mathbf{Cia}} (\neg A \vee \neg B)]$. By (ca1) we then conclude that $[\circ A, \circ B, \neg(A\wedge B) \vdash_{\mathbf{Cia}} (\neg A \vee \neg B)]$, but we also have, from LEMMA 3.43(ii), that $[\vdash_{\mathbf{Cia}} (\neg A \vee \circ A)]$, and from that we obtain $[\circ B, \neg(A\wedge B) \vdash_{\mathbf{Cia}} (\neg A \vee \neg B)]$. By a similar reasoning, from $[\neg B \vdash_{\mathbf{Cia}} (\neg A \vee \neg B)]$, we finally arrive at our goal, $[\neg(A\wedge B) \vdash_{\mathbf{Cia}} (\neg A \vee \neg B)]$. □

COROLLARY 3.65 (IpE) cannot hold in the calculi C_n, or in any extension of them.
Proof: Just recall THEOREM 3.51(iv). □

The FACT 3.64 also suggests some further information about the plausibility of calling either C_ω or C_{min} 'limits' for the hierarchy C_n, $1 \leq n < \omega$. For, as we have seen in THEOREM 3.18, these new forms of De Morgan rules that we now have in each C_n

---

[23] A supposedly general proof of this fact, dating still from the 'syntactical period', when no semantics had yet been presented to those calculi, appears for instance in da Costa's [49], pp.17–9, and once again in Alves's [2], pp.17–9, and is credited to Ayda Arruda. There is surely some mistake, however, in their attempt to prove the independence of each axiom $[A^{(n)}, A, \neg A \vdash B]$ with respect to the axioms of C_{n+1}, given that this very axiom assumes non-distinguished values in all matrices $\mathbf{T}_n$ thereby presented, if one only picks 1 as the value of A and picks for B any value in between 1 and $n+2$.

are *not* present even in C_{min}. Also, by a combination of FACT 3.40(i) and COROLLARY 3.56, we know that $((A\wedge\neg A)\rightarrow\neg\neg(A\wedge\neg A))$ is valid in **Cia**, and so in each C_n, while we also know, from the matrices of $\mathbf{P}^1$, as in THEOREM 3.15(ii), that this formula cannot be a theorem of neither C_{min} nor C_ω. Now, it is only compelling to think of a *deductive limit* for an infinite hierarchy of increasingly weaker calculi as the logic having as inferences exactly all sets of inferences common to the whole hierarchy![24] As we have shown in [39], it *is* possible to define such a logic, for each hierarchy of **dC**-systems as the one given above, by way of the useful tool of possible-translations semantics, obtaining, as a byproduct, some clear-cut and effective decision procedures, even though some other very interesting questions, such as how to finitely axiomatize this limit-calculi, or how to define a strong negation in them, if this is possible at all, were still left open (check also the paper [42]). Indeed, notice that when we go from each C_n to the following C_{n+1} we need in fact to add a further requirement in order to express the consistency of a formula A —while in C_n this was expressible by way of $A^{(n)}$, or, equivalently, by way of the set $\{A^1, A^2, \ldots, A^n\}$, in C_{n+1} that same set must be incremented by the formula A^{n+1}. So, ultimately, in C_{Lim}, the deductive limit of the hierarchy C_n, the consistency of A can evidently be expressed by an infinite number of formulas, and again we obtain a logic which is gently explosive, being thus an **LFI**. What we still do not know is if logics such as C_{Lim} can be alternatively characterized in such a way so as to also reveal themselves as *finitely* gently explosive, like all the other logics presented up to this point, based on the axiom (bc1). If this characterization is not possible, it is hard to see how a strong negation or a bottom particle could then be *defined* in such a logic,[25] so that this would make a

[24] This logic C_ω has puzzled people for too long as 'part' of the original hierarchy C_n. The existence of a logic as a real deductive limit to this hierarchy (see [39]) shows that it was clearly just a matter of coincidence that C_ω would appear as a kind of 'syntactical limit' to the original axiomatic formulation of the C_n, by the deletion of all the axioms and definitions involving the connectives $\circ$ and $\bullet$ ((bc1), (ci) and (cl)), and the 'deletion' also of (Min9) (actually, this last axiom was not in the original formulat i-on of these calculi, and could not even be proved from the other axioms of C_ω —see THEOREM 3.3). To put the matter in clear terms, C_ω *can* of course be studied in its own right, as a very weak paraconsistent (non-**LFI**) logic based on positive intuitionistic logic, but it *should not* be seen as part of the hierarchy C_n, $n\geq 1$, for it has no more right to occupy that position than C_{min} or many other logics that could substitute it would have!

This coincidence had also some harmful effects on the philosophical appreciation of the logics produced by da Costa. As da Costa himself has put it in his original piece on these systems (cf. [49], p.21), 'roughly speaking, we could say that human reason seems to attain the peak of its power the more it approaches the danger of trivialization'. This statement has been inspiring people to naively defend stances according to which, for instance, 'the more a theory is useful to found mathematics, the more easily it results to trivialize it; and the more difficult it is to trivialize it, the less it is useful to found mathematics' (see [26], p.243). There are good and bad points about these somewhat hasty conclusi-ons. First, as a general technical assertion about paraconsistent logics in general, da Costa's motto is certainly misleading, given the existence of maximal logics such as the three-valued *Pac* (subsection **2.4**), which is both as strong as a fragment of classical logic as it could be, and at the same time is not finitely trivializable at all. One could, then, restrict their attention to **LFI**s and repeat that motto in an environment in which it seems to make sense. In that case, of course, the second statement above would be affirming that no non-**LFI** could be useful to found mathematics —and *this* statement would be very likely to find its defensors (cf., for instance, Batens's attack [13] on Priest's [90]).

[25] Here, we really mean *defined* inside the logic, as a real *formula* of this logic. For instance, in Priest's [90] a logic containing no bottom particle is presented, but the author argues that such a propositional constant $\perp$ could be 'thought of informally as the conjunction of all formulas' (p.146), so that, for instance, a strong negation ~ would be obtainable from that in the usual way, by letting $\sim A$ be defined

case in which the Gentle Principle of Explosion does not coincide with the Supplementing one, or with *ex falso* (compare this with FACT 2.19, and the comments which follow that result).

Some other interesting theses of **Cila** are the following (see Urbas's [107]):

FACT 3.66 In **Cila** the schemas $\circ\bot$ and $\circ\dot{\sim}\dot{\sim}A$ are provable.
Proof: Recall first that $\dot{\sim}A$ was the classical negation defined inside of **bC** (in THEOREM 3.48) as $(A\to\bot)$, and $\bot$ is a bottom particle that can be defined, for instance, as $(\circ B\wedge(B\wedge\neg B))$, for some B. Now, by FACT 3.33 and COROLLARY 3.56, we know that both $\circ\circ B$ and $\circ(B\wedge\neg B)$ are theorems of **Cil**, and then we conclude, by the axiom (ca1), that $\circ(\circ B\wedge(B\wedge\neg B))$, and so $\circ\bot$ is a theorem of **Cila**. Recall also from FACT 3.49 that $\dot{\sim}A$ is equivalent, in **Ci**, to $\sim A$, and this last strong negation was defined as $(\neg A\wedge\circ A)$, and so we have in particular that $[(A\to\bot)\vdash_{\mathbf{Cila}}\circ A]$. So, from this last inference and from the fact that $\circ\bot$ is a theorem of **Cila**, as proved above, we use (ca3) and conclude that $[(A\to\bot)\vdash_{\mathbf{Cila}}\circ(A\to\bot)]$. As particular cases of the last two inferences, substituting A for $((A\to\bot)\to\bot)$ in the first case, and for $(A\to\bot)$ in the second case, we obtain, respectively, $[\dot{\sim}\dot{\sim}\dot{\sim}A\vdash_{\mathbf{Cila}}\circ\dot{\sim}\dot{\sim}A]$ and $[\dot{\sim}\dot{\sim}A\vdash_{\mathbf{Cila}}\circ\dot{\sim}\dot{\sim}A]$, and the version of proof by cases obtained for $\dot{\sim}$ leaves us at last with $[\vdash_{\mathbf{Cila}}\circ\dot{\sim}\dot{\sim}A]$. □

This last result gives us yet another reason for the failure of (IpE) in the C_n and in their extensions (COROLLARY 3.65) —now, it is THEOREM 3.51(i) that applies. As we shall see in subsection **3.12**, the failure of (IpE) makes it impossible for us to find a Lindenbaum-Tarski-like algebraization for these logics. In the case of the C_n the situation is actually worse: as Mortensen ([82]) has shown, no non-trivial congruence is definable for these logics, making these logics non-algebraizable even in a much more general sense, the one of Blok-Pigozzi (cf. THEOREM 3.83). There are, nevertheless, several extensions of the C_n in which non-trivial congruences *can* be defined, being thus much more receptive to algebraic treatments. We will be seeing many examples of these below.

Let us now investigate another way of propagating consistency, by liberalizing a little bit the conditions required by **Cila** (that is, C_1). Da Costa, Béziau & Bueno proposed, in [57], to substitute the above axioms, (ca1)–(ca3), by the following:

(co1) $(\circ A\vee\circ B)\vdash\circ(A\wedge B)$;
(co2) $(\circ A\vee\circ B)\vdash\circ(A\vee B)$;
(co3) $(\circ A\vee\circ B)\vdash\circ(A\to B)$.

We will call **Cilo** the logic obtained by the addition of (co1)–(co3) to **Cil**. It is very easy to see, using the positive axioms, that this logic, christened C_1^+ in [57], is a deductive extension of C_1. Requiring less assumptions in order to obtain consistency of a complex formula in terms of the consistency of its components, **Cilo** (or even **Cio**, already, without recourse to the axiom (cl)) gives us some interesting results such as:

THEOREM 3.67 $[\Gamma\vdash_{\mathbf{Cio}}\circ A]$ whenever $[\Gamma\vdash_{\mathbf{Cio}}\circ B]$, for some subformula B of A.
Proof: Immediate, using (co1)–(co3). □

---

as $(A\to\bot)$. But such an 'informal' bottom particle is simply no formula of our language! (And if it were, then we would have been done: all paraconsistent logics would turn out to be **LFIs**, in one way or another). This idea of Priest has already been criticized in Batens's [13].

FACT 3.68 **Cio** makes some new additions to FACT 3.64 and to the rules displayed in THEOREM 3.18 about the interdefinability of the binary connectives, namely the following provable schematic rules:

(vi) $\neg(A\wedge\neg B)\vdash_{\mathbf{Cio}}(A\to B)$;

(vii) $\neg(A\to B)\vdash_{\mathbf{Cio}}(A\wedge\neg B)$;

(xi) $\neg(\neg A\vee\neg B)\vdash_{\mathbf{Cio}}(A\wedge B)$.

Proof: Go to [57] or [76], or else the section on semantics for **Cibo** (and **Ciboe**) in [42], to check this. The actual syntactical proofs are, in any case, structurally similar to the one presented in FACT 3.64. □

Of course, given THEOREM 3.51(v), we know that FACT 3.68(vii) gives us yet another reason for the failure of (IpE) from this point on. But that we already knew, from the case of **Cila**, in FACT 3.64. What is new in this case is only that **Cilo** *can*, differently from **Cila**, define non-trivial congruences, making it possible to algebraize it *à la* Blok-Pigozzi (see FACT 3.81).

Once again, the axioms (co1)–(co3) have equivalent versions in terms of •, instead of ∘, and it is an easy exercise to try to find them. The reader should remember that both **Cila** and **Cilo** have not only associated decreasing hierarchies, and can evidently define different calculi as their deductive limits, but they also can be structurally varied in terms of their inner definition of consistency, if we only change the axiom (cl) for axiom (cd), or (cb), or (cg), as we did in the subsection **3.8**, or if we add to them the axiom for expansion of negations, (ce), as in the subsection **3.9** (defining the logics **Cido, Cibo, Cito, Ciloe, Cidoe, Ciboe** and **Citoe**). In all these cases, we can show that the resulting logics are extended by the three-valued paraconsistent logic $\mathbf{P}^2$, as in the THEOREM 3.59.

There are, actually, an unlimited number of ways of propagating consistency.[26] Before proceeding to a general investigation of the 'extreme cases', in the next subsection, let us just briefly survey some propagation axioms which have already shown up in the literature so far, and some of their consequences. Consider, for instance, the following axioms, converse to (co1)–(co3):

(cr1) $\circ(A\wedge B)\vdash(\circ A\vee\circ B)$;

(cr2) $\circ(A\vee B)\vdash(\circ A\vee\circ B)$;

(cr3) $\circ(A\to B)\vdash(\circ A\vee\circ B)$.

Adding these to **Cibo** and to **Cio** (that is, **Cibo** minus the axiom (cb)) we build, respectively, the logics **Cibor** and **Cior** (and so on, *mutatis mutandis*, for **Cilo** and **Cido**). These give us yet more perspectives on the propagation (and back-propagation) of consistency, and some of the possible meanings which can be assigned to it). Suppose, on the other hand, that we, more simply, consider the following axioms in order to automatically guarantee the consistency of some complex propositions:

(cv1) $\vdash\circ(A\wedge B)$;

(cv2) $\vdash\circ(A\vee B)$;

[26] Working with monotonic logics, the addition of consistency-propagation to some logic always means a gain in deductive strength. However, in a non-monotonic environment in which consistency is pressuposed by default, given that propagating consistency in one direction can mean propagating inconsistency the other way, the addition of such axioms for propagation can either be innocuous or in some cases even have a weakening effect on the resulting system (see [79], and [15]).

(cv3) $\vdash \circ(A \to B)$;
(cw) $\vdash \circ(\neg A)$.

Let's add **v** to the name of a logic that contains the axioms (cv1)–(cv3), and add **w** to the name of a logic containing (cw). Let's also recall the axiom (ce): $[A \vdash \neg\neg A]$, from which we obtained the FACT 3.60 (backward propagation of consistency through negation). There are now several new possible combinations to be considered. Evidently, any logic having (cv1)–(cv3) proves not only (co1)–(co3) but also (ca1)–(ca3). So we have, for instance, the logic **Cibv**, and have at least two immediate ways of enrichening it with respect to the behavior of negation, obtaining the logics **Cibve** and **Cibvw**. As it happens, **Cibvw** axiomatizes the three-valued maximal paraconsistent logic $\mathbf{P}^1$ (matrices in THEOREM 3.15(ii)), that has the peculiarity of admitting inconsistency only at the atomic level, and **Cibve** axiomatizes $\mathbf{P}^2$ (matrices in THEOREM 3.59), a logic that admits inconsistency only at the level of atomic propositions, or of propositions of the form $(\neg^n p)$, where p is atomic and $\neg^n$ denotes n applications of negation. If, on the other hand, one considers the logic **Ciborw**, once more a three-valued paraconsistent logic pops up, namely the one given by the following matrices:

∧	**1**	**½**	**0**
1	1	1	0
½	1	½	0
0	0	0	0

∨	**1**	**½**	**0**
1	1	1	1
½	1	½	1
0	1	1	0

→	**1**	**½**	**0**
1	1	1	0
½	1	½	0
0	1	1	1

	¬	∘
1	0	1
½	1	0
0	1	1

where 1 and ½ are both distinguished. We shall here call this logic $\mathbf{P}^3$. All these three logics, $\mathbf{P}^1$, $\mathbf{P}^2$ and $\mathbf{P}^3$, are in fact **dC**-systems, and can ultimately dispense with the axiom (cb), proving it from the other axioms. If, on the other hand, we consider the logic **Ciore**, the result is a maximal three-valued logic again, **LFI2** (investigated in [44]) whose matrices differ from those of $\mathbf{P}^3$ only in the matrix of negation, assigning ½ instead of 1 as the value of $\neg(½)$.

All the above logics have some kind of 'non-structural' propagation of consistency, that is, a propagation that does not really depend on the particular connective in focus. Alternatively, one can propose other forms of propagation which do depend on the connectives being considered. Now, some reasonable symmetry conditions on inconsistency and its behavior with respect to the different connectives could suggest to us, for instance, the consideration of the following forms:

(cj1) $\bullet(A \wedge B) \dashv\vdash ((\bullet A \wedge B) \vee (\bullet B \wedge A))$;
(cj2) $\bullet(A \vee B) \dashv\vdash ((\bullet A \wedge \neg B) \vee (\bullet B \wedge \neg A))$;
(cj3) $\bullet(A \to B) \dashv\vdash (A \wedge \bullet B)$.

The logic **Cij**, built from the addition of (cj1)–(cj3) to **Ci**, can now be enriched with (ce) in order to give us **Cije**, an axiomatization for the above many times mentioned maximal three-valued paraconsistent logic **LFI1** (see its matrices in FACT 3.29, and consult [44] again). It is interesting enough to note that neither **LFI1** nor **LFI2** are **dC**-systems, that is, they *cannot* define the consistency operator by way of the other connectives. Let us now just summarize, give some references and mention some properties of the five above mentioned maximal paraconsistent three-valued logics, before we proceed to show, in the next subsection, that these are just the top of the iceberg:

THEOREM 3.69 The matrices of $\mathbf{P}^1$ (in THEOREM 3.15(ii)) are axiomatized by **Civw**; the matrices of $\mathbf{P}^2$ (in THEOREM 3.59) by **Cive**; the matrices of $\mathbf{P}^3$ (above) by **Ciorw**; the matrices of **LFI2** (above) by **Ciore**; the matrices of **LFI1** (in FACT 3.29) by **Cije**.

Proof: As a general reference for all these logics and all the other ones in the next section, consult [78]. As specific references for some of them, go to [103] and [84] for $\mathbf{P}^1$ (or [10], where it appeared under the name PI^v); notice that $\mathbf{P}^2$ has also appeared in [10], under the name PI^m, but was then redefined in [84] (where it actually was wrongly supposed to be characterizable using just one distinguished value, invalidating the soundness proof therein presented —note 13 of that paper shows that the author had even been informed about that) and later rediscovered in [76]; go to [44] for **LFI2** and **LFI1** (but the reader should bear in mind that this last logic is in fact equivalent to the logic called $\mathbf{J}_3$ in [60], and to the propositional fragment of the logic called **CLuNs** in [15], which is in fact identical to the logic Φ_v presented in [101], a logic that has been reappearing quite often in the literature). □

FACT 3.70 Of the rules displayed in THEOREM 3.18 on the interdefinability of the binary connectives, these are the instances validated by each of the five three-valued logics above:

(i) in $\mathbf{P}^1$, $\mathbf{P}^2$, $\mathbf{P}^3$ and **LFI2**: parts (i), (iv), (vi), (vii), (ix), (xi);

(ii) in **LFI1**: parts (i), (iii), (iv), (v), (vii), (viii), (ix), (x), (xi), (xii).

Also, formulas such as $\neg(A \wedge \neg A)$ and $\neg(\neg A \wedge A)$ may be easily seen to hold in **LFI1** and **LFI2**, and rules such as $(A \wedge \neg A) \vdash \neg\neg(A \wedge \neg A)$ hold in $\mathbf{P}^2$, **LFI1** and **LFI2**.

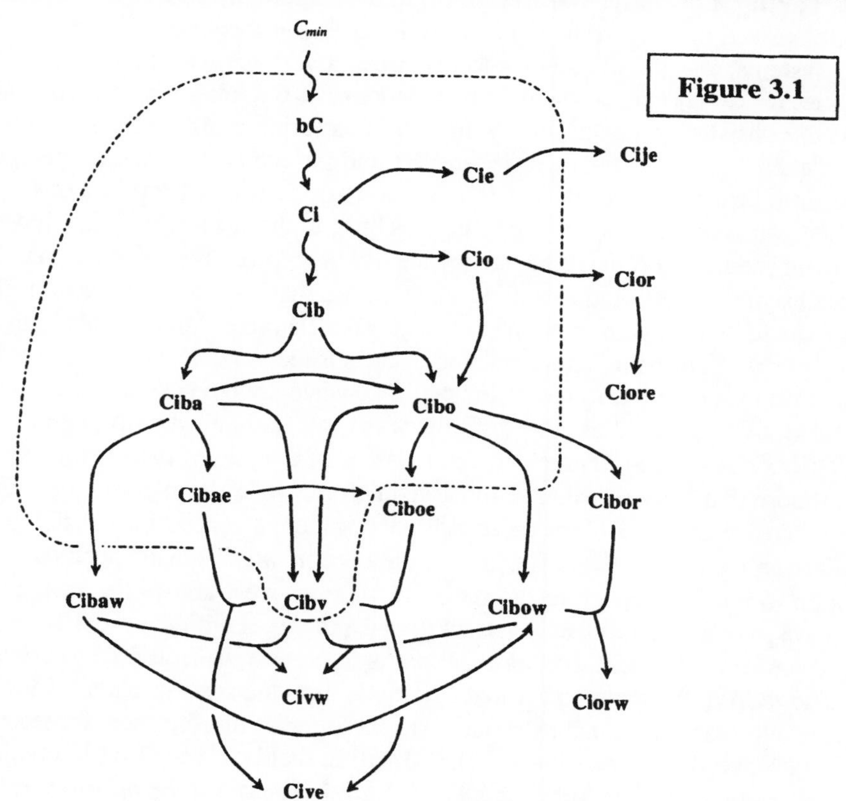

Figure 3.1

Proof: Just use the corresponding matrices to check this. Notice from FACT 3.68 that part (iv) was the only addition made by the first four logics above to the rules already validated by **Cilo**. □

This last result, of course, supplies us with still some further justifications for the failure of (IpE) in all these logics: parts (iii) and (vi) of THEOREM 3.51 applies to **LFI1**, parts (ii) and (vii) of 3.51 apply to both **LFI1** and **LFI2**, parts (iv) and (v) of 3.51 apply to all of the five logics; and finally we will see in FACT 3.76 that part (i) of 3.51 also applies to all of them.

We can, at this point, try our hand at sketching a very thin slice of the great number of **C**-systems introduced so far. Doing that, something like **Figure 3.1** might eventually be obtained. In that figure, an arrow leading from a logic **L**1 into a logic **L**2 says that **L**2 deductively extends **L**1. The logic C_{min}, at the upper end, is the only one that does not constitute a **C**-system; the logics at the lower ends are the three-valued ones appearing in THEOREM 3.69. The logics inside the dotted lines are some of those which we can prove to be *not* many-valued, by adapting the results in [76], pp.213–216 or, better, by checking [42]. The other logics not contemplated by these results, namely **Cior**, **Ciboe**, **Cibor**, **Cibaw**, and **Cibow**, are also conjectured to be not many-valued, but we must at this moment leave the proof of this fact in the hands of our clever readers. Do remember to have a look, however, at the elegant possible-translations semantics offered to **Ciboe** in [42] (also originating from [76], section 5.3).

3.11 Taking it literally: the Brazilian plan completed. The sagacious reader will have observed that all we have been doing so far, in this section on axiomatization of **C**-systems, was to basically try to explore at a very general level some of the possibilities for the formalization and understanding of the relationship between the concepts of consistency, inconsistency and contradictoriness. In particular, this research line makes it possible for us to reconsider and pursue, in an abstract perspective, a specific interpretation of da Costa's method and requisites on the construction of his first paraconsistent calculi (see **dC[i]**–**dC[iv]**, in the subsection **3.8**). Indeed, starting from the intuition that consistency should be expressible inside some classes of paraconsistent logics, and assuming furthermore that the consistency of a given formula would be enough to guarantee its explosive character (that is, assuming a Gentle Principle of Explosion, as formulated in the subsection **2.4**), we have arrived at the definition of an **LFI**, a Logic of Formal Inconsistency (see (D20), in the same subsection). To realize that (in a finitary way), we have above proposed the axiom (bc1): $[\circ A,\ A,\ \neg A \vdash B]$ for particular classes of **C**-systems based on classical logic. Even more than that, as we have remarked before, while **dC[ii]** simply establishes the non-explosive character of the paraconsistent negation, a general formulation of **dC[i]** is realized in a subclass of the **C**-systems, the ones in which the connectives '$\circ$' and '$\bullet$' happen to be definable from the remaining connectives, and to the members of this class we gave the name of **dC**-systems. Now, putting **dC[iii]**, the problem of providing higher-order versions of these logics, aside for a moment, we still need to provide an answer to **dC[iv]**, the requirement that 'most schemas and rules of classical logic' should hold in our logics. And that's the point we will ruminate in the present subsection.

Our proposed interpretation for **dC[iv]** will in fact be a very simple one, involving the following notion of 'maximality'. A logic **L**2 is said to be *maximal relative to* a logic **L**1 if: (i) both are written in the same language (so that they can be deductively compared); (ii) all theorems of **L**2 are provable by **L**1; (iii) given a theorem D of **L**1 which is not a theorem of **L**2, if D is added to **L**2 as a new schematic axiom, then all theorems of **L**1 turn to be provable. The idea, of course, is that any deductive extension of **L**2 contained in **L**1 and obtained by adding a new axiom to **L**2 would turn out to be identical to **L**1. We will call *maximal*, to simplify, any logic **L**2 which is maximal relative to some logic **L**1, previously introduced. Examples of maximal logics abound in the literature. It is widely known, for instance, that each Łukasiewicz's logic $Ł_m$ is maximal relative to **CPL**, the classical propositional logic, if and only if $(m-1)$ is a prime number. We also know that **CPL** is maximal relative to a 'trivial logic', in which all formulas are provable, but on the other hand it is also well-known that intuitionistic logic is *not* a maximal fragment of **CPL**, as the existence of an infinite number of *intermediate logics* promptly attests. As to the **C**-systems which have been introduced this far, only the five three-valued ones that were collected in the THEOREM 3.69 are maximal relative to **CPL**, or else relative to **eCPL**, the extended version of **CPL** introduced in the subsection **3.7** (so that, in particular, the calculus C_1, that we have presented as **Cila**, the strongest calculus introduced by da Costa on his first hierarchy of paraconsistent calculi, or the even stronger calculus C_1^{+}, that we presented as **Cilo**, proposed by da Costa and his collaborators much later, readily fail to be maximal, and to respect **dC[iv]**).

Let us explore then the idea that underlies the five three-valued maximal **C**-systems above. Suppose we are faced with this problem of finding models to contradictory, and yet non-trivial, theories. We might then intuitively start looking for non-trivial interpretations under which both some formula A and its negation $\neg A$ would be simultaneously validated. A very simple such interpretation would be found in the domain of the many-valued. Suppose we try to depart from classical logic as little as possible, so that the interpretation of our connectives will still be classical if we remain inside the classical domain, and suppose we just introduce a third value (½), besides true (1) and false (0), so that this third value will also be seen as a modality of *trueness*, that is, ½ will also be a distinguished value, together with 1, while 0 will be the only non-distinguished value. There are then two possible negations which are such that there is a model for both A and $\neg A$ being true, for some formula A (see the table to the right).

	$\neg$
1	0
½	½ or 1
0	1

One of these negations, the one that takes ½ into ½, is exactly the negation of **LFI1** and of **LFI2**, the other negation, taking ½ into 1, is exactly the negation of $\mathbf{P}^1$, $\mathbf{P}^2$ and $\mathbf{P}^3$. What about the other connectives? Let us again try to keep them as classical as possible (we want to keep on investigating **C**-systems *based on classical logic*), even at the level of the third value, that is, let us add to the requirement of coincidence of classical outputs for classical inputs the further higher-level 'classical' requirements to the effect that:

(C∧) $v(A\wedge B)\in\{½, 1\} \Leftrightarrow v(A)\in\{½, 1\}$ and $v(B)\in\{½, 1\}$;

(C∨) $v(A\vee B)\in\{½, 1\} \Leftrightarrow v(A)\in\{½, 1\}$ or $v(B)\in\{½, 1\}$;

(C→) $v(A\rightarrow B)\in\{½, 1\} \Leftrightarrow v(A)\notin\{½, 1\}$ or $v(B)\in\{½, 1\}$.

This leaves us then with the following options:

∧	1	½	0
1	1	½ or 1	0
½	½ or 1	½ or 1	0
0	0	0	0

∨	1	½	0
1	1	½ or 1	1
½	½ or 1	½ or 1	½ or 1
0	1	½ or 1	0

→	1	½	0
1	1	½ or 1	0
½	½ or 1	½ or 1	0
0	1	½ or 1	1

Thus, we have, theoretically, 2^3 options of 'conjunctions', 2^5 options of 'disjunctions', 2^4 options of 'implications', and, as we saw above, 2^1 options of 'negations', making a total of 2^{13}(=8,192, or 8K) possible 'logics' to play with. To remove the scare quotes of the previous passage we just have to show that these logics make some sense, and are worthy of being explored. To such an end, and to complete the definition of our 8K logics as **LFIs**, we will just also add to these logics the connectives for consistency and for inconsistency, implicitly assuming that the consistent models are the ones given by classical valuations, and only those (see matrices to the right). Evidently, all these 8K logics will be fragments of **eCPL**, the Extended Classical Propositional Logic (recall the subsection **3.7**). It is also clear that the logic *Pac* (subsection **2.4**) is not one of these, for it cannot define the connectives ∘ and •, though its conservative extension **LFI1** can (and it is one of the 8K).

	∘	•
1	1	0
½	0	1
0	1	0

Evidently, the five three-valued logics we discussed earlier are but special cases of the above outlined 8K logics, and we already know (from THEOREM 3.69 and **Figure 3.1**) that those five are axiomatizable by way of the addition of suitable axioms to the axiomatization of **Ci**, one axiom for each connective. In fact, as shown in [78], this idea can be extended to all the 8K logics above:

THEOREM 3.71 All the 8K three-valued sets of matrices above are axiomatizable as extensions of **Ci**.

Proof: In each case, one just has to add, for the negation, either the axiom $(A \to \neg\neg A)$ or the axiom $\circ\neg A$, depending respectively if the negation of ½ goes to ½ or to 1. And, for each other binary connective, $\#(\in \{\wedge, \vee, \to\})$, one just has to add either $\circ(A\#B)$ or else $(\bullet(A\#B) \leftrightarrow \sigma(A, B))$, where $\sigma(A, B)$ is a schema depending only on A and on B —these last axioms will evidently depend on the specific matrices of each #, and act in order to describe how inconsistency (or consistency) propagates back and forth for each binary connective. Full details on how to define these axioms may be found in [78]. □

Moreover, one can also prove that:

THEOREM 3.72 All the 8K three-valued logics above are distinct from each other, and they are all maximal relative to **eCPL**.

Proof: Again, we refer to [78] for the general proofs. The basic idea behind the proof of *distinctness* is the following: choosing any two of these 8K logics (without repetition), there will be some connective about which they differ, one of them giving 1 as an output for the same input(s) that the other one gives ½. But then the negations of such matrices will not be equivalent, and all we must do then is write down a formula which describes that situation in such a way that this formula will be a theorem of one of these logics, but a non-theorem of the other (again, see [78] on how to do it). For the *maximality* proofs, the reader might mind to be informed that for at least five of those logics (the ones referred to in THEOREM 3.69) the specific proofs were already presented elsewhere. It is also interesting to remark that the connective $\circ$ (or $\bullet$) plays a fundamental role in the general maximality proof exhibited in [78]. □

Now, how do these 8K logics compare with the other **C**-systems that have been studied this far? It is this simple: *every* logic investigated so far either coincides or is extended by some of the above 8K three-valued logics. So that now we have a very interesting class of (extended) solutions to the problem posed by da Costa's requirement **dC[iv]**! Furthermore, it is straightforward to check that all the above matrices do not only extend **Ci**, but also extend **Cia**, so that the original hunch by da Costa for the propagation axioms is a kind of minimal condition obeyed by every one of our 8K maximal three-valued logics. The only limitative point of the original proposal, under this approach, really rests in **dC[i]**, which is of course *not* verified by all those matrices, and in fact imposes a very restricted interpretation for the notion of consistency, limiting our sample space to only a *very* selective class of **dC**-systems, which is, however, larger than the reader might initially imagine (recall, in any case, that logics such as **LFI1** and **LFI2** are *not* **dC**-systems). Indeed:

FACT 3.73 All the 8,192 logics above are **C**-systems extending **Cia**. Of these, 7,680 are in fact **dC**-systems, being able to define $\circ$ and $\bullet$ in terms of the remaining connectives (and being maximal, thus, relative to **CPL**, and not only to **eCPL**). Of these, 4,096 are able to define $\circ A$ as $\neg(A\wedge\neg A)$, and so all of these do extend C_1 (that is, **Cila**). Of the 7,680 logics which are **dC**-systems, 1,680 extend **Cio**, the stronger alternative to **Cia**, and 980 of these are able to define $\circ A$ as $\neg(A\wedge\neg A)$, so that these 980 logics do extend C_1^+ (that is, **Cilo**).

Proof: This is just a combinatorial exercise on the above matrices, and we shall leave it for the reader to check. □

It might well be that not all of the above 8K three-valued maximal logics will be interesting as logics. Some of them, for instance, do not have symmetric matrices for the conjunction or for the disjunction (but notice that some such logics have had their use in results such as 3.21, 3.26, 3.35 and 3.57, or in 3.58), though any conjunction / disjunction is evidently equivalent to any other conjunction / disjunction (the negations of these conjunctions / disjunctions are what may differ). The fact that all the 8K three-valued logics do extend **Cia** (FACT 3.73) informs us, as a corollary to FACT 3.64, that:

FACT 3.74 (IpE) cannot hold in any of the 8K logics above.

Proof: Again, just recall THEOREM 3.51(iv). □

Now, if, in the next subsection, this failure of the replacement theorem will be seen to constitute a negative answer for the possibility of obtaining a Lindenbaum-Tarski-style algebraization for these logics (as already occurred for the calculi C_n and all of their extensions —see COROLLARY 3.65), the following result will help us to show in the following a positive answer for the possibility of obtaining a Blok-Pigozzi-like algebraization to each one of them (as already hinted for some extensions of C_n, such as C_1^+, see FACT 3.81 and FACT 3.82):

FACT 3.75 The following matrices of *classical negation* and *congruences* can be defined in each one of the above 8K logics:

	~
1	0
½	0
0	1

≡	**1**	**½**	**0**
1	1	0	0
½	0	½ or 1	0
0	0	0	1

Proof: To define the classical negation ~ one just has first to define $\bot$ either as $(B\wedge(\neg B\wedge\circ B))$ or as $(\circ B\wedge\neg\circ B)$, for some formula B, and then define $\sim A$ either as $(\neg A\wedge\circ A)$ or as $(A\rightarrow\bot)$. To define one of the above congruences one just has to set $(A\equiv B)$ as $((A\leftrightarrow B)\wedge(\circ A\leftrightarrow\circ B))$. If one wants to make sure that $v(A\equiv B)=1$ when both $v(A)=½$ and $v(B)=½$, this is also possible: just set some $(A\overset{1}{\equiv}B)$ as $\sim\sim(A\equiv B)$. □

In fact, it is not difficult to see that the above classical negation is indeed the *one and only* matrix of a strong negation that can be defined inside of these 8K three-valued logics (the paper [42] will also come back to this question). The reader will notice that this negation is indeed, in a sense, a 'highly' classical one. Indeed, it comes as a corollary, for instance, that:

FACT 3.76 The schema $\circ\sim A$ is provable in all of the above 8K three-valued logics.

This last result is more than what one needs to confirm, by way of THEOREM 3.51(i), the fact that (IpE) cannot hold in these logics (as in FACT 3.74).

A noteworthy expressibility result that can be proved for these 8K three-valued logics is the following:

FACT 3.77 (i) The matrices of $\mathbf{P^1}$ can be defined inside of any of the 8K three-valued logics above. (ii) All the matrices of all the 8K logics above can be defined inside of **LFI1**.

Proof: To check part (i), let $\wedge$, $\vee$, $\rightarrow$, $\neg$, $\circ$ and $\bullet$ be the connectives of any of the 8K logics above, and let ~ be the classical negation, defined inside this logic as in FACT 3.75. Then, the $\mathbf{P^1}$'s negation of a formula A can be defined as $\sim\sim\neg A$, the $\mathbf{P^1}$'s conjunction of some given formulas A and B, in this order, can be defined either as $\sim\sim(A\wedge B)$ or as $(\sim\sim A\wedge\sim\sim B)$, and the same we did for conjunction applies to both disjunction and implication, *mutatis mutandis*. The matrices for the connectives $\circ$ and $\bullet$ already coincide in all of these logics. Part (ii) is a particular consequence of the expressibility result that we have proven in [44], Theorem 3.6. In that result we showed, in fact, that the matrices definable in **LFI1** are all those, and exactly those, n-ary matrices that have classical (1 or 0) outputs for classical inputs (and that can have any output value if non-classical inputs are considered). Of course, all the above matrices, on these 8K three-valued logics, are, by definition, just 1-ary and 2-ary examples of such **LFI1**-definable *hyper-classical* matrices, as we have called them. □

COROLLARY 3.78 (i) The logic $\mathbf{P^1}$ can be conservatively translated inside any of the 8K three-valued logics above. (ii) Any of the 8K logics above can be conservatively translated inside of **LFI1**.

Are there other interpretations, besides *maximality*, of da Costa's requisite **dC[iv]** leading to yet some other classes of solutions to the problem of finding **C**-systems containing 'most rules and schemas of classical logic'? Are there non-many-valued (monotonic) solutions to that problem, or perhaps some other n-valued ones, for $n>3$? And, this is an important first step and probably an easier problem to solve, are there other interesting **C**-systems based on classical logic which are *not* extended by any of the above 8K three-valued logics? We must leave these questions open at this stage. It is interesting to notice, at any rate, that this problem has already been addressed here and there, in the literature. Besides [78], from which we drew the results in the subsection **3.12**, one could also recall, for instance, the adaptive programme for the confection of paraconsistent logics aiming to represent (non-monotonically) the dynamics of scientific reasoning and of argumentation (see [15]). Roughly speaking, the basic idea behind adaptive logics is that of working in between two boundary logics, classical logic often being one of them and a paraconsistent logic being the other one, so that consistency is pressuposed by default and we try to keep on reasoning (i.e. making inferences) inside of classical logic up to the point in which an 'abnormality' (an inconsistency?) pops out, a situation in which we had better descend to the level of the complementary paraconsistent logic, and go on reasoning over there. Indeed, the ancestral motivations of this programme (see [10]) seem to have been, as it is reasonable to conceive, yet another attempt to originate logics maintaining as much of classical logic as possible, so that paraconsistency will only be needed at limit cases.

As the reader will see in [42], when we go on to provide possible-translations semantics (as in [36], [39] or [76]) to some of the above non-three-valued logics, as for instance **Ci** or **Ciboe**, the intuition behind the construction of the previous three-valued paraconsistent logics can be pushed much farther, since it can be shown that some infinite-valued logics can also be *split* in terms of suitable combinations of clusters of three-valued logics.

3.12 Algebraic stuff. You may think, perhaps, that logic has 'too many formulas'. There is nothing unreasonable in supposing, however, that some of these formulas can in fact be identified, and indistinctly used in all contexts. If we consider classical logic, for instance, we will promptly see that there is no reason to distinguish between any two given theorems (or top particles) with respect to their relation to the other formulas of the classical language —even though they may very well still be understood as 'expressing' quite different facts, somehow conveying different bits of information. In the classical case, also, and for the very same reason, one does not really need to distinguish between any two given bottom particles (or two different pairs of contradictory formulas); even more than that, any two formulas A and B which turn out to be provably equivalent (that is, such that [$A \dashv\vdash B$], or, what in many logics amounts to be just the same, to put it in terms of a bi-implication, such that [$\vdash (A \leftrightarrow B)$]) are, in a certain sense, indistinguishable, and can be indistinctly employed in the same contexts, to attend similar purposes. The action of putting the glasses through which some 'contingent' properties of formulas are hidden and only those features related to their general behavior in relation to other formulas are exposed is the task of *algebraization*. Whenever a given logic turns out to be algebraizable, so that the logical problems can be faithfully and conservatively translated into some given well behaved algebra, then it will be possible for us to use the powerful (universal) algebraic tools to tackle those problems, so that, in the next and final step, we will be able to translate the results back into logic.

Being the above remarks all too informal, we had better strive to put them in more precise terms. If the whole activity of mathematics and logic involves 'forgetting' some things (and calling attention to others), and identifying what could otherwise, at the first look, have seemed just different, their tools by excellence, in such respect, are the key notions of *equivalence*, *congruence*, *isomorphism*, and so on. Once the very definition of a logic, as we have proposed it at the start of section **2**, can immediately be seen as some sort of algebra having the set of formulas as its domain (indeed, in the structural and propositional case, it is exactly the free algebra generated by the primitive connectives of the languages —here understood as operators— over the set of atomic propositions), the quest for dividing these formulas into disjunct packages of equivalent and indiscernible ones can easily be accomplished if one is able to define a *congruence* relation over these formulas, that is, an *equivalence* (reflexive, symmetric and transitive) relation such that any two equivalent formulas (with respect to this relation) can be just justifiably and indistinctly used in all and the same contexts. So, if some given formula A appears as a component of some other formula G, then any formula B which is congruent to A should be able to do the same job, with no loss or increase in expressibility or generality. What we implicitly mean with this is that the new *quotient* algebra obtained by dividing the original algebra of formulas by this congruence relation should preserve the original 'operations', existing thus an obvious homomorphism from the original algebra into the quotient algebra. So, in dividing the formulas this way into classes of congruent ones, one can go on to work and dialogue with just the (arbitrary) representants of these classes, once they are supposed to behave exactly the same as any of their congruent colleagues with respect to any operation of (any isomorph of) the quotient algebra. Any two congruent formulas are 'the same up to a congruence', and can play exactly the same roles in some specific dramas.

It comes perhaps as no surprise the confirmation that the most easy and standard way of algebraizing a given logic is obtained by way of the relation of provable equivalence induced by its underlying consequence relation. Indeed, the so-called *Lindenbaum-Tarski algebraization* sets two formulas A and B as congruent if $A \dashv\vdash B$ —let's denote this fact by writing $A \approx B$. Such 'congruence' relation $\approx$ is evidently an equivalence relation, and to confirm that any two so congruent formulas 'work the same in all contexts', one has to check if they can be intersubstituted everywhere, that is, one has to prove a *replacement* theorem, to the effect that the *intersubstitutivity of provable equivalents*, (IpE), holds (recall its definition in the subsection **2.3**, and check [111]). Many logics have Lindenbaum-Tarski-like algebraizations, as it is the case for classical logic, intuitionistic logic, several normal modal logics, several many-valued logics, and so on. But not all algebraizable logics are algebraizable in the sense of Lindenbaum-Tarski, not being able for instance to prove replacement with respect to provable equivalence (or provable bi-implication). In the case of many non-normal modal logics, for example, what one needs is *strict* (that is, *necessary*) provable equivalence (that is, strict bi-implication). In the case of the paraconsistent logics studied here, frequent negative results on what concerns the validity of (IpE) —and so, on what concerns the possibility of obtaining an algebraization *à la* Lindenbaum-Tarski— have been met: In fact, *all* of the above **C**-systems have been shown at some point to lack (IpE) (recall the results 3.22, 3.35, 3.58, 3.65, and 3.74). Yet, the possibility of obtaining some positive results within some extensions of those **C**-systems was not ruled out (recall 3.53, but confront it with 3.51).

An immediate result about algebraizations that may come quite handy is the following one (cf. [25], Corollary 4.9):

FACT 3.79 Every deductive extension of an algebraizable logic is algebraizable.

A case study which was particularly well investigated is that of the logic **Cila** (the logic C_1 of da Costa's [49] —check the subsection **3.10**). Even though at least as early as in da Costa & Guillaume's [54] it had already been noticed that (IpE) does not hold for **Cila** (COROLLARY 3.65), so that no Lindenbaum-Tarski-like algebraization for this logic (or for any other of the weaker calculi C_n) can be available, several attempts have been made to find other kinds of algebraizations for this logic (check, for instance, da Costa's [48]). The intuitive idea underlying the search of other algebraizations, generalizing the idea of Lindenbaum-Tarski, has been quite often that of finding 'any' congruence on the set of formulas that could be used to produce a quotient algebra from the algebra of formulas of the logic. Furthermore, if such a congruence is no more necessarily supposed to be induced directly by way of the consequence relation associated to the logic, nor should this congruence be necessarily supposed to be expressible by way of a formula written in the very language of the logic (it may happen to be definable only metalinguistically —for instance, if you do need a metalinguistical 'and' to characterize it, but there is no adequate conjunction available to express it in the language of the logic), it is still reasonable to suppose as well that this congruence should put no distinguished and non-distinguished formulas inside the same class of equivalence (so, for instance, no class will simultaneously contain a theorem and a non-theorem), so that we will have no trouble in attributing a distinguished or a non-distinguished status to some class of equivalence (cf. [84]). The final blow to the search for congruences algebraizing the logic **Cila** was delivered by Mortensen's [82], where this author proved that:

THEOREM 3.80 No non-trivial quotient algebra is definable for **Cila**, or for any logic weaker than **Cila**.

It is never too late to remember that, for non-trivial logics, a *trivial quotient algebra* is an algebra defined by a congruence relation $\approx$ such that $A \approx B$ if, and only if, A and B are the same formula (so, all equivalence classes are singletons). Now, some authors have argued that the exclusive existence of trivial quotient relations for a given logic is a major 'defect' (cf. [84], section 3), while others do not think so (cf. [20]) —and this is the reason why we have used scare quotes in writing ''any' congruence', above. In any case, this last result can be easily remedied by extensions of **Cila**. Consider, for instance, the logic **Cilo** (the logic $C_1^{\dagger}$ of da Costa, Béziau & Bueno's [57] —check again the subsection**3.10**). A non-trivial congruence can be defined within this logic by requiring, for any two given formulas, that they are not only provably equivalent, but are also both provably consistent. This can be put in terms of a single formula, by defining $A \approx B$ if $\vdash ((A \leftrightarrow B) \wedge (\circ A \wedge \circ B))$. One can then immediately prove that:

FACT 3.81 There is a non-trivial quotient algebra for **Cilo** (and already for **Cio**).
Proof: The above defined connective $\approx$ clearly sets up an equivalence relation. We have to show that it is in fact a congruence, so that given a schema $G(A)$ depending on A as a component formula (and possibly on some other formulas as well), we have to show that $G(A) \approx G(B)$ whenever $A \approx B$, where $G(B)$ is obtained by replacing each occurrence of A in $G(A)$ by B. Now, given this supposition that $A \approx B$, and recalling

from THEOREM 3.67 that the consistency of any component of a complex formula, in **Cio**, is enough to guarantee the consistency of the complex formula itself, we may infer that $\vdash \circ G(A)$ and $\vdash \circ G(B)$. To check that $\vdash (G(A) \leftrightarrow G(B))$ just do a straightforward induction on the complexity of G. In the trivial case in which no other connectives or formulas intervene, but A, there is really nothing to prove. The case of conjunction, disjunction and implication is also immediate, from positive logic. For negation, just recall, as a consequence of FACT 3.17, that contraposition holds for provably consistent formulas, so that from $A \dashv\vdash B$ and both $\vdash \circ A$ and $\vdash \circ B$ one can infer $\neg A \dashv\vdash \neg B$. This concludes the proof (a similar semantical argument can already be found in [57], Theorem 3.21), and it is obvious that this congruence is non-trivial —we know for example from FACT 3.66 that $\circ\bot$ is a theorem of **Cila**, and thus of **Cilo**, so that all bottom particles will of course belong to the same equivalence class determined by $\approx$ over **Cilo**. In all other respects, except for this last particular example, the above proof is clearly valid not only in **Cilo** but also in **Cio** (that is, **Cilo** without the axiom (cl) that transforms this last **C**-system, **Cio**, into a **dC**-system). □

Various other extensions of **Cila** having non-trivial quotient algebras have been proposed in the literature. In [84], for instance, Mortensen has proposed an infinite number of them, all situated of course somewhere in between **Cila** and classical logic. They were called $C_{n/(n+1)}$, for $n>0$ (C_0 is the name traditionally reserved for classical logic), and axiomatized by the addition to **Cila** of the following axioms, for each fixed $n>0$:

(M1n) $\neg^{n-1}A \vdash \neg^{n+1}A$, where $\neg^n$, as usual, denotes n iterations of $\neg$;

(M2n) $\bigwedge_{i=1}^{n}(\neg^{i-1}A \leftrightarrow \neg^{i-1}B) \vdash \bigwedge_{i=1}^{n}(\neg^{i}(A\#C) \leftrightarrow \neg^{i}(B\#C)) \wedge \bigwedge_{i=1}^{n}(\neg^{i}(C\#A) \leftrightarrow \neg^{i}(C\#B))$, where # is any binary connective, and $\bigwedge$ abbreviates, as usual, a long conjunction.

In the section 4 of [84] the connective $\approx$ defined by letting $A \approx B$ hold whenever $\vdash \bigwedge_{i=0}^{n}(\neg^{i}A \leftrightarrow \neg^{i}B)$ is shown to constitute a non-trivial congruence, for each $n>0$. The reason for non-triviality is that, in general, each $C_{n/(n+1)}$ can be understood as providing us with n+1 'negations': for any formula A of this logic we have that $\neg^{m-1}A$ is congruent to $\neg^{m+1}A$ if, and only if, $m \geq n$, so that there are n+2 distinct equivalent classes (represented by $A, \neg A, \ldots, \neg^{n+1}A$) of the quotient algebra generated by $\approx$. Do any of these new **C**-systems coincide with any of the other above studied ones? Do they have any special interest in themselves (besides being equipped with a non-trivial congruence)?

How can one understand these more general algebras induced by more esoteric congruence relations, if they do not fit inside the 'classical' algebraization theory of Lindenbaum-Tarski? A neat and elegant solution to that can be found in the study of Blok & Pigozzi (cf. [25]), where a much more general theory of algebraization is developed, extending the work of other authors. Some terminology and definitions are needed to explain what is a *Blok-Pigozzi algebraization*. Fixing some logic **L**=<*For*, ⊩>, an **L**-*algebra* is any structure homomorphic to **L** (being *For* a structured set of formulas constructed over some set of connectives, the corresponding **L**-algebra will of course contain, for each connective, an operator of the same arity 'interpreting' it). An **L**-*matrix model* of an **L**-algebra **Alg** is any pair <**Alg**, **D**>, where **D** is a proper subset of the universe of **Alg**, of the so-called *distinguished elements*. Formally, let an *interpretation* of a set of formulas *For* be an assignment of terms of **Alg** to each element of *For* (an assignment which is usually defined over some primitive elements and then extended to the whole set of formulas by way of the interpretation of the building structural operators). The *semantic consequence relation* $\vDash_{\mathbf{M}}$ associated to an **L**-matrix model **M** is then defined, as usual, by setting $\Gamma \vDash_{\mathbf{M}} A$ whenever A is assigned a distinguished element for every assignment of distinguished elements to all members of Γ. Matrices of finite many-valued logics are simple practical examples of *sound* and *complete* matrix models (that is, models such that $[(\Gamma \Vdash A) \Leftrightarrow (\Gamma \vDash_{\mathbf{M}} A)]$) that can be associated to some logics. In general, by a result of Wójcicki (see [111]) it is known that every structural logic can be characterized by sound and complete matrix models, in fact by κ-valued matrices, where κ has at most the cardinality of the set of formulas of the logic.

Any pair of terms φ and ψ of the **L**-algebra will be said to constitute an *equation*, to be designated by writing $(\varphi \doteq \psi)$. Such equations are always schematic, as any usual mathematical equation, and their non-operational components are said to be its *variables*; we may accordingly write $\varphi(C) \doteq \psi(C)$ to designate an equation having C as its single variable, and similarly for any number of variables. Now, what an interpretation does is exactly assigning values to these variables. One may then define an *equational consequence relation* induced by a class of **L**-algebras **KA**, to be denoted as $\vDash_{\mathbf{KA}}$, as follows: $[\Gamma \vDash_{\mathbf{KA}} (\varphi \doteq \psi)]$, where Γ is a set of equations, whenever the equation $(\varphi \doteq \psi)$ is a semantic consequence of Γ for every **L**-matrix model **M** of each **L**-algebra in **KA**, that is, when all those matrix models are such that $[\Gamma \vDash_{\mathbf{M}} (\varphi \doteq \psi)]$. The relation $\vDash_{\mathbf{KA}}$ is said to constitute an adequate *algebraic semantics* for a given logic **L** whenever there is a finite set of equations $\delta_i(C) \doteq \varepsilon_i(C)$, for $i<n$, such that: $(\Gamma \Vdash A) \Leftrightarrow [\{\delta_i(B) \doteq \varepsilon_i(B)$: for all $i<n$ and all $B \in \Gamma\} \vDash_{\mathbf{KA}} (\delta_i(A) \doteq \varepsilon_i(A))]$. In this case, the equations $\delta_i(C) \doteq \varepsilon_i(C)$, for $i<n$, are called *defining equations* of **L**, and we shall write simply $\delta \doteq \varepsilon$ as an abbreviation of them. Finally, an algebraic semantics for a logic **L**, induced by a class of **L**-algebras **KA**, is said to be *equivalent* (or *congruential*) if there can be defined in **L** a finite set of connectives with two variables $\approx_j$, for $j<m$, such that, for every equation $\varphi \doteq \psi$, we have that $[\{(\varphi \approx_j \psi)$: for all $j<m\}$ $\dashv\vDash_{\mathbf{KA}} \{\delta(\varphi \doteq \psi) \approx_j \varepsilon(\varphi \doteq \psi)$: for all $j<m\}]$. This set of connectives $\approx_j$, for $j<m$, will be abbreviated simply as $\approx$ and called a *system of equivalence* (or *congruence*) connectives for **L** and **KA**. Now, a logic **L** is said to be (Blok-Pigozzi-)*algebraizable* if it has an equivalent algebraic semantics. Another way of stating this definition (in terms of the consequence relation of **L**) is by requiring, to call a logic **L** algebraizable, to have in hand a set of equations $\delta \doteq \varepsilon$ and a set of formulas $\approx$ such that: (i) $\approx$ constitutes an equivalence relation; (ii) $(A_1 \approx B_1), \ldots, (A_n \approx B_n) \Vdash \sigma(A_1, \ldots, A_n) \approx \sigma(B_1, \ldots, B_n)$, for each n-ary connective σ; and (iii) $A \dashv\Vdash \delta(A) \approx \varepsilon(A)$. It should by now be completely clear how this generalizes the idea of (proving (IpE) and) producing a congruence over a set of formulas.

Not all logics are algebraizable (even in this broader sense of Blok-Pigozzi). For example, most modal logics, and the system **E** of entailment are not algebraizable, though they do have non-congruential algebraic semantics. As to the **C**-systems that we study in this paper, it has already been shown or mentioned some lines above (FACT 3.81 and below), that the logics **Cilo** and $C_{n/(n+1)}$, extensions of **Cila**, do have non-trivial congruences defined by finite sets of equations, being thus algebraizable in the sense of Blok-Pigozzi (though they are not algebraizable in the traditional sense of Lindenbaum-Tarski).

Also, as hinted in the last subsection, one can now prove that all the 8K three-valued maximal paraconsistent logics there presented are algebraizable (making use of and extending an argument by Lewin, Mikenberg & Schwarze, who have proved in [71] that the three-valued logic $\mathbf{P}^1$ is algebraizable):

FACT 3.82 All the 8K three-valued logics from the last subsection are algebraizable.
Proof: Just consider any of the two connectives $\equiv$ or $\stackrel{1}{=}$ defined in the FACT 3.75, let $\delta(A)$ be defined as $((A\rightarrow A)\rightarrow A)$ and $\varepsilon(A)$ be defined as $(A\rightarrow A)$, and check that the conditions (i)–(iii) defining an algebraizable logic two paragraphs above do hold. □

It can also be shown, at this point, that some of our **C**-systems are not algebraizable. To such an intent, yet another characterization of algebraizable logics can come on handy. Let **L** be a logic and **M** be an **L**-matrix model. A *Leibniz operator* Λ is a mapping from each arbitrary subset S of **M** into the largest congruence $\approx$ of **M** compatible with S, where $\approx$ is *compatible* with S if whenever we have that $\varphi\in S$ and $\varphi\approx\psi$ we also have that $\psi\in S$. It can be proved that a logic **L** is algebraizable if, and only if: (iv) Λ is injective and (v) order-preserving on the collection CT(**L**) of all closed theories of **L**, (vi) Λ preserves unions of directed subsets of CT(**L**), where a subset of CT(**L**) is *directed* if there is a common upper limit to every finite collection of elements of CT(**L**). (At this point, we had better direct the reader to [25] for details and proofs.) In any case, one might observe that a consequence of these last observations is that, for every logic **L**, the Leibniz operator produces an isomorphism between the lattice of filters of each **L**-matrix model **M** and the lattice of congruences of **M**. So, if such an operator is not an isomorphism, for some **L**-matrix model **M**, then the logic **L** is not algebraizable. This was the idea used by Lewin, Mikenberg & Schwarze in [72] (and that we extend here) to refine THEOREM 3.80:

THEOREM 3.83 The logic **Cila** (that is, da Costa's C_1) is not algebraizable. The same holds even for the stronger logic **Cibaw** (see **Figure 3.1**, in the subsection **3.10**), or any weaker logics extended by **Cibaw**.
Proof: Consider the following set of truth-values, $\mathbf{V}=\{0, a, b, 1, u\}$, ordered as follows: $0\leq a$, $0\leq b$, $a\leq 1$, $b\leq 1$, $1\leq u$, and where u and 1 are the distinguished elements. Consider now the following matrices defined over them:

$\wedge$	***u***	**1**	***a***	***b***	**0**
u	u	1	a	b	0
1	1	1	a	b	0
a	a	a	a	0	0
b	b	b	0	b	0
0	0	0	0	0	0

$\vee$	***u***	**1**	***a***	***b***	**0**
u	u	u	u	u	u
1	u	1	1	1	1
a	u	1	a	1	a
b	u	1	1	b	b
0	u	1	a	b	0

$\rightarrow$	***u***	**1**	***a***	***b***	**0**
u	u	u	a	b	0
1	u	1	a	b	0
a	u	1	1	b	b
b	u	1	a	1	a
0	u	1	1	1	1

	$\neg$	$\circ$
u	1	0
1	0	1
a	b	1
b	a	1
0	1	1

All axioms of **Cibaw** are validated by these matrices, as the reader can easily check. Now, it is also easy to check that there are no non-trivial congruences over **V**. Suppose for instance that $u \approx x$, for some $x \neq u$. In this case, as we know that $\neg\neg u \doteq 0$, and $\neg\neg x \doteq x$, then the condition (ii) above will give us $\neg\neg u \approx \neg\neg x$ from $u \approx x$, and so we conclude that $0 \approx x$, and thus $0 \approx u$. But, as we have observed before, there can be no congruence class containing both distinguished and non-distinguished values (in any case, this will violate condition (iii) above). We leave to the reader the easy exercise of showing, using the above connectives, that for any $x \approx y$, with $x \neq y$, one gets trapped at a similar predicament, namely, that of a distinguished value getting grouped with a non-distinguished one inside the same congruence class. Now, it is clear that $<A, \wedge, \vee>$ is a lattice, and that $\{0, a, 1, u\}$ and $\{0, b, 1, u\}$ are two filters over **V**. But there is just one congruence over **V** (which is of course the largest one compatible with both the filters just mentioned), and so the Leibniz operator cannot be an isomorphism. Once the logic **Cibaw** is, as a consequence, not algebraizable, FACT 3.79 informs us that none of its fragments can be algebraizable. □

Now, even non-algebraizable logics can happen to be amenable to sensible algebraic investigation. Indeed, a class of *weakly algebraizable logics* is characterized by the validity of conditions (iv) and (v) above, two of the three clauses of the characterization of Blok-Pigozzi algebraizability in terms of the Leibniz operator, and condition (v), alone, defines the class of *proto-algebraizable* logics. This last class includes all normal modal logics and most non-normal ones, but there are still some other logics which are not protoalgebraizable: an example is **IPC***, the implicationless fragment of intuitionistic logic, is neither algebraizable nor protoalgebraizable (cf. [25], chapter 5). Which of our non-algebraizable **C**-systems are protoalgebraizable, and which not (if any)? We shall leave this question open at this stage. It is interesting to notice, at any rate, that some sort of algebraic counterparts to some of these non-algebraizable **C**-systems have been proposed and studied, for instance, in Carnielli & de Alcantara's [37] and Seoane & de Alcantara's [102], where a variety of 'da Costa algebras' for the logic **Cila** has been introduced and studied, and a Stone-like representation theorem was proved, to the effect that every da Costa algebra is isomorphic to a 'paraconsistent algebra of sets'. It would be interesting now not only to extend that approach to other **C**-systems, but also to check how it fits inside this more general picture given by (Blok-Pigozzi-)algebraizable and protoalgebraizable logics.

An interesting application of the above mentioned algebraic tools is the following. Consider again, for example, the FACT 2.10(ii), where strong negations were shown to be 'definable' from bottom particles. Now, it is completely clear how this definition can be stated in practice if, for instance, a suitable implication is available inside of a compact logic (this is the case in all our examples, but needs not to be). This illustrates in fact how intuitionistic negation may be defined from a bottom particle and intuitionistic implication. But is that *definition*, in the general case, an *implicit* or an *explicit* one? For example, in positive classical logic (plus bottom and top) the theory containing both the formulas $((A \wedge B) \leftrightarrow \bot)$ and $((A \vee B) \leftrightarrow \top)$ implicitly defines the formula B (as the 'classical negation of A'). Are all implicit definitions also explicit ones? Or do we have, in some cases, to explicitly add some more structure to a logic to make explicit definitions expressible even when implicit ones are available? If, whenever a logic can implicitly define something, it can also explicitly define it, then the logic is said to have the *Beth definability property*, (BDP). Now, consider any class of **L**-algebras **KA**, for some logic **L**, and pick up a set **HA** of homomorphisms between any two of these **L**-algebras. A homomorphism $f\colon \mathbf{Alg}_1 \to \mathbf{Alg}_2$ in **HA** is said to be an *epi* if every pair of homomorphisms $g, h\colon \mathbf{Alg}_2 \to \mathbf{Alg}_3$ in **HA** is such that $g \circ f = h \circ f$ only if $g = h$. Evidently, all surjective homomorphisms are epis; if the converse also holds, that is, if all epis are surjective, we say that **KA** has the property (ES). Now, by a result of I. Németi (cf. [4]), an algebraizable logic has (BDP) if, and only if, its class of algebras has (ES). This is a very interesting result, and constitutes, in fact, just one example of how algebraic approaches can help us to solve real logical problems, in this case the problem of definability. Extensions of such results to wider classes of algebraic structures associated to (wider classes of) logics are clearly desirable.

4 FUTUROLOGY OF C-SYSTEMS

> When you encounter difficulties and contradictions, do not try to break them, but bend them with gentleness and time.
> —Saint Francis de Sales.

This is *not* the end. The next and natural small step for a paper, giant leap for paraconsistency, is providing reasonable interpretations for **C**-systems. This is the theme of our [42], where semantics for **C**-systems are presented and surveyed, ranging from the already traditional *bivaluations* to the more recently proposed *possible-translations semantics*, traversing on the way a few connections to many-valued semantics (a theme that already intromitted in our subsection **3.11**), and to modal semantics. A quite diverse approach to paraconsistent logics (in general) from the semantical point of view is also soon to be found in Priest's [91] (the remarkable possible-translations semantics, according to which some complex logics are to be understood in terms of *combinations* of simpler ones, will nevertheless not be found there —see instead [76], [36], [39], and, of course, [42]—, in addition, its section on many-valued logics is unfortunately too poor to give a reasonably good idea on the topic).

In this last section of the present paper we want to point out some interesting open problems and research directions connected to what we have herein presented. For example, in the section **2** we have extensively investigated the abstract foundations of paraconsistent logic, and the possibility and interest of defining the so-called *logical principles* at a purely logical level. There is still a lot of stirring open space to work here, and we will feel happy to have stimulated the reader to try their own hand at the relations between all the alternative formulations of (PPS), that is, all the different forms of explosion (or, if they prefer, the various forms of *reductio*, as in the subsection **3.2** —recall that the *reductio* and the Pseudo-Scotus are not always equivalent, for instance, if you think about intuitionistic logics). Think about it: are there any *interesting* logics, in our sense, disrespecting the Pseudo-Scotus, while respecting *ex falso* or the Supplementing Principle of Explosion, but still disrespecting the Gentle Principle of Explosion as well? In other words, are there interesting paraconsistent logics having either bottom particles or strong negations which do not constitute **LFI**s?

Recall that our approach contributed a novel notion of *consistency*. This is the picture again: There are consistent and inconsistent logics. The inconsistent ones may be either paraconsistent or trivial, but not both. The paraconsistent ones may be either dialectical or not. The consistent logics are explosive and non-trivial. The paraconsistent logics are non-explosive, and the dialectical paraconsistent ones are contradictory as well. The trivial logics (or trivial logic, if you fix some language) are explosive and contradictory (if the underlying logic has a negation symbol). Negationless logics are trivial if and only if inconsistent. Let us say that a theory *has models* only if these are non-trivial (they do not assign distinguished values to all formulas). So, the theories of a consistent logic have models if and only if they are non-contradictory. Paraconsistent logics may have models for some of its contradictory theories, and in the dialectical case all models of all theories are contradictory. Trivial theories (of trivial logics, or of any other logics) have no models. The consistency of each formula A of a logic **L** is defined exactly as what else one must say about A in order to make it explosive, that is, as what one should add to an A-contradictory theory in order to make it trivial. If the answer is 'nothing', then A is already consistent in **L** (whether the theories that derive this formula are contradictory or not). So, consistent logics are, quite naturally, those logics that have only consistent formulas. The above study sharply distinguishes the notions of non-contradictoriness and of consistency, and the model-theoretic impact of this should obviously be better appraised!

We now also have a precise definition of a large and fascinating class of paraconsistent logics, the *logics of formal inconsistency*, **LFI**s, and an important subclass of that, the **C**-*systems*. This is important to stress: according to our proposal it should be no more the case that the **C**-systems will be identified simply with the calculi C_n of da Costa, or with some other logics which just happen to be axiomatized in a more or less similar way. A general idea was put forward to be explored, namely that of being able to express *consistency* inside of our paraconsistent logics, and this helped collecting inside one big single class logics as diverse as the C_n and $\mathbf{P}^1$, or even $\mathbf{J}_3$ (now rephrased as **LFI1**), whose close kinship to the C_n seemed to have passed unsuspected until very recently (recall the subsections **2.4**, and end of **3.11**). This is, we may suggest, a fascinating challenge that we propose to our readers: To show that many other logics in the literature on paraconsistent logics can be characterized as **C**-systems, or, in general, as **LFI**s. This exercise has been explicitly put forward in the subsection **2.6**, but even previous to that, in the end of the subsection **2.4**, we have already hinted to the fact that other logics, such as Jaśkowski's **D2**, a discussive paraconsistent logic with motivations and technical features completely different from the ones that we study here, could be recast as an **LFI** (based on the modal logic $S5$) —more precisely, it can be recast as an **LFI** if only it is presented having some necessity operator, $\Box$, among its primitive or definable connectives (see note 11). Another recent example of that is the paraconsistent logic **Z** (are we perhaps running out of names?) proposed by Béziau in [24], in which a paraconsistent negation $\neg$ is defined from a primitive classical negation ~ and a possibility operator $\Diamond$, by setting $\neg A \stackrel{\text{def}}{=} \Diamond{\sim}A$.[27] Again, it is easy to see that **Z** can also be seen as an **LFI** (in fact, a

[27] By the way, exactly the same logic was proposed by Batens in [16] under the name **A**, and appears on another of Béziau's paper, [23], section 2.8, under the appelation of 'Molière's logic'. Strangely enough, after longly attacking, in the section 2.5 of [23], those logics that he calls 'paraconsistent atomical logics', that is, those logics in which 'only atomic formulas have a paraconsistent behavior' (being

C-system based on $S5$), in which the consistency of a formula A is expressed by the formula ($\Box A \vee {\sim}A$). Which other paraconsistent logics constitute **C**-systems, or **LFI**s, and which *not*? Inverting the question, are there good reasons for one trying to *avoid* **LFI**s, that is, can the investigation of non-**LFI**s have good technical or philosophical justifications? And how would the **C**-systems based on intuitionistic logic (**I**-systems?) or on relevance logic (**R**-systems?) look like, and which interesting properties would they have? How would this improve our map and understanding of **C**-land? In general, how could one use the very idea of a **C**-system to build up some new interesting paraconsistent logics, what advantages would they bring and particular technical tools could they contribute to the general inquiry about **LFI**s? The point to insist here is on the remarkable *unification* of aims and techniques that **LFI**s can seemingly produce in the paraconsistency terrain!

Another related interesting route is the one of *upgrading* any given paraconsistent logic in order to turn it into an **LFI**. This is exactly what is done by the logic **LFI1** (or **CLuNs**, or $\mathbf{J}_3$) over the logics *Pac* and *LP* (see subsection **2.4**), for which the gain in expressive power should already be obvious to the reader. Now, consider, for instance, the three-valued *closed set logic* studied by Mortensen and collaborators in [85], whose matrices of conjunction and of disjunction coincide with those of **LFI1**, and whose matrix of negation coincides with that of $\mathbf{P}^1$, having, again, 0 as the only non-distinguished value. Now, it is easy to see that the addition of appropriate matrices of implication and of a consistency operator will turn the upgraded closed-set logic into one of our 8K three-valued maximal paraconsistent logics, discussed in the subsection **3.11** above. The motivation for such closed set logic is also to be found among some of the most striking features of the 'Brazilian approach' to paraconsistency, namely, the idea of studying paraconsistent logics which are in a sense *dual* to other *broadly intuitionistic* (also called *paracomplete*) logics. We have slightly touched on this issue at a few points above —see, for instance, subsections **3.1** and **3.2**— as this has been one of the preferred justifications used by da Costa, among other authors, for the constitution of many of his paraconsistent logics. Indeed, if classical logic is not rarely held by some authors as the 'logic of sets', particularly because of its Boolean algebraic counterpart, Heyting's Intuitionistic Logic is very naturally held, in a topological setting, as the 'logic of open sets'. The very same dualizing intuition that we have just mentioned can then lead one to study the 'logic of closed sets' as a very natural paraconsistent logic. This is done in [85], where this investigation is also lifted to the categorial space —again, if intuitionistic logic is very naturally thought of as the logic of a *topos*, the closed set logic can be thought of as the underlying logic of a categorial structure called *complemented topos*. The upgrade of

---

thus controllably explosive with respect to every complex, or 'molecular' formula), a class of logics that seems to comprise not many logics up to this moment (the logic $\mathbf{P}^1$ —recall THEOREM 3.15(ii) and THEOREM 3.69— being among them), Béziau presents the above mentioned logic **Z** as the logic enjoying 'the best paraconsistent negation' around. But even if one concedes an enlargement of this definition of atomical logics in order to comprise all paraconsistent logics which are only 'paraconsistent up to some level of complexity', this author would still have to deal with the *bourgeois* fact that the negation of his preferred logic **Z**, exactly as what occurred with $\mathbf{P}^1$'s negation, is such that $\{p, \neg p\}$ is not trivial, for atomic p, while $\{\neg p, \neg\neg p\}$ *is* trivial: why, in this largely analogous case, would that same phenomenon be 'philosophically justifiable', and not be reducible just to some more bits of 'formal nonsense', as he puts it? One interesting such a justification, we suggest, may be found in terms of the dualization obtained by logics such as Mortensen's closed set logic, also mentioned in the present section.

the closed set logic into an **LFI** may thus set up some interesting new space for the study of topological and categorial interpretations of the notion of consistency. This proposed duality has also often been pushed, in the literature, in the contrary direction, namely, into the study of paracomplete logics which are dual to some given paraconsistent logics. Some samples of this can be found, for instance, in our papers [39] (where a logic called D_{min} is presented as dual to C_{min}, mentioned above in the subsection **3.1**), and [43] (where logics dual to slightly stronger versions of the calculi C_n are studied), as well as in Marcos's [78] (where 1K three-valued maximal paracomplete logics are presented, in addition to the 8K paraconsistent ones above mentioned). A more thorough study of the **LFU**s, the *Logics of Formal Undecidability* (or *Logics of Incomplete Information*), following the standards of the investigation set up by the present paper rests yet to be done.

We of course do not, and cannot, claim to have included and studied above *all* 'interesting' **C**-systems based on classical logic. We cannot but offer here a very partial medium-altitude mapping of the region, but we strongly encourage the reader to help us expand our horizons. So, if, guided mostly by technical reasons, we have started our study from the basic logic **bC** (subsection **3.2**), constructing all the remaining **C**-systems as extensions of **bC**, this should *not* be a impediment for the reader to study still weaker and *more basic* logics, such as **mbC**, the logic axiomatized by deleting (Min11): $(\neg\neg A \rightarrow A)$ from the axiomatization of **bC**, and their extensions, **mCi** (a logic studied under the name *Ci* in [17]) and so on. So, if **bC** was presented as a quite natural conservative extension of the logic C_{min} ([39]), **mbC** can similarly be presented as an extension of the logic *PI* ([10]). Just keep in mind that starting your study from **mbC**, instead of **bC**, and avoiding the axiom (Min11), you will be allowing for the existence of a few more in principle uninteresting partially explosive paraconsistent extensions of your logics (THEOREM 3.13), and you may also lose a series of other results, as for instance 3.14(iii), half of 3.17 and of 3.26, and also 3.20, 3.36, 3.41, 3.51(iii), 3.56, 3.57, 3.66, as well as some derived results and comments. Notice that we do not say that the loss of some of these results cannot be positive, but some symmetry certainly seems to be lost if the logic **mbC** happens to be extendable, for instance, in such a way as to validate the schematic rule $[(A\rightarrow B) \vdash (\neg B\rightarrow\neg A)]$, though it can in no way be extended so as to validate the similar rule $[(\neg A\rightarrow B) \vdash (\neg B\rightarrow A)]$.

What effects can the **LFI**s have on the study of some general mathematical questions, such as *incompleteness results* in Arithmetic? Indeed, recall that Gödel's incompleteness theorems are based on the identification of 'consistency' and 'non-contradictoriness' —what then if we start from our present more general notion of consistency (see (D19), subsection **2.4**)? And what if we integrate these logics with such modal logics as the *logic of provability*? (See [27], where consistency is also intended as a kind of dual notion to provability —if you cannot prove the negation of a formula, it is consistent with what you can prove—, and in which the necessary environment for the study of Gödel's theorems is provided.) What effects could that have (if at all) on the investigation of (set-theoretical) paradoxes? In fact, the analogy of the logics of formal inconsistency with the logics of provability is rather striking and worthy of being further explored; connections with other powerful internalizing-metatheoretical-notions logics, such as *hybrid logics*, and *labelled deductive systems* in general, are also to be expected.

We are *not* trying to escape from da Costa's requisite **dC[iii]** (subsection **3.8**), according to which extensions of our logics to higher-order logics ought to be available. But it still seems that most interesting problems related to *paraconsistency* appear already at the propositional level! Moreover, more or less automatic processes to *first-ordify* some given propositional paraconsistent logics can be devised, by the use of combination techniques such as *fibring*, if only we choose the right abstraction level to express our logics (see [34], where the logic C_1 is given a first-order version —coinciding with the one it had originally received, in [49] or [50], but richer in expressibility power— by the use of the notion of *non-truth-functional rooms*). One interesting thing about first-order paraconsistent logics is that they might allow for inconsistencies at the level of its objects, opening a new panorama for ontological investigations. Another interesting thing that we can conceive about first-order versions of paraconsistent logics in general, and especially of first-order **LFIs**, and that seems to have been completely neglected in the literature up to this point, is the investigation of consistent yet ω-*inconsistent* structures, or theories (also related to Gödel's theorems). Again, a point about that is scored by the logics of provability, but studious of paraconsistency should definitely have something to say about this.

Let us further mention some more palpable specific points to which we have already drawn attention here and there, and on which more research is still to be done. For instance, is the logic **bC** (subsections **3.2** to **3.4**) *controllably explosive*? Recall from FACT 3.32 that in the case of its extension **Ci**, the formulas causing controllable explosion coincide with the consistent theorems, and that, as it happens, **bC** does not have consistent theorems (THEOREM 3.10). For **bC**, however, only one side becomes immediate: consistent theorems cause controllable explosion... Another question: Does this logic **bC** have an intuitive adequate modal interpretation? And are there also extensions of **Ci** in which (IpE) holds (see the end of the subsection **3.7**)? What about investigating some other extensions of **bC** which do not extend **Ci** as well, such as **bCe**, obtained by the direct addition to **bC** of the axiom (ce): $A \vdash \neg\neg A$, studied in subsection **3.9**? Remember that we have extended **bC** to **Ci**, in the subsection **3.5**, arguing that all paraconsistent logics in the literature *do* identify inconsistencies and contradictions, and then all the other logics that we have studied after that in fact extended **Ci**. But the logic **bCe** also seems interesting in its own right, being able to accept some simple extensions that can express dual inconsistency, differently of what had occurred in the subsection **3.4** with some other extensions of **bC**, and paralleling a result obtainable for **Ci** (subsection **3.6**). **bCe** would also presumably constitute a step further in the direction of obtaining full (IpE), but in any case the general search for interesting **C**-systems extending **bC** and not extending **Ci** can already be funny enough. As to other ways of fixing the non-duality of the consistency and inconsistency operators in **bC**, alternative to extending this logic into **Ci**, suggestions have been made, for instance, for the addition to **bC** of schematic rules such as $\circ A, \neg\circ A \vdash B$ or else $\vdash \circ\circ A$ (perhaps having FACT 3.32 in mind, and trying to carry it forward into **bC**). None of these will work, however, as it is easy to see if one just considers again the matrices in THEOREM 3.16, noticing that they also validate the two last rules, while still not validating schemas such as (ci1) or (ci2). But there may quite well exist some other way out of this quagmire (perhaps the reader will find it).

Now, this is a quickie: can you find any (grammatical) conservative translation from **eCPL** (the extended classical propositional logic obtained by the addition to classical logic of the then innocuous consistency operator) into **bC** (recall the subsection **3.7**), as the one we had for **Ci** (COROLLARY 3.47)? And what happens when one considers, in the construction of **dC**-systems, the addition of more general rules such as (cg) or (RG) (see the subsection **3.8**), which implement more inclusive definitions for the consistency connective? Do we obtain interesting logics from that, fixing some asymmetries observed on the calculi C_n and its relatives? What effects does this move have on the semantical counterparts of these logics? Moving yet farther, we may ask about the logic C_{Lim}, the deductive limit to the hierarchy C_n (see subsection **3.10**), whether it can be proved to be *not* finitely gently explosive. For if it is not finitely gently explosive, given that it is gently explosive *lato senso*, then it cannot be *compact*. Or are we rather obligated to abandon *strong* completeness, in this case? Would there be other interesting **LFI**s in which consistency is not *finitely* expressible?

Recall the subsection **3.11**, where da Costa's requisite **dC[iv]** (subsection **3.8**) is taken very seriously, and we search for *maximal* paraconsistent fragments of classical logic, that is, logics having 'most rules and schemas of classical logic', and a class of 8K three-valued logics is presented as a solution to this. Now, are there other (full) solutions to this 'problem of da Costa'? And are there other interpretations, besides maximality, of da Costa's requisite **dC[iv]** leading to yet some other solutions to that problem? Are there non-many-valued (monotonic) solutions to that problem, or perhaps some other n-valued ones, for $n>3$? And, this is an important first step and probably an easier problem to solve: are there other interesting **C**-systems based on classical logic which are *not* extended by any of the above 8K three-valued logics? We must leave these questions open at this moment. It is interesting to notice, at any rate, that this problem has already been directly addressed here and there, in the literature. Besides [78], from which we drew the results in the subsection **3.12**, one could also recall, for instance, the *adaptive* programme for the confection of paraconsistent logics aiming to represent (non-monotonically) the dynamics of scientific reasoning and of argumentation (see [15]). Roughly speaking, the basic idea behind adaptive logics is that of working in between two boundary logics, often classical logic constituting one of them and a paraconsistent logic constituting the other one, so that consistency is pressuposed by default and we try to keep on reasoning inside of classical logic up to the point in which an 'abnormality' (an inconsistency?) pops out, a situation in which we had better descend to the level of the complementary paraconsistent logic, and go on reasoning over there. Indeed, the ancestral motivations of this programme seem to have been, as it is reasonable to conceive, yet another attempt to maintain the most of classical logic as possible (see [10]), so that paraconsistency is only needed at limit cases, while (most of) classical logic is, in principle, maintained in 'normal' situations. The difficulty here, as far as da Costa's requisite **dC[iv]** is concerned, seems to be *measuring* how much some different adaptive logics will respond to that same requirement of closeness to classical logic (if there is any measurable difference at all). As it is appealing to think of adaptive logics as situations in which two logics are combined in order to produce a third one, it seems also interesting to investigate if possible-translations semantics can, after all, be applied to such an environment as well, or at least stretch the analogies there as far as we can.

Some open questions can also be drawn from the subsection **3.12**. Notice, for instance, that all of Mortensen's axioms, (M1n) and (M2n), for every $n>0$, are validated by the matrices of the three-valued paraconsistent logic $\mathbf{P}^2$ (recall THEO-

REM 3.69), as that author himself have observed, and the question was left open, in [84], (and it remains still so, as far as we know) whether the logic $C_{1/2}$, the stronger of the logics $C_{n/(n+1)}$, would in fact coincide with the three-valued logic $\mathbf{P}^2$. Once $\mathbf{P}^2$ is known to be a maximal paraconsistent logic (cf. THEOREM 3.72), to show this coincidence would amount to showing that all axioms of $\mathbf{P}^2$ (and especially axioms (ca1)–(ca3), in the subsection **3.10**, specifying the consistent behavior of binary connectives) are provable from the axioms of $C_{1/2}$. Alternatively, using the fundamental FACT 3.32, one could try to show that $C_{1/2}$ is controllably explosive (or not, if what one wants is to disprove the conjectured coincidence) in contact with any formula involving binary connectives. In one way or another, it is quite interesting to note that the 8K maximal three-valued logics in subsection **3.11** show that there are several other logics different from $\mathbf{P}^1$ and from $\mathbf{P}^2$ that are next to the classical propositional logic 'in the same kind of way' (half of them extending C_1, as we point out in FACT 3.73), a problem that was left open in the closing paragraph of the above mentioned paper.

Now, what about extending our investigations on the *algebraizability* of **C**-systems (again, see the subsection **3.12**)? Can these algebras solve yet some other categories of logical problems? Again, notice that the problem of finding extensions of our **C**-systems which are algebraizable in the 'classical sense' was also left open (though the plausibility of the existence of such extensions was hinted) in the end of subsection **3.7**. To be precise, what was open was the existence of such extensions as fragments of some version of classical logic —the reader will have seen in the present section, however, that modal logics such as **Z** do extend **bC** and have no problem on what concerns (IpE) (given that *S5* is algebraizable). And what to say of extending our general approach on section **2** to other 'kinds of logics' (that is, varying the *logic structure* that we defined there) so as to include other kinds of consequence relations, such as (multiple-conclusion) non-monotonic ones (see [5] and [12])? Under this new light shed by **C**-systems and **LFI**s, it is also interesting to see how one can also move on to improve our present (rather poor) proof-theoretical approach. Indeed, Hilbert-style systems, such as the ones we present here, often require too much ingenuity to be applied, leaving intuition or mechanization of proofs far behind. For some **dC**-systems we know that *sequent* systems have already been proposed (see for instance [97] and [17]), as well as *natural deduction* systems (see [58]), and *tableau* systems (see [38]). The *really* interesting cases, however, seem to be those of **C**-systems that are *not* **dC**-systems, so that the consistency connective is, in a sense, 'ineliminable'! A first step towards such a general treatment of **C**-systems in terms of tableaux has already been offered by us in [41], where the logics **bC**, **Ci** and **LFI1** were all endowed with sound and complete tableau formulations.

There is so much yet to be done!

5 REFERENCES

[1] E. Agazzi. Il formale e il non formale nella logica. In: E. Agazzi, editor, *Logica filosofica e logica matematica*, pages 1119–1131. Brescia: La Scuola, 1990.

[2] E. Alves. *Logic and Inconsistency* (in Portuguese). Thesis, USP, Brazil, 137p, 1976.

[3] A. R. Anderson, N. D. Belnap. *Entailment*. Princeton: Princeton University Press, 1975.

[4] H. Andréka, I. Németi, and I. Sain. *Algebraic logic*. To appear in: D. Gabbay, editor, *Handbook of Philosophical Logic*, 2nd Edition.

[5] O. Arieli, and A. Avron. General patterns for nonmonotonic reasoning: From basic entailments to plausible relations. *Logic Journal of the IGPL*, 8:119–148, 2000.

[6] A. I. Arruda. A survey of paraconsistent logic. In: A. I. Arruda, R. Chuaqui, and N. C. A. da Costa, editors, *Mathematical Logic in Latin America:* Procedings of the IV Latin American Symposium on Mathematical Logic, Santiago, Chile, 1978, pp.1–41. Amsterdam: North-Holland, 1980.

[7] A. Avron. On an implication connective of *RM*. *Notre Dame Journal of Formal Logic*, 27:201–209, 1986.

[8] A. Avron. Natural 3-valued logics – characterization and proof theory. *The Journal of Symbolic Logic*, 56(1):276–294, 1991.

[9] A. Avron. On negation, completeness and consistency. To appear in: D. Gabbay, editor, *Handbook of Philosophical Logic*, 2nd Edition.

[10] D. Batens. Paraconsistent extensional propositional logics. *Logique et Analyse*, 90–91:195–234, 1980.

[11] D. Batens. Dialectical dynamics within formal logics. *Logique et Analyse*, 114: 161–173, 1986.

[12] D. Batens. Dynamic dialectical logics. In: [95], pp.187–217, 1989.

[13] D. Batens. Against global paraconsistency. *Studies in Soviet Thought*, 39: 209–229, 1990.

[14] D. Batens. Paraconsistency and its relation to worldviews. *Foundations of Science*, 3:259–283, 1999.

[15] D. Batens. A survey of inconsistency-adaptive logics. In: D. Batens, C. Mortensen, G. Priest, and J.-P. van Bendegem, editors, *Frontiers in Paraconsistent Logic:* Proceedings of the I World Congress on Paraconsistency, Ghent, 1998, pp.49–73. Baldock: Research Studies Press, King's College Publications, 2000.

[16] D. Batens. On the remarkable correspondence between paraconsistent logics, modal logics, and ambiguity logics. This volume.

[17] J.-Y. Béziau. Nouveaux résultats et nouveau regard sur la logique paraconsistante C_1. *Logique et Analyse*, 141–142:45–58, 1993.

[18] J.-Y. Béziau. Théorie legislative de la négation pure. *Logique et Analyse*, 147–148:209–225, 1994.

[19] J.-Y. Béziau. *Research on Universal Logic: Excessivity, negation, sequents* (French). Thesis, Univérsité Denis Diderot (Paris 7), France, 179p, 1995.

[20] J.-Y. Béziau. Logic may be simple. Logic, congruence, and algebra. *Logic and Logical Philosophy*, 5:129–147, 1997.

[21] J.-Y. Béziau. De Morgan lattices, paraconsistency and the excluded middle. *Boletim da Sociedade Paranaense de Matemática* (2), 18(1/2):169–172, 1998.

[22] J.-Y. Béziau. Idempotent full paraconsistent negations are not algebraizable. *Notre Dame Journal of Formal Logic*, 39(1):135–139, 1998.

[23] J.-Y. Béziau. Are paraconsistent negations negations? This volume.

[24] J.-Y. Béziau. The paraconsistent logic **Z** (a possible solution to Jaśkowski's problem). To appear in: *Logic and Logical Philosophy*, 7–8 (Proceedings of the Jaśkowski's Memorial Symposium), 1999/2000.

[25] W. J. Blok, and D. Pigozzi. *Algebraizable Logics*. Memoirs of the American Mathematical Society 396, 1989.

[26] A. Bobenrieth-Miserda. *Inconsistencias ¿Por qué no? Un estudio filosófico sobre la lógica paraconsistente*. Santafé de Bogotá: Tercer Mundo, 1996.

[27] G. Boolos. *The Logic of Provability*. Cambridge University Press, 1996.

[28] O. Bueno. Truth, quasi-truth and paraconsistency. In: W. A. Carnielli, and I. M. L. D'Ottaviano, editors, *Advances in Contemporary Logic and Computer Science:* Proceedings of the XI Brazilian Conference of Mathematical Logic, Salvador, 1996, pp.275–293. Providence: American Mathematical Society, 1999.

[29] M. W. Bunder. A new hierarchy of paraconsistent logics. In: A. I. Arruda, N. C. A. da Costa, and A. M. Sette, editors, *Proceedings of the* III *Brazilian Conference on Mathematical Logic*, Recife, 1979, pp.13–22. São Paulo: Soc. Brasil. Lógica, 1980.

[30] M. W. Bunder. Some definitions of negation leading to paraconsistent logics. *Studia Logica*, 43(1/2):75–78, 1984.

[31] M. W. Bunder. Some results in some subsystems and in an extension of C_n. *The Journal of Non-Classical Logic*, 6(1): 45–56, 1989.

[32] M. Bunge. *Racionalidad y Realismo*. Madrid: Alianza, 1995.

[33] H. Burkhardt, and B. Smith (editors). *Handbook of metaphysics and ontology* (2v). Munich: Philosophia Verlag, 1991.

[34] C. Caleiro, and J. Marcos. Non-truth-functional fibred semantics. In: H. R. Arabnia, editor, *Proceedings of the International Conference on Artificial Intelligence* (IC-AI'2001), v.II, pp.841–847. CSREA Press, USA, 2001.

[35] R. Carnap. *The Logical Syntax of Language*. London: Routledge & Kegan Paul, 1949.

[36] W. A. Carnielli. Possible-translations semantics for paraconsistent logics. In: D. Batens, C. Mortensen, G. Priest, and J.-P. van Bendegem, editors, *Frontiers in Paraconsistent Logic:* Proceedings of the I World Congress on Paraconsistency, Ghent, 1998, pp.149–163. Baldock: Research Studies Press, King's College Publications, 2000.

[37] W. A. Carnielli, and L. P. de Alcantara. Paraconsistent Algebras. *Studia Logica* 43(1/2):79–88, 1984.

[38] W. A. Carnielli, and M. Lima-Marques. Reasoning under inconsistent knowledge. *Journal of Applied Non-Classical Logics*, 2(1):49–79, 1992.

[39] W. A. Carnielli, and J. Marcos. Limits for paraconsistency calculi. To appear in: *Notre Dame Journal of Formal Logic*, 40(3), 1999.

[40] W. A. Carnielli, and J. Marcos. Ex contradictione non sequitur quodlibet. To appear in *Proceedings of the Advanced Reasoning Forum Conference*, held in

Bucharest, Rumania, July 2000. Edited by Mircea Dumitru, under the auspices of the New Europe College of Bucharest.

[41] W. A. Carnielli, and J. Marcos. Tableau systems for logics of formal inconsistency. In: H. R. Arabnia, editor, *Proceedings of the International Conference on Artificial Intelligence* (IC-AI'2001), v.II, pp.848–852. CSREA Press, USA, 2001.

[42] W. A. Carnielli, and J. Marcos. Semantics for **C**-systems. Forthcoming.

[43] W. A. Carnielli, and J. Marcos. Possible-translations semantics and dual logics. Forthcoming.

[44] W. A. Carnielli, J. Marcos, and S. de Amo. Formal inconsistency and evolutionary databases. To appear in: *Logic and Logical Philosophy*, 7–8 (Proceedings of the Jaśkowski's Memorial Symposium), 1999/2000.

[45] H. Curry. The inconsistency of certain formal logics. *The Journal of Symbolic Logic*, 7(3):115–117, 1942.

[46] N. C. A. da Costa. Nota sobre o conceito de contradição. *Anuário da Sociedade Paranaense de Matemática* (2), 1:6–8, 1958.

[47] N. C. A. da Costa. Observações sobre o conceito de existência em matemática. *Anuário da Sociedade Paranaense de Matemática* (2), 2:16–19, 1959.

[48] N. C. A. da Costa. Opérations non monotones dans les treillis. *Comptes Rendus de l'Academie de Sciences de Paris* (A–B), 263:A429–A423, 1966.

[49] N. C. A. da Costa. *Inconsistent Formal Systems* (in Portuguese). Thesis, UFPR, Brazil, 1963. Curitiba: Editora UFPR, 68p, 1993.

[50] N. C. A. da Costa. On the theory of inconsistent formal systems. *Notre Dame Journal of Formal Logic*, 15(4):497–510, 1974.

[51] N. C. A. da Costa. *Essay on the Foundations of Logic* (in Portuguese). São Paulo: Hucitec, 1980. (Translated into French by J.-Y. Béziau, under the title *Logiques Classiques et Non-Classiques*, 1997, Paris: Masson.)

[52] N. C. A. da Costa. The philosophical import of paraconsistent logic. *The Journal of Non-Classical Logic*, 1(1):1–19, 1982.

[53] N. C. A. da Costa, and E. Alves. A semantical analysis of the calculi C_n. *Notre Dame Journal of Formal Logic*, 18(4):621–630, 1977.

[54] N. C. A. da Costa, and M. Guillaume. Sur les calculs C_n. *Anais da Academia Brasileira de Ciências*, 36:379–382, 1964.

[55] N. C. A. da Costa, and D. Marconi. An overview of paraconsistent logic in the 80s. *The Journal of Non-Classical Logic*, 6(1):5–32, 1989.

[56] N. C. A. da Costa, and R. G. Wolf. Studies in paraconsistent logic I : The dialectical principle of the unity of opposites. *Philosophia (Philosophical Quarterly of Israel)*, 9:189–217, 1980.

[57] N. C. A. da Costa, J.-Y. Béziau, and O. A. S. Bueno. Aspects of paraconsistent logic. *Bulletin of the IGPL*, 3(4):597–614, 1995.

[58] M. A. de Castro, and I. M. L. D'Ottaviano. Natural Deduction for Paraconsistent Logic. *Logica Trianguli*, 4:3–24, 2000.

[59] I. M. L. D'Ottaviano. On the development of paraconsistent logic and da Costa's work. *The Journal of Non-Classical Logic*, 7(1/2):89–152, 1990.

[60] I. M. L. D'Ottaviano, and N. C. A. da Costa. Sur un problème de Jaśkowski. *Comptes Rendus de l'Academie de Sciences de Paris* (A–B), 270:1349–1353, 1970.

[61] R. L. Epstein. *Propositional Logics: The semantic foundations of logic*, with the assistance and collaboration of W. A. Carnielli, I. M. L. D'Ottaviano, S. Krajewski, and R. D. Maddux. Belmont: Wadsworth-Thomson Learning, 2nd edition, 2000.

[62] G. Gentzen. Untersuchungen über das logische Schliessen. *Mathematische Zeitschrift*, 39:176–210/405–431, 1934.

[63] H.-J. Glock. A Wittgenstein Dictionary. Blackwell, 1996.

[64] L. Goldstein. Wittgenstein and paraconsistency. In: [95], pp.540–562.

[65] G.-G. Granger. *L'irrationnel*. Paris: Odile Jacob, 1998.

[66] D. Hilbert. Mathematische Probleme: Vortrag, gehalten auf dem Internationalen Mathematiker Kongress zu Paris 1900. *Nachrichten von der Königlichen Gesellschaft der Wissenschaften zu Göttingen*, pages 253–297, 1900.

[67] S. Jaśkowski. Propositional calculus for contradictory deductive systems (in Polish). *Studia Societatis Scientiarum Torunensis*, sectio A–I:57–77, 1948. Translated into English: *Studia Logica*, 24:143–157, 1967.

[68] I. Johánsson. Der Minimalkalkül, ein reduzierter intuitionistischer Formalismus. *Compositio Mathematica*, 4(1):119–136, 1936.

[69] A. N. Kolmogorov. On the principle of excluded middle. In: Van Heijenoort, editor, *From Frege to Gödel*, pp.414–437. Cambridge: Harvard University Press, 1967. (Translation from the Russian original, from 1925.)

[70] W. Lenzen. Necessary conditions for negation-operators (with particular applications to paraconsistent negation). In: Ph. Besnard, and A. Hunter, *Reasoning with Actual and Potential Contradictions*, pp.211–239. Dordrecht: Kluwer, 1998.

[71] R. A. Lewin, I. F. Mikenberg, and M. G. Schwarze. Algebraization of paraconsistent logic $\mathbf{P}^1$. *The Journal of Non-Classical Logic*, 7(1/2):79–88, 1990.

[72] R. A. Lewin, I. F. Mikenberg, and M. G. Schwarze. C_1 is not algebraizable. *Notre Dame Journal of Formal Logic*, 32(4):609–611, 1991.

[73] C. I. Lewis, and C. H. Langford. *Symbolic Logic*. New York: Dover, 1st edition, 1932.

[74] A. Loparić, and E. H. Alves. The semantics of the systems C_n of da Costa. In: A. I. Arruda, N. C. A. da Costa, and A. M. Sette, editors, *Proceedings of the III Brazilian Conference on Mathematical Logic*, Recife, 1979, pp.161–172. São Paulo: Soc. Brasil. Lógica, 1980.

[75] D. Marconi. Wittgenstein on contradiction and the philosophy of paraconsistent logic. *History of Philosophy Quarterly*, 1(3):333–352, 1984.

[76] J. Marcos. *Possible-Translations Semantics* (in Portuguese). Thesis, Unicamp, Brazil, xxviii+240p, 1999.
URL = ftp://www.cle.unicamp.br/pub/thesis/J.Marcos/

[77] J. Marcos. *(Wittgenstein & Paraconsistência)*. Chapter 1 in [76], presented at the VIII National Meeting on Philosophy (VIII ANPOF Meeting), Caxambu, Brazil, 1998. Submitted to publication.

[78] J. Marcos. 8K solutions and semi-solutions to a problem of da Costa. Forthcoming.

[79] A. T. C. Martins. *A Syntactical and Semantical Uniform Treatment for the* **IDL** *&* **LEI** *Nonmonotonic System*. Thesis, UFPE, Brazil, xvi+225p, 1997.
URL = http://www.lia.ufc.br/~ana/tese.ps.gz

[80] F. Miró Quesada. Nuestra lógica. *Revista Latinoamericana de Filosofía*, 8(1): 3–13, 1982.

[81] F. Miró Quesada. Paraconsistent logic: some philosophical issues. In: [95], pp.627–652, 1989.

[82] C. Mortensen. Every quotient algebra for C_1 is trivial. *Notre Dame Journal of Formal Logic*, 21(4):694–700, 1980.

[83] C. Mortensen. Aristotle's thesis in consistent and inconsistent logics. *Studia Logica*, 43:107–116, 1984.

[84] C. Mortensen. Paraconsistency and C_1. In: [95], pp.289–305, 1989.

[85] C. Mortensen. *Inconsistent mathematics*, with contributions by P. Lavers, W. James, and J. Cole. Dordrecht: Kluwer, 1995.

[86] D. Nelson. Negation and separation of concepts in constructive systems. In: A. Heyting, editor, *Constructivity in Mathematics:* Proceedings of the colloquium held at Amsterdam, 1957, pp.208–225. Amsterdam: North-Holland, 1959.

[87] J. J. O'Connor, and E. F. Robertson. Girolamo Cardano. In: J. J. O'Connor, and E. F. Robertson, editors, *The Mac Tutor History of Mathematics archive* (March 2001 edition).
URL=http://www-history.mcs.st-andrews.ac.uk/history/Mathematicians/Cardan.html

[88] L. Peña. Graham Priest's 'dialetheism': Is it altogether true? *Sorites*, 7:28–56, 1996.

[89] K. R. Popper. *Conjectures and Refutations: The growth of scientific knowledge*. London: Routledge and Kegan Paul, 3rd edition, 1969.

[90] G. Priest. *In Contradiction. A Study of the Transconsistent*. Dordrecht: Nijhoff, 1987.

[91] G. Priest. Paraconsistent logic. To appear in: D. Gabbay, editor, *Handbook of Philosophical Logic*, 2nd Edition.

[92] G. Priest, and R. Routley. Introduction: Paraconsistent logics. *Studia Logica* (special issue on 'Paraconsistent Logics'), 43:3–16, 1984.

[93] G. Priest, and R. Routley. Systems of paraconsistent logic. In: [95], pp.151–186, 1989.

[94] G. Priest, and K. Tanaka. Paraconsistent Logic. In: Edward N. Zalta, editor, *The Stanford Encyclopedia of Philosophy* (August 2000 edition).
URL=http://plato.stanford.edu/entries/logic-paraconsistent/

[95] G. Priest, R. Routley, and J. Norman (editors). *Paraconsistent Logic: essays on the inconsistent*. Munich: Philosophia Verlag, 1989.

[96] W. V. O. Quine. *From a Logical Point of View.* Cambridge: Harvard University Press, 1953.

[97] A. R. Raggio. Propositional sequence-calculi for inconsistent systems. *Notre Dame Journal of Formal Logic*, 9(4):359–366, 1968.

[98] N. Rescher, and R. Brandom. *The Logic of Inconsistency.* Oxford: Basil Blackwell, 1980.

[99] G. Restall. *Laws of non-contradiction, laws of the excluded middle and logics.* Typescript.
URL= ftp://www.phil.mq.edu.au/pub/grestall/lnclem.ps

[100] R. Routley, and R. K. Meyer. Dialectical logic, classical logic and the consistence of the world. *Studies in Soviet Thought*, 16:1–25, 1976.

[101] K. Schütte. *Proof Theory.* Berlin: Springer, 1977. (Translated from German original version, from 1960.)

[102] J. Seoane, and L. P. de Alcantara. On da Costa algebras. *The Journal of Non-Classical Logic* 8(2):41–66, 1991.

[103] A. M. Sette. On the propositional calculus $\mathbf{P}^1$. *Mathematica Japonicae*, 18:181–203, 1973.

[104] B. H. Slater. Paraconsistent Logics? *Journal of Philosophical Logic* 24(4):451–454, 1995.

[105] R. Sylvan. Variations on da Costa C systems and dual-intuitionistic logics. I. Analyses of C_ω and CC_ω. *Studia Logica*, 49(1):47–65, 1990.

[106] A. Tarski. The semantic conception of truth and the foundation of semantics. *Philosophy and Phenomenological Research*, 4:341–378, 1944.

[107] I. Urbas. Paraconsistency and the **C**-systems of da Costa. *Notre Dame Journal of Formal Logic*, 30(4):583–597, 1989.

[108] I. Urbas. Paraconsistency. *Studies in Soviet Thought*, 39:343–354, 1990.

[109] L. Wittgenstein. *Bemerkungen über die Grundlagen der Mathematik.* 3rd revised edition. Suhrkamp: 1984. (In English as: *Remarks on the Foundations of Mathematics.* G. H. von Wright, R. Rhees, and G. E. M. Anscombe, editors, 3rd revised edition. Oxford: Basil Blackwell, 1978.)

[110] L. Wittgenstein. *Wittgenstein's Lectures on the Foundations of Mathematics:* Cambridge, 1939. C. Diamond, editor. The University of Chicago Press, 1989.

[111] R. Wójcicki. *Theory of logical calculi.* Dordrecht: Kluwer, 1988.

Paraconsistent Classical Logic*

BRYSON BROWN University of Lethbridge, Lethbridge, Alberta, Canada.
brown@uleth.ca

Abstract

The main aim of this paper is to show that the apparent oxymoron of the title is really a correct description of the logic I present here. This logic is paraconsistent, in the widely accepted sense that its consequence relation is not trivial for all inconsistent premise sets. It is also not trivial in the dual sense—not all conclusion sets that must have a true member follow trivially from every premise set. But it is a classical logic, both in the sense that its semantics is fully classical, and in the more significant sense that it results from a well-motivated type-raising of a fully classical sentence-sentence consequence relation to a set-set consequence relation. This logic is *weakly aggregative*: in general (using '$\Vdash$' for the consequence relation) $\{p,q\} \nVdash \{(p \wedge q)\}$ and $\{(p \vee q)\} \nVdash \{p,q\}$. The weakened aggregational principles that do hold have a direct and interesting connection to graph and hyper-graph colourings. Along the way we will encounter some of the main results that have emerged from work on this and related logics.

1 Consequence Relations

In what follows we will use lower case Greek letters for sentences, upper case for sets. As usual, $\alpha \vdash \gamma$ says that $<\alpha,\gamma>$ is in the syntactic sentence-sentence consequence relation. (This holds iff every consistent conjunction with α as a conjunct can be consistently extended by adding γ as a further conjunct.) Similarly, $\alpha \vDash \gamma$ says that $<\alpha,\gamma>$ is in the semantic sentence-sentence consequence relation. (This holds iff every classical model satisfying α also satisfies γ.) Thus we have a completely familiar picture of the starting point for our inquiry.

We will use $A \vdash \Gamma$ to say that $<A,\Gamma>$ is in the syntactic set-set consequence relation and we will use $A \vDash \Gamma$ to say that $<A,\Gamma>$ is in the semantic set-set consequence relation. Thus we will be using '$\vdash$' and '$\vDash$' ambiguously, to represent relations between sentences and relations between sets of sentences.

* Thanks are due to the SSHRC of Canada for their support of this work, and to two anonymous referees for corrections and suggested improvements. Of course the responsibility for any remaining errors rests solely with me.

But the ambiguity will not trouble us, since the relata will always allow us to tell which relation is meant. We take it for granted that we are dealing with a standard sort of formal language[1], and a standard recursive definition of truth-in-a-model. Since the details of the language do not matter for our purposes, we will forgo specifying them.

The classical semantic sentence-sentence consequence relation is often expressed in a way that treats it strictly as a matter of truth preservation, where formal satisfaction is our standard of truth-in-a-model:

$$\alpha \models \gamma \text{ iff every classical model satisfying } \alpha \text{ also satisfies } \gamma.$$

This way of putting it makes an important assumption: It assumes that our formal account of satisfaction is an adequate formal account of truth, in the sense that when and only when the classical models satisfying a given sentence also satisfy some other sentence, the truth of the first guarantees the truth of the latter.[2] Of course we can regard the formal apparatus of satisfaction as simply defining what truth-in-a-model is. But this simply dodges the issue, if our original intent was to capture some pre-existing, informal notion of truth and truth preservation.

The classical syntactic sentence-sentence consequence relation, similarly, can be based strictly on consistency. But again, if we mean by this some pre-theoretical notion of compatibility, this claim needs careful examination. On the other hand, if we simply mean that our formal notion of consistency provides the means to specify an interesting consequence relation, there can be no objection other than that the result is uninteresting or unilluminating. As to specifying our syntactic sentence-sentence relation, I suggest you pick your favourite classical deductive system, and restrict the premise rule to allow exactly one premise. Then we can say, as usual, that

$$\alpha \vdash \gamma \text{ iff } \gamma \text{ is derivable from } \alpha.$$

But if we define consistency in the usual way, as

$$\mathrm{Con}(\alpha) \text{ iff } \exists \gamma\ \alpha \nvdash \gamma$$

Then we can also say that

---

[1] The issues we will be dealing with here can be adequately captured in a propositional language, but they also apply, with some reservations about decidability, to first-order languages.

[2] We know, in fact, that this is false, since classical models ignore the meanings of predicates in favour of simply (and freely) assigning extensions to predicates ad libitum. So classical models ignore the constraints that arise from, for example, the incompatibility of colour properties. This is widely recognized, of course. More radical paraconsistent logics propose models "satisfying" sentences that are classically unsatisfiable; if these logics notion of truth is a correct account of our informal notion, then this assumption is false in both directions.

$$\alpha\vdash\gamma \text{ iff for all } \delta, \text{ if } \alpha\wedge\delta \text{ is consistent, then so is } \alpha\wedge\delta\wedge\gamma.$$

That is, $\alpha\vdash\gamma$ if and only if γ is a consistent conjunctive extension of every consistent conjunctive extension of α. It's also worth defining tautologousness at this point:

$$\text{Taut}(\delta) \text{ iff } \forall\alpha, \alpha\vdash\delta$$

We arbitrarily select some inconsistent sentence and give it the special and convenient name '$\perp$', and similarly select some sentence provable from no premises and name it '$\top$'.

The important thing to recognize at this point is that even given these consequence relations at the sentential level, it is not immediately obvious how we should define a corresponding set-set consequence relation. The problem, of course, is that it is sentences that are true (or false) and consistent or inconsistent. What properties of sets of sentences should be substituted for (or taken to correspond to) truth and consistency?

The standard approach to classical logic follows from adopting a straightforward answer to this question: A set of sentences is true_{set} iff all its members are true, and a set is $\text{consistent}_{\text{set}}$ iff every sentence in its closure under conjunction is consistent. But this answer begs an important question, and (worse) produces a consequence relation that fails to preserve some desirable features of the sentence-sentence consequence relation we began with.

Our aim in making the transition from a logic of sentences to a logic of sentence sets is to preserve desirable features of the sentence logic. That truth on the left guarantees truth on the right is one of these; but there are others, as well. Two of these will be particularly important for us. First, all and only contradictions (inconsistent sentences) are trivial on the left. And second, all and only tautologies are trivial on the right. That is, $\forall\gamma(\alpha\vdash\gamma)$ is true iff α is a contradiction, and $\forall\alpha(\alpha\vdash\gamma)$ is true iff γ is a tautology.

2 THE QUESTION OF AGGREGATION

The question that the standard proposal for type-raising truth and consistency begs is the question of *aggregation*. We certainly want the truth of a set of sentences to have something to do with the truth of the set's members. Similarly, we want the consistency of a set of sentences to depend on the consistency of some sort of aggregation of the set's members. But as we have just seen, the usual set-set consequence relation makes a very strong assumption about how we must aggregate. And there is a substantial price for aggregating in the standard way, a price that reveals itself in the ease with which premise sets become inferentially explosive, and with which conclusion sets become inferentially trivial.

This ease contrasts sharply with the demanding requirements for trivialization of the sentence-sentence consequence relation. As we mentioned above, only two very special classes of sentences count as trivial for the sentence-sentence consequence relation. But given this policy of aggregation, many sets that do not contain contradictions are nevertheless trivial on the left, and many sets that do not contain tautologies are trivial on the right.

In what follows we will impose some constraints on the set-set consequence relation, and see what sort of set-logic emerges from them.

The minimal preservation principle that we insist on is:

$$\textbf{Pres:}\quad \frac{\exists\gamma\in\Gamma,\ \delta\in\Delta\ \ \gamma\vdash\delta}{\Gamma\vdash\Delta}$$

The question arises what logic (and in particular, what sort of aggregation) we get if we have Pres, but insist as well that trivialization arises on the left if and only if a contradiction appears on the left, and on the right if and only if a tautology appears there [1]:

Triv: i. On the left: $\forall\Delta(\Gamma\vdash\Delta)$ iff $\exists\gamma\in\Gamma, \forall\delta(\gamma\vdash\delta)$
ii. On the right: $\forall\Gamma(\Gamma\vdash\Delta)$ iff $\exists\delta\in\Delta, \forall\gamma(\gamma\vdash\delta)$

The standard (one might say trivial) approach to aggregation creates contradictions in sets that initially lack them by closing premise sets under conjunction. Similarly, it creates tautologies in sets on the right (conclusion sets) by closing them under disjunction. In fact, one simple way to characterize the standard set logic in terms of the sentence logic base is this:

Standard: $A\vdash\Gamma$ iff $\exists\alpha\in C(A,\wedge), \gamma\in C(\gamma,\vee) \mid \alpha\vdash\gamma$

This characterization of '$\vdash$' makes the impact of the usual forms of aggregation plain. Aggregating in this very strong way trivializes many sets that do not contain a trivial sentence. As P.K. Schotch and R.E. Jennings have long urged, there is a very clear sense in which doing so "makes things worse": It converts sets that contain no trivialities into sets that do.

Discomfort with just these features of classical logic has led to many forms of non-classical logic. One aim of this paper is to reiterate (and clarify) the claim that responding to this discomfort does not demand a large departure from classical logic. In fact, as the title suggests, a mere re-thinking of just how classical logic ought to work at the level of sets can assuage much (perhaps even all) of the legitimate discomfort here.

The simplest response would be to reject aggregation altogether. But that will not do either. We need some form of aggregation, if the logic of sentence sets is to be of any interest at all. Pres alone gives us a set logic that has no aggregation whatsoever. As a result, it tells us nothing about consequences that isn't already there in the sentence logic. But there are alternatives to the strong form of aggregation that characterizes the standard approach to classical set logic. And these alternatives, which make up an infinite sequence of increasingly weakened principles of aggregation, allow us to retain Triv while also retaining some real interest for the set-set consequence relation.

3 PRINCIPLES OF AGGREGATION

What we need are alternative principles of aggregation, weaker than closure under $\wedge$ on the left and $\vee$ on the right. The key contribution to this search is a proposal by Schotch and Jennings, that aggregation be measured by the number of cells characterizing a family of covering families of sets for the set in question. We can aggregate in accord with a single-celled family of covering families, placing all members of a set of sentences together into a single cell. Or we can aggregate using a two-celled family of covering families, placing each member of a set of sentences into at least one of two cells. We can aggregate, in general, by selecting some natural number n and considering the aggregative consequences of placing each member of the set to be aggregated in at least one of n cells. Since, in general, we can distribute the members of the set in many different ways across n cells, the aggregation *forced* by a given choice of n is settled by the set of all such distributions, where the force of each individual distribution is the set of sentences that follows from the closure under conjunction (on the left) or disjunction (on the right) of each cell's contents.
If we choose to aggregate in the single-celled way, we get classical aggregation. But for n=2 or more, we get a *weakly aggregative* logic. A logic is *aggregative* if it allows conclusion sets to follow from premise sets even when no element of the conclusion set follows (in the underlying sentence logic) from any element in the premise set, and *weakly aggregative* if it is aggregative, but does not close (all) premise sets under conjunction, and/or conclusion sets under disjunction.
How can we capture the logical force of such aggregation? A general graph-theoretical argument [2] shows that, on the left, the aggregation by which premises combine to produce consequences is completely characterized by closing the set under

$$2/n{+}1_L\colon\ \vee(\gamma_i \wedge \gamma_j),\ 1 \leq i \neq j \leq n+1$$

That is, by closing the set under the operation of forming the disjunction of pairwise conjunctions of all n+1 tuples of elements. This operation is equivalent, by the duality of '$\wedge$' and '$\vee$', to forming the conjunction of n-wise disjunctions amongst all n+1 tuples of elements. Thus aggregation on the left can also be expressed by closing the premise set under

$$n/n{+}1_L\colon\ \wedge(\gamma_{i1} \vee \ldots \vee \gamma_{jn})\ 1 \leq i_1 \neq j_n \leq n+1$$ [3]

[3] In fact, graph theoretical considerations show that completeness follows for any aggregative principle formulated on the left as a disjunction of conjunctions corresponding to a hypergraph of chromatic index n+1

A completeness proof for the K_n modal logics based on this operation can be found in [3]. Dually (as one would expect), n-wise aggregation on the right is characterized by

$$2/n+1_R: \wedge(\gamma_i \vee \gamma_j),\ 1 \leq i \neq j \leq n+1$$

or equivalently by

$$n/n+1_R: \vee(\gamma_{i1} \wedge \ldots \wedge \gamma_{jn})\ 1 \leq i_1 \neq \ldots \neq j_n \leq n+1$$

Forming the disjunction of pairwise conjunctions amongst any n+1 sentences gives us new sentences that, in general, do not follow from any of the n+1 individual sentences we began with. Dually, forming the conjunction of pairwise disjunctions amongst any n+1 sentences gives us new sentences which, in general, may follow from a sentence when none of the individual sentences we began with followed from it. For example,

$$2/3_L(\alpha,\beta,\gamma) = (\alpha \wedge \beta) \vee (\alpha \wedge \gamma) \vee (\beta \wedge \gamma)$$

But then obviously enough, in general, $\alpha \nvdash 2/3_L(\alpha,\beta,\gamma)$, and similarly for β and γ. Thus $2/3_L(\alpha,\beta,\gamma)$ represents a real *aggregative* consequence of $\{\alpha,\beta,\gamma\}$, in the sense that it is a consequence of the set[4] but not of any individual member of the set.

Similarly,

$$2/3_R(\alpha,\beta,\gamma) = (\alpha \vee \beta) \wedge (\alpha \vee \gamma) \wedge (\beta \vee \gamma)$$

And obviously enough, in general,

$$\exists\delta \mid \delta \vdash 2/3_R(\alpha,\beta,\gamma),\ \delta \nvdash \alpha,\ \delta \nvdash \beta,\ \delta \nvdash \gamma.$$

Thus $2/3_R(\alpha,\beta,\gamma)$ represents a real *aggregative* weakening of $\{\alpha,\beta,\gamma\}$, since it can follow classically from sentences that do not have any of α, β, or γ as consequences. (One such sentence is $2/3_L(\alpha,\beta,\gamma)$. In fact, $2/3_L$ is self-dual, i.e. $2/3_L$ is equivalent to $2/3_R$.)

(i.e. such that any n-colouring must produce a monochrome edge). Dually, completeness follows for any aggregative formulated on the left as a conjunction of disjunctions corresponding to a hypergraph with what I call universal colour number n+1 (i.e. such that any n-colouring of its vertices has at least one colour appearing on all edges).

[4] Both classically, and (assuming we are dealing with n=2) in this aggregatively weaker logic.

Given these weaker aggregation principles, we can define a hierarchy of consequence relations:

$$\Gamma^{n}\vdash^{m}\Delta \text{ iff } \exists\gamma\in Cl(\Gamma,2/n{+}1_{L}),\ \delta\in Cl(\Delta,2/m{+}1_{R}) \mid \gamma\vdash\delta$$

Obviously, $\Gamma^{n}\vdash^{m}\Delta$ is identical to the usual classical set-set consequence relation if we set n=m=1. For higher values of n, m, the resulting aggregation is strictly weaker, in general, so if $n \leq n^{\bullet}$, $m \leq m^{\bullet}$, then ${}^{n^{\bullet}}\vdash^{m^{\bullet}} \subseteq {}^{n}\vdash^{m}$. The lower n is, the more sets are trivial on the left; the lower m is, the more sets are trivial on the right.

These weaker forms of aggregation allow us to insist on triv without giving up on aggregation altogether. To explain how we will apply these weaker aggregation principles, we need to introduce a generalization of consistency (and a dual generalization of non-tautologousness) for sets. Following Schotch and Jennings in [1], we define

$$Con(\Gamma,n) \text{ iff } \exists\pi\in\Pi_{n}(\Gamma) \mid \forall c\in\pi, \forall C\in C(c,\wedge)\mid C\nvdash\bot$$

$$Nontaut(\Delta,n) \text{ iff } \exists\pi\in\Pi_{n}(\Delta) \mid \exists c\in\pi, \forall C\in C(c,\vee) \mid \top\nvdash C$$

Where $\Pi_{n}(\Delta)$ is the set of n membered covering families of Δ. With these in hand, we can define the levels of premise and conclusion sets:

$$\mathfrak{l}(\Gamma) = \min n \mid Con(\Gamma,n), \text{ if this limit exists}$$
$$= \infty, \text{ otherwise.}$$

$$\mathfrak{l}^{\bullet}(\Delta) = \min n \mid Nontaut(\Delta,n), \text{ if this limit exists}$$
$$= \infty, \text{ otherwise.}$$[5]

Finally, we can now define a set-set consequence relation, *forcing*,[6] that does not trivialize except when a trivial sentence appears in the premise set (a contradiction) or conclusion set (a tautology):

$$\Gamma \Vdash \Delta \text{ iff, for } n=\mathfrak{l}(\Gamma) \text{ and } m=\mathfrak{l}^{\bullet}(\Delta),\ \Gamma\ {}^{n}\vdash^{m}.\Delta$$

That this relation does not trivialize unless Γ or Δ contains a triviality is evident from the definition of $\mathfrak{l}$ and $\mathfrak{l}$'. Further, it also follows from our definition and the hierarchy of aggregation principles described above that the principles of

---

[5] Here I've used n as the number of cells in a consistent covering family, rather than follow Schotch and Jennings' shift to ξ to insist on the fact that we're dealing with a cardinal here. This is a bit inelegant, but I think it makes the presentation somewhat easier to follow.

[6] The term is due to Schotch and Jennnings.

aggregation we apply here to Γ and Δ are the strongest general principles of aggregation we can apply while respecting Triv.[7]
As for completeness of this consequence relation, we already have most of the result in the classical completeness of the underlying sentence logic. All that's needed in addition is that closure of Γ under $2/n{+}1_L$ should produce sentences from which we can prove any sentence that follows from some cell of every n-partition of the premises, and that closure of Δ under $2/n{+}1_R$ should produce every sentence that follows from any sentence that proves the disjunction of every cell of every n-partition of Δ. And both these results, as we remarked above, follow from a graph-theoretical result. This result shows that using the complete graph on n+1 vertices as a template for aggregating singleton sets of sets of vertices, together with the trivial operations of eliminating some vertices from any chosen edge(s) and adding edges, allows us to construct all the non-n-colourable hypergraphs, i.e. the hypergraphs that must have a monochromatic edge on every n-colouring of their vertices- see the appendix to [2]. This construction is completely parallel to the 2/n+1 operations, and its completeness shows that they completely capture the force of the n-cell based picture of aggregation (which corresponds directly to n-colouring the sentences in our premise or conclusion set).

4 WHY THIS LOGIC IS CLASSICAL

Classical logic is usually presented in the form of a set-sentence consequence relation, where the premise set is assumed to be aggregated by closing the premise set under conjunction. Moving to a set-set relation also requires aggregating conclusion sets by closing them under disjunction. It is an important conceptual step, as well, since it emphasizes symmetries that are not apparent in the usual presentation of classical logic—in particular, the symmetry between assertion and denial, which makes the denial of all members of the conclusion set require the denial of at least one element of the premise set just as the assertion of the premise set requires the assertion of at least one element of the conclusion set.
Another way to think of this is in terms of two sorts of maximal consistent sentence sets– maximal consistently assertable sets, and maximal consistently deniable sets. Every consistently assertable extension of the premise set of a correct implication can be consistently extended by adding some member of the conclusion set, and every consistently deniable extension of the conclusion set of a correct implication can be consistently extended by adding some member of the

[7] Anything stronger (applied in general) would produce as a consequence those sentences that correspond to a subclass of the n membered covering sets of Γ and/or Δ. But nothing guarantees that such a subclass will include a consistent (consistently deniable) covering set, i.e. a set all of whose members are classically consistent (or classically consistently deniable), since our definition of [simply requires there be such a consistent (consistently deniable) covering set.

premise set. An important characteristic of classical logic is that maximal consistent assertable sets are complements to maximal consistently deniable sets—in general, the denial of a set of sentences will carry exactly the same information/descriptive force as the assertion of a corresponding set, and there is no overlap between what can consistently be asserted, and what can, simultaneously, be consistently denied.

A paraconsistent logic, in general, extends the sets that may be "consistently" (or, more broadly, "acceptably") asserted (i.e. asserted without trivializing the set's closure under deduction) beyond the classical limits; to preserve this symmetry, it must also extend the sets that may be consistently (or acceptably) denied. This is what characterizing level on the left and right has allowed us to do here—and it is a beautiful structural feature that I think is all too often neglected in work on paraconsistent logics. And like the cost of classical aggregation that has been our focus, there are costs in moving too quickly to a paraconsistent logic that preserves "truth" (or satisfiability) from left to right, but ignores the parallel preservation of denial (or anti-satisfiability) that is also a feature of the classical consequence relation.

In the approach to logic we are following here, we build a logic of sentence sets by starting from a logic of sentences, and "type-raising" the sentence logic to a relation between sets of sentences. This seems to me to be a well-motivated stand. Sentences, after all, are the real bearers (or at worst, syntactic stand-ins for the real bearers) of the property whose preservation most logicians consider central to their enterprise, viz. *truth*. Further, it is sentences (or their semantic contents) that we accept, entertain, suppose, hypothesize, refute and sometimes believe. Given this, it seems reasonable to suppose that logic should begin with the logic of sentences. But when we type-raise a relation like this, the aim (again) is preservation: preservation of the features that made the consequence relation interesting or worthwhile in its original form.

The suggestion Schotch and Jennings offer is that in the transition from a logic of sentences to a logic of sentence sets, it is all to easy to leap to the obvious but very constraining assumptions that a set is true iff all its members are true, and a set contains truth iff at least one of its members is true. This, together with the suggestion that the consequence relation should hold iff the truth of the left side guarantees the containment of some truth on the right, leads immediately to the classical picture of aggregation, with all its trivializing power.

What they propose instead is a more sensitive approach to the issue of type-raising. What features of the original sentence logic are worth preserving? When we answer this question, it's important not to cast too narrow a net. We should preserve not just the sentence logic itself, as Pres proposes, but also the identification of special classes of sentences as the sources of triviality on the left and the right. That these sentences should be trivial in these ways seems harmless, though of course this is a controversial remark, and still more so in a volume on paraconsistent logic—it is a standard sort of remark for defenders of classical logic who, in the eyes of most paraconsistent logicians, are biting bullets when they make it.

But in this case I think it really is harmless. There is no bullet here to be bitten, *given* classical semantics and the usual take on consequences as truth-preserving. Trivial sentences on the left are those which *on the proposed semantics* are never, under any circumstances, true; trivial sentences on the right are those which *on the proposed semantics* are never, under any circumstances, false.[8] Commitment to the truth of a left-trivial sentence *on this semantics* is literally nonsensical, as is denial of the truth of a right-trivial sentence.[9] Truth- (dually, from right to left, falsehood-) preservation trivializes for both. For the first there is no truth on the left to preserve, and for the second there is no falsehood on the right. Furthermore, there is a straightforward constructive response to this trivialization, at least in a propositional language: If you have a contradiction on the left, get rid of it. And if you have a tautology on the right, again get rid of it. They aren't doing anything useful there.[10]

Of course it's always open to someone to propose a different semantics, and perhaps the new semantics could be justified as a better account of contradictions and tautologies as they appear in ordinary language. But this is a long row to hoe, and not our concern here.[11] The question before us, instead, is what logic should we use for sets of sentences *if* we accept the classical treatment (or any

---

[8] Many, in fact most paraconsistent consequence relations are defined in terms of the preservation of some non-classical sort of satisfiability. So this point holds for them as well (though it becomes moot if their notion of satisfiability is so unconstrained as to allow any set whatsoever to be satisfied).

[9] Since trivialization destroys any distinction between what we are committed to and what we are not committed to, and the loss of that distinction leaves us without any *useful* account of our commitments at all.

[10] In a first-order language, as one referee reminds me, things are more difficult, since whether or not as sentence is contradictory is not decidable. Here we must retreat to a pragmatic response: When trivialization threatens because we come to recognize that a contradiction can be derived from a set of cardinality 2 or higher, we can respond by weakening aggregation. When trivialization threatens because we recognize a contradiction is already present, we can respond by eliminating the contradiction; if the "contents" of the contradiction do useful work for us, then we can try to preserve them by replacing the contradiction with a set of sentences that are individually enough to do the work for us, and whose closure under conjunction is equivalent to the contradiction.

[11] These sentences do no useful work that I can see in our description of goals or beliefs. How could we use the claim that someone believes a contradiction in the course of predicting or explaining their behaviour? There seems to be no way to do so, except in a very specifically linguistic sort of context: one could try to explain the fact that someone *asserts* a contradiction by appeal to their belief in it. But beyond such linguistic cases, there seems to be no behaviour that would reflect belief in a contradiction, as opposed to belief in a set of sentences whose conjunction is/is equivalent to the contradiction (in particular there is none that would reflect the entire, specific content of some specific contradiction). This contrasts sharply with belief in a pair of contrary claims, which can help to explain behaviour that in some circumstances, reflects and is guided by reasoning in terms of belief in one, and in other circumstances, reflects and is guided by reasoning in terms of belief in the other. Similarly, I can't imagine how to use the claim that someone had a contradiction as a goal to predict or explain their behaviour. Again, a kind of linguistic level allows it (I do this, I say, because I believe that, if I do it, my desired contradiction will result). But I find it difficult to imagine any actual way of bringing a contradiction about, or of recognizing them to be the case. The usual semantic treatment of them captures this difficulty very straightforwardly, which I think is a pretty attractive feature of that treatment.

other treatment you like, in fact) of the logical words expressed in the sentence logic.

Given the aim of such a type-raising maneuver, the preservation of desirable features of the sentence logic must guide our endeavours here. And as we have seen, going beyond Pres in a more delicate fashion leads us to a type raising that, at the least, avoids adding to the trivialities already present in the sentence logic. Moreover, as Schotch and Jennings have pointed out, there is no straightforward and constructive response to the wider sort of trivialization that arises in classical set-set logic. A set can be inconsistent$_{set}$ when none of its members is a contradiction. So we have an open question before us, in such a case: where have we gone wrong, and how shall we fix it? It would be helpful to be able to reason from the set, in some non-trivial fashion, as we try to discover where the problem lies. This is one important purpose that weakly aggregative logic aims to serve. This logic is a well-motivated paraconsistent logic for sets of sentences, based on a purely classical sentence logic. It is at least as deserving of the label "classical logic" as the more familiar, and more easily trivialized, logic so widely accepted as standard.

In fact, I suggest that this preservationist approach to paraconsistency, though it appears here as part of a damage-control operation aimed at making classical logic better at coping with inconsistency, in fact provides a more radical and liberating proposal for the re-thinking of consequence relations than other, avowedly non-classical paraconsistent logics. For it dispenses with a powerful constraint that they all respect: the semantic conception of a consequence relation as founded on the preservation of "designated values" from left to right. Rather than avoid undesirable consequences "the old-fashioned way – by making the premises true and the conclusions false"[12], we propose consequence relations that preserve desirable semantic properties other than truth (and its formal semantic parallel, being assigned a designated value).

Thus we reject some classical consequences not because they fail to be truth-preserving, but because they fail to preserve level[13], (or, in other forms of preservationist logic, other measures of the degrees of acceptable assertability for premise sets, and acceptable deniability for conclusion sets—see [4] and [5]). Looking more widely, both at what properties are preserved in classical logic (from right to left as well as from left to right), and at what properties we might wish to preserve when satisfiability (or anti-satisfiability on the right) are lost, is a program for paraconsistency that is broad in scope and rich in philosophical insights.

Further, it's worth also pointing out that whatever standards for consistent assertability and deniability are adopted by a logic, we can apply the techniques used here to weaken aggregation, the better to deal with sets of sentences that

---

[12] R.K. Meyer, in conversation.

[13] That is, there will be consequence pairs such that no element of the consequence set can be added to the premise set without increasing ℓ, and such that no element of the premise set can be added to the conclusion set without increasing ℓ.

violate the standards. So the issue of aggregation is really independent of to the issue of what base logic we choose to begin with.

5 A CLOSING REMARK ON MONOTONICITY

Classical logic is strictly monotonic; forcing—the logic of [⊢—is not. When the level of a set, on either side, increases, the consequence relation weakens because the applicable principle of aggregation weakens. Thus, for instance, if $\Gamma = \{p,q\}$, then the set $\Delta=\{(p \wedge q)\}$ follows, while from a superset of Γ, $\Gamma' = \{p,q,\neg p\}$, Δ no longer follows. And if $\Delta=\{p,q\}$, then Δ follows from $\Gamma = \{(p \vee q)\}$, while if we take a superset of Δ, $\Delta' = \{p,q,\neg p\}$, then the weakened aggregation of this conclusion set (now with $\ell'=2$) prevents Γ from proving Δ'.

But monotonicity is sometimes seen as the hallmark of deductive (as opposed to inductive) reasoning. Thus the surrender of monotonicity is sometimes regarded as a grievous black mark against this logic. But in fact the benefits of monotonicity are shared by this non-monotonic logic, while its costs are borne only by the standard logic.

The point is that strict monotonicity is only maintained by means of trivialization. Whenever we add something to a set of premises that renders it inconsistent, the new set is trivial, and so of course includes as consequences all the consequences of the original set. Classical logic in its usual set-sentence form preserves the consistency, or satisfiability, of premise sets: every consequence of a set of sentences is a consistent (satisfiable) extension of every consistent (satisfiable) extension of the set. When the premise set is inconsistent or unsatisfiable, this condition obviously becomes trivial. In the present set-set form, the aim is to preserve features of the consequence relation, specifically that truth on the left ensures the presence of truth on the right. Some sentence on the right, then, must be a consistent/satisfiable extension of every consistent/satisfiable extension of the premise set. Again obviously, this trivializes if the premise set is not consistent or satisfiable. And it also trivializes if the sentences on the right guarantee automatically that every consistent extension of our premise set can be consistently extended with one or another of them, because in fact every consistent set can be so extended. The trivialization of these sets reflects, in effect, a kind of idealization on the part of those who use such logics, viz. the assumption that we never really do any work with such sets.

But forcing preserves all of classical logic so long as this working assumption holds. Forcing becomes non-monotonic when and only when it fails. So long as classical monotonicity does us any good, that is, so long as inconsistency does not arise, forcing is also monotonic. But when inconsistency does arise, classical logic retains monotonicity at the cost of trivialization. By contrast, forcing surrenders monotonicity instead, and thereby preserves a reasonable (or at least an informative) account of the consequences we might reasonably derive from such a set, or the premises from which such a set might follow. This is a much more graceful response to the failure of our hopes for consistency: The benefits of forcing are as clear as the cost of the scorched-earth policy pursued by the standard logic.

In fact, forcing provides us with an instance of a healthier form of monotonicity: So long as extensions of the premise or conclusion set preserve the desirable, to-be-preserved properties of those sets, the logic is monotonic. But when an extension lacks these properties, we reject monotonicity and look for new properties worth preserving, rather than allow the consequence relation to trivialize. Such a policy gives up strict monotonicity only when monotonicity does us no good anyway. And it provides us with a constructive response to sets of sentences that lack whatever properties we initially designed our logic to preserve: "Find something you like about your premises, and preserve it!"[14] But I would also say find something you like about your conclusions (some standard of acceptable deniability) and preserve that, too.

REFERENCES

[1] P.K. Scotch and R.E. Jennings. On Detonating. In G. Priest, R. Routley and J. Norman, editors, *Paraconsistent Logic*, pages 306-327. Philosophia Verlag, Munich, 1989.

[2] B. Brown and P.K. Schotch. Logic and Aggregation. *Journal of Philosophical Logic*, 28(3):265-287, 1999.

[3] T. Nicholson. A New Completeness Proof for the K_n Modal Logics. In J. Woods and B. Brown, editors, *Logical Consequences: Rival Approaches*, pages 253-260. Hermes Science Publications, London, 2000.

[4] R.E. Jennings and D. Johnston. Paradox-Tolerant Logic. *Logique et Analyse* 26: 291-308, 1983.

[5] B. Brown. Yes, Virginia, There Really are Paraconsistent Logics. *Journal of Philosophical Logic*, 28(5): 489-500, 1999.

[14] R.E. Jennings, in a paper presented to the Society for Exact Philosophy, Austin Texas, 1994.

The Logic of Opposition

F.G. ASENJO Mathematics Department, University of Pittsburgh, USA
fgasenjo@pitt.edu

1 The Argument for a Logic of Opposition

1.1 Motivation

There is a mistaken generalized idea that antinomies are essentially dependent on negation. In fact, many antinomies have nothing to do with negation but rather with some basic, more general notion of opposition relative to which negation is only a particular case. We present here a logic of opposition as an instrument with which to place antinomicity in its most general setting. In the process, several arguments will be made about the foundations of mathematics which go beyond the issue of antinomicity and which reflect the present and future changes that are and will be taking place in mathematics as a whole, mostly related to the enormous impact that computers are effecting on the way mathematics is practiced from day to day.

1.2 Oppositions

The concept of opposition is a primitive logical idea; it cannot be defined in general, but we recognize its relevance when we apply our common-sense understanding of meaning – in other words, the intuitive semantics that we all use prior to any kind of formalization. No opposition is automatically antinomic, contradictory in some sense, but every opposition can be made antinomic by manipulation. For example, the predicates "less than" and "greater than" are opposite predicates not necessarily in antinomic opposition by themselves; they become so when I assert that there are objects a and b such that a is less than b and greater than b. "To be a member of itself" and "not to be a member of itself" are neutral opposite predicates, but they become antinomic when we feed to each of them the set of all sets which are not members of themselves. This also applies to the more basic opposition of objects. "To take as a unity" and "to take as a multiplicity" are opposite logical operations, not necessarily leading each by itself to antinomicity, but to take a multiplicity of objects as a unity creates a new entity which does involve antinomic opposition.

1.3 Unity and multiplicity

The pairs of opposites "unity versus multiplicity" and "one versus many" are not synonymous. An enormous gamut of meanings has been attached to each of these four ideas throughout history – including synonymous usages. I think it is fair to say that, mathematically speaking, one and many have chiefly a connotation of cardinality, whereas unity and multiplicity are more often understood as attributes that can be attached to an entity at the discretion of the observer. Thus we can address a single object (i) neutrally, that is, as having nothing to do with either unity or multiplicity, or (ii) as being a unity unto itself, a relatively closed, fairly fastened little universe, or (iii) as a multiplicity by itself – that is, a relatively loose entity, ready to be joined by other objects, any objects, chosen arbitrarily. The same can be said of more than one entity: they can be looked at (i) neutrally, as an adventitious bunch of objects each having an independent existence of its own, (ii) as constituting a relatively closed unity, or (iii) as a plurality of coexisting objects free to accept additional companions. Hence, rather than referring to unity and multiplicity merely as primitive ideas (which, of course, they are), we shall be concerned here with using the expressions "taking as a unity" and "taking as a multiplicity" as ***primitive logical attitudes*** – that is, as operations that we can apply to change the nature of the object or objects thus taken from that of a neutral or any other condition to that of an internally transformed new entity.

1.4 Identity and difference

The important point to be made here is that in the process of taking an object as a unity or as a multiplicity the original object is preserved as the same object being transformed into something ***different***. "To take as a unity" and "to take as a multiplicity" are not the usual mathematical operations of single-valued correspondences; they are not "external" mappings but "internal" transformations that fasten to the original entity at the same time that the latter is being changed into something new. Unless the entity was already a unity or a multiplicity respectively, in all other cases these two transformations are antinomic operations. An external operation maps an object or several objects into other objects, casting off the given object or objects after obtaining a "value." An antinomic internal operation instead preserves the given objects throughout the process of change into its final outcome. Just as becoming is based on the antinomy of entities being constantly the same and yet different, internal transformations produce new objects out of old ones, keeping the latter in the end as the same and yet different.

1.5 Bhaskara

Identity and difference represent another basic undefinable pair of fundamental opposite ideas. Bhaskara argued in the ninth century, "Difference is an attribute of identity. Everything is innately one and manifold, neither wholly indivisible nor wholly divisible..."[1]. The effect is only a particular state of the material cause, different and yet not different..."[2] No one can ever show that there is only unity or

[1] J. Pereira ***Hindu Theology***, New Delhi: M. Banarsidass, 1991, p. 262.

[2] S. Satchidanandendra, ***The Method of the Vedanta***, trans. by A. J. Alston, London: Kegan Paul International, 1989, p. 491.

only plurality..."[3] The suggestion that a thing cannot be its opposite is meaningless, because it is so experienced..."[4]

We must, of course, make a distinction between difference-in-identity and identity-in-difference – that is, between change that flows in a given sameness and sameness that is present through ongoing change. Bhaskara is an early and forceful promoter of the need to understand this intimate antinomic coexistence of opposites both in mind and in world. The day somebody writes a history of antinomicity, a chapter on his ideas will be very much in order. Along the same line, puzzled by the contradictory consequences of quantum mechanics, Eddington reflected a la Bhaskara: "We used to think that if we knew one, we knew two, because one and one are two. We are finding that we must learn a great more about 'and.'"[5]

1.6 Gathering and choosing

Two basic operations are at the heart of the foundations of mathematics: collecting entities into a whole, and choosing entities from a whole, either by merely pointing at them or by fully picking them out for the purpose of examining and using them individually. Cantor defined a set informally as "a multiplicity taken as a unity."[6] We shall take this definition seriously and consider "to take as a unity" one version of the fundamental process of gathering. The opposite fundamental operation of choosing is seen in the act of "taking as a multiplicity." Thus, for example, the Axiom of Choice is antinomic in that it chooses a multiplicity of representatives from each of the members of a family of sets in order to *gather* them subsequently into what is usually called "the choice set." Concrete choosing, of course, differs from abstract choosing in that it often merely points at distinguishable entities in a whole, leaving them there because it is impossible to pick them out. To make this distinction part of the logic of opposition, let us introduce the notion of *tes* as a kind of mirror image of set. From a set it is always possible to pull out one or several members ("Let us take a and b from the set c.") and consider them as a separable multiplicity. From a tes one can distinguish without being able to separate what is distinguished in the tes from the tes itself – one chooses but is not able to take out. A *tes* is a logical reflection of the fact that in the real world identity and difference come inextricably together and are discernible at the same time in the same entity.

As an egregious example of the latter, we have the way in which space is inseparable from real objects. When we abstract space from observable entities, we end up with the unreal conception of space as a container, an originally empty box in which things can be placed. Space, however, is not a container but a property of physical entities as well as of their relative motion. From relativity theory we know beyond any doubt that each moving particle is the center of a system of reference which structures space and time in its own individual way. From moving particle

[3] D.H.H. Ingalls, "Bhaskara the Vedantin," *Philosophy East and West*, 1967, p. 66.

[4] S. Dasgupta, *A History of Indian Philosophy*, Cambridge: Cambridge University Press, 1922, vol. IV, p. 329.

[5] A. N. Jeffares and M. Gray, *A Dictionary of Quotations*, New York: Barnes and Noble, 1997, p. 235.

[6] This is a paraphrase of Kant's well-known characterization of totality as a multiplicity taken as a unity, in turn a version of Leibniz's assertion to the effect that a whole is the parts taken together with their union. Cf. G. W. Leibniz, *Philosophical Papers and Letters*, trans. by L. E. Loemker, Dordrecht-Holland: D. Reidel, 1969, p. 76.

to moving particle the universe is very differently shaped. There is really no absolute empty space, not only because the universe is everywhere made up of either solid bodies or very rarefied gas, but more fundamentally because space is intrinsic, not extrinsic to any actual entity. Although Husserl talks about things "placed in space," he has made some apt comments on the subject, in particular: "Space is an 'inner' constitutive determination of the thing and has indeed a determinate structure."[7] Also: "A given empty space is necessarily an empty space between given things or phantoms of things. If nothing spatial at all is given, then neither is any space."[8] And as for time: "Duration is what it is only as the duration of a thing – of something real in general – and, in a somewhat different sense, as the duration of a process, which in turn presupposes a thing undergoing the process, and a thing is inconceivable without duration."[9] Endless instances of inseparability of aspects in the real world – that is, of identity-in-difference – could be added indefinitely. Logic must reflect this concrete situation.

1.7 "To take as..."

If we are going to use the gist of Cantor's informal definition of set seriously, then the primitive notions of unity and multiplicity must play a basic role in our logic of opposition, but even more so that primitive attitude of "taking as...," that is, the taking of an object, an operation, a predicate, a statement – anything – as something or other. "To take as" is a universal operation in mathematics, one that calls for answers to the questions: Who is taking what, in what context, to do what with? In other words, the role of the subject is not to become lost in the process of transforming that which is given to the subject into something else.

The inescapable role of the subject in mathematics was one of Brouwer's tenets. His system, however, was extremely restrictive, and it was not until Errett Bishop developed his Constructive Analysis that it was evident for everybody to see that the basic ideas of intuitionism, appropriately transformed, are able to produce an extensive body of mathematical discipline. This body provides a more suitable foundation than classical analysis for those types of mathematics based on or impacted by the current growth of computer-oriented systems.

Before Bishop, Brouwer's intuitionism had been transformed into a formalized variety of nonclassical logic, a narrow one, with limited mathematical application. We are not interested in this version of intuitionism but rather in Brouwer's very valid comments on intuition itself and its uneliminatable role in mathematics. He derided the conception of mathematics as a "museum of immovable truths,"[10] a Platonist heaven which is really the cemetery of mathematical progress. He said that "the fundamental parts of mathematics can be ***built up*** from units of perception."[11] For him, mathematics is a mental construction founded on basic intuitions. For example: "the ***continuum as a whole*** was given to us by intuition; ... [to] create from [this] mathematical intuition 'all' its points as individuals is inconceivable and

[7] E. Husserl, *Thing and Space*, trans. by R. Rojcewicz, Dordrecht: Kluwer, 1997, p. 297.

[8] E. Husserl, *Thing and Space*, trans. by R. Rojcewicz, Dordrecht: Kluwer, 1997, p. 323.

[9] E. Husserl, *Thing and Space*, trans. by R. Rojcewicz, Dordrecht: Kluwer, 1997, p. 298.

[10] L.E.J. Brouwer, *Collected Works*, ed. by A. Heyting, Amsterdam: North-Holland Publishing Co., 1975, vol. 1, p. 475.

[11] L.E.J. Brouwer, *Collected Works*, ed. by A. Heyting, Amsterdam: North-Holland Publishing Co., 1975, vol. 1, p. xiii.

impossible."[12] And since intuition develops as a process, one must conclude that "the only a priori element of science is time."[13] The subject that thinks and the time in which the thought develops are inseparable from mathematical constructions; no proof is independent of the mind. In fact, there are "languageless constructions which arise from the self-unfolding of the basic intuition, [constructions which] are exact and true by virtue of their very presence in the memory."[14]

Without reference to Brouwer's other tenets, we can see that these quoted comments are well in line with the situation encountered as we talk about "taking as...." Just as the computation of a function defined by recursion is a personal, time-consuming process, "to take as..." implies (i) a subject to whom the object, function, predicate, sentence, or anything else is *given*. This "being given" goes together with its "being taken," even if merely received neutrally. But then (ii) the subject acts on what is given and takes it in one very definite way – as a unity or as a multiplicity, for example. Then again (iii) the act of "taking as..." is applied once more in another direction, etc., all of which occur in successive stages. The somebody who takes things one way or another affects them intimately, transforms them in the course of time, reaches ever new conclusions. This applies even to the simplest mathematical situations. As Whitehead said, "A prevalent modern doctrine is that the phrase 'twice-three' says the same thing as 'six'; so that no new truth is arrived at in the sentence. My contention is that the sentence considers a process and its issue. Of course, the issue of one process is part of the material for process beyond itself. But with respect to the abstraction 'twice-three is six,' the phrase 'twice-three' indicates a fluent process and 'six' indicates a characterization of the completed fact."[15] Time is indeed of the essence of mathematics.

1.8 The anthropic principle of mathematics

The constructive presence of the subject is already widely recognized in physics in the classic phrase "the observer is part and parcel of the observation." Indeed, the impact of the observer on the structure of what is observed is regularly taken into account in quantum mechanics, for example. But in addition, there is today a so-called "Anthropic Cosmological Principle," which in one of its forms asserts "that the observed structure of the Universe is restricted by the fact that we are observing this structure; by the fact that, so to speak, the Universe is observing itself,"[16] or, in other words, "Observers are necessary to bring the Universe into being." [17]

This is true of mathematics as well. There would be no mathematics without people ready and willing to think and make mathematics. But not only this, the thinking mathematician is the one corner of the world to which the mathematical

[12] L.E.J. Brouwer, *Collected Works*, ed. by A. Heyting, Amsterdam: North-Holland Publishing Co., 1975, vol. 1, p. 45.

[13] L.E.J. Brouwer, *Collected Works*, ed. by A. Heyting, Amsterdam: North-Holland Publishing Co., 1975, vol. 1, p. 61.

[14] L.E.J. Brouwer, *Collected Works*, ed. by A. Heyting, Amsterdam: North-Holland Publishing Co., 1975, vol. 1, p. 443.

[15] A. Whitehead, *Modes of Thought*, New York: The Free Press, 1968, p. 92.

[16] J. D. Barrow and F. J. Tipler, *The Anthropic Cosmological Principle*, Oxford: Oxford University Press, 1988, p. 4.

[17] J. D. Barrow and F. J. Tipler, *The Anthropic Cosmological Principle*, Oxford: Oxford University Press, 1988, p. 22.

givens are given, the one agent of the world that takes the given and turns it into something else, that is to say, the source of mathematical transformations – internal transformations, not the mere arrangement of external correspondences. This is the anthropic principle of mathematics, a principle on which the logic of opposition rests.

1.9 The paradox of subjectivity

To show how far antinomicity reaches, we must refer now to the fact that the subject itself brings about its own reflexive contradictoriness. Akira Kurosawa referred to Dostoyevsky as the "one who writes most honestly about human existence.... He seems terribly subjective, but then you come to the conclusion that there is no more objective writing."[18] Indeed, if one wants to be absolutely objective about the human mind – one's own and others' – there is no recourse but to immerse subjectively into it, to exercise reflection and empathy, to live deeply into real intersubjectivity. This is the antinomy implicit in the understanding of consciousness from which there is no escape.

Husserl looked at this matter from an even more fundamental point of view. He asserted that the portion of reality that we observe transcends the mind at the same time that it is directly present in the mind, not as mere appearance but "in itself."[19] He called this antinomic situation "the mystery of intentionality," which he accepted as an unsolvable riddle, a paradox that we cannot dissolve and hence must just learn to live with – a paradox of identity-in-difference. He even asserts, in line with the Anthropic Cosmological Principle, that consciousness is fundamental to the world,[20] part and parcel of the real; otherwise, perception becomes meaningless. The transcendental self is in the midst of the transcendent reality that it faces, at the same time that the latter is itself at the heart of the mind. In other words, we do not have two parallel universes, one of the physical reality and another of the mind; we have one universe in which mind and world are co-present in full concrete participation. This reflects the existence of an internal relation that makes mind and the observed world constitute one another as they meet. All of this is perfectly natural, hardly a mystery, because, as Nagarjuna said, "when something is not related to anything, how then can that thing exist?"[21] Relations, then – applied internal relations, that is – make the related terms: they are true transforming operations.

1.10 Proclus

The previous considerations do not imply that the real is created by the mind but that the mind is as real as anything else, and whenever mind and world meet they create together a new structure of reality, a special cosmological entity. This universal situation applies to mathematics as well. Brouwer was not the first to assert that

[18] G. Sadul, *Dictionary of Films*, trans. by P. Morris, Berkeley, CA: University of California Press, 1972, p. 144.

[19] E. Husserl, *Ideas Pertaining to a Pure Phenomenology*, First Book, trans. by F. Kersten, The Hague: M. Nijhoff Publishers, 1982, pp. 119-120.

[20] Cf. R. Sokolowski, *The Formation of Husserl's Concept of Constitution*, The Hague: M. Nijhoff, 1964, p. 137.

[21] Nagarjuna, *Master of Wisdom*, trans. by C. Lindtner, Oakland, CA: Dharma Press, 1986, p. 17.

the subject that makes or reads mathematics remains the place where mathematics evolves. Proclus also maintained that the body of mathematical knowledge is the result of the activity of the mind. He said, "Let us follow our head."[22] Imagination – which leads to take entities one way or another – is prior to understanding.[23] There are Ones that include a Many,[24] Wholes which are a part of their Parts,[25] acts of the mind that possess the "double character of indivisibility and divisibility,"[26] acts in which the self reaches beyond itself and finds things all in one another,[27] as well as paradoxical theorems in geometry that, for example, show that some lines exhibit a nonconvergent convergence.[28] All in all, mathematical propositions exhibit a double process: a movement of "progression," or going forth from a source, an upward construction from atoms to compound entities, say, and a process of "reversion" that unfolds the simple into its inherent complexities.[29] Mathematics is a creation that follows the mathematical imagination according to the relatively arbitrary volitions of the mind, the active agency that takes objects, functions, predicates, and anything else in one or another transforming way. Complex structures proceed from the simple, and simple entities develop a complex inner structure, a "descent" into plurality, but a descent which "must not be cut off from its original state, for it would lose its definition."[30]

1.11 The algorithmic character of mathematics

Lésniewski and Tarski have been very influential in creating the present pattern of treating definitions as logical equivalences.[31] They were in turn influenced by *Principia Mathematica* where all symbols defined by tautological equivalences are described as convenient but superfluous. This approach, which so much lends itself to essentialism, does not correspond to actual mathematical practice. Constructive definitions are instructions to produce objects, or functions, or formulas, in a step by step manner. This is especially and obviously the case in inductive definitions. Kant said it most appropriately: "Whereas mathematical definitions make their concepts, in philosophical definitions concepts are only *explained*."[32] Constructive definitions, then, have nothing to do with truth and falsity. They are building instruments of a special kind. In Bishop's words, for example: "A set is defined by describing exactly what must be done to construct an element of the set"[33]; the definition, therefore, consists in the making of the mathematical object. This

[22]Proclus, *A Commentary on the First Book of Euclid's Elements*, trans. by G. Morrow, Princeton, NJ: Princeton University Press, 1992, p. lvi.

[23]Proclus, *A Commentary on the First Book of Euclid's Elements*, trans. by G. Morrow, Princeton, NJ: Princeton University Press, 1992, p. 45.

[24]Proclus, *A Commentary on the First Book of Euclid's Elements*, trans. by G. Morrow, Princeton, NJ: Princeton University Press, 1992, p. 40.

[25]Proclus, *The Elements of Theology*, trans. by E. Dodds, Oxford: Oxford University Press, 1971, p. 65.

[26]Proclus, *A Commentary on the First Book of Euclid's Elements*, p. 78.

[27]Proclus, *A Commentary on the First Book of Euclid's Elements*, p. 113.

[28]Proclus, *A Commentary on the First Book of Euclid's Elements*, p. 139.

[29]Proclus, *A Commentary on the First Book of Euclid's Elements*, p. lxii.

[30]L. Siorvanes, *Proclus*, New Haven, CT: Yale University Press, 1996, p. 105.

[31]Cf. A. N. Prior, *Formal Logic*, Oxford: Oxford University Press, 1962, pp. 96-98.

[32]I. Kant, *Critique of Pure Reason*, trans. by N. Smith, New York: St. Martin's Press, 1933, p. 588.

[33]E. Bishop and D. S. Bridges, *Constructive Analysis*, Berlin: Springer Verlag, 1985, p. 9.

was also Rickert's point of view: "A definition must form the concepts in such a way that, based on them, a system of assertions can be constructed. It is then an instrument to make the materials with which to build science."[34]

This means that mathematics, which started long ago as a computational discipline, is even today essentially an algorithmic project. Something is given; then it is manipulated, gathered together with other objects, then chosen again, etc., all in the course of a time period. Take, say, the Gramm-Schmidt process to produce an orthonormal basis for a finite-dimensional vector space: it is a step by step program for the construction of such a basis. Or in mathematical logic take the stage of atomic predicate formulas "not yet," "about to be" filled with terms, $P(-,-)$, "prime predicate expressions or ions," as Kleene called them, expressions "before they *become* prime formulas."[35] Or take the stage of functions "not yet" being filled with terms, $f(-,-)$, "prime function expressions or mesons" – also Kleene's terminology.[36] These, as each application of a logical rule of formation, are stages in the process of step by step constructions. Examples could be added endlessly to simply testify to the fact that the conception of logic and mathematics as grandiose tautologies, as static items of an immutable pantheon, is contrary to the reality of such disciplines. Mathematics is unmistakably algorithmic, and patently becoming even more so with the advent and current impact of computers.

1.12 Against reductionism

The idea of definition as a logical equivalence, one in which technically the defined concept is disposable, is a particular form of the extended practice of reductionism, the tendency to eliminate everything that is deemed unnecessary. This is a reincarnation of Occam's razor which, following Karl Menger, can be stated as follows: "It is vain to do with more what can be done with less."[37] The exercise of this notion can be highly salutary, but it can also do a great deal of damage. Time, for example, is an irreducible factor deemed superfluous by mathematical Platonists, who believe only in eternity. Menger proposes a counterpart of Occam's razor which can be stated as follows: "It is vain to try to do with less what requires more," or "entities must not be suppressed below sufficiency."[38] He produces several interesting mathematical examples of this counter-Occam principle.[39] Here we should like to add two: (i) the reduction of relations to classes of n-tuples effected by Wiener has had the unfortunate consequence of identifying relations with external relations and henceforth dropping any attention to internal relations – that is, relations which essentially modify the terms related and which are indispensable to understand the world as it is and not as a simplified, reduced abstraction of it; and (ii) the prevalence of atomism in logic to the exclusion of other alternatives. Let us look closely at this second example.

---

[34] H. Rickert, *Zur Lehre von der Definition*, Tübingen: Mohr Verlag, 1929, I, 2.

[35] S. C. Kleene, *Mathematical Logic*, New York: John Wiley, 1967, p. 77.

[36] S. C. Kleene, *Mathematical Logic*, New York: John Wiley, 1967, p. 148.

[37] K. Menger, *Selected Papers in Logic and Foundations, Didactics, and Economics*, Dordrecht-Holland: D. Reidel Publishing Co., 1979, p. 105.

[38] K. Menger, *Selected Papers in Logic and Foundations, Didactics, and Economics*, Dordrecht-Holland: D. Reidel Publishing Co., 1979, p. 106.

[39] K. Menger, *Selected Papers in Logic and Foundations, Didactics, and Economics*, Dordrecht-Holland: D. Reidel Publishing Co., 1979, p. 108-135.

We have an innate tendency to build from the bottom up. It pacifies us. In fact, the very moment that we scribe individual linguistic symbols on paper we are already prejudicing our thinking toward atomism. "Logical atomism" is precisely the name that Russell gave to his philosophy, and generally people resist admitting systems which allow the possibility of infinite descents, systems like those of today's non-well-founded set theories.[40] From this atomistic point of view, antinomies must be analyzed into separate, nonantinomic meanings – explained away – or be harmlessly blended into a synthesis. The fact is that most antinomies are irreducible and should be kept as such. Further, many syntheses should be replaced by antinomies if we want to stay closer to reality. It is not only that, as Kant said, antinomies keep the mind from slumbering, it is also that there is a special rational energy in antinomic reasoning that gives thought momentum and concreteness, and pushes reason further in ways that reductionism is incapable of even beginning to imagine. Indeed, antinomies must be seen as a source of logical and mathematical creativity.

We should, finally, point out that, were reductionism taken to its ultimate consequences, even the acts of intelligent reading and writing would be impossible. If we reduce the written sentence to a discrete, well-formed sequence of individual symbols, then the continuity of mental meaning is lost. Polanyi has it right when he says, "We can know more than we can tell,"[41] whereupon he proceeds to develop the concrete notion of tacit knowledge, which is, of course, irreducible to linguistic atomism and very much a fact of everyone's daily conscious life. With the understanding that the written word is a stage in a process of inarticulate thinking, a stage that remains immersed in the ever-moving cloud of knowledge, let us continue with our written statements.

1.13 Various ways of "taking as..."

To take something in any specific way implies a subject that performs this operation with various degrees of arbitrariness somewhere between rigid rules and an instantaneous unrestricted volition. We are not now referring to external operations such as the addition of integers, but to true transformations of the entities in question into something else. These transformations do not cast off what is fed to their action; they preserve the identity of the given, while changing it into something different.

(i) We have already mentioned unity and multiplicity, but we should add that, whereas there is only one way of "taking as a unity," we must distinguish "taking as a separable multiplicity" from "taking as an inseparable multiplicity." A group of physical objects arbitrarily placed on a table forms a separable multiplicity; living organs in a living organism form an inseparable multiplicity. As always, we must beware here of essentialism: separable and inseparable are conditions which, like everything else, are subject to transformations that can change each of these two conditions into the other just as we can glue the objects on the table and excise an organ from a body, at the price, of course, of changing the nature of the entities involved. We should also point out that even a single entity by itself can be taken either as a unity or as a multiplicity: a unity when taken as the only member of a set-theoretic

[40] Cf. P. Aczel, *Non-Well-Founded Sets*, Stanford, CA: CSLI Publications, 1988.

[41] M. Polanyi, *The Tacit Dimension*, New York: Doubleday and Co., 1967, p. 4.

singleton, say, a multiplicity when merely taken in its uniqueness, different from other entities. Madhva said, "To perceive objects is to perceive their uniqueness, and it is this uniqueness which constitutes difference. Difference is as simple and analyzable as unity. In themselves, things are of the nature of difference."[42]

(ii) "To take as a term" is in opposition to "to take as a relation." By the latter we mean an internal relation, that is, an operation, not a predicate. As an internal relation relates entities, it shows its character of being an antinomic transformation in that the entities related remain the same identifiable items while being transformed into something different.

(iii) "To take as a singularity" is in opposition to "to take as a specimen." The first means to make of the entity a unique item, be the entity an object, a function, a predicate, a statement, or anything else. The entity becomes a universe unto itself and remains in such a state until another transformation is applied to it. To take an entity as a specimen is to do it as a representative of many similar entities.

(iv) "To take as a whole" is in opposition to "to take as a part." Each of these transformations can be antinomic in still another sense besides the one of identity-in-difference that applies to every internal transformation. As Proclus and others have said, a whole can be part of its parts – in other words, be both whole and part simultaneously, which creates the logical situation of a part being connected to another part by a detour through the whole, or, as Kurt Goldstein put it in his classic characterization of the organism: "Every part-event, be it physical, be it mental, refers to the whole. And only *by way of the whole* is it related to another event."[43]

(v) "To take as a discrete entity" is in opposition to "to take as a continuous entity." Thus we can take a stone as an object circumscribed by the local space it seems to occupy in terms of its ostensibly simple location, or we can see the stone continuing beyond itself as constituting only a spurious fragment of a larger system of objects united by the same gravitational forces: a continuous system, only abstractly fragmented.

This list of pairs of opposite transformations can be extended indefinitely with the understanding that to iterate the same transformation after it has already been applied to an entity does not produce anything new. Thus, for example, to take a unity as a unity makes the second application of "to take as a unity" superfluous.

Let us look now at the various entities to which transformations can be applied as well as at some additional transformations.

(i) Objects are first given as neutral entities. Such to be "given" implies, of course, to be "received" by some subject who accepts the given neutrally. To the neutral object transformations can then be applied in succession. A set is the outcome of taking a neutral bunch of objects as a multiplicity,

[42] S. Dasgupta, op. cit., vol. IV, pp. 78-79.

[43] K. Goldstein, *The Organism*, Cambridge, MA: The MIT Press, 1995, p. 301. The underlining is Goldstein's as it appears in the first edition of the book.

and then this multiplicity is taken as a unity. A tes is the outcome of a neutral object being taken as a unity, and then the latter as a multiplicity. All objects, neutral or already transformed, are amenable to be taken in any other way possible. In particular, objects can be taken as operations, just as numbers can be taken as functions. Thus, for example, internal relations can be taken by themselves as objects or as internal operations when applied to other objects. In some versions of category theory based on mappings as the ultimate primitive items, these mappings can be taken by themselves as objects and then be fed as such to functions, predicates, etc.

(ii) Objects can also be taken as predicates, just as nouns are regularly taken as adjectives. Let us explain with an example this "predicatization" of an object in the sense of having the object function as a predicate. It is well known that fuzzy logic assigns to each statement a "degree of truth." But any statement that has a degree of truth also has a degree of falsity. In other words, fuzzy logic is inescapably an antinomic logic, one in which statements have a degree of antinomicity. Consider a fuzzy predicate like "goodness." "Peter is good" can be seen as having a certain degree of truth and a certain degree of falsity. Let us now take this object Peter and turn it into a predicate as follows. Thinking of Peter as a model that other people should emulate, and knowing that John has been trying to do so, we ask, "Is John sufficiently 'Peterized'?" In other words, we feed John, the object, to the Peter-predicate, and then assess the degree of truth of the corresponding statement. Every object is potentially subject to such kind of predicatization.

(iii) Predicates are subject to transformations as well. The binary predicate of distinguishability can itself be taken in various ways: We can distinguish in toto or we can distinguish in part, etc.

(iv) As for the truth or falsity of sentences and predicate formulas, we rely on a minimalist semantics.[44] Just as opposition, especially antinomic opposition, is determined by our understanding of the meaning of what is said, the truth of a sentence is a matter of asserting the sentence to be true – with some semantic reason to do so, to be sure. The truth may be provisional, just as when we take any hypothesis as temporarily true for the sake of an argument. In other words, I take this sentence as true and interpret its semantic consequences from such an assumption. No rules of deduction are involved here; hence we are free to use Frege's symbol $\vdash \mathcal{A}$ to indicate that the sentence or predicate formula is asserted, that is to say, taken as true. In his famous *Begriffsschrift*, Frege introduces the sign "$\vdash$" to signify that the judgment that follows is to be interpreted as true; $\vdash \mathcal{A}$ is the assertion that such is the case. Frege says: "If we omit the vertical stroke at the left of the horizontal one, the judgement will be transformed into a mere combination of ideas of which the writer does not state whether he acknowledges it to be true or not."[45] Prior to $\mathcal{A}$, $-\mathcal{A}$ will not express a true judgment; to this effect, the subject must take the further

[44]For a discussion of minimalist semantics, see P. Horwich's Truth, 2nd Ed., Oxford: Oxford University of Press, 1998.

[45]*From Frege to Gödel*, ed. by J. van Heijenoort, Cambridge, MA: Harvard University Press, 1967, p. 11.

step of asserting $\mathcal{A}$ to be true. Nowadays, $\vdash$ has the syntactic role of asserting provability or deducibility. Here we go back to Frege's initial use. And let us emphasize that assertion makes truth, whatever the semantic reasons anyone can have to assert and no matter how briefly the status of truth lasts. In other words, truth, like everything else, is a process, not an eternal attribute. In a regular indirect proof, the opposite of what is being proved is temporarily taken as true; then, after reaching a contradiction, the latter is taken as false. Here, on the other hand, to take a statement as false that has already been taken as true is to take the statement as antinomic. In order to obtain simple falsity, it is necessary to begin again with the statement in its original state of being neither true nor false and then start a new sequence of transformations.

2 Foundation of a Logic and Mathematics of Opposition

2.1 Alphabet and Rules of Formation

1. x, y, z, x_1, x_2, x_3, ..., variable ***first objects***, that is, entities before being taken in any special way (for example, as unities or as separable or inseparable multiplicities). In the words of Charles Peirce, "First are those objects that present themselves as immediate and undifferentiated."[46]

2. o, o_1, o_2, ..., variable or fixed objects of ***any kind***, i.e., either variable or fixed first objects or ***compound objects*** in the sense of having been obtained from first objects after applying to them one or several operations in succession. "Fixed" should not be understood in the sense of a variable taking an arbitrary, unspecified value, a parameter standing for some unknown or even unknowable entity. A fixed object here is a specific one, the opposite of a variable in abeyance, one with suspended variability but without any effective assignment.

3. a, b, c, a_1, a_2, a_3, ..., fixed objects of any kind. We assume that an unlimited number of fixed first and compound objects exists.

4. f, f_1, f_2, ..., variable external operations, each with a fixed arity – that is, operations in the usual sense of associating a given n-tuple of objects of any kind with a single object of any kind, provided that the n-tuple consists of an ordered sequence of n separable objects.

5. g, g_1, g_2, ..., variable ***internal operations***, each not necessarily with a fixed arity, also referred to as ***transformations***, and applicable to separable objects. These are operations which, instead of merely associating a given n-tuple of objects with another object, retain the given object or objects at the same time that they alter its or their character in an essential way. "External" and "internal" have no topological connotations; they only point respectively to (i) the operation's mere translational movement from the given n-tuple of objects to another object without changing the nature of the given objects – a changeless referral – and to (ii) the operation's opposite effect of retaining the given object or objects for the purpose of introducing a special point of view from which to modify the character of what was given; such

[46] *Writings of Charles S. Peirce*, Bloomington, IN: Indiana University Press, 1993, vol. 5, pp. 294-295.

change in character transforms the given objects into something that they were not before.

6. $A, A_1, A_2, \ldots$, variable predicates, each with a fixed arity. Every predicate can be made to involve antinomic opposition – each predicate by itself or in pairs – when fed the appropriate objects; also, many predicates can be taken to be internal in the sense of being capable of changing the nature of the objects involved rather than merely attaching an external property or relation to the objects in question.

7. $=$ (equality), $\in$ (membership), ∂ (distinguishability), $<$ (less than), $>$ (greater than), $\equiv$ (identity), fixed binary predicates. Other fixed predicates with a fixed arity can be added.

8. The usual logical symbols: (i) the propositional connectives $\rceil, \wedge, \vee, \rightarrow, \leftrightarrow$, and (ii) the quantifiers $\forall o$ and $\exists o$, employed in the usual sense where o is a variable object, and $\underline{\exists} a$ where a is a fixed object. The existential quantifier $\underline{\exists} a$ differs from $\exists o$ in that, instead of merely asserting the existence of an unspecified object o "such that" etc., it points to the existence of a specific such object. This goes beyond what is called "existential instantiation," the mere saying "let o be any one object such that..." without specifying the object. $\underline{\exists} a$ actually selects one such specific object, that is, it adds a concrete choosing and picking to the abstract assertion of individual existence. Whitehead and Russell noted this distinction in *Principia Mathematica*. They call "first truth" the sort of truth applicable to $A(a)$ if a is a specific object that makes $A(a)$ true. In contrast, "if we denote by '$(\exists x)A(x)$' the proposition '$A(x)$ sometimes,' then $(\exists x)A(x)$ has second truth."[47]

(iii) Parentheses and commas: (,), [,], and " , " the latter employed both as a standard linguistic symbol and as an operator, as explained in item 16 of this section. The curved parentheses are used to indicate the order of construction of a given expression. The straight brackets are used to avoid ambiguities in the use of the commas.

9. Opp, a marker without a fixed arity that indicates that objects, operations, or predicates, respectively, are in pairwise opposition. Opp is not a predicate, only a symbol to make explicit that certain entities oppose one another. Opp$(<, >)$ simply marks the semantic fact that "less than" is opposed to "greater than"; it is not a regular sentence. Opp$(A_1, A_2, \ldots)$ indicates that Opp(A_i, A_j) is the case for all $i \neq j$. Opp(o), Opp(g), and Opp(A) are meaningful cases also.

10. When operations, external or internal, are applied to first objects, one obtains compound objects as already stated in item 2, above. All objects, first or compound, can be fed to any operation, predicate, or suitable marker, respecting the appropriate arity in all cases where the arity is predetermined. When objects are fed to predicates in the appropriate arity, one obtains ***atomic predicate formulas***. However, the use of the word "atomic" as opposed to "compound" in connection with predicate formulas refers only to the nonoccurrence of propositional connectives and quantifiers, not to any ultimate simplicity in the composition of the formula. As is to be stated below, predicates can occur in the composition of an object and thus become components of any atomic formula to which the object is fed. There is always open a possible non-well-founded descent of objects and predicates in unlimited succession that makes of atomic formulas not first foundations upon which to build upwards but just a stage in the middle in the process of constructing

[47] A. N. Whitehead and B. Russell, ***Principia Mathematica***, Cambridge: Cambridge University Press, 1962, v. 1, p. 42.

both objects and formulas.

11. *Compound predicate formulas* are those obtained by application of the propositional connectives and the quantifiers in the usual way. "∃!o" and "∃!a" mean, respectively, "there exists a unique object o such that..." and "there exists a unique specified object a such that...."

12. Infinitely long sequences of objects and infinitely long sequences of formulas shall be used as markers when, say, an infinite multiplicity of objects is distinguishable in a given unitary object, and, respectively, when an infinite number of predicates applies to the same object. A. Robinson already used the device of expanding the classical propositional calculus to include infinite disjunctions and infinite conjunctions in order to handle the logic of non-Archimedean linearly ordered structures.[48] But whereas Robinson's infinitely long formulas are introduced only as part of the formal language, here infinitely long sequences of formulas (or objects) are also markers which may occur in an object (or formula) as indicators to record the role that nonfinite sequences of either kind of entities can play in the generation and composition of both objects and formulas (see item 29, below). Infinite sequences of objects may also occur as formal expressions (see items 14 and 15, below).

13. In addition to the symbols for variable internal operations listed in item 5 above, the unary symbols $t_1, t_2, ..., t_k, ...$ represent *fixed internal operations* applicable to any object – that is, each has a fixed transformational meaning, with the additional requirement that any transformation with an odd index $2k-1$ is in opposition with the next transformation in the sequence: $\mathrm{Opp}(t_{2k-1}, t_{2k})$. The list begins as follows (in all cases o represents any object).

(i) $t_1(o)$ stands for "o taken as a unity"

(ii) $t_2(o)$ for "o taken as a multiplicity"

(iii) $t_3(o)$ for "o taken as a term"

(iv) $t_4(o)$ for "o taken as an internal relation"

(v) $t_5(o)$ for "o taken as a singularity"

(vi) $t_6(o)$ for "o taken as a specimen"

(vii) $t_7(o)$ for "o taken as a whole"

(viii) $t_8(o)$ for "o taken as a part"

(ix) $t_9(o)$ for "o taken as a discrete entity"

(x) $t_{10}(o)$ for "o taken as a continuous entity."

This sequence can be extended indefinitely, and fixed internal or external operations of any arity can also be added (see items 14 and 15, below).

As a consequence of the t_i's forming opposite pairs in succession, the composite objects $t_{2k-1}t_{2k}o$ and $t_{2k}t_{2k-1}o$ constitute each an *intrinsically antinomic* object: any two mutually opposite internal operations applied in succession transform the

[48] A. Robinson, *Introduction to Model Theory and to the Metamathematics of Algebra*, Amsterdam: North-Holland Publishing Co., 1965, pp. 11, 29.

given object o into an intrinsically contradictory entity. Thus a singleton, a single object taken as a multiplicity and then as a unity, is an intrinsically antinomic object, and so is a tes, a unity taken as an inseparable multiplicity of a finite or infinite number of objects. $t_i t_i o$, on the other hand, is identical to $t_i o$.

14. The comma "," indicates that the arbitrary objects placed at both sides of it constitute an unordered, separable multiplicity; $o_1, o_2, ..., o_n$ represents a finite unordered separable multiplicity of n separable objects, whereas $o_1, o_2, ...,$ represents an infinite unordered separable multiplicity of separable objects. In contrast with t_2, the comma is an external operation of arity greater than or equal to 2. As such, it does not change the character of any of the entities involved. Adding now the markers without fixed arity Id and Dif – "identity" and "difference," respectively – we can say that for any object o_1 not of the form $t_i o_2$ it is always the case that $\text{Id}(o, t_i o) \wedge \text{Dif}(o, t_i o)$. This regular antinomy does not extend to the transit from a bunch of dispersed objects to their being associated into a separable multiplicity by the comma operator, which does not generate antinomies.

15. The curved parentheses "{" and "}" are used to place a finite or infinite separable multiplicity between them in order to take it as a unity. These parentheses constitute together an internal operation of no fixed arity. Apart from the arity, the parentheses differ from t_1 in that the latter applies to any object, whereas they apply only to multiplicities. Thus, $\{o_1, ..., o_n\}$ reads "the finite separable multiplicity $o_1, ..., o_n$ taken as a unity," and $\{o_1, o_2, ...\}$ reads "the infinite separable multiplicity $o_1, o_2, ...$ taken as a unity." Each of these expressions is called a set, an intrinsically antinomic compound object whose antinomicity does not derive from any predicate, including membership, but merely from the successive application of two antinomically opposed internal operations. This kind of antinomicity does not involve truth and falsity; let us call it ***horizontal antinomicity*** – including any of the antinomic combinations $t_{2k-1} t_{2k}$ and $t_{2k} t_{2k-1}$ – to emphasize that it originates purely in the interplay of internal operations and thus remains at the level of objects. Any single neutral object without exception can then become horizontally antinomic, as well as any other object, provided that the rule $t_i t_i o \equiv t_i o$ is observed.

16. The use of the comma as an external operator in o_1, o_2 is to be distinguished from its use as a purely linguistic symbol in $f(o_1, o_2), A(o_1, o_2)$, $\text{Opp}(o_1, o_2)$, etc. The objects o_1 and o_2 in these last three expressions are arbitrary assignments for the open positions in $f(-,-), A(-,-)$, and $\text{Opp}(-,-)$; their placement in the expressions does not automatically make of o_1 and o_2 a multiplicity. If one turns the dispersed objects o_1 and o_2 into the separable multiplicity o_1, o_2 and wants to emphasize for any reason the role of the comma as an operator, then we place o_1, o_2 between square brackets $[o_1, o_2]$; in this last expression there is no ambiguity as to the role of the comma. On the other hand, the role of the comma as an operator in $\{o_1, o_2\}$ is totally unambiguous. Unities are only the result of applying the specific transformations t_1 and $\{-,-,...,-\}$ or $\{-,-,...\}$. Separable multiplicities are the result only of applying the specific transformations t_2 and $[-,-,...,-]$ or $[-,-,...]$. Neither unities nor multiplicities are unmodifiable, eternal conditions. We should remark further that, since the comma generates a separable multiplicity out of arbitrary objects, not necessarily first ones, a_1 and a_2 in $[a_1, a_2]$ may themselves be multiplicities obtained by means of the comma operator. Hence, by a reversed recursion that begins with $[a_1, a_2]$ it is possible to construct non-well-founded multiplicities without any recourse whatsoever to predicates – that is, entities which

are made up of unlimited descending sequences of objects having no atoms of any kind as components.

17. Reversing the order of application of the internal operations involved in the generation of the set $\{o_1, o_2\}$, let us now select a single first object x, take it as unity $t_1 x$, represented from now on by $\times$ to simplify the notation, and take the latter as an inseparable multiplicity. This creates a new kind of object which we will call a tes, to reflect the fact that to take a unity as an inseparable multiplicity is to a great extent the mirror act of taking a separable multiplicity as a unity. A set is the outcome of a gathering, whereas a tes is the outcome of choosing – gathering and choosing being antinomically opposed processes. The notion of a mirror image here, however, is not completely exact: there are implications from the nonintrusive kind of choosing involved in a tes which do not mirror anything that takes place in gathering, such as the inseparability of the distinguishable components.

It should be noted that $[o_1, o_2, ...]$ and $\{o_1, o_2, ...\}$ are infinitely long formal expressions; each is a compound object and can be fed as a single entity to an operation, a predicate, or a marker. The same applies to teses, as will be made clear below. Let us use the symbol $o_1 \times o_2$, where o_1 and o_2 are objects already taken as unities, to indicate that the unity $\times$ is taken as the inseparable multiplicity $o_1 \times o_2$; the role of $\times$ parallels that of the parentheses in $\{o_1, o_2\}$ – that is, it functions as an internal operator that in this case generates an inseparable multiplicity out of itself. There is a difference between the separable multiplicity created by the comma in o_1, o_2 from the inseparable one obtained by $\times$ used as an operator: o_1, o_2 must not be identified with $o_1 \times o_2$. The contrast lies in the fact that to distinguish entities in a tes is a fundamentally different act from that of singling out an object from a set of objects arbitrarily put together. Whereas it is possible to pick an element out of a set and take it all by itself in isolation from the set, it is not possible to pick a distinguishable component out of a tes without dragging also the whole tes in such a choosing: we can gather totally heterogeneous objects into a set and keep them "loose" within the set, but when we distinguish objects in a tes they stay fastened to the unity from which they come. Let us call this kind of clinging distinguishability *bound choosing*, in contrast with the free choosing involved in taking an object out of a set. Correspondingly, distinguishable objects in the tes $o_1 \times o_2$ – that is, $\times, o_1$, and o_2 – constitute an inseparable multiplicity, whereas objects in o_1, o_2 form a separable multiplicity. A separable multiplicity may or may not be an antinomic object, depending on its components or on where it is placed. It is so if o_1 and o_2 in $[o_1, o_2]$ are objects taken as unities; it is not so if o_1 and o_2 are first objects. Sets and teses are always antinomic objects. Both gathering into a set and choosing into a tes are always antinomic operations.

The single first object x taken as a unity can be taken as a finite or infinite inseparable multiplicity, written respectively $o_1 \times o_2 \times ... \times o_n$ and $o_1 \times o_2 \times ...$, the latter representing an infinitely long formal expression. The o_i's are unities. Also, whereas a multiplicity can be taken as a unity in only one way, $\times$ can be taken as a multiplicity in many ways: $o_1 \times o_2, o_1 \times o_2 \times ... \times o_n, o_1 \times o_2 \times ...$, all compatible possibilities, each a compound object which can be fed as a single entity to an operation, a predicate, or a marker.

18. Given the single neutral first object x, $t_1 x$ is a pure unity, i.e., the taking of x as a multiplicity or as anything else has not been done prior to the application of t_1. Similarly, applying the operator t_2 to the neutral first object x we obtain the pure

multiplicity t_2x. If instead of t_2 we apply the comma operator to several neutral first objects, we also obtain pure multiplicities, to wit: $[x_1, x_2]$, $[x_1, x_2, ..., x_3]$, or $[x_1, x_2, ...]$. If in turn we iterate the application of the comma operator to pure multiplicities, we then obtain objects such as the following one: $[(x_1, x_2), (x_3, x_4)]$, a pure multiplicity of the previously constructed multiplicities $[x_1, x_2]$, and $[x_3, x_4]$. In contrast, no combination of the operators so far described allows us to construct a pure unity of unities; that is, every unity of unities requires that the unities should be first taken as a multiplicity in order to be able then to take them as a unity. A unity of unities makes sense only as the unity of the multiplicity of those unities.

19. The fixed internal operations introduced so far generate, applied in the proper order, intrinsically antinomic objects. Apart from the antinomicity due to the application of two opposite transformations in the proper order, they all have the general antinomicity of either identity-in-difference or difference-in-identity. In the transit from o_1, o_2 to $\{o_1, o_2\}$ the separable multiplicity o_1, o_2 is preserved as such inside the set at the same time as being changed into a unity. In the transit from $\times$ to $o_1 \times o_2$ the unity $\times$ is simultaneously preserved as pervading the inseparable multiplicity $o_1 \times o_2$ while being changed itself into a multiplicity. The same argument applies to any objects of the kind $t_{2k-1}t_{2k}o$ and $t_{2k}t_{2k-1}o$, both also intrinsically antinomic compound objects.

20. The role of the object $\times$ in, say, $\{\times, o\}$ is what is usually called a term, a separable object in the set. Its role in $o_1 \times o_2$, on the other hand, is that of an internal relation, a relation not generated by any predicate (as external relations are generated by membership after Wiener), but a relation that is the outcome of, first, an internal operation applied to the object x, with $\times$ then functioning itself as an internal operation, all this at the level of objects. The object $\times$ functioning as an internal relation in $o_1 \times o_2$ makes itself internally attached to o_1 and o_2 in a way that makes the bound choosing of o_1 or o_2 be inevitably encumbered with $\times$ itself. "Term" and "internal relation" point respectively to the separability and inseparability of the entities involved. Internal relations are an example of objects themselves functioning as internal operations.

The transformations t_3 and t_4 represent respectively the specific acts of "taking an object as a term" and "taking an object as an internal relation," in the latter case turning the entity in order into a potential internal operator. It is usually understood that relations relate terms, but if we apply t_4 to t_3o_2, then the object $(t_4o_1)(t_4t_3o_2)(t_4o_3)$ represents a term in the act of relating relations – after the term t_3o_2 had been taken itself as an internal relation. Every term can be taken as an internal relation and every relation, in turn, can be taken as a term: t_4t_3o and t_3t_4o represent these two antinomic situations. Integers, for example, can be taken both as terms or internal relations, and can then, in turn, generate antinomic relations and antinomic terms respectively.[49] Any set taken as a term can also be taken as a relation, and so can any tes. Thus, for example, the set $\{o_1, o_2\}$ can be taken as an internal relation, $t_4\{o_1, o_2\}$, and can then internally relate its own members: $o_1t_4\{o_1, o_2\}o_2$, changing in the process the separability of o_1 and o_2 in $\{o_1, o_2\}$ into the opposite status of inseparability while related. As it is the case with antinomicity, separability and inseparability are stages in the life of entities, not immutable properties. t_4 applied to an object o makes o become more than

[49]F. G. Asenjo, "Relations Irreducible to Classes," *Notre Dame Journal of Formal Logic*, 1963, vol. IV, pp. 193-200.

a "sticky" component; it turns o into a transformation itself, applicable to other objects. $o_1(t_4o)o_2$ is an inseparable multiplicity, but note that t_4, in contrast with $\times$ in $o_1 \times o_2$, applies to any object o, not only to unities. t_4o is not a unity; however, just as in the case of the unity $\times$, it pervades the objects it relates.

We must at this point emphasize the difference between the internal relations just described, which are, in effect, internal operations, compared to internal relations which are predicates. Predicates can function as external or internal predicates depending on the objects they are fed. In the latter case, just as with internal operations, they change the nature of the objects involved, creating new objects out of old ones. A stone does not love another stone, but Romeo loves Juliet, and the moment they become enamored, Romeo becomes a new person, while still being the same Romeo; in addition, Romeo and Juliet become together a special inseparable multiplicity. The predicate of distinguishability ∂, to be dealt with below, is another potentially internal predicate.

21. "To take an object as a singularity" (t_5) and "to take it as a specimen" (t_6) are also opposite operations. t_5o means to take the object o as a singular entity, that is, as an individual considered as a universe unto itself regardless of size and of how many – finite or infinite – objects are distinguishable in it, and of how many other objects there are of which o may be a part. To be a specimen, on the other hand, means to be the representative of a family of objects sharing in a situation, or displaying a common characteristic, or being in a mutual special relationship. t_6t_5o represents the antinomic act of taking the singularity t_5o as a specimen, and t_5t_6o represents the equally antinomic act of taking the specimen t_6o as a singularity all by itself.

22. t_7o, "to take the object o as a whole," means that o is taken fully, i.e., in such a way that no part of o is left out. t_8o, "to take the object o as a part," has one or both of the following meanings: o is taken as a part of other objects or o is taken partially, that is, portions of o may be left out. This allows us to distinguish between being an object wholly part of another object – in whatever meaning "being a part of" is intended – and being an object partially part of another object. These alternatives are especially significant when "being a part of" is understood not as mere clear-cut set-theoretic membership or inclusion, but as an object being distinguishable in another object. It is a common sense observation that o_1 may be distinguished in *toto* in o_2, or o_1 may be partially distinguishable in o_2, as a presence that fades beyond o_2, such as in a resemblance. This merely reflects the fact that in the physical world the word "contains" all too often means "partially contains," "intersects," or "has a portion of," and the word "belongs" means only "being partially present." Now, t_8t_7o represents "taking the whole t_7o as a part," and t_7t_8o represents "taking the object o already taken as a part – perhaps leaving out parts of o – as a whole." Both are antinomic combinations.

23. t_9o, "taking o as a discrete entity," means taking o as completely circumscribable. $t_{10}o$, "taking o as a continuous entity," means taking o as spreading without gaps beyond its assigned boundaries. Discreteness and continuity are not absolute properties but a matter of point of view and context – hence, subject to transformations. It is possible to turn a discrete object into a continuum, $t_{10}t_9o$, and a continuum into a discrete entity, $t_9t_{10}o$, again both antinomic combinations.

24. All the preceding transformations are algorithmic in character in the sense that they start with a first object and keep generating new compound objects

in succession. They are "creative" algorithms, though, in that the path of their combinations is not predetermined. Yet, along any path, the original object remains recognizable as the same through all its metamorphoses.

25. To summarize up to this point, these are all the objects. (i) Variable and fixed first objects: $x, y, z, x_1, x_2, ..., a, a_1, a_2, ...$; (ii) Compound objects:

$$f_i(o_1, ..., o_k), g_i(o_1, ..., o_k), t_i o, [o_1, o_2, ..., o_k], [o_1, o_2, ...], \{o_1, o_2, ..., o_k\},$$

$$\{o_1, o_2, ...\}, o_1 \times o_2, o_1 \times o_2 \times ... \times o_k, o_1 \times o_2 \times ..., o_1(t_4 o_2)o_3.$$

26. The following are predicate formulas. (i) $A_i(o_1, ..., o_k), o_1 = o_2, o_1 \in o_2, o_1 \partial o_2, o_1 < o_2, o_1 > o_2, o_1 \equiv o_2$, the atomic predicate formulas; and (ii) the usual combinations of predicate formulas with the propositional connectives and the quantifiers, including $\exists a$, the compound predicate formulas.

27. If A_i is any unary predicate, some objects o_j fed to it may become antinomic if the semantics of A_i and the nature of o_j make of $A_i(o_j)$ a true and false statement. This simply generalizes the situation that we have with Russell's antinomy: some sets are not members of themselves and are not antinomic, but the set T of all sets which are not members of themselves, fed to the predicate $\notin$, is antinomic because $T \notin T$ is true and false. Let us call ***vertical*** this kind of antinomicity of the object to refer to the fact that it is the predicate $\notin$ and not any transformation that generates the antinomicity of T. This extends to the logic of vague statements in the sense that every vague predicate fed a suitable object generates a statement that has a degree of truth t and a degree of falsity f; that is, a complex degree of antinomicity $t + if$. Fuzzy logic is necessarily an antinomic logic, and any suitable object to which a vague predicate applies to produce an antinomic statement becomes itself an antinomic object. Ordinary predicates occurring in common sense statements very often display this semantic situation. "Peter is good" is an antinomic statement that makes of Peter an antinomic object – every person is good to some degree and bad to another degree. Yet, to say that "this stone is good" (in the moral sense, of course) is simply false: neither the statement nor the stone is antinomic. The same Peter fed as an object to the predicate "human being" is, in turn, simply true. Again, the antinomicity of an object or a sentence is a matter of semantic circumstance.

The previous reasoning applies also to binary and n-ary predicates, as well as to compound predicate formulas in general. If the semantics of $A_i(o_j, o_k)$ make it an antinomic statement, then the objects o_j and o_k stand in antinomic opposition. This is the case in particular of fuzzy inclusion. If $\mathcal{A}(o_1, ..., o_n)$ is any antinomic compound predicate formula, each o_i in the formula is in antinomic opposition to any other o_j. Thus, $(o_1 < o_2) \wedge (o_2 < o_1)$ can be made into an antinomic predicate formula with an appropriate semantics for $<$, and then so can the objects o_1 and o_2.[50]

28. Just as internal operations as well as the objects to which they apply can be respectively in opposition of one another – $\text{Opp}(t_1, t_2)$, $\text{Opp}(t_1 o, t_2 o)$ – so can some predicates be in pairwise opposition of one another: for example, $\text{Opp}(\in, \partial)$, $\text{Opp}(<, >)$, etc. The arity may vary, but $\text{Opp}(A_i, A_j)$ indicates in general that the

[50] F. G. Asenjo, "Toward an Antinomic Mathematics," Chapter 15 in R. Routley, G. Priest, and J. Norman (eds.), *Paraconsistent Logic*, Munich: Philosophia Verlag, 1989, pp. 394-414.

predicates A_i and A_j are in opposition of one another. Similarly, $\text{Opp}(A_1, ..., A_n)$ means $\text{Opp}(A_i, A_j)$ for all $i \neq j$ in the sequence $A_1, ..., A_n$.

Although opposing predicates can generate antinomic statements and make some objects become antinomic, opposition is not synonymous with antinomicity. While it is indeed the foundation of antinomicity in many cases, predicates that induce antinomicity when fed certain specific objects yield only simply true or simply false statements in other cases, as is obvious with $\notin$ in set theory: the empty set is not a member of itself but neither is it for this reason an antinomic object.

29. Departing from the usual pattern of building logic exclusively from objects to formulas, let us now recognize that objects may have not only other objects as components but also predicate formulas as well. Since these formulas may be fed objects also having in turn other predicate formulas as components, etc., this makes of the present logic of opposition a non-well-founded one, i.e., a logic that allows a downward as well as an upward construction of expressions – infinite descents as well as infinite ascents. Let us introduce the letter h as a marker to indicate that the occurrence of a predicate formula attached to the object or objects that follow h are part of the composition of the objects: $h(o, A(o))$ means that $A(o)$ is inseparable from o. For example, we mentioned that Peter becomes an antinomic object when fed to the predicate of moral goodness; in this case, "Peter is good" is asserted adventitiously. But if we want to assert that goodness is essential to a particular Peter, an indelible part of him without which this Peter is not really Peter, then to write $A(o)$ is not sufficient; we must write $h(o, A(o))$ to stand or use for Peter in any statement that we may want to assert about him. $h(o, A_1(o), ..., A_k(o))$ indicates that o cannot be dealt with apart from the predicates $A_1, ..., A_k$, injected into o as part of o's internal composition. $h(o, A_1(o), A_2(o), ...)$ indicates that an infinity of attributes is essential to o. If any one of the statements $A_i(o)$ is antinomic by itself, or if $A_i(o)$ and $A_j(o)$ are in antinomic opposition, then the object o is also antinomic by definition.

Thus we have as possible initial steps of potentially infinite descents:

(i) $f_1(h(o_1, o_2, A_1(o_2), A_2(o_1, o_2)))$,

(ii) $t_2(h(o, A(o)))$,

(iii) $g_1(o_1, h(o_1, o_2, A_2(o_1, o_2)))$,

(iv) $A_2(o_1, h(o_2, A(o_2), \text{Opp}(A_1, A_2)))$,

(v) $A_1(h(o_1 \times o_2, A_3(\times), A_4(o_1 \times o_2)))$,

(vi) $A_2(t_1 o_1, h(t_2 o_2, A_5(t_1 o_1, t_2 o_2), \text{Opp}(o_1, o_2), \text{Opp}(A_2, A_5)))$.

The arity in (i), (iii), (iv), and (vi) is two; the arity in (ii) and (v) is one.

30. The fixed binary predicates $=, \in, \partial, <, >, \equiv$ can lead to antinomic statements each by itself or in combination when fed the appropriate objects in a suitable semantic setting. Thus, $o_1 = o_2, o_1 \in o_2, o_1 \partial o_2, o_1 < o_2, o_1 > o_2, o_1 \equiv o_2$ can each be simply true, simply false, or true and false, the latter when, respectively, o_1 and o_2 are equal and unequal, o_1 a member and not a member of o_2, o_1 distinguishable in and identified with o_2, o_1 less than and greater than o_2, o_1 and o_2 identical and different. That two binary predicates can generate antinomic statements is indicated by $(A_1(o_1, o_2) \wedge A_2(o_1, o_2), \text{Opp}(A_1, A_2))$, but it should be clear that it is

the particular semantic interpretation of the objects o_1 and o_2 that can make the conjunction $(A_1(o_1, o_2) \wedge A_2(o_1, o_2))$ both true and false.

31. Script letters $\mathcal{A}, \mathcal{A}_1, \mathcal{A}_2, \ldots$ are abbreviations of predicate formulas such as the ones already given. Thus, $\mathcal{A}(o_1, \ldots, o_n)$ may stand for $(A_1(o_1) \wedge A_2(o_2, o_3) \wedge \ldots \wedge A_k(o_1, \ldots, o_n)$, $\mathrm{Opp}(A_1, A_2, \ldots, A_k))$. Some or all of objects $o_1, \ldots, o_n$ may become antinomic if being fed to $\mathcal{A}(o_1, \ldots, o_n)$ makes this formula true and false.

32. Internal operations apply to predicates as well. $t_7\partial$ means "distinguishing in toto" and $t_8\partial$ "distinguishing in part"; that is, $o_1(t_7\partial)o_2$ means "the object o_1 is distinguished as a whole in o_2," and $o_1(t_8\partial)o_2$ means "the object o_1 is distinguished partially in o_2." In the best portraits of Velázquez and Rembrandt one looks at the eyes of a living person – there is a full concrete presence in what is, on the face of it, an inanimate object. The traits of Peter's father in Peter's face make us recognize presences that spread beyond Peter, just as looking at the sun's light on the surface of the moon, etc. We must then differentiate between first predicates, that is, predicates to which no internal operation has been applied, and *compound predicates*, predicates of the form t_iA, t_it_jA, or $t_i \ldots t_kA$.

2.2 Axioms

An assertion is a statement taken as true, represented by $\vdash \mathcal{A}$. Assertions can be made for immediate reasonings or computations, only to be put aside afterwards. Other assertions are made for longer standing and are called axioms. There is no other distinction between passing hypotheses or conjectures and axioms other than the intended time of duration of their use: they are relative names. Axioms, then, are not eternal verities, something that should be obvious after the many kinds – all legitimate – of opposite geometrics, logics, mathematics in general, and now with the enormous inroads that computer activities are having into the ways in which mathematics is made and thought about. The need for axioms emerges from the fact that, contrary to strict constructivism, not all useful mathematical relationships can be effectively constructed. Some must be postulated, not for the purpose of engaging in permanent deductions but for establishing the meaning and connections of the objects and the predicate formulas in case. Axioms are to be used to derive formulas purely on the ground of their meaning and their semantic consequences. No rules of deduction are to be added: everything that has to be taken as true has to be asserted, just as in the application of a computer program new hypotheses can be injected at any time, with the program itself being potentially subject to change or even substitution.

Alternatively, a statement can be taken as false, represented by $\models A$. An antinomic statement is one taken both as true and as false after pertinent semantic considerations. The symbols $\vdash$ and $\models$ are then both algorithmic in the sense that there is a subject that does the taking in succession with a relevant criterion. If we assert "Peter is good," it does not take me long before I realize that both algorithms $\vdash$ and $\models$ are suitable for such a statement. Before $\vdash$ and $\models$ are applied, statements are neither true nor false.

An object of the form $o_1 \times o_2$, with o_1 and o_2 unities, is called a tes. Also, if $\vdash o_1 \partial o_2$, then o_2 is called a tes as well, and we say that o_1 is distinguishable in o_2. If $o_1 \times o_2$ has been formed, then o_1 and o_2 are inseparable from $\times$, and so is $\times$ from o_1 and o_2. If $\vdash o_1 \partial o_2$, then o_1 is inseparable from o_2 and vice versa. Each first object is separable, and so is any object of the form $t_i o$. If $\vdash\rceil o_1 \partial o_2$, then o_1 is separable from o_2 and so is o_2 from o_1. Separable and inseparable are neither permanent nor exclusive conditions: any separable object can be made inseparable and vice versa. Separable means, of course, to be able to be taken by itself; inseparable means to be glued to another entity. To distinguish o_1 in o_2 does not mean that o_1 is necessarily fully contained in o_2; o_1 may well spill out of o_2 despite its being present in o_2. As shall follow from the axioms, no tes is a set, and no set is a tes. However, every set can be turned into a tes, and every tes can be turned into a set.

(We should not identify $\models\equiv o_1 \partial o_2$ with $\vdash\rceil o_1 \partial o_2$. To say that a statement is false is not the same as asserting the negation of such a statement: assertion and negation are independent of truth and falsity.[51])

For an object o_1 to be a member of a set o_k first it has to be taken necessarily as a component of a separable multiplicity, either by itself or in a finite or infinite lot. Then, in turn, the multiplicity has to be taken as a unity, i.e., the set o_k. To write $o_1 \in o_k$ automatically implies always that o_k is a unity and o_1 is a component of a separable multiplicity; $o_1 \in o_k$ is meaningless otherwise. Rather than an essentialist entity, $o_1 \in o_k$ is then the record of a process that goes from the separable object o_1 to $o_1, o_2, ...$, say, and then to $\{o_1, o_2, ...\}$, or from o_1 to $t_2 o_1$, and then to $t_1 t_2 o_1$. A set or a tes, both compound objects, can be made members of another set but only after first having been taken as separable multiplicities by themselves or as components of a lot.

Similarly, for an object o_1 to be distinguishable in a tes o_k, o_k has to be first the outcome of having taken another object as a unity; then $o_1 \partial o_k$ is the record of the following process: having taken some object as the unity o_k, we then take o_k as an inseparable multiplicity which we represent by $o_1 o_k o_2$, or $o_1 o_k o_2 o_k ... o_k o_n$, or $o_1 o_k o_2 o_k ...$. These last expressions are compound objects that can be fed as such to operations and predicates, but that, in contrast with the tes $o_1 \times o_2$, which does not depend on any predicate, are justified by the corresponding assertions $\vdash o_1 \partial o_k, \vdash o_2 \partial o_k$, ..., which assume the above-mentioned process. The predicate ∂ allows us then to go beyond the scope of the internal operator $\times$: now every unity, not only $\times$ and even sets, can be taken as an inseparable multiplicity. However, in $o_1 o_k o_2$, $o_1 o_k o_2 o_k ... o_k o_n$, and $o_1 o_k o_2 o_k ...$, the o_i's are all unities. This is because it is impossible to distinguish any entity or aspect in a given unity without perceiving it as itself constituting a unity, whatever else we can make later of the entity or aspect. As a consequence, to write $o_1 \partial o_k$ always automatically implies that both o_1 and o_k are unities; $o_1 \partial o_k$ is meaningless otherwise. Iterating ∂, chains of distinguishability become possible: $o_1 \partial o_k, o_2 \partial o_1, o_3 \partial o_2$, etc., as well as $o_1 \partial o_2, o_2 \partial o_3, o_3 \partial o_4$, etc. Distinguishability, though, is not necessarily transitive, nor is it necessarily symmetric, or reflexive, or connected.

Summing up, a tes is a unity in which other unities are distinguishable, and any unity can be made into a tes using the predicate ∂. On the other hand, all nonunities are not teses: they are separable entities, although they may be made inseparable after applying to them the appropriate transformations in the right order. Nothing just is: everything becomes. Also, some unities are distinguishable in other unities forming finite or infinite ascending or descending chains of distinguishability. If

[51] For an elaboration of this issue see F. G. Asenjo, "Antinomicity and the Axiom of Choice: A Chapter in Antinomic Mathematics," *Logic and Logical Philosophy*, Torún: 1996, vol. 4, pp. 53-95.

$\vdash \rceil o_1 \partial o_1$, o_1 is inseparable from itself. Yet every tes is separable and inseparable – separable, because if $\vdash o_1 \partial o_2$, then o_2 was already the separable object $t_1 o_k$ for some o_k, or because it was a set, also a separable object; inseparable, because o_2 is taken as an inseparable multiplicity.

If $\vdash o_1 \partial o_2$, one can apply any transformation t_i to o_1 (except t_1) and then $t_i o_1$ becomes a separable entity, while o_1 remains inseparably attached to o_2. Hence, every inseparable entity o which was not made separable before can be made separable using an appropriate transformation while preserving o's inseparability at the same time.

In what follows, whenever $o_1 \in o_2$ or $o_1 \partial o_2$ occurs, it shall be understood that o_1 and o_2 have been taken previously in the sequences of correspondingly appropriate ways that have just been described; both expressions will be meaningless otherwise. We should, finally, point out that, whereas the separable multiplicity o_1, o_2 is included in the multiplicity o_1, o_2, o_3, etc., there is no parallel way of considering the tes $o_1 o_k o_2$ as included in $o_1 o_k o_2 o_k o_3$: these are two unrelated inseparable entities.

Axiom A1. $o_1 = o_2 \leftrightarrow (\forall o_3(o_3 \in o_1 \leftrightarrow o_3 \in o_2) \vee \forall o_4(o_4 \partial o_1 \leftrightarrow o_4 \partial o_2))$. Extensionality; valid if o_1 and o_2 are either both sets or both teses, because no object is both a set and a tes. Equality does not apply when other objects are involved; in these other cases, we use the identity predicate $\equiv$.

A2. Cantor's axiom scheme of comprehension for $\in$. $\bigvee$ represents the set of all sets; it follows that $\bigvee \in \bigvee$. "It follows" merely signifies here that we apply the age-old semantic meaning of "for all." It is not necessary to have a full-fledged predicate calculus nor, in particular, a theorem $\forall o \mathcal{A}(o) \to \mathcal{A}(a)$ to sanction the semantic consequence that "what is true for all objects is true for a specific object." The latter connection is prior to any predicate calculus; it is all that is needed to assert $\bigvee \in \bigvee$.

A3. Cantor's axiom of infinity for $\in$.

A4. $\underline{\exists}! a \forall o(\rceil o \partial a)$. There exists a unique empty object (in the sense of identity) exhibiting no distinguishable object including itself. It is represented by $\oint_\theta$ and satisfies $t_i \oint_\theta \equiv t_i t_i \oint_\theta \equiv \oint_\theta$. Also, $\oint_\theta$ is neither a set nor a tes, and neither a unity nor a multiplicity.

A5. $\forall o(\rceil \oint_\theta \partial o)$. The empty object $\oint_\theta$ is not distinguishable in any object: we cannot distinguish nothingness in something.

A6. $\underline{\exists}! a \forall o(o \partial a)$. There exists a unique universal tes (unique in the sense of identity) in which every unity is distinguishable. It is represented by B, a unity, and it satisfies $t_i B \equiv t_i t_i B \equiv B$. It follows that $B \partial B$. Teses other than B can also be either distinguished in themselves, $o \partial o$, or not distinguished in themselves, $\rceil o \partial o$, or both.

A7. $\forall o(B \partial o \wedge \rceil B \partial o)$. It follows that $\rceil B \partial B$ in particular. The universal tes is ∂-antinomic for every unity including itself.

A8. $\exists o(o \equiv o_1, o_2, \ldots \wedge o_1 \not\equiv o_2 \wedge o_1 \not\equiv o_3 \wedge \ldots \wedge o_i \not\equiv o_j \wedge \ldots \wedge o_1 \partial o_2 \wedge o_2 \partial o_3 \wedge \ldots \wedge o_k \partial o_{k+1} \wedge \ldots)$. There exists an infinitely ascending chain of distinguishability.

A9. $\exists o(o \equiv o_1, o_2, \ldots \wedge o_1 \not\equiv o_2 \wedge o_1 \not\equiv o_3 \wedge \ldots \wedge o_i \not\equiv o_j \wedge \ldots \wedge o_2 \partial o_1 \wedge o_3 \partial o_2 \wedge \ldots \wedge o_{k+1} \partial o_k \wedge \ldots)$. There exists an infinitely descending chain of distinguishability.

A10. $\mathcal{A}(o_1) \wedge \exists o_2(o_2 \not\equiv o_1 \wedge o_2 \partial o_1) \to \underline{\exists} a(o_1 \not\equiv a \wedge a \not\equiv B \wedge a \partial o_1 \wedge \mathcal{A}(a))$. Axiom of choice for ∂. If there are objects distinguishable in a tes o_1 not identical to o_1, not all these objects necessarily inherit every property that o_1 has. A10 asserts the existence of an object distinguishable in o_1 having one such property, and it further chooses one specific object as distinguishable in o_1 with the property in question. Infinitely descending chains of distinguishable specific objects a_1, a_2, ... such that $a_2 \partial a_1, a_3 \partial a_2$, ..., and $\mathcal{A}(a_1), \mathcal{A}(a_2)$, ..., are then feasible, with different properties engendering different chains.

A11. $\forall o_1(\mathcal{A}(o_1) \wedge o_1 \not\equiv B \rightarrow \exists o_2(o_2 \not\equiv B \wedge o_1 \not\equiv o_2 \wedge \rceil o_1 \partial o_2 \wedge \rceil o_2 \partial o_1 \wedge \mathcal{A}(o_2)))$. Anticomprehension axiom scheme for ∂. Given a property $\mathcal{A}$, it is not possible to ∂-gather all objects with that property in a single tes different from B.

A12. If $o_1 \times o_2$ is given and o_1 and o_2 are, therefore, unities, then $\vdash o_1 \partial \times$ and $\vdash o_2 \partial \times$ follow. There is a link between teses formed at the level of objects and corresponding statements involving distinguishability.

As is the case with most antinomic systems, no standard semantics can be applied to the preceding axioms in any really productive way. In order to be able to talk about the axioms' semantic consequences, we need a general model theory for inconsistent systems, a theory that does not exist today despite the many significant results that keep treading into such uncharted territory. One major problem is whether to try to integrate the disparate points of view now active, or to allow antinomic semantics to branch out into incompatible paths. In particular, should antinomicity itself be systematically extended to the metalanguage as a whole, only in part, or not at all? Clearly we need a nonstandard semantics, but difficult unanswered questions must be examined before we can know how to proceed.

2.3 The principle of continuous induction

So far, the chief difference between sets and teses is that sets are formed out of separable multiplicities and teses are inseparable multiplicities formed out of unities. There is another difference to be pointed out now. The generation of sets out of sets, upward or downward, is a discrete process: $...o_i \in o_j \in o_k...$; step by step we gather or descend into ever new sets, which is why sets allow themselves to be constructed inductively. The generation of teses, on the other hand, is not necessarily discrete: $...o_i \partial o_j \partial o_k...$ It can also be continuous, as the following existential axiom asserts.

A13. There exists a stretch S of objects linearly ordered without gaps by ∂. S may or may not have initial or terminal objects, but it is, of course, dense.

There are cases in which, as we distinguish o_i in o_j, we know that an infinity of other objects lies between o_i and o_j forming a continuous stretch linearly ordered by ∂. Axiom A13 reflects this perception.

We want to give now a sufficient condition to determine that a given property extends to a whole stretch of objects of the kind described in A13. Khintchine gave a principle of induction for the real numbers, a principle that was extended by Sierpiński to all linearly ordered stretches without gaps.[52] This second version can be modified as follows to apply specifically to ∂.

A14. Let us call S a stretch of objects linearly ordered without gaps by ∂. Let P be a property concerning teses. Assume that (i) there exists an object a in S such that P holds for every object o_i in S such that $o_i \partial a$, and assume also that (ii) if the object b is in S and P holds for every object o_i in S such that $o_i \partial b$, then there exists an object c in S such that $b \partial c$ and P holds for every object o_j in S such that $o_j \partial c$; then P holds for every object in S.

[52] W. Sierpiński, *Cardinal and Ordinal Numbers*, Warsaw: Polish Scientific Publishers, 1965, pp. 265-266.

3 Applications, Extensions, and Conclusions

3.1 The arithmetic of term-relation number theory

As an application of horizontal antinomicity to mathematics, we outline now with some minor changes number systems developed in previous papers.[53] Let us take the ordinary integers as fixed first objects; then let us take each integer alternatively as either a term or an internal relation, denoted respectively by $..., -1, 0, 1, 2, ...$ and $... - \bar{1}, \bar{0}, \bar{1}, \bar{2}, ...$, and in general by $a, a_1, a_2, ...$, and $\bar{a}, \bar{a}_1, \bar{a}_2, ...$ Then $a_1\bar{a}_2a_3$ is a term-relation number, and so is $\bar{a}_1a_2\bar{a}_3$. Term-relation numbers are inseparable multiplicities, neither simply terms nor simply relations, although of course $t_4(a_1\bar{a}_2a_3)$ is a relation and $t_3(a_1\bar{a}_2a_3)$ is a term. In the spirit of Kleene's metaphors, wherein he called "ions" predicate expressions not yet but about to be fed objects, $P(-,-)$, and "mesons" functional expressions not yet but also about to be fed objects, $f(-,-,)$, let us give the name "gluons" to every integer taken as a term or as a relation not yet but about to be fed integers taken as relations or taken as terms. This is to emphasize that $-\bar{2}-$ and $-3-$ are about to tie together two integers, to glue them, in effect, into an inseparable multiplicity. Every term-relation of the form $a_1\bar{0}a_1$ is identical to a_1, and every expression of the form $\bar{a}_10\bar{a}_1$ is identical to $\bar{a}_1$. Compound term-relation numbers are formed as in the following examples: $(a_1\bar{a}_2a_3)(\bar{a}_4a_5\bar{a}_6)a_7, (\bar{a}_1a_2\bar{a}_3)a_4(\bar{a}_5(a_6\bar{a}_7a_8)\bar{a}_9)$, etc. The parentheses indicate the order of generation of the numbers; abstraction made of the parentheses, a term is followed by a relation, and a relation by a term, and any well-formed term-relation number begins and ends with a term or begins and ends with a relation. Let us represent a number that begins with a term by o_i and a number that begins with a relation by $\bar{o}_j$; then $o_i\bar{0}o_i$ is identical to o_i, and $\bar{o}_j0\bar{o}_j$ is identical to $\bar{o}_j$. Thus, $4 \equiv 4\bar{0}4 \equiv (4\bar{0}(4\bar{0}4))\bar{0}4 \equiv 4\bar{0}(4\bar{0}(4\bar{0}4)), \bar{4}0\bar{4} \equiv \bar{4}$, etc. In particular, $0\bar{0}0 \equiv 0$.

Some differently expressed inseparable multiplicities are in this way identified. Term-relation numbers can then be written in reduced form by substituting any expressions of the form $o_1\bar{0}o_1$ or $\bar{o}_10\bar{o}_1$ by o_1 and $\bar{o}_1$, respectively. On the other hand, any term-relation number can be expanded using these two substitutions in the opposite direction in order to match the more complex parenthesis structure of other term-relation numbers. By this reducing and expanding, addition and multiplication of term-relation numbers can be defined in general, as well as a partial ordering. As a consequence, term-relation numbers of the form $o_1\bar{o}_2o_3$ – and dually term-relation numbers of the form $\bar{o}_1o_2\bar{o}_3$ – constitute a ring Z^* which is a nonstandard model of the integers, one that even contains finite infinitesimals.

Z^* can be extended to a special nonstandard model R^* made up of term-relation real numbers, and this, in turn, by adjoining the imaginary unit i to R^*, can be extended to a nonstandard model C^* of term-relation complex numbers. C^* is a field, but it is not algebraically closed, nor is any finite extension of C^*. This means that the so-called Weierstrass' Final Theorem of Arithmetic to the effect that it is impossible to extend the number system beyond C into new fields by algebraic field extensions is circumvented if we start the extensions with the ring

[53] In addition to the paper referred to in note 49, the following ones are also pertinent. F. G. Asenjo, "The Arithmetic of Term-Relation Number Theory," "Rings of Term-Relation Numbers as Non-Standard Models," "Weierstrass' Final Theorem of Arithmetic is not Final" (this last written with J. M. McKean), all published in *The Notre Dame Journal of Formal Logic*, 1965, vol. IV, pp. 223-228; 1967, vol. VIII, pp. 24-26; and 1972, vol. XIII, pp. 991-994, respectively.

Z^* of term-relation integers rather than with the standard ring Z.

The last conclusion is of arithmetic and algebraic interest. Let me emphasize the following. If, when constructing the number systems, we start with the ordinary integers, but, instead of building immediately the field of fractions to obtain the rational numbers the usual way, we take, rather, the integers as either terms or relations and their combinations, we then construct effectively the antinomic extension of integers Z^*. From Z^* we can obtain Q^*, R^*, and C^*, extensions of the rational, real, and complex numbers, respectively. This alternative chain of extensions circumvents the usual path that from Z ends in the field of standard complex numbers C, which is algebraically closed. It is mathematically significant that the sequence of algebraic field extensions that we can form from C^* can be continued indefinitely to produce ever new *hypercomplex fields*.

3.2 Antinomic nonstandard analysis

As an example of applications of vertical antinomicity to mathematical analysis, let us now refer to the use made of order as an antinomic predicate in a previous paper for the purpose of building a nonstandard antinomic model of the natural numbers.[54] In this model, for any two different standard natural numbers m and n either $m < n$ or $n < m$ but not both. This is to say that in the galaxy N of all the standard (finite) natural numbers the usual nonantinomic properties of order obtain. For all the nonstandard (infinite) natural numbers, however, both $m < n$ and $n < m$ hold, including $m < m$. The linear order is preserved, but the representation of this model on a straight line necessarily requires bilocation of these nonstandard numbers. As already mentioned, when in general objects fed to a predicate yield an antinomic sentence, then the predicate as well as the objects that cause the antinomicity become antinomic. In particular, the occurrence of infinite nonstandard natural numbers in expressions of the form $o_1 < o_2$ are all antinomic precisely because for any two of them $m < n$ and $n < m$ both obtain. If m is a standard natural number and n is nonstandard, still $m < n$ and $n < m$ both obtain as well: the company of a nonstandard number in the formula $m < n$ makes the occurrence of the standard number m become antinomic.

Each infinite natural number takes, then, two locations on an appropriately stretched line: one to the right of the entire galaxy N of standard natural numbers, and another to the left.

Every standard natural number has only one successor and one predecessor, with 0 having only one successor. Every nonstandard antinomic number has two successors and two predecessors: $a+1$ and $a-1$ are both successors and predecessors of a. Thus, $a+1$ is a successor of a to the right of N and a predecessor of N to the left of N. No standard number can be reached by adding any number to a nonstandard one, or by multiplying a nonstandard number by any factor different from zero, or by raising any nonstandard number to a power different from zero.

[54] See note 50, above.

Let us now apply the same assumptions to a nonstandard model of the real numbers. In the galaxy R of standard real numbers we have $r_1 < r_2$ or $r_2 < r_1$ but not both, if r_1 and r_2 are different. For the nonstandard real numbers, on the other hand, we have both $r_1 < r_2$ and $r_2 < r_1$. Further, if r_1 is standard and r_2 is nonstandard and infinite, $r_1 < r_2$ and $r_2 < r_1$ both obtain also, making r_1 have an antinomic occurrence in $(r_1 < r_2) \wedge (r_2 < r_1)$. The same applies if r_1 is a standard real number and r_2 is a finite nonstandard real number greater in absolute value than r_1. Again, the antinomicity of order manifests itself in the bilocation of each nonstandard real number, but now, instead of having a blank stretch to the left of zero in which to place the second location of nonstandard numbers, we find that this second location falls in a place already occupied by their negatives. That is, each pair of nonstandard numbers r_1 and r_2 such that $r_1 < r_2$ will have a second location at $-r_1$ and $-r_2$, respectively, so that $r_2 < r_1$. This also includes the infinitesimals, of course. In other words, each nonstandard number occupies two positions, and each position is occupied by two nonstandard numbers, r and $-r$.

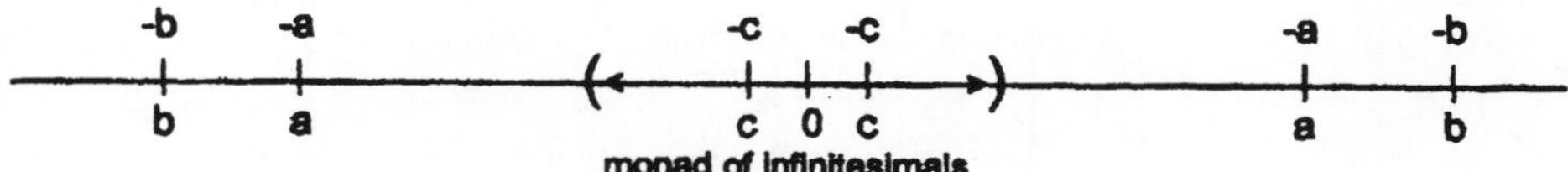

As with the antinomic model of the nonstandard natural numbers, using axioms paralleling those presented in the paper last referred to, the properties of arithmetic and order that obtain in a nonantinomic nonstandard model of the reals obtain also in the antinomic model. For example, the fact that c in $a + b = c$ has location in c and $-c$ does not mean that addition becomes a double-valued correspondence in the sense that $a+b = c$ and also $a+b = -c$. The bilocation of c does not imply that c is identical to $-c$. Further, while we have represented these antinomic models on a straight line, antinomicity can be extended to two, three, or n-dimensional spaces.

The constructions just described have definite clarifying applications to quantum mechanics, the most antinomic of all physical disciplines. Let us elaborate. According to Roger Penrose, to explain nonlocality, the instantaneous universality of some physical effects beyond the local space in which they can be observed, is the most serious and difficult problem that quantum physics will be faced with in the next decades, much more so than indeterminism. Nonlocality can be simply and positively described as ubiquity, the antinomy of here and there, the presence everywhere of local dynamic effects. Penrose says that, as things stand now with the theory, in experiments in which the same electron appears to pass through two different holes at the same time "we must accept that the particle is in two places at once! On this view, the particle *has actually passed through both slits at once.*"[55] Particles spread out in discontinuous slots of space – just as nonstandard numbers have two locations in the antinomic model of the real numbers.

As for two different entities sharing the same position, the so-called Bose-Einstein's condensate implies precisely the occupancy of the same space by two different particles. This characteristic of matter, the fact that "two atoms occupy the same volume of space, move at identical speed, scatter light of the same color,

[55] J. Gribbin, *Schrödinger's Kittens and the Search for Reality*, Boston: Little, Brown & Co., 1995, p. 172. Cf. also p. 159.

and so on,"[56] has been recently verified experimentally; the situation parallels that of infinitesimals and infinites sharing the same position with their respective negatives in the antinomic nonstandard model described above.

It is undoubtedly of physical interest to realize that there are definite mathematical models to represent some of the particular antinomic situations of quantum mechanics. Such is the case in general with "nonlocality" – a misnomer that should be appropriately replaced by "multiple location." Such is indeed the case, for instance, with the observation of having particles placed in two different locations at the same time, or the one of having two different particles sharing the same place. Both situations find respective clear-cut representations in mathematical models that extend to three dimensions the structures that we have just described, taking order to be an antinomic predicate.

3.3 Degrees of distinguishability

Fuzzy logic is based on fuzzy set theory, and set theory is the outcome of the subjective impulse to gather as thoroughly as possible. We want to open the alternative of bringing degrees to the act of choosing. Thus, just as fuzzy set theory allows degrees to the relation of membership, let us look at what it is that degrees of distinguishability do to teses. But first we must look at fuzzy set theory itself.

Peirce wrote about the urgent need to study vagueness, given that clear-cut exactness is a scientific abstraction and reality is full of graded transitions and diffuse contours. He said, "Vagueness is an ubiquitous presence and not a mark of faulty thinking, no more to be done away with in the world of logic than friction in mechanics."[57] In other words, fuzziness is not a matter of dealing with incomplete human thinking but a way of reflecting an essentially vague world, contrary to all efforts to satisfy our craving for clear-cut pictures. Fuzziness is not a shortcut to enable us to grapple with complex situations, or a means of covering up incomplete information; it is a way to mirror accurately essentially hazy, imprecise, graded facts. In connection with multiplicities, Husserl referred to the need for degrees of separability and inseparability.[58] Russell pointed out that "all language is vague."[59] But it was Karl Menger who introduced in 1951 the idea of "hazy sets," later developed by Lofti Zadeh in his theory of fuzzy sets. According to Zadeh, "the key elements in human thinking are not numbers, but labels of fuzzy sets, that is, classes of objects in which the transition from membership to nonmembership is gradual rather than abrupt. Indeed the pervasiveness of fuzziness in human thought processes suggests that much of the logic behind human reasoning is not the traditional two-valued or even multivalued logic, but a logic with fuzzy truths, fuzzy connectives, and fuzzy rules of inference."[60]

---

[56]E. A. Cornell and C. E. Wieman, "The Bose-Einstein Condensate," *Scientific American*, March 1998, p. 40.

[57]Quoted in D. McNeill and P. Freiberger, *Fuzzy Logic*, New York: Simon and Schuster, 1993, p. 28.

[58]E. Husserl, *Philosophie de l'arithmètique*, trans. by J. English, Paris: Presses Universitaires de France, 1992, p. 254.

[59]B. Russell, "Vagueness," *Australasian Journal of Psychology and Philosophy*, I, 1923, pp. 84-92.

[60]Quoted in D. Dubois, H. Prade, and R. Yager, *Readings in Fuzzy Sets for Intelligent Systems*, San Mateo, CA: Morgan Kaufman Publishers, 1993, p. 1.

In fuzzy set theory elements have a "degree of membership" to a set, a degree measured (as in the probability spectrum) by a real number taken from the interval $[0, 1]$: 1 corresponds to total membership, 0 to total nonmembership, and all other numbers between 0 and 1 to partial membership. Degrees of membership lead naturally to degrees of truth: if a belongs to b with degree p, it makes sense to say that the statement "a belongs to b" has a degree of truth p. But to say that a sentence s is partially true implies that it is also partially false, and its *degree of falsity* is, say q, determined in a manner parallel to but not necessarily dependent on the one by which p is determined – nor on negation, for that matter. But then s is partially antinomic, and its degree of antinomicity can be represented by the complex number $p + \mathbf{i}q$. In other words, what is not one hundred percent true must necessarily be partly false; hence, what has a degree of truth must also have a degree of falsity not mechanically tied to the degree of truth, all of which means that fuzzy logic is inescapably antinomic. Theoreticians in the field generally do not recognize this fact. Even such a recent book as Hajek's *Metamathematics of Fuzzy Logic* fails to acknowledge falsity as a primitive, primary factor and hence does not examine the implications of assigning to sentences both a partial truth and a partial falsity in respective degrees that can be determined in relatively independent ways.

In $a \in b$ we move from a to b; that is, we gather the entity a, together with other entities, into the set b. In contrast, in $a\partial b$ we move from b to a; that is, we choose the individual a from b, keeping it glued to b: if we want to separate a from b we need to apply additional transformations. In order to deal with degrees of distinguishability, set theory, fuzzy or not, is of no use. If anything, distinguishability leads us to reckon with individuals, including those which, although present in an object b, spill out of b, inseparable as they are from b. Further, the predicate ∂ does not require boundaries or capsules. We see in Mary's face her mother's features, but this is a hazy perception, and we cannot say that Mary "contains" her mother in any sense. On the other hand, the pain I feel in my stomach, although hazily localized, is fully contained in my body.

Let us say now that a is distinguishable in b with a degree of distinguishability measured by a real number p from the closed interval [0,1], 0 for fully indistinguishable, 1 for fully distinguishable, with any number in between providing an intermediate degree of partial distinguishability. Now, to have a degree of distinguishability p, with p between 0 and 1, implies that a also has a degree of indistinguishability q in b, with q also between 0 and 1. In other words, fuzziness implies a degree of intrinsic ∂-antinomicity that can be represented by $p + qi$. We encounter again the antinomy of difference-in-identity. Bhaskara could well have invented the fuzzy logic of distinguishability.

What operations can be defined for individuals distinguishable in a tes? Certainly not union, or intersection, or complementation. Individuals cannot be handled as classes. Two operations are to be defined here, but only for those cases in which if $a\partial c$ and $b\partial c$, then $a\partial b$ and $b\partial a$. That is to say, each individual distinguishable in c is distinguishable in any other individual distinguishable in c, but not necessarily with the same degree of distinguishability. Let us say that, on account of these different degrees, a and b have different perspectives of the entire tes c. Since degrees are linearly ordered real numbers, we now define the following operations.

(i) $a + b$ is an individual distinguishable in c whose degree of distinguishability is the maximum of the degrees that a and b have in c.

(ii) $a - b$ is an individual distinguishable in c whose degree of distinguishability is the minimum of the degrees that a and b have in c.

$a + b$ and $a - b$ have each a new perspective of c compared to those that a and b already had of c. Both operations are not external; they do not map the pair a, b into a necessarily pre-existent entity already distinguishable in c. They both effect an internal, creative transformation of c by constructing new individuals distinguishable in c. Just as $2+3i$ is a newly created arithmetical object vis-à-vis 2 and 3, so are $a+b$ and $a-b$, as defined in (i) and (ii), newly created individuals that generally might not have been distinguishable in c before. Let us add that there is no particular point in tes theory in gathering into a class the outcomes of applying these two internal operations to all the individuals distinguishable in a given tes. The calculus of individuals for teses is intended to remain in a step-by-step pattern. We require only a local kind of fuzzy logic, one in which fuzzy sets and subsets play no role.

3.4 Conclusions

Mathematics is going to be very different in the third millennium. Many classical mathematicians consider the inroads of computers as being no more than a passing fad. At the other extreme, computer enthusiasts consider classical mathematics all but dead. I believe that mathematics is to develop inevitably into a mixed world of universal statements of broad scope together with an immensely varied realm of local computer-generated disciplines. How will these two trends interact in depth? Who will provide a comprehensive program for such mixed enterprise? No less than another Hilbert is called for to think of a scheme, tentative as it may be, to encompass an overall foundational framework for such a brave new world.

Mathematics started as a codification of computation; only later did it develop into an even greater abstraction. Bishop complained that mathematics was not sufficiently computational in his time. Susan Landau, an editor of the American Mathematical Society Notices, describes the situation well: "When I was a mathematical child, abstraction was king. It was as if examples were the detritus, and proof and theorem the real thing. Computation was for those who could not think abstractly. [But] abstract understanding is like viewing terrain through a satellite map, while examples show what the land is under your feet. Computation can help uncover surprising connections, and it can uncover fruitful areas for study.... If for most of this century computation was held in low esteem, now the pendulum appears to be swinging back. It is doing so at a propitious time."[61] I believe that at present the situation has indeed changed and continues to change undeniably in directions away from the dominance of abstraction.

Whatever the shape of the new "Hilbert Program," whenever it is formulated, I believe that the following features should be part of it. First, there should be a bifurcation of semantics into global and local sections. Global semantics attempts to gather all the rules in a fixed code. Local semantics chooses the rules as it goes along in accordance with the meaning of the case. Global semantics is unbending; local semantics adjusts to the situation, it is creative. In global semantics we indulge in thinking how things are or should be; in local semantics an evolving situation sets the changing rules: things do not just exist, they also become, they are constantly

[61] S. Landau, "Compute and Conjecture," *American Mathematical Society Notices*, 1999, p. 189.

taken in one or another transforming way. It is a straitjacket approach to force semantics exclusively into a set-theoretic setting; computation, which with each new program is always beginning anew, is essentially local, generally informed by immediate concerns. Local semantics is ad hoc, minimalist, transformational, and proceeds either step by step or from circumscribed region to circumscribed region.

Second, as Brouwer realized, the subject that makes mathematics is to be preserved as part of the foundational picture. Brouwer's "units of perception" are, in fact, irreducible, nondisposable intuitive acts of transformation. The subject is the agent of internal operations, and it is the subject's motivation that provides the semantic reasons why a statement is being taken as true, or false, or both, or neither: we should "follow our head," as Proclus proposed. It is indeed a very shallow semantics that does not bring the subject to the heart of sense and meaning; the choices and transformations freely enacted by the mind with its own reasons not only keep the flow of intelligent signification moving but remain part of it: meaning is not a mere given. Schleiermacher maintained that there is a need to take account of the activity of the individual subject in the constitution of meaning, which is a constructive, creative process, not a passive reception. In his words, an "utterance can only arise as the action of an individual, and, as such, even if it is analytical in terms of its content, it still bears free synthesis within itself." Or, to put it another way, "the deliverances of receptivity already draw on capacities that belong to spontaneity, spontaneity being the activity of the mind which renders the world intelligible by linking together different phenomena."[62]

Third, if the subject is to be reckoned with as the agent that takes things one way or another with its own reasons, time is also an essential part of the mathematical picture. Whereas abstraction fosters essentialism, computation, on the other hand, is definitely engaged in real processes. Nobody who wrestles with down to earth computations has a mind for heavens of eternal ideas.

Finally, antinomicity should also be part of the general foundational picture. With myriads of computer programs starting with contradictory premises, with reality itself being eminently contradictory, with quantum physics producing paradox after paradox, the notion that mathematics can stay out of the overwhelming antinomicity of everything else is merely to cave in face to face to the sterile prejudice of centuries against all contradictions.

REFERENCES

[1] F. G. Asenjo, *In-Between: An Essay on Categories*, Washington, DC: University Press of America, 1988.

[2] F. G. Asenjo, "Verità, antinomicità e processi mentali," a chapter in *La Scommessa de la Verità*, Milano: Spirali Edizioni, 1981, pp. 40-69. English version: "Antinomicity as an Extension of Rationality," *International Journal of Communication*, 1994, pp. 13-37.

[3] F. G. Asenjo, *Antiplatitudes*, Valencia: University of Valencia Press, 1976.

[62] F. Schleiermacher, *Hermeneutics and Criticism*, trans. by A. Bowie, Cambridge: Cambridge University Press, 1998, pp. xxx and ix.

[4] F. G. Asenjo, "Formalizing Multiple Location," a chapter in *Non-Classical Logics, Model Theory, and Computability*, A. I. Arruda, N. C. A. da Costa, and R. Chuaqui (eds.), Amsterdam: North-Holland Publishing Co., 1977, pp. 25-36.

[5] F. G. Asenjo, "One and Many," *Philosophy and Phenomenological Research*, 1966, pp. 361-370.

[6] F. G. Asenjo, "Theory of Multiplicities," *Logique et Analyse*, 1965, pp. 104-110.

[7] F. G. Asenjo, *El Todo y las Partes: Estudios de Ontología Formal*, Madrid: Editorial Tecnos, 1962.

[8] F. G. Asenjo, "The General Concept of Antinomicity," *Foundations of Science*, Dordrecht: Kluwer, vol. 3, 1999, pp. 429-465.

[9] J. Barwise and L. Moss, *Vicious Circles*, Stanford, CA: CSLI Publications, 1996.

[10] R. L. Colie, *Paradoxia Epidemica*, Princeton, NJ: Princeton University Press, 1966.

[11] G. Deleuze, *The Logic of Sense*, trans. by M. Lester and C. Stivale, New York: Simon and Schuster, 1996.

[12] R. Farson, *Management of the Absurd*, New York: Simon and Schuster, 1996.

[13] J. Fletcher and K. Olwyler, *Paradoxical Thinking*, San Francisco: Berrett-Koehler Publishers, 1997.

[14] P. Hajek, *Metamathematics of Fuzzy Logic*, Dordrecht: Kluwer, 1998.

[15] B. Kosko, *Fuzzy Thinking*, New York: Hyperion, 1993.

[16] A. Mettinger, *Aspects of Semantic Opposition in English*, Oxford: Oxford University Press, 1994.

[17] C. K. Ogden, *Opposition*, Bloomington: Indiana University Press, 1967.

[18] A. Rollings, *The Logic of Opposites* (poems), Evanston, IL: Northwestern University Press, 1996.

[19] G. Weeks and L. L'Abate, *Paradoxical Psychotherapy*, New York: Brunner-Mazel Publishers, 1982.

[20] G. Weeks (editor), *Promoting Change through Paradoxical Therapy*, Rev. Ed., New York: Brunner-Mazel Publishers, 1991.

Categorical Consequence for Paraconsistent Logic

FRED JOHNSON Colorado State University
johnsonf@lamar.colostate.edu

PETER W. WOODRUFF University of California, Irvine
pwoodruf@uci.edu

Abstract
Consequence relations over sets of "judgments" are defined by using "overdetermined" as well as "underdetermined" valuations. Some of these relations are shown to be categorical. And generalized soundness and completeness results are given for both multiple and single conclusion categorical consequence relations.

Rumfitt [1] discusses multiple-conclusion consequence relations defined over sets whose members, if any, are assertions or rejections. The consequence relations are generated by sets of valuations whose members include those that admit truth value gaps – there are sentences that are neither true nor false on some valuations. Rumfitt shows that the consequence relations are categorical – that is, the consequence relations generated by distinct sets of valuations are distinct. Given Rumfitt's work, it is natural to ask whether categoricity holds for multiple-conclusion consequence relations generated by sets of valuations whose members include those that admit truth value gluts – there are sentences that are both true and false on some valuations. We will show that the answer is Yes.

Johnson [2] extends Rumfitt's work to obtain categoricity results for single conclusion consequence relations generated by valuations that allow truth-value gaps. We will extend his discussion by considering valuations that allow truth-value gluts.

We also will extend Shoesmith and Smiley's [3] generalized soundness and completeness results for consequence relations generated by valuations that admit neither gaps nor gluts to those that do.[1]

[1] Our discussion is heavily influenced by Shoesmith and Smiley [3] and Smiley [4]. The latter gives credit to Carnap [5] and [6] for initiating discussions of rejection and categoricity, respectively. This remark by Smiley [4, p.7] is especially noteworthy: 'If I am right, the practice of identifying a calculus with its consequence relation is only justified if specifying the consequence relation is sufficient to determine the entire sentential output of the semantics.' We endorse Smiley's view that the common practice of identifying calculi by using non-categorical consequence relations is a mistake.

No acquaintance with the literature mentioned above is required to understand the results that we present.

1 Multiple conclusion consequence

DEFINITION 1 Let S be a set of sentences. $+s$ is an ***assertion on*** $\{+,-\}/S$ iff $s \in S$. $-s$ is a ***rejection on*** $\{+,-\}/S$ iff $s \in S$. j is a ***judgment on*** $\{+,-\}/S$ iff j is an assertion on $\{+,-\}/S$ or a rejection on $\{+,-\}/S$.

So, given sentences a and b the judgments on $\{+,-\}/\{a,b\}$ are $+a$, $-a$, $+b$, and $-b$. Judgments on S are to be distinuished from sentences of S. And $-$ is not a symbol for negation.[2] Read $+a$ as 'a is asserted.' Read $-a$ as 'a is rejected.'

DEFINITION 2 Suppose J and K are (empty or non-empty) sets of judgments on $\{+,-\}/S$. Then $\langle J,K\rangle$ is an ***inference on*** $\{+,-\}/S$.

DEFINITION 3 Let S be a set of sentences. v is a ***pr-valuation on*** S iff v is a subset of $S \times \{t,f\}$.[3] ('pr' is short for 'partial relation.')

DEFINITION 4 Let v be a pr-valuation on S. Let s be a member of S. v ***satisfies*** $+s$ iff $\langle s,t\rangle \in v$, and v ***satisfies*** $-s$ iff $\langle s,f\rangle \in v$. Let $\langle J,K\rangle$ be an inference on $\{+,-\}/S$. v ***satisfies*** $\langle J,K\rangle$ iff v does not satisfy some member of J or v satisfies some member of K.

DEFINITION 5 Let V be a set of pr-valuations on S. Let R be the set of inferences on $\{+,-\}/S$ such that $\langle J,K\rangle \in R$ iff every pr-valuation in V satisfies $\langle J,K\rangle$. Then R is a ***pr-consequence relation on*** $\{+,-\}/S$ and R is the ***pr-consequence relation on*** $\{+,-\}/S$ ***generated by*** V.

DEFINITION 6 Suppose that R is the pr-consequence relation on $\{+,-\}/S$ generated by a set V of pr-valuations on S. R is a ***categorical pr-consequence relation on*** $\{+,-\}/S$ iff there is no set V' of pr-valuations on S such that $V \neq V'$ and R is the pr-consequence relation on $\{+,-\}/S$ generated by V'.

Example 1. Suppose a is a sentence. There are exactly 16 inferences on $\{+,-\}/\{a\}$, the subsets of $\{+a,-a\} \times \{+a,-a\}$. So there are exactly 2^{16} sets of inferences on $\{+,-\}/\{a\}$. There are exactly 4 pr-valuations on $\{a\}$, the subsets of $\{a\} \times \{t,f\}$. So there are exactly 16 sets of pr-valuations on $\{a\}$. So there are at most 16 pr-consequence relations on $\{+,-\}/\{a\}$.

We list the pr-valuations on $\{a\}$: $v_1 = \{\langle a,t\rangle, \langle a,f\rangle\}$; $v_2 = \{\langle a,t\rangle\}$; $v_3 = \{\langle a,f\rangle\}$; and $v_4 = \emptyset$. Each of these pr-valuations does not satisfy an inference that is satisfied by the others as indicated by the following pairs: $v_1, \langle\{+a,-a\},\emptyset\rangle$; $v_2, \langle\{+a\},\{-a\}\rangle$; $v_3, \langle\{-a\},\{+a\}\rangle$; and $v_4, \langle\emptyset,\{+a,-a\}\rangle$. So there are exactly 16 pr-consequence relations on $\{+,-\}/a$ and each of them is categorical.

[2]Smiley [4, p. 1] says 'asserting *not* P may be equivalent to rejecting P, but it is not the very same thing, any more than assserting P *and* Q is the same as the two assertions P and Q.

[3]The notion of a pr-valuation is taken from Dunn [7].

THEOREM 7 Given any set S of sentences, every pr-consequence relation on $\{+,-\}/S$ is categorical.

Proof: Let S be any set of sentences and let v be a pr-valuation on S. Let $\langle J, K\rangle$ be an inference on $\{+,-\}/S$ where $J \cup K$ is the universal set of judgments on $\{+,-\}/S$, $J \cap K = \emptyset$ and, for every sentence s in S, $+s \in J$ iff $\langle s,t\rangle \in v$ and $-s \in J$ iff $\langle s,f\rangle \in v$. Then v does not satisfy $\langle J,K\rangle$, but every valuation other than v satisfies $\langle J,K\rangle$. So distinct sets of pr-valuations on S generate distinct pr-consequence relations on $\{+,-\}/S$. □

DEFINITION 8 v is a *pf-valuation on a set S of sentences* iff v is a pr-valuation on S and for every sentence s in S either $\langle s,t\rangle \notin v$ or $\langle s,f\rangle \notin v$. v is a *tr-valuation on S* iff v is a pr-valuation on S and for every sentence s in S either $\langle s,t\rangle \in v$ or $\langle s,f\rangle \in v$. v is a *tf-valuation on S* iff v is a pf-valuation on S and a tr-valuation on S. ('pf,' 'tr,' and 'tf' are short for 'partial function,' 'total relation,' and 'total function,' respectively.)

Imitate Definitions 5 and 6 to define pf-, tr-, and tf- consequence relations and categorical pf-, tr-, and tf- consequence relations.

Example 2. Suppose a is the only sentence. Though $\{\langle a,t\rangle, \langle a,f\rangle\}$ is a tr-valuation it is not a pf-valuation and thus not a tf-valuation. Though $\emptyset$ is a pf-valuation it is not a tr-valuation and thus not a tf-valuation. There are exactly 8 sets of pf-valuations, exactly 8 sets of tr-valuations and exactly 4 sets of tf-valuations on $\{a\}$. Counts for the pf-, tr- and tf- consequence relations on $\{+,-\}/\{a\}$ are 8, 8, and 4, respectively. The pf- (tr-, tf-) consequence relations on $\{+,-\}/a$ are categorical pf- (tr-, tf-) consequence relations on $\{+,-\}/a$.

THEOREM 9 Every pf- (tr-, tf-) consequence relation on $\{+,-\}/S$ is a categorical pf- (tr-, tf-) consequence relation on $\{+,-\}/S$.[4]

Proof: Use Theorem 1. □

Suppose we exclude rejections or exclude assertions.

THEOREM 10 Every tf-consequence relation on $\{+\}/S$ is categorical, and every tf-consequence relation on $\{-\}/S$ is categorical.

Proof: Note that a tf-valuation on S satisfies $\langle +J \cup -K, +L \cup -M\rangle$ iff it satisfies $\langle +J \cup +M, +K \cup +L\rangle$ iff it satisfies $\langle -K \cup -L, -J \cup -M\rangle$. □

Rumfitt [1] points out for some S there are pf-consequence relations on $\{+\}/S$ that are not categorical. We show this by using a simple example. Let $S = \{a\}$. The pf-consequence on $\{+\}/a$ relation generated by the set $\{\{\langle a,t\rangle\}, \{\langle a,f\rangle\}\}$ of pf-valuations on $\{a\}$ is the pf-consequence relation on $\{+\}/a$ generated by the set $\{\{\langle a,t\rangle\}, \{\langle a,f\rangle\}, \emptyset\}$ of pf-valuations on $\{a\}$.

[4] Rumfitt[1] proves that pf-consequence relations on $\{+,-\}/S$ are categorical.

Note also that for some S there are tr-consequence relations on $\{+\}/S$ that are not categorical. Let $S = \{a\}$. The tr-consequence relation on $\{+\}/a$ generated by the set $\{\{\langle a,t\rangle\},\{\langle a,f\rangle\}\}$ of tr-valuations on $\{a\}$ is the tr-consequence relation on $\{+\}/a$ generated by the set $\{\{\langle a,t\rangle\},\{\langle a,f\rangle\},\{\langle a,t\rangle,\langle a,f\rangle\}\}$ of tr-valuations on $\{a\}$.

Given either of these results there are pr-consequence relations on $\{+\}/\{a\}$ that are not categorical.

By referring to the above sets of valuations one can easily show that there are pf-, tr- and pr-consequence relations on $\{-\}/\{a\}$ that are not categorical.

2 Multiple conclusion deducibility

DEFINITION 11 Let J be a set of judgments on $\{+,-\}/S$. $\langle J_1, J_2\rangle$ is a *partition of J* iff $J_1 \cup J_2 = J$ and $J_1 \cap J_2 = \emptyset$.

DEFINITION 12 A set R of inferences on $\{+,-\}/S$ is *an ODC-deducibility relation on $\{+,-\}/S$* iff it meets the following conditions:
i) (Overlap) If J and K are sets of judgments on $\{+,-\}/S$, then if $J \cap K \neq \emptyset$ then $\langle J,K\rangle \in R$;
ii) (Dilution) If J' and K' are sets of judgments on $\{+,-\}/S$, then if $J \subseteq J'$, $K \subseteq K'$, and $\langle J,K\rangle \in R$ then $\langle J',K'\rangle \in R$; and
iii) (Cut) If J, K and L are sets of judgments on $\{+,-\}/S$, then if, for every partition $\langle L_1, L_2\rangle$ of L, $\langle J \cup L_1, L_2 \cup K\rangle \in R$ then $\langle J,K\rangle \in R$.

So, for example, the universal set U of inferences on $\{+,-\}/\{a\}$ is an ODC-deducibility relation on $\{+,-\}/\{a\}$. But $U - \{\langle\emptyset,\emptyset\rangle\}$ is not. Though the latter set meets the Overlap and Dilution conditions it does not meet the Cut condition since $\langle\emptyset,\{+a\}\rangle$ and $\langle\{+a\},\emptyset\rangle$ belong to the set but $\langle\emptyset,\emptyset\rangle$ does not. $U - \{\langle\emptyset,\{+a,-a\}\rangle\}$ is not an ODC-deducibility relation on $\{+,-\}/\{a\}$. It meets the Overlap and Cut conditions but does not meet the Dilution condition since $\langle\emptyset,\emptyset\rangle$ belongs to it but $\langle\emptyset,\{+a,-a\}\rangle$ does not. $\{\langle\{+a,-a\},\{+a,-a\}\rangle\}$ is not an ODC-deducibility relation on $\{+,-\}/\{a\}$. It meets the Dilution and Cut conditions but does not meet the Overlap condition since $\langle\{+a,+a\}\rangle$ does not belong to it.

THEOREM 13 [Generalized soundness] Every pr-consequence relation on $\{+,-\}/S$ is an ODC-deducibility relation on $\{+,-\}/S$.

Proof: Suppose R is a pr-consequence relation on $\{+,-\}/S$ generated by a set V of pr-valuations on S. Suppose J and K are sets of judgments on $\{+,-\}/S$. *Overlap.* Suppose $J \cap K \neq \emptyset$. Then for some sentence a either $+a \in J \cap K$ or $-a \in J \cap K$. If the former holds then $\langle J,K\rangle \in R$ since, for every valuation v in V, $\langle a,t\rangle \in v$ or $\langle a,t\rangle \notin v$. If the latter holds then $\langle J,K\rangle \in R$ since, for every valuation v in V $\langle a,f\rangle \in v$ or $\langle a,f\rangle \notin v$. *Dilution.* Suppose J' and K' are sets of judgments on $\{+,-\}/S$, $J \subseteq J'$, and $K \subseteq K'$. Suppose $\langle J',K'\rangle \notin R$. Then there is a valuation v in V that does not satisfy $\langle J',K'\rangle$. Then v does not satisfy $\langle J,K\rangle$. *Cut.* Suppose L is a set of judgments on $\{+,-\}/S$. Given any pr-valuation v on S we construct a partition $\langle L_1, L_2\rangle$ of L such that v does not satisfy it. Suppose $+a \in L$. Put $+a$ in L_1 iff $\langle a, t\rangle \in v$. Suppose $-a \in L$. Put $-a$ in L_1 iff $\langle a, f\rangle \in v$. So, if $\langle J, K\rangle \notin R$ then $\langle J \cup L_1, L_2 \cup K\rangle \notin R$. □

THEOREM 14 [Generalized completeness] Every ODC-deducibility relation on $\{+,-\}/S$ is a pr-consequence relation on $\{+,-\}/S$.

Proof: Let R be an ODC-deducibility relation on $\{+,-\}/S$. We identify a pr-consequence relation $\models_V$ on $\{+,-\}/S$ such that $R = \models_V$. Let $v_{\langle U_1,U_2\rangle}$ be the pr-valuation that does not satisfy $\langle U_1, U_2\rangle$, where $\langle U_1, U_2\rangle$ is a partition of the universal set of judgments. Let $V = \{v_{\langle U_1,U_2\rangle} | \langle U_1, U_2\rangle \notin R\}$. Let $\models_V$ be the pr-consequence relation generated by V.

$R \subseteq \models_V$. Suppose $\langle J, K\rangle \notin \models_V$. Then, for some partition $\langle U_1, U_2\rangle$ of the universal set of judgments, $v_{\langle U_1,U_2\rangle}$ does not satisfy $\langle J, K\rangle$ and $\langle U_1, U_2\rangle \notin R$. Suppose $\langle J, K\rangle \in R$. If $J \subseteq U_1$ and $K \subseteq U_2$, by Dilution, $\langle U_1, U_2\rangle \in R$. Otherwise, $v_{\langle U_1,U_2\rangle}$ satisfies $\langle J, K\rangle$. So $\langle J, K\rangle \notin R$.

$\models_V \subseteq R$. Suppose $\langle J, K\rangle \notin R$. By Cut there is a partition $\langle U_1, U_2\rangle$ of the universal set of judgments such that $\langle J \cup U_1, U_2 \cup K\rangle \notin R$. By Overlap $J \subseteq U_1$ and $K \subseteq U_2$. So $\langle U_1, U_2\rangle \notin R$. So $v_{\langle U_1,U_2\rangle} \in V$. Since $v_{\langle U_1,U_2\rangle}$ does not satisfy $\langle U_1, U_2\rangle$, $v_{\langle U_1,U_2\rangle}$ does not satisfy $\langle J, K\rangle$. So $\langle J, K\rangle \notin \models_V$. □

DEFINITION 15 R is *an ODCE-deducibility relation on* $\{+,-\}/S$ iff R is an ODC-deducibility relation that meets this condition: (*ex falso quodlibet*) for every sentence s in S, $\langle\{+s, -s\}, \emptyset\rangle \in R$. R is *an ODCT-deducibility relation on* $\{+,-\}/S$ iff R is an ODC-deducibility relation that meets this condition: (*tertium non datur*) for every sentence s in S, $\langle\emptyset, \{+s, -s\}\rangle \in R$. R is *an ODCET-deducibility relation on* $\{+,-\}/S$ iff R is an ODCE-deducibility relation on $\{+,-\}/S$ and an ODCT-deducibility relation on $\{+,-\}/S$.

THEOREM 16 A set of inferences is a pf-consequence relation on $\{+,-\}/S$ iff it is an ODCE-deducibility relation on $\{+,-\}/S$.

Proof: *Soundness.* Every pf-consequence relation on $\{+,-\}/S$ is a pr-consequence relation on $\{+,-\}/S$. So, given Theorem 4, it suffices to show that every pf-consequence relation on $\{+,-\}/S$ meets the *ex falso quodlibet* condition. Suppose R is a pf-consequence relation on $\{+,-\}/S$ generated by a set V of pf-valuations on S. Suppose some member v of V does not satisfy $\langle\{+s, -s\}, \emptyset\rangle$ for some sentence s in S. Then $\langle s, t\rangle \in v$ and $\langle s, f\rangle \in v$. Then v is not a pf-valuation on S.

Completeness. Imitate the reasoning for Theorem 5. Let $\langle U_1, U_2\rangle$ range over partitions of the universal set of judgments on $\{+,-\}/S$ that meet this condition (E): for every sentence s in S either $+s \notin U_1$ or $-s \notin U_1$. □

THEOREM 17 A set of inferences is a tr-consequence relation on $\{+,-\}/S$ iff it is an ODCT-deducibility relation on $\{+,-\}/S$.

Proof: *Soundness.* Every tr-consequence relation on $\{+,-\}/S$ is a pr-consequence relation on $\{+,-\}/S$. So, given Theorem 4, it suffices to show that every tr-consequence relation on $\{+,-\}/S$ meets the ***tertium non datur*** condition. Suppose R is a tr-consequence relation on $\{+,-\}/S$ generated by a set V of tr-valuations on S. Suppose some member v of V does not satisfy $\langle\emptyset, \{+s, -s\}\rangle$ for some sentence s in S. Then $\langle s, t\rangle \notin v$ and $\langle s, f\rangle \notin v$. Then v is not a tr-valuation on S.

Completeness. Imitate the reasoning for Theorem 5. Let $\langle U_1, U_2\rangle$ range over partitions of the universal set of judgments on $\{+,-\}/S$ that meet this condition (T): for every sentence s in S either $+s \notin U_2$ or $-s \notin U_2$. □

THEOREM 18 A set of inferences is a tf-consequence relation on $\{+,-\}/S$ iff it is an ODCET-deducibility relation on $\{+,-\}/S$.

Proof: *Soundness.* Note that every tf-consequence relation on $\{+,-\}/S$ is a pf-consequence relation on $\{+,-\}/S$ and a tr-consequence relation on $\{+,-\}/S$.

Completeness. Imitate the reasoning for Theorem 5. Let $\langle U_1, U_2\rangle$ range over partitions of the universal set of judgments on $\{+,-\}/S$ that meet conditions E and T. □

Proofs for the following three theorems are given by modifying preceding reasoning.

THEOREM 19 A set of inferences is a tf-consequence relation on $\{+\}/S$ iff it is an ODC-deducibility relation on $\{+\}/S$.

THEOREM 20 A set of inferences is a tf-consequence relation on $\{-\}/S$ iff it is an ODC-deducibility relation on $\{-\}/S$.

Define consequence relations on $\emptyset/S$ in the natural way. Premise and conclusion sets are sets of sentences, not sets of judgments.

THEOREM 21 [Shoesmith and Smiley] A set of inferences is is a tf-consequence relation on $\emptyset/S$ iff it is an ODC-deducibility relation on $\emptyset/S$.

3 Single conclusion consequence

Following Johnson [2] we formalize two notions of assertion and two notions of rejection. In addition to the notions of "strong assertion" and "strong rejection" discussed above we recognize "weak assertion", which complements strong rejection, and "weak rejection", which complements strong assertion.[5]

[5]Rumfitt's [1] informal discussion of the the two rejection activities uses "internal" instead of "strong" and "external" instead of "weak." Rumfitt says that one who "rejects as *false*" the claim that the King of France is bald internally [strongly] rejects the claim, but one who "rejects as *not true*" the claim that the King of France is bald externally [weakly] rejects the claim. Rumfitt mentions the two kinds of rejection only to indicate that his focus is on what we are calling strong rejection. Since Rumfitt confined his discussion of categorical consequence relations to the multiple conclusion, partial function variety he had no need to formalize the two kinds of rejection and the two kinds of assertion.

DEFINITION 22 Let S be a set of sentences. Let s be a member of S. $+s$ ($\oplus s, -s, \ominus s$) is *a strong assertion (a weak assertion, a strong rejection, a weak rejection) on* $\{+,\oplus,-,\ominus\}/S$. j is *a judgment on* $\{+,\oplus,-,\ominus\}/S$ iff j is a strong assertion, a weak assertion, a strong rejection, or a weak rejection on $\{+,\oplus,-,\ominus\}/S$.

DEFINITION 23 Suppose J is a set of judgments and k is a judgment on $\{+,\oplus,-,\ominus\}/S$. Then $\langle J,k\rangle$ is an *sc-(single conclusion) inference on* $\{+,\oplus,-,\ominus\}/S$.

DEFINITION 24 Let v be a pr-valuation on S. Let s be a member of S. v satisfies $\oplus s$ iff $\langle s,f\rangle \notin v$ and v satisfies $\ominus s$ iff $\langle s,t\rangle \notin v$. Let $\langle J,k\rangle$ be an sc-inference on $\{+,\oplus,-,\ominus\}/S$. v satisfies $\langle J,k\rangle$ iff v does not satisfy some member of J or v satisfies k.

Extend the other notions defined for multiple conclusion inferences to single conclusion inferences in the natural way.

Example 3. Suppose $S = \{a\}$. There are exactly 256 sc-inferences on $\{+,\oplus,-,\ominus\}/\{a\}$. Let the pr-valuations on $\{a\}$ be defined as in Example 1. Each of these pr-valuations does not satisfy an sc-inference on $\{+,\oplus,-,\ominus\}/\{a\}$ that is satisfied by the others as indicated by the following pairs: $v_1, \langle\{+a\},\oplus a\rangle$, $v_2, \langle\{+a\},-a\rangle$, $v_3, \langle\{-a\},+a\rangle$, and $v_4, \langle\{\oplus a\},+a\rangle$. So there are exactly 16 sc-pr-consequence relations on $\{+,\oplus,-,\ominus\}/\{a\}$ and each of them is categorical.

THEOREM 25 Every sc-pr-consequence relation on $\{+,\oplus,-,\ominus\}/S$ is categorical.

Proof: For every pr-valuation v on S we construct an sc-inference $\langle J-\{k\},k\rangle$ on $\{+,\oplus,-,\ominus\}/S$ such that v does not satisfy it but every other pr-valuation on S does. Pick a sentence a that belongs to S. If $\langle a,t\rangle \in v$ let k be $\ominus a$, otherwise let k be $+a$. If $\langle a,f\rangle \in v$ put $-a$ in $J-\{k\}$, otherwise put $\oplus a$ in $J-\{k\}$. For every sentence b other than a in S: i) if $\langle b,t\rangle \in v$ put $+b$ in $J-\{k\}$, otherwise put $\ominus b$ in $J-\{k\}$; and ii) if $\langle b,f\rangle \in v$ put $-b$ in $J-\{k\}$, otherwise put $\oplus b$ in $J-\{k\}$. □

COROLLARY 26 Every sc-pf-, sc-tr-, and sc-tf-consequence relation on $\{+,\oplus,-,\ominus\}/S$ is categorical.[6]

Note that if F is a three-membered subset of $\{+,\oplus,-,\ominus\}$ we can choose S so that there are sc-pf- and thus sc-pr consequence relations on F/S that are not categorical. The pf-consequence relation on $\{-,\oplus,\ominus\}/\{a,b\}$ generated by $\{\{\langle a,t\rangle\},\{\langle b,t\rangle\}\}$ is the pf-consequence relation on $\{-,\oplus,\ominus\}/\{a,b\}$ generated by $\{\{\langle a,t\rangle\},\{\langle b,t\rangle\},\{\langle a,t\rangle,\langle b,t\rangle\}\}$. For $\{+,\oplus,\ominus\}/\{a,b\}$ use $\{\{\langle a,f\rangle\},\{\langle b,f\rangle\}\}$ and $\{\{\langle a,f\rangle\},\{\langle b,f\rangle\},\{\langle a,f\rangle,\langle b,f\rangle\}\}$. For $\{+,-,\ominus\}/\{a,b\}$ use $\{\{\langle a,f\rangle\}, \{\langle b,f\rangle\}\}$ and $\{\{\langle a,f\rangle\},\{\langle b,f\rangle\},\emptyset\}$. For $\{+,-,\oplus\}/\{a,b\}$ use $\{\{\langle a,t\rangle\},\{\langle b,t\rangle\}\}$ and $\{\{\langle a,t\rangle\},\{\langle b,t\rangle\},\emptyset\}$.

Likewise, if F is a three-membered subset of $\{+,\oplus,-,\ominus\}$ there are sc-tr-consequence relations on $F/\{a,b\}$ that are not categorical.

[6] Johnson [2] proves that every sc-pf-consequence relation on $\{+,\oplus,-,\ominus\}/S$ is categorical.

THEOREM 27 Let F be a two-membered subset of $\{+, \oplus, -, \ominus\}$ where one member is $+$ or $\oplus$ and the other member is $-$ or $\ominus$. Every sc-tf- consequence relation on F/S categorical.[7]

Proof: For every tf-valuation v on S we construct an sc-inference $\langle J - \{k\}, k\rangle$ on $\{+\ (\oplus), -\ (\ominus)\}/S$ such that v does not satisfy it but every other pr-valuation on S does. Pick a sentence a that belongs to S. If $\langle a, t\rangle \in v$ let k be $-a$ $(\ominus a)$, otherwise let k be $+a$ $(\oplus a)$. For every sentence b other than a in S if $\langle b, t\rangle \in v$ put $+b$ $(\oplus b)$ in $J - \{k\}$, otherwise put $-b$ $(\ominus b)$ in $J - \{k\}$. □

4 Single conclusion deducibility

DEFINITION 28 Let J be a set of judgments on $\{+, \oplus, -, \ominus\}/S$. J is an *fc (full and consistent) set of judgments relative to a set A of sentences ($A \subseteq S$)* iff: i) for each sentence s in A there are exactly two judgments with s as content that belong to J; ii) neither of these judgments is an opposite of the other; and iii) the only judgments in J are those whose content is a member of A. s is the content of $+s$, $\oplus s$, $-s$ and $\ominus s$. The opposite of $+s$ is $\ominus s$. In symbols $op(+s) = \ominus s$. $op(\ominus s) = +s$, $op(-s) = \oplus s$. And $op(\oplus s) = -s$.

DEFINITION 29 Let R be a set of sc-inferences on $\{+, \oplus, -, \ominus\}/S$. R is an *sc-ODCR deducibility relation on* $\{+, \oplus, -, \ominus\}/S$ iff R meets the following conditions:
i) (sc-overlap) If $k \in J$ then $\langle J, k\rangle \in R$;
ii) (sc-dilution) If $J \subseteq J'$ and $\langle J, k\rangle \in R$ then $\langle J', k\rangle \in R$;
iii) (sc-cut) If $\langle J \cup K, k\rangle \in R$ for every fc set K of judgments relative to A ($A \subseteq S$) then $\langle J, k\rangle \in R$; and
iv) (sc-reversal) If $\langle J \cup \{j\}, k\rangle \in R$ then $\langle J \cup \{op(k)\}, op(j)\rangle \in R$.

The following derived rule, Opposites, will be used below.

Let R be an sc-ODCR deducibility relation on $\{+, \oplus, -, \ominus\}/S$. *Opposites*: If $\{j, op(j)\} \subseteq J$ then, for every judgment k, $\langle J, k\rangle \in R$.

Proof: Assume the antecedent. Let $J = J' \cup \{j, op(j)\}$. By sc-overlap $\langle J' \cup \{j, op(j), op(k)\}\rangle j \in R$. By sc-reversal $\langle J' \cup \{j, op(j)\}, k\rangle \in R$. So $\langle J, k\rangle \in R$. □

THEOREM 30 Every sc-pr consequence relation on $\{+, \oplus, -, \ominus\}/S$ is an sc-ODCR deducibility relation on $\{+, \oplus, -, \ominus\}/S$.

Proof: Let v be a pr-valuation on S. *sc-overlap.* For each judgment on $\{+, \oplus, -, \ominus\}/S$ with content s, $\langle s, t\rangle$ does or does not belong to v and $\langle s, f\rangle$ does or does not belong to v. *sc-dilution.* Suppose $J \subseteq J'$. Note that if v satisfies every member of J' then v satisfies every member of J. *sc-cut.* Suppose v does not satisfy $\langle J, k\rangle$. Let A be any set of sentences. Let K be an fc set of judgments relative to A such that for each sentence s in A $+s$ $(-s, \oplus s, \ominus) \in K$ iff $\langle s, t\rangle \in v$ $(\langle s, f\rangle \in v$, $\langle s, f\rangle \notin v$, $\langle s, t\rangle \notin v)$, respectively. Then v does not satisfy $\langle J \cup K, k\rangle$. *sc-reversal.* Suppose v does not satisfy $\langle J \cup \{op(k)\}, op(j)\rangle$. Then v satisfies j but does not

[7] Smiley [4] proves that every sc-tf-consequence relation on $\{+, -\}/S$ is categorical.

satisfy k. So v does not satisfy $\langle J \cup j, k\rangle$.

□

THEOREM 31 Every sc-ODCR deducibility relation on $\{+, \oplus, -, \ominus\}/S$ is an sc-pr-consequence relation on $\{+, \oplus, -, \ominus\}/S$.

Proof: Let R be an sc-ODCR deducibility relation on $\{+, \oplus, -, \ominus\}/S$. Let $I = \langle U, m\rangle$ where $m \notin U$ and U is an fc set of judgments relative to S on $\{+, \oplus, -, \ominus\}/S$. Let v_I be the pr-valuation that does not satisfy I. Let $V = \{v_I | \; I \notin R\}$. Let $\models_V$ be the pr-consequence relation generated by V. Then $R = \models_V$.

$R \subseteq \models_V$. Suppose $\langle J, k\rangle \notin \models_V$. Then, for some v_I in V ($I = \langle U, m\rangle$), v_I does not satisfy $\langle J, k\rangle$. Since V_I satisfies U, $J \cup \{op(k)\} \subseteq U$. Suppose $\langle J, k\rangle \in R$. By sc-dilution $\langle J \cup U \cup \{op(m)\}, k\rangle \in R$. By sc-reversal $\langle J \cup \{op(k)\} \cup U\}, m\rangle \in R$. Then $\langle U, m\rangle \in R$. But $\langle U, m\rangle \notin R$. So $\langle J, k\rangle \notin R$.

$\models_V \subseteq R$. Suppose $\langle J, k\rangle \notin R$. By sc-cut $\langle J \cup U, k\rangle \notin R$, where U is an fc set of judgments relative to S on $\{+, \oplus, -, \ominus\}/S$. By sc-overlap $k \notin U$. By the derived rule Opposites, $J \subseteq U$. So $v_{\langle U,k\rangle} \in V$. Since $v_{\langle U,k\rangle}$ does not satisfy $\langle U, k\rangle$, $v_{\langle U,k\rangle}$ does not satisfy $\langle J, k\rangle$. So $\langle J, k\rangle \notin \models_V$.

□

DEFINITION 32 R is an *sc-ODCRE deducibility relation on* $\{+, -, \oplus, \ominus\}/S$ iff R is an sc-ODCR deducibility relation on $\{+, -, \oplus, \ominus\}/S$ that meets this condition: for every sentence s in S, $\langle\{+s\}, \oplus s\rangle \in R$. R is an *sc-ODCRT deducibility relation on* $\{+, -, \oplus, \ominus\}/S$ iff R is an sc-ODCR deducibility relation on $\{+, -, \oplus, \ominus\}/S$ that meets this condition: for every sentence s in S, $\langle\{\oplus s\}, +s\rangle \in R$. R is an *sc-ODCRET deducibility relation on* $\{+, -, \oplus, \ominus\}/S$ iff R is an sc-ODCRE deducibility relation on $\{+, -, \oplus, \ominus\}/S$ and an sc-ODCRT deducibility relation on $\{+, -, \oplus, \ominus\}/S$.

THEOREM 33 A set of inferences is an sc-pf- (sc-tr-, sc-tf-) consequence relation on $\{+, -, \oplus, \ominus\}/S$ iff it is an sc-ODCRE (sc-ODCRT, sc-ODCRET) deducibility relation on $\{+, -, \oplus, \ominus\}/S$.

Proof: Imitate the proofs of the preceding two theorems.

□

References

[1] I. Rumfitt. The categoricity problem and truth-value gaps. *Analysis*, 57:223–235, 1997.

[2] Fred Johnson. Categoricity of partial logics. In Timothy Childers, editor, *The Logica Yearbook 1998*, pages 194–202. Filosofia, Prague, 1999.

[3] D.J. Shoesmith and T. J. Smiley. *Multiple-conclusion logic*. Cambridge University Press, 1978.

[4] T. J. Smiley. Rejection. *Analysis*, 56:1–9, 1996.

[5] Rudolf Carnap. *Introduction to Semantics.* Harvard University Press, 1942.

[6] Rudolf Carnap. *Formalization of logic.* Harvard University Press, 1943.

[7] J.M. Dunn. Intuitive semantics for first degree entailment and 'coupled trees'. *Philosophical Studies*, 29:149–168, 1976.

Ontological causes of inconsistency and a change-adaptive, logical solution.*

GUIDO VANACKERE University of Ghent, Belgium
Guido.Vanackere@rug.ac.be

Abstract
This paper reveals an implicit ontological assumption that is presupposed in common thought. This assumption results in the fact that people usually do not make any distinction between 'the object a' and 'the object a at a given moment'. This laziness causes many inconsistencies. Several attempts to solve these inconsistencies are studied, and the most natural one is elaborated, namely the one obtained by applying Classical Logic to an ontological correct domain. This solution has a drawback with respect to communication, which is solved by the change-adaptive logic **CAL2**. This non-monotonic, paraconsistent logic, belongs to the family of ambiguity-adaptive logics. It has the special characteristic that it solves inconsistencies by the introduction of more precise names for objects, more exactly names that refer to objects at a moment. The dynamics of the logic captures the change in objects. **CAL2** has a nice proof theory, and an intuitive semantics. Interesting results and applications are commented upon, for instance those making use of the notion 'periods of invariance'. Of course, the philosophical background is discussed.

1 Background

As a logician I have been working on inconsistencies and ambiguities. As a philosopher I am mainly interested in the nature of individuals and their identity. This paper is the result of combining both topics.

1.1 Ambiguity-adaptive logics

The idea behind ambiguity-adaptive logics[1] is that we have to make an inference to identify two occurrences of one non-logical constant (sentential, predicative or individual constant, henceforth NLC). If, for instance, we meet two times the word "chair", we can either say that both occurrences mean the same, or we can suspect them to have a different meaning. In general there are two extreme positions: (1)

*Research for this paper was supported by the organizers of WCP2000, and by the the Fund for Scientific Research – Flanders. I would like to thank Bob Meyer for his philosophical support.

[1] See [1].

we identify all occurrences of one NLC; (2) we refuse to identify any two occurrences with one another. The first position can be handled by applying Classical Logic (henceforth **CL**) to the original reading of a set of premises Γ. The second position can be handled by applying **CL** to a maximally ambiguous reading of the set of premises Γ^I. This Γ^I is easily created by giving every occurrence of a NLC in Γ a different superscript. An ambiguity-adaptive logic starts from the maximally ambiguous reading, and reintroduces the original reading ***unless and until this would lead to the derivation of an inconsistency from the set of premises.*** The italic expression is typical for all adaptive logics, created by Diderik Batens, and the Ghent group.[2] The ambiguity-adaptive logic **ACL2** can be characterized by the theorems 1 and 2:[3]

THEOREM 1 $\Gamma \vdash_{\mathbf{CL}} \bot$ iff $\Gamma^I \vdash_{\mathbf{CL}} C_1^i \neq C_1^j \vee ... \vee C_n^k \neq C_n^l$ for some NLC C_m.

$C^i \neq C^j$ refers to either $\sim(C^i \equiv C^j)$, if C is a sentential letter, to $\sim C^i = C^j$ if C is a constant, or to $\sim(\forall x_1)...(\forall x_n)(C^i x_1...x_n \equiv C^j x_1...x_n)$ if C is a predicate of rank n.

THEOREM 2 $\Gamma^I \vdash_{\mathbf{ACL2}} A^I$ iff $\Gamma^I \vdash_{\mathbf{CL}} A^I \vee C_1^i \neq C_1^j \vee ... \vee C_n^k \neq C_n^l$ for some NLC C_m, none of which behaves ambiguously with respect to Γ^I.

A^I is a formula in which all NLC may be indexed. The importance of **ACL2** is straightforward: we assume that the possible inconsistency (with respect to **CL**) of a set of premises, is due to the ambiguity of some NLC. The logic **ACL2** is able to detect these ambiguities, and allows to reason completely classically outside the neighbourhood of ambiguities. If Γ is consistent, we have that $\Gamma \vdash_{\mathbf{CL}} A$ iff $\Gamma^I \vdash_{\mathbf{ACL2}} A^I$.

In this paper, we assume that we never have $C^i \neq C^j$ when C is a sentential letter or a predicate, and we interpret the superscripts of constants as time-indices. The result is a very interesting change-adaptive logic, with a very intuitive semantics and a dynamic proof theory.

There are some obvious reasons for creating the logic **CAL2**, I mean for interpreting the 'ambiguity-superscripts' as time-indices, and giving them to indivuals and not to predicates. Consider the following sentences. In 1999 John had a broken leg. In 1998 John did not have a broken leg. We can formalize them as (*i*) as $B^{99}j$ and $\sim B^{98}j$ (as in **CL**), or (*ii*) as Bj^{99} and $\sim Bj^{98}$ (as in the here presented approach). With approach (*i*) it seems like John remains the same, whereas the meaning of "has a broken leg" changes. This is quite unnatural.[4] Moreover, with

[2] See, for instance, [2] and [3].

[3] These theorems are proven in [1], resp. as Corollary 1 and Theorem 3.

[4] We can restrict the use of predicates to predicates that are defined sharply and independently of time. Especially in sciences this is the case. Hence, there is no need to suspect predicates to be ambiguous, or to have a different meaning on different times. Names for individuals however are not well-definable. (One can not give a clear and adequate definition of "Guido Vanackere".) Names often refer to changing individuals or to 'dead' or imaginary objects—which may cause 'stupid' inconsistencies. Two examples. (1) The sentences "John Smith is seven years old", "John Smith is twenty years old", and "If someone is seven, he is not twenty" allow us to derive an inconsistency. (2) In **CL**, we can derive from "August is a Roman emperor", "$(\exists x)x =$August"; which causes an inconsistency when we formalize the sentence "August is dead". Both inconsistencies can be avoided by not identifying the two occurrences of respectively "John Smith" and "August". Example (2) cannot be solved with approach (*i*).

approach (*ii*) combined with the change-adaptive logic **CAL2**, there is no need to make a non-logical distinction between "essential properties" where inconsistencies have to be avoided, and other properties where inconsistencies are allowed with respect to predication.[5] We can even say that the combination of approach (*ii*) and **CAL2** allows one to make a logical distinction between "essential" and "other" properties.[6]

In this paper it will become clear that the here presented approach is very easy and more adequate than the approach of time logic and free logics. Whereas these logics might allow for inconsistencies due to 'non-essential' predicates, they do not allow to make consistent extrapolations from sentences true at a moment (*e.g.* "In 1950 Georges has 9 children"), to sentences true in a long period (*e.g.* "From 1950 on, Georges has 9 children"). Indeed, the presented approach allows to define an interesting adaptive logic, which allows to derive all 'natural' consequences, and has some nice applications.[7]

To conclude this section, I want to mentioned that the here presented approach allows to change **CL** easily into a free logic.

1.2 Individuals and their identity

As an introduction to this section I mention the question: ***does the subject-predicate form of our sentences fit to the structure of reality?*** This question is profoundly treated by Whitehead in his [4]. We do not have to agree with Whitehead's answer to this question, but we can start with being humble, and taking into account that the structure of reality is not as simple as the structure of our descriptions. My suggestion is to start with being cautious with respect to the use of names for individuals.

In every day life, the perception of an individual is accompanied by the implicit assumption that the 3-dimensional 'thing' we see here and now ***is the individual.*** In fact, we only perceive ***the individual at this moment.*** We assume that an individual at a given moment represents ***the*** individual. By identifying "an individual a at a given moment" with "***the*** individual a", we implicitly assume that the individual a does not change. Everyone will agree that a lot of individuals do change. Due to this change, an individual a might meet a predicate P at moment i whereas it does not meet the predicate at moment j—which obviously causes inconsistencies if we do not take into account the change.

My proposal is to handle the fourth dimension of objects, namely the time-dimension, in the same way as the three space-dimensions. We will not pretend to have seen a 20 meter long object as soon as we have seen 50 cm of it. I think we should not pretend to have seen a 20 year lasting object as soon as we have seen half a day of it. I think we should not consider individuals as 3-dimensional things, but as 4-dimensional things. Therefore our description of individuals should be more

[5] In sections 3 and 4 it will become clear that a sentence like "John Smith 'throughout his life' is a human being" is **CAL2**-derivable from "John Smith 'at the age of seven' is a human being".

[6] See section 6.2.

[7] With approach (*ii*) we can formalize a sentence as "In 99, John wasn't the same anymore as in 95" as $\sim(j^{95} = j^{99})$. I do not know how to formalize it with approach (*i*) or within the language of time logic.

humble: we should talk about a on time i, a on time j, ... (notation: a^i, a^j,...), instead of talking about *the* individual a.

This humbleness allows us to introduce a unary connective that expresses that an individual a does not exist on a given time i (notation: $\dagger a^i$), which is not inconsistent with $(\exists x)x = a^i$. This 'existentially' quantified formula expresses that we can use the name a on time i, even if the thing it refers to, does not live anymore.

Our making of mental pictures of individuals goes along with *generalization*—we generalize from 'the individual as seen on some of its moments', to 'the individual during all of its life'— and goes along with *identification*—we identify the individual at time i with the individual at time j. Moreover, our pictures of individuals tend to survive the individuals—we still speak of the emperor August in the present tense; and Georges Staelens *is* my grandfather, though he died in 1984.

I think that the use of names for individuals is mainly due to a communicational wish. It is indeed easier to use one name for one individual throughout its history.[8] If it is true that we use names of individuals only because of a communicational wish, we might just as well talk about the individual a on time i (a^i) and a on time j (a^j), etc..., and assume that we can identify a^i and a^j unless this would lead to stupid inconsistencies. This way, the communicational wish is fulfilled, inconsistencies are avoided, and our descriptions are closer to reality.

An important question about individuals is whether they have an identity. Is there something real about 'the identity of an individual'? Does an individual have an identity independent of time? Is the identity more than a name? We can consider two extreme answers to these questions. The first one, I will call *Platonism*, the other one will be refered to with the name *Chaos*. Platonism says that an individual has an eternal identity, Chaos says that an individual has no identity; an individual is an ever-changing 'whole' of properties, which does not exist before its 'birth' or after its 'death'. Hence, if a name refers to the identity of an individual, then, in the Platonic view, we have *one name, one meaning*, and in the chaotic view we have *one name, ever-changing meaning*.

Both extreme views have serious drawbacks when we want to talk about change. Let us first consider the Platonic view:

There are a lot of situations in which time is not represented in our descriptions of the world, and hence, it often happens that we meet both Pa and $\sim Pa$ in our theories, for some predicate P and some object a. For example: when we look at a tomato at an early stage of its existence, it will be right to say

This tomato is green. (1)

When we look at the same tomato when it arrives in the shop, it will be right to say

This tomato is red. (2)

In view of the implicit rule

If something is green, it is not red. (3)

[8] One may even question whether we do need names for individuals at all. When we consider the domain of physics, for instance, we notice that one does not need names for individuals. We do not write $(\forall x)(Vx = Dx/Tx)$ or $Va = Da/Ta$, but we write $V = D/T$. (V stands for speed, D for distance, T for time.)

we meet an inconsistency. The main cause of this inconsistency is the implicit assumption that an object remains the same object throughout its history. In other words: the ontological assumption that there is something like "the time-independent individual tomato" is a cause of inconsistency. Notice that this kind of inconsistencies surfaces when a constant is linked to a predicate.

One might say: we can avoid these inconsistencies very easily by specifying our predicates. This approach would result in the following sentences (instead of (1) and (2)).

This tomato 'is green in its first week'. (4)

This tomato 'is red when it is ripe'. (5)

The analogous replacement of (3) would not be true anymore, and hence the inconsistency would be avoided. But this solution is merely an ad hoc solution. Actually, it is not a solution: it would become impossible to define the meaning of our predicates. Another problem of this 'solution' will become clear when you consider the following. Suppose the considered tomato has been eaten; thus it does not exist anymore. Should we create predicates like "does not exist after it has been eaten"? Is this a predicate or an existential quantifier? Or should we say: "If something has been eaten, it does not exist"?

$$Ea \supset {\sim}(\exists x)(x = a) \tag{6}$$

Formula (6) gives birth to an inconsistency in most common logics. (Even in time logic, Ea and ${\sim}(\exists x)x = a$ would be true on the same time.) The conclusion might be that the specification of the predicates is not a good solution to the problem. Anyway, this solution seems to assume that objects exist before their 'birth' and after their 'death', which is hard to maintain.

When we consider the Chaotic view, we meet another problem. Suppose we refer to the tomato with the name a. a will not have the same meaning on time i and time j. So, let us refer to these meanings with respectively a^i and a^j. If P stands for "is green", and Q for "is red", then we have Pa^i, Qa^j and the implicit rule $(\forall x)(Px \supset {\sim}Qx)$, from which we can derive both ${\sim}Pa^j$ and ${\sim}Qa^i$. Obviously, the inconsistency is avoided. However we meet another problem. Suppose time $i+1$ comes immediately after i. Then, in everyday life, we would infer from the fact Pa^i that also Pa^{i+1}, but this inference is not logically correct, at least not in **CL**.

$$Pa^i \nvdash_{\mathbf{CL}} Pa^{i+1} \tag{7}$$

Where the Platonic view has the drawback to lead to stupid inconsistencies, the Chaotic view has the drawback that predictions are no longer possible.

2 About the wanted logic

The logic we want should be as rich as possible—preferably as rich as **CL**—but it should avoid applications of ex contradictione quodlibet. Obviously, if we apply the "index-trick" of the Chaotic view, we can avoid inconsistencies. In order to obtain this, we apply **CL** to a domain of 'individuals at a moment'. I use the notation a^i to denote the (usual) individual a at moment i. With "moment", I mean the shortest period one wants to consider.[9] The individual variables—for instance x—refer to any individual at some moment. We can easily introduce the connective † (see AS† in section 2.1).

However, this is only one part of what we want. There are some drawbacks concerning 4-dimensional objects. For instance: if we replace an individual a by infinitely many a^n, we can no longer make predictions about the object. In order to have useful descriptions, we need to identify "a^i" and "a^j" in a lot of situations. The question is: when is it right to identify them and when is it wrong? A straightforward answer is: it is wrong to identify them if this would lead to the derivation of an inconsistency.

We may also consider the human wish to consider individuals as existing at the given moment unless and until it is proven that they do not exist at that moment, which can be 'translated' as "$\dagger a^i$" is false unless and until $\dagger a^i$ is derived from the premises.

We can introduce the wanted logic **CAL2** as 'the optimum' between a Chaotic logic (called **CHAOS**) and a Platonic logic (called Platonic Heaven or **PH**).

CHAOS and **PH** can be easily defined from **CL** extended with the unary connective †, and applied to a specific domain.

Let $\mathcal{C}$ be the set of individual constants 'at a moment', $\mathcal{V}$ the set of individual variables (referring to 'individuals at a moment'), and $\mathcal{P}^r$ the set of predicates of rank r.

$\mathcal{C}^{\mathcal{P}}$ is the set of (platonic forms of) individuals, *i.e.* the usual constants. $\mathcal{T}$ is the set of (shortest) periods. The relations $<$ and $=$ in $\mathcal{T}$ behave exactly as the relations $<$ and $=$ in the set of integers. Where $i \in \mathcal{T}$, $\langle \mathcal{C}^{\mathcal{P}}, i \rangle$ is a state of reality, and $(a,i) \in \langle \mathcal{C}^{\mathcal{P}}, i \rangle$.

The domain $\mathcal{D}$ is defined as follows:

$$\mathcal{D} = \bigcup_{n=-\infty \to +\infty} \langle \mathcal{C}^{\mathcal{P}}, n \rangle \tag{8}$$

$$\mathcal{D} = \mathcal{R} \cup \mathcal{I}; \ \mathcal{R} \cap \mathcal{I} = \emptyset \tag{9}$$

$\mathcal{R}$ stands for the set of real individuals on a given moment, $\mathcal{I}$ stands for the set of imaginary individuals on a given moment. $(a,i) \in \mathcal{R}$ iff a exists on time i. $(a,i) \in \mathcal{I}$ iff a does not exist on time i.

2.1 Proof theory of CHAOS

The syntax of **CHAOS** is obtained by extending the syntax of **CL** with:

$$\text{AS}\dagger : (\exists x_1)...(\exists x_{n-1})\Pi^n(\alpha^i) \supset \sim \dagger \alpha^i \tag{10}$$

in which Π^n is a predicate of rank n and $(\exists x_1)...(\exists x_{n-1})\Pi^n(\alpha^i)$ is a well-formed predicative expression, for instance Pa^1 or $(\exists x)(\exists y)Sxb^{45}y$.

Obviously, we have the following non-theorem:

$$\nvdash_{\text{CHAOS}} \alpha^i = \alpha^j \tag{11}$$

[9]Hence, there is a bijection between the natural numbers and the set of moments.

An example of a **CHAOS**-proof:[10]

1.	Pa^i	PREM
2.	Qa^j	PREM
3.	$\dagger a^k$	PREM
4.	$(\forall x)(Px \supset {\sim}Qx)$	PREM
5.	$Pa^i \supset {\sim}Qa^i$	4; UI
6.	$Pa^j \supset {\sim}Qa^j$	4; UI
7.	$Pa^k \supset {\sim}Qa^k$	4; UI
8.	${\sim}Qa^i$	1,5; MP
9.	${\sim}Pa^j$	2,6; MT
10.	${\sim}(Qa^i \equiv Qa^j)$	2,8; INE
11.	${\sim}(Pa^i \equiv Pa^j)$	1,9, INE
12.	${\sim}Pa^k$	3; E†
13.	${\sim}Qa^k$	3; E†
14.	${\sim}(Pa^i \equiv Pa^k)$	1,12; INE
15.	${\sim}(Qa^j \equiv Qa^k)$	2,13; INE
16.	${\sim}\dagger a^i$	1; I†
17.	${\sim}\dagger a^j$	2; I†

2.2 Semantics of CHAOS

Let v, the assignment-function, be as follows:

$$\mathsf{v} : \mathcal{C} \cup \mathcal{V} \longrightarrow \mathcal{D} \tag{12}$$

$$\mathsf{v} : \mathcal{P}^r \longrightarrow \mathbf{P}(\mathcal{R}^r) \tag{13}$$

$$\mathsf{v} : \mathcal{S} \longrightarrow \{0,1\} \tag{14}$$

The valuation function $\mathsf{v_M}$ is classical plus:

$$\mathsf{v_M}(\dagger\alpha^i) = 1 \text{ iff } \mathsf{v}(\alpha^i) \in \mathcal{I} \tag{15}$$

From (15) it follows that, if $\mathsf{v_M}(\dagger a^i) = 1$, then $\mathsf{v_M}(Pa^i) = 0$. The fact that the assignment function maps $\mathcal{P}^r$ on the powerset of $\mathcal{R}^r$ (and not on the power set of $\mathcal{D}^r$) has no influence on the definition of the valuation function with respect to the quantifiers. Notice that "$(\exists x)$" stands for "there is a real or imaginary individual on a given moment". Of course, as soon as we have $\mathsf{v_M}(\dagger a^i) = 1$, we have $\mathsf{v}(a^i) \in \mathcal{I}$, in view of (15), and hence, in view of (13) and the usual valuation for Pa^i, we have $\mathsf{v_M}(Pa^i) = 0$. Hence, $\mathsf{v_M}((\forall x)Px) = 0$ as soon as there is a constant 'that does not exists at a given time'. Nevertheless, formulas of the form $(\forall x)(Px \supset Qx)$ will get the same valuation in **CHAOS**-models as in **CL**-models.

2.3 Proof theory of PH

The syntax of **PH** is obtained by extending the syntax of **CL** with:

$$\mathrm{AS}\dagger' \ : \ {\sim}\dagger\alpha^i \tag{16}$$

[10] The rules applied in these proof are classical or easy derivable. "INE" stands for introduction of the negation of equivalence; "I†" and "E†" are easily derivable from AS†.

$$\mathrm{AS}=' : \alpha^i = \alpha^j \tag{17}$$

It is easy to see that the addition of AS=′ results in the fact that time-indices become meaningless. AS†′ makes every formula of the form $\dagger\alpha^i$ equivalent to $\perp$. Hence, we can say that **PH** and **CL** are equivalent.

2.4 Semantics of PH

Let v, the assignment-function, be as follows: (o is an arbitrary chosen moment).

$$\mathsf{v} : \mathcal{C} \cup \mathcal{V} \longrightarrow \langle \mathcal{C}^{\mathcal{P}}, o \rangle \tag{18}$$

$$\mathsf{v} : \mathcal{P}^r \longrightarrow \mathbf{P}(\langle \mathcal{C}^{\mathcal{P}}, o \rangle^r) \tag{19}$$

$$\mathsf{v} : \mathcal{S} \longrightarrow \{0, 1\} \tag{20}$$

The valuation is completely classical, plus:

$$\mathsf{v_M}(\dagger\alpha^i) = 0 \tag{21}$$

From this it follows:

$$\mathsf{v_M}(\alpha^i = \alpha^j) = 1 \text{ for every } i, j \tag{22}$$

$$\text{If for no } \alpha^i \in \mathcal{C},\ \Gamma \models_{\mathbf{PH}} \dagger\alpha^i, \text{ then } \Gamma \models_{\mathbf{PH}} A \text{ iff } \Gamma \models_{\mathbf{CL}} A \tag{23}$$

This result gives us a way to formulate an ontological assumption behind the normal use of **CL**, namely: ***CL** treats individuals as if they were platonic forms.*

2.5 The change-adaptive logic CAL2

CAL2 can be briefly characterized as follows. $\Gamma \vdash_{\mathbf{CAL2}} P\alpha^i \equiv P\alpha^j$, $\Gamma \vdash_{\mathbf{CAL2}} {\sim}\dagger\alpha^i$ and $\Gamma \vdash_{\mathbf{CAL2}} \alpha^i = \alpha^j$ *unless and until such an assumption leads to an inconsistency derivable from* Γ. In what follows, I have to translate the expression *'unless and until an assumption leads to an inconsistency derivable from* Γ*'* in terms of proof theory and semantics. People familiar with the adaptive-logic programme know that I only need to define an adaptive logic, that oscillates between a lower limit logic (in this case **CHAOS**), and an upper limit logic (in this case **PH**).

Every adaptive logic is characterized by abnormalities. These abnormalities are formulas which are tolerated in the lower limit logic whereas the (classical) negation is valid in the upper limit logic. For instance: $p\&{\sim}p$ is tolerated in the lower limit logic **CLuN** of the inconsistency-adaptive logics **ACLuN1** and **ACLuN2**, whereas ${\sim}(p\&{\sim}p)$ is valid in their upper limit logic **CL**. The abnormalities of the change-adaptive logic **CAL2** are formulas of the form $\dagger\alpha^i$, ${\sim}(\alpha^i{=}\alpha^j)$, or ${\sim}\forall(\Pi(\alpha^i) \equiv \Pi(\alpha^j))$ in which Π is a predicate of rank n and $\forall$ is an abrevation of $(\forall x_1)...(\forall x_m)$ $(0 \leq m \leq n-1)$, a universal quantification of all free variables in $\Pi(\alpha^i) \equiv \Pi(\alpha^j)$, in some preferred order. For instance ${\sim}\forall(P\alpha^i \equiv P\alpha^j)$ stands for ${\sim}(P\alpha^i \equiv P\alpha^j)$ (where P is a predicate of rank 1); ${\sim}\forall(Rxa^{56} \equiv Rxa^{59})$ stands for ${\sim}(\forall x)(Rxa^{56} \equiv Rxa^{59})$ (where R is a predicate of rank 2), ... In general, an abnormality will be written as $?C$. $\mathcal{A}$ is the set of all abnormalities.

A disjunction of abnormalities is what it says, and, where $?C_1, ..., ?C_n$ are abnoramilities, a disjunction of these abnormalities is written as $\mathsf{DA}\{?C_1, ..., ?C_n\}$ (a DA-formula). In general we write $\mathsf{DA}(\Sigma)$, in which Σ is a finite set of abnormalities. As for the lower and upper limit logic of ambiguity-adaptive logics, we have the following theorem:

THEOREM 3 $\Gamma \vdash_{\mathbf{PH}} A$ iff for some abnormilities $?C_1, ..., ?C_n$, $\Gamma \vdash_{\mathbf{CHAOS}} A \vee$ $\mathsf{DA}\{?C_1, ..., ?C_n\}$.

The proof is analogous to the proof of Theorem 1 in [1] . Remember that Γ is a set of premises in which all constants have a time-index.

As the semantics are very easy and intuitive, I will start with the semantics of the logic **CAL2**.

3 Semantics of CAL2

The logic **CAL2** is based on the idea: (*i*) it is right to identify a^i and a^j for different moments i, j unless and until an inconsistency is derivable from this identification; (*ii*) it is right to assume that a is real on time i, *i.e.*, it is right to assume $\sim \dagger a^i$ unless and until an inconsistency is derivable from this. These expressions bring us to adaptive logics, as they are developed by Diderik Batens and his Ghent Group. All these adaptive logics allow to make derivations of a rich intended logic, unless and until an inconsistency is derivable.[11] The most credulous adaptive strategy is the minimal abnormality strategy.[12] I will make use of this strategy in this paper.

DEFINITION 4 Where M is a **CHAOS**-model, $\mathsf{AC}(\mathsf{M}) = \{?C \mid ?C \in \mathcal{A}$ and $\mathsf{v_M}(?C) = 1\}$.

DEFINITION 5 A **CHAOS**-model M is *minimally abnormal with respect to* Γ iff M is a **CHAOS**-model of Γ, and there is no **CHAOS**-model M′ of Γ such that $\mathsf{AC}(\mathsf{M}') \subset \mathsf{AC}(\mathsf{M})$.

DEFINITION 6 M is a **CAL2**-*model of* Γ iff M is minimally abnormal with respect to Γ.

DEFINITION 7 $\Gamma \models_{\mathbf{CAL2}} A$ iff A is true in all **CAL2**-models of Γ.

All **CAL2**-models of Γ are **CHAOS**-models of Γ. Hence, $\Gamma \models_{\mathbf{CAL2}} A$ if $\Gamma \models_{\mathbf{CHAOS}} A$. Except for border cases, the **CHAOS**-consequenceset of Γ, will be a real subset of the **CAL2**-consequenceset of Γ. (In general, the set of **CAL2**-models of Γ is a real subset of the **CHAOS**-models of Γ.)

If for any **CAL2**-model of Γ, and for any DA-formula $\mathsf{DA}(\Sigma)$, $\mathsf{v_M}(\mathsf{DA}(\Sigma)) = 0$, then $\Gamma \models_{\mathbf{CAL2}} A$ iff $\Gamma \models_{\mathbf{PH}} A$.

The definitions show that **CAL2** interprets a set of premises as normal as possible: no more $?C \in \mathcal{A}$ are true than is required by the premises.

[11] For instance, the well-known inconsistency-adaptive logics allow to assume that a formula behaves consistently, unless and until it is proven that it does not behave consistently (with respect to the premises). See, for instance [2].

[12] For adaptive logics based on the minimal abnormality strategy, the Ghent Group usually adds the number 2 to the name of the logic. The number 1 is used for the reliability strategy. In [5], two more strategies are developed.

4 Proof theory of CAL2

The logic **CAL2** can capture change in objects. In order to obtain this, the logic has a dynamic proof procedure with unconditional and conditional rules. The unconditional rules are the rules of a lower limit logic, the conditional rule is a rule of the upper limit logic. The lower limit logic at hand is **CL** applied to an *"every object is completely different at each of its moments"*-interpretation, extended with † (namely the logic **CHAOS**). The upper limit logic is **CL** applied to the interpretation in which $\alpha^i = \alpha^j$ for all individuals α and all real numbers i and j, namely the logic **PH**. **PH** also contains †, but it functions as $\bot$, namely: $\vdash_{\mathbf{PH}} \dagger\alpha^i \supset A$. Where we have

$$\vdash_{\mathbf{CHAOS}} (Pa^i \equiv Pa^j) \vee {\sim}(Pa^i \equiv Pa^j) \tag{24}$$

$$\vdash_{\mathbf{CHAOS}} {\sim} \dagger a^i \vee \dagger a^i \tag{25}$$

$$\vdash_{\mathbf{PH}} Pa^i \equiv Pa^j \tag{26}$$

$$\vdash_{\mathbf{PH}} {\sim} \dagger a^i \tag{27}$$

we have $\Gamma \vdash_{\mathbf{CAL2}} Pa^i \equiv Pa^j$ on condition that $Pa^i \equiv Pa^j$ is normal with respect to Γ, and $\Gamma \vdash_{\mathbf{CAL2}} {\sim} \dagger a^i$ on condition that ${\sim} \dagger a^i$ is normal with respect to Γ.

The axiom scheme of **CAL2** is the one of **CHAOS** extended with a conditional rule RC and a marking rule RM. Here is the conditional rule:

RC: From $A \vee \mathsf{DA}(\Sigma)$ to derive A on condition Σ.

In order to do this, we change the format of **CAL2**-proofs: it is obtained from the format of **CHAOS**- or **CL**-proofs, by adding a fifth element to each line, which contains the conditions on which we rely in order for the formula in the second element to be derivable by the rule of inference mentioned in the fourth element, from the formulas of the lines enumerated in the third element.

If we apply a **CHAOS**-rule to lines the fifth element of which is not empty, the fifth element of the new line is the union of the fifth element of the used lines. Now, we have to define the **CAL2**-consequence relation.

DEFINITION 8 *A occurs unconditionally* at some line of a proof iff the fifth element of that line is empty.

DEFINITION 9 *?C is abnormal at a stage of a proof* iff $?C$ occurs unconditionally in the proof at that stage.

Suppose that A is derived on one or more lines the fifth element of which is not empty. A is considered as derived at a stage of a proof and the lines become a full part of the proof if A comes out true under a maximally normal 'interpretation' of the DA-formulas (at that stage). "Interpretation" should refer to formal properties of the formulas that occur in the proof.

The role of DA-formulas is crucial in **CAL2**-proofs. Clearly, when a DA-formula occurs unconditionally at some line of a proof, at least one of the disjuncts of that DA-formula is true. Some DA-formulas occurring unconditionally in a proof may be disregarded. As a first step, we only have to consider *minimal* DA-*formulas*. Where Σ and Θ are sets of abnormalities, let us stipulate that

DEFINITION 10 $\mathsf{DA}(\Sigma)$ is a *minimal* DA-*formula at a stage of a proof* iff (*i*) it occurs unconditionnally in the proof at that stage, and (*ii*) there is no $\Theta \subset \Sigma$ for which $\mathsf{DA}(\Theta)$ occurs unconditionally in the proof at that stage, and (*iii*) there is no $\mathsf{DA}(\Theta)$ that occurs unconditionally in the proof at that stage such that $\mathsf{DA}(\Theta) \vdash_{\mathbf{CHAOS}} \mathsf{DA}(\Sigma)$, whereas $\mathsf{DA}(\Sigma) \nvdash_{\mathbf{CHAOS}} \mathsf{DA}(\Theta)$.[13]

Next, let Φ^*_s be the set of all sets that contain one factor out of each minimal DA-formula at stage s of the proof. Φ^*_s may contain redundant elements: the same factor may occur in different minimal DA-formulas. If $\mathsf{DA}\{?C_1, ?C_2\}$ and $\mathsf{DA}\{?C_1, ?C_3\}$ are minimal DA-formulas, then $\Phi^*_s = \{\{?C_1\}, \{?C_1, ?C_2\}, \{?C_1, ?C_3\}, \{?C_2, ?C_3\}\}$. Of these $\{?C_1, ?C_2\}$ and $\{?C_1, ?_3\}$ are redundant. Both $\mathsf{DA}\{?C_1, ?C_2\}$ and $\mathsf{DA}\{?C_1, ?C_3\}$ are true if $?C_1$ is true; there is no need that also $?C_2$ and $?C_3$ be true. So, let Φ_s be obtained from Φ^*_s by eliminating elements from it that are proper supersets of other elements. Hence, the members of Φ_s are sets of formulas, such that, if all members of such a set are true, then all DA-formulas that occur unconditionally in the proof at stage s are true.

DEFINITION 11 Where Φ_s is as defined above and A is the second element of line j, line j *fulfils the integrity criterion at stage* s iff (i) the intersection of some member of Φ_s and of the fifth element of line j is empty, and (ii) for each $\varphi \in \Phi_s$ there is a line k such that the intersection of φ and of the fifth element of line k is empty and A is the second element of line k.

Now we can introduce the marking rule:

RM: A line is marked OUT at a stage iff it does not fulfil the integrity criterion.

If the fifth element of a line is empty, *i.e.*, when the formula in its second element is a **CHAOS**-consequence, the intigrity criterion is obviously fulfilled. All formulas in lines with an empty fifth element are **CAL2**-consequences.

DEFINITION 12 A is *finally derived* at some line in an **CAL2**-proof iff it is the second element of that line and any (possibly infinite) extension of the proof can be further extended in such way that the line is unmarked.

DEFINITION 13 $\Gamma \vdash_{\mathbf{CAL2}} A$ (A is **CAL2**-*finally derivable* from Γ) iff A is finally derived at some line of an **CAL2**-proof from Γ.

The proofs of the following theorems are analogous to those of Theorems 3 and 4 in [1]. Φ_Γ is obtained in the same way as Φ_s, taking into account all minimal DA-formulas derivable from Γ.

THEOREM 14 $\Gamma \vdash_{\mathbf{CAL2}} A$ iff there are one or more (possibly empty) finite sets $\Sigma_1, \Sigma_2, \ldots$ of abnormalities, such that $\Gamma \vdash_{\mathbf{CAL2}} A \vee \mathsf{DA}(\Sigma_1)$, $\Gamma \vdash_{\mathbf{CAL2}} A \vee \mathsf{DA}(\Sigma_2)$, ..., and for any $\varphi \in \Phi_\Gamma$, one of the Σ_i is such that $\Sigma_i \cap \varphi = \emptyset$.

THEOREM 15 If $\Gamma \vdash_{\mathbf{CAL2}} A$, then it is possible to extend any proof from Γ into a proof in which A is finally derived.

[13] Condition (*iii*) excludes that, *e.g.*, $\sim(a^i = a^j)$ can be a minimal DA-formula if $\sim(Pa^i \equiv Pa^j)$ is one, for $\sim(Pa^i \equiv Pa^j) \vdash_{\mathbf{CHAOS}} \sim(a^i = a^j)$, whereas $\sim(a^i = a^j) \nvdash_{\mathbf{CHAOS}} \sim(Pa^i \equiv Pa^j)$.

I skip the soundness and completeless theorems. They are analogous to those of [1].

An example of a **CAL2**-proof might make the matter more transparant. The last element of every line mentions the set of conditions on which the formula in this line is derived. The example is a continuation of the proof on page 7; in order to make a **CAL2**-proof of the mentioned **CHAOS**-proof, we have to add a fifth element to lines 1—17, all of which contain $\emptyset$.[14]

18.	$(Pa^i \equiv Pa^m) \vee \sim(Pa^i \equiv Pa^m)$	THEO	$\emptyset$
19.	$Pa^i \equiv Pa^m$	18; RC	$\{Pa^i \equiv Pa^m\}$
20.	Pa^m	1,19; EQ	$\{Pa^i \equiv Pa^m\}$
21.	$(Pa^j \equiv Pa^m) \vee \sim(Pa^j \equiv Pa^m)$	THEO	$\emptyset$
22.	$Pa^j \equiv Pa^m$	21; RC	$\{Pa^j \equiv Pa^m\}$
23.	$\sim Pa^m$	9,22; EQ	$\{Pa^j \equiv Pa^m\}$
24.	$\perp$	20,21; I$\perp$	$\{Pa^i \equiv Pa^m, Pa^j \equiv Pa^m\}$
25.	$\sim(Pa^i \equiv Pa^m) \vee \sim(Pa^j \equiv Pa^m)$	24; IDA	$\emptyset$
26.	$Pb^i \equiv Pb^j$	$\star$	$\{Pb^i \equiv Pb^j\}$
27.	$Ra^i \equiv Ra^j$	$\star$	$\{Ra^i \equiv Ra^j\}$

At stage 27 of the proof, Φ_{27} contains two sets φ_1 and φ_2. $\varphi_1 = \{\dagger a^k, \sim(Pa^i \equiv Pa^j), \sim(Pa^i \equiv Pa^k), \sim(Pa^i \equiv Pa^m), \sim(Qa^i \equiv Qa^j), \sim(Qa^j \equiv Qa^k)\}$; $\varphi_2 = \{\dagger a^k, \sim(Pa^i \equiv Pa^j), \sim(Pa^i \equiv Pa^k), \sim(Pa^j \equiv Pa^m), \sim(Qa^i \equiv Qa^j), \sim(Qa^j \equiv Qa^k)\}$; and hence, in view of the integrity criterion the formulas in lines 1—17, 18, 21, and 25—27 are **CAL2**-consequences, and the formulas in lines 19, 20, 22—24 are not, because they have to be marked out.

5 Applications

In this section I mention two kinds of applications. The first one[15] is based on (safe) generalizations of the information we have about time. For instance: if the only thing we know concerning an object a and a property P, is that a has property P at time i, and does not have property P at time $j > i$, then it is often safe to assume that Pa^n is true, for all n such that $n \leq i$, and that Pa^m is false for all m such that $j \leq m$. If we also happen to have information concerning the 'birth' (time h) and the 'death' of a (time k), we even can be more precise: $h < n \leq i$ and $j \leq m < k$.

The second kind of application mentioned here[16] is useful when we do not have information about time. In this case, **CAL2** is still more useful than, *e.g.*, **CL**. If we give meaningless but different indices to every occurrence of a constant anyway, we can interpret them afterwards for the interesting cases, *viz.* those cases that would give birth to an inconsistency when we use the original reading of the constants.

[14] "EQ" stands for the equivalence rule; IDA stands for the derivable rule introduction of DA-formulas, which is the 'reverse' rule of RC. $\star$ in lines 26 and 27, means that these lines are derived in an analogous way as the formulas in lines 19 and 22. I think the notation of the conditions in the fifth element becomes clearer, if I write, for instance, $Pa^i \equiv Pa^m$ instead of $\sim(Pa^i \equiv Pa^m)$.

[15] See section 5.1.

[16] See section 5.2.

5.1 Periods of invariance

In this section I give an example of how **CAL2** may become even more useful when we go beyond the limits of logic, or when we introduce non-logical hypotheses. We start from a **CAL2**-proof, at a certain stage, and introduce non-logical hypotheses as premises. I continue the proof on page 12. Let $0 < m \leq i$, $j \leq n < k$ and $k < l$. H1 can be read as "a meets predicate P until time i". H3 can be read as "from time k on, a does not exist anymore".

H1.	$(0 < z \leq i) \supset (Pa^z \equiv Pa^i)$	PREM	
H2.	$(j \leq z < k) \supset (Pa^z \equiv Pa^j)$	PREM	
H3.	$(k \leq z) \supset \dagger a^z$	PREM	
28.	$Pa^m \equiv Pa^i$	H1, MP	$\emptyset$
29.	Pa^m	H1,1,28; EQ	$\emptyset$
30.	$Pa^m \supset {\sim}Qa^m$	4; UI	$\emptyset$
31.	${\sim}Qa^m$	29,30; MP	$\emptyset$
32.	$\perp$	23,29; I†	$\{Pa^j \equiv Pa^m\}$
33.	${\sim}(Pa^j \equiv Pa^m)$	32; IDA	$\emptyset$
34.	$Pa^j \equiv Pa^n$	H2; MP	$\emptyset$
35.	${\sim}Pa^n$	9, 34; EQ	$\emptyset$
36.	$(Pa^i \equiv Pa^n) \vee {\sim}(Pa^i \equiv Pa^n)$	THEO	$\emptyset$
37.	$Pa^i \equiv Pa^n$	36; RC	$\{Pa^i \equiv Pa^n\}$
38.	Pa^n	1,37; EQ	$\{Pa^i \equiv Pa^n\}$
39.	$\perp$	35,38; I†	$\{Pa^i \equiv Pa^n\}$
40.	${\sim}(Pa^i \equiv Pa^n)$	39; IDA	$\emptyset$
41.	$\dagger a^l$	H3	$\emptyset$
42.	${\sim}\dagger a^m$	29; I†	$\emptyset$
43.	$(Qa^j \equiv Qa^n) \vee {\sim}(Qa^j \equiv Qa^n)$	THEO	$\emptyset$
44.	$Qa^j \equiv Qa^n$	43; RC	$\{Qa^j \equiv Qa^n\}$
45.	Qa^n	2,44; EQ	$\{Qa^j \equiv Qa^n\}$
46.	${\sim}\dagger a^n$	45; I†	$\{Qa^j \equiv Qa^n\}$

At stage 41 of the proof, the DA-formula in line 25 is no longer minimal (in view of the DA-formula derived in line 33), and hence Φ_{41} contains only one set $\varphi_1 = \{\dagger a^k$, $\dagger a^l$, ${\sim}(Pa^i \equiv Pa^j)$, ${\sim}(Pa^i \equiv Pa^k)$, ${\sim}(Pa^i \equiv Pa^n)$, ${\sim}(Pa^j \equiv Pa^m)$, ${\sim}(Qa^i \equiv Qa^j)$, ${\sim}(Qa^j \equiv Qa^k)\}$, and hence the formulas in lines 1—21, 25—31, 33—36, and 40—46 are **CAL2**-consequences from the original premises plus the 'hypotheses', whereas the formulas in lines 22—24, 32, and 37—39 are not. The introduction of H1 and H2 results in the fact that $Pa^i \equiv Pa^m$ and Pa^m are derivable. As one sees in lines 42 and 46, it can be easily proven that a exists until moment k; line 41 shows that a does not exist after k.

5.2 A rich solution of inconsistent theories

Let Γ be a set of premises in which every occurrence of a constant gets a different superscript. Let A be a formula in the same language. Let Γ^o and A^o be obtained from Γ and A respectively, by omitting all superscripts.[17] Consider the following consequence relation:

[17]Hence Γ^o and A^o are written in the original, classical language.

DEFINITION 16 $\Gamma^o \vdash_{\mathbf{CAL2^o}} A^o$ iff $\Gamma \vdash_{\mathbf{CAL2}} A$.

Clearly, **CAL2°** is paraconsistent. For instance: if $\Gamma = \{Pa^i, {\sim}Pa^j\}$, the **CAL2°**-consequenceset of Γ^o is not trivial. Moreover, whenever we have, for some formula A:

$$\Gamma^o \vdash_{\mathbf{CAL2^o}} A^o \& {\sim} A^o \tag{28}$$

there must be some i and j such that

$$\Gamma \vdash_{\mathbf{CAL2}} A \& {\sim} A^{j/i} \tag{29}$$

in which $A^{j/i}$ is the formula we obtain by replacing an occurrence of i in A by j.

This gives us an easy way to solve inconsistent theories, in a rich way. If some set of premises is inconsistent with respect to **CL**, and we suspect the change of some individuals to be the cause of this inconsistency, then an application of **CAL2** and the consequence relation $\vdash_{\mathbf{CAL2^o}}$ detects the individuals that have changed with respect to some predicate or with respect to their existence. Hence, we are able to replace the original inconsistent theory by a more adequate consistent theory.

Of course, if not a single individual (occurring in Γ^o) changes, applying **CAL2°** to Γ^o gives us exactly the same result as **CL**.

6 Concluding remarks

6.1 Technical

I think the logic **CAL2** proves to be very useful. The semantics are very intuitive and the proof theory is really not too complicated. Anyway it is at least as good as **CL** and in all interesting situations, it is even much better.

It may be interesting to compare the present results with results of time logics and results of update systems. I apologize for postponing this comparison.

6.2 Philosophical

Platonic philosophers attach great value to the (eternal) identity of individuals. As a result of their influence many people implicitly assume that names for individuals refer to something real and independent of time. I think that this assumption does not only lead to stupid inconsistencies, but also to a bad kind of egoism, *viz.*, it makes people too worried about the importance of their own identity. I hope this paper helps undermining this importance. Names are useful with respect to communication, and they help us to get a grip on the chaos, but there is no need of 'an identity' for these purposes. Anyway, who needs an identity for being glad to be alive and meeting people?

An analogous remark can be made about the philosophical distinction between "essential properties" and "other properties" (see section 1.1). Essential properties seem to refer to the identity of the considered objects. I think the question whether a property is essential or not, is relative to a given situation. The logic **CAL2** allows one to make a logical distinction between essential and other properties with respect to a given situation (it is: with respect to a given set of premises).

DEFINITION 17 P is an essential property of a with respect to Γ if and only if for all i, $\Gamma^I \vdash_{\mathbf{CAL2}} Pa^i \vee \dagger a^i$.

REFERENCES

[1] Guido Vanackere. Ambiguity-adaptive logic. *Logique & Analyse*, 159:261–280, 1997.

[2] D. Batens. Inconsistency-adaptive logics. In Ewa Orlowska, editor, *Logic at Work. Essays Dedicated to the Memory of Helena Rasiowa.*, pages 445–472. Heidelberg, New-York, Physica, Verlag, Springer, 1998.

[3] D. Batens. Extending the realm of logic. the adaptive-logic programme. In P. Weingartner, editor, *Alternative logics. Do sciences need them?* Springer Verlag, in print.

[4] Alfred North Whitehead. *Process and Reality, an essay in cosmology.* The Free Press, New York — London, 1978.

[5] Kristof De Clercq. Two new strategies for inconsistency-adaptive logics. *Logic and Logical Philosophy*, in print.

An Adaptive Logic for Pragmatic Truth *

JOKE MEHEUS Centre for Logic and Philosophy of Science, Ghent University, Belgium
Joke.Meheus@rug.ac.be

Postdoctoral Fellow of the Fund for Scientific Research—Flanders (Belgium)

Abstract
This paper presents the new adaptive logic **APT**. **APT** has the peculiar property that it enables one to interpret a (possibly inconsistent) theory Γ 'as pragmatically as possible'. The aim is to capture the idea of a partial structure (in the sense of da Costa and associates) that adequately models a (possibly inconsistent) set of beliefs Γ. What this comes to is that **APT** localizes the 'consistent core' of Γ, and that it delivers all sentences that are compatible with this core. For the core itself, **APT** is just as rich as Classical Logic. **APT** is defined from a modal adaptive logic **APV** that is based itself on two other adaptive logics. I present the semantics of all three systems, as well as their dynamic proof theory. The dynamic proof theory for **APV** is unusual (even within the adaptive logic programme) in that it incorporates two different kinds of dynamics.

1 Introduction

In [1], Mikenberg, da Costa and Chuaqui presented a mathematical analysis of the concept of pragmatic truth based on the notion of partial structures. This analysis gave rise to an impressive number of papers in which the concept was further elaborated and applied to important problems in a variety of domains, such as logic (see, for instance, [2]), the foundations of the theory of probability ([3]), philosophy of science (see, for instance, [4], [5], [6], and [7]), and the philosophy of mathematics (see [8]).

Intuitively, a partial structure is a relational structure that models one's knowledge about a certain domain. As this knowledge may be (and usually is) incomplete, the relations that make up such a structure are partial, in the sense that they are not necessarily defined for every element in the domain. Formally, a partial structure

*Research for this paper was carried out during a stay in the *Division of History and Philosophy of Science* at the University of Leeds which was financed by the Fund for Scientific Research – Flanders. I am greatly indebted to Steven French and Otávio Bueno for the many discussions on the topic, to Diderik Batens for his comments on an earlier version of this paper, and to the referees for several corrections and clarifications.

$\mathcal{A}$ is a set-theoretical structure of the form $\langle A, R_i, \mathcal{P}\rangle_{i\in I}$, where A is a non-empty set (the universe of the structure), R_i a partial relation defined on A, and $\mathcal{P}$ the set of ***true primary sentences***. Given a partial relation R of arity n and an n-tuple $\langle \alpha_1, \ldots, \alpha_n\rangle \in A^n$, there are three possibilities: $\langle \alpha_1, \ldots, \alpha_n\rangle$ is in the relation R, it is not in R, or it is undetermined whether it is or not. The set $\mathcal{P}$ consists of the sentences of $\mathcal{L}$ (the language that is used to talk about $\mathcal{A}$) that are accepted as true. A structure $\mathcal{B} = \langle B, R'_i, \mathcal{P}\rangle_{i\in I}$ is a ***total extension*** of a partial structure $\mathcal{A} = \langle A, R_i, \mathcal{P}\rangle_{i\in I}$ if and only if (i) $\mathcal{B}$ is a total structure (all R'_i of arity n are defined for all n-tuples of B), (ii) $A = B$, (iii) $R_i \subseteq R'_i$ for every $i \in I$, (iv) every individual constant of $\mathcal{L}$ is interpreted by the same element in both $\mathcal{A}$ and $\mathcal{B}$, and (v) for every $\phi \in \mathcal{P}$, $\mathcal{B} \models \phi$ (where $\mathcal{B} \models \phi$ means that ϕ is true in $\mathcal{B}$ in the usual Tarskian sense). A sentence ϕ is said to be ***pragmatically true*** in a partial structure $\mathcal{A}$ if and only if there exists a total extension $\mathcal{B}$ of $\mathcal{A}$ such that $\mathcal{B} \models \phi$.[1]

In [9], da Costa, Bueno and French presented the logic **PT** to study the notion of pragmatic truth. Their starting point is that the total extensions of a partial structure $\mathcal{A} = \langle A, R_i, \mathcal{P}\rangle_{i\in I}$ can be considered as worlds of a Kripke structure for **S5**. In line with this, $\mathcal{L}$ is extended to the standard modal language $\mathcal{L}^M$; $\Box A$ is interpreted as "A is pragmatically valid" and $\Diamond A$ as "A is pragmatically true". Where $\mathcal{A}$ is a partial structure in which $\mathcal{L}^M$ is interpreted, f a (total) function from the set of individual variables of $\mathcal{L}^M$ to the universe of $\mathcal{A}$, $\mathcal{B}$ a total extension of $\mathcal{A}$, and where **PV** stands for **S5** with quantification and necessary equality, we have:

> A is ***pragmatically valid*** in some partial structure $\mathcal{A}$ if and only if $\langle \mathcal{A}, f, \mathcal{B}\rangle \models_{\mathbf{PV}} \Box A$ for all total extensions $\mathcal{B}$ of $\mathcal{A}$.

and

> A is ***pragmatically true*** in some partial structure $\mathcal{A}$ if and only if $\langle \mathcal{A}, f, \mathcal{B}\rangle \models_{\mathbf{PV}} \Diamond A$ for all total extensions $\mathcal{B}$ of $\mathcal{A}$.

As is clear from these definitions, A is pragmatically true in a partial structure $\mathcal{A}$ if and only if $\langle \mathcal{A}, f, \mathcal{B}\rangle \not\models_{\mathbf{PV}} \Box{\sim}A$ for all total extensions $\mathcal{B}$ of $\mathcal{A}$. In other words, A is pragmatically true in $\mathcal{A}$ if and only if it is ***compatible*** with all sentences that are pragmatically valid in $\mathcal{A}$.[2]

PT is a discussive logic defined from **PV** in the following way:

> $\vdash_{\mathbf{PT}} A$ if and only if $\vdash_{\mathbf{PV}} \Diamond A$

and

> $\Gamma \vdash_{\mathbf{PT}} A$ if there are $B_1, \ldots, B_n \in \Gamma$, such that $\vdash_{\mathbf{PT}} (\Diamond B_1 \wedge \ldots \wedge \Diamond B_n) \supset \Diamond A$ or, equivalently, $\vdash_{\mathbf{PV}} \Diamond((\Diamond B_1 \wedge \ldots \wedge \Diamond B_n) \supset \Diamond A)$.

---

[1] I follow here the definition of "partial structure" as presented in most papers on the matter—see, for instance, [9]. In [1], $\mathcal{P}$ is not considered as an element of the partial structure itself. In line with this, a distinction is made between "total extensions" and "total $\mathcal{P}$-extensions". The former satisfy (i)–(iv); the latter moreover satisfy (v). The notion of "pragmatically true in a partial structure (relative to $\mathcal{P}$)" is then defined with respect to total $\mathcal{P}$-extensions.

[2] I say that A is compatible with some Γ if and only if ${\sim}A$ is not derivable from Γ by Classical Logic. From a semantic point of view, this means that A is true in some classical model of Γ. Hence, in terms of **PV**, A is compatible with Γ if and only if $\{\Box A \mid A \in \Gamma\} \not\models_{\mathbf{PV}} \Box{\sim}A$.

In view of these definitions, a *pragmatic theory* for a set of beliefs Γ may be defined as $Cn_{\mathbf{PT}}(\Gamma)$—compare [9, p. 612].[3]

As $\Diamond((\Diamond B \wedge \Diamond{\sim}B) \supset \Diamond A)$ is not a theorem of **PV**, **PT** is clearly paraconsistent ($B, {\sim}B \nvdash_{\mathbf{PT}} A$). For similar reasons, **PT** does not allow for the derivation of contradictions ($A, {\sim}A \nvdash_{\mathbf{PT}} A \wedge {\sim}A$). Both characteristics are in line with the partial structures approach. A may be true in some total extension of a partial structure $\mathcal{A}$, whereas ${\sim}A$ is true in another total extension of $\mathcal{A}$. Hence, both A and ${\sim}A$ may be pragmatically true in $\mathcal{A}$ without everything being pragmatically true in it. And, as $A \wedge {\sim}A$ is false in all total extensions of $\mathcal{A}$, $A \wedge {\sim}A$ cannot be pragmatically true.

As a logic of pragmatic truth, **PT** also exhibits two shortcomings. First, for sentences of $\mathcal{L}$, **PT** invalidates all (genuine) multiple-premise rules (Adjunction, Modus Ponens, Modus Tollens, Disjunctive Syllogism, ...). This seems highly undesirable. If A and B behave consistently on Γ, then they will come out pragmatically valid in the partial structures that adequately model Γ, and hence, also $A \wedge B$ will come out pragmatically valid in them. Next, **PT** does not enable one to derive all sentences that are *compatible* with the 'consistent core' of Γ.[4] Also this seems undesirable. Suppose, for instance, that $\Gamma = \{p \vee q, r, {\sim}r\}$. In that case, both p and q are pragmatically true in the partial structures for Γ, but neither of these is derivable from Γ by **PT**.

It is the aim of this paper to present a logic for pragmatic truth, **APT**, that solves both shortcomings. Like **PT**, **APT** enables one to define a pragmatic theory for a possibly inconsistent set of beliefs Γ. However, it does so in a way that is better in line with the partial structures approach.

APT has the peculiar property that it localizes the consistent core of a possibly inconsistent set Γ.[5] For this core, **APT** is just as rich as Classical Logic (henceforth **CL**). However, for sentences that do not belong to the consistent core, **APT** behaves paraconsistently and moreover does not allow for the derivation of contradictions. For example, although $q \wedge r$ is an **APT**-consequence of $\Gamma = \{p \wedge q, {\sim}p, r, t \vee s\}$, $p \wedge {\sim}p$ is not. In addition to this, **APT** delivers all sentences that are compatible with the consistent core of Γ. Thus, in our example, s as well as ${\sim}s$ are **APT**-consequences of Γ.

This last property has an important consequence, namely that **APT**, unlike **PT**, is non-monotonic. For example, as both r and ${\sim}r$ are compatible with $\{r \vee s\}$, both formulas are **APT**-consequence of $\Gamma = \{p, {\sim}p, r \vee s\}$. However, as ${\sim}r$ is not compatible with $\{r \vee s, {\sim}s\}$, ${\sim}r$ is not an **APT**-consequence of $\Gamma \cup \{{\sim}s\}$.

APT will be defined as a discussive logic from the modal logic **APV**. $\Gamma^{\Diamond}$ will denote $\{\Diamond A \mid A \in \Gamma\}$.

The techniques that led to **APV** derive from the adaptive logic programme—see, for instance, [10], [11], [12] and [13]. Putting it very generally, adaptive logics inter-

---

[3]In [9], da Costa, Bueno and French present an axiomatization for **PT** that is based on the discussive implication "$\supset_d$" and the discussive conjunction "$\wedge_d$" ($A \supset_d B =_{df} \Diamond A \supset B$; $A \wedge_d B =_{df} \Diamond A \wedge B$). As these connectives are definable within **PT**, and as they are not needed in the present paper, I shall not discuss them.

[4]Let $\Gamma^{\dagger}$ be the set of all formulas of the form $A_1 \vee \ldots \vee A_n$ such that every A_i is a primitive formula and $\Gamma \vdash_{\mathbf{PT}} A_1 \vee \ldots \vee A_n$. I shall say that a formula B belongs to the consistent core of Γ if and only if B follows by Classical Logic from every maximally consistent subset of $\Gamma^{\dagger}$.

[5]By this, I mean that **APT** enables one to determine whether or not some sentence belongs to the consistent core of Γ.

pret 'abnormal theories' (see below) 'as normally as possible'. In an inconsistency-adaptive logic, for instance, an abnormality is an inconsistency. In interpreting a set Γ, the logic presupposes that a sentence behaves normally (that is, consistently) with respect to Γ unless and until proved otherwise.

As we shall see below, the present enterprise requires a logic that 'adapts' itself to two different kinds of abnormalities. On the one hand, it should presuppose that a consequence A is pragmatically ***valid***—that $\Box A$ is derivable from $\Gamma^{\Diamond}$—unless and until proved otherwise. On the other hand, it should presuppose that A is pragmatically ***true***—that $\Diamond A$ is derivable from $\Gamma^{\Diamond}$—unless and until $\Box{\sim}A$ is derived from it. The latter comes to the same as presupposing (until proved otherwise) that A is compatible with the consequences of $\Gamma^{\Diamond}$ that are pragmatically valid. In order to meet these requirements, **APV** will be constructed in terms of two other adaptive logics, **AJ** and **COM**. Whereas **AJ** localizes the consequences of $\Gamma^{\Diamond}$ that should come out pragmatically valid, **COM** delivers all consequences that are compatible with these.[6] Both **AJ** and **COM** are based on **S5**.

I shall proceed as follows. In Section 2, I explain the basic ideas and techniques of adaptive logics. A first intuitive characterization of **APV** and **APT** is presented in section 3. In Section 4, I briefly discuss the semantics of **S5** that I shall rely on. In the following two sections, I discuss the adaptive logics **AJ** and **COM**. How the two logics should be combined to obtain **APV** is discussed in Section 7. In Section 8, I present the discussive adaptive system **APT**. The paper ends with some conclusions and open problems (Section 9).

For reasons of space, the presentation of all systems will be restricted to the propositional level.[7]

2 Some Basics of Adaptive Logics

Adaptive logics were designed to deal with reasoning patterns that are non-monotonic and/or dynamic. A reasoning pattern is called dynamic if the mere analysis of the premises may lead to the withdrawal of previously derived conclusions.

One of the main characteristics of such reasoning processes is that a specified set of logical presuppositions is followed 'as much as possible', that is, *unless and until* they are explicitly violated. When modelling one's knowledge about a certain domain, for instance, one of the presuppositions may be that whatever is not known to be false is pragmatically true. Whenever this presupposition is violated—for instance, when one's knowledge is extended—conclusions that were previously derived may be withdrawn. Thus, in our example, sentences that were at some point considered as pragmatically true may later be rejected.

What makes adaptive logics special is that they 'adapt' themselves to specific violations of presuppositions. Where a presupposition is violated, the rules of in-

[6] Both systems are interesting in themselves. As is shown in [14], **AJ** can be used to define an adaptive logic for Jaśkowski's **D2** as presented in [15]. This adaptive logic is non-adjunctive where inconsistencies are involved, but nevertheless allows for the application of Adjunction to all 'consistent parts' of the premises. **COM** is the adaptive logic for (classical) compatibility that was first presented in [16].

[7] The full version of **COM** can be found in [16], that of **AJ** in [14]. In view of these results, upgrading **APV** to the predicative level is rather straightforward.

ference are restricted in order to avoid triviality.[8] However, where this is not the case, the rules can be applied in their full strength. What this comes to is that adaptive logics do not invalidate a set of rules of inference, but invalidate specific *applications* of rules of inference.[9]

All currently available adaptive logics are defined in terms of three elements: an upper limit logic, a lower limit logic, and an 'adaptive strategy'. The upper limit logic is an extension of the lower limit logic. The former thus introduces a set of presuppositions on top of those of the latter. These additional presuppositions are the ones that are defeasible: they are followed 'as much as possible', but are abandoned when necessary to avoid triviality. The third element, the adaptive strategy, determines the interpretation of the ambiguous phrase "as much as possible".

When a set of premises violates one of the presuppositions of the upper limit logic, it will be said to behave abnormally with respect to the upper limit logic. It is important to note that "abnormality" does not refer to the purported standard of reasoning, say **CL**. It refers to properties of the application context—to presuppositions that are considered desirable, but that may be overruled.

Semantically, an adaptive logic is obtained by selecting a subset of the models of the lower limit logic. The selection is determined by the adaptive strategy. The Minimal Abnormality Strategy, for instance, selects those models that are minimally abnormal (in a set-theoretical sense) with respect to the upper limit logic. If some theory Γ behaves normally with respect to the upper limit logic, the adaptive models of Γ coincide with the models of the upper limit logic that validate Γ. Thus, an inconsistency-adaptive logic with **CL** as its upper limit logic leads to exactly the same consequence set as **CL** for every consistent Γ.

Syntactically, an adaptive logic is obtained by taking the rules of the lower limit logic as unconditional (as unconditionally valid), and the rules of the upper limit logic as conditional. The adaptive strategy determines a marking rule (see below). The proof theory of adaptive logics is *dynamic* in a strong sense. Sentences derived conditionally at some stage in a proof may at a later stage be withdrawn. The mechanism by which this is realized is quite simple. If a formula is added by the application of a conditional rule, a 'condition' (set of formulas) that is specified by the rule, is written to the right of the line. The members of this set have to behave normally for the formula to be derivable. At each stage of the proof—with each formula added—the marking rule is invoked: for each line that has a condition attached to it, it is checked whether the condition is fulfilled or not. If it is not, the line is marked and considered as not (any more) belonging to the proof. At a still later stage, a line may be unmarked again. The formulas derived at a stage are those that, at that stage, occur on lines that are not marked.

---

[8]It is typical of *monotonic* logics that the violation of presuppositions leads to triviality. **CL**, for instance, presupposes consistency as well as negation-completeness, and turns any theory that violates these presuppositions into the trivial one.

[9]Adaptive logics thus differ in important respects from non-classical logics that are obtained by dropping some **CL**-presuppositions and by restricting the *rules of inference* accordingly. As is argued in [17] and [18], such logics are inadequate to deal with dynamic reasoning processes. Not only do they fail to capture the dynamics involved, they are usually too poor to make sense of actual reasoning processes.

3 Intuitive Characterization of APT and APV

As I explained in Section 1, **APT** should enable one to define a pragmatic theory from a possibly inconsistent set of beliefs Γ. By a theory, I mean a couple $\langle \Gamma, \mathbf{L} \rangle$, where Γ is a set of non-logical axioms, and **L** is a logic. I say that A is a theorem of some theory $\langle \Gamma, \mathbf{L} \rangle$ if and only if A is derivable from Γ by **L**. A central requirement for **APT** is that A is a theorem of $\langle \Gamma, \mathbf{L} \rangle$ if and only if A is pragmatically true in the partial structures that adequately model Γ.[10]

As I explained in Section 1, this requirement will be met by defining **APT** from a modal logic **APV** that is itself based on two adaptive logics (**AJ** and **COM**). As **APV** is meant to serve as the basis of a discussive logic, it is only applied to sets of premises of the form $\Gamma^{\diamond}$. In order to explain how **APV** is obtained from **AJ** and **COM**, I first have to discuss the basic intuitions behind these two systems.

The idea behind **AJ** is to presuppose that some sentence is pragmatically valid—that $\Box A$ is derivable from $\Diamond A$—unless and until proved otherwise. Let us call this the ***pragmatic validity presupposition***. Where this presupposition is violated, **AJ** behaves like **S5**. So, the lower limit logic of **AJ** is **S5**. Its upper limit logic is the Trivial system **Triv**—compare p. 65 of [19]. Syntactically, **Triv** can be obtained by extending **S5** with $\vdash \Diamond A \supset \Box A$. Of course, **Triv** is nothing but **CL** (in that $\Box A$, $\Diamond A$, A are logically equivalent to one another). As we shall see below, this has the advantage that, if Γ is consistent, all **CL**-consequences are derivable from Γ by **APT**.

To semantically characterize **AJ**, we need a criterion for selecting a set of the **S5**-models of $\Gamma^{\diamond}$. In order to do so, we first need to specify which formulas are abnormal with respect to the upper limit logic. Next, we have to define the 'abnormal part' of a model: the set of abnormalities that are verified by the model. Finally, we have to choose an adaptive strategy to interpret the ambiguous phrase "as normally as possible".

In the case of **AJ**, the formulas that behave abnormally with respect to the upper limit logic are those that violate the pragmatic validity presupposition. As a first approximation, we may say that a formula A violates this presupposition if and only if $\Diamond A \wedge \Diamond{\sim}A$ is **S5**-derivable from the premises. Suppose, however, that $\Gamma^{\diamond} = \{\Diamond(p \vee q), \Diamond{\sim}p, \Diamond{\sim}q\}$. In that case, neither $\Diamond p \wedge \Diamond{\sim}p$ nor $\Diamond q \wedge \Diamond{\sim}q$ is **S5**-derivable from $\Gamma^{\diamond}$, but $(\Diamond p \wedge \Diamond{\sim}p) \vee (\Diamond q \wedge \Diamond{\sim}q)$ is. What this comes to is that the abnormal behaviour of p is *connected* to that of q: it can be inferred that at least one of them behaves abnormally with respect to $\Gamma^{\diamond}$. This suggests that a formula A violates the pragmatic validity presupposition if and only if $\Diamond A \wedge \Diamond{\sim}A$ is a disjunct of a 'minimal' disjunction of abnormalities that is **S5**-derivable from $\Gamma^{\diamond}$. A disjunction of abnormalities that is **S5**-derivable from $\Gamma^{\diamond}$ will be called a "*Dab*-consequence"; a *Dab*-consequence will be called "minimal" if and only if no result of dropping some disjunct from it is an **S5**-consequence of $\Gamma^{\diamond}$. For reasons that are explained in [14], the abnormalities in **AJ** have to be restricted to ***primitive*** formulas. The abnormal part of an **S5**-model $\mathcal{M}$ is the set of primitive formulas A, such that $\Diamond A \wedge \Diamond{\sim}A$ is verified by $\mathcal{M}$.

Having defined the formulas that behave abnormally with respect to the upper

[10] I shall assume that a partial structure $\mathcal{A}$ adequately models a set of beliefs Γ about some domain D if and only if all sentences about D that are considered as true and that behave consistently on Γ come out pragmatically valid in $\mathcal{A}$.

limit logic, and the abnormal part of a model, we should now turn to the choice of the adaptive strategy. As the abnormalities in **AJ** may be connected to each other, we have to choose between the Minimal Abnormality Strategy and the Reliability Strategy.[11] A model M is minimally abnormal if and only if there is no model M' such that the abnormal part of M' is a proper subset of the abnormal part of M; a model M is reliable if and only if its abnormal part contains only formulas that behave abnormally with respect to the premises. **AJ** is based on the Reliability Strategy. This choice is motivated by the fact that, within the partial structures approach, it is not demanded that the partial structures for a set of beliefs Γ are 'minimally abnormal'. For instance, if $\Gamma = \{p \vee q, \sim p, \sim q\}$, then there is no partial structure for Γ such that both p and q behave normally in it (that is, come out pragmatically valid). However, it is not demanded that in all of them either p or q should behave normally.[12]

Let us now turn to **COM**. For reasons that will immediately become clear, **COM** is only applied to sets of premises of the form $\Gamma^{\Box} = \{\Box A \mid A \in \Gamma\}$. The basic idea behind **COM** is to presuppose that some sentence is compatible with a (consistent) set of sentences Γ—that $\Diamond A$ is derivable from $\Gamma^{\Box}$— unless and until $\Box\sim A$ is derivable from it. In terms of pragmatic truth, **COM** presupposes that some sentence A is pragmatically true unless and until $\sim A$ is proved to be pragmatically valid. Let us call this the ***pragmatic truth presupposition.***

The lower limit logic of **COM** is **S5**. So, for sentences that violate the pragmatic truth presupposition (those of the form $\Box\sim A$), **COM** behaves like **S5**. The upper limit logic of **COM** is called $\mathbf{S5^P}$. Syntactically, it is obtained by extending **S5** with the rule "If $\nvdash_{\mathbf{S5}} \sim A$, then $\vdash_{\mathbf{S5^P}} \Diamond A$".[13]

The criterion for selecting the **COM**-models from $\Gamma^{\Box}$ is easily defined. A formula A behaves **COM**-abnormally with respect to $\Gamma^{\Box}$ if and only if $\Box\sim A$ follows from it by **S5**. The **COM**-abnormal part of an **S5**-model $\mathcal{M}$ consists of those formulas A such that $v_{\mathcal{M}}(\Box\sim A) = 1$. As the **COM**-abnormal behaviour of one formula is never connected to the **COM**-abnormal behaviour of another formula,[14] **COM** is based on the so-called Simple Strategy. According to this strategy, a model is selected if and only if its abnormal part coincides with the formulas that behave abnormally with respect to the premises. Thus, an **S5**-model is not a **COM**-model of $\Gamma^{\Box}$ if and only if it verifies $\Box\sim A$ for some A that behaves **COM**-normally with respect to $\Gamma^{\Box}$.

As **COM** is applied to sets of premises of the form $\Gamma^{\Box}$, it presupposes that Γ is consistent. Hence, in order to obtain **APV** from **AJ** and **COM**, we have to proceed in two steps. First, we shall use **AJ** to define the set of consequences of $\Gamma^{\Diamond}$ that are pragmatically valid—the latter is consistent even if Γ is inconsistent. Next, we shall apply **COM** to this set to obtain all compatible consequences.

---

[11] For a discussion of the different adaptive strategies, see [12].

[12] The Reliability Strategy proceeds a bit more cautiously than the Minimal Abnormality. For instance, where $\Gamma^{\Diamond} = \{\Diamond(p \vee q), \Diamond\sim p, \Diamond\sim q, \Diamond(p \vee r), \Diamond(q \vee r)\}$, $\Box r$ is verified by all minimally abnormal models of $\Gamma^{\Diamond}$, but is falsified by some reliable models.

[13] Whenever Γ contains well-formed formulas that are not **CL**-theorems, $Cn_{\mathbf{S5^P}}(\Gamma^{\Box})$ is trivial. This is as expected. Just as **CL** presupposes consistency and turns inconsistent theories into trivial ones, $\mathbf{S5^P}$ presupposes that any well-formed formula that is not logically true is possibly false, and hence turns $\Gamma^{\Box}$ into triviality whenever Γ contains a well-formed formula that is not a **CL**-theorem. As is explained in [16], $\mathbf{S5^P}$ is not closed under Uniform Substitution.

[14] As is proved in [16], $\Gamma^{\Box} \models_{\mathbf{S5}} \Box\sim A_1 \vee \ldots \vee \Box\sim A_n$ if and only if $\Gamma^{\Box} \models_{\mathbf{S5}} \Box\sim A_1$ or ... or $\Gamma^{\Box} \models_{\mathbf{S5}} \Box\sim A_n$.

4 The Lower and Upper Limit Logics

Both **AJ** and **COM** have **S5** as their lower limit logic. In view of the specific purpose of this paper, I shall rely on a rather unusual, but nevertheless very natural semantics for **S5** that was first presented in [16].

Let $\mathcal{L}$ be the standard language of **CL**, and $\mathcal{L}^M$ the standard modal language. Let $\mathcal{S}$ be the set of sentential letters, $\mathcal{W}$ the set of well-formed formulas of $\mathcal{L}$, and $\mathcal{W}^a$ the set of atoms of $\mathcal{L}$ (sentential letters and their negations). An **S5**-model is a couple $\mathcal{M} = \langle \Sigma_\Delta, M_0 \rangle$, where $\Delta \subset \mathcal{W}$, Σ_Δ is the set of **CL**-models of Δ, and $M_0 \in \Sigma_\Delta$.

The valuation determined by an **S5**-model $\mathcal{M}$ is defined by the following clauses:

C1 where $A \in \mathcal{S}$, $v_\mathcal{M}(A, M_i) = v_{M_i}(A)$
C2 $v_\mathcal{M}(\sim A, M_i) = 1$ if and only if $v_\mathcal{M}(A, M_i) = 0$
C3 $v_\mathcal{M}(A \wedge B, M_i) = 1$ if and only if $v_\mathcal{M}(A, M_i) = v_\mathcal{M}(B, M_i) = 1$
C4 $v_\mathcal{M}(\Box A, M_i) = 1$ if and only if $v_\mathcal{M}(A, M_j) = 1$ for all $M_j \in \Sigma_\Delta$.

The other logical constants are defined as usual.[15] Semantic consequence and validity are as usual (in terms of truth in M_0). Henceforth, "**S5**-model" will always refer to a model as defined here.

The above semantics is rather peculiar. Each of the **S5**-models corresponds to a standard worlds-model, but not *vice versa*. Nevertheless, it is proved in [16] that the semantics is adequate with respect to **S5**:

THEOREM 1 $\Gamma \models_{\mathbf{S5}} A$ if and only if $\Gamma \vdash_{\mathbf{S5}} A$.

The following theorems illustrate some interesting properties of the **S5**-semantics. The proof of the first is obvious in view of the definition of an **S5**-model.

THEOREM 2 $\mathcal{M} = \langle \Sigma_\Delta, M_0 \rangle$ is an **S5**-model of $\Gamma^\Box$ if and only if there is a $\Theta \subset \mathcal{W}^a$ and $\Sigma_\Delta = \Sigma_\Theta$.

THEOREM 3 If $\mathcal{M} = \langle \Sigma_\Delta, M_0 \rangle$ is an **S5**-model of $\Gamma^\Box$, then $\Sigma_\Delta \subseteq \Sigma_\Gamma$.

Proof: Suppose that $\Sigma_\Delta \not\subseteq \Sigma_\Gamma$. It follows that some $M \in \Sigma_\Delta$ does not verify Γ. But then $\mathcal{M} = \langle \Sigma_\Delta, M_0 \rangle$ is not an **S5**-model of $\Gamma^\Box$. □

THEOREM 4 If an **S5**-model $\mathcal{M} = \langle \Sigma_\Delta, M_0 \rangle$ does not verify $\Diamond A$, then $\Delta^\Box \models_{\mathbf{S5}} \Box{\sim}A$.

Proof: Suppose that $\mathcal{M} = \langle \Sigma_\Delta, M_0 \rangle$ does not verify $\Diamond A$. In that case, for every $M \in \Sigma_\Delta$, $v_M(A) = 0$, and hence, $v_M(\sim A) = 1$. But then, $\langle \Sigma_\Delta, M_0 \rangle$ verifies $\Box{\sim}A$. Hence, for every **S5**-model $\langle \Sigma_\Theta, M_0 \rangle$ of $\Delta^\Box$, $\Sigma_\Theta \subseteq \Sigma_\Delta$ (in view of Theorem 3). But then, all **S5**-models of $\Delta^\Box$ verify $\Box{\sim}A$. □

The above **S5**-semantics is easily adjusted to characterize the semantics of the upper limit logics. The $\mathbf{S5^P}$-models are the **S5**-models $\mathcal{M} = \langle \Sigma_\Delta, M_0 \rangle$ where Δ is empty—remark that $\Sigma_\emptyset$ contains *all* **CL**-models. The **Triv**-models are the **S5**-models $\mathcal{M} = \langle \Sigma_\Delta, M_0 \rangle$, such that $\Sigma_\Delta = \Sigma_\Theta$, and Θ is a maximally consistent subset of $\mathcal{W}^a$.

[15] For the extension to the predicative level, see [16]

DEFINITION 5 An **S5**-model $\mathcal{M}$ is $\Box$-complete if and only if $v_{\mathcal{M}}(\Diamond A \wedge \Diamond{\sim}A) = 0$, for all $A \in \mathcal{W}$.

The easy proofs of the following theorems are left to the reader:

THEOREM 6 An **S5**-model is $\Box$-complete if and only if it is a **Triv**-model.

THEOREM 7 $\Gamma^{\Diamond}$ has $\Box$-complete **S5**-models if and only if Γ is consistent.

5 The Adaptive Logic AJ

Semantically, **AJ** is obtained from the **S5**-models of $\Gamma^{\Diamond}$ by the Reliability Strategy. The idea is that any $\Gamma^{\Diamond}$ defines a set of unreliable formulas, and that the **AJ**-models of $\Gamma^{\Diamond}$ are those in which only unreliable formulas behave abnormally. If no formula is unreliable with respect to $\Gamma^{\Diamond}$, the **AJ**-models of $\Gamma^{\Diamond}$ are its **Triv**-models.

Let $Dab(A_1, \ldots, A_n)$ refer to $(\Diamond A_1 \wedge \Diamond{\sim}A_1) \vee \ldots \vee (\Diamond A_n \wedge \Diamond{\sim}A_n)$ *provided each* $A_i \in \mathcal{S}$. $A_1, \ldots, A_n$ will be called the *factors* of $Dab(A_1, \ldots, A_n)$. As permutations of the factors result in equivalent formulas, I shall use sets to refer to any of these permutations. $Dab(A_1, \ldots, A_n)$ is a *minimal Dab-consequence* of $\Gamma^{\Diamond}$ if and only if $Dab(A_1, \ldots, A_n)$ is an **S5**-consequence of $\Gamma^{\Diamond}$ and $Dab(\Delta)$ is not an **S5**-consequence of $\Gamma^{\Diamond}$, for any $\Delta \subset \{A_1, \ldots, A_n\}$.

I first define the **AJ**-abnormal part of a model:

DEFINITION 8 $Ab_{\mathbf{AJ}}(\mathcal{M}) =_{df} \{A \mid A \in \mathcal{S}; v_{\mathcal{M}}(\Diamond A \wedge \Diamond{\sim}A) = 1\}$.

The set of well-formed formulas that are unreliable with respect to $\Gamma^{\Diamond}$ is defined as:

DEFINITION 9 $U(\Gamma^{\Diamond}) = \{A \mid A$ is a factor of a minimal DAB-consequence of $\Gamma^{\Diamond}\}$.

DEFINITION 10 An **S5**-model $\mathcal{M}$ of Γ is reliable if and only if $Ab_{\mathbf{AJ}}(\mathcal{M}) \subseteq U(\Gamma^{\Diamond})$.

DEFINITION 11 $\Gamma^{\Diamond} \models_{\mathbf{AJ}} A$ if and only if all reliable models of $\Gamma^{\Diamond}$ verify A.

Let $sent(A)$ refer to the members of $\mathcal{S}$ that occur in A.

LEMMA 12 For any **S5**-model $\mathcal{M}$, if there is no B, such that $B \in sent(A)$ and $v_{\mathcal{M}}(\Diamond B \wedge \Diamond{\sim}B) = 1$, then there is a $\Box$-complete **S5**-model $\mathcal{M}'$, such that $v_{\mathcal{M}}(A) = v_{\mathcal{M}'}(A)$.

Proof: Suppose that the antecedent is true for some $\mathcal{M}$. Let $\mathcal{M}'$ be $\langle \Sigma_{\Delta}, M_0 \rangle$, where Δ is a maximally consistent subset of $\mathcal{W}^a$, and $\Delta \models_{\mathbf{CL}} A$ if and only if $v_{\mathcal{M}}(A) = 1$. It is easily checked that $\mathcal{M}'$ is an **S5**-model, that it is $\Box$-complete, and that $v_{\mathcal{M}}(A) = v_{\mathcal{M}'}(A)$. $\Box$

The following theorem and corollary are crucial for the proof theory of **AJ**:

THEOREM 13 If $\models_{\mathbf{Triv}} A$, then $\models_{\mathbf{S5}} Dab(sent(A)) \vee A$.

Proof: If $\models_{\mathbf{S5}} A$, the theorem obviously holds. So, suppose that $\models_{\mathbf{Triv}} A$ and that $\not\models_{\mathbf{S5}} A$. In view of Theorem 6, $v_{\mathcal{M}}(A) = 1$ in all $\Box$-complete **S5**-models. As $sent(A)$ is finite, $Dab(sent(A)) \vee A \in \mathcal{W}$. To prove that $Dab(sent(A)) \vee A$ is **S5**-valid, we have to consider two cases. If $v_{\mathcal{M}}(\Diamond C \wedge \Diamond{\sim}C) = 1$, for some $C \in sent(A)$, then $v_{\mathcal{M}}(Dab(sent(A)) = 1$. If $v_{\mathcal{M}}(\Diamond C \wedge \Diamond{\sim}C) = 0$, for all $C \in sent(A)$, then $v_{\mathcal{M}}(A) = 1$ in view of Lemma 12. □

COROLLARY 14 For every $A \in \mathcal{W}$, $\models_{\mathbf{S5}} Dab(sent(A)) \vee (\Diamond A \supset \Box A)$.

The following **S5**-valid formulas illustrate Theorem 13:

(1) $Dab(p,q) \vee (\Diamond(p \vee q) \supset \Box(p \vee q))$.
(2) $Dab(p,q) \vee (\Diamond(p \supset (q \vee {\sim}r)) \supset \Box(p \supset (q \vee {\sim}r)))$.
(3) $Dab(p,q,r) \vee ((\Diamond(p \vee r) \wedge \Diamond(q \vee {\sim}r)) \supset \Box(p \vee q))$.

In [20], Priest discusses an important property for adaptive logics which he calls Reassurance. Where **AL** is an adaptive logic and **L** is its lower limit logic, Reassurance holds in **AL** if and only if **AL** does not lead to triviality for any set of premises that has **L**-models. In [21], Batens discusses the related property Strong Reassurance: for every **L**-model M, there is a **AL**-model M' such that $Ab(M') \subseteq Ab(M)$. As is proved in [14], Strong Reassurance, and hence Reassurance, hold for **AJ**:[16]

THEOREM 15 If $\mathcal{M}$ is an **S5**-model of $\Gamma^{\Diamond}$ but not an **AJ**-model of $\Gamma^{\Diamond}$, then there is an **AJ**-model $\mathcal{M}'$ of $\Gamma^{\Diamond}$ such that $Ab(\mathcal{M}') \subset Ab(\mathcal{M})$. (Strong Reassurance)

COROLLARY 16 If $\Gamma^{\Diamond}$ has **S5**-models, it also has **AJ**-models. (Reassurance)

The proof theory for **AJ** is as for other adaptive logics. Lines of a proof have five elements: (i) a line number, (ii) the formula A that is derived, (iii) the line numbers of the formulas from which A is derived, (iv) the rule by which A is derived, and (v) the set of formulas that should behave normally in order for A to be so derivable. In line with the generic proof format for adaptive logics from [22], the proof theory is defined in terms of an unconditional rule RU, a conditional rule RC-J, and a marking rule MARK-J.[17] The unconditional rule is determined by the lower limit logic, the conditional one by the upper limit logic, and the marking rule by the strategy. A well-formed formula is said to be derived unconditionally if and only if it is derived on a line the fifth element of which is empty.

$Dab(A_1, \ldots, A_n)$ is a minimal Dab-consequence of Γ at stage s if and only if it is derived unconditionally in the proof at that stage and $Dab(\Delta)$ is not derived unconditionally at that stage for any $\Delta \subset \{A_1, \ldots, A_n\}$.

DEFINITION 17 $U_s(\Gamma) = \{A \mid A$ is a factor of a minimal Dab-consequence of Γ at stage $s\}$.

---

[16] The proof is analogous to, but much simpler than, the proof for Theorem 1 in [21].

[17] The systems **COM** and **APV** will be defined in terms of the same kinds of generic rules. The unconditional rule will be the same for all three systems, but the conditional rule and the marking rule will be different. I shall use the label "J" to refer to the rules that are specific for **AJ**, and the label "C" for those that are specific for **COM**. No labels will be used in the case of **APV**.

PREM At any stage of a proof one may add a line consisting of (i) an appropriate line number, (ii) a premise, (iii) a dash, (iv) "PREM", and (v) "$\emptyset$".

RU If $A_1, \ldots, A_n \vdash_{\mathbf{S5}} B$, and $A_1, \ldots, A_n$ occur in the proof on the conditions $\Delta_1, \ldots, \Delta_n$ respectively, then one may add to the proof a line that has B as its second element and $\Delta_1 \cup \ldots \cup \Delta_n$ as its fifth element.

RC-J If $A_1, \ldots, A_n \vdash_{\mathbf{S5}} B \vee Dab(C_1, \ldots, C_m)$, and $A_1, \ldots, A_n$ occur in the proof on the conditions $\Delta_1, \ldots, \Delta_n$ respectively, then one may add to the proof a line with an appropriate line number, B as its second element, and $\{C_1, \ldots, C_m\} \cup \Delta_1 \cup \ldots \cup \Delta_n$ as its fifth element.

MARK-J A line is marked at a stage iff, where Δ is its fifth element, $B \notin U_s(\Gamma)$, for any $B \in \Delta$.

DEFINITION 18 A is finally derived in an **AJ**-proof from $\Gamma^\diamond$ if and only if A is derived on a line that is not marked and, any extension of the proof in which A is marked, may be further extended in such a way that A becomes unmarked.

DEFINITION 19 $\Gamma^\diamond \vdash_{\mathbf{AJ}} A$ (A is finally derivable from $\Gamma^\diamond$) if and only if A is finally derived in an **AJ**-proof from $\Gamma^\diamond$.

I shall now show that the semantics of **AJ** is adequate with respect to final derivability, for all formulas of the form $\Box A$. Whether Adequacy obtains in general ($\Gamma \models_{\mathbf{AJ}} A$ if and only if $\Gamma \vdash_{\mathbf{AJ}} A$) is immaterial for the present paper. All we are interested in here is the set $\{\Box A \mid \Gamma^\diamond \models_{\mathbf{AJ}} \Box A\}$—this is the set to which **COM** will be applied in order to obtain **APV**.

LEMMA 20 If, in an **AJ**-proof from $\Gamma^\diamond$, $\Box A$ occurs as the second element and $\{C_1, \ldots, C_m\}$ $(0 \leq m)$ occurs as the fifth element of line i, then $\Gamma^\diamond \vdash_{\mathbf{S5}} \Box A \vee Dab(C_1, \ldots, C_m)$.

Proof: The proof proceeds by induction on the number of the line at which $\Box A$ occurs. The lemma obviously holds if $i = 1$, for then, the second element of line i is either not of the form $\Box A$ or $\vdash_{\mathbf{S5}} \Box A$. Suppose now that the Lemma holds for all lines that precede i. As all premises are of the form $\Diamond B$, we only have to consider those lines that are written down by application of RU or RC-J.

Case 1: The third element of line i is empty. Suppose that the fifth element of i is $C_1, \ldots, C_m$ $(m \geq 0)$. In that case, both RU and RC-J warrant that $\vdash_{\mathbf{S5}} \Box A \vee Dab(C_1, \ldots, C_m)$.

Case 2: The third element of line i is not empty. Suppose that the third element of i is $j_1, \ldots, j_n$ $(n \geq 1)$ and that $B_1, \ldots, B_n$ are the second elements of lines $j_1, \ldots, j_n$. Both RU and RC-J warrant that $\vdash_{\mathbf{S5}} ((B_1 \wedge \ldots \wedge B_n) \supset \Box A) \vee Dab(C_1, \ldots, C_m)$ $(m \geq 0)$. As the union of the fifth elements of lines $j_1, \ldots, j_n$ is a subset of $\{C_1, \ldots, C_m\}$, the supposition warrants that $\Gamma^\diamond \vdash_{\mathbf{S5}} (B_1 \wedge \ldots \wedge B_n) \vee Dab(C_1, \ldots, C_m)$. Hence, $\Gamma^\diamond \vdash_{\mathbf{S5}} \Box A \vee Dab(C_1, \ldots, C_m)$.

□

LEMMA 21 If $\Gamma^\diamond \vdash_{\mathbf{AJ}} \Box A$, then there are $C_1, \ldots, C_m$ $(m \geq 0)$ such that $\Gamma^\diamond \vdash_{\mathbf{S5}} \Box A \vee Dab(C_1, \ldots, C_m)$, and none of $C_1, \ldots, C_m$ is a member of $U(\Gamma^\diamond)$.

Proof: Suppose that $\Gamma^\diamond \vdash_{\mathbf{AJ}} \Box A$. In that case, $\Box A$ is finally derived at some line j of an **AJ**-proof from $\Gamma^\diamond$. Hence, where $\{C_1, \ldots, C_m\}$ $(m \geq 0)$ is the fifth element of j, $\Gamma^{\Diamond} \vdash_{\mathbf{S5}} \Box A \vee Dab(C_1, \ldots, C_m)$ in view of Lemma 20. Suppose now that $C_i \in U(\Gamma^{\Diamond})$. In that case, C_i is a factor of some minimal *Dab*-consequence $Dab(\Delta)$ of $\Gamma^{\Diamond}$. As **S5** is compact, there is an extension of the proof in which $Dab(\Delta)$ occurs unconditionally. But then, line j is marked in that extension, and will remain marked in any further extension. This contradicts the supposition that $\Box A$ is finally derived from $\Gamma^{\Diamond}$. □

THEOREM 22 If $\Gamma^{\Diamond} \vdash_{\mathbf{AJ}} \Box A$, then $\Gamma^{\Diamond} \models_{\mathbf{AJ}} \Box A$.

Proof: Suppose that $\Gamma^{\Diamond} \vdash_{\mathbf{AJ}} \Box A$. In view of Lemma 21, there are $C_1, \ldots, C_m$ ($m \geq 0$) such that $\Gamma^{\Diamond} \vdash_{\mathbf{S5}} \Box A \vee Dab(C_1, \ldots, C_m)$, and none of $C_1, \ldots, C_m$ is a member of $U(\Gamma^{\Diamond})$. Hence, $\Gamma^{\Diamond} \models_{\mathbf{S5}} \Box A \vee Dab(C_1, \ldots, C_m)$, in view of Theorem 1. As no C_i is a member of $\Gamma^{\Diamond}$, $Dab(C_1, \ldots, C_m)$ is false in all **AJ**-models of $\Gamma^{\Diamond}$. But then, $\Box A$ is true in all **AJ**-models of $\Gamma^{\Diamond}$, and $\Gamma^{\Diamond} \models_{\mathbf{AJ}} \Box A$. □

THEOREM 23 $\Gamma^{\Diamond} \models_{\mathbf{AJ}} \Box A$, then $\Gamma^{\Diamond} \vdash_{\mathbf{AJ}} \Box A$.

Proof: Suppose that $\Gamma^{\Diamond} \nvdash_{\mathbf{AJ}} \Box A$. Where $D_1, D_2, \ldots$ is a list of all members of $\mathcal{W}^a - U(\Gamma^{\Diamond})$, define:

$$\Delta_0 = \emptyset$$

$$\Delta_{i+1} = \Delta_i \cup \{D_{i+1}\}$$

if $\Gamma^{\Diamond} \cup \Delta_{i+1}^{\Box} \nvdash_{\mathbf{AJ}} \Box A$, and

$$\Delta_{i+1} = \Delta_i \cup \{\sim D_{i+1}\}$$

otherwise. Finally,

$$\Delta = \Delta_0 \cup \Delta_1 \cup \Delta_2 \cup \ldots$$

Suppose now that $\mathcal{M} = \langle \Sigma_\Delta, M_0 \rangle$ is not an **S5**-model of $\Gamma^{\Diamond}$. In that case, for some $\Diamond B \in \Gamma^{\Diamond}$, $v_{\mathcal{M}}(\Diamond B) = 0$. But then, in view of Theorem 4, $\Delta^{\Box} \models_{\mathbf{S5}} \Box \sim B$, and hence, $\Delta^{\Box} \cup \Gamma^{\Diamond} \models_{\mathbf{S5}} \Diamond B \wedge \sim \Diamond B$ as well as $\Delta^{\Box} \cup \Gamma^{\Diamond} \vdash_{\mathbf{S5}} \Diamond B \wedge \sim \Diamond B$. In view of $\Diamond B \wedge \sim \Diamond B \vdash_{\mathbf{S5}} \Box A$ and the rule RU, this would mean that $\Delta^{\Box} \cup \Gamma^{\Diamond} \vdash_{\mathbf{AJ}} \Box A$ which is impossible by the definition of Δ. Hence, $\mathcal{M} = \langle \Sigma_\Delta, M_0 \rangle$ is an **S5**-model of $\Gamma^{\Diamond}$. In view of the definition of Δ and of Definition 10, $\mathcal{M}$ is a *reliable* **S5**-model of $\Gamma^{\Diamond}$ that does not verify $\Box A$. Hence, in view of Definition 11, $\Gamma^{\Diamond} \not\models_{\mathbf{AJ}} \Box A$. □

COROLLARY 24 $\Gamma^{\Diamond} \models_{\mathbf{AJ}} \Box A$ if and only if $\Gamma^{\Diamond} \vdash_{\mathbf{AJ}} \Box A$.

6 The Adaptive Logic COM

Where **AJ** is obtained from the **S5**-models of $\Gamma^{\Diamond}$, **COM** is obtained from the **S5**-models of $\Gamma^{\Box}$. The **S5**-models of $\Gamma^{\Box}$ that are not **COM**-models of $\Gamma^{\Box}$ are those that verify $\Box \sim A$ for some A such that $\Gamma \not\models_{\mathbf{CL}} \sim A$. In other words, some well-formed formulas that are compatible with Γ come out impossible in those models.

The **COM**-abnormal part of a model is defined by:

DEFINITION 25 $Ab_{\mathbf{COM}}(\mathcal{M}) = \{A \mid A \in \mathcal{W};\ \mathcal{M} \text{ verifies } \Box{\sim}A\}$.

The abnormalities that are unavoidable in view of $\Gamma^{\Box}$ are:

DEFINITION 26 $Ab_{\mathbf{COM}}(\Gamma^{\Box}) = \{A \mid A \in \mathcal{W};\ \Gamma^{\Box} \models_{\mathbf{S5}} \Box{\sim}A\}$.

In view of these definitions,[18] **COM** is obtained by means of the Simple Strategy:

DEFINITION 27 An **S5**-model $\mathcal{M} = \langle \Sigma_{\Delta}, M_0 \rangle$ is a **COM**-model of $\Gamma^{\Box}$ if and only if $Ab_{\mathbf{COM}}(\mathcal{M}) = Ab_{\mathbf{COM}}(\Gamma^{\Box})$.

The proof of the following Theorem proceeds by an inspection of the **S5**-semantics:

THEOREM 28 Where $A \in \mathcal{W}$, and where $\Gamma^{\Diamond}$ has **S5**-models, $\Gamma^{\Box} \models_{\mathbf{COM}} \Diamond A$ if and only if $\Gamma^{\Box} \not\models_{\mathbf{S5}} \Box{\sim}A$.

The proof format for **COM** is like that for **AJ**, and the proof theory is defined in terms of the same four kinds of generic rules. Moreover, as **AJ** and **COM** have **S5** as their lower limit logic, the unconditional rule RU is the same for both proof theories. The only differences concern the unconditional rule and the marking rule:

RC-C At any stage in a proof, one may, for any formula $A \in \mathcal{W}$, add $\Diamond A$ on the condition $\{A\}$.

MARK-C A line is marked at a stage iff, where Δ is its fifth element, $\Box{\sim}B$ has not been unconditionally derived for any $B \in \Delta$.

The dynamics in a **COM**-proof is rather restricted. Once a line is marked, it remains marked at all future stages of the proof. As a consequence, the definition of final derivability is much simpler than that for **AJ**:

DEFINITION 29 A is finally derived in a **COM**-proof from $\Gamma^{\Box}$ if and only if A is derived on a line that is not marked and will not be marked in any extension of the proof.

DEFINITION 30 $\Gamma^{\Box} \vdash_{\mathbf{COM}} A$ (A is finally derivable from $\Gamma^{\Box}$) if and only if A is finally derived in a **COM**-proof from $\Gamma^{\Box}$.

As is proved in [16], the **COM**-semantics is adequate with respect to final derivability in **COM**, for all formulas of the form $\Diamond A$:

THEOREM 31 Where $A \in \mathcal{W}$, $\Gamma^{\Box} \vdash_{\mathbf{COM}} \Diamond A$ if and only if $\Gamma^{\Box} \models_{\mathbf{COM}} \Diamond A$.

[18] I use a simplified version of the semantics. Normally, the definitions should be restricted to those formulas that are not **S5**-valid. However, as **S5**-valid well-formed formulas hold true in all **S5**-models of $\Gamma^{\Diamond}$, the present definitions lead to the same selection of **S5**-models. In [16], it is shown that this simplified version is important for the dynamic proof theory.

7 The Adaptive Logic APV

Semantically, **APV** is obtained by applying **COM** to some Γ^* that is obtained from $\Gamma^\Diamond$ by **AJ**.

DEFINITION 32 $\Gamma^* = \{\Box A \mid \Gamma^\Diamond \models_{\mathbf{AJ}} \Box A\}$.

DEFINITION 33 $\Gamma^\Diamond \models_{\mathbf{APV}} A$ if and only if $\Gamma^* \models_{\mathbf{COM}} A$.

As will become clear below, this two-step semantics is helpful to understand the proof theory of **APV**. It is possible, however, to obtain **APV** directly from the **S5**-models by a rather unusual (but nevertheless very natural) strategy. In [23], I present this more direct semantics for **APV** and show that the two are equivalent.

The following theorem shows that $\Diamond A$ is an **APV**-consequence of $\Gamma^\Diamond$ if and only if it is compatible with the **AJ**-consequences of $\Gamma^\Diamond$. In other words, A comes out pragmatically true if and only if it is compatible with the consequences of $\Gamma^\Diamond$ that are pragmatically valid. In view of the aim to capture the notion of pragmatic truth as studied in the partial structures approach, this is as it should be.

THEOREM 34 Where $\Gamma^\Diamond$ has **S5**-models, $\Gamma^\Diamond \models_{\mathbf{APV}} \Diamond A$ if and only if $\Gamma^\Diamond \not\models_{\mathbf{AJ}} \Box{\sim}A$.

Proof: For the left-right direction, suppose that $\Gamma^\Diamond$ has **S5**-models, and that $\Gamma^\Diamond \models_{\mathbf{AJ}} \Box{\sim}A$. In that case, $\Box{\sim}A \in \{\Box B \mid \Gamma^\Diamond \models_{\mathbf{AJ}} \Box B\}$, and hence, $\{\Box B \mid \Gamma^\Diamond \models_{\mathbf{AJ}} \Box B\} \models_{\mathbf{S5}} \Box{\sim}A$. But then, $\{\Box B \mid \Gamma^\Diamond \models_{\mathbf{AJ}} \Box B\} \not\models_{\mathbf{COM}} \Diamond A$ (in view of Theorem 28). Hence, in view of Definition 33, $\Gamma^\Diamond \not\models_{\mathbf{APV}} \Diamond A$. The right-left direction proceeds analogously. □

As may be expected, the proof theory for **APV** consists of a combination of the proof theories for **AJ** and **COM**. What makes the proof theory interesting, however, is the interplay between two different kinds of dynamics.

As **APV** incorporates two adaptive strategies, there will be two conditional rules and two marking rules. As compared to the proof theories of **AJ** and **COM**, there are three changes. First, in order to avoid circularities, the conditional rule related to **AJ** (see RC1) will only be applicable to well-formed formulas that are *unconditionally derived.* Next, the marking rule related to **COM** (see MARK2) will refer not only to lines that are unconditionally derived (as is usual for adaptive logics), but also to lines that are conditionally derived. In this way, the marking rule takes into account lines that are derived on the basis of RC1. Finally, the conditions for lines added by the conditional rule related to **COM** (see RC2) will be written in a different way. This will enable us to distinguish the two kinds of conditions.

The proof format as well as the rules PREM and RU are as for **AJ** and **COM**. The other rules are:

RC1 If $A_1, \ldots, A_n \vdash_{\mathbf{S5}} B \vee Dab(C_1, \ldots, C_m)$, and $A_1, \ldots, A_n$ occur in the proof *on the condition* $\emptyset$, then one may add to the proof a line with an appropriate line number, B as its second element, and $\{C_1, \ldots, C_m\}$ as its fifth element.

RC2 At any stage in a proof, one may, for any formula $A \in \mathcal{W}$, add $\Diamond A$ on the condition $\{\Box{\sim}A\}$.

MARK1 A line is marked$_1$ at a stage iff, where Δ is its fifth element, $B \notin U_s(\Gamma)$, for any $B \in \Delta$.

MARK2 A line is marked$_2$ at a stage iff, where Δ is its fifth element, $\Box B$ has not been derived—*conditionally or unconditionally*—for any $\Box B \in \Delta$.

Note that if $\Box A$ occurs on a line that is marked$_1$, this indicates that A is (at that stage) not pragmatically valid. If $\Diamond A$ occurs on a line that is marked$_2$, then A is (at that stage) not pragmatically true. The following rule is obviously derivable and leads to proofs that are more interesting from a heuristic point of view:

RD If $\Diamond A$ occurs in the proof on the condition $\emptyset$, then one may add a line with an appropriate line number, $\Box A$ as its second element, and $sent(A)$ as its fifth element.

DEFINITION 35 A is finally derived in an **APV**-proof from $\Gamma^\Diamond$ if and only if A is derived on a line that is not marked$_{1/2}$ and any extension of the proof in which A is marked$_{1/2}$ may be further extended in such a way that A becomes unmarked$_{1/2}$.

DEFINITION 36 $\Gamma^\Diamond \vdash_{\mathbf{APV}} A$ (A is finally derivable from $\Gamma^\Diamond$) if and only if A is finally derived in an **APV**-proof from $\Gamma^\Diamond$.

A simple example of a proof follows below. At line 5, it is assumed that ${\sim}r$ is pragmatically true. At line 6, it is assumed that the formula on line 1 behaves **AJ**-normally, and hence that p is pragmatically valid. Analogously for line 7. On the basis of these assumptions, it is derived on line 8 that r is pragmatically valid. Hence, line 5 has to be marked$_2$ (${\sim}r$ is no longer compatible with the interpretation of the premises at that stage). At line 10, it is derived that p behaves **AJ**-abnormally. Hence, lines 6–8 have to be marked$_1$, and, as a consequence, line 5 is now unmarked$_2$. Lines 12–14 are finally derived. Note especially that $\Box q$ can be finally derived, despite the fact that, in the premises, q is conjoined to a formula that behaves **AJ**-abnormally.

1	$\Diamond p$	–; PREM	$\emptyset$	
2	$\Diamond({\sim}p \wedge q)$	–; PREM	$\emptyset$	
3	$\Diamond({\sim}p \vee r)$	–; PREM	$\emptyset$	
4	$\Diamond({\sim}q \vee s)$	–; PREM	$\emptyset$	
5	$\Diamond{\sim}r$	–; RC2	$\{\Box r\}$	marked$_2$ at stage 8; unmarked$_2$ at stage 10
6	$\Box p$	1; RD	$\{p\}$	marked$_1$ at stage 10
7	$\Box({\sim}p \vee r)$	3; RD	$\{p, r\}$	marked$_1$ at stage 10
8	$\Box r$	6,7; RU	$\{p, r\}$	marked$_1$ at stage 10
9	$\Diamond{\sim}p$	2; RU	$\emptyset$	
10	$\Diamond p \wedge \Diamond{\sim}p$	1,9; RU	$\emptyset$	
11	$\Diamond q$	2; RU	$\emptyset$	
12	$\Box q$	11; RD	$\{q\}$	
13	$\Box({\sim}q \vee s)$	4; RD	$\{q, s\}$	
14	$\Box s$	12, 13; RU	$\{q, s\}$	

The proofs of the following lemmas proceed as those for Lemma 20 and Lemma 21, and are left to the reader:

LEMMA 37 If, in an **APV**-proof from $\Gamma^\diamond$, $\diamond A$ occurs as the second element of line i and $\{C_1, \ldots, C_m, \Box{\sim}D_1, \ldots, \Box{\sim}D_n\}$ $(0 \leq m; 0 \leq n)$ as its fifth element, then $\Gamma^\diamond \vdash_{\mathbf{S5}} A \vee Dab(C_1, \ldots, C_m) \vee \Box{\sim}D_1 \vee \ldots \vee \Box{\sim}D_n$.

LEMMA 38 If $\Gamma^\diamond \vdash_{\mathbf{APV}} \diamond A$, then there are $C_1, \ldots, C_m \in \mathcal{S}$ $(m \geq 0)$ and $D_1, \ldots, D_n \in \mathcal{W}$ $(n \geq 0)$ such that $\Gamma^\diamond \vdash_{\mathbf{S5}} \Box A \vee Dab(C_1, \ldots, C_m) \vee \Box{\sim}D_1 \vee \ldots \vee \Box{\sim}D_n$, none of $C_1, \ldots, C_m$ is a member of $U(\Gamma^\diamond)$, and none of $\Box{\sim}D_1, \ldots, \Box{\sim}D_n$ is **APV**-derivable from $\Gamma^\diamond$.

The following lemma is easily established by an inspection of the generic rules for **APV**:

LEMMA 39 $\Gamma^\diamond \vdash_{\mathbf{APV}} \Box A_1 \vee \ldots \vee \Box A_n$ if and only if $\Gamma^\diamond \vdash_{\mathbf{APV}} \Box A_1$ or ...or $\Gamma^\diamond \vdash_{\mathbf{APV}} \Box A_n$.

THEOREM 40 Where $Cn_{\mathbf{S5}}(\Gamma^\diamond)$ is not trivial, $\Gamma^\diamond \vdash_{\mathbf{APV}} \diamond A$ if and only if $\Gamma^\diamond \nvdash_{\mathbf{AJ}} \Box{\sim}A$.

Proof: The right-left direction is evident in view of RC2 and MARK2.

For the left-right direction, suppose first that $Cn_{\mathbf{S5}}(\Gamma^\diamond)$ is not trivial, and that $\Gamma^\diamond \vdash_{\mathbf{APV}} \diamond A$. In that case, $\diamond A$ is finally derived in an **APV**-proof from $\Gamma^\diamond$. But then, in view of Lemma 38, there is a $\Theta \subset \mathcal{S} - U(\Gamma^\diamond)$, such that $\Gamma^\diamond \vdash_{\mathbf{S5}} \diamond A \vee Dab(\Theta) \vee \Box{\sim}D_1 \vee \ldots \vee \Box{\sim}D_n$, and none of $\Box{\sim}D_1, \ldots, \Box{\sim}D_n$ is **APV**-derivable from $\Gamma^\diamond$. Suppose next that $\Gamma^\diamond \vdash_{\mathbf{AJ}} \Box{\sim}A$. In view of Lemma 21, there is a $\Theta' \subset \mathcal{S} - U(\Gamma^\diamond)$ such that $\Gamma^\diamond \vdash_{\mathbf{S5}} \Box{\sim}A \vee Dab(\Theta')$. But then, in view of both suppositions, $\Gamma^\diamond \vdash_{\mathbf{S5}} Dab(\Theta') \vee Dab(\Theta) \vee \Box{\sim}D_1 \vee \ldots \vee \Box{\sim}D_n$. Hence, by RC1, $\Box{\sim}D_1 \vee \ldots \vee \Box{\sim}D_n$ occurs on a line in some **APV**-proof from $\Gamma^\diamond$ the fifth element of which is $\{\Theta \cup \Theta'\}$. As no member of $\Theta \cup \Theta'$ is unreliable, this line will not be marked in any extension of the proof. But then, by Lemma 39, $\Gamma^\diamond \vdash_{\mathbf{APV}} \Box{\sim}D_1$ or ...or $\Gamma^\diamond \vdash_{\mathbf{APV}} \Box{\sim}D_n$. This contradicts the supposition that none of them is **APV**-derivable. □

The following theorem follows immediately from Corollary 24 and the Theorems 34 and 40:

THEOREM 41 Where A is a well-formed formula of $\mathcal{L}$, $\Gamma^\diamond \vdash_{\mathbf{APV}} \diamond A$ if and only if $\Gamma^\diamond \models_{\mathbf{APV}} \diamond A$.

8 The Discussive Adaptive Logic APT

The inference relation $\vdash_{\mathbf{APT}}$ of **APT** is non-modal, and is defined from the modal inference relation $\vdash_{\mathbf{APV}}$ of the adaptive logic **APV** in the following way:

DEFINITION 42 $\Gamma \vdash_{\mathbf{APT}} A$ if and only if $\Gamma^\diamond \vdash_{\mathbf{APV}} \diamond A$ and $A \in \mathcal{W}$.

As I mentioned already in Section 1, **APT** is in two respects richer than **PT**. First, **APT** validates Adjunction (and hence, Modus Ponens, Modus Tollens, ...)

for all sentences that belong to the consistent core of Γ. Next, **APT** delivers all consequences that are compatible with the consistent core of Γ.

The following simple example nicely illustrates the differences with **PT**. Suppose that $\Gamma = \{p \wedge r, \sim p \wedge (r \supset s)\}$. In that case, the only atoms that follow from Γ by **PT** are $\{p, \sim p, r\}$. By **APT**, the same atoms are derivable. However, in addition to this, also s follows (as $r \supset s$ and r behave consistently on Γ, Modus Ponens can be applied to them) as well as all members of $\mathcal{W}^a$ that do not contain the letters p, r, s (all of these are compatible with the consistent core of Γ). Note that as $\sim s$ and $\sim r$ are not compatible with the consistent core of Γ, neither of them is **APT**-derivable. Note also that adding inconsistencies leads in general to a richer set of **APT**-consequences. For instance, whereas $\sim q$ is not an **APT**-consequence of $\{p \wedge r, p \supset q\}$, it is an **APT**-consequence of $\{p \wedge r, p \supset q, \sim p\}$. This should not surprise us. The more inconsistencies we add, the smaller the consistent core becomes, and hence, the more sentences are compatible with this core.

In view of these properties, **APT** enables us to define the notion of a pragmatic theory in a way that is in line with the partial structures approach:

DEFINITION 43 T is the pragmatic theory for a set of beliefs Γ if and only if $T = \langle \Gamma, \mathbf{APT} \rangle = Cn_{\mathbf{APT}}(\Gamma)$.

The proof of the following theorem proceeds by an inspection of the **APV**-semantics and Definition 42:

THEOREM 44 If Γ is consistent, then $Cn_{\mathbf{CL}}(\Gamma) \subseteq Cn_{\mathbf{APT}}(\Gamma)$.

THEOREM 45 If Γ contains no **CL**-contingent formulas, then every **CL**-contingent formula is a member of $Cn_{\mathbf{APT}}(\Gamma)$.

Also this last theorem is entirely in line with the partial structures approach. If our knowledge about a certain domain consists only of logical truths, then every logically contingent sentence is compatible with our knowledge.

9 Conclusion and open problems

In this paper, I presented the discussive adaptive logic **APT** that is based on the modal adaptive logic **APV**. I showed that **APT** captures the notion of pragmatic truth as studied within the partial structures approach of Newton da Costa and his associates. One of the central characteristics of **APT** is that $A \in Cn_{\mathbf{APT}}$ if and only if A is pragmatically true in the partial structures that adequately model Γ. What this comes to is that **APT** localizes the consistent core of a (possibly inconsistent) theory, and that it delivers all consequences that are compatible with this core.

One of the main results of this paper concerns the proof theory for **APV**. The notion of partial structures adequately defines the concept of pragmatic truth. However, the latter did not yet have a proof theory that does justice to this definition. I showed that the proof theory for **APV** involves two kinds of dynamics. One is related to assuming that some sentence is pragmatically valid (until proved otherwise), the other to assuming that some sentence is pragmatically true (until proved otherwise).

In this paper, I restricted the presentation of **APV** to the propositional level. Evidently, it should be upgraded to the predicative level. Another important problem concerns the design of a direct dynamic proof theory for **APT** (that proceeds directly in terms of Γ instead of $\Gamma^{\diamond}$). Also, it would be interesting to study some further relations between adaptive logics and partial structures. One such relation concerns the dynamics of partial structures. In view of results on other adaptive logics, a 'block semantics' for **APV** may be formulated—see [24]—that captures the dynamics of the proof theory. Given the relation between partial structures and **S5**-models, this block semantics would straightforwardly lead to a definition of 'dynamic' partial structures. These would have the enormous advantage that they could be used to represent changes in one's beliefs that are due to the mere analysis of these beliefs. Another important relation concerns the different 'adaptive strategies'. As far as I can see, partial structures as studied today are (implicitly) based on the Reliability Strategy. It would be interesting to examine whether this strategy can be changed (as is the case in adaptive logics) and, if so, what the implications would be.

REFERENCES

[1] Irene Mikenberg, Newton C.A. da Costa, and Rolando Chuaqui. Pragmatic Truth and Approximation to Truth. *Journal of Symbolic Logic*, 51:201–221, 1986.

[2] Newton C.A. da Costa and Steven French. Pragmatic Truth and the Logic of Induction. *The British Journal for the Philosophy of Science*, 40:333–356, 1989.

[3] Newton C.A. da Costa. Pragmatic Probability. *Erkenntnis*, 25:141–162, 1986.

[4] Newton C.A. da Costa and Steven French. Towards an Acceptable Theory of Acceptance. Partial Structures, Inconsistency and Correspondence. In Steven French and Harmke Kamminga, editors, *Correspondence, Invariance and Heuristics*, pages 137–158. Kluwer, 1993.

[5] Steven French. Partiality, Pursuit, and Practice. In Maria Luisa Dalla Chiara, Kees Doets, Daniele Mundici, and Johan van Benthem, editors, *Structures and Norms in Science*, pages 35–52. Kluwer, 1997.

[6] Otávio Bueno. Empirical Adequacy: A Partial Structures Approach. *Studies in History and Philosophy of Science*, 28:585–610, 1997.

[7] Newton C.A. da Costa and Steven French. Inconsistency in Science. A Partial Perspective. In Joke Meheus, editor, *Inconsistency in Science*. Kluwer, to appear.

[8] Otávio Bueno. Mathematical Change and Inconsistency: A Partial Structures Approach. In Joke Meheus, editor, *Inconsistency in Science*. Kluwer, to appear.

[9] Newton C.A. da Costa, Otávio Bueno, and Steven French. The Logic of Pragmatic Truth. *Journal of Philosophical Logic*, 27:603–620, 1998.

[10] Diderik Batens. Dynamic Dialectical Logics. In Graham Priest, Richard Routley, and Jean Norman, editors, *Paraconsistent Logic. Essays on the Inconsistent*, pages 187–217. Philosophia Verlag, 1989.

[11] Diderik Batens. Inconsistency-Adaptive Logics. In Ewa Orłowska, editor, *Logic at Work. Essays Dedicated to the Memory of Helena Rasiowa*, pages 445–472. Physica Verlag (Springer), 1999.

[12] Diderik Batens. A Survey of Inconsistency-Adaptive Logics. In Diderik Batens, Chris Mortensen, Graham Priest, and Jean Paul Van Bendegem, editors, *Frontiers of Paraconsistent Logic*, pages 49–73. Research Studies Press, 2000.

[13] Joke Meheus. An Extremely Rich Paraconsistent Logic and the Adaptive Logic Based on It. In Diderik Batens, Chris Mortensen, Graham Priest, and Jean Paul Van Bendegem, editors, *Frontiers of Paraconsistent Logic*, pages 189–201. Research Studies Press, 2000.

[14] Joke Meheus. An Inconsistency-Adaptive Logic Based on Jaśkowski's **D2**. In preparation.

[15] Stanisław Jaśkowski. Propositional Calculus for Contradictory Deductive Systems. *Studia Logica*, 24:143–157, 1969. Originally published in Polish in: *Studia Scientarium Torunensis*, Sec. A II, 1948, pp. 55–77.

[16] Diderik Batens and Joke Meheus. The Adaptive Logic of Compatibility. *Studia Logica*, 66:327–348, 2000.

[17] Joke Meheus. Adaptive Logic in Scientific Discovery: The Case of Clausius. *Logique et Analyse*, 143–144:359–389, 1993. Appeared 1996.

[18] Joke Meheus. Inconsistencies in Scientific Discovery. Clausius's Remarkable Derivation of Carnot's Theorem. In print.

[19] G.E. Hughes and M.J. Cresswell. *A New Introduction to Modal Logic*. Routledge, 1996.

[20] Graham Priest. Minimally Inconsistent **LP**. *Studia Logica*, 50:321–331, 1991.

[21] Diderik Batens. Minimally Abnormal Models in Some Adaptive Logics. *Synthese*, 125:5–18, 2000.

[22] Diderik Batens, Kristof De Clercq, and Guido Vanackere. Simplified Dynamic Proof Formats for Adaptive Logics. To appear.

[23] Joke Meheus. Discussive Adaptive Logics. In preparation.

[24] Diderik Batens. Blocks. The Clue to Dynamic Aspects of Logic. *Logique et Analyse*, 150–152:285–328, 1995. Appeared 1997.

A Multiple Worlds Semantics for a Paraconsistent Nonmonotonic Logic *

ANA TERESA MARTINS Federal University of Ceará, P.O.Box 12166, Fortaleza, CE, Brazil 60455-760
ana@lia.ufc.br

MARCELINO PEQUENO Federal University of Ceará, P.O.Box 12166, Fortaleza, CE, Brazil 60455-760
marcel@lia.ufc.br

TARCÍSIO PEQUENO Federal University of Ceará, P.O.Box 12166, Fortaleza, CE, Brazil 60455-760
tarcisio@lia.ufc.br

Abstract

Semantics have often been a problem with relation to both nonmonotonic and paraconsistent logics. In previous works, a system of logics called IDL & LEI was developed to formalize *practical reasoning*, i.e., reasoning under incomplete knowledge. This system combines the features of *nonmonotonicity* to model inferences on the basis of partial evidence, with *paraconsistency* to deal with the inconsistencies introduced by extended inferences. The aim of this paper is to present a semantics that is, as much as possible, uniform with respect to both the deductive and nonmonotonic part of this system. This semantics is presented in a possible worlds style and is intended to reflect the basic intuitions assumed in designing IDL & LEI. A possible worlds semantics to the paraconsistent logic LEI is first given and it is extended to encompass the nonmonotonic features formalized by the logic IDL. The resulting semantics is shown to be sound and complete with respect to the whole IDL & LEI system.

1 Introduction

The formalization of reasoning under conditions of incomplete knowledge has been a stimulating problem for Artificial Intelligence researchers. By incomplete knowledge, it is meant the property of not being able to decide some especially relevant questions that 'ought' to be decided for practical purposes.

*We acknowledge financial support from FAR/CNPq, VDL/ProTeM/CNPq and CNPq.

The field of Artificial Intelligence is often challenged to represent useful knowledge and to perform fruitful inferences in spite of the lack of information. One usual alternative to treating this problem is to simply forget about logic and formal methods altogether and implement particular ad hoc solutions which are computationally feasible, hand waving any theoretical or even systematic justification of the inferences to be done. Still, the alternative remains in trying to expand the scope of formal methods by providing a logical apparatus able to work under these mundane circumstances.

Our work can be grouped with those who take the second path and is part of a research program which started a decade ago. During the course of time we came to notice that although this problem has been motivated by its relevance to Artificial Intelligence, it is of interest to many other fields, both in science and philosophy.

Nonmonotonic logics were introduced in the field of Artificial Intelligence as a formal tool to deal with this problem [1]. These logics are characterized by the fact that an inference once done on the basis of partial evidence could possibly be withdrawn by the acquisition of new information.

By the time we started dealing with this problem [2, 3] the attitude prevailing, with respect to the formalization of nonmonotonic reasoning, was a very conservative one. Classical logic was taken as the basis for the deductive part of nonmonotonic reasoning, while some extra inferential mechanism was added on top of it to deal with the nonmonotonic inferences. This attitude was ultimately supported by the assumption that what we will call the superdeductive part of the reasoning is completely separable and does not interfere with the deductive part.

This assumption, no matter how convenient and simplified it might seem, showed itself to be false even for practical purposes, without mentioning its philosophical uneasiness. Intriguing malfunctions of all the pioneer nonmonotonic logics [1] have been detected in works such as [4]. The reasons for these malfunctions have been considered, among others, in studies [5, 6, 7, 8]. One of the insights that emerged from our work was the realization that classical deduction could not remain uncontaminated in its association with non-tautological forms of reasoning. The disposition to perform this kind of inference provokes the unavoidable migration of non-classical features to the deductive part of reasoning. This is the basic observation which founds the work on this subject which we have been doing since then.

The picture emerging from the reasoning under incomplete knowledge is described by the authors in [2, 3, 9, 10, 11] in the following way. The knowledge available provides items of evidence that clash with one another, which gives rise to contradictions. The role of reasoning should be, then, to perform a global analysis of the current evidence, in order to resolve those contradictions. Often, though, the available knowledge does not allow the resolution of all occurring contradictions. If we have to go on making inferences and taking decisions anyway, as frequently happens in the practical world, these contradictions must be dealt with, even at the deductive level of reasoning.

There are some other alternatives to dealing with contradictions. One of them is to incorporate all non-resolvable contradictions as part of the current knowledge, exactly the same way as is done with any other partial conclusion. This alternative has been described in the literature as a credulous attitude. Its formalization should require a paraconsistent logic. LEI, Logic of Epistemic Inconsistency, was especially designed to fulfil this function. It is intended to work in association with a nonmonotonic logic tolerant to contradictions: the Inconsistent Default Logic, IDL.

As previously mentioned, the logics presented in [2, 3] are among the first attempts at explicitly dealing with the inconsistencies arising from reasoning by combining paraconsistency with nonmonotonicity. It is true that there have been concerns about the damages that contradiction could do in a system in face of the inability of classical logic to deal with it, but these concerns came from the casualty of contradictions being encoded in a knowledge base. No matter how restricted these contradictions may be, when this base was treated as a large set of axioms to originate deductions, the whole system collapsed. To handle this problem, the common suggestion was to find some strategy to confine the range of contradiction in the system, in order to keep the rest of it uncontaminated.

In the last decade, this attitude has changed remarkably to a more tolerant and understanding one towards contradiction. Although this tolerance is still very far from becoming the main trend in the field of Artificial Intelligence, it is now possible to find studies that treat the incoming of contradictions as a natural and expected side effect of nonmonotonic reasoning [12, 13, 14, 15, 16]. It is especially worth mentioning a line of work started by Belnap [17, 18], which proposed a four-valued semantics as a suitable model for automated reasoning. This proposal was further developed by Ginsberg in [19, 20] who introduced algebraic structures called bilattices that generalize the four-valued structure of Belnap. In [21, 22], Fitting argued that bilattice structures were also suitable for providing semantics for logic programs. Following these lines, Arieli and Avron, in a valuable work [13], used a four-valued semantics together with a notion of logical consequence based on minimal preferential models to obtain a logic that is both nonmonotonic and paraconsistent. This logic was developed following the semantics tradition in nonmonotonic logic, started in McCarthy's circumscription, while our IDL & LEI system has come from the inferential tradition, in which the more well-known representative is Reiter's default logic. However, both approaches have the same aim: to provide a suitable model for reasoning under circumstances of incomplete knowledge and uncertain information, and they both take the stand of handling inconsistencies explicitly and formally. A comparison between these two logics in terms of their power of inference and expression is a task still to be performed.

Having started from inference rules and a syntactically conceived logic, our greatest challenge is to provide a semantics that seems natural for capturing our main intuitions. This is precisely the aim of the present paper. Although these logics have existed for some time now, the system as a whole was lacking a semantical presentation capable to express the intuition behind its design, in particular the interplay between both phases of reasoning. In order to attain this goal, a semantical framework able to support and combine the paraconsistency of LEI and the nonmonotonicity of IDL was required. This framework has been found in a multiple world construction that works as a simplified possible world semantics for LEI. This construction can be suitably generalized, in the form of a multiplicity of LEI worlds, in order to provide a semantics for the whole system. The contribution of this paper is, thus, to present semantics for LEI and IDL in this multiple worlds framework, proving their soundness and completeness with respect to a presentation of IDL&LEI.

In section 2, a possible world semantics for the paraconsistent logic LEI is given. This semantics has been developed from the original semantics given for LEI in [2], which is based on the idea of multiple observers having a possibly diverging perception about a certain state of affairs. It was constructed by computing the interchanging among the multiple views of these observers through two mutually recursive functions, designed to capture the behaviour of LEI connectives. This idea of a plurality of views is replaced by the one of a plurality of worlds, here called plausible worlds, in order to remain faithful to the intuition of LEI.

In section 3 the logic IDL is presented through its inference rule able to introduce uncertain knowledge in the course of reasoning. The basic definitions concerning IDL are given here. In section 4, we work on the concept of well-presented IDL theories. This concept is fundamental for the subsequent development of the paper because it expresses the conditions of an IDL theory to have a single extension. This is what is expected from an IDL theory. As a matter of fact, the semantics to be given in section 5 works and makes sense just for well-presented theories. Soundness and completeness is proved in section 6 and comparisons to other nonmonotonic semantics are discussed in section 7.

2 A Possible Worlds Semantics for LEI

The paraconsistent logic LEI will be introduced here through its semantics. This logic was specially designed to deal with inconsistencies arising from the disposition of realizing inferences on the basis of partial, inconclusive evidences. The conclusions coming from these inferences express a kind of precarious knowledge, subject to revision by a better judgement under the light of new evidence. The difference of epistemic status between irrevocable statements and plausible ones must be reflected in the way they are reasoned out in LEI. A key issue in LEI is that contradictions among plausible formulas are tolerated but not among irrevocable ones, or even among an irrevocable and a plausible one. In order to do so, the plausible formulas in LEI are syntactically distinguished by the use of a question mark suffixing them.

So, formulae in LEI are written in the language $\mathcal{L}_?$. $\mathcal{L}_?$ extends a classical first-order language $\mathcal{L}$, defined as usual, by introducing the new unary post-fixed operator '?' for plausibility. Formulae in $\mathcal{L}_?$ are called **?-formulae**, or simply **formulae**, and they are defined as:

DEFINITION 1 [?-Formulae: $\alpha, \beta, \gamma, \ldots$] (i) if $\alpha \in \mathcal{L}$, then α is a ?-formula. They will be called **?-free formulae** and Roman capital letters $A, B, C, \ldots$ will be used as meta-variables for them. (ii) if α is a ?-formula then $(\alpha?)$ and $(\neg\alpha)$ are ?-formulae; (iii) if α and β are ?-formulae then $(\alpha \wedge \beta)$, $(\alpha \vee \beta)$ and $(\alpha \rightarrow \beta)$ are ?-formulae; (iv) if α is a ?-formulae and x is a variable then $(\forall x\alpha)$ and $(\exists x\alpha)$ are ?-formulae; (v) nothing else is a ?-formula.

Greek capital letters Γ, Δ, ... will be used for a set of ?-formulae and, whenever it does not lead to ambiguities, parenthesis in a formula will be omitted. The notion of closed formulae is taken as usual.

A Kripke-like model for LEI, which is an alternative version of its original semantics presented in [2], will be presented here. Whereas in the original semantics multiple observers are used to represent possible distinct visions of the state of affairs, the idea of a plurality of views is here replaced by the one of a plurality of worlds, called ***plausible worlds***, in order to remain faithful to the intuition of LEI. LEI (modal) semantics will be defined through a set $\mathcal{W}_{LEI}$ of plausible worlds where each plausible world w within this set is intended to index the set V, $V = \{v_w \mid w \in \mathcal{W}_{LEI}\}$, of valuation functions for atomic formulae [23]. The set V of valuation functions v_w may be thought of as representing points of view of different observers or different points of view of the same observer.

A formula without '?' will be true if it is a consensus, that is, if it is true for all valuations. A plausible formula 'α?' may be analysed from two angles: a credulous and a skeptical one. Under the credulous view, 'α?' is true if 'α' is plausible for some observer (i.e., true under v_w, for some plausible world w in $\mathcal{W}_{LEI}$). Under the skeptical perspective, 'α' must be plausible to all observers (i.e., true under v_w, for all plausible worlds w in $\mathcal{W}_{LEI}$) in order to affirm that 'α?' is true.

The semantics of LEI might be regarded as a possible worlds semantics where '?' may be interpreted as the possibility modality with an important difference: satisfaction is defined based on two intermingled auxiliary relations: maximal and minimal satisfaction. These relations, which are recursively defined, are necessary to capture the paraconsistent behaviour of negation. Formally, LEI semantics is presented through the following definitions.

DEFINITION 2 [Plausible World Structure $\mathcal{M}$] $\mathcal{M} =<\mid \mathcal{A} \mid, \sigma, \mathcal{W}_{LEI}, V>$ is a plausible world structure for $\mathcal{L}_?$ where:

1. $\mid \mathcal{A} \mid$ is a nonempty universe of discourse;
2. σ is an interpretation mapping for terms which assigns to each constant symbol c a member $c^{\mathcal{A}}$ of the universe $\mid \mathcal{A} \mid$ and to each n-place function symbol f an n-place operation $f^{\mathcal{A}}$ on $\mid \mathcal{A} \mid$, i.e., $f^{\mathcal{A}} :\mid \mathcal{A} \mid^n \rightarrow \mid \mathcal{A} \mid$;
3. $\mathcal{W}_{LEI}$ is a non-empty set (possibly finite) of plausible worlds w;
4. V is an indexed set of valuation functions $\{v_w \mid w \in \mathcal{W}_{LEI}\}$. Any valuation function $v_w \in V$ assigns to each n-place predicate symbol P its interpretation $P^{\mathcal{A}}_w$ on $\mid \mathcal{A} \mid$, that is, $P^{\mathcal{A}}_w \subseteq \mid \mathcal{A} \mid^n$. $P^{\mathcal{A}}_w$ is intended to represent a relation whose elements are n-tuples of members of the universe $\mid \mathcal{A} \mid$.

In the following definitions, let $\mathcal{M} =<\mid \mathcal{A} \mid, \sigma, \mathcal{W}_{LEI}, V>$ be a Plausible World Structure for $\mathcal{L}_?$ under consideration.

DEFINITION 3 [Assignment $\bar{s}$ for Terms in $\mathcal{L}_?$]:
Let $s : \mathcal{V} \rightarrow \mid \mathcal{A} \mid$ be an assignment function for variables in $\mathcal{L}_?$. $\bar{s} : Terms \rightarrow \mid \mathcal{A} \mid$ is defined as usual:

1. for each variable x, $\bar{s}(x) = s(x)$;
2. for each constant symbol c, $\bar{s}(c) = \sigma(c) = c^{\mathcal{A}}$;
3. if $t_1, \ldots, t_n$ are terms and f is an n-place function symbol, then $\bar{s}(f(t_1, \ldots, t_n)) = \sigma(f)(\bar{s}(t_1), \ldots, \bar{s}(t_n)) = f^{\mathcal{A}}(\bar{s}(t_1), \ldots, \bar{s}(t_n))$.

As previously mentioned, in order to define the satisfaction of a formula α in a structure $\mathcal{M}$, two auxiliary notions of maximal and minimal satisfaction $\mathcal{M} \models^w_{max}$

and $\mathcal{M} \models^w_{min}$, respectively, are defined. They are intended to capture the behaviour of the paraconsistent negation and its relation to the other connectives. The function s is also used as the assignment function for variables.

DEFINITION 4 [$\mathcal{M} \models^w_{max} \alpha$ [s] and $\mathcal{M} \models^w_{min} \alpha$ [s]]: Let $w \in \mathcal{W}_{LEI}$. Then:

1. $\mathcal{M} \models^w_{max} P(t_1, \ldots, t_n)$ [s] iff $\mathcal{M} \models^w_{min} P(t_1, \ldots, t_n)$ [s] iff $< \overline{s}(t_1), \ldots, \overline{s}(t_n) > \in v_w(P) = P^{\mathcal{A}}_w$ where $P(t_1, \ldots, t_n)$ is an atomic formula in $\mathcal{L}_?$. For other formulae in $\mathcal{L}_?$, the definition is the following:
2. $\mathcal{M} \models^w_{max} \alpha?$ [s] iff for some $w' \in \mathcal{W}_{LEI}$, $\mathcal{M} \models^{w'}_{max} \alpha$ [s];
3. $\mathcal{M} \models^w_{min} \alpha?$ [s] iff for all $w' \in \mathcal{W}_{LEI}$, $\mathcal{M} \models^{w'}_{min} \alpha$ [s];
4. $\mathcal{M} \models^w_{max} \neg\alpha$ [s] iff $\mathcal{M} \not\models^w_{min} \alpha$ [s];
5. $\mathcal{M} \models^w_{min} \neg\alpha$ [s] iff $\mathcal{M} \not\models^w_{max} \alpha$ [s];
6. $\mathcal{M} \models^w_{max} \alpha \rightarrow \beta$ [s] iff $\mathcal{M} \not\models^w_{max} \alpha$ [s] or $\mathcal{M} \models^w_{max} \beta$ [s];
7. $\mathcal{M} \models^w_{min} \alpha \rightarrow \beta$ [s] iff $\mathcal{M} \not\models^w_{max} \alpha$ [s] or $\mathcal{M} \models^w_{min} \beta$ [s];
8. $\mathcal{M} \models^w_{max} \alpha \wedge \beta$ [s] iff $\mathcal{M} \models^w_{max} \alpha$ [s] and $\mathcal{M} \models^w_{max} \beta$ [s];
9. $\mathcal{M} \models^w_{min} \alpha \wedge \beta$ [s] iff $\mathcal{M} \models^w_{min} \alpha$ [s] and $\mathcal{M} \models^w_{min} \beta$ [s];
10. $\mathcal{M} \models^w_{max} \alpha \vee \beta$ [s] iff $\mathcal{M} \models^w_{max} \alpha$ [s] or $\mathcal{M} \models^w_{max} \beta$ [s];
11. $\mathcal{M} \models^w_{min} \alpha \vee \beta$ [s] iff $\mathcal{M} \models^w_{min} \alpha$ [s] or $\mathcal{M} \models^w_{min} \beta$ [s];
12. $\mathcal{M} \models^w_{max} \forall x\alpha$ [s] iff $\mathcal{M} \models^w_{max} \alpha$ [$s(x \mid d)$], for every $d \in \mid \mathcal{A} \mid$;
13. $\mathcal{M} \models^w_{min} \forall x\alpha$ [s] iff $\mathcal{M} \models^w_{min} \alpha$ [$s(x \mid d)$], for every $d \in \mid \mathcal{A} \mid$;
14. $\mathcal{M} \models^w_{max} \exists x\alpha$ [s] iff $\mathcal{M} \models^w_{max} \alpha$ [$s(x \mid d)$], for some $d \in \mid \mathcal{A} \mid$;
15. $\mathcal{M} \models^w_{min} \exists x\alpha$ [s] iff $\mathcal{M} \models^w_{min} \alpha$ [$s(x \mid d)$], for some $d \in \mid \mathcal{A} \mid$;

where $s(x \mid d) = s(y)$, if $y \neq x$ or $s(x \mid d) = d$, if $y = x$.
$\mathcal{M}$ **satisfies** α **at** w **with** s if $\mathcal{M} \models^w_{max} \alpha$ [s].

In order to capture the intended meaning of the paraconsistent negation and its relation with implication, $\mathcal{M} \models^w_{max} \alpha \rightarrow \beta$ and $\mathcal{M} \models^w_{min} \alpha \rightarrow \beta$ are not dually defined. As a consequence $(\alpha \rightarrow \beta) \rightarrow (\neg\alpha \vee \beta)$, but not the converse, is a valid formula (see definition 7).

DEFINITION 5 [$\mathcal{M} \models \alpha$][1] α is satisfied in a structure $\mathcal{M}$ or $\mathcal{M}$ is a **Plausible World Model** of α, represented by $\mathcal{M} \models \alpha$, iff for all worlds $w \in \mathcal{W}_{LEI}$ and all functions s, $\mathcal{M} \models^w_{max} \alpha$ [s].

DEFINITION 6 [$\mathcal{M} \models \Gamma$] $\mathcal{M}$ is a plausible world model of Γ iff it is a plausible world model of each formula in Γ.

DEFINITION 7 [$\Gamma \models \alpha$]: $\Gamma \models \alpha$ iff every plausible world model $\mathcal{M}$ of Γ is also a plausible world model of α. If the set Γ of formulae is empty, then α is said to be a **valid formula**.

Some examples of valid formulae are:

[1] In [10] the paracomplete logic LSR, the *Logic of Skeptical Reasoning*, is presented, which is dual to LEI. Its satisfaction relation is based on the auxiliary notion of minimal satisfaction.

- $\alpha \rightarrow \alpha?$
- $\alpha?? \rightarrow \alpha?$
- $(\neg\alpha)? \leftrightarrow \neg(\alpha?)$
- $\neg(\alpha \wedge \neg\alpha)$
- $\neg\neg\alpha \leftrightarrow \alpha$
- $\neg(\alpha \wedge \beta) \leftrightarrow (\neg\alpha \vee \neg\beta)$
- $\neg(\alpha \vee \beta) \leftrightarrow (\neg\alpha \wedge \neg\beta)$
- $\neg(\alpha \rightarrow \beta) \leftrightarrow (\alpha \wedge \neg\beta)$
- $(\alpha \rightarrow B) \rightarrow ((\alpha \rightarrow \neg B) \rightarrow \neg\alpha)$[2]

The first valid formula expresses that, from any formula α, its plausibility is also obtained. The converse is not possible in general, but just for formulae already suffixed by at least two question marks '?'.

All other valid formulae are about the paraconsistent negation. The third formula shows the relationship between the plausibility and the negation: if the negation of a formula α is plausible then it is an indication that *alpha* is not plausible, and vice versa. '$\neg(\alpha\wedge\neg\alpha)$' is a valid formula. However, it does not represent in LEI language the *Non Contradiction* law, which is indeed not valid as a meta principle. Double negation and De Morgan's equivalences do hold in LEI, but not ***reduction ad absurdum*** stated as '$(\alpha \rightarrow \beta) \rightarrow ((\alpha \rightarrow \neg\beta) \rightarrow \neg\alpha)$'. In fact, just the restricted formulation of ***reduction ad absurdum***, expressed by '$(\alpha \rightarrow B) \rightarrow ((\alpha \rightarrow \neg B) \rightarrow \neg\alpha)$', is possible involving contradiction among ?-free formulae.

Observe that LEI negation behaves classically for ?-free formulae and paraconsistently for ?-formulae:

THEOREM 8 [2] Let A be a ?-free formula, Γ be a set of ?-free formulae and let $\models_C$ be the classical consequence relation. Then, $\Gamma \models A$ iff $\Gamma \models_C A$.

DEFINITION 9 [LEI consequence operator Cn]: $\alpha \in Cn(\Gamma)$ iff $\Gamma \models \alpha$.

Finally, a definition of a consistent set under Cn is given:

DEFINITION 10 [Consistent Set]: Γ is a consistent set under Cn iff $\Gamma \not\models A \wedge \neg\alpha$, for any ?-formula α in $\mathcal{L}_?$, where the ?-free formula A comes from α by deleting all (possibly none) occurrences of the question mark '?' in α.

As an example, suppose that $\Gamma \models R(a) \wedge \neg(R(a)?)$, for some predicate symbol R and constant symbol a. By the definition 10, Γ is considered an inconsistent set. '$A \wedge \neg\alpha$' is thought to be a (strong) contradiction.

[2]Note that, according to the notation adopted in definition 1, B is a ?-free formula.

3 A Paraconsistent Nonmonotonic Logic

As it is usual for default theories, IDL theories will be represented by a pair $<\mathcal{W},\mathcal{D}>$ where $\mathcal{W}$ is a set of closed ?-free formulae, which represent irrefutable facts about the world and $\mathcal{D}$ is a set of defaults used to represent rules with exceptions. However, the definition of IDL default rules and IDL extension differ substantially from other traditional approaches [24, 25]. First of all, the plausibility operator '?' is introduced to allow the distinction between irrefutable formulae, the ones that come from the set of facts, and plausible formulae, the ones which are derived using defaults. Moreover, conflicts between plausible formulae, e.g. '$A? \wedge \neg(A?)$', will be paraconsistently tolerated.

An IDL default rule will have the following pattern: $\frac{:A(\overline{x})\rightarrow B(\overline{x});\ \neg C(\overline{x})}{A(\overline{x})?\rightarrow B(\overline{x})?}$, where $A(\overline{x}), B(\overline{x}), \neg C(\overline{x}), A(\overline{x})?$ and $B(\overline{x})?$ are formulae whose free variables are among those of $\overline{x} = (x_1, \ldots, x_n)$. (Note that $A(\overline{x})$, $B(\overline{x})$ and $\neg C(\overline{x})$ are ?-free formulae, following the notation introduced in definition 1.) $A(\overline{x}) \rightarrow B(\overline{x})$ will be called the ***normal justification***; $\neg C(\overline{x})$ the ***seminormal justification*** (sometimes omitted) used to represent the negation of the exception condition $C(\overline{x})$ for the application of this rule; and $A(\overline{x})? \rightarrow B(\overline{x})?$ will be called the ***conclusion*** of this default. The seminormal justification is treated differently from the normal justification and it plays a central rôle in detecting not well-presented or cyclic theories. Whenever $A(\overline{x})$ is a valid formula (see definition 7), an IDL default rule $\frac{:A(\overline{x})\rightarrow B(\overline{x});\ \neg C(\overline{x})}{A(\overline{x})?\rightarrow B(\overline{x})?}$ will be simply written as the equivalent one $\frac{:B(\overline{x});\ \neg C(\overline{x})}{B(\overline{x})?}$.

A default rule $\frac{:A(\overline{x})\rightarrow B(\overline{x});\ \neg C(\overline{x})}{A(\overline{x})?\rightarrow B(\overline{x})?}$ is open if there exists a free variable in at least one of its component formulae $A(\overline{x}), B(\overline{x}), \neg C(\overline{x}), A(\overline{x})?$ and $B(\overline{x})?$. Otherwise it is closed. An IDL default theory is closed if all defaults in $\mathcal{D}$ are closed. Otherwise it is open.

A ground instance of a default rule $\frac{:A(\overline{x})\rightarrow B(\overline{x});\ \neg C(\overline{x})}{A(\overline{x})?\rightarrow B(\overline{x})?}$ is obtained by uniformly replacing ground terms of the language $\mathcal{L}_?$ by the free variables in $A(\overline{x})$, $B(\overline{x})$, $\neg C(\overline{x})$, $A(\overline{x})?$ and $B(\overline{x})?$. An open default rule will be written as a schema representing all its ground instances. Thus, associated to an IDL theory $<\mathcal{W},\mathcal{D}>$, a closed theory $<\mathcal{W},\mathcal{D}^*>$ exists, where D^* is the set of all ground instances of all default rules in $\mathcal{D}$. (Recall that $\mathcal{W}$ was already taken as a set of closed ?-free formulae.)

Henceforth, all IDL default theories $<\mathcal{W},\mathcal{D}>$ are taken as closed and a default will be simply written as $\frac{:A\rightarrow B;\ \neg C}{A?\rightarrow B?}$.

The set $\mathcal{E}$ of all theorems of an IDL default theory, known as **IDL extension**, is inferred from the set $\mathcal{W}$ of facts and the set of all plausible conclusions introduced by those defaults rules $\frac{:A\rightarrow B;\ \neg C}{A?\rightarrow B?}$ in $\mathcal{D}$ whose application restrictions are respected, that is, '$A? \rightarrow B?$' is inferred if neither '$\neg(A \rightarrow B)$' nor '$C?$' is ***provable*** from $<\mathcal{W},\mathcal{D}>$. The formal definition of extension will make clear what it means to be provable from an IDL default theory and what is an applicable IDL default rule. Note that the restriction '$\neg(A \rightarrow B)$' must be derived from the irrefutable facts, while the exception condition '$C?$' needs only to be plausible.

DEFINITION 11 [IDL extension] Let $<\mathcal{W},\mathcal{D}>$ be an IDL default theory defined over $\mathcal{L}_?$ with all formulae in $\mathcal{W}$ being ?-free. Let $\mathcal{C}n$ be the LEI consequence operator and let $\mathcal{C}$ be an operator over any set Γ of formulae in $\mathcal{L}_?$. Then $\mathcal{C}(\Gamma)$ is the minimal set of formulae in $\mathcal{L}_?$, which satisfies the following properties:

1. $\mathcal{W} \subseteq \mathcal{C}(\Gamma)$;
2. $Cn(\mathcal{C}(\Gamma)) = \mathcal{C}(\Gamma)$;
3. If $\frac{:A\to B;\ \neg C}{A?\to B?} \in \mathcal{D}$, $\neg(A \to B) \notin \Gamma$ and $C? \notin \Gamma$ Then $A? \to B? \in \mathcal{C}(\Gamma)$.

A set $\mathcal{E}$ of formulae is an extension to $< \mathcal{W}, \mathcal{D} >$ iff $\mathcal{C}(\mathcal{E}) = \mathcal{E}$. $\mathcal{C}$ is the IDL consequence operator.

DEFINITION 12 [$< \mathcal{W}, \mathcal{D} > \mid\!\sim \alpha$] A formula '$\alpha$' is **provable** from $< \mathcal{W}, \mathcal{D} >$, that is, $< \mathcal{W}, \mathcal{D} > \mid\!\sim \alpha$ if $\alpha \in \mathcal{E}$ where $\mathcal{E}$ is some extension for $< \mathcal{W}, \mathcal{D} >$.

DEFINITION 13 [Applicable Default] A default $\frac{:A\to B;\ \neg C}{A?\to B?}$ is **applicable** in $< \mathcal{W}, \mathcal{D} >$ if $\neg(A \to B) \notin \mathcal{E}$, $C? \notin \mathcal{E}$ and $A? \to B? \in \mathcal{E}$ where $\mathcal{E}$ is some extension for $< \mathcal{W}, \mathcal{D} >$.

Semantics will be provided herein for just a subset of the IDL theories, the **well-presented IDL theories**. They will be characterized (see section 4) by imposing some constraints which will avoid an inadequate use of the IDL system [26].

In contrast to Reiter's default logic [27], which splits conflicting default conclusions into several extensions, a well-presented IDL theory always has a single extension [26]. All default conclusions, even conflicting ones, can be joined into the single IDL extension by the combined use of the plausibility operator and the paraconsistent negation. In this sense, IDL is considered to be a credulous logic, since conclusions from alternative extensions are taken altogether.

Paraconsistency plays a key role in this logical system whenever the aim is to formalize situations where conflicts between default conclusions are equally believed. In such a case, as previously stated, the conflict should be paraconsistently tolerated. As an example, suppose that a person who lives in a city X is plausibly rich and a person that comes from a city Y is plausibly not rich. However, if a is a person that lives in X and comes from Y then it is plausible that a is rich and it is plausible that a is not rich. This is exactly the case in which being paraconsistent is important to tolerate this weak contradiction. It will be formalized in a well-presented IDL theory as (use 'L' for 'Lives at X', 'C' for 'Comes from Y' and 'R' for 'Rich'.) :

EXAMPLE 14

$\mathcal{W} = \{L(a) \wedge C(a)\}$,
$\mathcal{D} = \{\frac{:C(a)\to\neg R(a);}{C(a)?\to(\neg R(a))?}, \frac{:L(a)\to R(a);}{L(a)?\to R(a)?}\}$.

Since $(\neg\alpha)? \leftrightarrow \neg(\alpha?)$ and $\alpha \to \alpha?$ are valid formulae in LEI, the extension is $\mathcal{E} = Cn(\mathcal{W} \bigcup \{R(a)?, \neg(R(a)?)\})$ and it is a consistent set. However, if $R(a)$ is later included as a fact, the first default rule will be no more applicable since $R(a)$ and $\neg(R(a)?)$ derives a (strong) contradiction.

Our next example emphasizes the reason why a distinct treatment is given to the normal and the seminormal part of the justification of the default rule. To inhibit the application of an IDL rule as $\frac{:A\to B;\neg C}{A?\to B?}$ you must either derive a ?-free formula '$\neg(A \to B)$', or merely a plausible formula '$C?$'. Such different treatment induces a priority (a preferential order) on defaults: a default has priority over another if the former may be used to prove the exception of the latter. This priority reflects the ***exception first*** principle presented in [9]. The ***exception first*** principle is important to correctly perform nonmonotonic reasoning. Pioneer nonmonotonic logics, for instance, McCarthy's Circumscription [28] and Reiter's default logic do not comply with this principle and they produce unsound inferences in some epistemic situations. This is the case regarding the famous Yale Shooting problem [4], which was intensively discussed in the late eighties [5, 6, 7, 8]. The next example, which may be regarded as a simpler version of the Yale Shooting problem, will be represented in Reiter's default logic and IDL to illustrate this issue.

First, consider the following example in Reiter's default logic.

EXAMPLE 15

$$\mathcal{W} = \{\forall x(W(x) \to F(x)), \forall x(B(x) \to A(x)), B(t)\},$$
$$\mathcal{D} = \{\frac{:A(t) \to \neg F(t), \neg W(t)}{A(t) \to \neg F(t)}, \frac{:B(t) \to W(t)}{B(t) \to W(t)}\}.$$

where 'A' stands for 'Animal', 'F' for 'Fly', 'W' for 'Winged', 'B' for 'Bird' and 't' for 'Tweety'.

In this case, two extensions are inferred, one where Tweety is a bird and flies and another one where Tweety is a bird but does not fly, that is:

$\mathcal{E}_1 = Cn_c(\mathcal{W} \bigcup \{W(t), F(t)\})$ and,

$\mathcal{E}_2 = Cn_c(\mathcal{W} \bigcup \{\neg W(t), \neg F(t)\})$

where Cn_c is the classical consequence operator.

Notice that the second default 'supports' the exception of the first. Thus, according to the exceptions-first principle, it should have priority over it. Therefore, this theory should have only one extension, namely $\mathcal{E}_1$. The second extension, $\mathcal{E}_2$, is anomalous since from the simple fact that Tweety is a bird, one should not infer that it does not fly.

Consider now the same example written in IDL.

EXAMPLE 16

$$\mathcal{W} = \{\forall x(W(x) \to F(x)), \forall x(B(x) \to A(x)), B(t)\},$$
$$\mathcal{D} = \{\frac{:A(t) \to \neg F(t); \neg W(t)}{A(t)? \to (\neg F(t))?}, \frac{:B(t) \to W(t);}{B(t)? \to W(t)?}\}.$$

$\mathcal{E} \;=\; Cn(\mathcal{W} \bigcup \{W(t)?, F(t)?\})$ is the only extension.

Therefore, IDL only infers the intended extension as it takes into account the exception-first principle already mentioned.

4 WELL-PRESENTED IDL THEORIES

As IDL has the ability to accommodate contradiction, it is expected that IDL theories have one and only one extension. The criteria presented in this section guarantee that indeed this is the case for well-presented theories, as proved in theorem 28.

4.1 Acyclic theories

The first condition required is the non-existence of cycles among defaults. Cycles reflect two problems in the presentation of the theory: mutual exclusion and self-exclusion. The first occurs whenever one represents defaults which 'support' (meaning it may be used in the derivation) exceptions to one another, whereas the second occurs if a default that 'supports' its own exception is included in the theory. Both cycles are defective. Potentially, self-exclusion may cause *incoherence*, the non-existence of extensions, and mutual exclusion may give rise to multiple extensions.

Exceptions play a key role in nonmomotonic reasoning as it is pointed out in [9, 29]. The relations based on the derivation of exceptions defined in the sequel are useful to detect cycles among defaults.

DEFINITION 17 [conclusions of a set R of defaults]:
conc(R) $= \{A? \to B? \mid \frac{:A \to B;\neg C}{A? \to B?} \in R, R \subseteq \mathcal{D} \text{ in } < \mathcal{W}, \mathcal{D} >\}$.

DEFINITION 18 [Support Set $R_{\alpha?}$]: Let $< \mathcal{W}, \mathcal{D} >$ be an IDL default theory, $\alpha?$ a ?-formula and $R_{\alpha?} \subseteq \mathcal{D}$. $R_{\alpha?}$ is a support set of $\alpha?$ iff:

1. $\mathcal{W} \cup conc(R_{\alpha?})$ is a consistent set of formulae;
2. $\alpha \in Cn(\mathcal{W} \cup conc(R_{\alpha?}))$;
3. $R_{\alpha?}$ is minimal with respect to $\subseteq$, that is, if there is an $R'_{\alpha?}$, $R'_{\alpha?} \subseteq R_{\alpha?}$, which satisfies the conditions above, then $R'_{\alpha?} = R_{\alpha?}$.

$R_{\alpha?}$ is thought to be a minimal set of defaults, which is used to prove $\alpha?$. Observe that a formula $\alpha?$ may have several associated support sets.

DEFINITION 19 [$\ll$, $\prec_i$ and $\prec$]: Let $< \mathcal{W}, \mathcal{D} >$ be an IDL default theory. Let also $\delta, \delta', \delta''$ be any default in $\mathcal{D}$ and $\alpha?$ be any ?-formula in $\mathcal{L}_?$. Then:

1. $\delta' \ll \alpha?$ iff $\delta' \in R_{\alpha?}$, for some support set $R_{\alpha?}$;
2. $\delta' \prec_i \delta''$ iff $\delta'' = \frac{:A \to B;\neg C}{A? \to B?}$ and $\delta' \ll C?$;
3. $\delta' \prec \delta''$ iff (a) $\delta' \prec_i \delta''$ or (b) there is a default δ such that $\delta' \prec \delta$ and $\delta \prec \delta''$.

Intuitively $\delta \ll \alpha$ implies that the conclusion of δ may be used to prove $\alpha?$; $\delta' \prec_i \delta''$ means that the conclusion of δ' may be used to directly prove the plausibility of the exception of δ''; and relation $\prec$ is the transitive closure of $\prec_i$. Note that δ'' may also be blocked if the irrefutable fact $\neg(A \to B)$ is derived but a default cannot derive a fact.

An IDL default theory $< \mathcal{W}, \mathcal{D} >$ is *acyclic* iff there is no default δ such that $\delta \prec \delta$. Hence, for acyclic theories, $(\mathcal{D}, \prec)$ is a partially ordered set and $\prec$ is a strict order.

EXAMPLE 20 [Self Exclusion]:
$\mathcal{W} = \emptyset$ and $\mathcal{D} = \{\frac{:\alpha,\neg\beta}{\alpha?}, \frac{:\beta,\neg\gamma}{\beta?}, \frac{:\gamma,\neg\alpha}{\gamma?}\}$

$\frac{:\alpha,\neg\beta}{\alpha?} \prec \frac{:\beta,\neg\gamma}{\beta?} \prec \frac{:\gamma,\neg\alpha}{\gamma?} \prec \frac{:\alpha,\neg\beta}{\alpha?}$. This is a case of self-exclusion for $\frac{:\alpha,\neg\beta}{\alpha?}$ indirectly supports $\frac{:\gamma,\neg\alpha}{\gamma?}$ (it blocks $\frac{:\beta,\neg\gamma}{\beta?}$ that blocks $\frac{:\gamma,\neg\alpha}{\gamma?}$) and $\frac{:\gamma,\neg\alpha}{\gamma?}$ blocks $\frac{:\alpha,\neg\beta}{\alpha?}$. After all, $\frac{:\alpha,\neg\beta}{\alpha?}$ indirectly blocks itself.

This theory has no IDL extension.

EXAMPLE 21 [Mutual Exclusion]:
$\mathcal{W} = \emptyset$ and $\mathcal{D} = \{\frac{:\beta,\neg\gamma}{\beta?}, \frac{:\gamma,\neg\beta}{\gamma?}\}$

$\frac{:\beta,\neg\gamma}{\beta?} \prec \frac{:\gamma,\neg\beta}{\gamma?} \prec \frac{:\beta,\neg\gamma}{\beta?}$. This is a case of mutual exclusion for one default supporting the derivation of the exception of the other.

This theory has two IDL extensions: $\mathcal{E}_1 = Cn(\{\beta?\})$ and $\mathcal{E}_2 = Cn(\{\gamma?\})$.

4.2 Stoppered theories

An IDL default theory $< \mathcal{W}, \mathcal{D} >$ is ***stoppered*** iff for all default $\delta \in \mathcal{D}$, there are at most finitely many defaults δ' such that $\delta' \prec \delta$. A theory is not stoppered if either there is an infinite descending chain of defaults — a sequence $\delta_0, \delta_1, \delta_2, \ldots$ such that $\delta_{n+1} \prec \delta_n$ for all n — or there is an exception to a default with infinitely many different supporting sets. Notice that by the compactness theorem of LEI [2, 10] each supporting set is finite. The class of stoppered theories is not too restrictive. For instance, closed IDL theories, where $\mathcal{W}$ and $\mathcal{D}$ are finite and the Herbrand Universe of the language $\mathcal{L}_?$ is finite [30], are stoppered.

Non-stoppered theories may have no extension. As an example, consider the one shown in [31] which is presented here in IDL formalism:

EXAMPLE 22
$\mathcal{W} = \{P_2 \rightarrow Q_1, P_3 \rightarrow Q_2, P_4 \rightarrow Q_3, \ldots\}$ and $\mathcal{D} = \{\frac{:P_1;\neg Q_1}{P_1?}, \frac{:P_2;\neg Q_2}{P_2?}, \frac{:P_3;\neg Q_3}{P_3?}, \ldots\}$

Such theory has the following descending chain:
$\ldots \frac{:P_1;\neg Q_1}{P_1?} \prec \frac{:P_2;\neg Q_2}{P_2?} \prec \frac{:P_1;\neg Q_1}{P_1?}$. This theory is acyclic but it does not have extension.

An enumeration of a set S is an one-to-one map from $C \subseteq \mathbf{N}$ onto S. In an acyclic IDL default theory, an enumeration h of the set $\mathcal{D}$ *conforms to the ordering* $\prec$ iff for all i, j, $i < j$ implies $h(j) \prec h(i)$ does not hold, i.e., either $h(i) \prec h(j)$ or they are incomparable. Theorem 25 assures that for stoppered acyclic theories, there exists an enumeration which conforms to the ordering $\prec$. In order to prove it, the following lemma is proven.

DEFINITION 23 Let $< \mathcal{W}, \mathcal{D} >$ be a stoppered acyclic IDL default theory and $\prec$ an ordering on $\mathcal{D}$, according to definition 19. A default $\delta \in \mathcal{D}$ is *reachable* in a set $S \subseteq \mathcal{D}$ iff for all $\delta' \in \mathcal{D}$, if $\delta' \prec \delta$, then $\delta' \in S$.

LEMMA 24 Let $< \mathcal{W}, \mathcal{D} >$ be a stoppered acyclic IDL default theory and $S \subseteq \mathcal{D}$, then for all $\delta \in \mathcal{D}$, either δ is reachable in S or there is $\delta' \in \mathcal{D}$, such that $\delta' \prec \delta$ and δ' is reachable in S.

Proof:

Let $\delta_0 \in \mathcal{D}$. If δ_0 is reachable in S, the proof is finished. Otherwise, there is $\delta_1 \in \mathcal{D}$, $\delta_1 \prec \delta$, $\delta_1 \notin S$. Now, if δ_1 is reachable in S, the proof is finished. If it is not, then there exists $\delta_2 \in \mathcal{D}$, $\delta_2 \prec \delta_1$, $\delta_2 \notin S$. Continuing in this way, if we stop after finitely many steps, we find a reachable default δ_m, $\delta_m \prec \delta$. Otherwise, there is an infinite descending sequence, $\delta_0, \delta_1, \delta_2, \ldots$, such that $\delta_{n+1} \prec \delta_n$, for all n. But, this is impossible since $< \mathcal{W}, \mathcal{D} >$ is a stoppered theory.

□

THEOREM 25 Let $< \mathcal{W}, \mathcal{D} >$ be a stoppered acyclic IDL default theory, then there is an enumeration of $\mathcal{D}$ that conforms to the ordering $\prec$.

Proof:

Let f be any enumeration of $\mathcal{D}$ (recall that $\mathcal{D}$ is countable). Define the following map h from $C \subseteq \mathbf{N}$ into $\mathcal{D}$.

$\mathcal{D}_0 = \emptyset$

$h(0) = f(k)$, where $k = \min\ \{i \in \mathbf{N}; f(i)$ is reachable in $\mathcal{D}_0\}$

For $n \geq 0$,

$\mathcal{D}_{n+1} = \mathcal{D}_n \bigcup h(n)$

$h(n+1) = f(k)$, where $k = \min\{i \in \mathbf{N}; f(i)$ is reachable in $\mathcal{D}_n$ and $f(i) \notin \mathcal{D}_n\}$.

Recursion theorem guarantees that h is a function, since the minimum element of any $\emptyset \neq S \subseteq \mathbf{N}$ exists and it is unique. Lemma 24 assures that the sets on the right hand side of the foregoing definition are not empty. Moreover, h is one-to-one since in each step $h(n+1) \neq h(i)$, for all $i \leq n$. To show that that h is onto lemma 26 is proven.

LEMMA 26 Let h and $\mathcal{D}_n$, for $n \geq 0$, defined as above and let $\delta \in \mathcal{D}$. If δ is reachable in $\mathcal{D}_{k0}$, for some k_0, and $\delta \notin \mathcal{D}_{k0}$, then $h(k) = \delta$, for some $k > k_0$.

Proof:

Let $\delta \in \mathcal{D}$, $\delta \notin \mathcal{D}_{k0}$, δ reachable in $\mathcal{D}_{k0}$. Then $f(i_0) = \delta$, for some i_0. So, there are at most i_0 defaults δ' also reachable in $\mathcal{D}_{k0}$, such that $f^{-1}(\delta') < i_0$. Therefore, $h(k) = \delta$, for some $k_0 < k \leq k_0 + i_0 + 1$.

□

Hence, by lemma 26, to show that h is onto, it is enough to show that every $\delta \in \mathcal{D}$ is reachable in some $\mathcal{D}_k$. So, let $\delta_0 \in \mathcal{D}$ and let $k_0 \in \mathbf{N}$. If δ_0 is reachable in $\mathcal{D}_{k0}$, the proof is finished. Otherwise, by lemma 24, there is $\delta_1 \prec \delta_0$, such that δ_1 is reachable in $\mathcal{D}_{k0}$. So, by lemma 26, $h(k_1) = d_1$, for some $k_1 \geq k_0$, hence $\delta_1 \in \mathcal{D}_{k1+1}$. If δ_0 is still not reachable in $\mathcal{D}_{k1+1}$, then there is $\delta_2 \prec \delta_0$, δ_2 is reachable in $\mathcal{D}_{k1+1}$. So, once again by lemma 26, $h(k_2) = \delta_2$, for some $k_2 \geq (k_1 + 1)$ and $\delta_2 \in \mathcal{D}_{k2+1}$. As there are only finitely many defaults $\delta_m \prec \delta_0$, after at most m steps, δ_0 will be reachable in $\mathcal{D}_{km}$, for some k_m, so $h(k) = \delta_0$, for some k.

Notice that h conforms to the ordering $\prec$, for if $h(k+1) = \delta$, then δ is reachable in $\mathcal{D}_k$, so for all $\delta' \prec \delta$, $\delta' \in \mathcal{D}_k$, therefore, $h^{-1}(\delta') < k$.

□

4.3 Consistency of facts

The third condition for theories to be well-presented is quite obvious, it requires the set of irrefutable facts $\mathcal{W}$ to be (classically) consistent, that is, for no ?-free formula A, $\mathcal{W} \models A \wedge \neg A$. Otherwise, the theory would have an inconsistent trivial extension.

The three conditions above characterize ***well-presented theories.***

DEFINITION 27 [Well-Presented IDL Theories]: An IDL Theory $< \mathcal{W}, \mathcal{D} >$ is well-presented if it is: (i) acyclic; (ii) stoppered and (iii) $\mathcal{W}$ is a consistent set of ?- free formulae.

The next theorem, proved in [26], vindicates the restrictions imposed on well-presented theories. It asserts that, indeed, they have a single extension.

THEOREM 28 [Uniqueness]: A well-presented IDL theory $< \mathcal{W}, \mathcal{D} >$ has a unique extension $\mathcal{E}$.

As for Reiter's default logic, multiple extensions do not pose a problem, and if one is only interested in criteria to guarantee the existence of extensions, mutual exclusion cycles should be allowed. However, in [32], Etherington gives a sufficient condition to the existence of extensions that is too restrictive, since it unnecessary rules out mutual exclusion cycles. He uses an order on defaults following an intuition similar to ours, although formalized in a completely different way. In [11], Martins defines an order on defaults separating self- exclusion from mutual exclusion cycles and M. Pequeno [33] proves that acyclic default theories, where only self-exclusion cycles are considered, have extensions. Still, there is a reason for considering mutual exclusion cycles defective even for default logic. Splitting of extensions should be caused by the conclusions of defaults not by their exceptions, but this is very polemical and this is not the appropriate place for such a debate.

Verification of whether a given theory is well-presented is computationally untractable, since it involves consistency tests and derivation of formulae, which is ultimately undecidable. In fact, nonmonotonic logics, in general, are undecidable, since nonmonotonicity stems from an appeal to consistency. In practice, restrictions to decidable and tractable theories are necessary.

Before finishing this section, a constructive way to calculate the single extension for a well-presented IDL theory will be presented. It will be useful to prove the completeness theorem in section 6.

DEFINITION 29 Let $< \mathcal{W}, \mathcal{D} >$ be an IDL default theory and let $\delta_0, \delta_1, \ldots$ be an enumeration for all default in $\mathcal{D}$ which conforms to the order among them. Then:

$\mathcal{E}_0 = Cn(\mathcal{W})$

and for $i \geq 0$,

$\mathcal{E}_{i+1} = Cn(\mathcal{E}_i \cup \{A? \rightarrow B? \mid \delta_i = \frac{:A \rightarrow B; \neg C}{A? \rightarrow B?}, \neg(A \rightarrow B) \notin \mathcal{E}_i \text{ and } C? \notin \mathcal{E}_i\})$.

THEOREM 30 Let $< \mathcal{W}, \mathcal{D} >$ be an IDL default theory and let $\mathcal{E}_i$, for $i \geq 0$, be defined as in definition 29. Then $\mathcal{E}$ is the extension for $< \mathcal{W}, \mathcal{D} >$ iff $\mathcal{E} = \bigcup_{i=0}^{\infty} \mathcal{E}_i$.

The proof of this theorem is detailed in appendix A.

5 IDL SEMANTICS

IDL semantics will be defined as an extension of LEI semantics. An IDL model for a well-presented IDL theory $< \mathcal{W}, \mathcal{D} >$ will satisfy all formulae in $\mathcal{W}$ and all plausible conclusions that comes from applicable defaults (see definition 13). From a class of all plausible world models $\mathcal{M}$ for $\mathcal{W}$, a subset, which satisfies exactly all applicable defaults is found. Such a set is named the IDL model $\mathcal{U}$.

In the following definitions, let $< \mathcal{W}, \mathcal{D} >$ be the well-presented IDL default theory under consideration. Then:

DEFINITION 31 [$MOD(\mathcal{W})$]: $MOD(\mathcal{W})$ is the set of all plausible world models $\mathcal{M}$ for $\mathcal{W}$.

DEFINITION 32 [IDL Structure $\mathcal{U}$] An IDL Structure $\mathcal{U}$ is any non-empty subset of MOD($\mathcal{W}$).

The satisfaction of a formula α, of a set of formulae Γ and of a default will be now defined. For the sake of simplicity, the same symbol of satisfaction will be used for all notions, but the identification of which notion it will refer to is straightforward.

DEFINITION 33 [$\mathcal{U} \models\!\approx \alpha$]: $\mathcal{U} \models\!\approx \alpha$ iff for all $\mathcal{M} \in \mathcal{U}$, $\mathcal{M} \models \alpha$.

DEFINITION 34 [$\mathcal{U} \models\!\approx \Gamma$]: $\mathcal{U} \models\!\approx \Gamma$ if $\mathcal{U} \models\!\approx \alpha$ for all formula α in Γ.

DEFINITION 35 [$\mathcal{U} \models\!\approx \frac{:A \to B;\neg C}{A? \to B?}$]: $\mathcal{U} \models\!\approx \frac{:A \to B;\neg C}{A? \to B?}$ iff $\mathcal{U} \not\models\!\approx \neg(A \to B)$ and $\mathcal{U} \not\models\!\approx C?$.

However, how could a model for $< \mathcal{W}, \mathcal{D} >$, which satisfies exactly all applicable defaults be defined? In order to do this, the enumeration defined at the end of the previous section will be used. Therefore, the satisfaction of a default δ_j will be tested in a structure that already satisfies all applicable defaults δ_i, $i < j$. The definition of the IDL model $\mathcal{U}$ for $< \mathcal{W}, \mathcal{D} >$ is then:

DEFINITION 36 [IDL Model $\mathcal{U}$ for $< \mathcal{W}, \mathcal{D} >$] Let $< \mathcal{W}, \mathcal{D} >$ be an IDL default theory and let $\delta_0, \delta_1, \ldots$ be an enumeration for all defaults in $\mathcal{D}$, which conforms to the order among them. Then:

$\mathcal{U}_0 = MOD(\mathcal{W})$

and for $i \geq 0$,

$\mathcal{U}_{i+1} = \mathcal{U}_i \cap MOD(\{A? \to B? \mid \delta_i = \frac{:A \to B;\neg C}{A? \to B?},\ \mathcal{U}_i \models\!\approx \frac{:A \to B;\neg C}{A? \to B?}\})$. [3]

The **IDL Model** for $< \mathcal{W}, \mathcal{D} >$ is the IDL Structure $\mathcal{U}$ such that $\mathcal{U} = \bigcap_{i=0}^{\infty} \mathcal{U}_i$. In this case, $\mathcal{U}$ satisfies $< \mathcal{W}, \mathcal{D} >$ or, equivalently, $\mathcal{U} \models\!\approx < \mathcal{W}, \mathcal{D} >$.

Note that $\mathcal{U}$ is well defined for $\bigcap_{i=0}^{\infty} \mathcal{U}_i$ is invariant for any enumeration which conforms to the order among defaults.

DEFINITION 37 [$< \mathcal{W}, \mathcal{D} > \models\!\approx \alpha$]: $< \mathcal{W}, \mathcal{D} > \models\!\approx \alpha$ iff the IDL Model $\mathcal{U}$ for $< \mathcal{W}, \mathcal{D} >$ satisfies α, that is, if $\mathcal{U} \models\!\approx < \mathcal{W}, \mathcal{D} >$ then $\mathcal{U} \models\!\approx \alpha$.

[3] Observe that if $\mathcal{U}_i \not\models\!\approx \frac{:A \to B;\neg C}{A? \to B?}$ then $\mathcal{U}_{i+1} = \mathcal{U}_i$.

In order to illustrate these definitions, the examples of section 3 will be used again:

EXAMPLE 14 revisited:
$\mathcal{W} = \{L(a) \wedge C(a)\}$,
$\mathcal{D} = \{\frac{:C(a)\rightarrow\neg R(a);}{C(a)?\rightarrow(\neg R(a))?}, \frac{:L(a)\rightarrow R(a);}{L(a)?\rightarrow R(a)?}\}$.

Since both default rules are normal ones, any enumeration is acceptable. Consider, for instance, $\delta_0 = \frac{:C(a)\rightarrow\neg R(a);}{C(a)?\rightarrow(\neg R(a))?}$ and $\delta_1 = \frac{:L(a)\rightarrow R(a);}{L(a)?\rightarrow R(a)?}$. Thence:

$\mathcal{U}_0 = MOD(\mathcal{W})$

$\mathcal{U}_1 = \mathcal{U}_0 \cap MOD(\{C(a)? \rightarrow (\neg R(a))?\})$
since $\mathcal{U}_0 \approx \frac{:C(a)\rightarrow\neg R(a);}{C(a)?\rightarrow(\neg R(a))?}$, that is, since there is a plausible world model $\mathcal{M} \in \mathcal{U}_0$ such that $\mathcal{M} \not\models \neg(C(a) \rightarrow \neg R(a))$. For instance, consider $\mathcal{M}$ satisfying $L(a), C(a)$ and $\neg R(a)$.

$\mathcal{U}_2 = \mathcal{U}_1 \cap MOD(\{L(a)? \rightarrow R(a)?\})$
since $\mathcal{U}_1 \approx \frac{:L(a)\rightarrow R(a);}{L(a)?\rightarrow R(a)?}$, that is, since there is a plausible world model $\mathcal{M} \in \mathcal{U}_1$ such that $\mathcal{M} \not\models \neg(L(a) \rightarrow R(a))$. For instance, consider $\mathcal{M}$ satisfying $L(a), C(a)$ and $R(a)$.

EXAMPLE 16 revisited:
$\mathcal{W} = \{\forall x(W(x) \rightarrow F(x)), \forall x(B(x) \rightarrow A(x)), B(t)\}$,
$\mathcal{D} = \{\frac{:A(t)\rightarrow\neg F(t);\neg W(t)}{A(t)?\rightarrow(\neg F(t))?}, \frac{:B(t)\rightarrow W(t);}{B(t)?\rightarrow W(t)?}\}$.

The enumeration should be $\delta_0 = \frac{:B(t)\rightarrow W(t);}{B(t)?\rightarrow W(t)?}$ and $\delta_1 = \frac{:A(t)\rightarrow\neg F(t);\neg W(t)}{A(t)?\rightarrow(\neg F(t))?}$ since $\frac{:B(t)\rightarrow W(t);}{B(t)?\rightarrow W(t)?} \ll W(t)?$ and $\frac{:B(t)\rightarrow W(t);}{B(t)?\rightarrow W(t)?} \prec \frac{:A(t)\rightarrow\neg F(t);\neg W(t)}{A(t)?\rightarrow(\neg F(t))?}$.

$\mathcal{U}_0 = MOD(\mathcal{W})$

$\mathcal{U}_1 = \mathcal{U}_0 \cap MOD(\{B(t)? \rightarrow W(t)?\})$
since $\mathcal{U}_0 \approx \frac{:B(t)\rightarrow W(t);}{B(t)?\rightarrow W(t)?}$, that is, since there is a plausible world model $\mathcal{M} \in \mathcal{U}_0$ such that $\mathcal{M} \not\models \neg(B(t) \rightarrow W(t))$. For instance, consider $\mathcal{M}$ satisfying $B(t)$ and $W(t)$.

$\mathcal{U}_2 = \mathcal{U}_1$
since $\mathcal{U}_1 \not\approx \frac{:A(t)\rightarrow\neg F(t);\neg W(t)}{A(t)?\rightarrow(\neg F(t))?}$, that is, since for all plausible world models $\mathcal{M} \in \mathcal{U}_1$, $\mathcal{M} \models W(t)?$ as it satisfies $B(t)$ and $B(t)? \rightarrow W(t)?$.

6 Soundness and Completeness

LEMMA 38 Let $< \mathcal{W}, \mathcal{D} >$ be a well-presented IDL theory and consider the enumeration $\delta_0, \delta_1, \ldots$ for defaults as defined in the previous section. Let $\mathcal{E}$ be its associated IDL extension $\mathcal{E} = \bigcup_{i=0}^{\infty} \mathcal{E}_i$ (see theorem 30) and the IDL model $\mathcal{U}$ for $< \mathcal{W}, \mathcal{D} >$ defined as $\mathcal{U} = \bigcap_{i=0}^{\infty} \mathcal{U}_i$. Then: for all $i \geq 0$, $\mathcal{U}_i = MOD(\mathcal{E}_i)$.

Proof: By induction on i.

(Basis): $\mathcal{U}_0 = MOD(\mathcal{E}_0)$.

$\mathcal{W} \models \alpha$ iff $\alpha \in Cn(\mathcal{W}) = \mathcal{E}_0$. Then, $\mathcal{W} \models \mathcal{E}_0$. Furthermore, as $\mathcal{W} \subseteq Cn(\mathcal{W})$, then $\mathcal{E}_0 \models \mathcal{W}$.

Therefore, $\mathcal{M} \in \mathcal{U}_0 = MOD(\mathcal{W})$ iff $\mathcal{M} \models \mathcal{W}$ iff $\mathcal{M} \models \mathcal{E}_0$ iff $\mathcal{M} \in MOD(\mathcal{E}_0)$.

(Hypothesis): $\mathcal{U}_k = MOD(\mathcal{E}_k)$.

(Step): $\mathcal{U}_{k+1} = MOD(\mathcal{E}_{k+1})$.

$\mathcal{M} \in MOD(\mathcal{E}_{k+1})$ iff

$\mathcal{M} \in MOD(Cn(\mathcal{E}_k \cup \{A? \rightarrow B? \mid \delta_k = \frac{:A \rightarrow B; \neg C}{A? \rightarrow B?}, \neg(A \rightarrow B) \notin \mathcal{E}_k, C? \notin \mathcal{E}_k\}))$ iff

$\mathcal{M} \in MOD(\mathcal{E}_k \cup \{A? \rightarrow B? \mid \delta_k = \frac{:A \rightarrow B; \neg C}{A? \rightarrow B?}, \neg(A \rightarrow B) \notin \mathcal{E}_k, C? \notin \mathcal{E}_k\})$ iff

$\mathcal{M} \in MOD(\mathcal{E}_k) \cap MOD(\{A? \rightarrow B? \mid \delta_k = \frac{:A \rightarrow B; \neg C}{A? \rightarrow B?}, \neg(A \rightarrow B) \notin \mathcal{E}_k, C? \notin \mathcal{E}_k\})$ iff

(since $MOD(\mathcal{E}_k) = \mathcal{U}_k$, by hypothesis)

$\mathcal{M} \in \mathcal{U}_k \cap MOD(\{A? \rightarrow B? \mid \delta_k = \frac{:A \rightarrow B; \neg C}{A? \rightarrow B?}, \mathcal{U}_k \not\approx \neg(A \rightarrow B), \mathcal{U}_k \not\approx C?\})$ iff

$\mathcal{M} \in \mathcal{U}_k \cap MOD(\{A? \rightarrow B? \mid \delta_k = \frac{:A \rightarrow B; \neg C}{A? \rightarrow B?}, \mathcal{U}_k \approx \frac{:A \rightarrow B; \neg C}{A? \rightarrow B?})$ iff

$\mathcal{M} \in \mathcal{U}_{k+1}$.

□

THEOREM 39 [Soundness and Completeness for IDL]: Let $< \mathcal{W}, \mathcal{D} >$ be a well-presented IDL theory and α a formula in $\mathcal{L}_?$. Then:

$< \mathcal{W}, \mathcal{D} > \mid\!\sim \alpha$ iff $< \mathcal{W}, \mathcal{D} > \approx \alpha$.

By the use of definition 33, definition 36 and theorem 30, in order to prove soundness and completeness it is just necessary to prove: $\alpha \in \mathcal{E}$ iff $\mathcal{U} \approx \alpha$, where $< \mathcal{W}, \mathcal{D} >$ is the well-presented IDL theory under consideration, $\mathcal{E}$ is its associated IDL extension, $\mathcal{U}$ is its associated IDL model and α is a formula in $\mathcal{L}_?$. In fact, to prove that $\alpha \in \mathcal{E}$ iff $\mathcal{U} \approx \alpha$ one just needs to prove that $\mathcal{U} = MOD(\mathcal{E})$.

Proof:

(1th Case): $\mathcal{U} \subseteq MOD(\mathcal{E})$: Let $\mathcal{M} \in \mathcal{U}$. Then $\mathcal{M} \in \mathcal{U}_i$, for all $i \geq 0$ since $\mathcal{U} = \bigcap_{i=0}^{\infty} \mathcal{U}_i$. Let α be any formula in $\mathcal{E}$. So $\alpha \in \mathcal{E}_i$, for some $i \geq 0$, since $\mathcal{E} = \bigcup_{i=0}^{\infty} \mathcal{E}_i$. Hence, by lemma 38, $\mathcal{U}_i \approx \alpha$. As $\mathcal{M} \in \mathcal{U}_i$ then $\mathcal{M} \models \alpha$. Therefore, $\mathcal{M} \models \mathcal{E}$ and $\mathcal{M} \in MOD(\mathcal{E})$.

(2nd Case): $MOD(\mathcal{E}) \subseteq \mathcal{U}$: Let $\mathcal{M} \in MOD(\mathcal{E})$. Then, $\mathcal{M} \models \mathcal{E}$. Since $\mathcal{E} = \bigcup_{i=0}^{\infty} \mathcal{E}_i$, $\mathcal{M} \models \mathcal{E}_i$ for all $\mathcal{E}_i$, $i \geq 0$. Hence, $\mathcal{M} \in MOD(\mathcal{E}_i)$, for all $i \geq 0$. By lemma 38, $\mathcal{M} \in \mathcal{U}_i$, for all $i \geq 0$. Thus, since $\mathcal{U} = \bigcap_{i=0}^{\infty} \mathcal{U}_i$, $\mathcal{M} \in \mathcal{U}$.

□

7 Comparison to Other Works

In recent years, several semantics for nonmonotonic logics that follow an altogether different approach from the one presented in this paper have been proposed. Some approaches develop a semantics based on possible worlds [34], [35]; others translate the operational meaning of a default rule to modal operators in an epistemic logic [36]; and there are those which formalize default reasoning through algebraic notions [37]. In this section, we make comparisons with semantics related to our approach, namely: stable models for default logics; preferential semantics for preferential logics and semantics for systems that combines nonmonotonicity and paraconsistency.

7.1 Stable Semantics

Stable models semantics for Reiter's default logic was presented by Etherington in [32] and [38]. He characterized his semantics by observing that defaults can be viewed as extending the first-order knowledge about an incompletely specified world. Default select restricted subsets of the models of the underlying first-order theory. Lukaszewics [39] formalized this intuition for default theories with just normal defaults [4]. His semantics requires the property of semimonotonicity which allows each normal default to be treated independently. However, for defaults with seminormal justifications, semimonotonicity does not always hold. Etherington's semantics generalizes his work to cover the entire class of default theories.

By the fact that semimonotonicity is not valid for arbitrary default theories, the application of one default after another does not, in general, lead to extensions. It is necessary to ensure that the application of each default does not violate the justifications of the already applied defaults. Etherington's semantics defines an order of preference $\geq_\delta$ among models indexed by a default δ in order to capture more specialized world descriptions, starting with the class of all classical models for $\mathcal{W}$.

For normal default theories, it is sufficient to take into account the maximal elements of this order. A stability condition is introduced as a pruning mechanism to properly deal with the failure of semimonotonicity in arbitrary default theories. As a consequence, the satisfiability of all justifications of the applied default rules must be ensured in the resulting class of models. Stable set of models provides a semantical interpretation for a default theory. Results of soundness and completeness in relation to Default Logic are stated in [38].

Another stable semantics for Reiter's default logic was presented by Guerreiro and Casanova in [40]. In this formalism, the preference criterion is based upon set inclusion not directly indexed by defaults as in the Etherington's proposal. Both semantics find the same set of first-order classical models to default theories but Etherington's semantics begins with models of $\mathcal{W}$ and then moves to smaller sets whereas Guerreiro and Casanova's semantics starts with smaller sets (D-models) and moves to larger ones (stable D-models).

There are several differences and similarities between IDL semantics and stable semantics for Reiter's default logic. The first difference is that IDL is based on the set of all plausible world models for $\mathcal{W}$, instead of classical models. This is an obvious consequence of the fact that the IDL monotonic basis is paraconsistent,

[4] Normal default rules do not have seminormal justifications.

and not classical, as in the case of Reiter's default logic. On the other hand, the stability condition is not imposed since it is assured, in the definition of the IDL model, by the use of the order $\prec$ among defaults.

Despite these differences, stable semantics has inspired IDL semantics. In fact, starting from the set of all LEI plausible world models for $\mathcal{W}$, a subset of it is found, and this subset satisfies exactly all applicable defaults of the IDL extension. Like Etherington, we move from a larger set of models in the direction of a smaller set, taking set inclusion as the underlying order as do Guerreiro and Casanova. An enumeration of defaults which conforms to the order $\prec$ among them is also used to eliminate models that do not satisfy the conclusion of the correspondent applicable defaults in the IDL extension. In IDL formalism, the order $\prec$ is defined among defaults whereas in Etherington's semantics the order $\geq_\delta$ is defined among models that satisfy a default δ.

7.2 Preferential Models

Preferential models were developed by Shoham [41], [42] as a generalisation of the notion of circumscription [28]. The essential idea of circumscription is to select models from those that satisfy a set Γ of formulae, which are in some respect 'minimal'. Shoham's work generalizes the idea of minimality to any strict partial order over the models.

In its most general form ([43] and [44], [45]), a *preferential model* may be defined to be a triple $\mathcal{M} = (M, \models, <)$ where:

1. M is an arbitrary set. Its elements are called ***states***.
2. $\models$ is an arbitrary relation between elements of M and formulae of the language $\mathcal{L}$ under consideration, i.e. $\models \subseteq M \times \mathcal{L}$. It is called the ***satisfaction relation*** of the model.
3. $<$ is an arbitrary relation between elements of M, i.e. $< \subseteq M \times M$. It is called the ***preference relation*** of the model.

Given a preferential model $\mathcal{M} = (M, \models, <)$, the notion of preferential satisfaction is defined: if Γ is a set of formulae and $m \in M$, then m ***preferentially satisfies*** Γ, written $m \models_< \Gamma$, iff $m \models \Gamma$ and there is no $n \in M$ with $n < m$ such that $n \models \Gamma$. Given a preferential model $\mathcal{M} = (M, \models, <)$, a relation $\mid\!\sim_<$ of preferential entailment is defined as:

$\Gamma \mid\!\sim_< \alpha$ iff for all m $\in M$, if m $\models_< \Gamma$ then m $\models \alpha$

In contrast to the stable semantics for default logics, preferential models give an intuitive semantical view not attached to the syntax of those logics for which they provide semantics. They characterize a large class of nonmonotonic logics called preferential logics [43].

In [11, 46], it was proved that IDL satisfies the same advantageous properties that a preferential logic must obey, such as: reflexivity, cut, cautious monotonicity, left logical equivalence, right weakening and OR [43, 44, 45]. Hence, the IDL model $\mathcal{U}$ for a well-presented IDL theory $< \mathcal{W}, \mathcal{D} >$ may be thought of as the preferential model for this theory. The preferential entailment would be defined as: $< \mathcal{W}, \mathcal{D} > \models\!\!\!\approx$ α iff the *prefered* IDL Model $\mathcal{U}$ for $< \mathcal{W}, \mathcal{D} >$ satisfies α, that is, if $\mathcal{U} \models\!\!\!\approx < \mathcal{W}, \mathcal{D} >$ then for all $\mathcal{M} \in \mathcal{U}$, $\mathcal{M} \models \alpha$.

7.3 Semantics for Nonmonotonic and Paraconsistent Formalisms

In [15], a logical system of reasoning about truth is presented. The logic **LP** is obtained by relaxing the assumption that sentences cannot be simultaneously true and false. The semantical interpretation of formulae allows one to assign three truth values to atomic formulae, $\{1\}$, $\{0\}$ and $\{1,0\}$, which are intended to mean 'true and true only', 'false and false only' and 'both', respectively. A formula α is true (under an interpretation) iff 1 is in its truth value (under that interpretation) and α is false iff 0 is in its truth value. The semantical interpretation of **LP** connectives — conjunction, disjunction and negation — are classical. '$\alpha \rightarrow \beta$' is defined as '$\neg\alpha \vee \beta$'. The notions of model of a formula, of a set of formulae and of entailment is also taken as in the classical case.

LP is proved to be a paraconsistent logic. Therefore the *Ex contradiction quodlibet* principle fails. The disjunctive syllogism fails too. This seems to be a weakness since it is sometimes a reasonable inference. However, as pointed out in [15], there are situations where disjunctive syllogism must fail, the ones that deal with inconsistency. In order to properly formalize reasoning about truth, **LP** is extended to take consistency as a default assumption, that is, it is supposed that there is no more inconsistency than one is forced to suppose. A notion of minimal model with respect to inconsistency is characterized and the default consequence relation $\models_m$ is defined as: $\Gamma \models_m \alpha$ iff every minimal model of Γ is an **LP** model of α. This logic is called $\mathbf{LP}_m$ and it is shown to be paraconsistent and nonmonotonic.

Differently from **LP**, all connectives, including implication, are primitives. In fact, $(\alpha \rightarrow \beta) \rightarrow ((\neg\alpha) \vee \beta)$ is a LEI valid formula but not the converse. Moreover, Modus Ponens is a valid rule in LEI but not the disjunctive syllogism. In **LP**, Modus Ponens is not a valid inference since it is just an alternative way of representing the disjunctive syllogism.

Although paraconsistent and nonmonotonic, $\mathbf{LP}_m$ is distinct from IDL in several aspects. First of all, the underlying monotonic basis are different as just mentioned. Moreover, $\mathbf{LP}_m$ implements a default reasoning just in its semantical definition to complement the inferential power of the paraconsistent negation whereas IDL, on the other hand, uses default reasoning as a general rule in the language. IDL semantics is therefore more expressive than $\mathbf{LP}_m$ semantics in the sense that it allows one to model default rules with exceptions.

Some other systems of logics, which combine paraconsistency and nonomonotonicy, are the ones presented by Arieli and Avron in [13]. This is an interesting paper, which has already been briefly commented on in the first section of this paper. Some paraconsistent and nonmonotonic entailment relations are therein presented. They are based on a four-valued based semantics, which is proved to simulate $\mathbf{LP}_m$ entailment and also arbitrary multi-valued semantics proposed to model incompleteness and inconsistencies that arise in the course of practical reasoning. A four-valued semantics for IDL has also been investigated in [47]. A comparison with Arieli and Avron work is under investigation.

A A Constructive Definition of the IDL Extension

THEOREM 30: Let $<\mathcal{W},\mathcal{D}>$ be an IDL default theory and let $\mathcal{E}_i$, for $i \geq 0$, as in definition 29. Then $\mathcal{E}$ is the extension for $<\mathcal{W},\mathcal{D}>$ iff $\mathcal{E} = \bigcup_{i=0}^{\infty} \mathcal{E}_i$.

Proof:

($\Rightarrow$) if $\mathcal{E}$ is the extension for $<\mathcal{W},\mathcal{D}>$ then $\mathcal{E} = \bigcup_{i=0}^{\infty} \mathcal{E}_i$.

It is necessary to show that (a) $\bigcup_{i=0}^{\infty} \mathcal{E}_i \subseteq \mathcal{E}$ and (b)$\mathcal{E} \subseteq \bigcup_{i=0}^{\infty} \mathcal{E}_i$.

(a) $\bigcup_{i=0}^{\infty} \mathcal{E}_i \subseteq \mathcal{E}$

By induction on i.

$\mathcal{E}_0 = Cn(\mathcal{W}) \subseteq \mathcal{E}$, by conditions 1 and 2 in definition 3.1.

Suppose that $\mathcal{E}_j \subseteq \mathcal{E}$, for all $j \leq i$.

It will be proven that $\mathcal{E}_{i+1} \subseteq \mathcal{E}$.

$\mathcal{E}_{i+1} = Cn(\mathcal{E}_i \cup \{A? \to B? \mid \delta_i = \frac{:A \to B; \neg C}{A? \to B?}, \neg(A \to B) \notin \mathcal{E}_i \text{ and } C? \notin \mathcal{E}_i\})$.

By condition 2 in definition 3.1 and by the inductive hypothesis, it is enough to show that

$A? \to B? \in \mathcal{E}$, for $\delta_i = \frac{:A \to B; \neg C}{A? \to B?}$, $\neg(A \to B) \notin \mathcal{E}_i$ and $C? \notin \mathcal{E}_i$.

By condition 3 in definition 3.1, it is enough to show that $\neg(A \to B) \notin \mathcal{E}$ and $C? \notin \mathcal{E}$.

Remember that $\neg(A \to B) \notin \mathcal{E}_i$. However, since $\neg(A \to B)$ is ?-free, for $\neg(A \to B) \in \mathcal{E}$ it must belong to $Cn(\mathcal{W})$. But this is impossible since $Cn(\mathcal{W}) = \mathcal{E}_0 \subseteq \mathcal{E}_i$. Then, $\neg(A \to B) \notin \mathcal{E}$.

Now suppose that $C? \in \mathcal{E}$. As $C? \notin \mathcal{E}_i$, then there is at least a default $\delta' = \frac{:A' \to B'; \neg C'}{A'? \to B'?}$ such that $\delta' \ll C?$, $A'? \to B'? \in \mathcal{E}$ and $A'? \to B'? \notin \mathcal{E}_i$.

Hence, $\delta' \prec \delta_i$. Therefore, $\delta' = \delta_j, j < i$.

Thence, $\mathcal{E}_j \subseteq \mathcal{E}_i \subseteq \mathcal{E}$.

As $A'? \to B'? \in \mathcal{E}$, then $\neg(A' \to B') \notin \mathcal{E}$ and $C'? \notin \mathcal{E}$.

Therefore, $\neg(A' \to B') \notin \mathcal{E}_j$ and $C'? \notin \mathcal{E}_j$.

Thus, $A'? \to B'? \in \mathcal{E}_{j+1} \subseteq \mathcal{E}_i$, a contradiction!

(b) $\mathcal{E} \subseteq \bigcup_{i=0}^{\infty} \mathcal{E}_i$.

Since $\mathcal{E} = \mathcal{C}(\mathcal{E})$ and $\mathcal{C}(\mathcal{E})$ is the minimal set of formulae satisfying conditions 1, 2 and 3 in definition 3.1, it is enough to show that $\bigcup_{i=0}^{\infty} \mathcal{E}_i$ also satisfies these three conditions for $\Gamma = \mathcal{E}$. In effect,

1. $\mathcal{W} \subseteq \bigcup_{i=0}^{\infty} \mathcal{E}_i$, since $\mathcal{E}_0 = Cn(\mathcal{W})$;
2. $Cn(\bigcup_{i=0}^{\infty} \mathcal{E}_i) = \bigcup_{i=0}^{\infty} \mathcal{E}_i$ by definition of $\mathcal{E}_i$;
3. If $\delta = \frac{:A \to B; \neg C}{A? \to B?} \in \mathcal{D}$ and $\neg(A \to B) \notin \mathcal{E}$ and $C? \notin \mathcal{E}$ then it is necessary to show that $A? \to B? \in \bigcup_{i=0}^{\infty} \mathcal{E}_i$.

In effect, $\delta = \delta_j$, for some j in the enumeration.

As, $\bigcup_{i=0}^{\infty} \mathcal{E}_i \subseteq \mathcal{E}$ by item (a), $\neg(A \to B) \notin \mathcal{E}_j$ and $C? \notin \mathcal{E}_j$.

Hence, $A? \to B? \in \mathcal{E}_{j+1}$. Therefore, $A? \to B? \in \bigcup_{i=0}^{\infty} \mathcal{E}_i$.

($\Leftarrow$) If $\mathcal{E} = \bigcup_{i=0}^{\infty} \mathcal{E}_i$ then $\mathcal{E}$ is the extension for $< \mathcal{W}, \mathcal{D} >$.

It must be shown that $\mathcal{C}(\bigcup_{i=0}^{\infty} \mathcal{E}_i) = \bigcup_{i=0}^{\infty} \mathcal{E}_i$. Then, it is necessary to prove that (a) $\mathcal{C}(\bigcup_{i=0}^{\infty} \mathcal{E}_i) \subseteq \bigcup_{i=0}^{\infty} \mathcal{E}_i$ and (b) $\bigcup_{i=0}^{\infty} \mathcal{E}_i \subseteq \mathcal{C}(\bigcup_{i=0}^{\infty} \mathcal{E}_i)$.

(a) $\mathcal{C}(\bigcup_{i=0}^{\infty} \mathcal{E}_i) \subseteq \bigcup_{i=0}^{\infty} \mathcal{E}_i$

As $\mathcal{C}(\bigcup_{i=0}^{\infty} \mathcal{E}_i)$ is the minimal set of formulae satisfying conditions 1,2,3 of definition 3.1, it is enough to show that $\bigcup_{i=0}^{\infty} \mathcal{E}_i$ satisfies these three conditions with $\Gamma = \bigcup_{i=0}^{\infty} \mathcal{E}_i$.

In effect,

1. $\mathcal{W} \subseteq \bigcup_{i=0}^{\infty} \mathcal{E}_i$, since $\mathcal{E}_0 = Cn(\mathcal{W})$;
2. $Cn(\bigcup_{i=0}^{\infty} \mathcal{E}_i) = \bigcup_{i=0}^{\infty} \mathcal{E}_i$, by definition of $\mathcal{E}_i$;
3. If $\delta = \frac{:A \to B; \neg C}{A? \to B?} \in \mathcal{D}$ and $\neg(A \to B) \notin \bigcup_{i=0}^{\infty} \mathcal{E}_i$ and $C? \notin \bigcup_{i=0}^{\infty} \mathcal{E}_i$, it must be shown that $A? \to B? \in \bigcup_{i=0}^{\infty} \mathcal{E}_i$.

In effect, $\delta = \delta_j$, for some j in the enumeration. Since $\neg(A \to B) \notin \mathcal{E}_j$ and $C? \notin \mathcal{E}_j$, thence $A? \to B? \in \mathcal{E}_{j+1}$. Therefore, $A? \to B? \in \bigcup_{i=0}^{\infty} \mathcal{E}_i$.

(b) $\bigcup_{i=0}^{\infty} \mathcal{E}_i \subseteq \mathcal{C}(\bigcup_{i=0}^{\infty} \mathcal{E}_i)$.

By induction on i:

$\mathcal{E}_0 = Cn(\mathcal{W}) \subseteq \mathcal{C}(\bigcup_{i=0}^{\infty} \mathcal{E}_i)$, by conditions 1 and 2 in definition 3.1.

Suppose that $\mathcal{E}_j \subseteq \mathcal{C}(\bigcup_{i=0}^{\infty} \mathcal{E}_i)$, for all $j \leq i$.

It must be proven that $\mathcal{E}_{i+1} \subseteq \mathcal{C}(\bigcup_{i=0}^{\infty} \mathcal{E}_i)$.

$\mathcal{E}_{i+1} = Cn(\mathcal{E}_i \cup \{A? \to B? \mid \delta_i = \frac{:A \to B; \neg C}{A? \to B?}, \neg(A \to B) \notin \mathcal{E}_i \text{ and } C? \notin \mathcal{E}_i\})$.

By condition 2 in definition 3.1 and the inductive hypothesis it is enough to show that

$A? \to B? \in \mathcal{C}(\bigcup_{i=0}^{\infty} \mathcal{E}_i)$, for $\delta_i = \frac{:A \to B; \neg C}{A? \to B?}$ such that $\neg(A \to B) \notin \mathcal{E}_i$ and $C? \notin \mathcal{E}_i$.

By condition 3 in definition 3.1, it is enough to show that $\neg(A \to B) \notin \bigcup_{i=0}^{\infty} \mathcal{E}_i$ and $C? \notin \bigcup_{i=0}^{\infty} \mathcal{E}_i$.

Recall that $\neg(A \to B) \notin \mathcal{E}_i$. However, since $\neg(A \to B)$ is ?-free, for $\neg(A \to B) \in \bigcup_{i=0}^{\infty} \mathcal{E}_i$ it must belong to $Cn(\mathcal{W})$. But it is impossible since $Cn(\mathcal{W}) = \mathcal{E}_0 \subseteq \mathcal{E}_i$. Then, $\neg(A \to B) \notin \bigcup_{i=0}^{\infty} \mathcal{E}_i$.

Now suppose that $C? \in \bigcup_{i=0}^{\infty} \mathcal{E}_i$. As $C? \notin \mathcal{E}_i$, then there is at least a formula $A'? \to B'? \in \mathcal{E}_j, j > i$ such that $\delta_j = \frac{:A' \to B'; \neg C'}{A'? \to B'?} \in \mathcal{D}$ and $\neg(A' \to B') \notin \mathcal{E}_j$ and $C? \notin \mathcal{E}_j$.

Notice that $\delta_j \ll C?$. Hence $\delta_j \prec \delta_i$. Therefore, $j < i$ in the enumeration. A contradiction ! $\square$

REFERENCES

[1] D.G. Bobrow. Special issue on nonmonotonic logic. *Artificial Intelligence*, 13, 1980.

[2] T. Pequeno and A. Buchsbaum. The Logic of Epistemic Inconsistency. In *2nd International Conference on Principles of Knowledge Representation and Reasoning*, pages 453–460, Boston, 1991.

[3] T. Pequeno. A Logic for Inconsistent Nonmonotonic Reasoning. Technical Report 90/6, Department of Computing, Imperial College, London, 1990.

[4] S. Hanks and D. McDermott. Nonmonotonic Logic and Temporal Projection. *Artificial Intelligence*, 33:27–39, 1987.

[5] M.L. Ginsberg, editor. *Readings on Nonmonotonic Reasoning*. Morgan Kaufmann Publishers, Inc., Los Altos, 1987.

[6] F. Brown, editor. *The Frame Problem in Artificial Intelligence: Proceedings of the 1987 Workshop*. Morgan Kaufmann Publishers, Inc., Los Altos, 1987.

[7] C.G. Morris. The Anomalous Extension Problem in Default Reasoning. *Artificial Intelligence*, 35:383–99, 1988.

[8] G. Zaverucha. On Cumulative Default Logic with Filters. In *Sixth International Workshop on Nonmonotonic Reasoning*, Timberline, June, 10-12 1996.

[9] M. Pequeno. *Defeasible Logic with Exception First*. PhD thesis, Imperial College, London, 1994. Supervisor: D. Gabbay.

[10] A. Buchsbaum. *Lógicas da Inconsistência e Incompletude: Semântica, Axiomatização e Automatização*. PhD thesis, Departamento de Informática, Pontifícia Universidade Católica, Rio de Janeiro, 1995. Supervisor: T. Pequeno.

[11] A.T. Martins. *A Syntactical and Semantical Uniform Treatment for the IDL & LEI Nonmonotonic System*. PhD thesis, Departamento de Informática, Universidade Federal de Pernambuco, Recife, 1997. Supervisor: T. Pequeno.

[12] K. M. Sim. Reasoning Tractably About Explicit Belief: A Model-Theoretic Approach. *International Journal of Intelligent Systems*, 15:811–848, 2000.

[13] O. Arieli and A. Avron. The Value of the Four Values. *Artificial Intelligence*, 102:97–141, 1998.

[14] E. L. Lozinskii. Resolving Contradictions: A Plausible Semantics for Inconsistent Systems. *Journal of Automated Reasoning*, 12:1–31, 1994.

[15] G. Priest. Reasoning about Truth. *Artificial Intelligence*, 39:231–244, 1989.

[16] D. Gabbay. What is a Negation in a System? In F.R. Drake and J.K. Truss, editors, *Logic Colloquium '86*, pages 95–112, Amsterdam, 1988. Elsevier.

[17] N.D. Belnap. A useful four-valued Logics. In G. Epstein and J.M. Dunn, editors, *Modern Uses of Multiple-Valued Logics*, pages 7–73. Reidel Publishing Company, Boston, 1977.

[18] N.D. Belnap. How a computer should think. In G. Ryle, editor, *Contemporary Aspects of Philosophy*, pages 30–56. Oriel Press, 1977.

[19] M.L. Ginsberg. Multi-valued Logics. In M.L. Ginsberg, editor, *Readings in Nonmonotonic Reasoning*, pages 251–258. Morgan Kaufmann, Los Altos, CA, 1987.

[20] M.L. Ginsberg. Multivalued Logics: a uniform approach to reasoning in AI. *Computer Intelligence*, 4:256–316, 1988.

[21] M. Fitting. Bilattices in logic programming. In G. Epstein, editor, *20th International Symposium on Multiple-Valued Logic*, pages 238–246. IEEE Press, 1990.

[22] M. Fitting. Bilattices and the semantics of logic programming. *J. Logic Programming*, 11(2):91–116, 1991.

[23] A.T. Martins and T. Pequeno. Paraconsistency and Plausibility in the Logic of Epistemic Inconsistency. In *Proceedings of the 1st World Congress on Paraconsistency* , Belgium, Jul 31, Ago, 1-2 1997. University of Ghent. Abstract.

[24] G. Antoniou. A Tutorial on Default Logics. *ACM Computing Surveys*, 31(3):337–359, 1999.

[25] G. Antoniou. *Nonmonotonic Reasoning.* MIT Press, Cambridge, 1997.

[26] A.T. Martins, M. Pequeno, and T. Pequeno. Well-Behaved IDL Theories. In *Lecture Notes in Artificial Intelligence, 1159:11-20.* Springer-Verlag, Curitiba, Oct 1996. Proceedings of the 13th Brazilian Symposium on Artificial Intelligence.

[27] R. Reiter. A Logic of Default Reasoning. *Artificial Intelligence*, 13:81–132, 1980.

[28] J. McCarthy. Circumscription - A Form of Nonmonotonic Reasoning. *Artificial Intelligence*, 13(171–172):27–39, 1980.

[29] J.L. Pollock. Defeasible Reasoning. *Cognitive Science*, 11:481–518, 1987.

[30] C.L. Chang and R.C. Lee. *Symbolic Logic and Mechanical Theorem Proving.* Academic Press, New York, 1973.

[31] P. Besnard. *Problems with Default Logic.* Springer Verlag, 1989.

[32] D.W. Etherington. Formalizing Nonmonotonic Reasoning Systems. *Artificial Intelligence*, 31:41–85, 1987.

[33] M. Pequeno. A new criterion to determine the coherence of default theories. Technical report, 2000. Unpublished.

[34] P. Besnard and T. Schaub. Possible worlds semantics for default logics. In J. Glasgow and B. Hadley, editors, *Proceedings of the Canadian Artificial Intelligence Conference*, pages 148–155. Morgan Kaufmann Publishers Inc., 1992.

[35] T. Schaub. *Considerations on Default Logics.* PhD thesis, Technische Hochschule Darmstadt, Darmstadt, 1992.

[36] F. Lin and Y. Shoham. A Logic of Knowledge and Justified Assumptions. *Artificial Intelligence*, 57:271–89, 1992.

[37] S. Ben-David and R. Ben-Eliyahu. A Modal Logic for Subjective Default Reasoning. In *IEEE*, 1994.

[38] D.W. Etherington. A semantics for default logic. In *Proceedings of the International Joint Conference on Artificial Intelligence*, pages 495–8, 1987.

[39] W. Lukaszewicz. Two Results on Default Logic. In *Proc.IJCAI-9*, pages 459–61, Los Angeles, 1985.

[40] R.A. de Guerreiro and M.A. Casanova. An Alternative Semantics for Default Logic. In K. Konolidge, editor, *Third International Workshop on Nonmonotonic Reasoning*, pages 141–157, South Lake Tahoe, 1990.

[41] Y. Shoham. A semantical approach to nonmonotonic logics. In *Proceedings of the Tenth International Joint Conference on Artificial Intelligence*, pages 388–392, 1987.

[42] Y. Shoham. *Reasoning About Change.* MIT Press, Cambridge, 1988.

[43] S. Kraus, D. Lehmann, and M. Magidor. Nonmonotonic reasoning, preferential models and cumulative logics. *Artificial Intelligence*, 44:167–207, 1990.

[44] D. Makison. General theory in cumulative inference. In Reinfrank, de Kleer, Ginsberg, and Sandewall, editors, *Nonmonotonic Reasoning, Lecture Notes in Artificial Intelligence, 346.* Springer-Verlag, January 1989.

[45] D. Makison. *General patterns in nonmonotonic reasoning*, volume II of *Non-Monotonic and Uncertain Reasoning, Handbook of Logic in Artificial Intelligence and Logic Programming*, chapter 2. Oxford University Press, Oxford, 1992.

[46] A.T.C. Martins and T. Pequeno. Some Characteristics of the Inconsistent Default Logic Reasoning Style. *Journal of the Interest Group in Pure and Applied Logics*, 4(3):517–519, June 1996. Also in the Proceedings of the 3th Workshop on Logic, Language, Information and Computation, Organized by a join effort of the Federal University of Bahia and Pernambuco, Salvador, May, 8-10.

[47] M. Pequeno and J-Y. Beziau. A new look at Vasiliev's paraconsistent and paracomplete logic. Technical report. Draft.

An inductive annotated logic *

NEWTON C. A. DA COSTA Department of Philosophy, University of São Paulo and Paulista University (UNIP)
ncacosta@usp.br

DÉCIO KRAUSE Department of Philosophy, Federal University of Santa Catarina
dkrause@cfh.ufsc.br

Abstract
The evolution of techniques in AI has motivated the investigation of various forms of reasoning, such as non-monotonic and defeasible reasoning, which in some cases have been associated with non-classical logics. But in general, in such developments, attention has been given only to non-doxastic states of inputs and outputs, that is, to those forms of reasoning which are performed without explicitly using degrees of belief, or confidence, about the states of the data (or premisses and conclusions of the inference rules). In this paper we outline the use of a certain kind of paraconsistent logic, termed annotated logic, for dealing with propositions which are vague in a sense but that, despite their vagueness, can be 'believed' with a certain degree of confidence.

1 Introduction

Computer programs are written in formal languages, having precise and well defined grammatical structures. Despite the fact that deductions in computer programs should not be identified with *proofs* in formal systems, the cannons of inference of such programs, by means of which it is possible to reach to a certain conclusion starting from certain premisses, are generally based on inference rules of some kind of logic. In other words, when inferences are made, all happens as in

*Partially supported by CNPq. This paper was to be presented at the Second World Congress on Paraconsistency, held in Juquehy, São Sabastião, Brazil, 8-12 May 2000, but the authors decided to discuss some of its parts in more detail before presenting it "officially". Anyway, the subject was discussed during the Congress with other people, particularly with Renato Lewin and his students from the Pontifical Catholic University of Chile, and for this reason we decided to present it to the Proceedings of the Congress, with the agreement of the organizers. We would like to thank Prof. Renato Lewin for comments and criticisms of a first version of our work, as well as the organizers of the meeting for having accepted the submission of this paper for the Proceedings. We would also like to thank the two anonymous referees for useful criticism and suggestions.

standard deductive logics, that is, by using Imre Lakatos' words, "[the conclusion] follows downwards [from the premisses] through the deductive channels of truth-transmission (*proofs*)" [1, p. 2].

But, if computer programs will evolve and approach the human ways of reasoning (independently of whether they will reach this or not), their softwares will need to make inferences involving various forms of reasoning, other than mere deductive ones and, in particular, those involving vague assertions (or propositions), as we, humans, usually do. For instance, we usually also 'make inferences' from vague propositions, to which we in general cannot attribute, with certainty, one of the two truth-values *true* or *false*. Frequently, our reasoning is performed by attributing only some degree of confidence as either the involved propositions are true or not. This is what may happen, for instance, when we are visiting a foreign country and someone gives us (vague) information about the location of a certain place (perhaps because the native is also not sure about the right geography of the city), and we 'believe' in the information with a certain degree of confidence and 'decide' the way to be taken.

So, it is interesting to ask for the ways by means of which computers (or expert systems) could 'reason' in this sense, that is, we should also investigate how to provide computer programs with an apparatus which may enable them to make inferences in the presence of vague propositions by 'believing' them with certain degrees of confidence. The attribution of degrees of belief (or of 'confidence') to the propositions links the subject with inferences which do not follow strictly deductive channels, being closer to defeasible forms of reasoning, since (using the above example) a further information provided by a tourist guide may change our conclusion about the direction we should take.

The literature on non-deductive ways of reasoning presents various systems of computational tools devoted to non-monotonic and defeasible reasoning ([2], [3], to mention two of them). Here, without revising these approaches, we propose a different process involving vagueness which we believe should also be seriously considered by all AI researchers. Our main motivation was of course to handle vague information mechanically, but in this paper we shall limit ourselves to the description of a vague inductive logic which we hope can be useful in the mechanical treatment of inductive information. Further developments should provide a way of elaborating, for example, expert systems based on our scheme.

2 Vagueness

Let us recall that the discussion on vagueness is old in the philosophical tradition. The question 'Are there vague objects?' is usually answered with a 'yes', and examples are provided by mentioning mountains, heaps of grains of wheat, hairs in a head and even ships (the case of Theseu's ship is well known). For instance, the so-called *sorites paradoxes* arise as a result of the indeterminacy involving the limits of application of the involved predicates. If we agree that a single grain of wheat is not a heap, it is reasonable to suppose that we also cannot accept that two grains of wheat form a heap. But, in adding grains of wheat to them, we must admit that we will be in the presence of a heap sooner or later, so, there seems to be a drawn line.

But, where is it?[1] Usually, philosophers tend to accept that the indeterminacy is not in the world but in our ways of describing things; that is, according to a certain tradition, vagueness infects our descriptions only, and not the things we describe. We will not enter into this discussion here, but only accept that there may be vague propositions in our usual ways of speech. Examples of vague predicates are "to be old", "to be intelligent", and so on. So, in general the accepted situation is that we have a 'sharp' object, say John, and we may have some doubt in describing him as old; in other words, the predicate "to be old" is vague, but not John.[2]

3 Two case studies

Let us give some examples of situations which motivate the system to be presented below. Suppose we are working in an insurance company and need to provide a way of classifying people into disjointed classes by age, since the company needs to distinguish young people from old people in order to differentiate among several prices for the insurance premiums. For example, the company could guide its interviewers that they should classify people in (say) four different categories: (C_1) "young" (less or equal than 25 years), (C_2) "not very young" (less or equal than 35 years and more than 25), (C_3) "not very old" (between 35 and 45 years) and (C_4) "old" (more than 45 years). Of course this is just an example and more detailed and accurate descriptions could be required. But suppose the Government would like to extend (and to pay for) the insurance services for the people who don't have birth certificates, and who sometimes don't precisely know their age, and their hard and poor conditions of life may confound the interviewers in what concerns attributing them precise ages.[3]

So, the information provided by the interviewers may be sometimes considered as not completely precise. For instance, interviewer A may account to the company an information which represents his belief that John looks not very old (that is, it seems to the interviewer that John has an age of between 35 and 45 years) and similarly, he may tell them that "Paul does not look very young" (that is, John and Paul were classified as satisfying respectively the predicates C_3 and C_2 above, but the interviewer is not completely sure about the correctness of such a classification). If John and Paul have their birth certificates, then apparently there is no problem regarding the adequate manipulation of the information. But suppose they don't have their certificates. How do we deal with this situation? Since the attribution of ages cannot be done with precision, the interviewer, in the case of doubt, may adopt one of the following alternatives, depending on several involved factors: (1) to classify John as "old" and Paul as "not very young" and (2) to classify John as "not very old" and Paul as "young". These mentioned 'factors' may be, for instance, the expertize of the interviewer (that is, his experience in the job), or his

[1] Some forms of the sorites paradox may be described in more mathematical terms; for instance, if we agree that someone with just one hair in his head is bald and (the induction hypothesis) if someone being bald is still bald if one more hair is added to his head, then it follows by induction that independently of the number of hairs someone has in his head, he is bald.

[2] See [4], were the possibility of vagueness *in the world* is taken into account, that is, the existence of actual vague objects is discussed. The motivation for such a supposition is given by quantum mechanics.

[3] This situation is common in developing and in poor countries.

interest in the defense of some kind of policy, or some guidance, such as for instance an hypothetical company's interest in classifying people in classes of ages so that the due premium prices are as great as possible, thus getting more money from the Government.[4]

Humans in general go around situations of this kind by fixing some *ad hoc* criterion in a more or less arbitrary way: they may "decide" what to do by fixing some rule, or by deciding case by case. But, what happens if the case is to be handled by an expert system? In other words, is it possible to keep the system with the capacity of dealing with propositions to which different "degrees of confidence" are attached to, or to follow, for instance, an insurance policy where it is necessary to express a "confidence" in the information provided by the interviewers?

Another situation may be the following. The manager of a supermarket would like to know if the quantity of sugar is sufficient for the next few days or if he needs to order a new load. He may tell his staff that they must verify the quantity of sugar at disposal, and the answers he gets can be classified into four categories (as above, other situations could be admitted): (C_1) "it is not lacking sugar" (the stock is full); (C_2) "it lacks some sugar"; (C_3) "there is some sugar" or, alternatively, "it lacks much sugar", and (C_4) "there is no sugar at all". Of course the predicate "it lacks sugar at the supermarket x" is vague. How to decide something taking into consideration such vague pieces of information? The manager may of course verify the stock by himself, but let us suppose that the manager uses an expert system (a robot, say), and so it must use the pieces of information as they are given to it, that is, despite their vagueness. Of course, this case can be dealt with by providing some criterion in the robot's software, for instance by fixing some rule for avoiding dealing with vague information (if possible),[5] but if we hope that one day a robot can be used to help the management of a supermarket, whatever decision he takes may depend on other possibilities which should be considered as well, as we usually do. Furthermore, we cannot specify in a program every situation a manager may find in his day-to-day activity, but we can introduce some very general guidance procedures which make the system as general as possible. In short, the software cannot be 'rigid' in classifying the bits of information (inputs) by attributing them the same 'value'; some degree of belief, here expressing the confidence in the people who are providing them is in order. That is, the robot should sometimes 'decide' among certain situations by attributing them certain weights, which can be understood as degrees of confidence which may influence the decisions. These 'decisions' of course should be incorporated into a software so flexible that it can take into account the possibilities that inputs provided by different sources can be 'believed' differently. For instance, the robot-manager, by following some previously specified criterion, gives to the supplier, who is another robot, the information "my supermarket A lacks some sugar", while the robot-manager of another supermarket says "in my supermarket B the stock of sugar is low". If we admit that due to a strike or due to the lack of sufficient stock the robot-supplier can provide just one load, then he could 'decide' that the load must go to B. This seems to be the 'natural' decision.

[4]Of course the situation could be precisely the opposite.

[5]As already suggested, the members of staff could be previously classified in different categories depending on their expertize, so that the collected information can be qualified with different degrees of confidence, putting more confidence in the information provided by the more experienced people.

The inference made from the two mentioned premisses is of course defeasible, since the pieces of information that the robot-supplier has received could be incorrect, for instance for the purpose of preserving the future stock of sugar, and an additional information saying that the previous account of robot-B was incorrect may modify the final decision.

In short, we are trying to take into account two aspects of a given information: (a) its vagueness, in the sense that inputs of the form 'it lacks sugar' or 'John is old' may be not true or false *tout court*, but 'partially true' (vague),[6] and (b) a degree of confidence in the truthfulness of such a vague proposition.[7] In other words, we aim to make inferences with vague propositions which are 'believed' with a certain amount of confidence. These degrees of belief or 'degrees of confidence', as we prefer to call them for avoiding paralleling them with the concept of subjective probability (but see Section 7), may be interpreted as fixing the confidence the agent may have in a proposition, or event, in order to use it for making inferences.

But there is still another difficulty. In the above examples, suppose that we have (among other details) the following premisses of an inference: "almost sure that supermarket A lacks sugar", and "it seems that supermarket A does not lack sugar". What kind of conclusion are we allowed to infer from these premisses? Classical propositional logic, of course, cannot help us, since the premisses are 'propositionally independent' from one another. Classical first order logic seems not to help us here as well. In order to deal with such situations, and *to link* premisses of this kind to one another, as well as with some conclusion obtained from them, we shall use annotated logic with an operator corresponding to the concept of degree of confidence. Furthermore, annotated logics enable us also to cope with contradictory situations which may appear due to vagueness, although this topic will be not treated here (but see the papers listed in the References).

So, we shall sketch a way of making sense of the possibility of dealing with inferences such that in both the premisses and in the conclusion we may admit that some kind of incomplete or vague information might be involved. We will not provide all the details in this paper, since we aim to deal with the basic underlying ideas of our approach only. Furthermore, we shall be restricted here to the propositional level.

We would like to recall that authors such as John Casti and Keith Devlin have pointed out that new forms of 'inductive rationality' seem to be essential for present day studies of complex processes, in particular for the full understanding of the human mind ([5, Chap. 3], [6]). In the same vein, John Pollock, who has been investigating several aspects of defeasible forms of reasoning, says that

> A common misconception about reasoning is that reasoning is deducing, and in good reasoning the conclusions follow logically from the premisses. It is now generally recognized both in philosophy and in AI that non-deductive reasoning is at least as common as deductive reasoning, and a reasonable epistemology must accommodate both. [3]

These ideas of course give an additional motivation for the present study.

---

[6]Of course it seems that fuzzy logics could be used instead, but we remark that we are trying to provide an alternative approach by using annotated logics. See also footnote 10 below.

[7]This is, of course, a way of saying that some degree of confidence is attributed to the proposition, be it true or false.

4 ANNOTATED LOGICS AND VAGUENESS

As already explained, our point is to extend the common ways of using arguments by accepting that a proposition may have a certain 'degree of vagueness'. For instance, 'Peter is smart' is a vague proposition. Furthermore, we still aim to attribute a degree of confidence in the truthfulness of these propositions, as suggested in the previous sections. In other words, we intend to suppose that we believe that Peter is smart with some degree of confidence. Peter's mother has a great confidence in such a proposition, but his teacher may be not so confident. This degree of confidence can be interpreted as an amount of confidence someone accepts in relation to a proposition.

In order to deal with these two concepts related to propositions, namely its vagueness and its degree of confidence, we make use of annotated logics ([7], [8], [9]), which are paraconsistent logics. We will be more precise in what follows.[8]

Let us call $\mathcal{I}_\tau$ a propositional logic whose language has the following categories of primitive symbols: a countable set of propositional letters, which stay for propositions (we use $P, Q, \ldots$ as syntactical variables for propositions); the elements $\mu, \ldots, \mu_1, \ldots$ of a complete lattice τ ordered by $\leq$, termed *the values of vagueness* and the usual logical connectives ($\neg$, $\wedge$, $\vee$, $\rightarrow$), as well as auxiliary symbols (parentheses).

The concept of *formula* of $\mathcal{I}_\tau$ is introduced in the following way:

(i) If P is a propositional letter and $\mu \in \tau$, then $P : \mu$ is a formula of $\mathcal{I}_\tau$ (atomic formula).

(ii) If α and β are formulas, then $\neg\alpha$, $\alpha \wedge \beta$, $\alpha \vee \beta$, $\alpha \rightarrow \beta$ are formulas.[9]

(iii) Every formula is obtained from just one of the two above clauses.

Furthermore, we employ a standard way of eliminating parentheses, and Greek capital letters for denoting collections of formulas. Intuitively speaking, $P : \mu$ means that P is true with degree of vagueness μ. Let us remark that we are attaching degrees of vagueness to atomic formulas only, and not to formulas in general; so, expressions like

$$((P : \mu_1) \vee (Q : \mu_2)) : \mu \tag{1}$$

are not well formed in our system.

DEFINITION 1

(i) If P is a propositional letter and $\mu \in \tau$, then:

(ii) $\neg^0 P : \mu$ means $P : \mu$

[8] We should remark that we are not considering here propositions with two annotations taken from different lattices, but with just one. Propositions with two annotations were first considered by Subrahmanian; da Costa and others, in several papers, have used this idea in interesting applications (see the References). Another approach to the subject is being developed by Renato Lewin, who is investigating algebraic aspects of annotated logics.

[9] $\alpha \leftrightarrow \beta$ is introduced in the standard way.

(iii) $\neg^1 P : \mu$ means $\neg(P : \mu)$

(iv) $\neg^k P : \mu$ means $\neg(\neg^{k-1}(P : \mu))$, with k a natural number, $k \neq 0$.

(v) Let $\sim: \tau \longrightarrow \tau$ be a fixed mapping.[10] We shall write $\sim \mu$ instead of $\sim (\mu)$ from now on. If $\mu \in \tau$, then:

(a) $\sim^0 \mu$ means μ

(b) $\sim^1 \mu$ means $\sim \mu$

(c) $\sim^k \mu$ means $\sim (\sim^{k-1} \mu)$, for $k \neq 0$ being a natural number.

Expressions like $P : \mu$ are called ***annotated atoms***, while $\neg^k(\alpha : \mu)$ are *hyper-literals* of order k ($k \geq 0$); the other formulas are called *complex*.

5 SEMANTICS

Let τ be the complete lattice above with least element $\perp$ and greatest element $\top$; let $h : \mathcal{P} \longrightarrow \tau$ be a mapping, called an ***interpretation*** of $\mathcal{I}_\tau$, where $\mathcal{P}$ is the collection of propositional letters of $\mathcal{I}_\tau$. The image of the proposition P by the mapping h shall be denoted $P : \mu$, where $\mu \in \tau$. Informally speaking, as we have said, $P : \mu$ means that P is true with degree of vagueness μ. To each interpretation h we associate a *valuation* $v_h : \mathcal{F} \longrightarrow \{0, 1\}$, where $\mathcal{F}$ is the above defined collection of formulas of $\mathcal{I}_\tau$. Intuitively speaking, 1 and 0 stand for 'true' and 'false' respectively.

Particular applications may demand appropriate choices of the complete lattice, as the papers in our References show. Here, to cope with the above mentioned case studies, we shall be concerned with a particular finite linearly ordered set $\tau = \{\mu_1, \ldots, \mu_4\}$ (with $\mu_1 \leq \cdots \leq \mu_4$) for expressing the distinct degrees of vagueness of a proposition, but of course the scheme presented here is quite general.

DEFINITION 2 If h and v_h are as above and P is a propositional letter and α and β denote formulas, then:

(i) $v_h(P : \mu) = 1$ iff $\mu \leq h(P)$.

(ii) $v_h(\neg^k(P : \mu)) = v_h(\neg^{k-1}(P :\sim \mu))$, where $k \neq 0$.

(iii) $v_h(\alpha \wedge \beta) = 1$ iff $v_h(\alpha) = v_h(\beta) = 1$.

(iv) $v_h(\alpha \vee \beta) = 1$ iff $v_h(\alpha) = 1$ or $v_h(\beta) = 1$.

(v) $v_h(\alpha \rightarrow \beta) = 1$ iff either $v_h(\alpha) = 0$ or $v_h(\beta) = 1$.

(vi) if α is a complex formula, then $v_h(\neg\alpha) = 1$ iff $v_h(\alpha) = 0$.

If $v_h(\alpha) = 1$, we say that v_h ***satisfies*** α, and that it does not satisfy α otherwise (that is, when $v_h(\alpha) = 0$). If Γ is a set of formulas, then we say that a formula α is a ***semantic consequence*** of (the formulas of) Γ, and write $\Gamma \models \alpha$, iff for every valuation v_h such that $v_h(\beta) = 1$ for each $\beta \in \Gamma$, then $v_h(\alpha) = 1$. A formula α is ***valid*** iff $\emptyset \models \alpha$, and in this case we write $\models \alpha$.

As usual, we say that a valuation v_h is a ***model*** for a set Γ of formulas iff $v_h(\beta) = 1$ for every $\beta \in \Gamma$. In particular, v_h is a model of α iff $v_h(\alpha) = 1$. The other concepts like maximal non-trivial sets of formulas and so on are defined like the standard ones.

---

[10]The specific definition of this mapping depends on the particular application. For instance, by taking τ to be the unit interval $[0, 1] \subseteq \Re$ and $\sim (x) := 1 - x$, the introduction of 'fuzzy' ways of reasoning can be performed within the scope of annotated logics (see [9]).

6 The postulates of $\mathcal{I}_\tau$

If α, β and γ are formulas and P is a propositional letter, then the postulates (axioms plus inference rules) of $\mathcal{I}_\tau$ are the following (adapted from [9], [10]):

(I1) All the postulates of classical positive logic.

(I2) If α and β are complex formulas, then the following is an axiom: $(\alpha \to \beta) \to ((\alpha \to \neg\beta) \to \neg\alpha)$.

(I3) If α is complex, then $\alpha \vee \neg\alpha$ is an axiom.

(I4) If α is complex and β is a formula whatsoever, then $\alpha \to (\neg\alpha \to \beta)$ is an axiom.

Then, classical logic holds for complex formulas. The presence of inconsistencies will be allowed at the level of atomic formulas only [9].

(I5) $P :\perp$ is an axiom. The technical motive for using this axiom is that $v_h(P :\perp) = 1$ iff $h(\alpha) \geq 0$, which is always true.

(I6) If $\lambda \leq \mu$, then $P : \mu \to P : \lambda$

(I7) $\neg^k(P : \mu) \leftrightarrow \neg^{k-1}(P :\sim\mu)$, if $k \neq 0$.

(I8) If α is a formula whatsoever, then if $\alpha \to (P : \mu_i)$, $i \in I$, then $\alpha \to (P : \bigsqcup_{i\in I} \mu_i)$. If τ is a finite lattice, then this axiom may be replaced by the following one (cf. [9]):

$$P : \mu_1 \wedge \ldots \wedge P : \mu_n \to P : \bigsqcup_{i=1}^{n} \mu_i \tag{2}$$

The syntactical concepts of $\mathcal{I}_\tau$ are introduced in the standard way, so as in particular, the symbol of deduction $\vdash$ (see [9]).

We can prove the soundness and completeness of the logic $\mathcal{I}_\tau$ with respect to the semantic described in the previous section, as we shall sketch below. Let us first introduce a definition:

DEFINITION 3 [Strong Negation]

$$\neg^*\alpha := \alpha \to ((\alpha \to \alpha) \wedge \neg(\alpha \to \alpha)) \tag{3}$$

It is easy to prove that $\neg^*$ has all the properties of classical negation, hence the classical laws hold when $\neg^*$ is used instead of $\neg$ in the formulas of our system. For instance, the reductio ad absurdum $(\alpha \to \beta) \to ((\alpha \to \neg^*\beta) \to \neg^*\alpha)$ is a theorem of $\mathcal{I}_\tau$, so as is the excluded middle law $\alpha \vee \neg^*\alpha$. Although we can show that if α is a complex formula, then $\neg\alpha \leftrightarrow \neg^*\alpha$ is valid (see the theorem below), this does not hold for formulas in general; for instance, if Q is a hyper-literal, then in general $\neg Q \leftrightarrow \neg^* Q$ is not valid.[11]

Other results are the following:

THEOREM 4

(i) If $\Gamma, \alpha \vdash \beta$, then $\Gamma \vdash \alpha \to \beta$ (the Deduction Theorem)

[11] See [9].

(ii) If $\Gamma \vdash \alpha$ and $\Gamma \vdash \alpha \to \beta$, then $\Gamma \vdash \beta$

(iii) $\alpha \wedge \beta \vdash \alpha$, $\alpha \wedge \beta \vdash \beta$, $\alpha, \beta \vdash \alpha \wedge \beta$

(iv) $\alpha \vdash \alpha \vee \beta$, $\beta \vdash \alpha \vee \beta$

(v) $\Gamma, \alpha \vdash \gamma$ and $\Gamma, \beta \vdash \gamma$, then $\Gamma, \alpha \vee \beta \vdash \gamma$ (Proof by Cases)

(vi) $\Gamma, \alpha \vdash \beta$ and $\Gamma, \alpha \vdash \neg^*\beta$, then $\Gamma \vdash \neg^*\alpha$ (Reductio ad Absurdum)

(vii) $\alpha, \neg^*\alpha \vdash \beta$, $\neg^*\neg^*\alpha \vdash \alpha$, $\alpha \vdash \neg^*\neg^*\alpha$

(viii) If α is complex, then $\neg^*\alpha \leftrightarrow \neg\alpha$

(ix) $(\alpha : \mu_i)_{i \in I} \vdash \alpha : \bigsqcup_{i \in I} \mu_i$

(x) If $\Gamma \vdash \alpha$, then $\Gamma \models \alpha$ (Soundness Theorem).

In order to prove the completeness theorem, we need a few definitions and results.

DEFINITION 5

(i) $\overline{\Gamma} := \{\alpha : \Gamma \vdash \alpha\}$

(ii) Γ is *trivial* iff $\overline{\Gamma} = \mathcal{F}$, where $\mathcal{F}$ is the set of formulas of $\mathcal{I}_\tau$; otherwise, Γ is *non-trivial*.

(iii) Γ is *inconsistent* iff there exists α such that both α and $\neg\alpha$ belong to $\overline{\Gamma}$. Otherwise, Γ is *consistent*.

(iv) Γ is *strongly inconsistent* iff there exists α such that both α and $\neg^*\alpha$ belong to $\overline{\Gamma}$. Otherwise, Γ is *strongly consistent*.

It is easy to see that Γ is strongly inconsistent iff it is trivial and that Γ is strongly consistent iff it is non-trivial. Furthermore, by an adequate choice of τ, we may prove that there exist inconsistent but non-trivial sets of formulas, which are still not strongly inconsistent [7]. So, the logic $\mathcal{I}_\tau$ is a paraconsistent logic. This means that there exist interpretations h and formulas α such that $v_h(\alpha) = v_h(\neg\alpha) = 1$. But we may also prove that for certain τ, there are formulas α and interpretations h such that $v_h(\alpha) = v_h(\neg\alpha) = 0$. So, $\mathcal{I}_\tau$ is also a paracomplete logic. All these results are treated in details in the papers listed in the References.

LEMMA 6 Every non-trivial set of formulas is a subset of some maximal non-trivial set of formulas.

Proof: See [9], [7].

□

The completeness theorem results from the following Lemma:

LEMMA 7 If Γ is a maximal non-trivial set of formulas, then its characteristic function $\chi_\Gamma : \mathcal{F} \longrightarrow \{0, 1\}$ is a model of Γ, that is, such a mapping is a valuation such that $\chi_\Gamma(\beta) = 1$ for every $\beta \in \Gamma$.

Proof: The trick is to define a valuation v_h, for a given interpretation h, in such a way so that the rules of $\mathcal{I}_\tau$ are 'preserved'. This means that, given Γ, we may define $h : \mathcal{P} \longrightarrow \tau$ such that for every proposition $P \in \mathcal{P}$,

$$h(P) := \bigsqcup_i \{\mu_i : \mu_i \in \Gamma\} \tag{4}$$

It is now not difficult to prove that the valuation generated by such an interpretation coincides with the characteristic function χ_Γ. □

As a consequence, we have the completeness theorem:

THEOREM 8 If $\Gamma \models \alpha$, then $\Gamma \vdash \alpha$.

Proof: See [7], [9]. □

7 Degrees of confidence

Now, we shall sketch a *theory of confidence*, which enables us to attribute degrees of confidence to propositions, even to vague ones.

Our degrees of confidence are, in general, only qualitative, characterized by the elements of an appropriate lattice with least and greatest elements. Abstractly speaking, degrees of confidence are elements of a lattice σ, and are attributed to the formulas of the language $\mathcal{I}_\tau$, when the propositional variables are interpreted as denoting specific vague statements as described above. In order to do so, let σ be a lattice with least and greatest elements denoted respectively by $\perp\prime$ and $\top\prime$. The algebraic lattice operations are represented by $\sqcap$ and $\sqcup$, and the corresponding partial order by $\leq$. If $\mathcal{F}$ is the set of formulas of the logic $\mathcal{I}_\tau$, let $\mathsf{C} : \mathcal{F} \longrightarrow \sigma$ be a mapping satisfying the postulates below, where α and β denote annotated propositions whatever:

(C1) $\mathsf{C}(\alpha \wedge \neg^* \alpha) = \perp\prime$

(C2) $\mathsf{C}(\alpha \vee \neg^* \alpha) = \top\prime$

(C3) $\mathsf{C}(\bigvee_{i\in I} \alpha_i) \geq \bigsqcup_{i\in I} \mathsf{C}(\alpha_i)$, for I finite.

(C4) $\mathsf{C}(\bigwedge_{i\in I} \alpha_i) \leq \sqcap_{i\in I} \mathsf{C}(\alpha_i)$, for I finite.

(C5) If $\vdash \alpha \leftrightarrow \beta$, then $\mathsf{C}(\alpha) = \mathsf{C}(\beta)$.

Any such a mapping C is called a *confidence function*. Depending on the applications, the above postulates can be extended by other convenient ones, for instance those expressing that if $\vdash \alpha \rightarrow \beta$, then $\mathsf{C}(\alpha) \leq \mathsf{C}(\beta)$, and that $\mathsf{C}(\alpha) \sqcup \mathsf{C}(\neg^* \alpha) = \top\prime$.

Our proposal is to combine the concept of vagueness with that of confidence, that is, the logic $\mathcal{I}_\tau$ with the operator C; this is similar to the introduction of probability measures in a classical system of probability calculus. Obviously, an "algebra of confidence", as we suggest, extends the classical case of subjective measure of probability, defined on Boolean algebras of propositions. On the other hand, our procedure also encompasses Zadeh's theory of possibility (cf. [11], [12]).[12]

When the degrees of confidence reduce to strict degrees of belief, *i.e.*, subjective probability, they can be handled in the standard way (following the rules of Bayesian probability calculus). Be such hypothesis satisfied or not, it is convenient to strengthen our system with extra rules such as the following Warning Rule (we write $P : \mu : \lambda$ for $\mathsf{C}(P : \mu) = \lambda$):[13]

[The Warning Rule]

$$\frac{P : \mu_i : \lambda_i\,,\ P : \mu_j : \lambda_j}{P : \mu_i \sqcap \mu_j : \lambda_i \sqcup \lambda_j}\quad \Gamma \tag{5}$$

where Γ stands for the set of side conditions which are the underlying base for the application of the rule, conditions that are accepted as true.

The above rule has an intuitive appeal and constitutes a tool for dealing simultaneously with degrees of confidence and of vagueness. Some inductive rules, such as those of induction by simple enumeration and analogy (cf. [14]), complemented by the degrees of vagueness and of confidence, are also useful to reinforce the logic we are trying to built; in the same vein, the methods of the theory of possibility may help here.

We designate the resulting system of inductive logic by $\mathbf{I}_\tau$. Our principal objective is to employ an appropriate $\mathbf{I}_\tau$ for the mechanization of inductive inference, say in robotics and in the theory of expert systems. Let us, then, outline how this can be done by considering the above mentioned case studies.

Considering the case of the insurance company (see Section 3), let τ be a linearly ordered set $\tau = \{\mu_1, \ldots, \mu_4\}$, which is a complete lattice (were $\mu_1(=\perp) \leq \ldots \leq \mu_4(=\top)$). Suppose that two interviewers (this example can be generalized) give as inputs $P : \mu_2$ and $P : \mu_3$ and suppose that P stands for 'John is old'. Then, according to that example, $P : \mu_2$ means 'John is not so young', while $P : \mu_3$ says that 'John is not so old'. More precisely, the first interviewer has classified John as having an age between 25 and 35 years, while the second interviewer admitted that he is between 35 and 45 years old.

An expert system, in order to attach a certain value to John's due taxes, may follow the following rule, being $C_1, \ldots, C_4$ the classes of ages in the considered example (see Section 3): *in the case of doubt if someone belongs to either class C_i or C_{i+1}, classify him/her as belonging to the class C_i.* So, in the exemplified case, John would be considered as 'not very young', and then his due taxes would be supposed to be smaller than if he were classified as being 'not very old'.

The choice of $P : \mu_2$ may be interpreted as resulting from the application of the Warning Rule, since it was preferred to put the greater degree of confidence in the vaguer proposition. In other words, according to the convention made in writing the Warning Rule, if we write $P : \mu_2 : \lambda_2$ for $\mathsf{C}(P : \mu_2) = \lambda_2$ and $P : \mu_3 : \lambda_3$ for $\mathsf{C}(P : \mu_3) = \lambda_3$, then we may say that from the 'premisses' $P : \mu_2$ and $P : \mu_3$ we have arrived at a 'conclusion' $P : \mu_2$ with a greater degree of confidence. This can

[12]The contents of [13] can also be dealt with within our scheme.

[13]We remark that, in writing the rule, we have used $\sqcap$ and $\sqcup$ to denote the algebraic operations in both lattices τ and σ by simplicity, but of course this will not cause any confusion.

be expressed by saying that the 'conclusion' is $P : \mu_2 \sqcap \mu_3 : \lambda_2 \sqcup \lambda_3$ (we remark that τ is a linear lattice in which $\mu_2 \sqcap \mu_3 = \mu_2$).

Similar remarks apply to the case of the supermarket.

Roughly speaking, the Warning Rule expresses that a possible expert system elaborated for dealing with vagueness and with degrees of confidence, when faced with situations like the aforementioned, one should opt for the more prudent situation. This is in accordance with a rational stance, for we may say that 'rationality' also means the tentative of optimizing our rational degrees of confidence in the propositions we are concerned with, but with the caution of not taking conclusions with degrees of confidence greater than those attributed to the premisses.

References

[1] I. Lakatos. Infinite regress and foundations of mathematics. In J. Worrall and G. Currie, editors, *Mathematics, science and epistemology*, pages 3–23. Cambridge Un. Press, 1978.

[2] T. Pequeno. A logic for inconsistent non-monotonic reasoning. Technical Report 90/6, Department of Computing, Imperial College, London, 1990.

[3] J. L. Pollock. Defeasible reasoning. *Cognitive Science*, 11:481–518, 1987.

[4] S. French and D. Krause. Vague identity and quantum non-individuality. *Analysis*, 55(1):20–26, 1995.

[5] J. L. Casti. *Would-be worlds*. John Wiley, 1997.

[6] K. Devlin. *Goodbye Descartes: the end of logic and the search for a new cosmology of the mind.* John Wiley & Sons, 1997.

[7] N. C. A. da Costa and V. S. Subhramanian. Paraconsistent logics as a formalism for reasoning about inconsistent knowledge bases. *Artifical Intelligence in Medicine*, 1:167–174, 1989.

[8] N. C. A. da Costa, L. J. Henschen, J. J. Lu and V. S. Subrahmanian. Automatic theorem proving in paraconsistent logics: theory and implementation. In M. E. Stickel, editor, *Proceedings of the 10th International Conference on Automated Deduction*, Lecture Notes in Artificial Intelligence 449, pages 72–86. Springer-Verlag, 1990.

[9] N. C. A. da Costa, V. S. Subrahamanian and C. Vago. The paraconsistent logics P τ. *Zeitschrift fur Mathematische Logik und Grundlagen der Mathematik*, 37:139–148, 1991.

[10] N. C. A. da Costa, V. S. Subrahamanian and J. M. Abe. Remarks on annotated logic. *Zeitschrift fur Mathematische Logik und Grundlagen der Mathematik*, 37:561–570, 1991.

[11] L. A. Zadeh. Fuzzy sets as a basis for a theory of possibility. *Fuzzy sets and systems*, 1:3–28, 1978.

[12] D. Dubois and H. Prade. *Possibility theory: an approach to computerized processing of uncertainty.* Plenum Press, 1988.

[13] N. C. A. da Costa and S. French. On Russell's principle of induction. *Synthese*, 86:285–295, 1991.

[14] N. C. A. da Costa. *Lógica indutiva e probabilidade.* Hucitec-EdUSP, 2nd. ed., 1993.

On NCG_ω: a paraconsistent sequent calculus

JOSÉ EDUARDO DE A. MOURA Universidade Federal do Rio Grande do Norte
jemoura@ufrnet.br

ITALA M. LOFFREDO D'OTTAVIANO Universidade Estadual de Campinas
itala@reitoria.unicamp.br

Abstract
This paper introduces the system NCG_ω, a sequent-calculus formulation for da Costa's system C_ω that explores Raggio's intuitions in a paper of 1968, where the *cut* elimination theorem is proved in a system stronger than C_ω, WG_ω. The new logic, here presented, is equivalent to C_ω and is suitable for the application of Gentzen techniques, and it also allows us to prove the *cut* elimination, the decidability and the consistency theorems for C_ω.

1 INTRODUCTION

The system here introduced, NCG_ω, is a presentation of da Costa's C_ω (in [1]) in the form of a sequent calculus ([2]), derived from the system CG_ω (of [3]).

1.1 The System C_ω

The paraconsistent propositional calculus C_ω is built up from a usual propositional language with the logical symbols $\neg$ (negation), & (conjunction), $\vee$ (disjunction) and $\supset$ (conditional) and a countable set of propositional variables. Formulae are defined as usual and are denoted by uppercase Latin letters. The ***postulates*** that characterize C_ω, as introduced by da Costa in [1], are the following:

Postulate 1. $A \supset (B \supset A)$

*We would like to thank the anonymous referees for useful comments and Prof. Luiz Carlos P.D. Pereira, who suggested and discussed these ideas in the 80's, for substantial suggestions and corrections incorporated in the final version of this paper.

Postulate 2. $(A \supset B) \supset ((A \supset (B \supset C)) \supset (A \supset C))$

Postulate 3. $\dfrac{A, A \supset B}{B}$

Postulate 4. $A\&B \supset A$

Postulate 5. $A\&B \supset B$

Postulate 6. $A \supset (B \supset A\&B)$

Postulate 7. $A \supset A \vee B$

Postulate 8. $B \supset A \vee B$

Postulate 9. $(A \supset C) \supset ((B \supset C) \supset (A \vee B \supset C))$

Postulate 10. $A \vee \neg A$

Postulate 11. $\neg\neg A \supset A$

This system was presented as a kind of limit system of a hierarchy of calculi, $C_n, 1 \leq n \leq \omega$, with the following defined symbols and, for every n, the corresponding Postulates 12(n) and 13(n).

DEFINITION 1.1.1. $A^o =_{df} \neg(A\&\neg A)$.

DEFINITION 1.1.2. $A^n =_{df} A^{\overbrace{oo\ldots o}^{n\,times}}$.

DEFINITION 1.1.3. $A^{(1)} =_{df} A^o$.

DEFINITION 1.1.4. $A^{(n)} =_{df} A^o\&A^{oo}\&\ldots\&A^n$.

DEFINITION 1.1.5. $\neg^* A =_{df} \neg A\&A^o$.

DEFINITION 1.1.6. $\neg^{(n)} A =_{df} \neg A\&A^{(n)}$.

Postulate 12(n). $B^{(n)} \supset ((A \supset B) \supset ((A \supset \neg B) \supset \neg A))$,

Postulate 13(n). $A^{(n)}\&B^{(n)} \supset (A \supset B)^{(n)}\&(A\&B)^{(n)}\&(A \vee B)^{(n)}$.

$\neg^*$ has all the properties of classical negation. In $C_n, 1 \leq n < \omega$, this can be easily proved through the *reduction ad absurdum* theorem demonstration.

Adopting the notions of demonstration, deduction, (formal) theorem and the properties of the $\vdash$ symbol as in [4], Elias H. Alves, in [5], presents the following results about C_ω.

1.1.7 In C_ω all the theorems and deduction rules of positive intuitionistic logic are provable. In particular,

(a) $A\&B \vdash B\&A$,
(b) $A \vee B \vdash B \vee A$,
(c) $(A\&B)\&C \vdash A\&(B\&C)$,
(d) $(A \vee B) \vee C \vdash A \vee (B \vee C)$.

1.1.8 "Peirce's Law", $((A \supset B) \supset A) \supset A$, does not hold in C_ω.

1.1.9 The following formulae are not valid in C_ω :

(a) $\neg(A\&B) \supset (\neg A \vee \neg B)$,
(b) $\neg(A \vee B) \supset (\neg A\&\neg B)$,
(c) $\neg A \supset (A \supset B)$.

1.1.10 C_ω is not finitely trivializable.

1.1.11 The Deduction Theorem is valid in C_ω. In particular,

(a) $A \supset (B \supset C) \vdash (A\&B) \supset C$,
(b) $(A\&B) \supset C \vdash A \supset (B \supset C)$.

1.1.12 The Substitution of Equivalents Theorem does not apply to C_ω.

1.1.13 C_ω's postulates are independent.

1.2 The System CG_ω

When the decidability of the hierarchy $C_n, 1 \leq n \leq \omega$, was still an open problem, Raggio, in "Propositional sequence-calculi for inconsistent systems" (*Notre Dame Journal of Formal Logic* IX (1968), p. 359-366), presented the calculi $CG_n, 1 \leq n \leq \omega$, built up from a standard propositional language. Besides formulae, these sequent calculi have expressions of the form $\Gamma \to \Delta$, called sequents, where Γ (antecedent) and Δ (succedent) are sequences of formulae. An inference rule of the calculi is a figure of the form

$$\frac{S_1 \ldots S_n}{S},$$

where $S_1, \ldots, S_n$ ($n = 1$ or $n = 2$) and S are sequents, and a derivation is a chain of applications of inference rules, in tree form, where the root is the endsequent and the topmost leaves are initial sequents of the form $A \to A$.

The rules of inference of the calculus CG_ω which are the rules of the propositional part of Gentzen's *LK* ([2]), with two modifications, are the following.

Structural Rules

Thinning:

$$\text{at} \rightarrow \frac{\Gamma \rightarrow \Theta}{A, \Gamma \rightarrow \Theta} \qquad \rightarrow \text{at}\frac{\Gamma \rightarrow \Theta}{\Gamma \rightarrow \Theta, A}$$

Contraction:

$$\text{cont} \rightarrow \frac{A, A, \Gamma \rightarrow \Theta}{A, \Gamma \rightarrow \Theta} \qquad \rightarrow \text{cont}\frac{\Gamma \rightarrow \Theta, A, A}{\Gamma \rightarrow \Theta, A}$$

Interchange:

$$\text{perm} \rightarrow \frac{\Gamma, A, B, \Delta \rightarrow \Theta}{\Gamma, B, A, \Delta \rightarrow \Theta} \qquad \rightarrow \text{perm}\frac{\Gamma \rightarrow \Theta, B, A, \Delta}{\Gamma \rightarrow \Theta, A, B, \Delta}$$

Cut:

$$\text{cut}\frac{\Gamma \rightarrow \Theta, D \quad D, \Delta \rightarrow \Lambda}{\Gamma, \Delta \rightarrow \Theta, \Lambda}.$$

Operational Rules

Conjunction:

$$\& \rightarrow \frac{A, \Gamma \rightarrow \Theta}{A\&B, \Gamma \rightarrow \Theta} \qquad \rightarrow \& \frac{\Gamma \rightarrow \Theta, A \quad \Gamma \rightarrow \Theta, B}{\Gamma \rightarrow \Theta, A\&B}$$

$$\& \rightarrow \frac{B, \Gamma \rightarrow \Theta}{A\&B, \Gamma \rightarrow \Theta}$$

Disjunction:

$$\vee \rightarrow \frac{A, \Gamma \rightarrow \Theta \quad B, \Gamma \rightarrow \Theta}{A \vee B, \Gamma \rightarrow \Theta} \qquad \rightarrow \vee \frac{\Gamma \rightarrow \Theta, A}{\Gamma \rightarrow \Theta, A \vee B}$$

$$\rightarrow \vee \frac{\Gamma \rightarrow \Theta, B}{\Gamma \rightarrow \Theta, A \vee B}$$

Conditional:

$$\supset \rightarrow \frac{\Gamma \rightarrow \Theta, A \quad B, \Delta \rightarrow \Lambda}{A \supset B, \Gamma, \Delta \rightarrow \Theta, \Lambda}$$

FES'

$$\frac{A, \Gamma \rightarrow B}{\Gamma \rightarrow A \supset B}$$

NEA'

$$\frac{A, \Gamma \rightarrow \Theta}{\neg\neg A, \Gamma \rightarrow \Theta}$$

Negation:

$$\rightarrow \neg \frac{A, \Gamma \rightarrow \Theta}{\Gamma \rightarrow \Theta, \neg A}$$

We observe that, in CG_ω, Gentzen's rule for introduction of conditional in the succedent, "FES"

$$FES\frac{A, \Gamma \to \Theta, B}{\Gamma \to \Theta, A \supset B}$$

is restricted to intuitionistic form, stated as FES'. Gentzen's rule for introduction of negation in the antecedent, "NEA"

$$NEA\frac{\Gamma \to \Theta, A}{\neg A, \Gamma \to \Theta}$$

is appropriately modified as NEA'.

The systems $CG_n, 1 \leq n < \omega$, have all the rules of CG_ω and also the following rule NEA'', with the corresponding value of n :

$$NEA''\frac{A_1^{(n)}, A_2^{(n)}, ..., A_p^{(n)}, \Gamma \to \Theta, \beta(A_1, A_2, ..., A_p)}{\neg\beta(A_1, A_2, ..., A_p), A_1^{(n)}, A_2^{(n)}, ..., A_p^{(n)}, \Gamma \to \Theta}$$

where $\beta(A_1, A_2, ..., A_p)$ is any schema built from A_1, A_2, ... ,A_p using only $\neg, \vee, \&$ and $\supset$.

DEFINITION 1.2.1 The **length** of a derivation Π, $l(\Pi)$, is defined as follows:

i) If Π is an initial sequent, $l(\Pi) = 0$;

ii) If Π is obtained from Π_1 by means of a rule, $l(\Pi) = l(\Pi_1) + 1$;

iii) If Π is obtained from Π_1 and Π_2 by means of a rule, $l(\Pi) = max(l(\Pi_1), l(\Pi_2))+ 1$.

Some fundamental results can be shown:

THEOREM 1.2.2 In CG_ω there is no derivation of any sequent of type $\Gamma \to$.

Proof: By induction on the length of the derivation Π of $\Gamma \to$, in CG_ω.

Let Π be a derivation such that $l(\Pi) = 0$. There is nothing to prove, for $\Gamma \to$ is not an initial sequent.

Suppose the theorem holds for $l(\Pi) = n$.

If $l(\Pi) = n + 1$, $r(\Pi)$, i.e. the rule that generates the endsequent of Π, cannot be any operational rule.

So, by induction hypothesis there is no derivation of at least one of its premisses, the same occurring if $r(\Pi)$ is *cut*, or any structural rule.

□

COROLLARY 1.2.3 In CG_ω, $\nvdash \to$.

THEOREM 1.2.4 **a)** If $\vdash_{C_\omega} A$, then $\vdash_{CG_\omega} \to A$;

b1) If $\vdash_{CG_\omega} A_1, A_2, ..., A_n \to B_1, B_2, ..., B_p$, then $\vdash_{C_\omega} A_1 \& A_2 \& ... \& A_n \supset B_1 \vee B_2 \vee ... \vee B_p$;

b2) If $\vdash_{CG_\omega} \rightarrow B_1, B_2, ..., B_p$, then $\vdash_{C_\omega} B_1 \vee B_2 \vee ... \vee B_p$.

Proof: The usual one for (a). The derivations of (b1) and (b2) are by induction on the length of the derivations in CG_ω. □

The necessity of the intuitionistic restriction over the rule FES' is explicitly presented by showing that the formula

$$\rightarrow (C \supset (A \vee B)) \supset (A \vee (C \supset B))$$

is provable using the rule FES (without the intuitionistic restriction) and that this same formula is not valid in C_ω (see [3], p. 361-362).

Raggio says that although equivalent to the $C_n, 1 \leq n \leq \omega$, the $CG_n, 1 \leq n \leq \omega$, are not suitable for *Hauptsatz*, for FES'

> "contains the typical intuitionistic restriction that the succedent should have at most one formula. But the other rules of the CG_n are free from this restriction and this prevents the application of Gentzen's proof." ([3], p. 362)

Raggio does not make explicit either the difficulties he reached in developing Gentzen's proof, from case 3.232.1 (see [2]), or the difficulties corresponding to conditioned introduction of negation in the antecedent (NEA'').

In order to avoid these difficulties, Raggio proposes a new hierarchy of systems, the $WG_n, 1 \leq n \leq \omega$. By substituting FES' for FES, he introduces the system WG_ω, and by substituting NEA''' for NEA''

$$NEA''' \qquad \frac{\Gamma \rightarrow \Theta, A^{[n]}}{\neg A^{[n]}, \Gamma \rightarrow \Theta}$$

he constructs, for the corresponding n, the system $WG_n, 1 \leq n < \omega$, where $A^{[1]} =_{df} A \& \neg A$ and $A^{[n+1]} =_{df} A^{[n]} \& \neg A^{[n]}$.

The *Cut* Elimination Theorem was proved for $WG_n, 1 \leq n \leq \omega$, but the calculus WG_ω is not equivalent to C_ω, for it has a rule that is not admissible in C_ω, as can be seen.

So, by analyzing the most important properties of Raggio's systems, we conceived the possibility of developing the ideas concerning the introduction of the calculus C_ω in sequents, *à la* Gentzen. This sequent-calculus, the system NCG_ω, is presented here.

2 THE SYSTEM NCG_ω

2.1 Language and Rules of Inference

The system NCG_ω is defined from a standard propositional language with a countable set of propositional variables and the logical symbols $\neg$ (negation), $\supset$ (conditional), $\&$ (conjunction) and $\vee$ (disjunction), from which the formulae are constructed.

A **sequent** is an expression of the form $\Gamma \to \Theta$, where Γ (*antecedent*) and Θ (*succedent*) are finite sequences of formulae, but Θ has at most one formula. A sequent of the form $A \to A$, where A is a formula, is an **initial sequent**.

The **rules of inference** of NCG_ω, with the restriction of the succedents of the sequents having at most one formula, are the following.

Structural Rules

Thinning:
$$at \to \frac{\Gamma \to \Theta}{A, \Gamma \to \Theta}$$

Contraction:
$$cont \to \frac{A, A, \Gamma \to \Theta}{A, \Gamma \to \Theta}$$

Interchange:
$$perm \to \frac{\Gamma, A, B, \Delta \to \Theta}{\Gamma, B, A, \Delta \to \Theta}$$

Mix:
$$mix \frac{\Gamma \to D \qquad D, \Delta \to \Theta}{\Gamma, \Delta^* \to \Theta}$$

Operational Rules

Conjunction:
$$\& \to \frac{A, \Gamma \to \Theta}{A\&B, \Gamma \to \Theta} \qquad \to \& \frac{\Gamma \to A \qquad \Gamma \to B}{\Gamma \to A\&B}$$
$$\& \to \frac{B, \Gamma \to \Theta}{A\&B, \Gamma \to \Theta}$$

Disjunction:
$$\vee \to \frac{A, \Gamma \to \Theta \qquad B, \Gamma \to \Theta}{A \vee B, \Gamma \to \Theta} \qquad \to \vee \frac{\Gamma \to A}{\Gamma \to A \vee B}$$
$$\to \vee \frac{\Gamma \to B}{\Gamma \to A \vee B}$$

Conditional:
$$\supset \to \frac{\Gamma \to A \qquad B, \Delta \to \Theta}{A \supset B, \Gamma, \Delta \to \Theta} \qquad \to \supset \frac{A, \Gamma \to B}{\Gamma \to A \supset B}$$

Negation:
$$\neg\neg \to \frac{A, \Gamma \to \Theta}{\neg\neg A, \Gamma \to \Theta}$$

Neg:
$$\frac{A, \Gamma \to \Theta \qquad \neg A, \Delta \to \Theta}{\Gamma, \Delta \to \Theta} neg$$

D is the *mix* **formula** and Δ^* is equal to Δ, except that it has no occurrence of D.

The **degree** of an application of *mix* is the degree of D. The **degree of a proof** Π, $g(\Pi)$, is the maximum degree of the *mix* occurring in Π, or 0 (zero), if Π has no *mix*. If $g(\Pi) = 0$ one says that Π is **normal**.

When necessary, **"str"** will indicate the application of Structural Rules in the combination required to get the sequent indicated.

The rule *neg* has characteristic properties and cannot be classified as a rule of antecedent or of succedent. It encodes the postulate of excluded middle, which destroys the intuitionistic features present through the restriction imposed on the succedents.

The definitions and conventions used here are those of [4] and [6].

THEOREM 2.1 In NCG_ω there is no proof of any sequent of type $\Gamma \rightarrow$.

Proof: By induction on length of proofs.

Let Π be a proof of $\Gamma \rightarrow$, such that $l(\Pi) = 0$. There is nothing to prove, for $\Gamma \rightarrow$ is not an initial sequent.

If $l(\Pi) = n + 1$ and $r(\Pi)$, i.e. the rule that generates the endsequent of Π, is any operational or structural rule, then, by induction hypothesis, there is no proof of at least one of its premisses. If $r(\Pi)$ is *mix* or *neg*, also by induction hypothesis, there is no proof of at least one of its premisses.

□

2.2 The equivalence between NCG_ω and C_ω

We prove, here, the equivalence between the sequent calculus NCG_ω and C_ω.

LEMMA 2.2.1 If $\vdash_{C_\omega} A$, then $\vdash_{NCG_\omega} \rightarrow A$.

Proof: C_ω's Postulates 1, 2 and 4 to 9 are positive intuitionistic provable formulae, so usual proofs are clearly applicable. Postulate 3, *Modus Ponens*, corresponds, in NCG_ω, to applications of $\supset\rightarrow$ and *mix* rules and, in order to prove that this rule is admissible in NCG_ω, we have to prove that $\rightarrow B$ can be obtained from the proofs of $\rightarrow A$ and $A \supset B$:

$$\begin{array}{ccccl}
 & & \underline{B \rightarrow B} & & \text{at, perm} \\
 & A \rightarrow A & B, A \rightarrow B & & \supset\rightarrow \\
 & & \overline{A \supset B, A, A \rightarrow B} & & \text{cont} \\
 & \Sigma_1 & \overline{A \supset B, A \rightarrow B} & & \text{perm} \\
\Pi_1 & \rightarrow A & \overline{A, A \supset B \rightarrow B} & & \text{mix(1)} \\
\underline{\rightarrow A \supset B} & & \overline{A \supset B \rightarrow B} & & \text{mix(2)} \\
 & \overline{\rightarrow B} & & &
\end{array}$$

Following, direct applications of *neg* and *mix* demonstrate Postulates 10 and 11.□

LEMMA 2.2.2 If $\vdash_{NCG_\omega} A_1, A_2, \ldots, A_n \rightarrow B$, then $\vdash_{C_\omega} A_1 \& A_2 \& \ldots \& A_n \supset B$.

Proof: By induction on the length of the proof Π of $A_1, A_2, \ldots, A_n \rightarrow B$ in NCG_ω. Base: $\vdash_{C_\omega} A \supset A$.

Induction step: a) structural rules correspond to properties of "$\vdash$" and idempotent and commutative laws of "&" and "$\vee$", in C_ω; b) operational rules are admissible (or derived) rules of C_ω, according to Theorem 1 of [1].

An application of *mix* is handled thus: by induction hypothesis we have that $\vdash_{C_\omega} \wedge\Gamma \supset D$ and $\vdash_{C_\omega} (D\&\wedge\Delta) \supset C)$, where $\wedge\Gamma$ and $\wedge\Delta$ stand for the conjunction of the formulae of Γ and Δ, respectively. By using the properties of commutativity and idempotence of "&", we obtain that $\vdash_{C_\omega} (D\&\wedge\Delta^*) \supset C)$ and, by the Deduction Theorem, $\vdash_{C_\omega} (D \supset (\wedge\Delta^* \supset C)$. By Modus Ponens, from Postulates 1 and 2, we have that $\vdash_{C_\omega} \wedge\Gamma \supset (\wedge\Delta^* \supset C)$, and so, $\vdash_{C_\omega} (\wedge\Gamma\& \wedge \Delta^*) \supset C$.

An application of *neg* is handled thus: by induction hypothesis we have that $\vdash_{C_\omega} A\&\wedge\Gamma \supset C$ and $\vdash_{C_\omega} \neg A\&\wedge\Delta \supset C$. By using properties of "$\vdash$" (the Deduction Theorem), we obtain that $\vdash_{C_\omega} (A\& \wedge\Gamma\& \wedge\Delta) \supset C$ and $\vdash_{C_\omega} (\neg A\& \wedge\Gamma\& \wedge\Delta) \supset C$ and by applying Postulates 9, 10 and 3 we have the desired $\vdash_{C_\omega} (\wedge\Gamma\& \wedge \Delta) \supset C$.

□

LEMMA 2.2.3 If $\vdash_{NCG_\omega} \rightarrow B$, then $\vdash_{C_\omega} B$.

Proof: This is a particular case of Lemma 2.2.2.

□

THEOREM 2.2.4 The system NCG_ω is equivalent to C_ω.

Proof: From Lemmas 2.2.1-2.2.3.

□

2.3 The equivalence between NCG_ω and CG_ω

By using the proof of the equivalence between CG_ω and C_ω, presented in [3] we have that NCG_ω is equivalent to C_ω.

COROLLARY 2.3 NCG_ω is equivalent to CG_ω.

2.4 Elimination of *Mix*

We define the rank of a *mix* in the same way as Gentzen:

> "We shall call the *rank* of the derivation the sum of its rank on the left and its rank on the right. These two terms are defined as follows:
>
> The *left rank* is the largest number of consecutive sequents in a path so that the lowest of these sequents is the *left-hand* upper sequent of the mix and each of the sequents contains the mix formula in the *succedent.*
>
> The *right rank* is (correspondingly) the largest number of consecutive sequents in a path so that the lowest of these sequents is the *right-hand* upper sequent of the mix and each of the sequents contains the mix formula in the *antecedent.*" ([2], p. 89-90)

LEMMA 2.4.1 All *mix* that have a negation as *mix* formula and with left rank equal to 1, have an initial sequent as left upper sequent.

Proof: There is no rule of introduction of negation in succedent.

□

LEMMA 2.4.2 Let Π be a proof in NCG_ω such that $r(\Pi)$ is the unique application of *mix*. Then Π can be transformed into a proof Π', of the same endsequent, without application of *mix*.

Proof: By induction on the degree of the *mix*, with a subsidiary induction on its rank. It follows the standard proof, as developed in [2], with the necessary adaptations. □

The difficulties previewed by [3] are bypassed, following the very observations of Gentzen:

> "In order to transform an LJ-derivation into an LJ-derivation ***without cuts***, we apply ***exactly the same procedure*** as for LK-derivations. (...) We have only to convince ourselves that with every transformation step an LJ-derivation becomes another LJ-derivation, i.e., that the D-sequent of the transformed derivation does not contain more than one S-formula in the succedent, given that this was the case before." ([2], p. 101)

And

> "**3.232.1.** For the cases 3.121 it holds generally that Σ^* is empty, since in $\Pi \rightarrow \Sigma$, Σ must contain only *one* formula, and that formula must be equal to $\mathcal{M}$." ([2], p. 102)
>
> "In the cases 3.111, 3.113.1, 3.113.31, 3.113.35 and 3.113.36, only such formulae occur in each succedent of the sequent of a new derivation as had already occurred in the succedent of the sequent of the original derivation." ([2], p. 102)

Nonetheless, losing the perfect symmetry of the proofs in LK, NCG_ω admits *cut* (*mix*) elimination.

THEOREM 2.4.3 Every proof Π, in NCG_ω, can be transformed into a normal proof of the same endsequent.

Proof: By reiterated applications of the former result over each subderivation Π' of Π such that $r(\Pi')$ is *mix*. □

COROLLARY 2.4.4 Every normal proof without application of *neg*, contains, only, subformulae of the formulae of its endsequent.

Proof: By induction on $l(\Pi)$.

The examination of the form of the rules shows the result.

□

In order to establish that the rule *neg* preserves the subformula property, we go back to the origin of the negated formulae that occur in the premisses of *neg*, and define "cluster sequence" [1], to express the fact that the side formulae remain untouchable in a proof.

DEFINITION 2.4.5 A side formula that is the ith formula occurrence in Γ (or Δ, or Θ) in the upper sequent of an inference rule is **clustered** with the ith formula occurrence in Γ (or Δ, or Θ) in the lower sequent of this inference rule.

DEFINITION 2.4.6 The premisses of an application of contraction rule are clustered with the principal formula, and the premiss A (B) in an application of interchange rule is clustered with the principal formula A (B).

DEFINITION 2.4.7 A **cluster sequence**, or a simply a **cluster**, of a formula occurrence B in a proof Π is a sequence of formulae $B_1, B_2, \ldots, B_n$ in Π such that:

1. B_n is B;
2. each B_i stands immediately above and is clustered with B_{i+1}, for $1 \leq i < n$; and
3. B_1 is either a formula occurring in an initial sequent, the principal formula of an operational inference, or the principal formula of an application of thinning.

The main result of this paper is the following.

LEMMA 2.4.8 Let Π be a normal proof such that $r(\Pi)$ is the unique application of *neg*. Then, Π is of the form

$$\dfrac{\begin{matrix}\Pi_1 & & \Pi_2 \\ A, \Gamma_1 \to C & & \neg A, \Gamma_2 \to C\end{matrix}}{\Gamma_1, \Gamma_2 \to C}$$

and:

I) If every cluster sequence of $\neg A$ begins with $at \to$, then there exists a proof, without *neg*, of $\Gamma_1, \Gamma_2 \to C$;

II) If there is at least one cluster sequence of $\neg A$ beginning with an initial sequent, then $\neg A$ is a subformula either of C or of some formula of Γ_2; and

III) If at least one of the cluster sequences of $\neg A$ begins with an operational rule, then there exists a proof of $\Gamma_1, \Gamma_2 \to C$ such that all cluster sequences of the premisses of *neg* begin with $at \to$ or by an initial sequent.

Proof: I) By eliminating all applications of $at \to$ that introduce $\neg A$ in Π_2, we obtain the normal proof

$$\dfrac{\begin{matrix}\Pi_2' \\ \Gamma_2' \to C\end{matrix}}{\Gamma_1, \Gamma_2 \to C}\, est$$

[1] According to [6], p. 19-20.

without *neg*, where Π_2' e Γ_2' are different from Π_2 and Γ_2, respectively, since they contain no $\neg A$. Observe that $\neg A$ is not a principal formula in Π_2.

II) Consider the proof

$$\dfrac{\begin{matrix} & \neg A \to \neg A \\ \Pi_1 & \Pi_2 \\ A, \Gamma_1 \to C & \neg A, \Gamma_2 \to C \end{matrix}}{\Gamma_1, \Gamma_2 \to C}\, neg,$$

where any other cluster sequence of $\neg A$ can begin with thinning or an operational rule. By hypothesis, the following derivation Σ

$$\begin{matrix} \neg A \to \neg A \\ \Pi_2 \\ \neg A, \Gamma_2 \to C \end{matrix}$$

is normal and contains no application of *neg*.

If $l(\Sigma) = 1$, trivially, $\neg A$ is a subformula of C. If $l(\Sigma) > 1$, and Π_2 has no application of $\supset\to$, the occurrence of $\neg A$ in the consequent of $\neg A \to \neg A$ is a subformula of C; otherwise, if there is an application of $\supset\to$, in Π_2, this occurrence of $\neg A$ is transformed into a subformula of a formula belonging to Γ_2.

III) By induction on the degree of A.

If $g(A) = 1$, the lemma does not apply.

If $g(A) > 1$, and A is not of the form $\neg B$, there is no cluster sequence of $\neg A$ that begins with an operational rule. On the other hand, if A is of the form $\neg B$, there is a cluster sequence of $\neg A$ beginning with an operational rule, and Π has the form

$$\dfrac{\begin{matrix} & \Pi_2 \\ & \dfrac{B, \Gamma_3 \to D}{\neg\neg B, \Gamma_3 \to D}\,\neg\neg\to \\ \Pi_1 & \Pi_3 \\ \neg B, \Gamma_1 \to C & \neg\neg B, \Gamma_2 \to C \end{matrix}}{\Gamma_1, \Gamma_2 \to C}\, neg,$$

which can be transformed into Σ':

$$\dfrac{\begin{matrix} \Pi_2 & \\ B, \Gamma_3 \to D & \\ \Pi_3' & \Pi_1 \\ B, \Gamma_2 \to C & \neg B, \Gamma_1 \to C \end{matrix}}{\Gamma_1, \Gamma_2 \to C}\, neg.$$

The derivation Π'_3 differs from Π_3 only because it contains no occurrence of $\neg\neg B$, and it is a correct derivation, for every rule applied over $\neg\neg B$ can be applied over B.

The degree of A is reduced, hence, by induction hypothesis, this derivation Σ' can be transformed into a derivation where every cluster sequence of $\neg A$ begins either with *at* $\rightarrow$, or with with initial sequent.

We observe that no inference rule allows us to introduce the simple negation $\neg A$. Then, any application of contraction in two cluster sequences of $\neg A$, whose origins are distinct, leads the discussion to the previously analysed cases.

□

The rule *neg* is not eliminable in NCG_ω. In fact, we claim that exactly this rule allows us to dominate the intuitionistic characteristic determined by the set of rules of NCG_ω-$\{neg\}$.

The previous result means that the formula $\neg A$, independently of its origin in the proof, remains in the conclusion of the application of *neg*. It implies that *neg* has the subformula property and allows us to directly prove the consistency and decidability of the system NCG_ω.

DEFINITION 2.4.9 An **operational neg** is an application of *neg* that has at least one premiss whose cluster sequences begin with a conclusion of $\neg\neg\rightarrow$.

COROLLARY 2.4.10 All operational *neg* can be transformed into applications of *neg* such that its premisses have cluster sequences beginning with an initial sequent.

Proof: By Lemma 2.4.8. □

COROLLARY 2.4.11 Every normal proof Π, in which the applications of *neg* have premisses with clusters beginning with an initial sequent, contains only subformulae of formulae of its endsequent.

Proof: By Lemma 2.4.8. □

COROLLARY 2.4.12 Any provable sequent of NCG_ω has a proof with the subformula property.

Proof: By Lemma 2.4.8, and Corollaries 2.4.10 and 2.4.11. □

COROLLARY 2.4.13 In NCG_ω there is no proof of a sequent of type $\rightarrow \neg A$.

COROLLARY 2.4.14 NCG_ω is consistent.

THEOREM 2.4.15 NCG_ω is decidable.

Proof: The standard proof of decidability applies for NCG_ω. □

References

[1] N. C. A. da Costa. Calculs propositionnels pour les systèmes formels inconsistants. *C.R. Acad. Sc. Paris*, 257:3790–3792, 1963.

[2] G. Gentzen. Investigations into logical deduction. In G. Szabo, editor, *Collected papers of G. Gentzen*, pages 69–131, Amsterdam, 1969. North-Holland.

[3] A. Raggio. Propositional sequence-calculi for inconsistent systems. *Notre Dame Journal of Formal Logic*, 9:359–366, 1968.

[4] S. C. Kleene. *Introduction to metamathematics*. D. Van Nostrand Company, New York, 1972.

[5] Elias H. Alves. Lógica e inconsistência: um estudo dos cálculos $C_n, 1 \leq n < \omega$. Master's thesis, Universidade de São Paulo, 1976.

[6] L. C. P. D. Pereira. *On the Estimation of the Length of Normal Derivations*. Akademilitteratur, Stockholm, 1982.

A, Still Adorable

ROBERT K. MEYER Australian National University, Canberra, ACT, AUSTRALIA
rkm@arp.anu.edu.au

JOHN K. SLANEY Australian National University, Canberra, ACT, AUSTRALIA
jks@arp.anu.edu.au

To Professor Newton C. A. da Costa to mark his 70th birthday, who has recalled, "The Qu'ran has a thousand interpretations and the last one I tell you is true"

Abstract
This paper expands on an earlier paper, "Abelian Logic (from A to Z)", by the same authors. The main result is that the Abelian logic A is rejection-complete; i. e., that each formula B is either a theorem of A, or else leads in the style of Lukasiewicz to a proof of the variable p. Meredith's single axiom formulation is introduced for the implicational fragment of A and the finite model property proved. An interesting normal form is shown for A, using distribution laws not normally available to relevant logics.[1]

INTRODUCTION

One score and one years ago, your authors brought forth upon this planet a logic new to them, conceived in Integers and dedicated to Inferential Negation. We beheld this logic, called it **A** [2] and showed **Z** [3] characteristic for it. Having announced preliminary results in [1], commercial confidentiality was served when we entrusted them to paraconsistent publication in [2]. Sure enough, the secrets of **A** remained secret for another decade. Even the TV rights that we signed away [4] are not (yet) used.

But we are now on the threshold of a new millennium. New occasions teach new systems. Breaking our old vow of silence, we commend **A**, still adorable, to you **now**. We are all the more emboldened to do so because the collection [3] in which [2] appeared is sadly out of print. The title is suggested by an old popular song, which wanders through the alphabet from "A, you're adorable." Here goes.

---

[1] Thanks, mainly, to one of the referees for this abstract.
[2] Abelian (meaning *commutative*), alternatively Australian, maybe ANU
[3] Familiarly, the Integers, as a lattice-ordered Abelian group
[4] Yes, we really did. Imagine Bill Gates doing something that stupid and unprofitable!

1 CHASING GROUPS WITH J. M. DUNN

In his dissertation [4], Dunn announced his invention of **DeMorgan semigroups** (ever after, DeMorgan **monoids**),[5] devised to correspond algebraically to the relevant logic **R**. Clearly [4] yearned after lattice-ordered groups (henceforth, l-groups). The ingredients were there, namely

Particle	Current Name	Mathematical properties	
o	Fusion	Associative, commutative, (square-increasing)	
→	Implication	Residual of o	
t	Ackermann **t**	Identity for **o**, left identity for →	
⊃	Conditional	Enthymematic implication, B ⊃ C is B∧**t** → C	
∧	Conjunction	Lattice meet, extensional conjunction	
∨	Disjunction	Distributive lattice join, extensional disjunction	
~	Negation	DeMorgan *not,* a dual automorphism	
≤	Entailment	Partial order relation	
+	Fission	DeMorgan dual of **o**, intensional disjunction	
f	Ackermann **f**	~**t**, may define ~B inferentially as B → **f**	
F	Church **F**	The *Absurd,* **F** ≤ *all* B, lattice zero	
T	Church **T**	The *Trivial, all* B ≤ **T**, lattice unit	

There is some rewriting of history here (but not much). Note the *paired* couples {**t**,**f**} and {**T**,**F**} of truth-values. Contrast also the *intensional* pair {**o**, +} with their *extensional* analogues {∧,∨} of counterparts for 'and' and 'or'. Finally *enthymematic implication* ⊃, which has an *intuitionist* character for **R**, turns out in [2] to capture infinite-valued Łukasiewicz logic for **A**.

Clearly Dunn was yearning after l-group theory in his assault on relevant algebras. But that assault could only prosper on some amendment of logic. For although ~ is already in [4] an *involution,* just like group *inverse,* there are elements in relevant implication that resist identification with group-theoretic ideas. The mountain of group theory, we confessed in [2], is not coming to the Muhammed of **R**. But nothing prevents Muhammed from going to the mountain.

2 SLANEY SUPPLIES A TRUE INFERENTIAL NEGATION

Meanwhile Slaney had the crazy idea that, if negation were really to be defined inferentially, then fundamental principles like full double negation should already be present in the implicational part of logic. This requires new theses not even bought by classicists suckled in creeds outworn, like the axiom scheme (of Relativity),

AxRel. ((A→B)→B) → A

[5] As in Dunn's contribution to [5].

And it requires the *excision* of long cherished theorems, such as the *contraction* axiom scheme

AxW. (A→(A→B)) → (A →B)

It will be immediately clear why AxRel is ineluctable for **A**, while AxW is inadmissible. We alluded above to the standard inferential definition of ~ (due to Peirce and then Russell, whence usually credited to Johansson), via a suitable proposition **f** and

D~. ~B =df B → **f**

AxRel is needed that the theory of → should deliver full double negation principles, *without appeal* to special properties of **f**. In its presence not only AxW but also *any theorem scheme* of the form (B→C)→C is acceptable only if B is already acceptable. Conversely one step of **→E** with AxRel as major premiss and AxW as minor premiss yields A itself as a theorem scheme, trivializing logic.

We turn next to some *old* and *new* pictures of inferential negation.

2A. Belnap's M_0

Consider the 8 element Boolean algebra $\mathbf{M_0}$, with the following Hasse diagram:

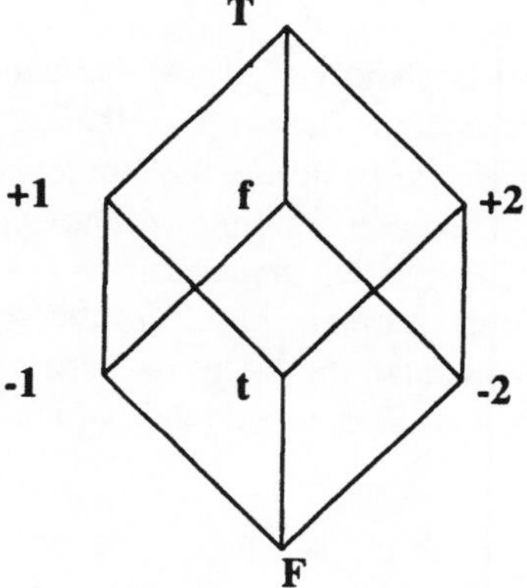

This is a rendition of $\mathbf{M_0}$ (Squint until you see a *cube*). The *atoms* are **t**, **-1**, **-2** (which may be identified respectively with the triples of truth-values 100, 010, 001). The *counteratoms* are **f**, **+1**, **+2** (similarly identifiable with 011, 110, 101). The lattice connectives ∧ and ∨ may be defined *pointwise* (truth-functionally, whence **T** and **F** turn out respectively to be 111 and 000). On the (Boolean) plan just parenthetically suggested, elements with a 1 in the *initial* position (**t**, **+1**, **+2**, **T**) will count as *designated (i. e., true) values*, whence the others (**f**, **-1**, **-2**, **F**) are *false*.

Note that there are *two* candidate false constants in $\mathbf{M_0}$, namely **f** and **F**. Intuitionists, classicists and other partisans of the outworn creeds are likely to opt for **F** as *the* false, since **F** is that than which nothing falser can be conceived.

(Nothing, let us hope, is *that false.*) Meanwhile **f** looks like a considerably more plausible candidate to be *the false*. Instead of the *conjunction* of all falsehoods, **f** is rather their *disjunction.* And it suffices, we think with [5], for B to be false that it entails *something* false, without throwing in the supernumerary requirement that it entail *absolutely everything*.

$\mathbf{M_0}$ was proposed by Belnap (and is used in [5]) to show that relevant logics like **R** (and all its subsystems, including **E** and **C**) have the *relevance* property; i. e., if A→B is a theorem, then there is some propositional variable p which occurs in *both* A and B.[6] To show this it is necessary to define the implication → on $\mathbf{M_0}$. The following will do:

→	F	-2	-1	f	t	+1	+2	T
F	T	T	T	T	T	T	T	T
-2	F	+2	F	+2	F	F	+2	T
-1	F	F	+1	+1	F	+1	F	T
f	F	F	F	t	F	F	F	T
*t	F	-2	-1	f	t	+1	+2	T
*+1	F	F	-1	-1	F	+1	F	T
*+2	F	-2	F	-2	F	F	+2	T
*T	F	F	F	F	F	F	F	T

We have defined → on $\mathbf{M_0}$. And we already have given the (Boolean) recipe for defining ∧ and ∨ on it as well. But what of the DeMorgan negation ~? As a Boolean algebra, we may of course define the Boolean negation ¬ on $\mathbf{M_0}$.[7] But in view of p∧¬p → q, the relevance principle will not survive that definition.

On the other hand, the *inferential definition* by D~ above will do just fine, taking **f** as the so-called counteratom of $\mathbf{M_0}$. Put otherwise–the *column* headed by **f** in the → table just above is also the table for DeMorgan ~. (Or, as Belnap has put it in conversation, $\mathbf{M_0}$ is a Boolean algebra which is a little mixed up about complementation.)

2B. Meyer's *C6*

From a paraconsistent viewpoint, $\mathbf{M_0}$ is pretty dull, splitting its (DeMorgan monoid) elements univocally into the *true* ones and the *false* ones. Where, readers may wonder, do the *true contradictions* go? We shall exhibit *another* Hasse diagram, for the 6 element *crystal lattice* **C6**, which is equally efficient (at least) in showing that **R** and its subsystems have the Belnap relevance property. Like $\mathbf{M_0}$, **C6** is a DeMorgan monoid.[8] Here is its → table.

---

[6] Brady in [7] has in fact *axiomatized* $\mathbf{M_0}$, as an *extension* of **R**.

[7] It is also *interesting* to have *both* DeMorgan ~ and Boolean ¬, getting the *conservative extension* of **R** (which we now call **CR**) discussed in [9].

[8] And, yes, **C6** also has been axiomatized by Brady.

→	F	t	1	2	f	T
F	T	T	T	T	T	T
*t	F	t	1	2	f	T
*1	F	F	1	F	1	T
*2	F	F	F	2	2	T
*f	F	F	F	F	t	T
*T	F	F	F	F	F	T

And now, here is the Hasse diagram.[9]

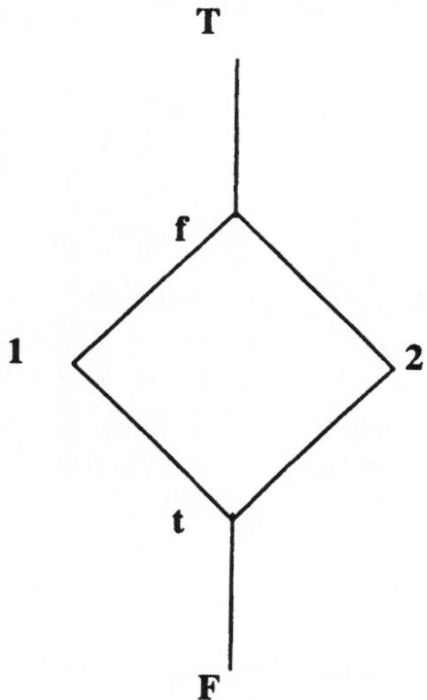

Again we apply D~ to define ~b as b→**f**, for all elements b in **C6**. This has the effect that each of **1, 2** are *fixed points* with respect to negation. Lattice connectives may be read off the Hasse diagram in the usual way. (E. g., **1** ∧ **2** = **t**, etc.) Note moreover that **F** is the *only* undesignated element. Nonetheless there is an elementary argument that **R** (and any other logic of which **C6** is a model) has the relevance property. Suppose the contrary. Then there are formulas A and B such that A→B is a theorem of **R**; but, for *reductio*, that A and B share no variable. Simply assign 1 to all variables of A, and 2 to all variables of B. It is immediate that, on this assignment, A→B takes the value 1→2 = **F** in **C6**. So **R** has the variable-sharing property.

[9] Elements of **C6** also have a Boolean representation. See Birkhoff [6]. The simplest is **F** = 0000, **t** = 1000, 1 = 1100, 2 = 1010, **f** = 1110, **T** = 1111. Such facts prove useful in the Brady axiomatizations of **C6** and $\mathbf{M_0}$ noted above, and in his algebraic work recorded in [7].

2C. Inferential negation in *A*

We have looked in the past two subsections at a couple of propositional interpretations for the relevant logic **R**–the semantically normal $\mathbf{M_0}$ of 2A, and the paraconsistent **C6** of 2*B*. We turn now to **A**, with its affinity for **Z**, linearly ordered. We recall from our motivating discussion above that **f** can be anything that we want it to be. We might want a large number of *undecided propositions,* to be taken as *neither true nor false.* Since the *truths,* in **A** perspective, are pretty well fixed in **Z** as the *non-negative integers* from 0 (= **t**) on, it is the *falsehoods* with respect to which we get a choice. So let us simply take **f** as some negative number of staggering absolute value, such as -10^{123}. We then end up with a picture like the following:

```
                    -10^123          .. -3  -2  -1  0 +1 +2 +3 ...
...Stinkers ...        |                            |
                       f   ...Undecided ...         t  ...Good as gold ...
```

On another scenario there are lots of propositions (as in the **C6** case, perhaps) which there is *evidence for,* and also *evidence against.* We can allow for this situation also, choosing **f** as a *positive integer*. Asserting each of A, ~A comes intuitively to much the same thing as failing to assert either (truth to tell, we don't know about A, save on the fantasies of standard logics on which, if you've got one contradiction, you've got them all). Here's a picture, with $\mathbf{f} = +10^{79}$.

```
      ... -3  -2  -1   0 +1 +2 +3 ...            +10^79
                       |                          |
  ... Bad guys  ...    t ...A bit mixed up here...  f   ...No worries, mate! ...
```

We wish once more to stress that it is *up to us* where we put **f**. But there are a couple of places that commend themselves especially. For we might, in classical or at least $\mathbf{M_0}$ mode, wish to split the propositions *exactly* into the definite truths and the definite falsehoods. In the integers, the choice $\mathbf{f} = -1$ will have that effect, with the picture

```
          ...  -3  -2   f   t  +1  +2  +3 ...
     False to the Left |   |  True to the Right
```

This picture may seem consonant with the New World Order–bad guys on the *left,* good guys on the *right,* and never the twain shall meet. Casari in [10] has a somewhat similar idea–save that he contemplates the addition of a new element – 0 to serve as the negation of 0.[10] But we like better the integers as *originally served,* to wit

[10] Interestingly [5] also labels the **f** of $\mathbf{M_0}$ as –0, taking **t** as +0. When will they ever learn that 0 is *its own* negate?

t ... Truths begin here
|
-3 -2 -1 **0** +1 +2 +3 ...
|
Falsehoods end here ... **f**

On this picture, 0 is *both* the *Final Falsehood* **f** and the *First Truth* **t**. It enables us to launch a last catapult against the truth-functional logic **2**. Namely, **2** admits *too many truth-values,* whereas **A** has only the *single* truth-value 0. Granted, **Z** supplies many *other values* for sentences to take. But why call them *truth* values, in subservience to standard propaganda that it is the functional manipulation of **T** and **F** that Logic is *really* about? Finally, as we said in [2], 0 is a good spot to localize those paradoxes that turn upon themselves, like The Liar. If p, through some filthy[11] logical trick, turns out equivalent to not-p, **A** has 0 ready to be its value. And the choice of **f** as 0, which we will henceforth prefer, we call *canonical.* The resulting *canonical negation* defined by D~ makes the model theory of **A** particularly smooth, as we shall see in more detail below. Here come some *axioms.*

3 AXIOMS FOR A

Our policy in formulating **A** was then as follows: Helping ourselves to the Principle of Strength of Belnap's [11], we made the system as strong as was consistent with its purposes. Here, in Backus-Naur form, is our vocabulary:

Variables: $p \in \Phi$, Φ countable
Formulas: $A \in \text{FORM}(\Phi)$

$$A ::= p \mid \mathbf{t} \mid A \oplus A \mid A \wedge A \mid A \vee A \mid A \rightarrow A$$

Definitions:

D–. $-A =_{df} A \rightarrow \mathbf{t}$ D↔. $A \leftrightarrow B =_{df} (A \rightarrow B) \wedge (B \rightarrow A)$

D⊃. $A \supset B =_{df} A \wedge \mathbf{t} \rightarrow B$ D$\sim_B$. $\sim_B A =_{df} A \rightarrow B$

For ease in reading formulas we give unary connectives minimal scope. Binary ones are ranked

$\circ, \oplus, +, \wedge, \vee, \supset, \rightarrow, \leftrightarrow$

in order of *increasing* scope. We may also write $A \leq B$ to indicate that $A \rightarrow B$ is a *theorem* of **A**, and $A \cong B$ when $A \leq B$ and $B \leq A$. And we use ⇒ as a *metalogical if,* whence our *modus ponens* principle **Ru→E** below says that if A→B is a theorem then if A is a theorem then B is a theorem of **A.** Dropping only AxW from **R** and adding AxRel, henceforth usually **AxA**, here are axiom schemes and rules sufficient for a Hilbert formulation of **A.**

AxA. $(A \rightarrow B) \rightarrow B \leq A$

[11] A referee opines, "I do not think the Liar Paradox constitutes a 'filthy logical trick'. It follows by applying normal meanings of words to a justifiably self-referring expression." We admit that '*This sentence is false*' is grammatical English. Whether it is *good English* is another question—perhaps a *moot* one, since the sentence may nonetheless be good Formalese.

AxB. $B \rightarrow C \leq (A \rightarrow B) \rightarrow (A \rightarrow C)$
AxC. $A \rightarrow (B \rightarrow C) \leq B \rightarrow (A \rightarrow C)$
AxI. $A \leq A$
Ax∧E. $A \wedge B \leq A$, and $A \wedge B \leq B$
Ax→∧. $(A \rightarrow B) \wedge (A \rightarrow C) \leq A \rightarrow B \wedge C$
Ax∧∨. $A \wedge (B \vee C) \cong (A \wedge B) \vee (A \wedge C)$
Ax∨→. $(A \rightarrow C) \wedge (B \rightarrow C) \cong A \vee B \rightarrow C$
Ax⊕→. $A \oplus B \rightarrow C \cong A \rightarrow (B \rightarrow C)$
Axt→. $A \cong \mathbf{t} \rightarrow A$
Ru→E. $(A \leq B) \Rightarrow (A \Rightarrow B)$
Ru∧I. A and B $\Rightarrow A \wedge B$

Readers who recall that o is an intensional *and* in **R** and that + is a similar *or* will wonder at their absence in **A**. And what, they may also wonder, is that squiggle ⊕? Readers of [2], if any, will answer that ⊕ is *indifferently* whichever of o and + that we want it to be. I. e., at the *canonical* limit, intensional *and* and *or* are the same in **A**. In line with the Slaney motivation of section 2, *every* formula B gives rise to a corresponding *negation* $\sim_B$, which applied to A is $A \rightarrow B$. But we have stated our preference for – , which is the *canonical negation* $\sim_t$. Finally, since Abelian l-groups are automatically *distributive* lattices, the distribution axiom **Ax∧∨** is *redundant.*

4 ENTER THE INTEGERS Z

We admitted in [2] that, next to the two-valued logic **2**, **A** is the one logic that *absolutely everybody* can understand. The reason is that it has a *simple, clear* and *characteristic* model in the set **Z** of integers. It makes you wonder why Boole went to all that trouble to invent *another* calculus for logic in [12]. His idea was to use the symbolic methods of mathematics for *non-quantitative reasoning*. But why bother, if the methods of quantitative mathematics will already do the job?

First, we recall the modeling. An *interpretation* of FORM in **Z** shall be any function I from formulas to integers, subject to the following constraints:

C1. $I(\mathbf{t}) = 0$
C2. $I(A \oplus B) = I(A) + I(B)$
C3. $I(A \rightarrow B) = I(B) - I(A)$
C4. $I(A \wedge B) = MIN(A,B)$
C5. $I(A \vee B) = MAX(A,B)$

A formula A is *verified* on an interpretation I in **Z** iff $0 \leq I(A)$. A is **Z**-*valid* iff A is verified on all interpretations I in **Z**. [2] proved the following

Soundness and completeness theorem for A in Z. A necessary and sufficient condition for B to be a theorem of **A** is that B be **Z**-valid.

5 THE DELICACIES OF A

A has some absolutely lovely properties, which have been *insufficiently attended to.* We recall some of them immediately.

Infinite model property. All non-trivial models of **A** are infinite. For **A** admits *neither* a least proposition **F** nor a greatest proposition **T**. (If it did, A → **T** ≤ **T** and **AxRel** together validate *every* A. This is *bad.* Note also that, by ∨-semilattice properties, every *finite* model of **A** must have such a **T**.)

Many a logic was *made up for the purpose* to *lack* the finite model property. But only **A**, among the logics we know, lacks this property *because it has the infinite model property.* Next,

A is decidable. For the theorems are recursively enumerable (having been finitely presented), while the non-theorems also may be recursively enumerated by working our way systematically through interpretations in **Z**. Accordingly the theorems of **A** constitute a *recursive* set.

We drew in [2] the corollary that

Every interesting logic with the infinite model property is decidable.

6 A LINEAR LINEAR LOGIC

It is far from clear to us why Girard called the system introduced in his [13] *linear logic.* (Our name for it will be **LL**. Girard's terminology, we suppose, was suggested by *linear algebra.*) Rather more vertebrately linear is **A**. An easy verification in **Z** produces the *linear order theorem*

LOT. (A → B) ∨ (B → A)

In view of LOT, we may restrict our semantic attention to *chains* (like **Z**, on its natural order).

7 THE WAYS OF POST AND REJECTION-COMPLETENESS FOR A

A is Post complete. That is, the result of adding *any non-theorem* B as a new axiom to **A**, and closing under substitution and the rules of **A**, produces *all formulas* as theorems. Proved in [2].

The *only logic* which is *famously* Post-complete is the truth-functional logic **2**. The Post completeness of **A** also makes it possible for us to set out a *rejection formulation* of the system, similar to the analogous formulation of truth-functional logic.[12] We add to the basis of section 3 a *rejection* symbol ⊣, together with a single *counteraxiom*

Caxp. ⊣ p, where p is the first propositional variable,

and two rules. The first of these is a *modus tollens* rule, namely

CMT. ⊢ A→B ⇒ (⊣ B ⇒ ⊣ A),

and the second is a *converse rule of substitution*, namely

Csub. ⊣ A' ⇒ ⊣ A, where A' is a substitution instance of A.

We call a propositional logic **L** *rejection-complete* iff, for each formula A in its vocabulary, either ⊢ A or ⊣ A. The extension of a formulation of a logic **L** by just Caxp, CMT, and Csub above will be a *p-rejection* formulation of **L**. We call **L** so formulated *p-rejection-complete* if it is rejection-complete.

To set the stage for the p-rejection-completeness of **A,** we review a corresponding argument that **2** has this property. Where *n* is a natural number, we call a logic *n-reducible* provided that, for each formula B, ⊢ B iff, for all substitution instances B' of B in at most *n* variables, ⊢ B'. We observe first, by truth-table considerations,

1-reducibility lemma for 2. 2 is 1-reducible.

Now suppose that B is a non-theorem of **2.** By the lemma, there is an instance B' of B, in the single variable p, such that B' is also a non-theorem. There are 4 non-equivalent 1-variable formulas in **2**, represented by p∧~p, p, ~p, p∨~p. Clearly we may choose B' to be one of the first three (else B would be already a theorem). But ⊣ p, by Caxp. And ⊣ p∧~p, which implies p in **2**, by CMT. Finally by Csub, ⊣ ~p, since its instance ~~p implies the rejected p. In all cases B' is rejected, in virtue of classically implying one of the 3 *bad* 1-variable formulas.[13] And so ⊣ B, by Csub. We conclude

p-rejection theorem for 2. 2 is p-rejection complete.

We now wish to *adapt* the above argument to **A**. We observe first, again,

1-reducibility lemma for A. A is 1-reducible.

[12] Cf. Łukasiewicz [23].

[13] Truth-table considerations allow us to choose B' as materially equivalent to p∧~p. For B, as a non-tautology, comes out 0 on some line of a truth-table. Form B' by substituting p∧~p for each variable q assigned 0, and p∨~p for each variable r assigned 1. And we get a B' equivalent to p in *positive* classical logic, which [24] calls **HC**, substituting p for variables assigned 0 but p→p for those assigned 1.

On this occasion, *proof* is called for. Suppose that B is a non-theorem of **A**. By the completeness theorem (section 4), there is an interpretation I in **Z** such that I(B) < 0. We use I to define a 1-variable instance B' of B and an interpretation I' such that I'(B') < 0, which suffices. I is determined homomorphically by its value on the propositional variables q_i which occur in B. Without loss of generality, we assume that p itself does *not* occur in B.

We introduce integer coefficients into the vocabulary of **A** as follows: where C is a formula,

*0*C =df **t**

Where *(m-1)*C has been defined, for m > 0, define

*m*C =df C ⊕ *(m-1)*C
*(-m)*C =df –(*m*C)

Examples (near enough):
*3*q ≅ q ⊕ q ⊕ q
(-2)(p∨**t**) ≅ (**t**∧–p) ⊕ (**t**∧–p)

We now complete the proof of the 1-reducibility lemma for **A**. Our refutation I of B in **Z** assigns to each variable q in B some integer m, whence I(B) = –n for some n > 0. Define now a substitution ' by setting, for each q in B, q' = *m*p, where I(q) = m. We claim that B' ≅ –*n*p, where I(B) = –n and n > 0. The elementary argument is by structural induction on subformulas of B, and is left to the reader. Define I' by setting I'(p) = +1. Another elementary argument shows, for all subformulas C of B, I(C) = I'(C'). In particular, I'(B') = –n < 0, whence the 1-variable formula B' is an *unprovable* instance of the non-theorem B. Of course the class of theorems of the logic **A** is closed under substitution, an observation that ends the verification that **A** is 1-reducible.

Now we show **A** to be p-refutation-complete. The grounding intuition behind Post-completeness is that *nothing* can be added to a system, in conformity with its underlying principles, without getting absolutely *everything*, even p. It is this grounding intuition with which our characterization above of a p-refutation formulation was intended to conform.

So suppose that we try to add some non-theorem B as a new axiom. By the 1-reducibility lemma we find a substitution instance B' in the sole variable p, which is also a non-theorem. By Csub, it will suffice for ⊣ B if we can show ⊣ B'. We may assume, invoking the proof just concluded, that there is an interpretation I' such that I'(p)=+1 but I'(B') = –n, for some positive integer n. Define a substitution s by setting s(p) = p ∨ –p. Using I', it is clear that the formula B" = s(B') ≅ –*n*(p∨–p) ≅ *n*(p∧–p), where *n* is positive. But, for *positive n,* it is easily verified in **A** that ⊢ *n*(p∧–p) → p. This completes the daisy chain. Since ⊣ p we have ⊣ B", by CMT. But then ⊣ B', by Csub. And ⊣ B, by Csub again. Recapitulating,

p-rejection theorem for A. A is p-rejection complete.

8 GETTING AND SPENDING, WE LAY WASTE OUR POWERS

A motivational argument urged for **LL** is the following: The **use** of a premiss incurs a cost. If the premiss is used twice, the cost is doubled. Suppose, for example, a Coke costs $1. If one has a dollar, one may use 1 Coke. But if one wants to use 2 Cokes, one had better send out for an *extra* dollar. Loose classical (and even intuitionist and some relevant) thinking encourages the attitude that, if one has a premiss, one can use it as many times as one wants. But as our friend Robert Bull put it to us, "If I want to *use* a premiss repeatedly, I can *write it down* repeatedly."

9 MORE ON A→

The *implicational* (or *multiplicative,* or *intensional*) fragment **A→** of **A** is the part determined by our axiom schemes **A, B, C, I** above, together with the rule **→E** of *modus ponens.* It is immediately apparent that there is more in **A→** than one might have thought. For we may add the following to the definitions above:

> **Dt.** **t** =df p→p, where p is the *first* propositional variable
> D–. –A =df A → **t**
> D⊕. A ⊕ B =df –A → B

Recalling that ⊕ stands indifferently for an Abelian *and* or a (canonical) Abelian *or,* we have defined the *entire* intensional fragment of **A** in its pure implicational fragment **A→**. (Use **Dt** and **D⊕** to show, from the pure → axioms alone, that we have **Axt** and **Ax⊕** anyway as theorems.)

Speaking of Robert Bull, as we were in the last section, he was present for one of our first presentations of **A**, in 1979. He had, he said, come up with a *single axiom* formulation of the implicational *heart* **A→** of **A** in the 1950's, as a graduate student at Manchester. Alas, we *neglected* to get his axiom. Given that Bull has long since *retired,* it was a little late to repair that neglect for the present paper.

Single axiom formulations of logics are *grist for the mill* of the Automated Reasoning community. So we sent off e-mail to Ulrich, at Purdue, and to McCune, at Argonne. We received in immediate reply an e-note from Wisconsin's Fitelson, who had been reading McCune's mail. "Are not your implicational axioms *redundant?*" asked Fitelson. For, he went on to note, **AxI** follows by a single step of condensed detachment from **AxA**. (Oops! Though this was immediately confirmed by Rogerson, by Bunder, and by the FMO program of [14].) More than that, **AxC** is also redundant. (We set this as an exercise for readers, or their programs–e. g., Argonne's OTTER, which Fitelson used to derive **C** from **A** and **B**. Again, FMO confirms.)

Then we heard from Ulrich, who ended the quest for a single implicational axiom. Sufficient for this purpose with **Ru→E** is

> **AxM.** A→B ≤ ((C→B) → A) → C

The **M** is for C. A. Meredith, who published this axiom in [15] with Prior in 1968. Does this not leave priority with Bull? No (says Ulrich)–for the back of an old envelope, or whatever, on which Meredith jotted down his axiom was, according to [15], dated no later than 1952.

10 A→ AS ABELIAN GROUP LOGIC

We advertised the full system **A** in [2] as the logic of *lattice-ordered* Abelian groups. The (distributive) lattice order is imposed via the *lattice connectives* ∧ and ∨. But →, on its own, imposes no more than a *partial order.* Note next that under the *equality relation* =, every group counts moreover as a partially ordered group. It follows quickly that B is a theorem of **A→** iff I(B) = 0 on all interpretations I of *intensional* formulas in the integers Z. Unlike all of **A**, its intensional part **A→** has *the finite model property.*

For suppose I(B) = –m for some m > 0. Let h be the natural homomorphism from Z onto the integers *modulo* m+1, which we call Z'. We view Z' as a logical matrix, in which 0 is *the sole designated value.* And now let I'(C) = h(I(C)), for all formulas C in the vocabulary of **A→**. For our chosen non-theorem B we have I'(B) = h(–m) = 0–m = m+1–m = +1. Since +1 must be distinct from 0, we have refuted B in the finite matrix Z'.[14]

Put otherwise, B→C is a theorem of **A→** iff I(B) = I(C) for *all* interpretations I in **Z**. This fact led logicians at Monash to propose in [16] a (Lovelight) system **L→** *dual* to **A→**, switching our I(B→C) = I(C) – I(B) to the kindred I(B→C) = I(B) – I(C). The result, spelled out in [17], is that A is a theorem of **A→** iff the result A^r of systematically *reversing* all arrows is a theorem of **L→**. And *vice versa.*

A→ is also pleasant to *Gentzenize.* See Paoli [18] for a *consecution calculus.* Exercise: Extend [18], perhaps along the lines of (Belnap's) *display logics* in [19], to *all* of **A**.

We complete this section by recalling from [2] that, along with **Ru→E**, we have in **A→**

Ponens modus. A ≤ B ⇒ (⊢ B ⇒ ⊢ A)

Like the finite model property, the above admissible rule *no longer holds* in the *extensionally enriched* full system **A**.

[14] Hey! We have been insisting all paper long that *non-negative* integers count as *true*. So what is this "refuted at +1" stuff? Since m>0, it is, mod m+1, "refuted at –m". For the *nicest* harmony in the ideas, we should take the *representatives* of the integers mod m+1 as –m, 1–m, 2–m, ..., –1, 0.

11 NORMAL FORMS IN A

Consider the following classical equivalence:

Th→∨. $A \to B \vee C \;\cong\; (A \to B) \vee (A \to C)$

This is one of a family of classical distribution laws, which enable us to rearrange logical formulas so that certain particles occur *inside* others. Alas, outside of the friendly turf of **2**, Th→∨ *hardly ever holds.* It fails for the *intended implicational* → of the intuitionist logic **J**, the relevant logic **R**, the modal logic **S5** and *all their subsystems.* This deprives us, computationally, of a useful tool–the capacity to place *intensional* formulas equivalently inside *extensional* ones.[15] And it induces *fans* of the thus deprived logics (including, very often, us) to find *classes* of formulas A (e. g., the *Harrop* formulas of [20], the *normal* formulas of Brady [21] and the *HRSK formulas* of Kron [22]) for which some semblance of the above distribution law can be recovered.

We dwell on this hobbling of logical technique mainly to note that **A** is *not* so hobbled. Put otherwise, Th→∨ holds for **A** in full generality. This and related distributional principles enable us to drive *all* intensional particles inside extensional ones. (Imagine the sense of *freedom*–not to say *delicious* sin–conveyed by the capacity to make logical moves long inhibited by *relevant* propriety.) The resultant scheme of (Abelian) *normal forms* comes out thus in [2].

A *basic intensional formula* (henceforth, *bif*) is a formula B of the form,

$$(1)\quad m_1p_1 \oplus \ldots \oplus m_np_n\,,$$

for $n > 0$, distinct propositional *variables* p_i , and corresponding *integer coefficients* m_i.

Bif theorem. ([2]) Every formula A of **A→** is provably equivalent to some *bif* B in the same propositional variables, which may be made unique by taking the variables in alphabetical order.

Examples: (i) $\mathbf{t} =_{df} p\to p \cong -1p \oplus 1p \cong 0p.$

(ii) $(p\to(p\to q)) \to (p\to q) \cong 2p \oplus -1q \oplus -1p \oplus 1q \cong 1p \oplus 0q$

(i) is a template for *theoremhood* in **A→**. E. g., a formula A in variables p, q, r is a theorem of **A→** iff its equivalent *bif* B is $0p \oplus 0q \oplus 0r$. (ii) likewise is a template for *non-theoremhood,* in this case of the **A**-rejected *contraction* axiom **AxW** of **R**. In general, if after collecting terms and adding or subtracting coefficients by associativity and commutativity of ⊕ we find any variable p in the resultant *bif* B with a *non-zero* coefficient m, we refute B (and hence A) in **Z** by setting $I(p) = -1$ if $m > 0$ and $I(p) = +1$ if $m < 0$. In either case $I(mp) < 0$, which refutes B if we simply set $I(q) = 0$ for every q distinct from p.

---

[15] *Linear logicians* might enter the same complaint by saying that *multiplicatives* don't go inside *additives*.

Things get more complicated when the lattice connectives $\wedge$ and $\vee$ are also present. But they do not get *that much* more complicated. First, every formula B may be reduced to an equivalent (canonical) *negation-normal form* B', using DeMorgan laws and double negation for – to confine negation to propositional variables; at the same time, $\rightarrow$'s may be eliminated for – and $\oplus$. Next we confine $\oplus$'s *within* the lattice connectives, in view of the (distribution) laws

Th⊕∨. $A \oplus (B \vee C) \cong (A \oplus B) \vee (A \oplus C)$

Th⊕∧. $A \oplus (B \wedge C) \cong (A \oplus B) \wedge (A \oplus C)$

Why do these laws hold for **A**? Recall that $\oplus$ is *both* a fusion o and a fission +. In relevant logics generally o distributes over $\vee$, justifying Th⊕∨. And + distributes over $\wedge$, yielding Th⊕∧. So both hold, whence we may assume that there are no occurrences of the lattice connectives $\wedge$ and $\vee$ *within* the scope of $\oplus$ in the normal form B' of B.

A final ingredient in the construction of B' is the application of *distributive lattice laws* for $\wedge$ and $\vee$. These enable us to take our choice of *conjunctive* or *disjunctive* normal forms. We choose *conjunctive normal form.* I. e., we may take B' to be a *conjunction* of disjunctions of bifs. Specifically,

(a) D is a *disbif* if D is of the form $D_1 \vee \ldots \vee D_n$, where each D_i is a bif.

(b) C is a *condisbif* if C is of the form $C_1 \wedge \ldots \wedge C_n$, each C_j a disbif.

We sum up with the

Condisbif theorem. Every formula A of **A** is provably equivalent to some condisbif C.

Proof by the normal forming maneuvers set out above.

12 COMPLEXITY, MUNDICI AND A WALK ON THE BEACH

Daniele Mundici, responding to Meyer's presentation of this material at WCP'2000, commented as follows: It would appear that the decision problem for **A** is co-NP complete. Is this good news, or bad?

In the first place, it *has* to be good news. As relevant logicians, we have grown accustomed to systems of *frightening* levels of complexity. Even the system **LR**, mechanized in [26] with Thistlewaite and McRobbie, was shown by St. Alasdair Urquhart in [27] to have a decision procedure primitive recursive in the Ackermann function. And the hitherto *major* relevant logics, like **R** and **E**, were shown *undecidable* by Urquhart in [28]. So the thought that **A** is no worse than **2**, the original co-NP complete truth-functional logic, is not unwelcome.

But was Mundici's observation *true?* Meyer explored this with him briefly during a walk on the beautiful Juquehy beach. In the first place, deciding **A** must be at least as *hard* as truth-tables. For the *canonical* 0-degree fragment of **A**, in $\wedge$, $\vee$, – and propositional variables, consists *exactly* of the truth-functional tautologies in these variables.

The open possibility is that **A** is *harder to decide* than **2**. Mundici thought *not,* in view of the known result that the word problem for ordered Abelian groups is co-NP. We are *not yet persuaded* .It is a sensitive point to translate word problems for classes of algebras into the decision question for the corresponding logic, which depends on the *shape* of the deduction theorem for the logic.

Still, we expect that Mundici's observation was in fact correct. By the Condisbif theorem of the last section, a formula will be a theorem of **A** iff an equivalent condisbif C is provable. And C, as a conjunction of disbifs, will be provable iff each of its component disbifs $B_1 \vee \ldots \vee B_n$ is a theorem. This would seem, at worst, a question whether we have a generalized excluded middle.

It accordingly *appears* that none of the steps in the reduction lead *out* of co-NP. And so we vote, at least for now, *for* the claim that **A** is co-NP complete.

13 WHO DOES THE METALOGIC?

When we proposed **A** in [2] as the logic of Abelian l-groups, it occurred to us that generations of mathematicians had already been working out theoretical results, to which we could help ourselves. So far, this hope has a *poor* track record. True, our original completeness proof for **A** did use Hahn's theorem, which states that ordered Abelian groups can be embedded in certain products of the real numbers, lexicographically ordered.[16] But that was overkill, and Slaney found a simple appeal to mathematical induction that replaced our profound argument with the humdrum one of [2].[17]

And as we survey what we have learned since, we have learned it mainly from *philosophical* logicians. It was Ulrich who put us onto Meredith and Prior, all of them philosophers. It was Fitelson who exposed the redundancies in our original axiom set for **A**→. Finally and most significantly it has been Italian philosophers from Firenze and Pisa, led by Casari, who came up quite independently in his [10] not only with **A** but with the related systems mentioned above.

14 OPEN PROBLEMS AND CONCLUSION

A, we think, has been *unjustly neglected* in the research of the paraconsistent logical community. Indeed, we agree with Paoli's suggestion that the time has come for a conference *devoted to* Abelian logic. Those logicians who see **2** as a kind of *natural top* of the class of logics they find interesting should think the same thing of **A**. Whatever one thinks of the arguments *in favour of* **2**, they are widely accepted. We have rehearsed the properties that **A** shares with **2**, from its Post-completeness through its admission of *smooth* normal forms to its straightforward containment of well-known logics, including **2**. And while it has hitherto been fashionable to concentrate on logics that are *subsystems of* **2** (who knows why?), many of these logics are subsystems of **A** as well. We include in this family the system **C** (got by *subtracting* the contraction axiom AxW from **R**). Further subtractions produce (the additive-multiplicative fragment of) *linear logic* **LL** of [13], the system **S** (for syllogism) of [29] and the minimal relevant

[16] Cf. [6] or [25].

[17] Still, it was *fun* to bring down the gnat of A-completeness with modern mathematical missiles. Our eventual proof applied some high-school algebra. *Everything* one learns is useful *somewhere!*

logic **B** of [30] and [31]. All of these systems (and their fragments with interesting properties, like $\mathbf{S}_{\rightarrow}$ and **B+**) may be extended to **A** by adding theorems. Why not?

We close with some challenges for readers. The first is to *mobilize* the Łukasiewicz infinitely many-valued logic, henceforth **Lω+**, which is a well-defined (positive) fragment (in ∧, ∨, and the *enthymematic implication* ⊃ defined by A∧**t** → B) of **A**. Since, e.g., *modus ponens* for ⊃ is *admissible* in **A**, we might wish a reformulation of the axioms in which a primitive Ru⊃E replaces our Ru→E. We might then drop the (annoying) Ru∧I , appealing rather to the **A**-theorem A ⊃ (B ⊃ A∧B) and ⊃E to close the theorems under ∧.

There are other prospects for cleaning up the axioms. The redundant Ax∧∨ is certainly to be dropped. So, thanks to Fitelson, are our AxI and AxC. Ought we to go the whole hog, thanking Ulrich by trading in *all* the → axioms for Meredith's AxM? In that case, we think *not.* In the trade-off between perspicuity and efficiency, our judgment is that AxM *loses.* Both AxA and AxB make *motivational* sense; AxA, on our insistence that an inferential ~ give negation its *ordinary* properties; AxB, on the even more fundamental point that → should be *transitive.* But the *proof* of the logical pudding nonetheless remains in the *eating.* All else being *definable,* we may rest with → and ∧ as our sole logical particles. Will we get a *tastier* logic by simplifying the ∧ postulates? Is there a single axiom for *all* of **A**? These questions are *open.*

We referred in passing to Paoli's Gentzenization [18] of **A→**. There *must* [18] be some smooth way of admitting the lattice connectives ∧ and ∨ into that analysis. In fact our colleague Raje'ev Goré has stuck up for Belnap's Display Logic with the observation that, in humdrum mathematics, we often *isolate* (i. e., *display*) a particular term or formula *alone* on one side or the other of an equation. There is hardly anything *more humdrum* than the algebra of *the integers* **Z**, with special attention to *addition* +, (unary) *complement* – and the *natural order* ≤ , with its associated *lattice operations* ∧ and ∨ producing *minima* and *maxima* respectively. So if there is anywhere that the Display ideas may be applied with profit, we believe that **A** is the place for them![19]

---

[18] Method of Authority!

[19] We conclude with lots of thanks to many people. Our Australian colleagues (and Priest, Brady, Mortensen and Petersen were among those who joined us at WCP'2000), do keep us on our toes. Other participants–especially Asenjo, Batens, Lewin, Mundici, Sagastume, Vanackere and Van Bendegem–were helpful in ways that we particularly appreciate. The Brazilian organizers, led by d'Ottaviano , Carnielli and Coniglio, did a bang-up job. Marcos found a magnificent venue on the ocean. We presented this material again (and again near an ocean) to the 2000 conference of the Australasian Association for Logic, whose members have heard a great deal about A (from each of us). We are ever grateful for the support of our colleagues and students in the Automated Reasoning Group in the Computer Science Laboratory at ANU. Meyer was an ANU University Fellow while this paper was in preparation. Thanks to many folks for this support, but especially to former student and present co-author and boss John Slaney. This made it possible for Meyer to travel both to Italy and to Chicago for further discussion of these topics, especially with members of the Lambda Group at Torino and the Automated Reasoners gathered for a workshop at the Argonne National Laboratories. Among the latter, Ulrich from Purdue and Fitelson from Wisconsin expanded our horizons. Mariangiola Dezani, Larry Wos and Bill McCune are especially to be thanked for organizing these visits. Finally, we are ever indebted to our old mentors–especially Nuel Belnap, Mike Dunn and the founder of the ANU Logic Group, the late Richard Sylvan.

REFERENCES

[1] R. K. Meyer and J. K. Slaney. Abelian Logic. Abstract. *The Journal of Symbolic Logic* 45:425, 1980.

[2] R. K. Meyer and J. K. Slaney. Abelian Logic (from **A** to **Z**). In [3], 245-288.

[3] G. Priest, R. Routley and J. Norman, editors, *Paraconsistent Logic. Essays on the Inconsistent.* Analytica, Philosophia Verlag, München, 1989.

[4] J. M. Dunn, *The algebra of intensional logics*, U. of Pittsburgh PhD thesis, University Microfilms, Ann Arbor, 1966. Cf. Dunn's contributions to [5], esp. 352-371.

[5] A. R. Anderson and N. D. Belnap, Jr., *Entailment*, vol. I. Princeton, 1975.

[6] G. Birkhoff, *Lattice Theory*, 3rd edition. American Mathematical Society, Providence, 1967.

[7] R. T. Brady, with R. K. Meyer, C. Mortensen and R. Routley. Algebraic analyses of relevant affixing logics, and other Polish connections. Research paper no. 16, Logic group, Philosophy, RSSS, ANU, 1983. To appear in [8].

[8] R. Routley, R. T. Brady and others. *Relevant Logics and Their Rivals 2*, Ashgate, 2002?.

[9] R. K. Meyer and R. Routley. Classical relevant logics II. *Studia Logica*, 33:183-94, 1974.

[10] E. Casari, Comparative logic and Abelian l-groups. In R. Ferro *et al.*,editors, *Logic Colloquium '88*, pages 161-190. North Holland, 1989.

[11] N. D. Belnap, *A formal analysis of entailment*, ONR, New Haven, 1960

[12] G. Boole, *The mathematical analysis of logic.* Cambridge, 1847

[13] J-Y. Girard. Linear logic. *Theoretical computer science,* 50:1-102, 1987.

[14] R. K. Meyer, M. W. Bunder and L. Powers. Implementing the "fool's model" of combinatory logics. *Journal of Automated Reasoning,* 7:597-630, 1991.

[15] C. A. Meredith and A. N Prior. Equational logic. *Notre Dame Journal of Formal Logic,* 9:212-226, 1968.

[16] S. Rogerson and S. Butchart. Naïve comprehension and contracting implications. Abstract. Presented to the Australasian Association for Logic, July, 1999. *The Bulletin of Symbolic Logic* 5(2): , 2000.

[17] S. Rogerson. Investigations into a class of robustly contraction free logics. Abstract. Presented to the Australasian Association for Logic, June, 2001. To appear in *The Bulletin of Symbolic Logic.*

[18] F. Paoli. Logic and groups. In *Proceedings of 'JS98.* (Torun, Poland, 15-18 July 1998).

[19] A. R. Anderson, N. D. Belnap and J. M Dunn. *Entailment,* vol. II. Princeton, 1992. See esp. pp. 294-332, originally N. D. Belnap. Display logic. *Journal of Philosophical Logic,* 11:375-417, 1982.

[20] M. Dezani-Ciancaglini., R. K. Meyer and Y. Motohama. The semantics of entailment omega. Tentatively to appear in G. Restall's memorial volume for Richard Sylvan. Preprint obtainable from www.arp.edu.au.

[21] R. T. Brady. Simple Gentzenizations of the Normal Formulae of Contraction-less Logics. *The Journal of Symbolic Logic,* 61:1321-1346, 1996.

[22] A. Kron. Decidability and Interpolation for a First-Order Relevance Logic. In P. Schroeder-Heister and K. Došen, editors, *Substructural Logics,* pages 153-177. Oxford, 1993.

[23] J. Łukasiewicz. *Aristotle's syllogistic, from the standpoint of modern formal logic*, 2nd edition, enlarged, Clarendon, Oxford, 1957.

[24] H. B. Curry. *Foundations of mathematical logic*. McGraw-Hill, N. Y., 1963. Reprint Dover, 1977.

[25] L. Fuchs. *Partially ordered algebraic systems.* Pergamon Press, Oxford, 1963.

[26] P. B. Thistlewaite, M. A. McRobbie and R. K. Meyer, *Automated theorem-proving in non-classical logics.* Pitman (London) and Wiley (N. Y.), 1988. An update of Thistlewaite's ANU doctoral thesis, 1984.

[27] A. Urquhart. The complexity of decision procedures in relevance logics II, to appear. This sharpens results in A. Urquhart. The complexity of decision procedures in relevance logics. In J. M. Dunn and A. Gupta, editors, *Truth or consequences: essays in honor of Nuel Belnap*, pages 61-76. Kluwer, 1990.

[28] A. Urquhart. The undecidability of entailment and relevant implication. *The Journal of Symbolic Logic,* 47:1059-1073, 1984. Edited version with additions in [19], 348-375.

[29] E. P. Martin and R. K. Meyer. Solution to the P-W problem. *The Journal of Symbolic Logic,* 47:869-886, 1982.

[30] R. Routley and R. K. Meyer. The semantics of entailment III. *Journal of Philosophical Logic,* 1:192-208, 1972.

[31] R. Routley, with V. Plumwood, R. K. Meyer, and R. T. Brady, *Relevant logics and their rivals 1*, Ridgeview, Atascadero, California, 1982.

Fuzzy Relevant Logic

GRAHAM PRIEST University of Queensland, Australia
g.priest@unimelb.edu.au

Dedicated to Newton da Costa on the occasion of his 70th birthday.

Abstract
This paper raises the question of what a logic should be like if it is both fuzzy and relevant. Two strategies are considered for answering the question. In the first, standard world-semantics for relevant logics are "fuzzified". In the second, algebraic semantics for relevant logics are simply reinterpreted, showing that we can think of standard relevant logics as already fuzzy. The two strategies deliver a number of logics with different properties, especially concerning the conditional.

1 Relevance, Vagueness and Paraconsistency

The study of paraconsistent logic is now about 50 years old. A major pioneer of the subject, Newton da Costa, articulated many paraconsistent logics, showing the way to this rich and important field. There are now very many different kinds of paraconsistent logic; and they have been suggested with very many different applications in mind.[1] Two such applications, which will concern us in this paper, are vagueness and relevance.

Let us start with relevance. The thought that there must be some connection between the antecedent and consequent of a true conditional is an ancient and very natural one. A (propositional) relevant logic is one which respects this intuition in the following form: whenever $A \to B$ is a logical truth, A and B share a propositional parameter. In particular, then, $\not\models (p \wedge \neg p) \to q$. Strictly speaking, relevant logics need not be paraconsistent. For example, Ackermann's system Π' was relevant. However, one of its primitive rules was the disjunctive syllogism, and (interpreting this as a rule of derivability) it quickly gives $p, \neg p \vdash q$. Still, the same kind of intuition that rebels against the logical truth of conditionals of the form $(p \wedge \neg p) \to q$ rebels against the validity of inferences of the form $p, \neg p \vdash q$. It is not, therefore, surprising that most relevant logics are paraconsistent as well. Indeed, Ackermann's Π' was reworked by Anderson and Belnap into their favourite relevant logic, E, which is paraconsistent.

A quite different motivation for paraconsistency comes from the notion of vagueness. Most of the normal predicates we operate with are vague. Specifically, there

[1] For a survey, see Priest [6].

appear to be transition areas where their applicability fades out or fades in. Thus, for example, as a child grows into an adult, there would seem to be a transitional period around adolescence, where they would seem to be as much child as adult, or as little adult as child. What status does the claim that the person in question is a child have during this period, when they are symmetrically poised between childhood and adultery? The natural answers are also symmetric: neither a child nor not a child; both a child and not a child. Common sense seems to be comfortable with both possibilities, though most logicians have taken only the second seriously. If one does, though, one clearly needs a paraconsistent logic. For during the transition period it is not true, for example, that the person is a chicken. Hence, this cannot follow from the contradictory characterisation.

Actually, there seems to be more to vagueness than so far indicated. Any simple semantic dichotomy or trichotomy appears to be inadequate to characterise vagueness. As the child grows up, there seems to be no precise line between being a child and not being a child; or between being a child and neither being nor not being a child; or between being a child and both being and not being a child. In virtue of this, it is very natural to suppose that there are degrees of truth. A standard way of implementing this idea is by representing truth values as real numbers in the closed interval $[0,1]$. If one gives a natural semantics for negation then, again, a paraconsistent logic is obtained. For it is easy to arrange for the value of $p \wedge \neg p$ to be greater than that of q, making the inference from $p \wedge \neg p$ to q invalid. We will come back to the details of this in a moment.

All of this background will be familiar to most paraconsistent logicians. Now to the main issue I want to raise. Intuitively, at least, there is nothing incompatible about relevance and vagueness—quite the opposite: the conditional 'if John is a child and not a child then he is a chicken', seems intuitively quite rebarbative, even though the predicates in question are vague. Yet, though the studies of relevance and of degrees of truth have both given rise to paraconsistent logics, relevant logics and fuzzy logics are currently quite distinct. Standard relevant logics countenance only the truth values *true* and *false*—though sometimes they may be allowed to occur in combination. And standard fuzzy logics are certainly not relevant. How, then, to put these two ideas together? What should a fuzzy relevant logic be like? Surprisingly, that question seems scarcely to have been raised.[2] It is the issue I want to address in the rest of this paper. I will consider two strategies for producing a fuzzy relevant logic: "fuzzification" and reinterpretation. The considerations are purely technical: I shall not discuss the philosophical adequacy of any of the logics concerned here.

[2]There have been some near misses. Peña's fuzzy logic of [4] is an extension of the relevant system E, but it is not a relevant logic. Closer, Slaney's logic F of [8] is clearly in the same family as standard relevant logics, but it, too, is not relevant. Sylvan and Hyde [10] argue that a relevant logic without *modus ponens* is a suitable logic for vagueness, but the logic is not fuzzy, having just the usual two truth values. In a sense, Boričić [1] fuzzifies possible-world semantics for intuitionist and stronger logics. The construction adds operators like 'Is true to at least degree r' to the language itself; but it leaves the semantics two-valued. The same techniques can be applied to other possible-world semantics, including those for relevant logics.

2 STRATEGY 1: FUZZIFICATION

2.1 Fuzzy Logic

To see how fuzzification works, let us start with a clean statement of a fuzzy logic. There are many such logics. Standard ones differ as to how they give the truth conditions of connectives. Here, I will employ the connectives that are probably most familiar to philosophers, the Łukasiewicz truth conditions.[3] It should be clear that fuzzification could be performed in exactly the same way with others.

Truth values are represented by the closed interval, $[0, 1]$. If ν is an assignment of truth values to propositional parameters, this is extended to other formulas by the following clauses:

$$\nu(\neg A) = 1 - \nu(A)$$

$$\nu(A \wedge B) = Min(\nu(A), \nu(B))$$

$$\nu(A \vee B) = Max(\nu(A), \nu(B))$$

$$\nu(A \rightarrow B) = \nu(A) \ominus \nu(B)$$

where:[4]

$$\begin{array}{lll} a \ominus b & = 1 & \text{if } a \leq b \\ a \ominus b & = 1 - (a - b) & \text{if } a > b \end{array}$$

In many-valued logics validity is defined in terms of the preservation of designated values. In standard fuzzy logics, the set of designated values is taken to be $\{1\}$. We will be a little more general here. It is natural to think of the designated values as the values of those things that are true enough to be acceptable, and taking 1 to be the only such value seems a little over-zealous. Technically, the set of designated values could be any subset of $[0, 1]$. However, if we are thinking of designated values in the way just explained, it is natural to require the designated values to be closed upwards. Hence, if $0 \leq \varepsilon \leq 1$, any set of the form $[\varepsilon, 1]$ is a possible set of designated values, defining a corresponding notion of validity. Thus, we have $\Sigma \models_\varepsilon B$ iff $\forall \nu$(if $\forall A \in \Sigma, \nu(A) \geq \varepsilon$ then $\nu(B) \geq \varepsilon$).[5]

ε is the lower bound of those degrees of truth that are acceptable; and it is plausible to suppose that this is a contextual matter. In some contexts (for example, choosing a safe drug where someone's life is at stake), one would require a higher degree than others (for example, choosing a coloured paint where one is decorating a house). Hence, it makes sense to abstract from context, and define an absolutely valid inference as one that is valid, no matter what ε is; that is, $\Sigma \models B$ iff $\forall \varepsilon\, \Sigma \models_\varepsilon B$. It is not difficult to establish that this is equivalent to the following definition. $\Sigma \models B$ iff:

$$\forall \nu (Glb(\nu[\Sigma]) \leq \nu(B))$$

[3]Technically, standard fuzzy logics turn around the notion of a *t-norm*, which determines the behaviour of the conditional (and the conjunction of which it is the residuum). Apart from the Łukasiewicz *t*-norm, the best known are those of Gödel and Product logics. For details, see Hájek [2], esp. chs. 2-4.

[4]In Gödel and Product logics, the definiens in the second clause is replaced by b, and b/a, respectively.

[5]Alternatively, we could take the designated values to be the half-open interval, $(\varepsilon, 1]$, and replace the '$\leq$'s with '$<$'s.

where $\nu[\Sigma] = \{\nu(A): A \in \Sigma\}$, and $Glb(X)$ is the greatest lower bound of X (between 0 and 1).[6]

In particular, if $\Sigma = \{A_1, ...A_n\}$, then $Glb(\nu[\Sigma]) = Min(\nu(A_1), ..., \nu(A_1))$, so $\Sigma \models A$ iff $\nu(A_1 \wedge ... \wedge A_n) \leq \nu(B)$ for all ν; and if Σ is empty, then $Glb(\nu[\Sigma]) = 1$. Thus, $\models B$ (i.e., $\phi \models B$) iff $\forall \nu \, \nu(B) = 1$.[7]

One of the most distinctive features of fuzzy logic, thus formulated, is the failure of *modus ponens*—as one might expect, given its role in sorites paradoxes. To see this, set $\nu(p) = 0.75$, $\nu(q) = 0.5$. Then $\nu(p \to q) = 0.75$. Hence $\nu(p \wedge (p \to q)) = 0.75 > \nu(q)$. Note also that for any ν, $\nu(p \to (q \to q)) = 1$. Hence, $\models p \to (q \to q)$. As I observed in the introduction, Lukasiewicz' fuzzy logic is not a relevant logic.[8] It is also known not to be compact.[9] It follows that it has no sound and complete proof theory. It does have a proof theory sound and complete with respect to finite sets of premises, though.[10]

2.2 Fuzzy Modal Logic

Before we turn to relevant logic, let us see how fuzzification works in the slightly simpler case of modal logic. Specifically, let us see how to fuzzify the simplest normal modal logic, K. As is well known, an interpretation for K is a structure $\langle W, R, \nu \rangle$, where W is a set of worlds, R is an arbitrary binary relation on W, and for every $w \in W$, and propositional parameter, p, $\nu(w, p) \in \{0, 1\}$. (I will write $\nu(w, A)$ as $\nu_w(A)$.) The truth conditions for a standard set of connectives are as follows. For all $w \in W$:

$$\nu_w(\neg A) = 1 \text{ iff } \nu_w(A) = 0$$

$$\nu_w(A \wedge B) = 1 \text{ iff } \nu_w(A) = \nu_w(B) = 1$$

$$\nu_w(A \supset B) = 1 \text{ iff } \nu_w(A) = 0 \text{ or } \nu_w(B) = 1$$

$$\nu_w(\Box A) = 1 \text{ iff for all } w' \text{ such that } wRw',\ \nu_{w'}(A) = 1$$

A little thought shows that these may be expressed equivalently as:

$$\nu_w(\neg A) = 1 - \nu_w(A)$$

$$\nu_w(A \wedge B) = Min(\nu_w(A), \nu_w(B))$$

[6] *Proof*: Suppose that $\forall \nu(Glb(\nu[\Sigma]) \leq \nu(B))$. Now, suppose, for fixed ε, that $\forall A \in \Sigma, \nu(A) \geq \varepsilon$. Then $Glb(\nu[\Sigma]) \geq \varepsilon$. Hence $\Sigma \models_\varepsilon B$. Thus, $\Sigma \models B$. Conversely, suppose that for some ν, $Glb(\nu[\Sigma]) = \varepsilon > \nu(B)$. Then for all $A \in \Sigma$, $\nu(A) \geq \varepsilon > \nu(B)$. Thus, $\Sigma \not\models_\varepsilon B$, and hence, $\Sigma \not\models B$.

[7] So the logical truths of this logic coincide with the fuzzy logic in which 1 is taken to be the sole designated value.

[8] If one takes 1 to be the only designated value, it is not even paraconsistent, since $\alpha, \neg\alpha \models_1 \beta$.

[9] To see this, define $A \oplus B$ as $\neg A \to B$. Then it is not difficult to check that $\nu(A \oplus B) = Min(\nu(A) + \nu(B), 1)$. Let $1A$ be A, and $(n+1)A$ be $A \oplus nA$. Then if $\nu(A) > 0$, the sequence: $\nu(1A), \nu(2A), \nu(3A)...$ eventually takes the value 1. Let $\Sigma = \{p, \neg 1p, \neg 2p, \neg 3p, ...\}$. Then it follows that for any ν, there is a $B \in \Sigma$ such that $\nu(B) = 0$. Hence, $\Sigma \models q$. But there is no finite $\Sigma' \subseteq \Sigma$ such that $\Sigma' \models q$. For let m be the greatest n such that $nA \in \Sigma'$. Let $\nu(p) = 1/(m+1)$. Then it is easy to check that if $A \in \Sigma'$, $\nu(A) \geq 1/(m+1)$. Now let $\nu(q) = 0$, to see that $\Sigma' \not\models q$.

[10] Observe that $\{A_1, A_2, ..., A_n\} \models B$ iff $\models_1 (A_1 \wedge A_2 \wedge ... \wedge A_n) \to B$. Hence, an inference with a finite set of premises is valid iff the corresponding conditional is logically true in Lukasiewicz' continuum-valued logic with designated value 1. This is well known to be axiomatizable. See, for example, Hájek [2], ch. 3.

$$\nu_w(A \supset B) = \nu_w(A) \ominus \nu_w(B)$$

$$\nu_w(\Box A) = Glb\{\nu_{w'}(A); wRw'\}$$

In particular, for the modal operator: if $\nu_{w'}(A) = 1$ for all w' such that wRw', $Glb\{\nu_{w'}(A); wRw'\} = 1$; and if $\nu_{w'}(A) = 0$ for some w' such that wRw', $Glb\{\nu_{w'}(A); wRw'\} = 0$.

The standard definition of logical consequence for K is:

$\Sigma \models A$ iff for every $\langle W, R, \nu\rangle$ and $w \in W$, if $\nu_w(B) = 1$ for all $B \in \Sigma$, $\nu_w(A) = 1$

Again, a little thought shows that this may be expressed equivalently as a simple generalisation of the fuzzy definition of the previous section:

$\Sigma \models A$ iff for every $\langle W, R, \nu\rangle$ and $w \in W$, $Glb(\nu_w[\Sigma]) \leq \nu_w(A)$

Fuzzifying K is now completely routine. We simply take the above account, where the truth conditions and definition of validity are expressed in the equivalent terms, and replace $\{0, 1\}$ by $[0, 1]$. Let us call the result FK (*fuzzy* K).

The relationship between K and FK is not difficult to establish. For a start, any K counter-model is a special case of an FK counter-model (where everything takes the value 1 or 0). Hence if $\Sigma \models_{FK} A$, then $\Sigma \models_K A$. The converse is not true, however. For *modus ponens* in valid in K, but invalid in FK. (Just consider the one-world model corresponding to the counter-model of the last section.)

As should be clear, other modal logics with world semantics can be fuzzified in exactly the same way. Thus, for example, fuzzified $S4$ is the same as FK, except that R is required to be reflexive and transitive.[11] Similarly, we can fuzzify intuitionist logic by fuzzifying its world semantics. However, I will not go into any of this here, since this is not a paper about modal logic. The preceding material was just to illustrate the basic idea of fuzzification, which we will now apply to relevant logic.

2.3 Fuzzifying Relevant World Semantics

Perhaps the simplest and most natural semantics for relevant logics are world-semantics. Leave negation aside for a moment. A very simple world semantics for a positive relevant logic is a structure $\langle N, W, \nu\rangle$, where W is a set of worlds; N is a subset of W, the normal worlds—the rest being non-normal; and ν is an evaluation function that assigns a truth value, 0 or 1, to every propositional parameter at every world, and to every conditional at every non-normal world. Truth values are then assigned to other formulas by the recursive clauses:

For $w \in W$: $\nu_w(A \wedge B) = 1$ iff $\nu_w(A) = 1$ and $\nu_w(B) = 1$

For $w \in W$: $\nu_w(A \vee B) = 1$ iff $\nu_w(A) = 1$ or $\nu_w(B) = 1$

For $w \in N$: $\nu_w(A \rightarrow B) = 1$ iff for all $w' \in W$ such that $\nu_{w'}(A) = 1$, $\nu_{w'}(B) = 1$.

[11] Fuzzy versions of some modal logics, and in particular $S5$, are already known. See Hájek [2], 8.3, for discussion and references.

(The truth values of conditionals at non-normal worlds are already taken care of by ν.) The truth conditions may be expressed equivalently thus:

For all $w \in W$: $\nu_w(A \wedge B) = Min(\nu_w(A), \nu_w(B))$

For all $w \in W$: $\nu_w(A \vee B) = Max(\nu_w(A), \nu_w(B))$

For all $w \in N$: $\nu_w(A \to B) = Glb\{\nu_{w'}(A) \ominus \nu_{w'}(B);\ w' \in W\}$

Validity is defined as truth preservation at all *normal* worlds of all interpretations:

$\Sigma \models A$ iff for all $\langle W, N, R, \nu\rangle$ and $w \in N$, if $\nu_w(B) = 1$ for all $B \in \Sigma$, then $\nu_w(A) = 1$

Or equivalently:

$\Sigma \models A$ iff for all $\langle W, N, R, \nu\rangle$ and $w \in N$, $Glb\{\nu_w(B);\ B \in \Sigma\} \leq \nu_w(A)$

These semantics give the positive relevant logic H^+. One of its characteristic features is that it has no entailments that involve conditionals essentially. That is, any logical truth of the form $A \to B$ is a substitution instance of one of the form $A' \to B'$, where A' and B' contain no occurrences of $\to$. H^+ is, at any rate, a relevant logic.[12]

An axiom system for H^+ is as follows.

A1 $A \to A$

A2 $(A \wedge B) \to A \qquad (A \wedge B) \to B$

A3 $A \to (A \vee B) \qquad B \to (A \vee B)$

A4 $A \wedge (B \vee C) \to ((A \wedge B) \vee (A \wedge C))$

R1 $A, A \to B \vdash B$

R2 $A, B \vdash A \wedge B$

R3 $A \to B, B \to C \vdash A \to C$

R4 $A \to B, A \to C \vdash A \to (B \wedge C)$

R5 $A \to C, B \to C \vdash (A \vee B) \to C$

The axiom system is weakly complete (i.e., complete for the empty set of premises). For strong completeness (i.e., completeness for arbitrary sets of premises), the disjunctive forms of the rules of inference have to be added as well. (The disjunctive form of R1 is: $A \vee C, (A \to B) \vee C \vdash B \vee C$. The others are similar.)

To transform these semantics into a fuzzy logic, we simply replace the set of truth values $\{0, 1\}$ with the closed interval $[0, 1]$, as we did for K, taking the truth conditions and definition of validity in their equivalent forms. Let us call this system FH^+.

[12] For further details concerning this, and all the other facts about relevant logic referred to in this paper, see Priest [6].

As with K, all two-valued interpretations are fuzzy interpretations. It follows that if $\Sigma \models_{FH+} B$ then $\Sigma \models_{H+} B$. In particular, then, FH^+ is a relevant logic. Moreover, again as for K, the implication does not go in the opposite direction, since *modus ponens* fails in FK^+. (Take an interpretation with one world, w, which is normal, where $\nu_w(p) = 0.75$, and $\nu_w(q) = 0.5$.) This time, however, we can say a little more. It is easy to check that each of the axioms of H^+ is logically valid in FH^+, and that each of the rules of inference (including the disjunctive forms) preserves logical validity. It follows that if $\models_{H+} B$ then $\models_{FH+} B$. Thus, H^+ and FH^+ have the same logical truths.[13] Whether there is a complete proof theory for FH^+, or even just for finite sets of premises, is presently an open question.[14]

2.4 Negation

So much for a basic relevant fuzzy logic. How should this be extended to include negation? There are standardly two treatments of negation that may be joined to the semantics of H^+ to give a full relevant logic.

The first of these relaxes the condition that ν be a function. Instead, it is a relation, relating each formula to some, all, or none, of the truth values, 1 and 0. One can pursue this strategy in the fuzzy case, too, but it is problematic. Let $\nu_w[A] \subseteq [0,1]$ be the set of values to which A is related by ν_w. What now are the appropriate truth conditions for the connectives, and definition of validity? Perhaps the most natural truth conditions are the combinatorial ones. E.g.:

$$x \in \nu_w[A \wedge B] \text{ iff } \exists y \in \nu_w[A], \exists z \in \nu_w[B], x = Min(y, z)$$

Thus, the values of $A \wedge B$ are the values that one can get by combining all the values of A and B in the usual way. And $\Sigma \models B$ iff for all $\langle W, N, \nu\rangle$ and $w \in N$:

$$Glb\{Glb(\nu_w[A]);\ A \in \Sigma\} \ \leq Glb(\nu_w[B])$$

This certainly makes sense, but it is not the generalisation of the two valued case that one would expect. For if, say, $\nu_w[A] = \{0\}$ and $\nu_w[B] = \phi$ then $\nu_w[A \wedge B] = \phi$. Consequently, $A \wedge B \not\models A$. (In the non-fuzzy case, $A \wedge B \models A$, since if $\nu_w[A] = \{0\}$ and $\nu_w[B] = \phi$ then $\nu_w[A \wedge B] = \{0\}$. One false conjunct is sufficient to make the conjunction false.) Insisting that for all A and w, $\nu_w[A]$ be non-empty does not help, either. For then, for all w, $Glb(\nu_w[p \wedge \neg p]) \leq 0.5$ and $Glb(\nu_w[q \vee \neg q]) \geq 0.5$. Hence, $\models (p \wedge \neg p) \rightarrow (q \vee \neg q)$.

Whether there are other ways of generalising the relational case, with less untoward consequences, I do not know. But the other way of handling negation in relevant logic is more straightforward. In this, an interpretation is augmented with an operator on worlds, $*$ (the Routley star), such that for all $w \in W$, $w^{**} = w$. The truth conditions of negation are then:

$\nu_w(\neg A) = 1$ iff $\nu_{w^*}(A) = 0$

[13]This is not the same for K: $A \vee \neg A$ is logically valid in K, but not FK.

[14]Another option for creating a fuzzy relevant logic here (and in what follows) is to retain the definition of logical validity in its original form. That is, validity is defined in terms of preservation of the value 1 (at all normal worlds). This gives a stronger logic with the same set of logical truths. In particular, *modus ponens* is valid for this logic.

Or equivalently:

$\nu_w(\neg A) = 1 - \nu_{w^*}(A)$

Adding this machinery to H^+ gives the logic H.

To obtain an axiom system for H, we add the following rules and axioms to those for H^+:

A5 $A \leftrightarrow \neg\neg A$

R6 $A \rightarrow B \vdash \neg B \rightarrow \neg A$

where $A \leftrightarrow B$ is $(A \rightarrow B) \wedge (B \rightarrow A)$.

The generalisation of this to the fuzzy case is obvious. Formulate the semantics in the appropriate terms, and simply replace {0,1} with [0,1]. Call the system produced FH. The relation between H and FH is the same as that between H^+ and FH^+, and for exactly the same reason as before. The logical truths of the two are the same; but for deducibility, FH is a proper sublogic of H. Again, whether there is a complete proof theory for FH^+, or even just for finite sets of premises, is presently an open question.[15]

There remains, also, the question of what $*$ means, and why it should poke its nose into the truth conditions of negation. This is an unresolved question in relevant logics.[16] As far as I can see, fuzzification does nothing to help the matter; but neither does it seem to make the question any harder.[17]

2.5 Ternary Accessibility Relations

As relevant logics go, H is a relatively weak one. The standard way of making it stronger employs a ternary accessibility relation, R. This is added to interpretations, and the truth conditions for $\rightarrow$ *at non-normal worlds*, w, become:

$\nu_w(A \rightarrow B) = 1$ iff for all $y, z \in W$ such that $Rwyz$, if $\nu_y(A) = 1$ then $\nu_z(B) = 1$

A little thought suffices to show that this is equivalent to:

$\nu_w(A \rightarrow B) = Glb\{\nu_y(A) \ominus \nu_z(B)\text{: } y, z \in W \text{ and } Rwyz\}$

If we put no constraints at all on R, we have the relevant logic B. An axiom system for this is obtained from that for H by deleting R3-R5, and replacing them by the stronger:

[15] We cannot obtain a proof theory for the finite case as in fn.10 since in FH the equivalence:

$$\{A_1, ..., A_n\} \vDash_{FH} B \text{ iff} \vDash_{FH} (A_1 \wedge ... \wedge A_n) \rightarrow B$$

fails from left to right. (The left hand side constrains behaviour only at normal worlds.) On the other hand, if one defines $A \oplus B$ as $\neg A \rightarrow B$, then $\oplus$ does not have the monotone properties required to refute compactness as in fn. 9, either.

[16] The best account on the market is, I think, Restall [7], which defines $*$ in terms of a primitive notion of incompatibility.

[17] Other paraconsistent logics can be fuzzified in the same way that H is fuzzified here. Consider, for example, Da Costa's positive-plus logics. To fuzzify these, we start with a positive fuzzy logic, say Łukasiewicz'. We then add a non-truth-functional negation. The value of $\neg A$ is a number in $[0, 1]$ that is independent of the value of A (though one may put some constraints on the relationship between the two). The logics produced in this way are fuzzy and paraconsistent, but they are not relevant.

A6 $((A \to B) \wedge (A \to C)) \to (A \to (B \wedge C))$

A7 $((A \to C) \wedge (B \to C)) \to ((A \vee B) \to C)$

R7 $A \to B, C \to D \vdash (B \to C) \to (A \to D)$

The semantics for B can be fuzzified in the obvious way, proceeding as before. Call the logic obtained in this way FB. It is more complex in this case to check that the new axioms/rules for B are valid/validity-preserving in FB; but it is true. I leave it as an exercise for the committed reader.[18] Hence the relationship between B and FB is the same as that between H and FH. Similar remarks also apply to its proof-theory.

Stronger logics in the relevant family are obtained by adding constraints on the ternary relation R. A novelty here is that the natural correspondence between constraints and axioms in standard relevant logics breaks down in some of the fuzzy cases. For example, in the standard case, adding the constraint that for all $w \in W$, $Rwww$, suffices to verify the axiom $(A \wedge (A \to B)) \to B$. This is no longer the case once we fuzzify. For example, consider the interpretation $\langle W, N, R, *, \nu\rangle$, where $W = \{g, x\}$; $N = \{g\}$; for every $w \in W$, $Rwww$; $\nu_x(p) = 0.8$ and $\nu_x(q) = 0.6$. Then $\nu_x(p \to q) = 0.8$. So $\nu_x(p \wedge (p \to q)) = 0.8$, and $\nu_g((p \wedge (p \to q)) \to q) \leq 0.8$. A full study of the connection between conditions on R and the corresponding axioms still needs to be undertaken for the fuzzy case.

As for the Routley $*$, there is a philosophical problem concerning the meaning of the ternary R. And as for $*$, fuzzification does nothing, as far as I can see, to help with this matter or to hinder it.

3 Strategy 2: Reinterpretation

3.1 De Morgan Lattices

Let us now turn to the second approach to the construction of a fuzzy relevant logic; and let us start by returning to standard fuzzy logic. A familiar criticism of this is that degrees of truth do not seem to be linearly ordered. 'Russell was old when he died' might have a higher degree of truth than 'Wittgenstein was old when he died'. But how does the degree of truth of 'Australia has a small population' relate to either of these?

The natural suggestion in response to this criticism is to trade in a linear order of truth values for a partial order. The values are no longer real numbers, of course. In fact, we may not care too much what the members of the order are. Let a semantic structure, then, be a partial order $\langle D, \leq\rangle$.

If sentences take values in this order, how do the connectives function? Conjunction and disjunction are easy. The order $\langle [0,1], \leq\rangle$ of Łukasiewicz logic is a distributive lattice, and, in that lattice, Min and Max are the meet and join, respectively. Hence, the natural assumption is that the partial order is a distributive lattice, and that if $\nu(A)$ is the truth value of A then:

$\nu(A \wedge B) = \nu(A) \wedge \nu(B)$

[18] Details can be found in Priest [5], ch. 11.

$\nu(A \vee B) = \nu(A) \vee \nu(B)$

I write the lattice meet and join as $\wedge$ and $\vee$, respectively, context sufficing to disambiguate. I will employ the same convention for the algebraic operators corresponding to other connectives.

What of negation? In the semantics of Łukasiewicz logic, negation functions as in involution, that is, an order-inverting function of period two. We can generalise these aspects of its behaviour by supposing that our lattice comes with an operator, $\neg$, such that:

if $a \leq b$ then $\neg b \leq \neg a$

$\neg\neg a = a$

where $\nu(\neg a) = \neg\nu(a)$.

In summary, a natural generalisation of the semantic structure of a fuzzy logic for conjunction, disjunction and negation, once we drop the condition that it must be a linear order, is a distributive lattice with an involution. Such structures are known as De Morgan lattices. Moreover, if conjunction, disjunction and negation relate to the lattice in the way indicated, we have one of the well known semantics for a relevant logic. In particular, $A_1, ..., A_n \models B$ in First Degree Entailment iff $\nu(A_1 \wedge ... \wedge A_n) \leq \nu(B)$ for every De Morgan lattice, and every evaluation, ν, into the lattice.[19]

As a semantics for First Degree Entailment, the algebraic values would normally be thought of as propositions or Fregean senses, and $\leq$ would be thought of as some sort of containment relationship. But as we see, if we reconceptualise the interpretation of these notions, a fuzzy relevant logic falls straight out of the construction.[20]

3.2 De Morgan Groupoids

First Degree Entailment has no conditional connective. Can the preceding considerations be extended to cover such a connective? The most versatile algebraic semantics for relevant logic extends De Morgan lattices with new algebraic operators. A structure is now of the form $\langle \mathcal{D}, e, \circ, \rightarrow \rangle$, where $\mathcal{D}$ is a De Morgan lattice, $\circ$ and $\rightarrow$ are binary operators on the domain of the algebra, and e is a member of the domain. $\circ$ is standardly thought of as some sort of intensional conjunction, and e is thought of as the (value of the) conjunction of all truths. If the new components satisfy the following constraints:

$e \circ a = a$

$a \circ b \leq c$ iff $a \leq b \rightarrow c$

if $a \leq b$ then $a \circ c \leq b \circ c$ and $c \circ a \leq c \circ b$

$a \circ (b \vee c) = (a \circ b) \vee (a \circ c)$ and $(b \vee c) \circ a = (b \circ a) \vee (c \circ a)$

[19] This characterisation does not account for inferences with an infinite number of premises; but this is no loss, since the logic is compact.

[20] For good measure, we might note that classical logic can be seen as a fuzzy logic in this way too, since it is sound and complete with respect to the class of Boolean algebras.

then we have a structure called a De Morgan groupoid. And if we define $\Sigma \models B$ to mean that for all such groupoids, and all evaluations into its domain, ν, if $e \leq \nu(A)$ for all $A \in \Sigma$ then $e \leq \nu(B)$, we have a semantics sound and strongly complete with respect to the relevant logic B. In particular, A is a logical truth iff for all ν, $e \leq \nu(A)$. Stronger relevant logics in the family can be obtained by adding constraints on $\circ$.[21]

What sense can be made of this in fuzzy terms? For present purposes, $\circ$ can be thought of as an auxiliary notion. The crucial question therefore concerns e and $\rightarrow$. If we think of the members of the algebra as degrees of truth, then we may think of e as the lower bound of the things which are true enough to be acceptable. That is, $e \leq a$ iff something with value a is acceptable as true. What of $\rightarrow$? The algebraic postulates tell us that $e \leq a \rightarrow b$ iff $e \circ a \leq b$ iff $a \leq b$. Thus, a conditional is acceptable iff the truth value of the consequent is at least as great as that of the antecedent.

This is a plausible enough condition. One might have one's reservations about it, though. If, in a conditional, the truth value drops from antecedent to consequent, but only a very little, shouldn't it still be acceptable? (This is certainly how it works in standard fuzzy logics if $\varepsilon < 1$.) Perhaps not. On these semantics, we have to say that, strictly speaking, a conditional of this kind is not acceptable, though it might be as close to acceptable as one might like. Indeed, given the possibility of sorites arguments, and the validity of *modus ponens* in the logic B, it has to be like this. At any rate, we see that, if we are prepared to live with the validity of *modus ponens* and the consequences of this, we can think of the relevant logic B, and its strengthenings, as already fuzzy logics.

3.3 Defeasible Conditionals

If one cannot live with these things, there is another way to proceed.[22] We add a new one-place function, δ, to the algebra, governed by the condition:

$$\delta a \leq a$$

Intuitively, δa is a value a little below a, and represents the value to which an antecedent would have to drop to make a conditional acceptable in the strong sense of the previous section. Employing this, we can define a different notion of conditionality, $a \rhd b$, as $\delta a \rightarrow b$.

[21] It is worth noting that standard fuzzy logics (taking 1 as the only designated value) correspond to well know algebraic structures: MV algebras, G algebras, Π algebras, and BL algebras. The most basic of these are BL algebras, the collection of which characterises the logical truths common to all continuous t-norms. For details, see Hájek [2], chs. 3, 4. All these algebras are special cases of De Morgan groupoids. $\circ$ is, in fact, the connective whose truth conditions are given my the t-norm. In particular, a BL algebra is a groupoid in which $\circ$ is commutative and associative (as in some relevant logics), and which also satisfies the conditions:

$a \circ (a \rightarrow b) = a \wedge b$
$(a \rightarrow b) \vee (b \rightarrow a) = 1$

From a relevant point of view, the undesirablity of the second condition hardly needs to be pointed out. The first fails in all standard relevant logics as well. A logic called monoidal logic is given in Höhle [3]. This is characterised by the class of all residuated lattices. It is therefore a sublogic of B.

[22] Hinted at in Sylvan and Hyde [10], p.13.

The easiest way to handle the new conditional proof-theoretically is to add a corresponding monadic functor, δ, to the language, augment the axioms with:

A8 $\delta A \to A$

R8 $A \leftrightarrow B \vdash \delta A \leftrightarrow \delta B$

and define $A \rhd B$ as $\delta A \to B$. It is clear that the resulting axiom system is sound with respect to the semantics. A simple modification of the completeness proof for the algebraic semantics shows that it is complete also.[23] Let us call the logic obtained by extending B in this way DB (*defeasible* B). Stronger logics of the same kind can be obtained by modifying the algebras appropriate for stronger relevant logics in the same way.

The conditional $\rhd$ in DB is a relevant one. For suppose that A and C share no propositional parameter. Then $\nvDash_B A \to C$. Consider a De Morgan groupoid counter-model. Extend this by defining δa as a. This is a model for DB, and in it, $\rhd$ collapses into $\to$. Hence, $\nvDash_{DB} A \rhd C$. But *modus ponens* for $\rhd$ fails in DB and its extensions. To show this, we can construct a counter-model as follows. Let $\mathcal{B}$ be the Boolean algebra of all subsets of ω. Let $\circ$ and $\to$ collapse into the corresponding extensional connectives (so that $e = \omega$). This is a De Morgan groupoid. (In fact, it is an algebra appropriate for every relevant logic.) Let δa be a with its least member removed (or if $a = \phi$, $\delta a = \phi$). Augmented by δ, the algebra is appropriate for DB (and the defeasible logics produced by using stronger relevant logics). To show that $p, p \rhd q \nvDash q$, set the values of p and q as ω and $\omega - \{0\}$, respectively.

The conditional $\rhd$ is also a defeasible one. That is, $p \rhd q \nvDash (p \wedge r) \rhd q$. For a counter-model, take the same algebra as before, except that:

$$\delta a = \begin{array}{ll} a - \{0\} & \text{if } 1 \in a \\ a - \{2\} & \text{otherwise} \end{array}$$

Take the values of p, q, and r, to be ω, $\omega - \{0\}$, and $\omega - \{1\}$, respectively.

In fact, DB has all the marks of a relevant conditional logic. In the world semantics for these, $A \rhd B$ is true at a world, w, iff $s(w, [A]) \subseteq [B]$, where $[C]$ is the set of worlds where C is true, and s is a function selecting subsets of W. If we impose the constraint on s that $s(w, [A]) \subseteq [A]$, we get the relevant conditional logic obtained by adding the following proof-theoretic rules to B.[24]

ID $A \to B \vdash A \rhd B$

REA $A \leftrightarrow B \vdash (A \rhd C) \to (B \rhd C)$

RPC $(A \wedge B) \to C \vdash ((D \rhd A) \wedge (D \rhd B)) \to (D \rhd C)$

It is easy to check that all these rules are sound in DB. I suspect that the rules are also complete (for the fragment without δ), but I have not been able to prove this yet.

It is not an implausible thought that the conditional involved in sorites arguments is a conditional of the kind given by DB: it certainly does not seem to be an

[23] Specifically, construct the Lindenbaum algebra in the usual way. R8 tells us that the function δ on the algebra may be defined in the standard fashion, and A8 gives δ its appropriate property.

[24] See Sylvan [9].

entailment. At any rate, the construction that we have just been considering gives us another fuzzy relevant logic.[25]

4 Conclusion

In this paper, we have examined two semantic strategies for constructing fuzzy relevant logics. The first fuzzifies standard world-semantics for relevant logics, changing the discrete truth values to continuum-valued ones. This construction gives sublogics of the corresponding standard relevant logics. In particular, *modus ponens* is no longer valid. This is as one might expect in a fuzzy logic where 1 is not taken to be the only designated value. In the second strategy, we simply reinterpret the algebraic semantics for relevant logic, thinking of the algebraic values as degrees of truth. The upshot of this is that standard relevant logics can already be thought of as fuzzy logics. In particular, then, in these semantics *modus ponens* holds. These semantics can be extended by a "decrease in value" operator, δ, which can be used to define a defeasible relevant conditional, for which, again, *modus ponens* fails. Which, if any, of these logics is philosophically the best for their intended application is another matter. But at least we now have some fuzzy relevant logics to philosophise about.[26]

References

[1] B.Boričić. On the Fuzzification of Propositional Logics. Fuzzy Sets and Systems 108: 91-98, 1999.

[2] P.Hájek. Metamathematics of Fuzzy Logics. Dordrecht: Kluwer Academic Publishers, 1998.

[3] U.Höhle. Commutative Residuated 1-Monoids. In: U.Höhle and E.P.Klement, eds., Non-Classical Logics and their Applications to Fuzzy Subsets. Dordrecht: Kluwer Academic Publishers, 1995, pp. 53-106.

[4] L.Peña. A Chain of Fuzzy Strengthenings of Entailment Logic. In: S.Barro and A.Sobrino, eds., III Congreso Español de Tecnologías y Lógica Fuzzy. Spain: Universidad de Santiago, 1993, pp. 115-122.

[5] G.Priest. Introduction to Non-Classical Logic. Cambridge: Cambridge University Press, 2001.

[25] Another algebraic way of proceeding is by modifying the notion of a De Morgan groupoid. Everything is as in 3.2, except that the condition $e \circ a = a$ is weakened to $e \circ a \leq a$. In effect, $e \circ a$ now plays the role of δa. The logic obtained in this way is clearly a relevant logic, and, as may be checked, *modus ponens* for $\rightarrow$ fails. But $\rightarrow$ is not the same as $\rhd$ in DB, since $\rightarrow$ is not defeasible. For suppose that $e \leq a \rightarrow b$. Then $e \circ a \leq b$. But $a \wedge c \leq a$. So $e \circ (a \wedge c) \leq e \circ a$. It follows that $e \circ (a \wedge c) \leq b$, i.e., $e \leq (a \wedge c) \rightarrow b$.

[26] Versions of the paper were given at the Second Word Congress on Paraconsistency, Saõ Paulo, May 2000, and a meeting of the Australasian Association for Logic, Queensland, June 2000. I would like to thank those present for their helpful thoughts, and especially Daniele Mundici, Greg Restall, and Peter Woodruff. Helpful comments were also made by anonymous referees.

[6] G.Priest. Paraconsistent Logic. In: D.Gabbay and F.Guenthner, eds., The Handbook of Philosophical Logic. 2nd. edition. Dordrecht: Kluwer Academic Publishers, forthcoming.

[7] G.Restall. Negation in Relevant Logics (How I Stopped Worrying and Learned to Love the Routley Star. In: D.Gabbay and H.Wansing, eds., What is Negation? Dordrecht: Kluwer Academic Publishers, 1999, pp. 53-76.

[8] J.Slaney. Vagueness Revisited. Technical Report TR-ARP-15/88, Automated Reasoning Project, Australian National University, 1988.

[9] R.Sylvan. Relevant Conditionals, and Relevant Applications Thereof. In: S.Akama, ed., Logic, Language, and Computation. Dordrecht: Kluwer Academic Publishers, 1997, pp. 191-224.

[10] R.Sylvan and D.Hyde. Ubiquitous Vagueness without Embarrassment: Logic Liberated and Fuzziness Defuzzed (i.e., Respectabilized). Acta Analytica 10: 7-29, 1993.

On some Remarkable Relations between Paraconsistent Logics, Modal Logics, and Ambiguity Logics*

DIDERIK BATENS Centre for Logic and Philosophy of Science, Universiteit Gent, Belgium
Diderik.Batens@rug.ac.be

Abstract
This paper concerns some connections between paraconsistent logics, modal logics (mainly **S5**), and Ambiguity Logic **AL** (Classical Logic applied to a language in which all letters are indexed and in which quantifiers over such indices are present). **S5** may be defined from **AL**.

Three kinds of connections will be illustrated. First, a paraconsistent logic **A** is presented that has the same expressive power as **S5**. Next, I consider the definition of paraconsistent logics from **S5** and **AL**. Such definition is shown to work for some logics, for example Priest's **LP**. Other paraconsistent logics appear to withstand such definition, typically those that contain a detachable material implication. Finally, I show that some paraconsistent logics and inconsistency-adaptive logics serve exactly the same purpose as some modal logics and ampliative adaptive logics based on **S5**. However, they serve this purpose along very different roads and the logics cannot be defined from one another.

The paper intends to open lines of research rather than pursuing them to the end. It also contains a poor person's semantics for **S5** as well as a description of the simple but useful and powerful **AL**.

1 Aim of this Paper

The respectability assigned to a family of logics partly depends on their relations with other logics. Such relations include definability, embedding, and their converses. They also include the capability of a logic to serve a purpose served by another logic. The present paper supports the respectability of paraconsistent logics in this respect.

*The research for this paper was financed by the Fund for Scientific Research – Flanders, and indirectly by the Flemish Minister responsible for Science and Technology (contract BIL98/37). I am indebted to the referees and especially to João Marcos for several improvements.

Although some technicalities cannot be avoided, I shall try to reduce them to a minimum. I shall mainly display theorems that require no proof (or very simple proofs). I shall rely on published results without repeating them. And I shall try to keep an eye on the philosophical import and 'the story' behind the technical result.

The discussion will first concentrate on the relations between paraconsistent logics, **S5**, and the Ambiguity Logic **AL**. The latter is a very simple system (Classical Logic applied to a slightly enriched language) obtained by generalizing a result of Guido Vanackere in [1]. It will turn out that **S5** can be defined from a paraconsistent logic, here called **A**, and vice versa. **S5** can also be defined from **AL**, but not the other way around. In **AL**, a very natural consequence relation for formulas of the standard (object) language may be defined. Its identity-free fragment is Priest's **LP**; the same fragment of **LP** may be defined within **S5**, and full **LP** may be defined within **AL**. Several other paraconsistent logics seem to withstand such reduction.

In this paper, I am only interested in a specific strong sense of defining logic $\mathbf{L}_1$ from logic $\mathbf{L}_2$—intuitively, for any formula in the language of $\mathbf{L}_1$ there is a formula in the language of $\mathbf{L}_2$ that 'has the same meaning'. Let $\mathbf{L}_1$ and $\mathbf{L}_2$ be formulated respectively in the languages $\mathcal{L}_1$ and $\mathcal{L}_2$, with $\mathcal{F}_1$ and $\mathcal{F}_2$ the respective sets of formulas and $\mathcal{W}_1$ and $\mathcal{W}_2$ the respective sets of closed formulas.[1] I shall say that $\mathcal{W}_1^*$ *covers* $\mathcal{W}_1$ iff $\mathcal{W}_1^* \subseteq \mathcal{W}_1$ and, for any $A \in \mathcal{W}_1$, there is a $B \in \mathcal{W}_1^*$ such that, whenever D is obtained by replacing A by B in C or vice versa, then $C \vdash_{\mathbf{L}} D$ and $D \vdash_{\mathbf{L}} C$. Let $\mathcal{F}_1^{at} \subset \mathcal{F}_1$ be a set of *atoms* iff there is a set Σ of logical symbols (including quantifiers) such that (i) any formula of $\mathcal{W}_1^*$ is compounded from atoms by members of Σ and parentheses, and (ii) no member of $\mathcal{F}_1^{at}$ is compounded from other members of $\mathcal{F}_1^{at}$ by members of Σ (and parentheses). Where $\mathcal{F}_1^{at}$ is a set of atoms, $f : \mathcal{W}_1^* \mapsto \mathcal{W}_2$ will be called a *straight translation* iff $f(A)$ is obtained from A by replacing (i) members of $\mathcal{F}_1^{at}$ by members of $\mathcal{F}_2$ and (ii) logical symbols of $\mathcal{L}_1$ by logical symbols of $\mathcal{L}_2$. $\mathbf{L}_1$ can be *defined* from $\mathbf{L}_2$ iff there is a $\mathcal{W}_1^*$ that covers $\mathcal{W}_1$, and a straight translation $f : \mathcal{W}_1^* \mapsto \mathcal{W}_2$, such that, for all $\Gamma \subseteq \mathcal{W}_1^*$ and $A \in \mathcal{W}_1^*$,

$$\Gamma \vdash_{\mathbf{L}_1} A \text{ iff } \{f(B) \mid B \in \Gamma\} \vdash_{\mathbf{L}_2} f(A)\,.$$

In Sections 9 and 10, I discuss some couples of very different adaptive logics, one defined from a paraconsistent logic and the other from **S5**, that serve exactly the same purpose along different and apparently unrelated roads.

Several results are partial, and many suggest related questions that are not discussed here. My aim is to illustrate some lines of research, not to pursue them to the end. By way of retribution, I list a set of interesting open problems in the conclusion.

2 Notational Conventions

Let $\mathcal{L}$ be the language of Classical Logic—henceforth **CL**—with $\mathcal{S}$, $\mathcal{P}^r$, $\mathcal{C}$, $\mathcal{V}$, $\mathcal{F}$, and $\mathcal{W}$ the sets of, respectively, sentential letters, predicative letters of rank r (for any $r \geq 1$), letters for individual constants, letters for individual variables, (open and closed) formulas, and wffs (closed formulas). $\mathcal{L}$ will contain bottom ($\bot$), which is

[1] I shall throughout consider a style according to which premises and conclusions are closed formulas.

defined axiomatically by $\bot \supset A$ and semantically by $v_M(\bot) = 0$ and $v_M(\bot, w_i) = 0$ respectively.

To simplify the semantic metalanguage, I introduce a (non-denumerable) set of pseudo-constants $\mathcal{O}$, *requiring that any element of the domain D is named by at least one member of* $\mathcal{C} \cup \mathcal{O}$. Let $\mathcal{W}^+$ denote the set of wffs of $\mathcal{L}^+$ (in which $\mathcal{C} \cup \mathcal{O}$ plays the role played by $\mathcal{C}$ in the original language schema). The function of $\mathcal{O}$ is to simplify the clauses for the quantifiers, for example (in the **CL**-semantics):

C∀ $v_M((\forall\alpha)A(\alpha)) = 1$ iff $v_M(A(\beta)) = 1$ for all $\beta \in \mathcal{C} \cup \mathcal{O}$

The (standard) modal language will be called $\mathcal{L}^M$, its extension with pseudo-constants $\mathcal{L}^{M+}$.

In the language $\mathcal{L}^I$ every member of $\mathcal{S} \cup \mathcal{P}^r \cup \mathcal{C} \cup \mathcal{V}$ is *indexed* by some member of a set $\mathcal{I}$ of indices.[2] For example, if $A \in \mathcal{S}$, then $A^i \in \mathcal{S}^I$ for all $i \in \mathcal{I}$. The formulas in $\mathcal{F}^I$ and wffs in $\mathcal{W}^I$ will be composed from the thus indexed letters, but the quantifiers will be over the original members of $\mathcal{V}$—for example, $(\forall x)(P^4x^1 \supset Q^2x^3) \in \mathcal{W}^I$.

$\mathcal{L}^I$ contains quantifiers over index variables, as in $(\exists i)(\exists j)(\forall x)R^1a^ix^j$. To avoid clutter, I introduce two special conventions. $A^{(4)}$ will indicate that all letters in A have the same index, viz. 4. Similarly, $((B \wedge A^{(4)}) \vee C)^{(3)}$ indicates that all letters in A have index 4 and all occurrences of letters outside A (that is, in B and in C) have index 3. This notation is easily generalized to variable indices, as in $(\forall j)((B \wedge (\exists i)A^{(i)}) \vee C)^{(j)}$.

The second convention concerns cases where the indices are mixed. Let us first write A^i as $A^{\{i\}}$, $\sim A^i$ as $(\sim A)^{\{i\}}$, $(A^{\{i_1,\ldots,i_n\}} \wedge B^{\{i_{n+1},\ldots,i_{n+m}\}})$ as $(A \wedge B)^{\{i_1,\ldots,i_{n+m}\}}$, etc. Next, let $(\exists \mathbf{i})A^{\mathbf{i}}$ abbreviate $(\exists i_1, \ldots, i_n)A^{\{i_1,\ldots,i_n\}}$ in which all i_j are different from each other.

The extension of $\mathcal{L}^I$ with pseudo-constants will be called $\mathcal{L}^{I+}$. In it, all members of $\mathcal{O}$ are indexed; $\mathcal{W}^{I+}$ is the set of its wffs.

3 Defining Modal Logic from Paraconsistent Logic

One link between paraconsistent and modal logics was the source of Jaśkowski's approach—see [2]. The basic idea is summarized by:

$$\Gamma \vdash_{\mathbf{D2}} A \text{ iff } \Gamma^\diamond \vdash_{\mathbf{S5}} \diamond A$$

in which $\Gamma^\diamond = \{\diamond B \mid B \in \Gamma\}$. Some enthusiastically extended Jaśkowski's approach, for example in [3], [4], and [5]. Others objected to the non-adjunctive character of the approach—it is easily seen that $\Gamma \vdash_{\mathbf{D2}} A$ iff there is some $B \in \Gamma$ such that $B \vdash_{\mathbf{D2}} A$. I shall not enter this discussion here, but rather point to (what I found) a rather unexpected observation.

Consider a **S5**-structure (without accessibility relation), and the semantics consisting of the usual clauses for all logical constants of $\mathcal{L}$, except that the clause for negation is replaced by:

[2]If I is denumerable, $\mathcal{L}^I$ is isomorphic to $\mathcal{L}$ and the logic **AL**—see below—is just **CL** in disguise. If the **S5**-semantics allows for non-denumerable domains, its reconstruction in terms of **AL**—see Section 5—requires that I be non-denumerable.

C$\sim$ $v_M(\sim A, w_i) = 1$ iff $v_M(A, w_j) = 0$ for some $w_j \in W$

Call this paraconsistent logic **A**.[3] That **A** is defined here from a worlds-semantics, is rather immaterial. It could have been given axiomatically, for example by an axiomatization for positive **CL** ($\perp$ and hence classical negation $\neg$ included), with the following rules and axioms (chosen in honour of Georg Henrik von Wright) to characterize the paraconsistent negation:[4]

NR1 If $\vdash A$, then $\vdash \neg\sim A$
NR2 If $\vdash A \equiv B$, then $\vdash \sim A \equiv \sim B$
NA1 $\vdash A \vee \sim A$
NA2 $\vdash \sim(A \wedge B) \equiv (\sim A \vee \sim B)$
NA3 $\vdash \sim\sim A \supset \neg\sim A$

It is easily seen from the semantics that the definitions

$$\Box A \quad =_{df} \quad \neg\sim A \qquad (1)$$
$$\Diamond A \quad =_{df} \quad \sim\neg A \qquad (2)$$

define the **S5**-modalities in **A**. So, where $f(A)$ is the result of replacing the boxes and diamonds according to these definitions, and $f(\Gamma) = \{f(A) \mid A \in \Gamma\}$, we have

THEOREM 1 $\Gamma \vdash_{\mathbf{S5}} A$ iff $f(\Gamma) \vdash_{\mathbf{A}} f(A)$

As **A** may itself be defined in **S5** (by $\sim A =_{df} \Diamond\neg A$), **A** and **S5** have exactly the same expressive power.[5]

Jaśkowski's **D2** is defined directly in terms of **A** by

$$\Gamma \vdash_{\mathbf{D2}} A \text{ iff } \Gamma^{\sim\neg} \vdash_{\mathbf{A}} \sim\neg A$$

in which $\Gamma^{\sim\neg} = \{\sim\neg B \mid B \in \Gamma\}$.

Incidentally, this reformulation relates Jaśkowski's approach to the Rescher–Manor consequence relations (see Section 9). Indeed, each such consequence relation **X** may be defined by $\Gamma \vdash_{\mathbf{X}} A$ iff $\Gamma^{\sim\neg} \vdash_{\mathbf{ACLuNx}} A$, in which **ACLuNx** is an inconsistency-adaptive logic defined (by a specific adaptive strategy) from the (very weak) paraconsistent logic **CLuN**. I presented the result at the Jaśkowski Memorial Symposium—see [7]—and it is a pity I failed to see the connection with Jaśkowski's work at that time. The connection is not unexpected. Where Jaśkowski's approach is non-adjunctive, the Rescher–Manor consequence relations are *non-adjunctive beyond consistency*—that is, if $\{A_1, \ldots, A_n\} \subseteq \Gamma$ is inconsistent, no such consequence relation recognizes $A_1 \wedge \ldots \wedge A_n$ as a consequence of Γ.

4 Ambiguity Logics

In [1], Guido Vanackere presented the first ambiguity-adaptive logic. The underlying idea is simple but ingenious. The inconsistency of a text may derive from the

[3] João Marcos brought it to my attention that **A** is identical to the logic **Z** from [6].
[4] The converse of NA3 is obviously derivable, and so is, for example $\sim\sim A \supset A$.
[5] So, that **A** is Sound and Complete with respect to its semantics is immediate.

ambiguity of its non-logical constants. To take these possible ambiguities into account, one attaches *indices* to all occurrences of non-logical constants—see Section 2. To detect where the ambiguity is located, an ambiguity-adaptive logic interprets a set of premises Γ as unambiguously as possible. This is realized by presupposing that two occurrences of a non-logical constant have the same meaning 'unless and until' proven otherwise.

Given that Guido was not interested in the lower limit logic as such (in which all occurrences of non-logical constants receive a different meaning), but rather in the adaptive logic defined from it, he presented a rather crude version of the lower limit logic. As my aim is different, I offer a (rather heavy) reconstruction of the lower limit logic, viz. the application of **CL** to the indexed language $\mathcal{L}^I$.

The first step is a child's play: the same letter with two different indices is simply treated as two distinct letters. Individual variables require some care. While occurrences of members of $\mathcal{V}$ are indexed, the quantifiers that bind them are over the non-indexed members of $\mathcal{V}$—see Section 2. This warrants, for example, that $P^1a^1a^2 \vDash (\exists x)P^1x^1x^2$. Next, the matter becomes more attractive if quantification over indices is introduced. Indeed, the specific indices of the letters in the premises and conclusion are immaterial. Clearly, we have

$$(\exists i)(\exists j)(p^i \wedge {\sim}p^j) \vDash_{\mathbf{CL}} (\exists i)p^i$$

as well as

$$(\exists i)(\exists j)(p^i \wedge {\sim}p^j) \nvDash_{\mathbf{CL}} (\exists i)q^i\,.$$

¿From now on in this paper, I shall call the thus defined system the ambiguity logic **AL**. Ambiguity logics for the language $\mathcal{L}$ may be defined from **AL**, for example by the following recipe:

$$\Gamma \vDash A \text{ iff } f(\Gamma) \vDash_{\mathbf{AL}} f(A)$$

where f is some function that takes care of the indexing.

5 Ambiguity Logics and Modal Logics

There is an obvious connection between **AL** and a Kripke semantics. The former actually describes a Kripke semantics more accurately that any (first order) modal logic can possibly do. To see this, interpret the index of any letter in $\mathcal{S} \cup \mathcal{P}^r \cup \mathcal{C} \cup \mathcal{V}$ as referring to a member of W. Thus $v_M(p^1) = 1$ from the **AL**-semantics corresponds to $v_M(p, w_1) = 1$ in the **S5**-semantics. The idea is simple, but worth being spelled out a bit.

To avoid unnecessary complications, I consider a style of semantics that evaluates wffs of $\mathcal{L}^{I+}$ within a model, rather than within a world of that model. I also disregard $\sim$, as $\neg$ is available anyway. Let a model $M = \langle W, D, d, v \rangle$, in which W is a set of worlds, D a set, $d : w_i \mapsto \wp(D)$ a function that assigns to any $w_i \in W$ its particular domain,[6] and v an assignment defined by:

C1.1 $v : \mathcal{S} \times W \mapsto \{0, 1\}$

[6] I chose the more general approach that does not require all worlds to share the same domain and allows individual constants to have a different interpretation in different worlds. However, to keep things simple, I introduce the restriction that each individual constant has an interpretation in each world.

C1.2 $v : (\mathcal{C} \cup \mathcal{O}) \times W \mapsto D$
C1.3 $v : \mathcal{P}^r \times W \mapsto \wp(D^r)$

For all $w_i \in W$, $d(w_i)$ comprises the members of D that occur in the interpretation in w_i of a constant, pseudo-constant or predicate. Thus $v(\alpha, w_i) \in d(w_i)$ for all $\alpha \in \mathcal{C} \cup \mathcal{O}$, and $v(\alpha_1, w_i), \ldots, v(\alpha_r, w_i) \in d(w_i)$ for all $\alpha_1, \ldots, \alpha_r \in \mathcal{C} \cup \mathcal{O}$ and $\pi \in \mathcal{P}^r$ such that $\langle v(\alpha_1, w_i), \ldots, v(\alpha_r, w_i)\rangle \in v(\pi, w_i)$.

The valuation function $v_M : \mathcal{W}^{I+} \mapsto \{0, 1\}$, determined by M, is defined by:

C2.1 where $A \in \mathcal{S}$, $v_M(A^i) = v(A, w_i)$; $v_M(\bot) = 0$
C2.2 where $\pi \in \mathcal{P}^r$ and $\alpha_1 \ldots \alpha_r \in \mathcal{C} \cup \mathcal{O}$, $v_M(\pi^{i_0}\alpha_1^{i_1} \ldots \alpha_r^{i_r}) = 1$ iff $\langle v(\alpha_1, w_{i_1}), \ldots, v(\alpha_r, w_{i_r})\rangle \in v(\pi, w_{i_0})$
C2.3 where $\alpha, \beta \in \mathcal{C} \cup \mathcal{O}$, $v_M(\alpha^i = \beta^j) = 1$ iff $v(\alpha, w_i) = v(\beta, w_j)$
C2.4 $v_M(A \vee B) = 1$ iff $v_M(A) = 1$ or $v_M(B) = 1$
C2.5 $v_M((\forall \alpha)A(\alpha^i)) = 1$ iff $v_M(A(\beta^i)) = 1$ for all $\beta \in \mathcal{C} \cup \mathcal{O}$
C2.6 $v_M((\forall i)A(\xi^i)) = 1$ iff $v_M(A(\xi^i)) = 1$ for all $i \in \mathcal{I}$.

In C2.6, ξ^i is any member of $\mathcal{S} \cup \mathcal{P}^r \cup \mathcal{C} \cup \mathcal{O} \cup \mathcal{V}$. The other logical constants are explicitly defined in the usual way.[7] Where $A \in \mathcal{W}^I$, A is verified by M iff $v_M(A) = 1$.

Before we proceed, let us compare the present semantics with the semantics from Section 4. The latter was introduced informally because it simply is the **CL**-semantics for the language $\mathcal{L}^I$, with an obvious clause added to take care of the quantification over indices—that clause simply is C2.6. Any model from one semantics corresponds to an equivalent model from the other. Where expressions to the left refer to the semantics from Section 4, we have $v(A^i) = v(A, w_i)$ for $A \in \mathcal{S}$, $v(\alpha^i) = v(\alpha, w_i)$ for $\alpha \in \mathcal{C} \cup \mathcal{O} \cup \mathcal{V}$, ..., and $\{v(\alpha^i) \mid \alpha \in \mathcal{C} \cup \mathcal{O}\} = d(w_i)$. So, the semantics from Section 4 is simply a poor person's replacement for a worlds-semantics (without accessibility relation—but only a slight complication is required to introduce that). It is worth remarking that the poor person's replacement is just as good, accurate, suggestive, etc. as the original. Philosophically, the poor person's semantics may be easily connected to Carnap's approach (see [8]) and nicely suits people who want to avoid discussions on the ontological status of possible worlds.[8]

It remains to be shown that **S5** may be defined within **AL**. Let $m(A)$ be the result of replacing in A, $\Diamond B$ by $(\exists i)B^{(i)}$ and $\Box B$ by $(\forall i)B^{(i)}$.[9] Moreover, let $\Gamma^m = \{m(A)^{(1)} \mid A \in \Gamma\}$.

THEOREM 2 $\Gamma \vDash_{\mathbf{S5}} A$ iff $\Gamma^m \vDash_{\mathbf{AL}} m(A)^{(1)}$

Here are some hints for the easy proof. Consider a **S5**-semantics in which a model $M = \langle W, w_1, v\rangle$ verifies A iff $v_M(A, w_1) = 1$. Any **S5**-model that verifies Γ and falsifies A is easily transformed to an **AL**-model that verifies Γ^m and falsifies $m(A)^{(1)}$, and vice versa.[10]

[7] Obviously, $v_M(\neg A) = 1$ iff $v_M(A \supset \bot) = 1$ iff $v_M(A) = 0$, as expected.

[8] I am indebted to João Marcos for pointing out the relation between the present use of **AL** and hybrid logics.

[9] Thus, $\Diamond(\Box\Diamond p \supset q)) = (\exists i)((\forall i)(\exists i)p^{(i)} \supset q^{(i)})$, which is equivalent to $(\exists i)p^i \supset (\exists i)q^i$.

[10] Obviously, **A** too may be characterized in terms of **AL**: replace any $\sim A$ by $(\exists i)\neg A^{(i)}$; next add index 1 to all not indexed letters. Remark that $\sim\sim p \vee q$ becomes $(\exists i)\neg(\exists i)\neg(p^{(i)})^{(i)} \vee q^1$, i.e. $(\exists i)\neg(\exists i)\neg p^i \vee q^1$, which is equivalent to $(\forall i)p^i \vee q^1$.

The upshot is that **S5** is a tiny fragment of **AL**. So, any (paraconsistent or other) logic that may be defined in terms of **S5**, may also be defined in terms of **AL**. The converse does not hold because the expressive power of **AL** supersedes that of **S5**. This is obvious in view of the fact that such formulas as $a^1 = b^3$ and R^4a^7, which are respectively true iff $v(a, w_1) = v(b, w_3)$ and $v(a, w_7) \in v(R, w_4)$, cannot be expressed in $\mathcal{L}^M$.[11] In view of the expressive power of **AL**, it may be expected that more paraconsistent logics can be defined in **AL** than in **S5**. I give an example in Sections 6 and 7.

A final word on the interpretation of the semantics. In connection with ambiguity, the worlds or indices may be interpreted as viewpoints, conceptual systems, contexts, etc.—the connection with Jaśkowski's views are obvious. Interestingly, some viewpoints are themselves mixtures or confusions of several viewpoints. This is why we do not only need **AL** to get a grasp on the Babylonian situation, but often also to understand a single interlocutor, most often ourselves.

6 LP (Without Identity) Defined Within AL

Suppose that a post-modernist wants to apply **AL**, but wants to reduce its effects to $\mathcal{L}$. At the end of Section 4, I mentioned a recipe to do so; the only problem is to specify the function f. Clearly, the most straightforward specification, given the post-modernist's purposes, relies on the following idea: whatever the meaning of the non-logical constants in the premises, there is a meaning of the non-logical constants in the conclusion such that the conclusion follows from the premises. Formally,

$$\Gamma \models A \text{ iff } \Gamma^{\exists\mathbf{i}} \models_{\mathbf{AL}} (\exists\mathbf{i})A^{\mathbf{i}}$$

in which $\Gamma^{\exists\mathbf{i}} = \{(\exists\mathbf{i})B^{\mathbf{i}} \mid B \in \Gamma\}$.

This produces a nice logic, with Double Negation, De Morgan properties, etc. Disjunctive Syllogism is invalid in it, as $(\exists\mathbf{i}){\sim}A^{\mathbf{i}}, (\exists\mathbf{i})(A \vee B)^{\mathbf{i}} \not\models (\exists\mathbf{i})B^{\mathbf{i}}$, and material implication is not detachable. In general, every consequence of two premises is a consequence of one of them. There is one exception: $A, B \vdash A \wedge B$. But, again every consequence of $A \wedge B$ is equivalent to $D \wedge E$ in which D is a consequence of A (or empty) and E is a consequence of B (or empty). So, no detachment-like rule is valid.

Let $\mathcal{L}^{\setminus=}$ be the language $\mathcal{L}$ with identity removed. If restricted to $\mathcal{L}^{\setminus=}$, the just defined logic turns out to be equivalent to Graham Priest's **LP**. In the (outlined) proof of the following theorem, I consider Priest's preferred semantics of **LP**, listed, for example, in [9]. The style is slightly modified to simplify the proof.

THEOREM 3 Where A and the members of Γ are wffs of $\mathcal{L}^{\setminus=}$, $\Gamma \models_{\mathbf{LP}} A$ iff $\Gamma^{\exists\mathbf{i}} \models_{\mathbf{AL}} (\exists\mathbf{i})A^{\mathbf{i}}$.

Proof: For the first direction, suppose that $M = \langle D, v\rangle$ is an **AL**-model (from Section 4) that verifies $\Gamma^{\exists\mathbf{i}}$ and falsifies $(\exists\mathbf{i})A^{\mathbf{i}}$. From M, we define a **LP**-model $M' = \langle S, I\rangle$, in which S is the domain and I the interpretation or assignment.

[11] That a is a rigid designator can be expressed in **AL** by $(\forall i)a^i = a^1$, or equivalently by $(\forall i)(\forall j)a^i = a^j$ or $(\exists i)(\forall j)a^i = a^j$. If a is not a rigid designator, the *de re* reading of $\Box Pa$ can be expressed by $(\forall i)P^i a^1$, and its *de dicto* reading by $(\forall i)P^i a^i$.

Where $\alpha \in \mathcal{C} \cup \mathcal{O}$, let $\widehat{\alpha} = \{v(\alpha^i) \mid i \in \mathcal{I}\}$ (the elements of D that are assigned to some α^i).

(1) $S = \{\widehat{\alpha} \mid \alpha \in \mathcal{C} \cup \mathcal{O}\}$,
(2) Where $A \in \mathcal{S}$, $1 \in I(A)$ iff $v(A^i) = 1$ for some i and $0 \in I(A)$ iff $v(A^i) = 0$ for some i,
(3) Where $\alpha \in \mathcal{C} \cup \mathcal{O}$, $I(\alpha) = \widehat{\alpha}$,
(4) Where $\pi \in \mathcal{P}^r$, $\langle I(\alpha_1), \ldots, I(\alpha_r)\rangle \in I^+(\pi)$ iff $\langle v(\alpha_1^{i_1}), \ldots, v(\alpha_r^{i_r})\rangle \in v(\pi)$ for some $i_1, \ldots, i_r$, and $\langle I(\alpha_1), \ldots, I(\alpha_r)\rangle \in I^-(\pi)$ iff $\langle v(\alpha_1^{i_1}), \ldots, v(\alpha_r^{i_r})\rangle \notin v(\pi)$ for some $i_1, \ldots, i_r$.[12]

Consider some primitive formula A. Clearly, $I(A) = \{1\}$ iff $v_M(A^{\mathbf{i}}) = 1$ for all $\mathbf{i}$; $I(A) = \{0\}$ iff $v_M(A^{\mathbf{i}}) = 0$ for all $\mathbf{i}$; and $I(A) = \{0,1\}$ iff $v_M(A^{\mathbf{i}}) = 1$ for some $\mathbf{i}$ and $v_M(A^{\mathbf{i}}) = 0$ for some $\mathbf{i}$.

This result is generalized to all members of $\mathcal{W}^+$ by a straightforward induction. For negation, suppose that $I(A) = \{1\}$, and hence that $v_M(A^{\mathbf{i}}) = 1$ for all $\mathbf{i}$. But then $v_M(\sim A^{\mathbf{i}}) = 0$ for all $\mathbf{i}$, which agrees with $I(\sim A) = \{0\}$. Similarly for the other values of $I(A)$. The proof is just as straightforward for the other logical constants—the only further specific bit needed is that $(\exists \mathbf{i})A^{\mathbf{i}} \wedge (\exists \mathbf{i})B^{\mathbf{i}}$ is obviously equivalent to $(\exists \mathbf{i})(A \wedge B)^{\mathbf{i}}$.[13]

As M verifies $(\exists \mathbf{i})B^{\mathbf{i}}$ for all $B \in \Gamma$, $1 \in I(B)$ for all $B \in \Gamma$, which means that M' verifies Γ; as M falsifies $(\exists \mathbf{i})A^{\mathbf{i}}$, $1 \notin I(A)$, which means that M' falsifies A.

For the second direction, suppose that $M' = \langle S, I\rangle$ is a **LP**-model that verifies Γ and falsifies A. From M', we define an **AL**-model $M = \langle D, v\rangle$. Consider some set S' that has the same cardinality as S and is such that $S \cap S' = \emptyset$, and let R be a one-one mapping from S on S'. Let $o_1, o_2, \ldots$ be (some) variables for members of S. Below, each $[o_i]$ stands for either o_i or $R(o_i)$; thus $\langle [o_1], [o_2]\rangle \in v(P)$ abbreviates $\langle o_1, o_2\rangle, \langle R(o_1), o_2\rangle, \langle o_1, R(o_2)\rangle, \langle R(o_1), R(o_2)\rangle \in v(P)$.[14]

(1) $D = S \cup S'$,
(2) Where $A \in \mathcal{S}$, (i) if $I(A) = \{1\}$, then $v(A^i) = 1$ for all i, (ii) if $I(A) = \{0\}$, then $v(A^i) = 0$ for all i, and (iii) if $I(A) = \{0,1\}$, then $v(A^{2i}) = 1$ and $v(A^{2i-1}) = 0$ for all i,
(3) Where $\alpha \in \mathcal{C} \cup \mathcal{O}$, $v(\alpha^{2i}) = I(\alpha)$ and $v(\alpha^{2i-1}) = R(I(\alpha))$ for all i,
(4) Where $\pi \in \mathcal{P}^r$, (i) if $\langle o_1, \ldots, o_r\rangle \in I^+(\pi)$ and $\langle o_1, \ldots, o_r\rangle \notin I^-(\pi)$ then $\langle [o_1], \ldots, [o_r]\rangle \in v(\pi^i)$ for all i, (ii) if $\langle o_1, \ldots, o_r\rangle \notin I^+(\pi)$ and $\langle o_1, \ldots, o_r\rangle \in I^-(\pi)$ then $\langle [o_1], \ldots, [o_r]\rangle \notin v(\pi^i)$ for all i, and (iii) if $\langle o_1, \ldots, o_r\rangle \in I^+(\pi) \cap I^-(\pi)$ then $\langle o_1, [o_2], \ldots, [o_r]\rangle \in v(\pi^i)$ for all i and $\langle R(o_1), [o_2], \ldots, [o_r]\rangle \notin v(\pi^i)$ for all i.

These definitions warrant that, for any primitive formula A, (i) $v_M(A^{\mathbf{i}}) = 1$ for all $\mathbf{i}$ iff $I(A) = \{1\}$, (ii) $v_M(A^{\mathbf{i}}) = 0$ for all $\mathbf{i}$ iff $I(A) = \{0\}$, and (iii) $v_M(A^{\mathbf{i}}) = 1$ for some $\mathbf{i}$ and $v_M(A^{\mathbf{i}}) = 0$ for some $\mathbf{i}$ iff $I(A) = \{0,1\}$. This result too is easily generalized by a straightforward induction to all members of $\mathcal{W}^+$. As M' verifies

[12] $I(\pi)$ is a couple of sets of r-tuples. To simplify the notation, $I^+(\pi)$ refers to the first set (the positive extension of π) and $I^-(\pi)$ to the second set (the negative extension of π).

[13] Remark that $(\exists \mathbf{j})A^{\mathbf{j}} \wedge (\exists \mathbf{k})B^{\mathbf{k}}$ can be written as $(\exists \mathbf{i})A^{\mathbf{i}} \wedge (\exists \mathbf{i})B^{\mathbf{i}}$.

[14] **AL** is so rich that, in each of (1)–(4) below, its possibilities are used only to a very limited extent.

Γ, $1 \in I(B)$ for all $B \in \Gamma$ and hence $v_M(((\exists\mathbf{i})B^{\mathbf{i}}) = 1$ for all $B \in \Gamma$, which means that M verifies $\Gamma^{\exists\mathbf{i}}$. As M' falsifies A, $1 \notin I(A)$ and hence $v_M((\exists\mathbf{i})A^{\mathbf{i}}) = 1$, which means that M falsifies $(\exists\mathbf{i})A^{\mathbf{i}}$.

□

Why does this result not generalize to *identity*? Consider again $\Gamma \models A =_{df} \Gamma^{\exists\mathbf{i}} \models_{\mathbf{AL}} (\exists\mathbf{i})A^{\mathbf{i}}$. This delivers all **CL**-*theorems*, including those on identity: $\models a = a$, $\models a = b \supset (b = c \supset a = c)$, etc. It also delivers $a = b \models b = a$. However, we also obtain $a = b, Pa \nvDash Pb$, including $a = b, b = c \nvDash a = c$. So, identity comes out nearly useless for inferential purposes. A post-modernist might be happy with this. After all, if $p, p \supset q \nvDash q$ because the two occurrences of p in the premises might have a different meaning, why then should the two occurrences of a in $a = b, Pa$ have the same meaning? If they have not, $a = b, Pa \nvDash Pb$.

It is worth expanding on the precise difference between Jaśkowski's plot and Priest's plot. Both are in a sense non-adjunctive, but Jaśkowski's plot is worse in this respect. Translating the definiens of

$$\Gamma \models_{\mathbf{D2}} A \text{ iff } \Gamma^{\Diamond} \models_{\mathbf{S5}} \Diamond A$$

in terms of **AL** gives

$$\Gamma \models_{\mathbf{D2}} A \text{ iff } \Gamma^{\exists(i)} \models_{\mathbf{AL}} (\exists i)A^{(i)}$$

in which $\Gamma^{\exists(i)} = \{(\exists i)B^{(i)} \mid B \in \Gamma\}$. Compare this to

$$\Gamma \models_{\mathbf{LP}} A \text{ iff } \Gamma^{\exists\mathbf{i}} \models_{\mathbf{AL}} (\exists\mathbf{i})A^{\mathbf{i}}$$

What is the difference? On the Jaśkowski plot, all letters in a premise receive the same index, and similarly for all letters in the conclusion. The index itself is arbitrary, and the index that goes to all letters in one premise may be different from the one that goes to all letters of another premise. On the Priest plot, all letters in a premise (or in the conclusion) receive an arbitrary index. The crucial difference shows in examples as the following:

$$(\exists i)(p \wedge \sim p)^{(i)}, \text{ that is } (\exists i)(p^i \wedge \sim p^i))$$

is logically false in **AL**, whereas

$$(\exists\mathbf{i})(p \wedge \sim p)^{\mathbf{i}}, \text{ that is } (\exists i)(\exists j)(p^i \wedge \sim p^j))$$

is logically contingent in **AL**. So, Priest can handle contradictory formulas, whereas Jaśkowski cannot (they have no models). As explained above, this allows **LP** to be adjunctive, but the presence of Adjunction is hardly consequential: no detachment-like rule (Modus Ponens, Modus Tollens, Disjunctive Syllogism, Dilemma, ...) is validated by **LP**; everything derivable from two formulas or from their conjunction, is derivable from one of the formulas (or is a conjunction of consequences of the separate formulas).[15] It is highly questionable whether such logics have any use for people facing inconsistency—see for example [10], [11], and [12] for some relevant evidence from the history of the sciences.

[15] From *this* point of view, that **LP** validates the Replacement of Identicals does not contribute to its coherence: it is the only case of interacting premises (from $A(\alpha)$ and $\alpha = \beta$ to derive $A(\beta)$).

Two further comments seem useful. The first is that I did *not* claim that **LP** is just an Ambiguity Logic. And I did *not* claim that its inconsistency-tolerance derives from the ambiguity of the non-logical constants, rather than from the meaning of the logical constant $\sim$. All I showed is that a person who would allow for ambiguous non-logical constants, would end up with (exactly) **LP** if identity is disregarded. The second comment is that, in [13], Graham Priest stated his adherence to the inconsistency-adaptive view: consistency is the normal situation, and is presupposed until and unless shown otherwise. Still, some post-modernists would accept that communication requires one to consider ambiguities as exceptional—interpret a text as unambiguously as you can—and would end up with exactly the same formal machinery. (I do not enter here the discussion on adaptive strategies; see [14] and [15].)

7 Full LP Defined Within AL

One wonders whether it is possible to rescue the Replacement of Identicals by a definition that resembles the one from Section 6. To do so, the post-modernist line has to be phrased in a more complex way. Still restricting attention to $\mathcal{L}^{\setminus=}$, define the function g and its 'complement' g^*, by the following clauses:

(Cg^{prim}) If A is a primitive formula, $g(A) = (\exists \mathbf{i})A^{\mathbf{i}}$ and $g^*(A) = (\forall \mathbf{i})A^{\mathbf{i}}$
(C$g^{\sim}$) $g(\sim A) = \neg g^*(A)$ and $g^*(\sim A) = \neg g(A)$
(C$g\vee$) $g(A \vee B) = g(A) \vee g(B)$ and $g^*(A \vee B) = g^*(A) \vee g^*(B)$
(C$g\forall$) $g((\forall\alpha)A) = (\forall\alpha)g(A)$ and $g^*((\forall\alpha)A) = (\forall\alpha)g^*(A)$

and eliminate the other logical constants by their definitions. Next let $g(\Gamma) = \{g(B) \mid B \in \Gamma\}$.

It is easily seen that $g(A)$ is **AL**-equivalent to $(\exists \mathbf{i})A^{\mathbf{i}}$ (and is the result of pushing, for each i, the quantifier over i inside until it occurs in front of the primitive formula in which the index i occurs). This invites to a small digression.

Let g' and g'^* be as g and g^* except that

(Cg'^{prim}) If A is primitive formula, $g'(A) = \Diamond A$ and $g'^*(A) = \Box A$

Given the definition of the **S5**-modalities from Section 5, Theorem 3 immediately translates to:[16]

THEOREM 4 Where A and the members of Γ are wffs of $\mathcal{L}^{\setminus=}$, $\Gamma \vDash_{\mathbf{LP}} A$ iff $g'(\Gamma) \vDash_{\mathbf{S5}} g'(A)$.

Here is a simpler way of phrasing the same result. Consider in turn A as well as each member of Γ. Locate each primitive formula in it. If the formula occurs within the scope of an even number of negations (including zero), add $\Diamond$ in front of it; otherwise add $\Box$ in from of it. A is a **LP**-consequence of Γ iff the thus transformed A is a **S5**-consequence of the thus transformed Γ. Simple propositional examples: $p, \sim p \nvDash_{\mathbf{LP}} q$ because $\Diamond p, \sim\Box p \nvDash_{\mathbf{S5}} \Diamond q$—in other words $\Diamond p, \Diamond{\sim}p \nvDash_{\mathbf{S5}} \Diamond q$—and $p, \sim p \vDash_{\mathbf{LP}} p \wedge \sim p$ because $\Diamond p, \sim\Box p \vDash_{\mathbf{S5}} \Diamond p \wedge \sim\Box p$—in other words $\Diamond p, \Diamond{\sim}p \vDash_{\mathbf{S5}}$

[16] Similarly, **LP** may be defined within **A** from Section 3.

$\Diamond p \wedge \Diamond{\sim}p$.[17] It is also useful to see what happens to negations of theorems. Take, for example, the **LP**-theorem $p \vee {\sim}p$. While $g'(p \vee {\sim}p) = \Diamond p \vee {\sim}\Box p$, $g'({\sim}(p \vee {\sim}p)) = {\sim}(\Box p \vee {\sim}\Diamond p)$. The former is an **S5**-theorem, while the latter is **S5**-contingent (and equivalent to $\Diamond p \wedge \Diamond{\sim}p$).

But let us return to identity. **LP** with identity cannot possibly be defined within **S5**. As we have seen, $\Diamond a = b$ is too weak and $\Box a = b$ is too strong in that it rules out $a = b \wedge {\sim}a = b$.

With the present plot, however, a special treatment for identity formulas may be devised. Restricting ($\mathrm{C}g^{\mathrm{prim}}$) to members of $\mathcal{S}$ and primitive predicative formulas (as intended above), we add:

$$(\mathrm{C}g{=}) \quad g(\alpha = \beta) = (\forall i)(\exists j)\alpha^i = \beta^j \wedge (\forall j)(\exists i)\alpha^i = \beta^j$$
$$\text{and } g^*(\alpha = \beta) = (\exists i)(\forall j)\alpha^i = \beta^j \wedge (\exists j)(\forall i)\alpha^i = \beta^j$$

It is easily seen that these definitions warrant Replacement of Identicals: any a^i is identical to some b^j and vice versa. They nevertheless allow for inconsistent identities such as $a = b \wedge {\sim}a = b$. Indeed, $a = b$ is true *only* iff all a^i and b^j are identical. But of course, they should give us more.

THEOREM 5 Where A and the members of Γ are wffs of $\mathcal{L}$, $\Gamma \vDash_{\mathbf{LP}} A$ iff $g(\Gamma) \vDash_{\mathbf{S5}} g(A)$.

The proof of this theorem is nearly obvious in view of the **LP**-semantics from the appendix of [15]. Here is a summary of that semantics.

Consider a classical structure $S = \langle D, \{R_i \mid i \in I\}\rangle$, with D a non-empty set, I a set of indices, and each R_i a relation with a certain adicity. A relation of adicity 0 may serve as the interpretation of a sentential letter.

Where $\alpha \in \mathcal{C} \cup \mathcal{O}$, $v(\alpha) \subseteq D$. Where $\pi \in \mathcal{P}^n$, $v(\pi) \subseteq \{R_i \mid i \in I; R_i$ has the same adicity as $\pi\}$. Here is how the valuation function handles primitive formulas:

- if the adicity of π is 0, then
  - $v_M^+(\pi) = 1$ iff $R_i = 1$ for some $R_i \in v(\pi)$
  - $v_M^-(\pi) = 1$ iff $R_i = 0$ for some $R_i \in v(\pi)$
- if the adicity of π is $n > 0$, then
  - $v_M^+(\pi\, \alpha^1 \ldots \alpha^n) = 1$ iff $\langle \mathsf{a}_1, \ldots, \mathsf{a}_n\rangle \in R_i$ for some $R_i \in v(\pi)$, $\mathsf{a}_1 \in v(\alpha_1)$, ..., $\mathsf{a}_n \in v(\alpha_n)$
  - $v_M^-(\pi\, \alpha^1 \ldots \alpha^n) = 1$ iff $\langle \mathsf{a}_1, \ldots, \mathsf{a}_n\rangle \notin R_i$ for some $R_i \in v(\pi)$, $\mathsf{a}_1 \in v(\alpha_1)$, ..., $\mathsf{a}_n \in v(\alpha_n)$
- for identity:
  - $v_M^+(\alpha = \beta) = 1$ iff $v(\alpha) = v(\beta)$
  - $v_M^-(\alpha = \beta) = 1$ iff $v(\alpha) \neq v(\beta)$, or $\mathsf{a} \neq \mathsf{b}$ for some $\mathsf{a}, \mathsf{b} \in v(\alpha)$.

[17] Recall the difference with Jaśkowski's approach: $p, {\sim}p \nvDash_{\mathbf{D2}} p \wedge {\sim}p$ because $\Diamond p, \Diamond{\sim}p \nvDash_{\mathbf{S5}} \Diamond(p \wedge {\sim}p)$.

Except for the quantifiers, the logical constants are handled as in the usual **LP**-semantics. Thus $v^+_M(\sim A) = v^-_M(A)$, etc. For the quantifiers, let A^β_α be the result of replacing any free occurrence of the individual variable α by the individual constant or pseudo-constant β.

- $v^+_M((\forall\alpha)A) = 1$ iff $v^+_M(A^\beta_\alpha) = 1$ for all $\beta \in \mathcal{C} \cup \mathcal{O}$
- $v^-_M((\forall\alpha)A) = 1$ iff $v^-_M(A^\beta_\alpha) = 1$ for some $\beta \in \mathcal{C} \cup \mathcal{O}$
- $v^+_M((\exists\alpha)A) = 1$ iff $v^+_M(A^\beta_\alpha) = 1$ for some $\beta \in \mathcal{C} \cup \mathcal{O}$
- $v^-_M((\exists\alpha)A) = 1$ iff $v^-_M(A^\beta_\alpha) = 1$ for all $\beta \in \mathcal{C} \cup \mathcal{O}$

Semantic consequence and validity are defined as usual.

In view of this **LP**-semantics and the proof of 3, Theorem 5 readily follows. Remark that $v(g(a = b)) = 1$ just in case $\{v(a^i) \mid i \in \mathcal{I}\} = \{v(b^i) \mid i \in \mathcal{I}\}$, and that $v(g(\sim a = b)) = 1$ iff these sets are different *or* are not singletons. If they are not singletons, then the model also verifies $\sim a = a$ and $\sim b = b$. It is then obvious that an **AL**-model of $g(\Gamma)$ may be 'collapsed' into a **LP**-model (from the above semantics) of Γ, and that this model verifies, respectively falsifies, A just in case the original model verifies, respectively falsifies, $g(A)$.

That the semantics from the previous paragraphs characterizes exactly the same logic as Priest's semantics is merely a technical point. The existence of both semantic systems, however, has also a philosophical import. Suppose that there are strong arguments for considering some contradiction as true. This may be the effect of an 'inconsistency' in the structure of the world, but it may just as well be an effect of the fact that the interpretation of the language is ambiguous—see [16] for a related (but largely informal and not completely correct) criticism.

8 Detachable Material Implication

LP is a fragment of the logic (called in Ghent) **CLuNs**, which is the predicative version of Schütte's $\mathbf{\Phi_v}$ from [17]—I refer the reader to [18] for the axiomatization, some semantic systems, and a metatheoretic study.[18] **LP** is obtained from **CLuNs** by removing the detachable implication and equivalence (and defining a non-detachable implication and equivalence in terms of conjunction, disjunction and the paraconsistent negation). In **CLuNs**, one may also define the very remarkable paraconsistent system **AN** from [20] (which validates Disjunctive Syllogism and in general all 'analysing' rules of **CL**). So, this raises the question whether the detachable **CLuN**-implication may be defined in terms of **AL**.

Apparently this is impossible. An obvious suggestion would be to handle $A \supset B$ as follows:

(Cg⊃) $\quad g(A \supset B) \;=\; g(A) \supset g(B)$ and $g^*(A \supset B) \;=\; g^*(A) \supset g^*(B)$

[18] An equivalent system (without identity) is studied in [19]. João Marcos informs me that the same system is studied in I.M.L. D'Ottaviono, The completeness and compactness of a three-valued first-order logic (In *Proceedings of the V Latin American Symposium on Mathematical Logic*. Bogotá, 1981).

On the earlier definition, $g(p \supset q)$ is equivalent to $\sim(\forall i)p^i \vee (\exists i)q^i$, and from this together with $(\exists i)p^i$ does not follow $(\exists i)q^i$. The new definition makes $g(p \supset q)$ identical to $\sim(\exists i)p^i \vee (\exists i)q^i$, and from this together with $(\exists i)p^i$ follows $(\exists i)q^i$ as desired. However, the definition runs into trouble if an implication occurs in the scope of a negation.

There is a further reason to be pessimistic about defining **CLuNs** within **AL**. If we again restrict our attention to $\mathcal{L}^{\setminus=}$, then all clauses, including (C$g\supset$), lead to a definition in **S5**—see the paragraph before Theorem 4. In **CLuNs**, we may define strict equivalence as follows: $A \leftrightarrow B =_{df} (A \equiv B) \wedge (\sim A \equiv \sim B)$. In Figure 1, I list, on the left hand side, the formulas in one sentential letter that are not strictly equivalent in **CLuNs** (not naming conjunctions and disjunctions of named formulas). The two highest middle nodes are theorems but are not strictly equivalent. The middle bottom node is logically false. On the right hand side, I list the formulas that are not equivalent in **S5**. The conclusion is rather obvious.[19]

Figure 1: Not strictly equivalent expressions built from p.

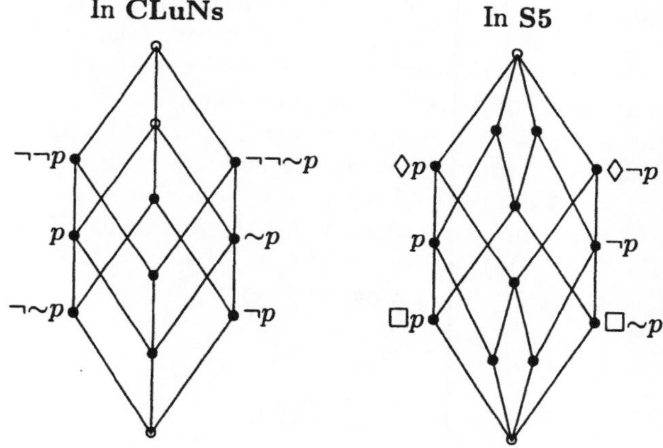

This apparently negative result may presumably be extended to other paraconsistent logics that contain a detachable implication, such as da Costa's $\mathbf{C}_n$-systems and my favourite 'basic' system **CLuN**. As paraconsistent logics without detachable implication have a rather restricted use—remember the discussion on forms of non-adjunctivity—the upshot seems to be that interesting paraconsistent logics cannot be defined (in the sense of Section 1) from modal logics or ambiguity logics. Still, the systems that may be defined within **AL** by means of such clauses as (C$g\supset$) seem worth attention and study.

[19]Still, João Marcos has shown (see [21]) that the propositional fragment of **CLuNs** may be defined in **S5** in a weaker sense than the one described in Section 1. According to that translation, there are A, B, C and D such that $f(A) = f(B)$, D is obtained by replacing A in C by B, but $C \nvdash_{\mathbf{CLuNs}} D$.

9 The Rescher–Manor Puzzle

Rescher–Manor Consequence Relations are defined in terms of **CL**-consequences of the maximal consistent subsets of a (possibly inconsistent) set of formulas. The mechanism is implicitly present in [22] and is spelled out in [23]. Extensions and applications appear in [24], [25], and elsewhere. Later, further consequence relations were defined within the same approach. For a survey and comparative study of all those relations, I refer to [26] and [27]. In [7], I showed that all 'flat' such relations (those not relying on non-logical preferences) are characterized by an inconsistency-adaptive logic defined from **CLuN**: where **X** is such a consequence relation, $\Gamma \vdash_{\mathbf{X}} A$ iff $\Gamma^{\sim\neg} \vdash_{\mathbf{ACLuNx}} A$, in which **ACLuNx** is an inconsistency-adaptive logic defined (by some adaptive strategy) from the (very weak) paraconsistent logic **CLuN**.

We have seen before that there is a correspondence between the modal logic expression $\Diamond A$ and the paraconsistent logic expression $\sim\neg A$. This suggests that each of the Rescher–Manor consequence relations may be characterized by an adaptive logic defined from **S5**. And indeed, this turns out possible.

Actually, several approaches are possible, and I can present only one here. Moreover, I only mention the Weak and Strong Consequence relations—anyone familiar with the matter can extend the result to the other Rescher–Manor consequence relations. Finally, I skip the dynamic proof theory, and confine my attention to the semantics. The lower limit logic is **S5** and the adaptive logic will be called $\mathbf{AS5^c}$. Semantically, the latter is obtained from **S5** by two selections of models. First, one selects the **S5**-models of $\Gamma^{\Diamond}$ in which all worlds w are minimally abnormal with respect to Γ—no world w' of any **S5**-model is such that $Ab_{\Gamma}(w') \subset Ab_{\Gamma}(w)$, where $Ab_{\Gamma}(w) = \{A \in \Gamma \mid v_M(A, w) = 0\}$. From these one selects the models M that are themselves minimally abnormal with respect to Γ—there is no M' such that $Ab_{\Gamma}(M') \subset Ab_{\Gamma}(M)$, where $Ab_{\Gamma}(M) = \{\{A_1, \ldots, A_n\} \mid n > 1; A_1, \ldots, A_n \in \Gamma; M \nvDash \Diamond(A_1 \wedge \ldots \wedge A_n)\}$. It is provable that:[20]

THEOREM 6 $\Gamma \vdash_{Weak} A$ iff $\Gamma^{\Diamond} \vDash_{\mathbf{AS5^c}} \Diamond A$.

THEOREM 7 $\Gamma \vdash_{Strong} A$ iff $\Gamma^{\Diamond} \vDash_{\mathbf{AS5^c}} \Box A$.

There are several remarkable aspects to this result. The Weak and Strong Consequence relations are inconsistency handling mechanisms and are characterized by the inconsistency-adaptive logics $\mathbf{ACLuN^s}$ and **ACLuN2** respectively—see [7]. Inconsistency-adaptive logics are *corrective*; their consequence set is in general a subset of the **CL**-consequence set of the premises (and is identical to the latter iff the premises are normal, viz. consistent). However, $\mathbf{AS5^c}$ is an *ampliative* adaptive logic: its consequence set is in general a superset of the **S5**-consequence set of the premises. The same holds for the other Rescher–Manor consequence relations. In a sense, the situation is similar to the relation between a paraconsistent logic and a modal logic in Jaśkowski's approach. As an inconsistency-adaptive logic (non-monotonically) extends the paraconsistent consequences of a set of premises in the direction of the **CL**-consequences (a corrective step) the translation of this in modal terms leads to an adaptive logic that extends the (in our case) **S5**-consequences of a set of premises (in the direction of its $\mathbf{S5^c}$-consequences), and hence is ampliative.

[20] The following theorems are proved as Theorems 2 and 4 in [28]. The notation there is slightly different in view of the aim of that paper.

More puzzling is that the inconsistency-adaptive characterizations of the Weak and Strong Consequence relations requires two different adaptive logics, whereas their characterization in terms of a modal adaptive logic requires two definitions but only one adaptive logic. A further, and really puzzling, aspect is that there is absolutely no clear correspondence between **CLuN** and **S5**.[21] The two characterizations of the Weak and Strong Consequence relations lead to the same result, as required (and both provide a dynamic proof theory), but they do so along completely different roads. Even the role played by the lower limit logic and the adaptive logic (the unconditional and the conditional steps in the dynamic proofs) is different in both cases. In view of the easy correspondence between some paraconsistent logics and **S5**, the absence of any such correspondence in the present case is puzzling.

As mentioned before, there are several adaptive characterizations of the Rescher–Manor consequence relations in terms of modal logics. This, however, does not remove the puzzle. The problem is that the inconsistency-adaptive characterizations cannot apparently be translated in modal terms, and that the translation in the opposite direction (in terms of an adaptive logic that has **CLuN** as its lower limit logic) seems to fail just as well.

10 The Compatibility Puzzle

The point I want to make here is that paraconsistent logics are excellent instruments for some purposes that have in themselves no relation at all with inconsistency. To illustrate this, I consider compatibility. This simple and transparent notion is obviously undecidable and lacks a positive test—a reliable sign that an adaptive logic is able to take care of it.

In [30], "A is compatible with Γ" is defined by: $\Gamma^{\Box}$ has **COM**-models and all of them verify $\Diamond A$—here $\Gamma^{\Box} = \{\Box B \mid B \in \Gamma\}$ and **COM** is an (ampliative) adaptive logic based on **S5**. Semantically, the idea is that $Ab(M) = \{A \mid A \in \mathcal{W};\ M \models \neg\Diamond A\}$, that $Ab(\Gamma^{\Box}) = \{A \mid A \in \mathcal{W};\ \Gamma^{\Box} \models_{\mathbf{S5}} \neg\Diamond A\}$, and that a **S5**-model M of $\Gamma^{\Box}$ is a **COM**-model of $\Gamma^{\Box}$ iff $Ab(M) = Ab(\Gamma^{\Box})$.[22] This approach is proved adequate, and a sound and complete dynamic proof theory is defined from it. The latter is simple, but requires more space than available.

Compatibility may also be characterized in terms of an inconsistency-adaptive logic defined from the lower limit logic **CLuN**,[23] and this approach leads to a dynamic proof theory as well. I shall call the adaptive logic **ACLuNc**.

On this approach, "A is compatible with Γ" (where the only negation occurring in A and in the members of Γ is $\neg$) is defined by: $\Gamma^{\neg\sim}$ has **ACLuNc**-models and all of them verify $\sim\neg A$. Let me characterize **ACLuNc** by its semantics. $Ab(\Gamma^{\neg\sim}) = \{A \mid A \in \mathcal{W};\ \Gamma^{\neg\sim} \models_{\mathbf{CLuN}} \neg\sim\neg A\}$, $Ab(M) = \{A \mid A \in \mathcal{W};\ M \models \neg\sim\neg A$ or $M \models A \wedge \sim A\}$, and a **CLuN**-model M of $\Gamma^{\neg\sim}$ is a **CLuN**-model of $\Gamma^{\neg\sim}$ iff $Ab(M) = Ab(\Gamma^{\neg\sim})$. Here too, I have to skip the dynamic proof theory.

[21] **CLuN** can be *embedded* within **CL**, and hence within **S5**—see [29]—but the embedding does not provide any relevant information for the present point.

[22] Never mind that all **S5**-theorems are abnormalities. For further details the original paper should be consulted.

[23] See [31] for the study of the two most interesting inconsistency-adaptive logics defined from **CLuN**, and [32] for a survey of the domain.

As A and the members of Γ do not contain any occurrence of $\sim$, they themselves behave fully classically—for example, if Γ is inconsistent, then it has no **ACLuNc**-models and hence nothing is compatible with it. The use of the inconsistency-adaptive logic **ACLuNc** is that it enables one to express that both A and $\neg A$ are compatible with Γ—where $A \in \mathcal{W}$, $\sim\neg A \wedge \sim\neg\neg A$ is not an **ACLuNc**-inconsistency.

Here comes the puzzling bit. That $\Box$ corresponds to $\neg\sim$ and $\Diamond$ to $\sim\neg$ raises more questions than it resolves. To suppose that **CLuN** might be defined within **S5** or vice versa seems utterly absurd.[24] Moreover, while $\Box p$ is **S5**-derivable from $\Box(p \wedge q)$, $\neg\sim p$ is not **CLuN**-derivable from $\neg\sim(p \wedge q)$. Only the inconsistency-adaptive logic **ACLuNc** makes sure that $\neg\sim p$ is derivable from $\neg\sim(p \wedge q)$. So, although the ampliative adaptive logic **COM** and the corrective adaptive logic **ACLuNc** ultimately lead to the same result, they do so along very different roads. Both semantically and proof-theoretically, the adaptive logic based on **S5** is extremely different from the one based on **CLuN**.

The puzzle is not that a modal and a paraconsistent approach may be made to serve the same purpose. In view of the results from Sections 3 and 5, the adaptive logic **COM** may be converted to an adaptive logic based on **A** or **AL**. The puzzle derives from the fact that **S5** and **CLuN** have very different structures, and that the adaptive logics **COM** and **ACLuNc** follow extremely different roads, both semantically and proof-theoretically. The puzzle may be rephrased about paraconsistent logics only: the **ACLuNc** approach to compatibility is radically different from the adaptive approach based on **A**, which is easily 'translated' from **COM**.

11 In Conclusion

As promised, I presented some correspondences, some of them rather unexpected, between paraconsistent logics, modal logics, and the ambiguity logic **AL**. Many of these results are partial; related and analogous questions have not been investigated. Moreover, I presented two puzzles the solution of which might greatly advance our insight in paraconsistent and inconsistency-adaptive logics and in their relation with modal and ambiguity systems.

Here is a list of (to the best of my knowledge) open problems:

(1) To study the properties of the paraconsistent logic **A** from Section 3 (a simple task in view of insights in modal logics—start from [6]), to find a sensible paraconsistent interpretation for it, and to find alternative semantics for it.

(2) To study variants of **A** obtained from worlds-semantics with different properties than that for **S5**. Somewhat unusual modal logics should not be overlooked here, as they might lead to poor (and hence interesting) paraconsistent logics.

(3) To study ambiguity logics with an accessibility relation defined on the indices. This will lead to predictable reconstructions of modal logics. However, the surplus expressive power of the ambiguity logics will presumably result in interesting characterizations of paraconsistent (and other) logics.

(4) In the proof of Theorem 3, an **AL**-model is reduced to a **LP**-model, and a **LP**-model is expanded to an **AL**-model. However, **AL** is so rich that many such

[24] As in the previous section, the embedding from [29] does not provide any relevant information for the present point.

reductions and expansions would do the job. Are some of them more natural than others in delivering a nicer interpretation of **LP**?

(5) To spell out the proof of Theorem 5.

(6) To investigate the systems obtained from **AL** by such clauses as ($Cg\supset$).

(7) To vary on (6) with respect to the variants for **AL** obtained from (3).

(8) To resolve the puzzle from Section 9. One approach would start by investigating other inconsistency-adaptive logics that allow for an adequate characterization of the Rescher–Manor consequence relations. The most obvious first candidate, the adaptive logic defined from the system **A** from Section 3, is obtained by literally rephrasing the results from Section 9 in terms of **A**. However, one should look for the performance of variants of **A** and of **CLuN**.[25]

(9) To resolve the puzzle from Section 10. Again, an obvious first step might be to literally rephrase the results from [30] in terms of the system **A**. Next, starting with inconsistency-adaptive logics defined from variants of **A**, one might investigate which other inconsistency-adaptive logics allow for an adequate definition of compatibility.

Each of these seem worth being tackled, but some might turn out to be dead ends. For example, the results from (2) will only be interesting if they lead to paraconsistent logics that are worth attention, either from a theoretical point of view or with respect to applications.

Let me add a final warning. We have seen that some paraconsistent logics are connected to modal logics and to ambiguity logic. While this is a sign of the technical maturity of the paraconsistent enterprise, it does not follow that those paraconsistent logics are superior to their alternatives. It turns out easy enough to find (or define) an operator $\sharp$ such that some models verify both A and $\sharp A$. A very different question is whether such an operator constitutes the right instrument for understanding (or regulating) the way in which people proceed when facing inconsistency—a situation that is not uncommon in both science and everyday practice. For reasons expounded before, paraconsistent logics that reduce negation to possibility (or to indices) should be mistrusted.[26]

REFERENCES

[1] Guido Vanackere. Ambiguity-adaptive logic. *Logique et Analyse*, 159:261–280, 1997. Appeared 1999.

[2] Stanisław Jaśkowski. Propositional calculus for contradictory deductive systems. *Studia Logica*, 24:243–257, 1969.

[3] J. Kotas and Newton C.A. da Costa. Problems of modal and discussive logics. In Graham Priest, Richard Routley, and Jean Norman, editors, *Paraconsistent Logic. Essays on the Inconsistent*, pages 227–244. Philosophia Verlag, München, 1989.

[25] It seems advisable to first study [33]. There a direct dynamic proof theory (in terms of **CL**) is derived from the inconsistency-adaptive one.

[26] Most unpublished papers in the reference section (and many others) are available from the internet address `http://logica.rug.ac.be/centrum/writings/`.

[4] Irene Mikenberg, Newton C. A. da Costa, and Rolando Chuaqui. Pragmatic truth and approximation to truth. *Journal of Symbolic Logic*, 51:201–221, 1986.

[5] Newton C.A. da Costa, Otávio Bueno, and Steven French. The logic of pragmatic truth. *Journal of Philosophical Logic*, 27:603–620, 1998.

[6] Jean Yves Béziau. The paraconsistent logic Z (A possible solution to Jaskowski's problem). *Logic and Logical Philosophy*, to appear.

[7] Diderik Batens. Towards the unification of inconsistency handling mechanisms. *Logic and Logical Philosophy*, in print.

[8] Rudolf Carnap. *Meaning and Necessity.* University of Chicago Press, Chicago, 1947.

[9] Graham Priest. *In Contradiction. A Study of the Transconsistent.* Nijhoff, Dordrecht, 1987.

[10] Joel Smith. Inconsistency and scientific reasoning. *Studies in History and Philosophy of Science*, 19:429–445, 1988.

[11] Joke Meheus. Adaptive logic in scientific discovery: the case of Clausius. *Logique et Analyse*, 143–144:359–389, 1993. Appeared 1996.

[12] Newton C.A. da Costa and Steven French. Inconsistency in science. A partial perspective. In Joke Meheus, editor, *Inconsistency in Science.* Kluwer, Dordrecht, in print.

[13] Graham Priest. Minimally inconsistent **LP**. *Studia Logica*, 50:321–331, 1991.

[14] Diderik Batens. Minimally abnormal models in some adaptive logics. *Synthese*, 125:5–18, 2000.

[15] Diderik Batens. Linguistic and ontological measures for comparing the inconsistent parts of models. *Logic and Logical Philosophy*, to appear.

[16] David Lewis. Logic for equivocators. *Noûs*, 16:431–441, 1982.

[17] Kurt Schütte. *Beweistheorie.* Springer, Berlin, 1960.

[18] Diderik Batens and Kristof De Clercq. A rich paraconsistent extension of full positive logic. *Logique et Analyse*, in print.

[19] Walter Carnielli, João Marcos, and Sandra de Amo. Formal inconsistency and evolutionary databases. *Logic and Logical Philosophy*, 2001. In print.

[20] Joke Meheus. An extremely rich paraconsistent logic and the adaptive logic based on it. In Batens et al. [34], pages 189–201.

[21] João Marcos. **CLuNs** as modal logic. Forthcoming.

[22] Nicholas Rescher. *Hypothetical Reasoning.* North-Holland, Amsterdam, 1964.

[23] Nicholas Rescher and Ruth Manor. On inference from inconsistent premises. *Theory and Decision*, 1:179–217, 1970.

[24] Nicholas Rescher. *The Coherence Theory of Truth.* Clarendon, Oxford, 1973.

[25] Nicholas Rescher. *Plausible Reasoning. An Introduction to the Theory and Practice of Plausibilistic Inference.* Van Gorcum, Assen/Amsterdam, 1976.

[26] Salem Benferhat, Didier Dubois, and Henri Prade. Some syntactic approaches to the handling of inconsistent knowledge bases: A comparative study. Part 1: The flat case. *Studia Logica*, 58:17–45, 1997.

[27] Salem Benferhat, Didier Dubois, and Henri Prade. Some syntactic approaches to the handling of inconsistent knowledge bases: A comparative study. Part 2: The prioritized case. In Orłowska [35], pages 473–511.

[28] Diderik Batens. A strengthening of the Rescher–Manor consequence relations. To appear.

[29] Diderik Batens, Kristof De Clercq, and Natasha Kurtonina. Embedding and interpolation for some paralogics. The propositional case. *Reports on Mathematical Logic*, 33:29–44, 1999.

[30] Diderik Batens and Joke Meheus. The adaptive logic of compatibility. *Studia Logica*, 66:327–348, 2000.

[31] Diderik Batens. Inconsistency-adaptive logics. In Orłowska [35], pages 445–472.

[32] Diderik Batens. A survey of inconsistency-adaptive logics. In Batens et al. [34], pages 49–73.

[33] Diderik Batens and Timothy Vermeir. Direct dynamic proofs for the Rescher–Manor consequence relations: The flat case. To appear.

[34] Diderik Batens, Chris Mortensen, Graham Priest, and Jean Paul Van Bendegem, editors. *Frontiers of Paraconsistent Logic.* Research Studies Press, Baldock, UK, 2000.

[35] Ewa Orłowska, editor. *Logic at Work. Essays Dedicated to the Memory of Helena Rasiowa.* Physica Verlag (Springer), Heidelberg, New York, 1999.

The Dialogical Dynamics of Adaptive Paraconsistency *

SHAHID RAHMAN University of Saarland, Germany
s.rahman@mx.uni-saarland.de

JEAN PAUL VAN BENDEGEM University of Brussels, Belgium
jpvbende@vub.ac.be

Abstract
The dialogical approach to paraconsistency as developed by Rahman and Carnielli ([1]), Rahman and Roetti ([2]) and Rahman ([3], [4] and [5]) suggests a way of studying the dynamic process of arguing with inconsistencies.
In his paper on *Paraconsistency and Dialogue Logic* ([6]) Van Bendegem suggests that an adaptive version of paraconsistency is the natural way of capturing the inherent dynamics of dialogues. The aim of this paper is to develop a formulation of dialogical paraconsistent logic in the spirit of an adaptive approach and which explores the possibility of eliminating inconsistencies by means of logical preference strategies.

INTRODUCTION

The dialogical approach to paraconsistency as developed by Rahman and Carnielli ([1]), Rahman and Roetti ([2]) and Rahman ([3]), ([4]) and ([5]) suggests a way of studying the dynamic process of arguing with inconsistencies.
In his paper on *Paraconsistency and Dialogue Logic* ([6]) Van Bendegem, who supports the dialogical approach to paraconsistency, argues however that Rahman-Carnielli's formulation defeats the aim of a general dynamic structure of dialogical paraconsistent logics because their formulation uses a paraconsistency rule which applies not only to arguments containing inconsistent formulae but also to arguments where no inconsistent formulae occur. Van Bendegem suggests that an adaptive version of paraconsistency is the natural way of capturing the

* We would like to thank Diderik Batens (Ghent), Gerhard Heinzmann (Nancy), for helpful philosophical and historical comments on the subject of this paper, João Marcos (Unicamp) and Helge Rückert (Saarbrücken / Leiden) for critical comments on an earlier version, and Graham Priest (Queensland) for sending us unpublished material on the topic. We would also wish to thank Marcelo Coniglio (Unicamp) for his thorough proof-reading. The first author would like to thank the CNPq for financial support of position as a visiting-researcher at the University of Campinas where he had the opportunity to discuss with high-level researchers about the issues developed in this paper.

inherent dynamics of dialogues. In fact, Diderik Batens proposed in his paper *A survey of inconsistency-adaptive logics* ([7]) that the study of adaptive logic should be carried out in the context of argumentation.
The aim of this paper is to develop a formulation of dialogical paraconsistent logic in the spirit of an adaptive approach and which explores the possibility of eliminating inconsistencies by means of logical preference strategies.

1 THE DIALOGICAL LITERAL APPROACH TO PARACONSISTENCY AND ITS PROBLEMS

1.1 Literal Paraconsistency

One way to formulate paraconsistent logic within the dialogical approach as developed in Rahman and Carnielli ([1]), Rahman and Roetti ([2]) and Rahman ([4]) can be achieved in the following way. Assume that to the structural rules of the standard dialogical logic[1] we add the following:

- *Negative Literal Rule*:
  The Proponent is allowed to attack the negation of an atomic (propositional) statement (the so called *negative literal*) if and only if the Opponent has already attacked the same statement before.

This structural rule can be considered in analogy to the formal rule for positive literals. The idea behind this rule is the following: An inconsistency of the Opponent may be tolerated by using a type of charity principle. The inconsistency might involve different semantic contexts in which, say, a and $\neg a$ have been asserted. Now, if the Opponent attacks $\neg a$ with a he concedes thereby that there is some common context between $\neg a$ and a which makes an attack on $\neg a$ possible. This allows the Proponent to attack the corresponding negation of the Opponent.
In Rahman and Carnielli ([1]) the logics produced by this rule were called *Literal Dialogues*, or shorter: L-D. In order to distinguish between the intuitionistic and the classical version Rahman and Carnielli wrote L-D^{i} (for the intuitionistic version) and L-D^{c} (for the classical version). To be precise we should call these logical systems literal dialogues with classical structural rules and literal dialogues with intuitionistic structural rules respectively.

In L-D the (from a paraconsistent point of view) dangerous formulae $(a\wedge\neg a)\rightarrow b$, $a\rightarrow(\neg a\rightarrow b)$ and $(a\rightarrow b)\rightarrow((a\rightarrow\neg b)\rightarrow\neg a)$ are not valid. Let us see the corresponding literal dialogues in L-D^{c} for the first one:

---

[1] For a short introduction to standard dialogical logic see the appendix.

Example 1

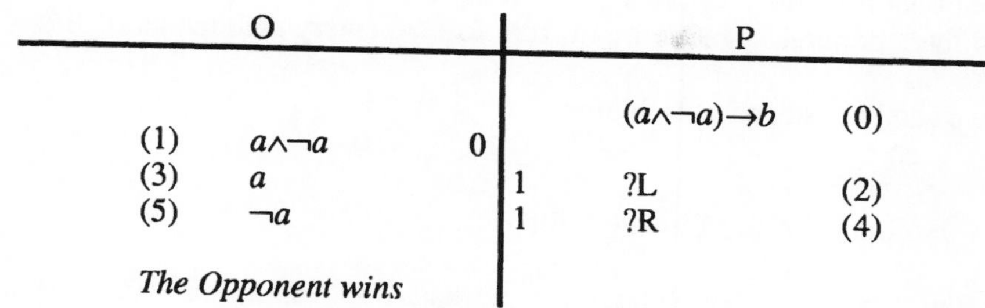

O				P	
				$(a\wedge\neg a)\rightarrow b$	(0)
(1)	$a\wedge\neg a$	0			
(3)	a		1	?L	(2)
(5)	$\neg a$		1	?R	(4)
The Opponent wins					

The Proponent loses because he is not allowed to attack the move (5) (see negative literal rule). In other words the Opponent may have contradicted himself, but the semantic context of the negative literal is not available to the Proponent until the Opponent starts an attack on the same negative literal — an attack which in this case will not take place.

All classically valid formulae without negation are also valid in L-D^c. All intuitionistically valid formulae without negation are also valid in L-D^i. As in da Costa's system C_1 neither of the following is valid in L-D^c:

$(a\wedge\neg a)\rightarrow b$
$(a\wedge\neg a)\rightarrow\neg b$
$\neg(a\wedge\neg a)$
$a\rightarrow\neg\neg a$
$(a\rightarrow b)\rightarrow((a\rightarrow\neg b)\rightarrow\neg a)$
$((a\rightarrow b)\wedge(a\rightarrow\neg b))\rightarrow\neg a$
$((\neg a\rightarrow b)\wedge(\neg a\rightarrow\neg b))\rightarrow a$

$\neg a\rightarrow(a\rightarrow b)$
$\neg a\rightarrow(a\rightarrow\neg b)$
$a\rightarrow(\neg a\rightarrow b)$
$a\rightarrow(\neg a\rightarrow\neg b)$
$((a\rightarrow\neg a)\wedge(\neg a\rightarrow a))\rightarrow b$

$(a\rightarrow(b\vee c))\rightarrow((a\wedge\neg b)\rightarrow c)$
$((a\rightarrow\neg a)\wedge(\neg a\rightarrow a))\rightarrow\neg b$
$((a\wedge b)\rightarrow c)\rightarrow((a\wedge\neg c)\rightarrow\neg b)$
$((a\vee b)\wedge\neg a)\rightarrow b$
$(a\vee b)\rightarrow(\neg a\rightarrow b)$
$(a\rightarrow b)\rightarrow(\neg b\rightarrow\neg a)$
$(a\rightarrow b)\vee(\neg a\rightarrow b)$

$(\neg a\vee\neg b)\rightarrow\neg(a\wedge b)$
$(\neg a\wedge\neg b)\rightarrow\neg(a\vee b)$
$(\neg a\vee b)\rightarrow(a\rightarrow b)$
$(a\rightarrow b)\rightarrow\neg(a\wedge\neg b)$
$a\rightarrow((\neg a\vee b)\rightarrow b)$

In L-D^i all the intuitionistically non-valid formulae have to be added to the list, for example:

$\neg\neg A\rightarrow A$
$A\vee\neg A$
$((A\rightarrow B)\rightarrow A)\rightarrow A$

$A\vee(A\rightarrow B)$
$A\vee((A\vee B)\rightarrow B)$
$\neg(A\rightarrow B)\rightarrow A$

Validity is defined in dialogical logic via winning strategies for the Proponent. A systematic description of the winning strategies for these logics can be obtained

from appropriate tableau systems. [2] The extension of literal dialogues for propositional logic to first-order quantifiers is straightforward: We only have to extend the structural negative literal rule to elementary statements of first-order logic. [3]
Now we shall consider the problems.

1.2 Problems of the Literal Approach

The large number of non-valid formulae listed above suggests that the literal approach presented is too restrictive. There are many formulae in this list we would like to have as valid. A further problem is what to do with inconsistencies between complex propositions. One could argue that contradictions which cannot be carried on at the literal level should be released of paraconsistency restrictions — this was defended in Rahman ([3], [4] and [5]) where a distinction between internal or *de re* and external or *de dicto* negation was introduced. One can also follow another complementary strategy and introduce a device combining two types of restrictions, one applying to the literal case and the other to the complex case. Thus, it looks like more has to be done than just introducing a structural rule restricting the use of negations. One first idea is to introduce an adaptive extension of the literal negative rule making use of the dynamics of argumentation. That is, one argues consistently as long as possible and when necessary adapts to the specific inconsistencies that occur. In other words, the consistency of all sentences is presupposed unless and until proven otherwise (cf. Batens [7], 1). This fits nicely with the problem of complex inconsistencies: If one analyses the complex inconsistency until the literal level has been reached and no inconsistency occurs, then one continues using a non-paraconsistent framework. But as remarked by Max Urchs the adaptive logic of Batens has another deep and appealing contribution to paraconsistency: What is really dangerous is not *inconsistency* as such, but *contradictions* (the latter are also called *explosive inconsistencies*) ([8]). Overlooking such a distinction leads to a uniform treatment of some types of formulae which should be kept apart — e.g. the inconsistency which can occur in a dialogue for disjunctive syllogism or de Morgan should be distinguished from the inconsistency produced by ex falso sequitur quodlibet. The idea is that if you have an inconsistency in the premises you can either isolate it and continue with the rest or take that part of the inconsistency which is necessary for the proof — this strategy has been called by Batens the *logical preferential mechanism* — in contrast to the non-logical preferential mechanisms of standard non-monotonic logics. Now what distinguishes inconsistencies from contradictions? Our proposal is the following: While contradictions (i.e. explosive inconsistencies) in the concessions of an argument make at least one defensive or aggressive move redundant, non-explosive inconsistencies do not. That is why contradictions lead in this approach

[2] See details on how to build the tableau systems in Rahman and Carnielli ([1]) and Rahman ([3] and [5]).
[3] João Marcos (Unicamp) observed that the classical literal system corresponds to Sette's P1.

to the paraconsistent non-validity of those formulae in which they occur. Moreover, some parts of the (non-explosive) inconsistencies can be used for the defence of the argument (the other parts can be dismissed). Thus, an appropriate concept of redundancy should be introduced here. This is the line we are going to follow in this paper.

2 THE DIALOGICAL WAY TO ADAPTIVE PARACONSISTENCY AT THE GAME LEVEL

As mentioned above, we introduce here a redundancy concept which can be applied in order to distinguish between non-explosive inconsistencies and contradictions. Clearly, this procedure should be understood adaptively. That is, if no inconsistencies are present, use the standard definition of dialogical validity (see the appendix). If inconsistencies appear, restrict the definition of validity by means of the redundancy principle.

2.1 Formulae Which are Paraconsistently Free of Redundancies: Dialogical Adaptivity

We will present here one concept of redundancy which is a variation of some ideas on dialogical relevance logic developed in Rahman and Rückert [9]. In fact other versions based on relevance considerations introduced in the above-mentioned paper could also be applied. We have chosen this version because it seems more inherent to the paraconsistent point of view in the sense that it stresses the point of having inconsistency without triviality.

The idea behind our dialogical approach to adaptivity is that in the case an inconsistency occurs the Proponent can try one of the following procedures:

i) He may try to win by isolating the inconsistency. That is, he will just ignore it and continue with the rest. If he can answer all the attacks of the Opponent anyway the inconsistency was redundant and he is through.

ii) He may try to win using only one part of the inconsistency and continue with the rest just as in i.

iii) He may try to win using the inconsistency and ignoring the rest of the moves of the Opponent.

We will say that the inconsistency is explosive and the thesis not paraconsistently valid iff the third case is the only successful procedure available. Shorter, we call explosive such inconsistencies which make a defensive move redundant.
Now, what happens in the case of a negation occurring in the then-part of a conditional, as in $(a\wedge\neg a)\rightarrow\neg r$? Well, if nothing is added we obtain a minimal

logic; if we want more than minimality then here again various options are available. We will discuss an option which tolerates inconsistencies, does not lead to triviality but renders the principle of non-contradiction as valid. The idea is that a non-explosive inconsistency should not make any defensive move redundant, that is, the Proponent has to use all his defensive moves. (Counterattacks resulting from an attack on negations are considered to be defensive moves.) Suppose the Proponent states the thesis $\neg(a\wedge\neg a)$. The Opponent will attack with move $(\mu)(a\wedge\neg a)$. The Proponent should now use this move. Now, this leads to an awkward formulation where two types of structural rules have to be distinguished: one applying to attacks and the other to counterattacks. A more general approach results from the following considerations: Suppose, on one hand, that the Proponent notices that he can win without using (or with only one part of) the inconsistency which appears in a conjunction stated by the Opponent, then he should be allowed to attack only one part of this conjunction. Suppose, on the other hand, that the Proponent notices that he can win without using (or with only one part of) the inconsistency which appears in a disjunction stated by himself, then he should be allowed to defend only one part of this disjunction
Thus, if an inconsistency in the concessions occurs, the Proponent must attack each formula at least once, use all the atomic formulae of the Opponent and defend himself against all attacks. It is time to be a little more precise:

DEFINITION 1*: O-Variants:* Any possible development of a dialogue as determined by the choice of the Opponent constitutes a variant. Thus, the developments of a dialogue as determined by the choice between attacks ?L and ?R, or ?R after L? on a conjunction, the different possible defences of a disjunction (left or right side) and the choice between defence and counterattack while reacting to an attack on a conditional are said to be different *O-variants*.

DEFINITION 2*: O-Inconsistency:* It is said that an O-inconsistency occurs iff

1. The Opponent stated $(\nu)A$, $\neg(\mu)\ A$, where ν and μ signalise the numbers of the respective moves and A is a formula.
2. The Opponent stated a negative refutable formula (like $\neg(A\vee\neg A)$; $\neg(A\rightarrow A)$)

Notice that here we make paraconsistency for complex formulae possible. Clearly, a more restrictive definition applying to literals is possible.

If we want to put the idea of adaptivity into practice we need to precise the notion of *used*

1. An atomic O-formula has been used iff the Proponent states this formula in order to state an aggressive (=attack) or a defensive move.
2. A complex formula A has been used iff all the aggressive and defensive moves related to A have been stated according to the conditions stated by definition 3 (below).

DEFINITION 3: A formula is said to be *free of paraconsistent redundancies* iff the Proponent wins under the standard structural rules and he *can* win under the following conditions:

1. He can attack every O-formula at least once (e.g. he can win with only one of the possible attacks on a conjunction) in any of the possible O-variants (not necessarily in the same O-variant).
2. He can defend himself at least once against all attacks (e.g. he can win defending only one part of the disjunction) in any of the possible O-variants (not necessarily in the same O-variant)
3. He can use at least one occurrence of any atomic O-formula in any of the possible O-variants (not necessarily in the same O-variant).

DEFINITION 4: A formula is said to contain paraconsistent redundancies iff the Proponent cannot win without dismissing at least one of the conditions stated before.

We will tag all attacked O-formulae and all atomic O-formulae used. We do not need to tag the defences because they can be recognised by the *closed rounds* – see appendix.
Now the definition of formulae which are valid by adaptation:

DEFINITION 5: A formula in the dialogical proof of which an inconsistency occurs is *valid by adaptation* if this formula is valid by the standard definition of validity and is free of paraconsistent redundancies. Inconsistencies occurring in the dialogical proof of formulae not paraconsistently free of redundancies are called *explosive inconsistencies* or *contradictions*.

Let us study some examples. In example 1 of chapter 1.1 (played in standard dialogical logic) it is clear that the inconsistency occurring in the concessions (O-formulae) is explosive: The Proponent wins according to the standard definition of validity, but there is no variant where the Proponent can win by stating the (redundant) defence *b*. In the next two examples the inconsistencies do not collapse into contradictions.

Example 2

	O			P	
				$\neg a\rightarrow(a\rightarrow\neg(a\wedge b))$	(0)
(1)	$\neg a$	0√		$a\rightarrow\neg(a\wedge b)$	(2)
(3)	a	2		$\neg(a\wedge b)$	(4)
(5)	$a\wedge b$	4√		⊗	
(7)	a	√	5	?L	(6)
	⊗		1	a	
	The Proponent wins				

Because of the inconsistency occurring in 1, 3 the Proponent plays adaptively. In doing so, in a first stage he ignores the inconsistency. Then he uses move 7 (or move 3) to attack $\neg a$ and wins. Moreover, since there is no other variant to be considered the formula is valid.
It should not be difficult for the reader to check that neither $\neg a \rightarrow (a \rightarrow b)$ nor $(\neg a \wedge a) \rightarrow b$ hold and that $\neg(\neg a \wedge a)$, $\neg a \rightarrow (a \rightarrow \neg a)$, $(\neg a \wedge a) \rightarrow a$ and $(a \rightarrow \neg\neg a)$ hold.

The next example will lead us to determine how to check whether a given formula is redundant. In this example two variants have to be considered. In order to keep track of the possible redundancies at the game level we will introduce subdialogues (the tableau system due to be introduced later will include all the variants, but at the strategy level).[4] The idea is quite simple. Suppose that at a given stage of the dialogue, the Opponent, who stated an inconsistency realises that he loses by the standard rules but he realises too that one defensive move has been dismissed. He then can ask something like: Well, I will lose in this way, but what about, say, the defensive move *b*? (shorter ?*b*) You did not show me that I can win using that. In this case the Proponent will open a subdialogue where he will have to show that he can win with the move challenged. Now, it can be the case that the Proponent says: I can win using *b*, but I will use it only if in your variant (i.e., the O-variant) you choose, say, the right part — notice that this is a consequence of condition 3 of definition 3 which states that the Proponent can use one of the defensive answers in only one of the O-variants. This can be formulated with a new structural rule:

> R5 (subdialogues): The Opponent may at a given stage μ of the dialogue ask for some unused formula which has been dismissed so far and which should be used according to definition 3. The Proponent will have to defend this challenge by asking the Opponent to develop all of his O-variants by means of subdialogues. Those formula of the initial dialogue which define the O-variant can be used in the subdialogue.

Example 3

	O			P	
				$((a\vee b)\wedge\neg a)\rightarrow b$	(0)
(1)	$(a\vee b)\wedge\neg a$	0√			
(3)	$a\vee b$	√	1	?L	(2)
(5)	a	√	3	?	(4)
(7)	$\neg a$	√	1	?R	(6)
			7	a	(8)

[4] A systematic formulation of subdialogues for relevant aims can be found in Rahman and Rückert ([10] and [11]) and Rückert ([12] and [13]).

At this stage, where the Opponent is going to lose he realises that round 1 is still open. O then asks for a response to 1 and P will ask O to open a subdialogue for the right-side variant of the defence to the attack 4. In this subdialogue P will show that *b* is not redundant because if the Opponent chooses *b* at move 5 he will answer the attack of move 1 and win. Thus the dialogue now looks as follows:

	O			P	
				$((a \vee b) \wedge \neg a)) \rightarrow b$	(0)
(1)	$(a \vee b) \wedge \neg a$	0√		b	(12)
(3)	$a \vee b$	√	1	?L	(2)
(5)	a	√	3	?	(4)
(7)	$\neg a$	√	1	?R	(6)
	⊗		7	a	(8)
(9)	?1	√			
			3	?b	(10)

O	P
(11) b	

At move 9 the Opponent asks for the round opened at move 1. The Proponent counterattacks by asking for the O-variant *b* in the defence of the disjunction. After this counterattack the Proponent uses *b* to close the opened round. Actually we could rewrite all the variant again but we dispense with doing so by allowing jumping from the subdialogue to the initial dialogue.

Notice that a clever Proponent will always try to play classically because the adaptive rules restrict his own moves and not the moves of the Opponent. Thus, the Opponent will try to state an inconsistency as fast as he can in order to play paraconsistently. For example, in the dialogue for $((\neg a \vee b) \wedge a) \rightarrow a$ the Proponent will not try to produce the O-inconsistency $\neg a$, a. He will instead attack the right part of the conjunction obtaining *a* and then he will immediately close the first round using precisely this *a* and win.

2.2 Quantifiers and Ontological Adaptivity

The quantifiers do not in this approach present any special problems in relation to inconsistencies. But there are some interesting considerations about the ontological status of inconsistencies. Actually there are two main interpretations of paraconsistency possible. The one, called the *compelling interpretation*, stresses that paraconsistent theories are ontologically committed to inconsistent objects. The other, called the *permissive interpretation*, does not assume this ontological commitment of paraconsistent theories. Da Costa formulates this latter approach in the following way:

> *I suggest to address the inconsistency issue differently. We may well explore the rich representational devices allowed by the use of paraconsistency in inconsistent domains, but withholding any claim to the effect that there are 'inconsistent objects' in reality.* (Da Costa [14], 33).
> *This allows for the accommodation of inconsistency by acknowledging that it is not a permanent feature of reality to which theories must correspond, but is rather a temporary aspect of such theories* [...]. *In this view, to accept a theory is to be committed, not to believing it to be true per se, but to holding it as if it were true, for the purposes of further elaboration, development and investigation.* (Da Costa, Bueno and French [15], 616-617).

In Rahman ([4] and [5]) a logic was developed in which the temporary acceptance of inconsistencies will be abandoned as soon as the singular terms occurring in the elementary propositions producing these inconsistencies are ontologically committed. In other words, the Proponent concedes the Opponent's inconsistency $\boldsymbol{A}_\tau \wedge \neg \boldsymbol{A}_\tau$ iff the constant τ carries no ontological commitment. Actually we are not only differentiating between explosive and non-explosive inconsistencies: we propose here to distinguish between inconsistencies which are ontologically committing and those which are not. This distinction can be considered in the context of Da Costa's epistemology of quasi truths. The idea is the following. It could well be that, say, B_τ has somehow been verified in reality and thus we commit ontologically with it and reject $\neg B_\tau$ But it could also happen that there is no verification either of A_τ or of $\neg A_\tau$ and according to our theory it seems we should concede A_τ and $\neg A_\tau$ In this case, though we will concede it, we will not take this concession as being ontologically committed.

The introduction of ontologically committing quantifiers can be achieved through the following definition and a new structural rule:

DEFINITION 6: A constant τ is said to be *introduced with ontological commitment by* X if

(1) X states a formula $A[\tau/x]$ to defend $\vee_x A$ or

(2) X attacks a formula $\wedge_x A$ with $?\vee_{/\tau}$, and τ has not been used in the same way before.

R6 (formal introduction of constants): P may not introduce constants with ontological commitment: any such constant must first be introduced by O according to definition 6. (cf. [16])

The intuitive idea behind the ontologically committing quantifiers should be clear: The Proponent is allowed to use a constant for a defence (of an existential

quantifier) or an attack (on a universal quantifier) iff the Opponent has already conceded that this constant has ontological commitment by an attack (on a universal quantifier) or by a defence (of an existential quantifier).
This logic also contains the quantifiers $\exists$ and $\forall$ for which neither definition 6 nor R6 hold. Since the quantifiers of $\exists$ and $\forall$ are ruled by the standard and structural rules it looks like these quantifiers work as in the standard logic. Actually they are very different: here they work as quantifiers *without any ontological commitment at all.* In the standard logic the ontological commitment is presupposed by the use of these quantifiers. Here this is not the case.

Our adaptive rule for quantified adaptive paraconsistent logic is the following:

> R7 (ontological adaptivity): If an O-inconsistency occurs and its constants were not introduced according to definition 6 and R6, replace these quantifiers with $\exists$ and $\forall$ and run an adaptive dialogue as described in the definitions 1 to 5. Proceed classically otherwise.

Example 4

	O			P	
				$(\bigwedge x((Ax \vee Bx) \wedge \neg Ax)) \rightarrow \bigvee x(\neg\neg Bx \vee Cx)$	(0)
(1)	$\bigwedge x((Ax \vee Bx) \wedge \neg Ax)$	0√		$\bigvee x(\neg\neg Bx \vee Cx)$	(2)
(3)	?	2√			
(5)	$(A\tau \vee B\tau) \wedge \neg A\tau$		1	$?\tau$	(4)
(7)	$A\tau \vee B\tau$		5	?L	(6)
(9)	$A\tau$		7	?	(8)
(11)	$\neg A\tau$		5	?R	(10)

At moves 9 and 11 an O-inconsistency occurs. The Proponent will concede the inconsistency and thus, run an adaptive dialogue, but he will not take it as ontologically committing:

	O			P	
				$(\forall x((Ax \vee Bx) \wedge \neg Ax))) \rightarrow \exists x (\neg\neg Bx \vee Cx)$	(0)
(1)	$\forall x((Ax \vee Bx) \wedge \neg Ax))$	0√		$\exists x (\neg\neg Bx \vee Cx)$	(2)
(3)	?	2√		$\neg\neg B\tau \vee C\tau$	(16)
(5)	$(A\tau \vee B\tau) \wedge \neg A\tau$		1	$?\tau$	(4)
(7)	$A\tau \vee B\tau$		5	?L	(6)
(9)	$A\tau$		7	?	(8)
(11)	$\neg A\tau$		5	?R	(10)
	$\otimes$		7	$A\tau$	(12)
(13)	?3	√			
			7	$? B\tau$	(14)

O			P		
(15)	$B\tau$	√			
17)	?	16√		$\neg\neg B\tau$	(18)
(19)	$\neg B\tau$	18√		⊗	
	⊗		19	$B\tau$	(20)
The Proponent wins					

Now it could well be that we do not want to replace every quantifier automatically. It could be the case that some formulae in which these quantifiers occur are known to have an empirical verification and thus should be read as ontologically committing. In fact, the different quantifiers allow us to think about different possible combinations between types of quantifiers. We could, as already mentioned, study the possibility that not all of the quantifiers lose their ontological commitment. This would allow one part to have ontological commitment and the other not. In this case we could, for example, proceed with an anti Meinongian point of view and give priority to the existent. That is, we could understand that if one part of the inconsistency carries ontological commitment the inconsistency cannot hold anymore because this part will be considered to be verified and we should proceed classically. This can be made explicit by copying the quantifiers with ontological commitment in the P-column at the start of the dialogue:

Example 5
In this example we will not go into the process of replacing. We will instead replace at the start what has not been conceded as ontologically committing:

O			P		
				[Not committing quantifer: $\wedge x\neg(Ax\wedge Bx)$]	
				$\forall x(\neg Ax\wedge Ax)\rightarrow \wedge x\neg(Ax\wedge Bx)$	(0)
(1)	$\forall x(\neg Ax\wedge Ax)$	0√		$\wedge x\neg(Ax\wedge Bx)$	(2)
(3)	$?\tau$	2√		$\neg(A\tau\wedge B\tau)$	(4)
(5)	$A\tau\wedge B\tau$	4√		⊗	
(7)	$A\tau$	√	5	?L	(6)
(9)	$\neg A\tau\wedge A\tau$√		1	$?\tau$	(8)
(11)	$\neg A\tau$	√	9	?L	(10)
	⊗		11	$A\tau$	(12)
The Proponent wins					

Another possibility due to be explored is to introduce many quantifiers with different grades of ontological commitment or verification. Furthermore it could be that we have a hierarchy of quantifiers corresponding to grades of confirmation. We could in this case require that if a constant of level ν has been introduced with the help of quantifiers of level μ<ν then this constant should be considered as having ontological commitment in μ. More precisely, think of the first pair of quantifiers as having the upper index 0 and add new pairs of quantifiers with higher indices, as many as we need to express every type of (verified) reality (or fiction) that could possibly appear. The extended set of quantifiers requires a new notion of introduction.

- A constant τ is said to be *introduced with ontological commitment as belonging to the type* i iff it is used to attack a universal quantifier of type i or to defend an existential quantifier of type i and has not been used in the same way before.

- For each type of quantification the following rule holds: Constants may only be introduced by O.

Now the hierarchy:

- P may introduce a constant τ with ontological commitment on a level μ iff O has introduced τ in the same way on some level ν with ν<μ before.

We leave two examples as an exercise for the reader:

1. $(\vee^1{}_x\boldsymbol{A}_x \vee \vee_x{}^2\boldsymbol{A}_x) \rightarrow \exists_x\boldsymbol{A}_x$
2. $(\vee^1{}_x\boldsymbol{A}_x \wedge \wedge_x{}^2(\boldsymbol{A}_x \rightarrow \boldsymbol{B}_x)) \rightarrow \vee^1{}_x\boldsymbol{B}_x$

Thus, in our paraconsistent context we could think of applying the following rule:

- if an O-inconsistency occurs at level μ, check first whether the corresponding constants have been introduced at level ν<μ.

If constants have been introduced in ν two procedures are possible:

1. stay at level μ, ignore this, and use an adaptive paraconsistency,
2. stay at level μ, proceed classically for the inconsistency with constants introduced at level ν but paraconsistently with the rest of the inconsistencies of level μ. After doing so jump to another level.

The jumps might be regulated by subdialogues. The choice between procedures 1 and 2 could be part of the agreements either at the very start or, more dynamically, during the dialogue. Here many details have to be fixed, for example: what happens if one constant is ontologically committed but for a given predicate different from those defining the O-inconsistency? We will leave this and other related problems for a future research where we will study the combination of the hierarchy of quantifiers with Newton da Costa's theory of partial truth.

Now although we believe that all this is congenial to the spirit of adaptive logic our approach does not seem to agree exactly with the present adaptive systems developed in Gent. The first difference is that adaptive logic concentrates mainly on how to derive from a given set of formulae. We can emulate this by considering the set of premises as a set of concessions at the very start of the dialogue. Such dialogues are known in the literature as *Hypotheses-dialogues* (*H-dialogues* for short). The approach we have presented above seems to agree more with Joke Meheus' system ANA ([17]), as remarked in a personal e-mail by Diderik Batens, for example in that q does but r does not follow from the H-dialogue with initial concessions $\{p, \neg p, p\vee q\}$ – or in the language of formulae of the standard dialogues $((a\vee b)\wedge a\wedge\neg a)\rightarrow b$ holds but $((a\vee b)\wedge a\wedge\neg a)\rightarrow r$ does not. In the first case the O-inconsistency produced by the conjunction $a\wedge\neg a$ can be dismissed without hindering the defence b. In the second case the round which should be closed with r will remain for ever open. The precise relations between our approach and Joke Meheus' should be studied thoroughly in a future research, but nevertheless we claim that our formulations deserve the adjective *adaptive*.

We will now describe the tableau system which serves as a basis for checking validity.

3 WINNING STRATEGIES

As already mentioned, validity is defined in dialogical logic via winning strategies of P, i.e. the thesis A is logically valid iff P can succeed in defending A against all possible allowed criticism by O. In this case, P has a *winning strategy* for A. A systematic description of the winning strategies available can be obtained from the following considerations:

If P is to win against any choice of O, we will have to consider two main different situations, namely the dialogical situations in which O has stated a (complex) formula and those in which P has stated a (complex) formula. We call these main situations the O-cases and the P-cases respectively.

In both of these situations another distinction has to be examined:

1. P wins by *choosing* an attack in the O-cases or a defence in the P-cases, iff he can win *at least one* of the dialogues he can choose.
2. When O can *choose* a defence in the O-cases or an attack in the P-cases, P can win iff he can win *all of the* dialogues O can choose.

The closing rules for dialogical tableaux are the usual ones: a branch is closed iff it contains two copies of the same atomic formula, one stated by O and the other one by P. A tableau for (P)*A* (i.e. starting with (P)*A*) is closed iff each branch is closed. This shows that strategy systems for classical and intuitionistic logic are nothing other than the very well known tableau systems for these logics.

For the intuitionistic tableau system, the structural rule about the restriction on defences has to be considered. The idea is quite simple: the tableau system allows all the possible defences (even the atomic ones) to be written down, but as soon as determinate formulae (negations, conditionals, universal quantifiers) of P are attacked all other P-formulae will be deleted — this is an implementation of the structural rule R_I4 for intuitionistic logic. Clearly, if an attack on a P-statement causes the deletion of the others, then P can only answer the last attack. Those formulae which compel the rest of P's formulae to be deleted will be indicated with the expression "$\Sigma_{[O]}$Ó which reads: in the set Σ save O's formulae and delete all of P's formulae stated before.

3.1 Classical Tableaux

(O)-*Cases*	(P)-*Cases*
$\Sigma, (O)A\vee B$ -------------------- $\Sigma, <(P)?>(O)A \mid \Sigma, <(P)?>(O)B$	$\Sigma, (P)A\vee B$ -------------------- $\Sigma, <(O)?>(P)A$ $\Sigma, <(O)?>(P)B$
$\Sigma, (O)A\wedge B$ -------------------- $\Sigma, <(P)?L>(O)A$ $\Sigma, <(P)?R>(O)B$	$\Sigma, (P)A\wedge B$ -------------------- $\Sigma, <(O)?L>(P)A \mid \Sigma, <(O)?R>(P)B$
$\Sigma, (O)A\rightarrow B$ -------------------- $\Sigma, (P)A \ldots \mid <(P)A>(O)B$	$\Sigma, (P)A\rightarrow B$ -------------------- $\Sigma, (O)A; \Sigma,(P)B$
$\Sigma, (O)\neg A$ -------------------- $\Sigma, (P)A; \otimes$	$\Sigma, (P)\neg A$ -------------------- $\Sigma, (O)A; \otimes$
$\Sigma, (O)\wedge_x A$ -------------------- $\Sigma, <(P)?_\tau>(O)A_{[\tau/x]}$	$\Sigma, (P)\wedge_x A$ -------------------- $\Sigma, <(O)?_\tau>(P)A_{[\tau/x]}$ τ *is new*
$\Sigma, (O)\vee_x A$ -------------------- $\Sigma, <(P)?>(O)A_{[\tau/x]}$ τ *is new*	$\Sigma, (P)\vee_x A$ -------------------- $\Sigma, <(O)?>(P)A_{[\tau/x]}$

By a dialogically signed formula we mean (P)X or (O)X where X is a formula. If Σ is a set of dialogically signed formulae and X is a single dialogically signed formula, we will write Σ, X for $\Sigma \cup \{X\}$. Observe that the formulae below the line always represent pairs of attack and defence moves. In other words, they represent rounds. Note that the expressions between the symbols "<" and ">", such as <(P)?> or <(O)?> are moves – more precisely they are attacks – but not statements.

3.2 Intuitionistic Tableaux

(O)-*Cases*	(P)-*Cases*
Σ, (O)$A \vee B$ -------------------- Σ, <(P)?>(O)A | Σ, <(P)?>(O)B	Σ, (P)$A \vee B$ -------------------- Σ, <(O)?>(P)A Σ, <(O)?>(P)B
Σ, (O)$A \wedge B$ -------------------- Σ, <(P)?L>(O)A Σ, <(P)?R>(O)B	Σ, (P)$A \wedge B$ -------------------- Σ, <(O)?L>(P)A | Σ, <(O)?R>(P)B
Σ,(O)$A \rightarrow B$ -------------------- Σ, (P)A ... | <(P)A>(O)B	Σ, (P)$A \rightarrow B$ -------------------- $\Sigma_{[O]}$, (O)A; (P)B
Σ, (O)$\neg A$ -------------------- Σ, (P)A; $\otimes$	Σ, (P)$\neg A$ -------------------- $\Sigma_{[O]}$, (O)A; $\otimes$
Σ, (O)$\bigwedge_x A$ -------------------- Σ, <(P)$?_\tau$>(O)$A_{[\tau/x]}$	Σ, (P)$\bigwedge_x A$ -------------------- $\Sigma_{[O]}$, <(O)$?_\tau$>(P)$A_{[\tau/x]}$ *τ is new*
Σ, (O)$\bigvee_x A$ -------------------- Σ, <(P)?>(O)$A_{[\tau/x]}$ *τ is new*	Σ, (P)$\bigvee_x A$ -------------------- Σ, <(O)?>(P)$A_{[\tau/x]}$

Let us look at two examples, namely one for classical logic and one for intuitionistic logic. We use the tree shape of the tableau made popular by Smullyan ([18]) and omit the expressions between < and >:

Example 6

(P) $\bigwedge_x \neg P_x \rightarrow \neg P_\tau$
(O) $\bigwedge_x \neg P_x$
(P) $\neg P_\tau$
(O) P_τ
(O) $\neg P_\tau$
(P) P_τ
The tableau closes.

The following intuitionistic tableau makes use of the deletion rule:

Example 7

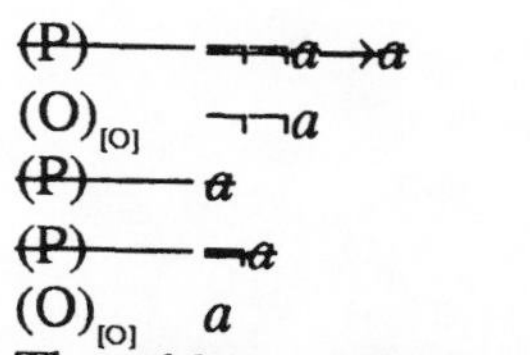

The tableau remains open.

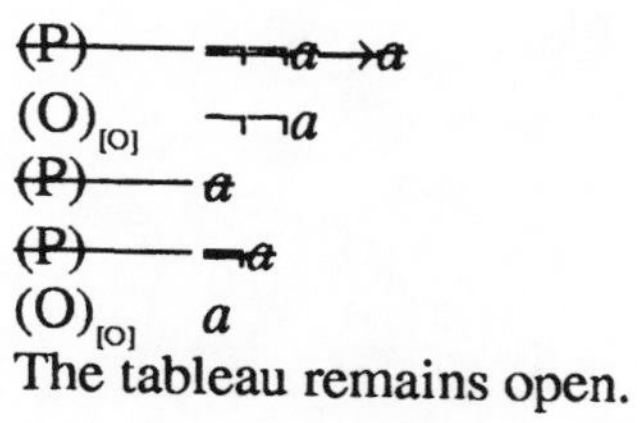

3.3 Winning Strategies for Adaptive Paraconsistency as Tableaux

In order to introduce a tableau system for dialogical adaptively paraconsistent logic which captures the idea of redundancy described above the rule for closing branches has to be reformulated.
For this aim we will use a notational system which makes use of the device of tagging those formulae which have been used in the sense of our definition of adaptive paraconsistency. More precisely:

Tagging Rule:

1. We will tag any O-atomic formula at least one occurrence of which has been used to close at least one branch (and delete the unused occurrences).
2. We will tag any O-conjunctive formula iff at least one of the conjuncts has been used in at least one branch (and delete the unused conjunct).
3. We will tag any O-complex formula other than a conjunction which has been used in at least one branch.
4. We will tag any P-atomic formula at least one occurrence of which has been used to close at least one branch.
5. We will tag any P-disjunctive formula iff at least one of the disjuncts has been used in one branch (and delete the unused occurrences).

6. We will tag any P-complex formula other than a disjunction which has been used in at least one branch.

Now the instructions for developing an adaptive tableau:

- If it is possible to close the tableau without producing an inconsistency of the forms described in definition 6 use the standard closing rules. If an inconsistency occurs develop the tableau completely, tag the formulae according to the tagging rule and apply the following adaptive rule.

- A *tableau* for (P)*A* (i.e. starting with (P)*A*) is *adaptively closed* iff each branch is closed by the standard closing rule and there are no untagged formulae.

Example 8

i)	$(P)(a\wedge\neg a)\rightarrow\neg(a\wedge b)$
ii)	$(O)(a\wedge\neg a)$
iii)	$(P)\neg(a\wedge b)$
iv)	$(O)a\wedge b$
v)	$<(P)?L>\ (O)a$
vi)	$<(P)?R>\ (O)\mathrm{b}$
vii)	$<(P)?R>\ (O)\neg a$

First we use the standard rules and develop the tableau. In doing so we realise that lines v and vii state an inconsistency. We then develop an adaptive strategy applying the tagging rule. In doing so the Proponent deletes the redundant formula *b* of the conjunction at move iv and the redundant conjunct *a* of the conjunction at move ii. The Proponent then uses the resulting atomic formulae to close the branch and the tree:

i)	$(P)(a\wedge\neg a)\rightarrow\neg(a\wedge b)$	√	
ii)	$(O)(a\wedge\neg a)$	√	
iii)	$(P)\neg(a\wedge b)$	√	
iv)	$(O)a\wedge b$		√
v)	$<(P)?L>\ (O)a$		√
vi)	$<(P)?R>\ (O)\neg a$		√
vii)	$(P)a$		√

Example 9

i)	$(P)((a\vee b)\wedge\neg a)\rightarrow(\neg\neg b\vee c)$	√
ii)	$(O)((a\vee b)\wedge\neg a)$	√
iii)	$(P)\ \neg\neg b\vee c$	√
iv)	$<(O)?>\ (P)\neg\neg b$	√
v)	$<(P)L?>(O)(a\vee b)$	√

vi)	<(P)R?> (O)¬a)				√
vii)	<(P)?>(O)a	√	viii)	<(P)?>(O)b	√
ix)	(P)a	√	x)	(O)¬b	√
			xi)	(P)b	√

Each branch is closed. Moreover every formula has been tagged. Thus, everything that should be used has been used and the tableau is adaptively closed.

As far as quantification is concerned we will only suggest how to produce tableau systems with ontologically committed quantifiers and leave the rest of the work to the reader:

Rewrite the rules for the quantifiers in the following way:

(O)-*Cases*	(P)-*Cases*
Σ, (O)$\wedge^{i}_{x}A$ -------------------- Σ, <(P)?$_{\tau*}$>(O)$A_{[\tau/x]}$ *τ has been labelled with an asterisk before*	Σ, (P)$\wedge^{i}_{x}A$ -------------------- $\Sigma_{[O]}$, <(O)?$_{\tau*}$>(P)$A_{[\tau*/x]}$ *τ is new*
Σ, (O)$\vee^{i}_{x}A$ -------------------- Σ, <(P)?>(O)$A_{[\tau*/x]}$ *τ is new*	Σ, (P)$\vee^{i}_{x}A$ -------------------- Σ, <(O)?>(P)$A_{[\tau*/x]}$ *τ has been labelled with an asterisk before*

DEFINITION 7: A constant τ is said to be *introduced with ontological commitment belonging to the type* i iff it has been labelled with an asterisk by the use of quantifiers of the type i.

4 FINAL REMARKS

As suggested in the text, several logics which we did not describe may be introduced by changing the notions of redundancy, namely, not allowing any occurrence of an atomic formula to be left unused, not allowing any O-formula to be left unused and not allowing any P-formula to be left unused. This could be combined with the option literal versus complex inconsistencies. Several others result from combining these logics, with the idea of ontological adaptivity.

Neither did we explore how to combine the logical way of eliminating inconsistencies by redundancy with a material strategy for establishing preferences. We will leave this for future research.

5 APPENDIX: A BRIEF INTRODUCTION TO DIALOGICAL LOGIC

Dialogical logic, suggested by Paul Lorenzen in 1958 and developed by Kuno Lorenz in several papers from 1961 onwards,[5] was introduced as a pragmatic semantics for both classical and intuitionistic logic.

The dialogical approach studies logic as an inherently pragmatic notion using an overtly externalised argumentation formulated as a *dialogue* between two parties taking up the roles of an *Opponent* (O in the following) and a *Proponent* (P) of the issue at stake, called the principal *thesis* of the dialogue. P has to try to defend the thesis against all possible allowed criticism (*attacks*) by O, thereby being allowed to use statements that O may have made at the outset of the dialogue. The thesis *A* is logically valid if and only if P can succeed in defending *A* against all possible allowed criticism by O. In the jargon of game theory: P has a *winning strategy* for *A*. We will now describe an intuitionistic and a classical dialogical logic.

Suppose the elements and the logical constants of first-order language are given with small italic letters (*a*, *b*, *c*, ...) for elementary formulae, capital italic letters for formulae that might be complex (*A*, *B*, *C*, ...), capital italic bold letters (***P***, ***Q***, ***R***, ...) for predicators and τ_i for constants. A dialogue is a sequence of formulae of this first-order language that are stated by either P or O.[6] Every move — with the exception of the first move through which the Proponent states the thesis — is an aggressive or a defensive act. In dialogical logic the meaning in use of the logical particles is given by two types of rules which determine their *local* and their *global* meaning (*particle* and *structural rules* respectively).

The particle rules specify for each particle a pair of moves consisting of an attack and (if possible) the corresponding defence. Each such pair is called a *round*. A round is *opened* by an attack and is *closed* by a defence if one is possible.

---

[5] Cf. [19]. Further work has been done for example by Rahman ([20]).

[6] Sometimes we use **X** and **Y** to denote **P** and **O** with **X**≠**Y**.

PARTICLE RULES

$\neg, \wedge, \vee, \rightarrow, \bigwedge, \bigvee$	ATTACK	DEFENCE
$\neg A$	A	⊗ *(The symbol '⊗' indicates that no defence, but only counter-attack is allowed)*
$A \wedge B$	?L ------ ?R *(The attacker chooses)*	A ------ B
$A \vee B$	?	A ------ B *(The defender chooses)*
$A \rightarrow B$	A	B
$\bigwedge_x A$	$?_\tau$ *(The attacker chooses)*	$A[\tau/x]$
$\bigvee_x A$	?	$A[\tau/x]$ *(The defender chooses)*

The first column contains the form of the formula in question, the second one possible attacks against this formula, and the last one possible defences against those attacks. (The symbol " ⊗" indicates that no defence is possible.) Note that for example "?L" is a move — more precisely it is an attack — but not a formula. Thus if one partner in the dialogue states a conjunction, the other may initiate the attack by asking either for the left-hand side of the conjunction ("show me that the left-hand side of the conjunction holds", or "?L" for short) or the right-hand side ("show me that the right-hand side of the conjunction holds", or "?R"). If, on the other hand, one partner in the dialogue states a disjunction, the other may initiate the attack by asking to be shown *any* side of the disjunction ("?").

Next, we fix the way formulae are sequenced to form dialogues with a set of structural rules (orig. *Rahmenregeln*):

R0 *(starting rule)*:

Moves are alternately uttered by P and O. The *initial formula* is uttered by P. It provides the topic of argument. Every move below the initial formula is either an attack or a defence against an earlier move stated by the other player.

R1 *(formal rule for atomic formulae)*:
P may not introduce atomic formulae: any atomic formula must be stated by O first. Atomic formulae can not be attacked.

R_I2 *(intuitionistic rule)*:
In any move, each player may attack a (complex) formula asserted by his partner or he may defend himself against *the last not already defended* attack. Only the latest open attack may be answered: if it is X's turn at position ν and there are two open attacks μ, λ such that $\mu < \lambda < \nu$, then X may not defend against μ.[7]

These rules define an intuitionistic logic. To obtain the classical version simply replace R_I2 by the following rule:

R_C2 *(classical rule)*:
In any move, each player may attack a (complex) formula asserted by his partner or he may defend himself against *any attack* (including those which have already been defended).

Before stating the next rule we need the following definition (the observation about *dialogical contexts* in the following definition is only relevant for relevance logic, modal logic and linear logic with exponentials):

DEFINITION 8: We speak of the strict repetition of an attack iff

a. A move is being attacked although the same move (from the same dialogical context) has already been attacked with the same attack before (notice that ?L and ?R are in this context different attacks).
   In the case of moves where a universal quantifier has been attacked with a new constant, the following type of move has to be added to the list of strict repetitions:
b. A universal-quantifier-move is being attacked using a new constant, although the same move (from the same dialogical context) has already been attacked before with a constant which was new at the time of that attack.

DEFINITION 9: We speak of the strict repetition of a defence iff

c. An aggressive move (=attack) ν which has been already defended with the defensive move μ (=defence) before, is being defended from the challenge at ν once more with the same defensive formula (notice that the left part and the right part of a disjunction are in this context two different defences)

[7] Notice that this does not mean that the last open attack was the last move.

In the case of moves where an existential quantifier has been defended with a new constant, the following type of move has to be added to the list of strict repetitions:

d. An attack on an existential quantifier is being defended using a new constant although the same quantifier (from the same dialogical context) has already been defended before with a constant which was new at the time of that defence.

R3 *(no delaying tactics rule)*:

While playing with the classical structural rule (see R_C2) P may perform a strict repetition of a defence stating *a* (atomic) twice (or more) if and only if O has conceded *a* twice (or more). No other strict repetitions are allowed.

While playing with the intuitionistic structural rule P may perform a strict repetition of an attack (see R_I2) if and only if O has introduced a new atomic formula (see R1 below) which can now be used by P (or iff O has introduced a new dialogical context which is now accessible for P).

(Notice that according to the definitions leading to this rule neither the new defence of a existential quantifier nor a new attack on a universal quantifier using a constant different (but not new) from the one used in the first defence (in the first attack) represents a strict repetition)

R4 *(winning rule)*:

X wins iff it is Y's turn but he cannot move (whether to attack or defend).

As already mentioned, validity is defined in dialogical logic via winning strategies of P:

DEFINITION 10: Validity:

In a certain dialogical system a formula is said to be valid iff P has a (formal) winning strategy for it, i.e. P can in accordance with the appropriate rules succeed in defending *A* against all possible allowed criticism by O.[8]

It is possible to build tableau systems for winning strategies, which correspond to the well-known semantic tableau methods. We present them in the text.[9]

Example 10 (either with the classical or the intuitionistic structural rule: it makes no difference):

---

[8] See consistency and completeness theorems in Krabbe ([21]) and Rahman ([20]).

[9] See more details on how to build tableau systems from dialogues in Rahman ([20]), Rahman and Rückert ([22]). Alternative systems with the corresponding proofs have been given by Felscher ([23]) and by Krabbe ([21]).

	O			P	
				$((a\to b)\wedge a)\to b$	(0)
(1)	$(a\to b)\wedge a$	0		b	(8)
(3)	$a\to b$		1	?L	(2)
(5)	a		1	?R	(4)
(7)	b		3	a	(6)
	P wins				

Example 12 (classical):

	O		P	
			$\bigwedge_x(P_x\vee\neg P_x)$	(0)
(1)	$?_\tau$	0	$P_\tau\vee\neg P_\tau$(2)	
(3)	?	2	$\neg P_\tau$	(4)
(5)	P_τ	4	⊗	
(3')	?	2	P_τ	(6)
	P wins			

Remarks:
Notation: Moves are labelled in (chronological) order of appearance. They are not listed in the order of utterance, but in such a way that every defence appears on the same level as the corresponding attack. Thus, the order of the moves is labelled by a number between brackets. Numbers without brackets indicate which move is being attacked.

Example 12 shows how the classical structural rule works: the Proponent may, according to the classical structural rules, defend an attack which was not the last one. This allows the Proponent to state P_τ in move (6). For notational reasons we repeated the attack of the Opponent, but actually this move does not take place. That is why, instead of tagging the attack with a new number, we repeated the number of the first attack and added an apostrophe.

The quite simple structure of the dialogue in this and the following examples should make it possible to recognise with the help of only one dialogue whether P has a winning strategy or not.

The next rule shows how the non-delaying rule works in classical logic:[10]

Example 13 (classical):

	O		P	
			$((a\to b)\to a)\to a$	(0)
(1)	$(a\to b)\to a$	0	a/a	(4)/(6)
(3)	a		$a\to b$	(2)
(5)	a	2		
	P wins			

[10] João Marcos (Unicamp), used this example in a seminar offered by Rahman at the Unicamp in order to show that earlier versions of the non-delaying rule should be reformulated with more precision.

Here, the clever Opponent, who does not give up so easily, after the Proponents s defence 4, attacks move 2 stating *a* once more. But, the Opponent's twofold use of *a* allows the Proponent to repeat his defence.

The next two examples deal with dialogues with hypotheses. In these dialogues the Proponent states his thesis under the condition of some hypotheses. Any such hypothesis is to be included as a concession of the Opponent at the very start of the dialogue. The hypotheses will be formulated with schematic letters and the Proponent can at any stage of the dialogue make use of these hypotheses asking first for an adequate instantiation (chosen by the Proponent) of the according schematic letters.[11] In the first of the next two examples we show how to obtain tertium non datur from an adequate instantiation of Peirce's Law:

Example 14 (intuitionistic):

O			P		
H: $((\mathfrak{R}\to\mathfrak{I})\to\mathfrak{R})\to\mathfrak{R}$				$a\vee\neg a$	(0)
(1)	?	0			
(3)	$(((a\vee\neg a)\to\neg(a\vee\neg a))\to(a\vee\neg a))\to(a\vee\neg a)$		H:	? $(a\vee\neg a)/\mathfrak{R}$, $\neg(a\vee\neg a)/\mathfrak{I}$	(2)
			3	$((a\vee\neg a)\to\neg(a\vee\neg a))\to(a\vee\neg a)$	(4)
(5)	$(a\vee\neg a)\to\neg(a\vee\neg a)$	4		$a\vee\neg a$	(6)
(7)	?	6	3	$\neg a$	(8)
(9)	a	8		⊗	
(11)	$\neg(a\vee\neg a)$		5	$a\vee\neg a$	(10)
	⊗		11	$a\vee\neg a$	(12)
(13)	?	12		a	(14)

P wins

With move 3, the Proponent asks for an adequate instantiation of the schematic letters. The reader can verify that if at move the Opponent instead of attacking move 4 he defends himself from move 4 he can loose even faster.

Our last example show how to obtain tertium non datur from an adequate instantiation of double negation:

[11] Actually in the case of dialogues with hypotheses we need to extend the non-delaying rule in such a way that unnecessary repetitions of instantiations should be avoided. This extension is not difficult if we think that an aggressive instantiation-move works in a similar way as an attack on a universal quantifier: the Proponent may repeat instantiation-attacks until all propositional variables occurring in the thesis have been used for an instantiation. We leave the details for the reader.

Example 15 (intuitionistic):

O			P		
H: $\neg\neg\mathfrak{R}\rightarrow\mathfrak{R}$				$a\vee\neg a$	(0)
(1)	?	0			
(3)	$\neg\neg(a\vee\neg a)\rightarrow(a\vee\neg a)$		H:	? $(a\vee\neg a)/\mathfrak{R}$	(2)
			3	$\neg\neg(a\vee\neg a)$	(4)
(5)	$\neg(a\vee\neg a)$	4		$\otimes$	
	$\otimes$		5	$a\vee\neg a$	(6)
(9)	?	6		$\neg a$	(8)
(11)	a	8		$\otimes$	
	$\otimes$		5	$a\vee\neg a$	(10)
(13)	?	10		a	(12)
	P wins				

Here the Proponent can repeat an attack on move 5 because the Opponent introduced a new atomic formula at move 11.

REFERENCES

[1] Rahman, S. and Carnielli, W.A. The Dialogical Approach to Paraconsistency. In D. Krause, editor, *The work of Newton da Costa*, special issue of *Synthese*, 2001. In press.

[2] Rahman, S. and Roetti, J. A. Dual intuitionistic paraconsistency without ontological commitments. In *Proceedings of the International Congress: Analytic Philosophy at the turn of the Millenium in Santiago de Compostela (Spain)*, 1-4 December 1999, pages 120- 126, 1999.

[3] Rahman, S. Ways of Understanding Hugh MacColl's Concept of Symbolic Existence. *Nordic Journal of Philosophical Logic*, volume 3 (1-2): 35-58, 1998.

[4] Rahman, S. Argumentieren mit Widersprüchen und Fiktionen. In K. Buchholz, S. Rahman and I. Weber, editors, pages 131-148, Campus, New York-Frankfurt a.M., 1999.

[5] Rahman, S. On Frege's Nightmare. In H. Wansing, editor, *Essays on non-classical logic*, King's College, London, 2001. In press.

[6] Van Bendegem, J. P. Paraconsistency and Dialogue Logic. Critical Examination and Further Explorations. In Rahman and Rückert, editors, 2000. In press.

[7] Batens, D. A survey of inconsistency-adaptive logics. In press, 2000.

[8] Urchs, M. Recent trends in paraconsistent logics. Konstanz, Technical Report of the DFG-Group *Logik in der Philosophie*, volume 46, 1999,

[9] Rahman, S. and Rückert, H. Dialogische Logik und Relevanz. *FR 5.1 Philosophie, Universität des Saarlandes*, volume 27 (Memo), 1998.

[10] Rahman, S. and Rückert H. Dialogical Modal Logic. *Logique et Analyse*, 2001, volume March 2001. In press.

[11] Rahman, S. and Rückert, H. Eine neue dialogische Semantik für die lineare Logik. In C.F. Gethmann, G. Kamp, *Neue Entwicklungen der Konstruktiven Logik*, 2001. In press.

[12] Rückert, H. Why Dialogical Logic? In H. Wansing, editor, *Essays on non-classical logic*, King's College: London, 2001. In press.

[13] Rückert, H. Dialogue games and connexive logic. In A. Baltag, M. Pauly, editor, *Workshop Logic and Games*, pages 35-42, ILLC, Amsterdam, 1999.

[14] Da Costa, N. C. A. Paraconsistent Logic. In *Stanislaw Jáskowski Memorial Symposium. Paraconsistent Logic, Logical Philosophy, Mathematics & Informatics at Torún,*, pages 29-35, 1998.

[15] Da Costa, N. C. A; Bueno, O. and French, S. The Logic of Pragmatic Truth. *Journal of Philosophical Logic*. volume 26: 603-620, 1998.

[16] Rahman, S., Rückert, H. and Fischmann, M.. *On Dialogues and Ontology. The dialogical approach to free logic. Logique et Analyse*, 1997, volume 160, 327-374.

[17] Meheus, J. Rich paraconsistent logics: the three-valued logic AN and the adaptive logic ANA that is based on it. In print.

[18] Smullyan, R.. *First-Order Logic*. Heidelberg, Springer, 1968.

[19] Lorenzen, P. and Lorenz, K. *Dialogische Logik*. Wissenschaftliche Buchgesellschaft, Darmstadt, 1978.

[20] Rahman, S. *Über Dialoge, protologische Kategorien und andere Seltenheiten*. Peter Lang, Frankfurt a. M.-NewYork-Paris, 1993.

[21] Krabbe, E. C. W. Formal Systems of Dialogue Rules. *Synthese*, volume 63(3): 295-328, 1985.

[22] Rahman, S. and Rückert, H. Die pragmatischen Sinn- und Geltungskriterien der Dialogischen Logik beim Beweis des Adjunktionssatzes. *Philosophia Scientiae*, volume 3(3): 145-170, 1999.

[23] Felscher, W. Dialogues, strategies and intuitionistic provability. *Annals of Pure and Applied Logic*, volume 28: 217-254, 1985.

[24] Rahman, S. and Rückert, H., editors. *New Perspectives in Dialogical Logic*. Special issue of *Synthese*, with contributions of P. Blackburn, D. Gabbay and J. Woods, J. Hintikka, E. Krabbe, K. Lorenz, U. Nortmann, H. Prakken, S. Rahman and H. Rückert, G. Sandu, J. P. Van Bendegem, G. Vreeswijk. Volume 127 (1-2), 2001. In press

An inconsistency-adaptive proof procedure for logic programming *

TIMOTHY VERMEIR Centre for Logic and Philosophy of Science, Ghent University
timothy@logica.rug.ac.be

Abstract
It is the goal of this paper to define a paraconsistent proof procedure that has the best of two mechanisms, in casu logic programming and inconsistency-adaptive logics. From logic programming we will maintain the ease of computing, and from adaptive logics their paraconsistency, dynamics and non-monotonicity. This will be done by combining the notion of competitor from logic programming with the conditionallity that is common in all adaptive proofs.

1 INTRODUCTION

The primary aim of this article is to show that adaptivity, as it is already known in some non-monotonic logics (most of them paraconsistent, cfr. Diderik Batens' [1], [2] (the oldest paper), [3] and others), is also applicable in other fields, in casu logic programming. Moreover, our claim is that moving to an (inconsistency-) adaptive strategy brings along many benefits for logic programming.

The adaptive strategy has proven to be a very useful tool in all areas where abnormalities make classical logic inappropriate (see for instance Guido Vanackere's *Ambiguity-adaptive logic* [4]). Central notions in such adaptive mechanisms are reliability and the conditional derivation of formulas. However, inconsistency-adaptive logics have a proof theory which is not that straightforward.[1]

It is my aim to devise a way of handling logic programs that is *paraconsistent, non-monotonic and dynamic.* The paraconsistency lies in the fact that an inconsistency does not lead to the derivation of everything, i.e. the *ex contradictione quodlibet* is abandoned. The system is non-monotonic according to the classical definition, but what is more, it is also dynamic, i.e. in one and the same proof

*I would like to thank Dirk Vermeir (VUB) for many fruitful discussions that lead to the marriage of adaptive logic and logic programming. Research for this paper was supported by the Fund for Scientific Research – Flanders.

[1]Two simplified proof theories have been proposed for inconsistency-adaptive logics, namely by Batens, De Clerq and Vanackere in [5] and by Batens and Vermeir in [6]. The latter is specifically for the so-called Rescher-Manor mechanisms.

there is a dynamics in the set of derived formulas. What at one stage is considered as a consequence of a program, can at a later stage be deleted again: as such, a dynamic proof procedure is capable of representing one's knowledge at a certain time (knowledge that can be updated when new information becomes available).

On the other hand, we want to incorporate some central notions of logic programming into the adaptive strategy. Hence, one will find (be it slightly modified) the notion of competitor (common in almost all non-monotonic extensions of logic programming and mechanisms for argumentation) central in the construction of the proof procedure.[2] Also, instead of a set of premises from which to reason, we start with a logic program, consisting of rules. Hence, the system we will construct here, does not have any derivation rules like ordinary logics, but solely depends on the program given. As such, the ease of proofs can be brought to adaptive logics.

The rest of this paper is organized as follows. In the next section, we shall introduce the basic ingredients of adaptive logic programming: in subsection 2.1 an informal introduction to inconsistency-adaptive logics, and adaptivity in general, is formulated. Subsection 2.2 then gives the basic definitions of logic programming (readers familiar with logic programming are nevertheless advised to read this subsection, since there is some slight deviation in terminology).

In section 3 we come to the essence of this article, namely the combination of adaptivity with logic programming. We here define the actual proof theory of our system *ALP*. At the end of this section, we list a few examples in order to shed some light on the theory. Some possible extensions (with respect to the types of negation allowed) of adaptive logic programming are summed up in section 4, together with a brief account on how these extensions can actually be integrated in the system constructed in the previous section. As is common, the last section (section 5) formulates a conclusion, and points towards directions for further research.

2 Basic definitions

2.1 Introducing adaptivity

Conditional derivation and reliability of formulas are central in inconsistency-adaptive logics. As such, I think both deserve some explanation.[3]

Conditional derivations occur when a derivation rule is used that is "sensitive" to inconsistencies, for instance disjunctive syllogism. As an example, consider $p \vee q$ and $\sim p$ as premises. It is clear that we can only derive q if p does not behave inconsistently (is not true together with it's negation). More formally, we know that if $\sim p$ is true, then two possibilities arise in a paraconsistent logic: either p is false (and hence p behaves consistently), or p is true also. In this last case, p behaves inconsistently, which is formalized by $p \& \sim p$. Hence, we can unconditionally derive, without many difficulties, $q \vee (p \& \sim p)$, which we write as $q_{\{p\}}$: "q is derived on the condition that p behaves consistently".

[2]The undercutting competitors of a rule, as defined here, are also central in the semantics for our system, which will be the subject for another paper.

[3]The introduction given here is a very informal one, and I gladly refer the interested reader to [1], among others, for the details.

A well formed formula (hereafter abbreviated as wff) A is said to be *reliable*[4] at a certain stage of a proof, if there is, at that stage, no evidence of A behaving inconsistently. The possible inconsistent behaviour of a formula A is formalized by it being a factor in a minimal (with respect to set-inclusion) DEK-consequence. A DEK-consequence is an unconditionally derived disjunction of existentially quantified contradictions:

$$\exists x(A_1x \& \sim A_1x) \vee \ldots \vee \exists x(A_nx \& \sim A_nx)$$

abbreviated as $DEK(A_1, \ldots, A_n)$ or $DEK(\Delta)$, where $\Delta = \{A_1, \ldots, A_n\}$. Each of the A_i is called a *factor* of the DEK-formula. When the DEK-consequence is minimal, we say that the consistent behaviour of each such A_i is *connected* to the consistent behaviour of the rest of the factors of the DEK-formula (see also definition 4.3 in [1]).

From an unconditionally derived minimal DEK-consequence we learn that at least one of the factors (in the above example at least one A_i where $1 \leq i \leq n$) has to behave inconsistently in order to make the DEK-formula true, but we do not know which. Hence, we call all factors "unreliable".

From the above paragraphs, it should be clear that adaptive logics are non-monotonic. E.g., from the premises $p \vee q$ and $\sim p$, one can derive q. However, if p were added to the premises, we can no longer derive q, not even conditionally, since the consistency of p is required for the derivation of q.

Also, these logics are *dynamic*: they present those formulas we can believe, *given the knowledge we have at that time.* If at a certain point in time (or stage) we know that a formula A holds, and that A and B can not hold together, we shall naturally conclude *at that time* that B does not hold. If we would, however, later learn that $\sim A$ holds (from an independent source for instance), then we shall have to revise *at that later stage* our conclusion B.

Furthermore, inconsistency-adaptive logics have a richer consequence set than ordinary paraconsistent logics: most paraconsistent logics do for instance not allow for the application of Disjunctive Syllogism.

2.2 Introducing adaptive logic programs

The basic entity of the language of logic programs is the *term*, which is either a variable or a constant. An *atomic formula*, or *atom*, is of the form $P(t_1, \ldots t_r)$,[5] where P is a predicate symbol of arity r ($r \geq 1$), and all t_i ($1 \leq i \leq r$) are terms. A *literal* is either an atom, or the negation of an atom. In this paper, we restrict ourselves to just one type of negation, namely the paraconsistent negation $\sim$ (we shall, towards the end of this paper, suggest a hint towards including classical negation $\neg$, and negation as failure *not* into adaptive logic programming). A term, atom or literal is *ground* iff it is variable-free.

A *rule* is of the form $\Delta \rightarrow \Gamma$ or $\Delta \Rightarrow \Gamma$, where Δ and Γ are sets of literals[6]. Rules of the latter form (with $\Rightarrow$) denote *defeasible* rules, while the $\rightarrow$-rules are said

[4]The term "reliable" is originally restricted to one specific inconsistency-adaptive logic, namely **ACLuN1** (see [1]). We use the term here in a wider context.

[5]Brackets shall be omitted where they are not necessary for readability.

[6]This is a deviation from traditional logic programming, where Δ and Γ are required to be sets of *atoms*.

to be ***strict***: they can not be defeated by other rules[7].

We also write, omitting the braces for ease, (where $n, m \geq 0$)

$$A_1, \ldots, A_n \rightarrow B_1, \ldots, B_m$$

for strict rules (which can be read as "if all A_i hold, then at least one of the B_i does too"), and

$$A_1, \ldots, A_n \Rightarrow B_1, \ldots, B_m$$

for defeasible ones. This last rule has the intuitive meaning "if all the A_i hold, then at least one of the B_j does too (unless something is wrong with one or more of the A_i)".[8]

REMARK 1 The difference between strict and defeasible rules is in the "real world" not as obvious as it is above. Nute [9] formulates it as follows: "The point is, of course, that we often conclude that something has or will happen because somebody tells us that it has or will happen. These conclusions are risky, and we often must abandon them in the light of definite evidence to the contrary." Defeasible rules are thus rules without certainty. Strict rules, on the other hand, can be those which represent definitions, classifications, lawlike rules from science[9] etcetera.

To use a classic example, it is like the difference between the rules "All penguins are birds" and "All birds fly": the first is definitely strict, while the latter is defeasible in the sense that what should be understood is "Typically, all birds fly". In general, a rule is defeasible if a counterexample is not unthinkable. Using the above example, it is not hard to come across a non-flying bird, whereas if we would discover a penguin that is not a bird, biology would be in serious trouble. As defeasible rules, we could also consider abduction, induction and default reasoning.

We denote the set $\{A_1, \ldots A_n\}$ as the ***body*** of the rule, and $\{B_1, \ldots, B_m\}$ as the ***head*** of the rule. Given a rule r, $H(r)$ and $B(r)$ denote the head and the body of the rule respectively. A rule is ***ground*** iff it is variable-free. A rule r is ***definite*** iff $H(r)$ is a singleton.[10]

Where α is a literal, by $\tilde{\alpha}$ we denote (a) $\sim\alpha$ if α is an atom, and (b) β if α is of the form $\sim\beta$ (where β is an atom). $\tilde{\alpha}$ is said to be the ***complement*** of α. Note that ***being the complement of*** is a symmetric relation.

By an ***adaptive logic program***, we will denote a set of (strict and/or defeasible), definite rules. A definite rule r for which $B(r) = \emptyset$ is said to denote a ***fact***. The ***set of facts*** of a program P is denoted $\mathcal{F}_P$ and is defined by

$$\{f \mid \exists r \in P \text{ such that } H(r) = \{f\} \text{ and } B(r) = \emptyset\}$$

i.e. the set of literals that are the head of a bodyless, definite rule.

---

[7] Donald Nute also makes a distinction between these two types of rules (cfr. [7], [8], [9] among others), but he uses the terms in a slightly different manner. Nevertheless, it were these and other articles by Nute that inspired the approach presented here.

[8] The formalization of the "something is wrong with" is given in the next section.

[9] However, several scientific "certainties" have in the past been proven not to be as certain as some may have thought.

[10] Again, this is a deviation from the traditional definition, where a rule is definite when its head contains exactly one *atom*.

Given a program P, the subsets $\{r|r \in P$ and r is defeasible$\}$ and $\{r|r \in P$ and r is strict$\}$ are the defeasible and strict part of P, respectively denoted $D(P)$ and $S(P)$. Clearly, $D(P) \cup S(P) = P$ and $D(P) \cap S(P) = \emptyset$.

By the ***Herbrand Universe*** of a program P, written as $\mathcal{H}_P$, we denote the set of all possible ground terms of P. The ***Herbrand Base*** of P, denoted $\mathcal{B}_P$ is the set of all possible ground atoms whose predicate symbols occur in P and whose arguments are elements of $\mathcal{H}_P$. By $\sim\mathcal{B}_P$, we denote the set $\{\sim\varphi|\varphi \in \mathcal{B}_P\}$. A ground instance of a rule r in P is a rule obtained from r by replacing every variable x occurring in r by $\psi(x)$, where ψ is a mapping from the set of all variables occurring in P to $\mathcal{H}_P$. By $ground(P)$ we denote the set of all ground instances of all rules in P. In the rest of this paper, we shall, where no confusion is possible, simply speak of ***program*** where we mean an adaptive logic program.

EXAMPLE 2 Let us, to clarify the terminology, consider a very simple example. Let P_1 be given by

$$
\begin{aligned}
P_1 \ = \{ \ & Px, Qx \rightarrow Rx, \\
& \sim Px \Rightarrow Qx, \\
& \rightarrow \sim Pa, \\
& \Rightarrow Rb, \\
& Rb \rightarrow Pa \\
\} &
\end{aligned}
$$

The Herbrand Universe of P is the set $\mathcal{H}_{P_1} = \{a, b\}$, and the Herbrand Base is the set $\mathcal{B}_{P_1} = \{Pa, Qa, Ra, Pb, Qb, Rb\}$. From this it is obvious that $ground(P_1)$ is the following program:

$$
\begin{aligned}
ground(P_1) \ = \{ \ & Pa, Qa \rightarrow Ra, \\
& Pb, Qb \rightarrow Rb, \\
& \sim Pa \Rightarrow Qa, \\
& \sim Pb \Rightarrow Qb, \\
& \rightarrow \sim Pa, \\
& \Rightarrow Rb, \\
& Rb \rightarrow Pa \\
\} &
\end{aligned}
$$

Furthermore, $D(P_1) = \{\sim Px \Rightarrow Qx, \ \Rightarrow Rb\}$, and $D(ground(P_1)) = \{\sim Pa \Rightarrow Qa, \ \sim Pb \Rightarrow Qb, \ \Rightarrow Rb\}$.

The defeasible rule $\sim Px \Rightarrow Qx$ can be understood as abbreviating the set of ground instances of it, and can hence intuitively be read as "for all $\alpha \in \mathcal{H}_{P_1}$, if $P\alpha$ holds, then so does $Q\alpha$, unless something is wrong with $P\alpha$." The reader can already see that something ***is*** wrong with $\sim Pa$, since also Pa is derivable. However, Pb does not seem to share this problem.

2.3 Domain restriction and a note on defeasible rules

The domain of logic programming is a large one, and hence, we need to restrict it. A first restriction is that we will only consider logic programs consisting solely of definite rules. Also, we do, for the moment, only allow literals built up of an atom and a paraconsistent negation. It seems obvious that the claims made here can be generalized in a trivial manner to programs with two kinds of negations.

Secondly, we restrict ourselves in this paper to the *proof procedure* of adaptive logic programming. Although the model-theoretic semantics can quite easily be defined through the Herbrand base of a program, this would lead us too far here. These semantics will be the subject of another paper.[11]

Before defining the *ALP* proof procedure, there is one remark to be made. Above, we distinguished between strict and defeasible rules. In an inconsistency-adaptive setting, the difference between the two, is like the difference between disjunctive syllogism and modus ponens. When $p \rightarrow q$ and $\rightarrow p$ are given, q follows, no matter what contradictions are derived. Hence, this is "MP-like". On the other hand, if the rule were to be defeasible, i.e. $p \Rightarrow q$, then the derivation of q is like in the case of disjunctive syllogism: the consistency of p is required to ensure the derivation of q.[12] The major difference with implications and disjunctions, is that rules are by definition only applicable in one direction: rules are not "contraposable". See below for further details.

3 The proof procedure of ALP

When dealing with inconsistencies, a classical approach within logic programming is to define the set of *competitors* of a rule r. Intuitively, $\hat{r}$ is a competitor of r iff $H(\hat{r}) = \neg H(r)$, i.e. when r and $\hat{r}$ have strict contradicting heads. This definition leads to proofs in which no inconsistencies can be derived. Here we do not use this classical notion of competitor, but define a similar set of ***undercutting competitors***. The main result is that, although inconsistencies are derivable, they do not lead to triviality.

DEFINITION 3 The **set of undercutting competitors** of a rule $r \in ground(P)$ is the set

$$Ucomp_P(r) \;=\; \{\hat{r} | \hat{r} \in ground(P) \text{ and } \exists \ell \in B(r) : \ \ell = \widetilde{H(\hat{r})}\}$$

We say that each such $\hat{r}$ **attacks** r, denoted $\hat{r} \leadsto r$.

Intuitively, $\hat{r}$ attacks r iff the application of $\hat{r}$ leads to the derivation of one or more literals contradicting $B(r)$, i.e. contradicting the antecedent of r. Note that, while the classical '*is-a-competitor-of*'-relation is symmetric, $\hat{r} \leadsto r$ does not necessarily imply $r \leadsto \hat{r}$.

EXAMPLE 4 Let us return to the example we started above (example 2). As a reminder, P_1 is the program consisting of the following rules

[11] A hint towards paraconsistent semantics of logic programs is already given in [10]. See also [11] and [12].

[12] Of course, for facts it does not matter whether the rule is strict or defeasible, since bodyless rules can not be defeated.

$$\begin{array}{lll} P_1 & = \{ & Px, Qx \rightarrow Rx, \\ & & {\sim}Px \Rightarrow Qx, \\ & & \rightarrow {\sim}Pa, \\ & & \Rightarrow Rb, \\ & & Rb \rightarrow Pa \\ & \} & \end{array}$$

From definition 3, it follows immediately that the only rule in $ground(P_1)$ that has an undercutting competitor, is ${\sim}Pa \Rightarrow Qa$, namely

$$Ucomp_{P_1}({\sim}Pa \Rightarrow Qa) \;=\; \{Rb \rightarrow Pa\}$$

EXAMPLE 5 Consider the very simple program $P_2 = \{p \Rightarrow q \;,\; {\sim}q \rightarrow r\}$. Here, clearly $(p \Rightarrow q) \;\rightsquigarrow\; ({\sim}q \rightarrow r)$. This only to show that rules need not be defeasible to be attacked, nor do the attackers need to be strict. A dwarf can attack a giant by all means, but that does not bring along that the giant shall be defeated, as we will see in a moment.

We will construct proofs from programs by means of directed, finite trees, consisting of nodes which are labeled by elements of $\mathcal{B}_P \cup {\sim}\mathcal{B}_P$. We denote by $[n]$ the labeling of node n.

DEFINITION 6 Let P be a logic program, $r \in ground(P)$, and T a finite, directed tree with nodes labeled with elements of $\mathcal{B}_P \cup {\sim}\mathcal{B}_P$ (i.e. the set $\{[n] | n \text{ is a node of } T\} \subseteq \mathcal{B}_P \cup {\sim}\mathcal{B}_P$). Nodes may be marked by means of a †. We say r is

- **applicable** in T iff there are nodes $n_1, \ldots, n_i$ such that $\{[n_1], \ldots, [n_i]\} = B(r)$. These nodes are said to **denote** $B(r)$.
- **applied** in T iff there is a node n such that $[n] = H(r)$.
- **defeated** in T iff r is defeasible, and there is a $\hat{r} \in Ucomp_P(r)$ such that $\hat{r}$ is applied in T and the node n, such that $[n] = H(\hat{r})$, is not marked.
- **cut off** in T iff there is a literal $b \in B(r)$ such that for all applicable rules r' with $H(r') = b$, r' is defeated or cut off in T.

DEFINITION 7 Let P be a program, and T a directed, finite tree with nodes labeled with elements of $\mathcal{B}_P \cup {\sim}\mathcal{B}_P$. Then let $\mathcal{A}$ be built up as follows:

- $\mathcal{A} = \langle \mathcal{A}_0, \ldots, \mathcal{A}_n \rangle$ be an $n+1$-tuple of lists of rules, where $n \geq 0$.
- for all i $(0 \leq i \leq n)$, $\mathcal{A}_i$ is a list of m^i (for all such i, $m^i \geq 1$) rules $r_{i,1}, \ldots, r_{i,m^i}$ such that $r_{i,1} \in D(ground(P))$ and for all j $(1 \leq j \leq m^i)$ $r_{i,j} \in ground(P)$.
- for all j $(1 \leq j < m^i)$, $H(r_{i,j}) \in B(r_{i,j+1})$.

Then, $\mathcal{A}$ is an **attack-cycle** iff for all i $(0 \le i \le n)$:

$$B(r_{i,1}) \to H(r_{i,m^i}) \rightsquigarrow B(r_{i\oplus 1,1}) \to H(r_{i\oplus 1,m^{i\oplus 1}})$$

where $\oplus$ is addition modulo $n+1$.
A rule $r \in \mathit{ground}(P)$ is said to be **involved in an attack-cycle** $\mathcal{A}$ iff there is an i, j such that $r = r_{i,j}$ (for simplicity we write $r \in \mathcal{A}$).
The **body of an attack-cycle** $\mathcal{A}$, abbreviated $B(\mathcal{A})$, is the union of all sets $B(r_{i,1})$, $0 \le i \le n$. By the **head of an attack-cycle** $\mathcal{A}$, denoted $H(\mathcal{A})$, we mean the set $\{H(r) | r \in \mathcal{A}\}$.
The attack-cycle $\mathcal{A}$ is said to be **applicable** in T iff (1) there is for every $r \in B(\mathcal{A})$ a set of unmarked nodes $\{n_1, \ldots, n_k\}$ such that $\{[n_1], \ldots, [n_k]\} = B(r)$, and (2) no other attack-relations occur between two rules involved in $\mathcal{A}$ other than those specified above. That is, the attack-cycle should be minimal: to be applicable the cycle $\mathcal{A}$ may not contain a sub-cycle which is an applicable attack-cycle (see also example 8).
The attack-cycle $\mathcal{A}$ is said to be **applied** in T iff $\mathcal{A}$ is applicable in T and, for every i $(0 \le i \le n)$, for every j $(1 \le j \le m^i)$ there are nodes denoting $B(r_{i,j})$, and there is a node denoting $H(r_{i,j})$ in T.

EXAMPLE 8 A simple example of an attack-cycle involves two rules of the form $p \Rightarrow \sim q$ and $q \Rightarrow \sim p$. These two attack each other, but the fight can not be won (which is the essence of the notion of attack-cycle). A more complicated example is

$$\begin{array}{lll} \mathcal{A}_1 & = \langle & \langle p \Rightarrow q;\ q \to \sim r \rangle, \\ & & \langle r \Rightarrow s \rangle, \\ & & \langle \sim s \Rightarrow t;\ u, t \to \sim p \rangle \\ & \rangle & \end{array}$$

The smallest attack-cycle possible only involves one rule, of the form $p \Rightarrow \sim p$. It should also be noted that a rule can be involved in more than one attack-cycle. Consider the cycle

$$\begin{array}{lll} \mathcal{A}_2 & = \langle & \langle p \Rightarrow q;\ q \Rightarrow r;\ r \Rightarrow s \rangle, \\ & & \langle \sim s \Rightarrow \sim r;\ \sim r \Rightarrow \sim q;\ \sim q \Rightarrow \sim p \rangle \\ & \rangle & \end{array}$$

Clearly, $\mathcal{A}_2$ can never be applied, since it is not applicable (independent of the program and the tree already available). One can easily see that there is a "sub-cycle" in $\mathcal{A}_2$ that is also an attack-cycle (involving r and q). So this particular cycle is not applicable since it does not meet the second condition in the definition of applicability of attack-cycles. We can say that $\mathcal{A}_2$ is *not minimal.*

Intuitively, an attack-cycle can be understood as a circle of sets of rules, such that each such set attacks its right neighbour. So, every set of rules itself attacks another set, but is at the same time itself under attack. This situation reminds us clearly of the problems dining philosophers tend to have.

We can in such a situation not have only one winner, since this would imply that the winners right neighbour is defeated, and hence the right neighbour of this defeated set of rules is no longer under attack. Furthermore, the choice of a sole winner is arbitrary, since there are no logical reasons to prefer one such set.[13] Also, we could be generous, and declare it a tie, with only winners. This will however lead to an unnatural situation in which several defeasible rules are attacked, but not defeated. Hence, we shall choose for a third option, namely a tie with only losers.

DEFINITION 9 Let P be a program, and T a finite, directed tree with nodes labeled with elements of $\mathcal{B}_P \cup \sim\mathcal{B}_P$. Then for all nodes n of the tree T, we say n is

- **undercut** in T by $r_1, \ldots, r_n$ iff for all such r_i: $H(r_i) = [n]$, r_i is defeated or cut off in T; and there is no $r \in \textit{ground}(P)$ such that $H(r) = [n]$ and r is applicable, not defeated, and not cut off in T.
- **deadlocked** in T iff there is an applied attack-cycle $\mathcal{A}$ such that (i) $[n] \in H(\mathcal{A})$, and (ii) $\forall \ell \in B(\mathcal{A}) \cup H(\mathcal{A})$, ℓ is not the head of a rule r that is undercut by rules not involved in $\mathcal{A}$.

A node n is thus undercut in a tree for a program P iff all paths between the facts of P, i.e. the set $\mathcal{F}(P)$ which is represented in the tree by the set of those nodes that do not have any ancestors, and n are "interrupted", meaning that each such path includes at least one edge representing a defeated rule. The node n is deadlocked in the cases where it is involved in an applied attack-cycle (see example 8).

DEFINITION 10 Let P be a program and T a finite, directed tree with nodes labeled with elements of $\mathcal{B}_P \cup \sim\mathcal{B}_P$. We say T is a **prooftree for** P iff

- if there is a rule $r \in \textit{ground}(P)$ such that r is applicable, then the nodes denoting $B(r)$ have a common child n such that $[n] = H(r)$.
- for all nodes n, n is marked iff n is undercut or deadlocked in T.

Intuitively, the above definition is a "tree-version" of traditional adaptive proofs, where lines are marked when they rely on unreliable formula's. Here, we mark nodes when they are the result of defeated rules. A prooftree for a program P gives us a clear view of what is going on in P: the edges show the derivations that can be made, and the markings tell us what is abnormal in P. We can define the consequences of a program as follows:

DEFINITION 11 Let P be a program, and $\varphi \in \mathcal{B}_P \cup \sim\mathcal{B}_P$. We say φ is an **ALP- consequence** of P iff there is a prooftree T for P containing an unmarked node n: $[n] = \varphi$. We write $P \vdash_{ALP} \varphi$. The **consequence set** of P is the set $Cn_{ALP}(P) = \{\varphi | P \vdash_{ALP} \varphi\}$.

Several things may be noted about the above definition. First of all, it is a *dynamic* proof-format. When constructing a prooftree for a program P, we constantly trust on the knowledge we have at the time, i.e. on the nodes already written. At a

[13] One could add extra-logical preferences—see for instance [13], [14], [15], [16] for some accounts on priorities and preferences in (non-monotonic) logic—to decide about winning and loosing. This is a possibility, but one we shall not investigate in this paper.

later stage, it can turn out that our knowledge was wrong, or incomplete, and certain nodes must be marked (indicating that they are not part of the consequences of P at that stage). Furthermore, markings are by no means definite (they are not deletions), and hence nodes can be "unmarked" at still a later stage (note the "iff" in the definition).

Secondly, the above proof procedure is definitely ***paraconsistent***, in the sense that a contradiction (which is formalized in the tree by two nodes with complementary labels) does not lead to triviality. Even more, the procedure is ***non-monotonic***, as the reader may easily verify.

Furthermore, the system incorporates the two central notions of adaptive proofs as they are mentioned above. The derivations are ***conditional*** since they depend on the ***reliability*** of all ancestor-nodes.

Apart from these properties from adaptive logics, the above system has also much in common with logic programming, as can easily be seen.

Let us look at some examples to illustrate the theory.

EXAMPLE 12 Let P_3 be the program

$$\begin{aligned} P_3 \; = \{ \; & \rightarrow s, \\ & s \rightarrow p, \\ & s \rightarrow q, \\ & p \rightarrow r, \\ & r \Rightarrow \sim q, \\ & q \Rightarrow t, \\ & t \rightarrow \sim r \\ \} & \end{aligned}$$

Clearly, there is a problem of inconsistency, namely concerning q and r. We can easily see that there is an attack-cycle in P_3:

$$\begin{aligned} \mathcal{A}_{P_3} \; = \langle \; & \langle r \Rightarrow \sim q \rangle, \\ & \langle q \Rightarrow t;\; t \rightarrow \sim r \rangle \\ \rangle & \end{aligned}$$

where, $H(\mathcal{A}_3) = \{\sim q, t, \sim r\}$ and $B(\mathcal{A}_3) = \{r, q\}$. The prooftree for P_3 is then given by figure 1. Note that if $\rightarrow \sim p$ would be added to P_3, then the attack-cycle $\mathcal{A}_{P_3}$ would not lead to any markings (since one of the literals would have been undercut by a not-involved rule).

EXAMPLE 13 Let P_4 be given by the following set of rules:

$$\begin{aligned} P_4 \; = \{ \; & \Rightarrow f \\ & f \Rightarrow p, \\ & f \Rightarrow r, \\ & f \Rightarrow s, \\ & s \Rightarrow \sim r, \\ & r \Rightarrow \sim p, \\ & p \Rightarrow q \\ \} & \end{aligned}$$

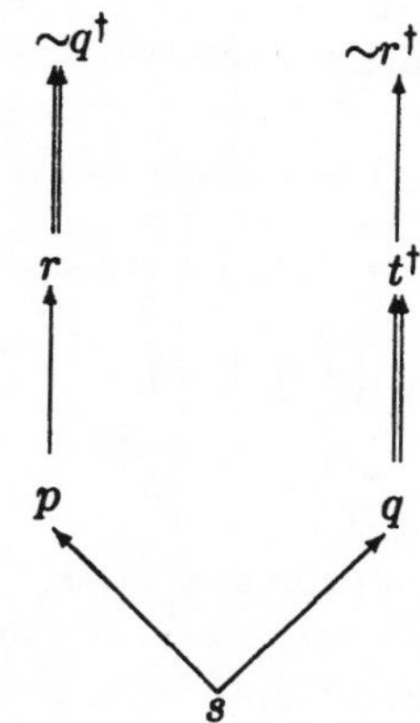

Figure 1: Prooftree of P_3

From the set of rules, we can easily see that $(s \Rightarrow \sim r) \rightsquigarrow (r \Rightarrow \sim p) \rightsquigarrow (p \Rightarrow q)$. This program will lead to the marking of the node labeled $\sim p$, since this is the only node that is the result of a rule that is undercut by a rule the head of which is not marked. The node labeled q will ***not*** be marked, since the competitor of $p \Rightarrow q$ is defeated. See figure 2.

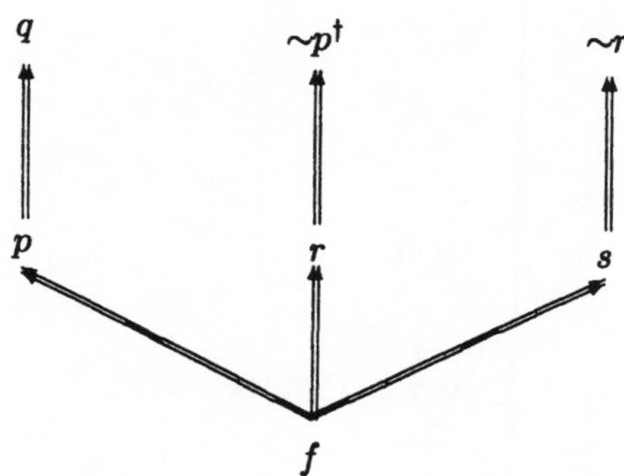

Figure 2: Prooftree of P_4

EXAMPLE 14 A third and last example. Let P_5 be given by

$$
\begin{aligned}
P_5 \; = \{ \; & \rightarrow p \\
& \rightarrow s \\
& p, s \rightarrow \sim r, \\
& p \rightarrow q, \\
& q \Rightarrow r, \\
& r \rightarrow \sim p, \\
& \sim r \rightarrow \sim q, \\
& \sim r, \sim q \rightarrow t, \\
\}
\end{aligned}
$$

At first glance it may appear that there are two attack-cycles in P_5, but this is not so. In order for

$$\langle\langle p \rightarrow q; q \Rightarrow r; r \rightarrow \sim p\rangle\rangle$$

to be an attack-cycle, the first rule must be defeasible, which it is not. For similar reasons

$$\begin{array}{ll}\langle & \langle q \Rightarrow r\rangle, \\ & \langle \sim r \rightarrow \sim q\rangle \\ \rangle & \end{array}$$

is not an attack-cycle. Hence, the prooftree for P_5 will lead to only two markings, namely of the nodes marked r and $\sim p$. The derivation of t on the other hand is not troublesome since only strict rules are applied to get from the set of facts $\mathcal{F}_{P_5}$ to the node labeled t. (Note that a line connecting two or more edges denotes a connected derivation, i.e. the two (or more) ancestor nodes *together* lead to the derivation of the child node.)

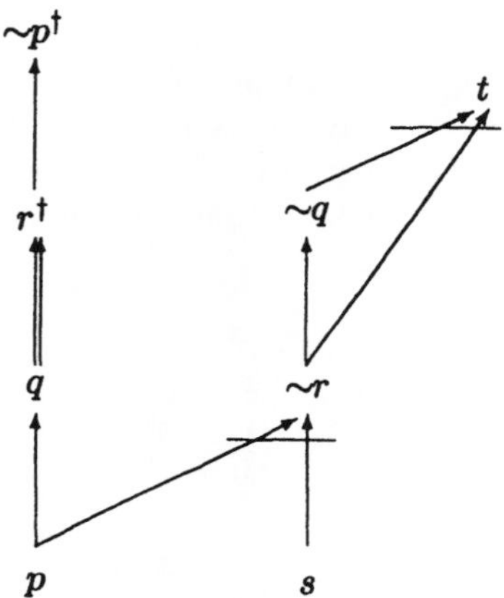

Figure 3: Prooftree of P_5

REMARK 15 At this point, we can shed some light on the relation between DEK-consequences, known from inconsistency-adaptive logics, and the attack-cycles presented in definition 7. Consider the premises a, b and $\sim a \vee \sim b$, which lead, when using an inconsistency-adaptive logic, to the DEK-consequences $DEK(a, b)$. If we were to write these three premises as a set of rules, the result would be $P = \{\rightarrow a;\ \rightarrow b;\ a \Rightarrow \sim b;\ b \Rightarrow \sim a\}$.[14] Clearly,

$$\begin{array}{lll}\mathcal{A} = & \langle & \langle a \Rightarrow \sim b\rangle, \\ & & \langle b \Rightarrow \sim a\rangle \\ & \rangle & \end{array}$$

is an applied attack-cycle in any prooftree of P. Note that, just as the disjunction $\sim a \vee \sim b$ is not troublesome in inconsistency-adaptive logics, so the attack-cycle itself

[14] More on transforming sets of premises into an equivalent set of rules can be found in [10].

is not of any worry when reasoning according to ALP. It is only when an attack-cycle is ***applied*** in a prooftree that trouble arises. The reader may easily verify that the applied attack-cycle in example 12 corresponds to the two DEK-consequences $DEK(q,r)$ and $DEK(q,t,r)$; that is, one DEK-consequence for each m^i-tuple of rules. It can be shown that this holds in general for all applied attack-cycles.

4 Extending ALP with two more types of negation

We have until now constructed a workable system, that combines inconsistency-adaptive logics with logic programming, with a relatively simple theory. However, in order to keep the above clean and simple, we made some restrictions, not incorporating the whole of our two starting points. From inconsistency-adaptive logics we left out classical negation, even though many benefits rise from its presence. And from logic programming we left negation as failure aside. Hence, one could argue that the claim made in the introduction of this article, that the aim is to combine the best of both worlds, is not achieved.

In this section I show that ALP is capable of incorporating both classical negation and negation as failure. However, this shall be done very briefly, for the details would lead us on a lengthy journey.

First, consider classical negation $\neg$, which is commonly defined as $\neg A \equiv (A \supset \bot)$ (where $\bot$ is the well-known ***falsum***).[15] Only having a paraconsistent negation is not always enough, for we can then not declare some formula simply as false (i.e. $\sim A$ does not guarantee us that A is false). Adding this second negation into ALP is not a problem. First note that if $\neg A$ holds, then so does $\sim A$, since it follows from "A is false" that "A is false or A behaves inconsistently".

We then have two types of complements in our system. Besides the "A is the complement of $\sim A$", we also have that A is the ***strong*** complement of $\neg A$ (and vice versa). Denote this by $A = \overline{\neg A}$. Naturally then, in view of the $\bot$-characterization of the strong negation, any proof in which both A and $\overline{A}$ appear, leads to triviality.[16]

A possibility for the proof theory, is to define a function ε on programs, in such a way that

$$\varepsilon(P) \;=\; P \cup \{\sim\alpha \Rightarrow \neg\alpha | \alpha \in \mathcal{B}_P\} \cup \{\neg\alpha \rightarrow \sim\alpha | \alpha \in \mathcal{B}_P\}$$

In words, $\varepsilon(P)$ is an extension of the program P with for every member of $\mathcal{B}_P$ a rule strictly deriving from the strict negation the paraconsistent one, and, in a defeasible manner, the strict from the paraconsistent negation. Proof trees are then constructed in a similar fashion as defined above, the only difference being that a tree containing two nodes n and m such that $[n] = \overline{[m]}$ implies triviality, unless one of both is a defeasible consequence of a node p such that $\widetilde{[p]}$ is equal to either $[n]$ or $[m]$.

[15] For some accounts on classical negation in logic programming, see among others Alferes and Pereira [17], Alferes, Pereira and Przymusinski [18], and Gelfond and Lifschitz [19].

[16] Note that the same holds in inconsistency-adaptive logics: only contradictions of the form $A\& \sim A$ are "manageable", hard contradictions, which can be read as "A is true and A is false", inevitably lead to triviality.

Another type of negation that is very common in logic programming (in particular database-theory) is the so-called ***negation as failure***, which in standard logic programming is the ***only*** negation available—see Clark [20] and Reiter [21]. The idea is that we presume our database or program to be closed, i.e. to contain all relevant information. This assumption (the ***closed world assumption***) allows us to assume every atom false, unless it is stated explicitly otherwise (or if the atom can be derived from explicit information). A standard example is that of the database of an airline company. In the database, one will find a list of all planes, schedules, ... For instance, it will have a record stating that flight F54 is a flight from Brussels to São Paulo, on a Boeing 747. What it will not contain, for obvious reasons, is that flight F54 is ***not*** a flight from Madrid to New York, and ***not*** a flight from Tokyo to London, and ... The database is said to be complete, so that we can assume that flight F54 is not a flight from A to B, where A is different from Brussels and B different from São Paulo. Obviously, negation as failure is a very welcome tool in such applications.

Formally, negation as failure is written as ***not***. As a result from the closed world assumption, a rule of the form $\mathbf{not}p \rightarrow q$ can by default be applied, unless p is derived or derivable. When such a rule is present in a program, and our aim is to prove q, we first need to show that p is not provable.

To incorporate negation as failure into ALP is not a hard task. First, we need to add a clause that says that a rule $A_1, \ldots, A_n, \mathbf{not}A_{n+1}, \ldots, \mathbf{not}A_m$ is applicable whenever $A_1, \ldots, A_n$ occur in the prooftree. Secondly, we add that whenever such a rule with *not*-negated atoms in its body is attacked in its *not*-part of its body by an applied, non-defeated rule, it is defeated. What is meant is that a rule $a, \mathbf{not}b \rightarrow c$ is defeated by every applied, non-defeated rule that has b in its head. A rule leading to $\sim a$ falls under the cases dealt with in this paper. Thirdly, a negation as failure version of an attack-cycle must be included (see the remark below). Further adjustments are minor, and follow immediately from the above and the previous sections.

REMARK 16 It is interesting to point out that ALP extended with negation as failure, is in fact an extension of classical general logic programming. A logic program only consisting of definite rules, with only negation as failure (and thus not the paraconsistent $\sim$) will yield the same consequences in ALP as it would in ordinary logic programming. A very simple example from Baral and Gelfond [22] will illustrate this. Consider $P = \{\mathbf{not}p \rightarrow q;\ \mathbf{not}q \rightarrow p\}$. ALP will lead to a tree with four nodes, with edges $\langle \mathbf{not}p, q\rangle$ and $\langle \mathbf{not}q, p\rangle$, and the marking of both the node labeled p as the node labeled q (this is the negation as failure version of the attack-cycle of definition 7). Baral and Gelfond give the two stable models $\{p\}$ and $\{q\}$, which is clearly the same result. What ALP adds to general logic programming is a way of sensibly dealing with further negations and defeasible rules.

5 Conclusion and hints towards further research on ALP

Clearly, we have done what we set out to do, namely combine the best of two worlds. Using logic programming as a starting point, we have successfully defined some of the most central notions of adaptivity. As such, we are a step closer to automated deduction in an adaptive manner.

But of equal importance is the successful combination of two different visions on logic. On the one hand, there is the logic as developed by philosophers, and on the other the more mechanical, computational view that is common in computer science and its applications. We have proven here that the gap between these two visions on logic is not as wide as some may think. It is perfectly well possible to build a bridge that is strong enough to carry the load it is designed for.

Let us briefly return to the distinction between strict and defeasible rules from remark 1. The choice for formalizing beliefs by either one of them, is highly a personal matter, which reflects the amount of trust one has in the available knowledge. If one chooses to reason with only strict rules, this is an indication of a high, if not to say infinite, confidence in the correctness of ones knowledge. As we have seen, strict rules are always applicable, never defeated, and can lead to strange and doubtful conclusions. At the other end of the spectrum, the choice of only allowing defeasible rules, is extremely cautious: it is the position of he who doubts all knowledge. But, whatever position one takes between these two extremes, the strategy for reasoning presented here, is workable, and highly intuitive.

Still, much work remains to be done in this area of research: first, a semantics for *ALP* must, and can, be defined. A semantics can be given by means of ***minimal inconsistent Herbrand interpretations***, which will be the subject of a forthcoming paper.

Secondly, in addition to the system presented here, we can extend the *ALP*-system to include the here proposed extensions of negation as failure and classical negation. Further extensions might include a way of ***doubting*** premises (see e.g. Guido Vanackere's [13]), non-logical preferences, non-definite rules, Skolem-functors etcetera. Also, more credulous or skeptical versions can be defined.[17]

REFERENCES

[1] Diderik Batens. Inconsistency–adaptive logics. In Orłowska [23], pages 445–472.

[2] Diderik Batens. Dynamic dialectical logics. In G. Priest, R. Routley, and J. Norman, editors, ***Paraconsistent Logic. Essays on the Inconsistent***, pages 187–217. Philosophia Verlag, Munchen, 1989.

[3] Diderik Batens. In defence of a programme for handling inconsistencies. In Joke Meheus, editor, ***Inconsistency in Science***. Kluwer, Dordrecht, 200x. To appear.

[4] Guido Vanackere. Ambiguity–adaptive logic. ***Logic et Analyse***, 159:261–280, 1997.

[5] Diderik Batens, Kristof De Clercq, and Guido Vanackere. Simplified dynamic proof formats for adaptive logics. To appear.

[17] All unpublished papers cited here, and many more, can be found on the WWW `http://logica.rug.ac.be/centrum/writings/`.

[6] Diderik Batens and Timothy Vermeir. Direct dynamic proofs for the Rescher–Manor consequence relations: The flat case. To appear.

[7] Donald Nute. Defeasible logic. In Dov Gabbay, C.J. Hogger, and J.A. Robinson, editors, *Nonmonotonic Reasoning and Uncertain Reasoning*, volume 3 of *Handbook of Logic in Artificial Intelligence*. Clarendon Press, 1993. Volume Coordinator: Donald Nute.

[8] Donald Nute. Defeasible prolog. Technical Report AI-1993-04, Artificial Intelligence Programs, University of Georgia, 1993. Presented at the AAAI Fall Symposium on Automated Deduction and Nonstandard Logics, Raleigh, NC, 1993.

[9] Donald Nute. Defeasible reasoning: A philosophical analysis in prolog. In James Fetzer, editor, *Aspects of Artificial Intelligence*, pages 251–288. Kluwer Academic Publishers, 1988.

[10] Timothy Vermeir. From wffs to clauses: transforming wffs in clauses without loss of meaning. To appear.

[11] Chiaki Sakama. Extended well-founded samantics for paraconsistent logic programs. In *Proceedings of the International Conference on Fifth Generation Computer Systems (FGCS'92)*, pages 592–599. Ohmsha LTD., 1992.

[12] Chiaka Sakama and Katsumi Inoue. Paraconsistent stable semantics for extended disjunctive programs. *Journal of Logic and Computation*, 5:265–285, 1995.

[13] Guido Vanackere. Preferences as inconsistency–resolvers: the inconsistency–adaptive logic PRL. To appear in *Logic and Logical Philosophy*.

[14] Dov Gabbay, Els Laenens, and Dirk Vermeir. Credulous vs. sceptical semantics for ordered logic programs. In *Proceedings of the Second International Conference on Principles of Knowledge Representation and Reasoning*, pages 208–217. Morgan Kaufmann, 1991.

[15] Salem Benferhat, Didier Dubois, and Henri Prade. Some syntactic approaches to the handling of inconsistent knowledge bases: A comparitive study. part 2: The prioritized case. In Orłowska [23], pages 473–511.

[16] Henry Prakken and Giovanni Sartor. A system for defeasible argumentation, with defeasible priorities. In *Proceedings of the International Conference on Formal and Applied Practical Reasoning. Bonn 1996*, volume 1085 of *Springer Lecture Notes in AI*, pages 510–524, Berlin, 1996. Springer Verlag.

[17] José Alferes and Luís Pereira. On logic program semantics with two kinds of negation. In K. Apt, editor, *Joint International Conference and Symposium on Logic Programming*, pages 574–588. MIT Press, 1992.

[18] José Alferes, Luís Pereira, and Teodor Przymusinski. "Classical" negation in non-monotonic reasoning and logic programming. *Journal of Automated Reasoning*, 20(1):107–142, 1998.

[19] Michael Gelfond and Vladimir Lifschitz. Classical negation in logic programs and disjunctive databases. *New Generation Computing*, 9(3/4):365–386, 1991.

[20] Keith Clark. Negation as failure. In H. Gallaire and J. Minker, editors, *Logic and Data Bases*, pages 293–322. Plenum, New York, 1978. Reprinted in *Readings in Nonmonotonic Reasoning*, M. Ginsberg, editor, Morgan Kaufmann, 1987.

[21] Raymond Reiter. On closed world databases. In Herve Gallaire and Jack Minker, editors, *Logic and Data Bases*, pages 119–140. Plenum Press, New York, 1978.

[22] Chitta Baral and Michael Gelfond. Logic programming and knowledge representation. *Journal of Logic Programming*, 12:1–80, 1994.

[23] Ewa Orłowska, editor. *Logic at Work. Essays Dedicated to the Memory of Helena Rasiowa*. Physica Verlag (Springer), Heidelberg, New York, 1999.

Referential and inferential many-valuedness *

GRZEGORZ MALINOWSKI University of Łódź, Poland
gregmal@krysia.uni.lodz.pl

Abstract

The development of the method of logical matrices at the turn of 19th century made it possible to define the concept of many-valued logic. However, the problem of interpretation of logical values other than truth and falsity is still among the most controversial questions of contemporary logic. The aim of this paper is to present two approches to the problem of many-valuedness, referential and inferential. In the former, many-valuedness may be received as the result of multiplication of semantic correlates of sentences, and not logical values. In the latter, many-valuedness (more precisely, three-valuedness) is the metalogical property of inference which takes from non-rejected assumptions to accepted conclusions.

The inferential framework provides a new perspective for some logical problems. In this paper, this is illustrated by two examples of its application: the Ł-modal logic and paraconsistency.

1 LOGICAL MATRICES AND TAUTOLOGICAL MANY-VALUEDNESS

A generic construction of a many-valued logic starts with the choice of the sentential language L which may be shown as an algebra $L = (For, F_1, \ldots, F_m)$ freely generated by the set of sentential variables $Var = \{p, q, r, \ldots\}$. Formulas, i.e. elements of For, are then built from variables using the operations $F_1, \ldots, F_m$ representing the sentential connectives. In most cases, either the language of the classical sentential logic

$$L_k = (For, \neg, \rightarrow, \vee, \wedge, \leftrightarrow)$$

with negation ($\neg$), implication ($\rightarrow$), disjunction ($\vee$), conjunction ($\wedge$), and equivalence ($\leftrightarrow$), or some of its reducts or extensions is considered. Subsequently, one

*I would like to acknowledge financial support from CNPq, CAPES and FAPESP (Brazil) for this research

defines a multiple-valued algebra A similar to L and chooses a non-empty subset of the universe of $A, D \subseteq A$ of designated elements. The interpretation structures

$$M = (A, D)$$

are called logical *matrices*.

Given a matrix M for a language L, the system $\mathrm{E}(M)$ of sentential logic is defined as the *content* of M, i.e. the set of all formulas taking a designated value for every valuation h (a homomorphism) of L in M. Thus,

$$\mathrm{E}(M) = \{\alpha \in For : \text{for every } h \in \mathrm{Hom}(L, A), h(\alpha) \in D\}.$$

In the case where M is the classical matrix based on $\{0, 1\}$, i.e.

$$M_2 = (\{0, 1\}, \neg, \rightarrow, \vee, \wedge, \leftrightarrow, \{1\})$$

with the connectives defined by the classical truth-tables, $\mathrm{E}(M_2)$ is the set of *tautologies* TAUT. Hence, in the sequel we shall sometimes refer to $\mathrm{E}(M)$ as the set of tautologies of a matrix M even if it is not classical.

When the content of a multiple-element matrix M does not coincide with TAUT, $\mathrm{E}(M) \neq$ TAUT, we say that the matrix in question defines *tautologically many-valued logic*. The examples of such logics are well known from the literature. Every scholar in logic knows the historic first construction of the three-valued logic by Łukasiewicz [3], Ł$_3$, whose matrix is

$$M_3 = (\{0, \tfrac{1}{2}, 1\}, \neg, \rightarrow, \vee, \wedge, \leftrightarrow, \{1\})$$

and the connectives are defined by the following tables:

x	¬ x
0	1
$\frac{1}{2}$	$\frac{1}{2}$
1	0

→	0	$\frac{1}{2}$	1
0	1	1	1
$\frac{1}{2}$	$\frac{1}{2}$	1	1
1	0	$\frac{1}{2}$	1

∨	0	$\frac{1}{2}$	1
0	0	$\frac{1}{2}$	1
$\frac{1}{2}$	$\frac{1}{2}$	$\frac{1}{2}$	1
1	1	1	1

∧	0	$\frac{1}{2}$	1
0	0	0	0
$\frac{1}{2}$	0	$\frac{1}{2}$	$\frac{1}{2}$
1	0	$\frac{1}{2}$	1

↔	0	$\frac{1}{2}$	1
0	1	$\frac{1}{2}$	0
$\frac{1}{2}$	$\frac{1}{2}$	1	$\frac{1}{2}$
1	0	$\frac{1}{2}$	1

Obviously, $\mathrm{E}(M_3) \neq$ TAUT, since e.g., $p \vee \neg p$ and $\neg(p \wedge \neg p)$ do not belong to the set of tautologies of Łukasiewicz logic, i.e. to $\mathrm{E}(M_3)$.

1.1. Consider the matrix

$$M_3 = (\{0, t, 1\}, \neg, \rightarrow, \vee, \wedge, \leftrightarrow, \{t, 1\})$$

for L_k with the operations defined by the following tables:

x	¬ x
0	1
t	0
1	0

→	0	t	1
0	1	t	1
t	0	t	t
1	0	t	1

∨	0	t	1
0	0	t	1
t	t	t	1
1	1	1	1

∧	0	t	1
0	0	0	0
t	0	t	t
1	0	t	1

↔	0	t	1
0	1	0	0
t	0	t	t
1	0	t	1

We claim that this three-valued matrix determines the system of tautologies of the classical logic. To verify this it suffices to notice that due to the choice of the set of designated elements $\{t, 1\}$, to each each $h\alpha \in Hom(L, A)$ there corresponds a classical valuation $h^* \in Hom(L, M_2)$ in a one-to-one way such that $h\alpha \in \{t, 1\}$ iff $h^*\alpha = 1$. Thus, the logic under consideration in neither sense is many-valued.

2 MATRIX CONSEQUENCE AND STRUCTURAL LOGICS

The notion of matrix consequence being a natural generalization of the classical consequence is defined as follows: relation $\models_M \subseteq 2^{For} \times For$ is a matrix consequence of M provided that for any $X \subseteq For, \alpha \in For$

$$X \models_M \alpha \text{ iff for every } h \in Hom(L, A)(h\alpha \in D \text{ whenever } hX \subseteq D).$$

Notice, that E(M) $= \{\alpha : \emptyset \models_M \alpha\}$. Therefore, if a matrix determines a non-classical set of tautologies, then obviously $\models_M$ does not coincide with the consequence relation of the classical logic, i.e. with $\models_{M_2}$. As before, we may consider the three-element matrix of Łukasiewicz as an example. In what follows, we note also that the consequence relation determined by the matrix in 1.1 is classical.

The next example of a three-element matrix logic is surprising and shows that there are matrices determining as its content the same set of classical tautologies, but its consequence relation is non-classical.

2.1. Consider the matrix

$$M_3 = (\{0, t, 1\}, \neg, \rightarrow, \vee, \wedge, \leftrightarrow, \{t, 1\})$$

for L_k with the operations defined by the following tables:

x	¬ x
0	1
t	1
1	0

→	0	t	1
0	1	1	1
t	1	1	1
1	0	0	1

∨	0	t	1
0	0	0	1
t	0	0	1
1	1	1	1

∧	0	t	1
0	0	0	0
t	0	0	0
1	0	0	1

↔	0	t	1
0	1	1	0
t	1	1	0
1	0	0	1

The above truth tables define the classical connectives of negation, implication, disjunction, conjunction and equivalence. To see this, just observe that 0 and t behave alike - they always give the same output in all tables. Since t and 1 are distinguished and no formula may take the value t, 1 is the only one to decide about logical properties of formulas and, therefore, we obtain that $E(M_3) = TAUT$.

Furthermore, it turns out that M_2 is the only two-element matrix which might determine $\models_{M_3}$. Simultaneously, $\models_{M_3} \neq \models_{M_2}$, since, for example,

$$\{p \to q, p\} \models_{M2} q \text{ while not } \{p \to q, p\} \models_{M3} q.$$

To verify this, it simply suffices to turn over a valuation h such that $hp = t$ and $hq = 0$

The example shows that it is reasonable to make a distinction between tautological and consequential many-valuedness. Accordingly, the classical system (of tautologies) was extended to a three-valued logic and the very property was assured by rules of inference rather than by its logical laws.

With every $\models_M$ an operation $Cn_M : 2^{For} \to 2^{For}$ may be uniquely associated such that

$$\alpha \in Cn_M(X) \text{ if and only if } X \models_M \alpha.$$

called a *matrix consequence operation* of M.

The concept of structural sentential logic is the ultimate generalization of the notion of the matrix consequence operation. A structural logic for a given language L is identified with Tarski's consequence $C : 2^{For} \to 2^{For}$,

(T_0) $X \subseteq C(X)$

(T_1) $C(X) \subseteq C(Y)$ whenever $X \subseteq Y$

(T_2) $C(C(X)) = C(X)$,

satisfying the condition of structurality,

(S) $eC(X) \subseteq C(eX)$ for every *substitution* of L.

Structural logics are characterized through their set of logical laws and schematic (sequential) rules of inference. The most important property of these logics is their matrix description: for every such C, a class of matrices $\underline{K}$ exists such that C is the intersection of $\{Cn_M : M \in \underline{K}\}$ i.e. for any $X \subseteq For$

$C(X) = \cap\{Cn_M(X) : M \in \underline{K}\}$, cf. [13].

The problem of their many-valuedness is thus reducible to the problem of many-valuedness of matrix consequence relations (or operations).

3 LOGICAL TWO-VALUEDNESS OF STRUCTURAL LOGICS

In the 1970's the investigations of logical formalizations bore several descriptions of many-valued constructions in terms of zero-one valuations. The interpretations associated with these descriptions shed new light on the problem of logical many-valuedness. One of the best justified and general approaches is that of R. Suszko, the author of non-Fregean logic, cf. [11].

The basis of R. Suszko's philosophy of logic was the distinction between semantic correlates of sentences and their logical values. In this perspective, the traditionally many-valued logics (tautologically or consequentially) may be regarded as only ***referentially many-valued*** and not necessarily ***logically many-valued.*** Recall that one of the central points of the Fregean approach in [1], the so-called ***Fregean Axiom,*** identified semantic correlates of sentences with their logical values.

Suszko [12] draws attention to the referential character homomorphisms associating sentences with their possible semantic correlates (i.e. referents or situations) and sets them against the ***logical valuations*** being zero-one-valued functions on For and, thus, differentiates the referential and the logical ***valuedness.*** Suszko claims that each sentential logic, i.e. a ***structural*** consequence relation, can be determined by a class of logical valuations and thus, it is ***logically two-valued.***

The argument supporting the Suszko thesis is the completeness of any structural consequence C with respect to a Lindenbaum bundle, which is the class of all Lindenbaum matrices of the form (L,C(X)) with $X \subseteq For$.

Given a sentential language L and a matrix $M = (A, D)$ for L, the set of valuations TV_M is defined as:

$$\mathrm{TV}_M = \{\mathrm{t}_h : h \in \mathrm{Hom}(L, A)\},$$

where

$$t_h(\alpha) = \begin{cases} 1 & \text{if} \quad h(\alpha) \in D \\ 0 & \text{if} \quad h(\alpha) \notin D. \end{cases}$$

Consequently, the matrix consequence operation $\models_M$ may be described using valuations as follows:

$$X \models_M \alpha \text{ iff if for every } t \in TV_M \ (t(\alpha) = 1 \text{ whenever } t(X) \subseteq \{1\}).$$

The definition of logical valuations may be simply repeated with respect to any structural consequence operation C using its Lindenbaum bundle. Thus, each structural logic (L, C) can be determined by a class of logical valuations of the language L or, in other words, ***it is logically two-valued.***

The justification of the thesis that states logical two-valuedness of an important family of logics lacks a uniform description of TV_C's, i.e. the respective classes of valuations. The task is, in each particular case, a matter of an elaboration. An example of a relatively easily definable and readable set of logical valuations is LV_3, the class adequate for the $(\neg, \rightarrow)$, a version of the three-valued Łukasiewicz logic, cf. [3]. LV_3 is the set of all functions $t : For \rightarrow \{0, 1\}$ such that for any $\alpha, \beta, \gamma \in For$ the following conditions hold:

(0) $\mathrm{t}(\gamma) = 0$ or $\mathrm{t}(\neg\gamma) = 0$
(1) $\mathrm{t}(\alpha \rightarrow \beta) = 1$ whenever $\mathrm{t}(\beta) = 1$
(2) if $\mathrm{t}(\alpha) = 1$ and $\mathrm{t}(\beta) = 0$, then $\mathrm{t}(\alpha \rightarrow \beta) = 0$
(3) if $\mathrm{t}(\alpha) = \mathrm{t}(\beta)$ and $\mathrm{t}(\neg\alpha) = \mathrm{t}(\neg\beta)$, then $\mathrm{t}(\alpha \rightarrow \beta) = 1$
(4) if $\mathrm{t}(\alpha) = \mathrm{t}(\beta) = 0$ and $\mathrm{t}(\neg\alpha) \neq \mathrm{t}(\neg\beta)$, then $\mathrm{t}(\alpha \rightarrow \beta) = \mathrm{t}(\neg\alpha)$
(5) if $\mathrm{t}(\neg\alpha) = 0$, then $\mathrm{t}(\neg\neg\alpha) = \mathrm{t}(\alpha)$
(6) if $\mathrm{t}(\alpha) = 1$ and $\mathrm{t}(\beta) = 0$, then $\mathrm{t}(\neg(\alpha \rightarrow \beta)) = \mathrm{t}(\neg\beta)$
(7) if $\mathrm{t}(\alpha) = \mathrm{t}(\neg\alpha) = \mathrm{t}(\beta)$ and $\mathrm{t}(\neg\beta) = 1$, then $\mathrm{t}(\neg(\alpha \rightarrow \beta)) = 0$,

cf. [12].

4 INFERENTIAL THREE-VALUEDNESS

In [8] a generalization of Tarski's concept of consequence operation based upon the idea that rejection and acceptance need not be complementary was proposed. The central notions of the framework are counterparts of the concepts of matrix and consequence relation - both distinguished by the prefix "q" which may be read as "quasi". Where L is a sentential language and A is an algebra similar to L, a *q-matrix* is a triple

$$M^* = (A, D^*, D),$$

where D^* and D are disjoint subsets of the universe A of A ($D^* \cap D = \emptyset$). D^* and D are then interpreted as sets of ***rejected*** and ***accepted*** values of M^*, respectively. For any such M^* one defines the relation $\models_{M*}$ between sets of formulae and formulae, a ***matrix q-consequence of*** M^* such that, for any $X \subseteq For$ and $\alpha \in For$

$$X \models_{M^*} \alpha \text{ iff for every } h \in Hom(L, A)(h\alpha \in D \text{ whenever } hX \cap D^* = \emptyset).$$

The relation of q-consequence was designed as a formal counterpart of reasoning, admitting rules of inference which from non-rejected assumptions lead to accepted conclusions. The q-concepts coincide with usual concepts of matrix and consequence only if $D^* \cup D = A$, i.e. when the sets D^* and D are complementary. Then, the set of rejected elements coincides with the set of non-designated elements.

For every $h \in Hom(L, A)$ let us define a three-valued function $k_h : For \to \{0, \frac{1}{2}, 1\}$ such that

$$k_h(\alpha) = \begin{cases} 0 & \text{if} \quad h(\alpha) \in D^* \\ \frac{1}{2} & \text{if} \quad h(\alpha) \in A - (D^* \cup D) \\ 1 & \text{if} \quad h(\alpha) \in D. \end{cases}$$

Given a q-matrix M^* for L, let $KV_M = \{k_h : h \in Hom(L, M^*)$; we get the following three-valued description of the q-consequence relation $\models_{M^*}$:

$$X \models_{M^*} \alpha \text{ iff for every } k_h \in KV_M(k_h(X) \cap \{0\} = \emptyset \text{ implies } k_h(\alpha) = 1).$$

It is worth emphasising that this description in general is not reducible to the two-valued description possible for the ordinary (structural) consequence relation. Since the latter property may be interpreted as logical two-valuedness of logics identified with the consequence, we may say that a q-matrix consequence is logically either two or three valued, see also next section. The example below shows that such q-logics exist.

4.1. (cf. [8]). Consider the three-element q-matrix

$$Ł_q 3 = (\{0, \tfrac{1}{2}, 1\}, \sim, \Rightarrow, \vee, \wedge, \equiv, \{0\}, \{1\}),$$

where the connectives are defined as in the Łukasiewicz three-valued logic. Then, for any $p \in Var$, it is not true that $\{p\} \models_{M^*} p$. To see this, it suffices to consider the valuation sending p into $\frac{1}{2}$.

The most striking remark is perhaps the fact that even logics generated by some two-element q-matrices may be three-valued. This is illustrated by our last example:

4.2. Let us consider the two-element algebra

$A_2 = (\{0,1\}, \neg, \rightarrow, \vee, \wedge, \leftrightarrow)$,

with the operations defined by the classical truth-tables of negation, implication, disjunction and equivalence. Next, let us consider the following two q-matrices:

$M_1 = (A_2, \emptyset, \{1\})$,
$M_0 = (A_2, \emptyset, \{0\})$.

The q-consequence relations of M_1 and M_0 are such that for any $X \subseteq For, \alpha \in For$

$X \models_{M_1} \alpha$ if and only if for every $h \in Hom(L, A_2) h\alpha = 1$,
$X \models_{M_0} \alpha$ if and only if for every $h \in Hom(L, A_2) h\alpha = 0$.

Thus, in the first case a formula α is a q-consequence of any set of formulas, whenever it is a tautology. In the second case α is a contradictory formula.

The standard description of $\models_{M_1}$ in terms of $\{0, \frac{1}{2}, 1\}$-valuations k_h is then defined in such a way that for every $\alpha \in For, k_h(\alpha) = 1$ iff $\alpha \in TAUT$, $k_h(\alpha) = \frac{1}{2}$ otherwise; for no formula k_h takes the value 0.

Similarly, $X \models_{M_0} \alpha$ whenever $k_h(\alpha) = 0$, where $k_h(\alpha) = 0$ iff α is contradictory and $k_h(\alpha) = \frac{1}{2}$ otherwise.

5 Tarski-like characterization

The inferential approach is based on the notions of q-matrix and the q-consequence relation. In Section 4 we have already described how these notions are related to usual concepts of a matrix and the matrix consequence. Let us note, in turn, that similarly to the usual consequence, with every q-matrix consequence relation $\models_{M^*}$ an operation $Wn_{M^*} : 2^{For} \rightarrow 2^{For}$ is uniquely associated such that

$$\alpha \in Wn_{M^*}(X) \text{ is and only if } X \models_{M^*} \alpha.$$

Now, to go deeply into the subject, let us first observe that

5.1. For any q-matrix $M^* = (A, D^*, D)$ and a corresponding matrix $M = (A, D), Wn_{M^*}(\emptyset) = Cn_M(\emptyset) = E(M)$.

This means that any logical system may equally well be extended to logically two-valued logic (L, Cn_M) or to a three-valued logic (L, Wn_{M^*}). Obviously, depending on the quality and cardinality of M, both kinds of extensions may take different shapes, thus defining different logics. Moreover, in several cases it is also possible to define two (or more) different inferential extensions of a given system thus receiving different q-logics. The idea was applied in [4] to get the framework permitting to make a distinction between two "indistinguishable" modal connectives of the four-valued modal system of Łukasiewicz.

5.2.(cf. [2]). The Łukasiewicz modal algebra has the form:

$\pounds = (\{1,2,3,4\}, \neg, \rightarrow, \Delta, \nabla)$, with the operations defined by the following tables:

$\to$	1	2	3	4
1	1	2	3	4
2	1	1	3	3
3	1	2	1	2
4	1	1	1	1

x	$\neg$ x
1	4
2	3
3	2
4	1

x	Δ x
1	1
2	1
3	3
4	3

x	∇ x
1	1
2	2
3	1
4	2

The system Ł of modal logic was defined on the language $L = (For, \neg, \to, \Delta, \nabla)$, with two modal connectives Δ and ∇ as the set of all formulas taking for every valuation of L in £ the distinguished value 1, thus

$$\text{Ł} = \{\alpha \in For : \text{for every } h \in Hom(L,\pounds), h(\alpha) = 1\}.$$

Therefore, the matrix $M = (\pounds, \{1\})$, whose content coincides with Ł is the natural candidate for the base of the Tarski extension to the structural consequence. However, Łukasiewicz claims in [2] that the two modal connectives are indistinguishable.

The q-consequence framework permits, in contrast, two inferential extensions to be defined, determined by the following q-matrices:

$$M_\Delta = (\pounds, \{3,4\}, \{1\}),$$
$$M_\nabla = (\pounds, \{2,4\}, \{1\}),$$

The choice of the sets of rejected and accepted elements in the two q-matrices and the very idea of considering inferential extensions of the system of modal logic are in a way connected with Łukasiewicz attempts to discern between the two operators of possibility. Note that in the first case are rejected those elements of the algebra of values which Δ "sends to" not designated values, i.e. different from 1.

The q-matrices M_Δ and M_∇ define two inferential extensions of £-modal logic, i.e. the following inferential calculi:

$$(L\ ,\ Wn_{M_\Delta}) \text{ and } (L\ ,\ Wn_{M_\nabla}).$$

Obviously, $Wn_{M_\Delta}(\emptyset) = Wn_{M_\nabla}(\emptyset) = \text{Ł}$.

5.3. The q-framework is more general and it can be related to the theory of the so-called q-consequence operation W : $2^{For} \to 2^{For}$ satisfying the following postulates

(W_1) $W(X) \subseteq W(Y)$ whenever $X \subseteq Y$

(W_2) $W(X \cup W(X)) = W(X)$,

and, possibly, the condition of structurality,

(S) $eW(X) \subseteq W(eX)$ for every *substitution* of L.

It may be easily seen that any Wn_{M^*} satisfies (W_1), (W_2) and (S). Thus, Wn_{M^*} is structural q-consequence operation. Q-consequence operations also satisfying Tarski's first postulate (T_0)

(W_0) $X \subseteq W(X)$

for every $X \subseteq T$, are consequence operations: since (W_2) is reduced to (T_2). Therefore, Tarski's consequence is a special case of a q-consequence.

The last property sets the new approach against the standard one. The unique features of the q-framework justify entirely its usefulness.

To prove this, let us remark that the Tarski's first postulate (T_0) for consequence unconditionally validates the repetition rule

(Rep) $\alpha \vdash_c \alpha$ for every $\alpha \in For$.

Note also, that no other rule is thus validated in Tarski's environment. Hence, (Rep) is the rule of any consequence operation exceptionally distinguished exclusively by a methodological tool. Contrary to this, (Rep) is not in general valid for a q-consequence, see Example 4.1. Moreover, it is particularly interesting that when the "full" repetition rule is not among the rules of a given q-consequence W, some of its instances may be. Hence, the limitations imposed by W onto (Rep) can play a role of filtration of formulas.

6 APPLICATION: INFERENTIAL PARACONSISTENCY AND ITS DUAL

Paraconsistent logics provide the basis for inconsistent but not trivial theories. They are usually constructed as systems of theorems and rules, defining not explosive consequence relations $\models$. A consequence $\models$ is explosive whenever from any set of contradictory formulas $\{\alpha\ ,\ \neg\alpha\}$ any other formula β follows, $\{\alpha\ ,\ \neg\alpha\} \models \beta$, cf. [9], [10]. Usually, paraconsistency is already present in the set of theorems or tautologies of a logic construed or rediscovered for that purpose. Thus, it reflects only an internal quality of a logical system.

It follows that paraconsistency can be generated outside of a system as a property of inference. Accordingly, using the q-approach it is possible to get an inferential paraconsistent version of some non-paraconsistent logics. In the sequel, we present a paraconsistent q-version of the three-valued Lukasiewicz propositional logic and its dual counterpart (cf. [6]).

6.1. The three-valued Lukasiewicz logic $\mathbf{L}_3$ is defined as the content of the three-valued matrix $M_3 = (A_3, \{1\})$ based on the algebra

$$A_3 = (\{0, \tfrac{1}{2}, 1\}, \neg, \rightarrow, \vee, \wedge, \leftrightarrow)$$

with the operations defined by the first set of tables in Section 1. The matrix consequence operation extending the Lukasiewicz system $E(M_3)$ is then defined as follows:

$$X \models_M \alpha \text{ iff for every } h \in Hom(L, A_3)(\text{if } hX \subseteq \{1\} \text{ then } h\alpha = 1)\ .$$

The formula $\alpha \rightarrow (\neg\alpha \rightarrow \beta)$ belongs to $E(M_3)$ which, in turn, is closed on *ModusPonens*, being also a rule of $\models_M$, i.e. for any α, $\beta \in For$

(MP) $\{\alpha, \alpha \rightarrow \beta\} \models_M \beta$.

Therefore, $\{\alpha , \neg\alpha\} \models_M \beta$ for any α, $\beta \in For$, which means that the logic in question is not paraconsistent.

Now, retaining 1 as the accepted value and taking 0 as the only rejected element we receive the following Łukasiewicz q-matrix

$$M_3^1 = (A_3, \{0\}, \{1\}) .$$

The q-consequence determined by M_3 is such that

$$X \models_{M^*} \alpha \text{ iff for every } h \in Hom(L, A_3)(\text{if } hX \subseteq \{\tfrac{1}{2}, 1\}, \text{ then } h\alpha = 1).,$$

if all premises are not rejected, i.e. not false, then the conclusion is accepted, i.e. true. It is easy to note that $\models_{M^*}$is paraconsistent. This holds true since there are α, β such that not $\{\alpha , \neg\alpha\} \models_M \beta$. To see this, one may simply take $\alpha = p$ and $\beta = q$ and conclude that any valuation sending both variables p, q into $\frac{1}{2}$ falsifies the inference.Obviously, this also means that (MP) is not a rule of $\models_{M_3^1}$.

6.2 Let us now consider the q-matrix

$$M_3^0 = (A_3, \{1\}, \{0\}) .$$

resulting from M_3^0 after the mutual exchange of the sets of accepted and rejected values. In this case, each set takes the role of the other and, for this reason, we may call M_3^0 the q-matrix *dual* to M_3^1. The resulting logic has the following properties: (1) the logical system $E(M_3^0)$ is the set of formulas false under every valuation in A, (2) the q-consequence $\models_{M_3^0}$ admits inferences from non-accepted premises to the rejected conclusions:

$$X \models_{M_3^0} \alpha \text{ iff for every } h \in Hom(L, A_3)(\text{if } hX \subseteq \{0, \tfrac{1}{2}\} \text{ then } h\alpha = 0) \quad .$$

Accordingly, for the resulting "negative" Łukasiewicz q-logic, one may also formulate a paraconsistency problem and find a satisfying answer to it. Also in this case some sets $\{\alpha , \neg\alpha\}$ are not explosive: as in the "positive" case, $\{p , \neg p\} \models_{M_3^0} q$ is not true.

6.3. The procedure of dualization of matrix quasi-logics is general and it may be applied to a wide range of the known propositional logics.

7 Final remarks

The logical two-valuedness in the sense of Section 3 is related to the division of the universe of a matrix into two subsets: designated and undesignated elements. Accordingly, the logical three-valuedness set out in Section 4 is clearly mirrored in the construction of q-matrices: the universe of any q-matrix is divided into three subsets. The common feature of both approaches is that logical valuations are characteristic functions of actual division, in the latter case the "generalized" function ranging over a set of three elements.

An extensive study of the framework given in [8] shows that the theory of q-matrices and structural q-consequence operation is very similar to the theory of matrices and structural consequence operations. First, the basic concepts of universal algebra extended in [13] in a straightforward way to matrices may be redefined similarly for q-matrices. Next, any structural q-consequence W is complete with respect to the Lindenbaum bundle of q-matrices of the form $(L, For - (X \cup W(X)), W(X))$ Consequently, for every such W, one may find a class of q-matrices K^*_W characteristic for it, i.e. such that

$$W(X) = \cap\{Wn_{M^*}(X) : M^* \in K^*_W\}.$$

Every matrix q-consequence operation Wn_{M^*} can be determined by two classes of two or three-element logical valuations (compare Section 4). The class of all valuations for the q-matrices in K^*_W uniquely determines W. We then arrive at the result corresponding to Suszko's thesis and we can say that **each q-logic**, i.e. structural q-consequence operation, **is logically two or three-valued,** cf. [5].

References

[1] Frege, G., **Über Sinn und Bedeutung.** Zeitschrift für Philosophie und philosophische Kritik C, 1892, 25–50.

[2] Łukasiewicz, J., A system of modal logic, **The Journal of Computing Systems**, 1, 1953, 111–149.

[3] Łukasiewicz, J., O logice trójwartościowej, **Ruch Filozoficzny,** 5, 1920, 170-171. English tr. On three-valued logic [in:] Borkowski, L. (ed.) Selected works, North- Holland, Amsterdam, 87–88.

[4] Malinowski, G., Inferential extensions of Łukasiewicz modal logic, (a lecture to the Conference "Łukasiewicz in Dublin" University College Dublin, Department of Philosophy, Dublin 7 - 10 July 1996), **Bulletin of the Section of Logic**, 26 (4), 1997, 220–224.

[5] Malinowski, G., Inferential many-valuedness [in:] Woleński, J. (ed.) **Philosophical logic in Poland**, Synthese Library, 228, Kluwer Academic Publishers, Dordrecht, 1994, 75–84.

[6] Malinowski, G., Inferential paraconsistency, **Paraconsistent logic, Logical Philosophy Mathematics & Informatics**, abstracts of Stanisław Jaśkowski Memorial Symposium, Toruń, 1998, 77–82.

[7] Malinowski, G., **Many-valued logics**, Oxford Logic Guides, 25, Clarendon Press, Oxford, 1993.

[8] Malinowski, G., Q-consequence operation, **Reports on Mathematical Logic**, 24, 1990, 49–59.

[9] D'Ottaviano, I. M. L., Da Costa, N. C. A., Sur un probleme de Jaskowski, **Comptes Rendus de l'Academie des Sciences de Paris, 270 A**, 1970, 1349–1353.

[10] Priest G., Routley R,. Systems of paraconsistent logic, [in:] **Paraconsistent logic: Essays on the inconsistent**, Philosophia, Munich, 1989, 151–186.

[11] Suszko, R., Abolition of the Fregean Axiom [in:] Parikh, R. (ed.) Logic Colloquium, Symposium on Logic held at Boston, 1972-73. **Lecture Notes in Mathematics**, vol. 453, 1972, 169–239.

[12] Suszko, R., The Fregean Axiom and Polish Mathematical Logic in the 1920s, **Studia Logica**, XXXVI (4), 1977, 377–380.

[13] Wójcicki R., **Theory of logical calculi. Basic theory of consequence operations**, Synthese Library, 199. Kluwer Academic Publishers, Dordrecht, 1988.

When is a Substructural Logic Paraconsistent?
Structural conditions for paraconsistency in ternary frames*

MARCELO FINGER Departamento de Ciência da Computação, Instituto de Matemática e Estatística, Universidade de São Paulo, 05508-900, São Paulo, Brazil
mfinger@ime.usp.br

Abstract
In this work, we study structural, model-theoretical conditions that support paraconsistency in Substructural Logics. The idea is to follow the notion of *Correspondence Theory* from Modal Logics and apply it to *Substructural Logics*.

Several logics in the family of Substructural Logics were initially defined with goals similar to those of Paraconsistent Logic. There are several possible ways of defining paraconsistency, but this work takes a neutral way towards all such definitions. We note that the formalization of such definitions vary according to the set of connectives present in the logical language, and also according to whether we view paraconsistency as the possibility to deny the principles of Non-contradiction or Trivialization. All this yields a number of possible definitions of *paraconsistency*. We propose a method that allows us to compute which effects a given definition may have upon the model theoretical structures of a Substructural Logic that adopt one such definition.

It has been known since the work of Routley and Meyer [RM73] that binary logical connectives can be seen as modalities interpreted over Kripke frames (W, R) with a ternary accessibility relationship $R \subseteq W \times W \times W$. More recently, a correspondence theory was developed for substructural logics in analogy to the usual modal correspondence theory.

In this a setting, we derive structural restrictions over ternary frames corresponding to the violation of a *consistency condition*, that is, an axiom. Such a process is performed on a fragment consisting of the connectives $\otimes$ (*tensor product*, also called *multiplicative conjunction*), $\rightarrow$(multiplicative implication), $\neg$ (classical negation), $\sim$ (intuitionistic negation) and $\wedge$ (classical conjunction).

*Marcelo Finger was partly supported by the Brazilian Research Council (CNPq), grants PQ 300597/95-5 and APQ 468765/2000-0.

1 INTRODUCTION

In this work, we study structural, model-theoretical conditions that support paraconsistency [dC74] in Substructural Logics [Res00]. One of the initial motivations for the proposal of Relevant Logics was to avoid the classical trivialization of theories, where from a formula A and its negation one can infer a formula B, even if A has nothing in common with B [AB75]. The way that Relevant and other Substructural Logics followed to achieve that goal was to restrict the set of classical structural rules in deductions; hence the name of the family of logics. With the elimination of structural rules, classical connectives unfolded into several others, so many new fragments were created for Substructural Logics. Actually, the family was unified as such only much later [Doš93], and for several years there were just several groups of logics (Relevant, Linear, Lambek, Intuitionistic, etc). A semantics for Relevant Logic based on ternary frames was proposed by Routley and Meyer [RM73], which was later extended to the whole family of Substructural Logics [Res00].

The way Paraconsistency is treated in da Costa's approach is different [dC74], and consists of weakening the notion of classical negation. Initial tentatives to create a semantics for paraconsistent logics tried to provide set theoretical constructions to accommodate the "inconsistent elements" present in most paraconsistent systems, with partial success [CA81]. Recent approaches to a semantics of paraconsistent logics have totally avoided the manipulation of the usual set theoretical structures, preferring to give a semantics based on the ***translation*** of a paraconsistent logics into a set of many-valued logics, plus some mechanism for the combination/interaction of these translations [Car98].

We do not deny that there are interesting aspects in these translation-based approaches to semantics, but since we are taking the substructural point of view, we will study the model theoretical conditions present at the intersection between Substructural and Paraconsistent Logics in the light of model theoretical constructions for substructural paraconsistency.

It is important to note that we do not mean that Paraconsistent Logics *are* Substructural Logics. Quite the opposite, we simply note that some Substructural Logics display a ***paraconsistent behaviour***, e.g. Relevant Logics as mentioned above. So some substructural logics do accept some paraconsistent theories, but some others do not. This does not rule out the possibility of existing other logics termed Paraconsistent that are not Substructural or vice versa.

In this way, we proceed with our study of model theoretical conditions that permit a substructural logic to accept paraconsistent theories.

1.1 Paraconsistency and Substructural Logics

Our approach here does not start with *the* definition of a Paraconsistent Logic, so that we can put forward a sound and complete semantics for it. We do not have a final definition for paraconsistency, nor do we think that one such definition is desirable.

In the literature, there are two basic notions related to paraconsistency, both involving a formula A and its negation $\mathsf{not}\, A$, both related to the violation of a logical principle:

- *Non-contradiction:* according to this principle, a theory should not derive a formula and its negation. Therefore, a ***paraconsistent theory*** that violates non-contradiction cannot validate an axiom of the form not(A and not A).

- ***Trivialization:*** according to this principle, a theory containing both a formula and its negation derives any formula. A ***paraconsistent theory*** that violates triviality must not validate an axiom of the form (A and not A) implies B.

In this explanation above, we have used the connectives not, and, implies to remain neutral as to their definition, for in substructural logics there may exist several possible connectives for negation, conjunction and implication. The present work is also neutral towards such definitions and we analyse structural conditions for several possible definitions of these connectives.

As stated earlier, our approach is based on the semantics. We start with a pure semantical structure for substructural logics, that is, a semantical structure free from any structural pressuposition. We then study what kind of properties should be imposed on that structure for each alternative definition of paraconsistency.

1.2 Paraconsistency and Correspondence Theory

The idea is to follow the notion of *Correspondence Theory* from Modal Logics [vB84]. In modal semantics we have the notion of a basic Kripke frame, $\mathcal{F} = (W, R)$, consisting of a set W of possible worlds with a binary relation R, called the ***accessibility relation***, which provides a sound and complete semantic basis for the minimal modal logic K. We know that by adding some property to the system, e.g. reflexivity, some formulas become valid in the class of all Kripke models obeying that property; e.g. the axiom T, $\Box p \rightarrow p$, is valid in all reflexive Kripke frames. Conversely, if we add an axiom to a modal axiomatization, we get completeness over some class of Kripke frames; e.g. logic K + axiom $\Box p \rightarrow p$ is complete over the class of reflexive Kripke frames [BS84, Che80].

In this way, the relationship between modal axioms and classes of Kripke frames can be studied without the need to define *the* modal logic.

We develop here a similar approach for ***substructural logics*** [Doš93, Res00], that is, the family of logics obtained by rejecting some of the structural rules used in classical logic deductions. The works of Roorda [Roo91] and Kurtonina [Kur94] have shown that, in the same way that monadic modalities are interpreted over binary accessibility relationships, binary connectives can be seen as modalities interpreted over Kripke frames with a ternary accessibility relationship. In particular, we may study the usual connectives (implication, conjunction, negation) as modalities.

In such a setting we can start asking what sort of properties corresponds to a given axiom, as is done in modal correspondence theory. In particular, some axiom may be taken as the definition of consistency in the system, so that we may investigate what structural properties correspond to each definition of consistency.

Note that it follows from the modal examples above that if we want to allow for the falsity of modal axiom T at some worlds, we may not have all worlds reflexive; that is, $\forall x Rxx$ must fail for some x. This is the way we are going to treat paraconsistency conditions, namely by falsifying the structural conditions imposed by ***consistency axioms*** on ternary frames.

1.3 Automated Methods

Recently, we have been able to find an automatic way to compute a first-order condition on ternary frames associated to an axiom [Fin00], in a manner analogous to the way that modal Sahlqvist formulas can computationally generate a restriction on traditional (binary) Kripke frames [vB84]. Such automatic computation is performed on a substructural fragment known as Categorial Grammar [Car97, Moo97], consisting of the connectives $\rightarrow$ (*right-implication*), $\leftarrow$ (*left-implication*) and $\otimes$ (*tensor product*, also called *multiplicative conjunction* or *fusion*).[1]

We claim that such techniques can be applied for the study of first-order condition on ternary frames that allows a logic to support paraconsistent theories.

The rest of the paper develops as follows. Ternary frames, and its relationship to first-order formulas are presented in Section 2, with an example on how to compute the first-order restriction associated with an axiom. Then in Section 3 we show that different definitions of what constitutes a consistency axiom lead to distinct structural constraints; in particular, we study consistency conditions based on:

- non-contradiction *vs.* trivialization principles;
- boolean *vs.* intuitionistic negation;
- boolean *vs.* multiplicative conjunction.

Finally, we analyse in Section 4, we apply those methods for relevant negation and in Section 5 we discuss several other possible negations which can be analysed by our method.

2 Ternary Frames

The idea of using ternary frame for the semantics of substructural logics goes back to [RM73], where it was used to provide a semantics for relevance logics. In a context free of structural pressuposition, that semantics has been used in, for example, [Kur94, DM97].

A ternary frame is a pair $\mathcal{F} = (W, R)$, where R is a any ternary relation on $W \times W \times W$. The set W is a set of possible worlds. We normally represent that a triple $\langle a, b, c\rangle \in R$ by writing $Rabc$. The elements of R are seen as a binary tree, with a being the root node, b its left daughter, and c its right daughter. To reinforce this point of view, $Rabc$ is sometimes written as Ra, bc.

Every model has a distinguished world $\mathbf{0} \in W$. Unlike modal Kripke models, a valid formula is not required to hold at all worlds of every model, but only at the distinguished world of every model. The distinguished $\mathbf{0}$ has the following properties:

$$Ra\mathbf{0}a \text{ and } Raa\mathbf{0}$$

The language fragment we work with in this section consists of a countable set of propositions, $\mathcal{P} = \{p_1, p_2, \ldots\}$, and the binary connectives $\rightarrow, \leftarrow, \otimes$. We use A, B, C as variables ranging over substructural formulas. The connectives $\Rightarrow$, $\wedge$ and $\neg$ are, respectively, the classical implication, conjunction and negation.

[1] These connectives also appear in the literature as /, \ and $\bullet$.

A model $\mathcal{M} = (W, R, V, \mathbf{0})$ consists of a ternary frame plus a valuation $V : \mathcal{P} \to 2^W$ that maps propositional variables into a set of possible worlds. Formulas are evaluated with respect to a possible world $a \in W$, so that $\mathcal{M}, a \models A$ reads that the formula A holds at a in model $\mathcal{M}$. The semantics of the binary connectives over a ternary model is given by:

$$\begin{array}{ll} \mathcal{M}, a \models p & \textit{iff } a \in V(p) \\ \mathcal{M}, a \models A \otimes B & \textit{iff } \exists b \exists c (Rabc \wedge \mathcal{M}, b \models A \wedge \mathcal{M}, c \models B) \\ \mathcal{M}, a \models A \to B & \textit{iff } \forall b \forall c (Rcab \wedge \mathcal{M}, b \models A \Rightarrow \mathcal{M}, c \models B) \\ \mathcal{M}, a \models B \leftarrow A & \textit{iff } \forall b \forall c (Rcba \wedge \mathcal{M}, b \models A \Rightarrow \mathcal{M}, c \models B) \end{array}$$

A formula is ***valid*** if it holds at $\mathbf{0}$ in all models. It is easy to see that a formula of the form $A \to A$ or $A \leftarrow A$ is valid at ternary formulas.

A ternary model $\mathcal{M} = (W, R, V)$ can be seen as a first-order model structure over $\mathcal{M}_{FO} = (W, R, P_1, P_2, \ldots)$, where each unary predicate P_i corresponds to a propositional letter $p_i \in \mathcal{P}$. A substructural formula can thus be translated into a first-order one, with respect to a world a, in the following way:

$$\begin{array}{lcl} FO_a(p_i) & = & P_i(a) \\ FO_a(A \otimes B) & = & \exists b \exists c (Rabc \wedge FO_b(A) \wedge FO_c(B)) \\ FO_a(A \to B) & = & \forall b \forall c (Rcab \wedge FO_b(A) \Rightarrow FO_c(B)) \\ FO_a(B \leftarrow A) & = & \forall b \forall c (Rcba \wedge FO_b(A) \Rightarrow FO_c(B)) \end{array}$$

It is straightforward to see that $\mathcal{M}, a \models A$ iff $\mathcal{M}_{FO} \models FO_a(A)$.

Like in usual modal correspondence theory, if we want to make a formula A valid over all models, this means that A should be true in all models, for all valuations; this translates into a second-order formula, obtained by the universal closure of $FO_a(A)$ over a and over all the predicate symbols occurring in it, that is:

$$\forall P_1 \ldots \forall P_n \forall a FO_a(A).$$

Such a formula provides a second-order constraint over the ternary relation R. It is particularly interesting here (as in modal logic) to know whether this second-order formula is equivalent to a first-order formula. However, it is not always possible to find such a first-order equivalent to a second-order frame constraint. We illustrate next a case where it is possible.

EXAMPLE 1 Consider the formula $A = (p \to q) \to (q \leftarrow p)$. We want to know what restrictions should be imposed on ternary frames for it to be a valid formula. For that, we compute $FO_a(A)$:

$$\begin{array}{l} FO_a((p \to q) \to (q \leftarrow p)) = \\ \quad = \forall bc (Rcab \wedge FO_b(p \to q) \Rightarrow FO_c(q \leftarrow p)) \\ \quad = \forall bc (Rcab \wedge \forall de (Rebd \wedge P(d) \Rightarrow Q(e)) \Rightarrow \forall fg (Rgfc \wedge P(f) \Rightarrow Q(g))) \\ \quad = \forall bcfg \exists de (Rcab \wedge (Rebd \wedge P(d) \Rightarrow Q(e)) \wedge (Rgfc \wedge P(f) \Rightarrow Q(g))) \end{array}$$

At this point we know that for A to be a valid formula, the ternary frame has to obey the second-order restriction $\forall P \forall Q \forall a (FO_a(A))$. To obtain a first-order equivalent to this formula, an appropriated valuation for P and Q must be provided; this is equivalent to finding a valuation for p and q in the modal context. Finding such a valuation is the crucial point of this method. Although we have a way of computing one [Fin00], if one exists, for the substructural fragment, here we just present one:

$$\begin{array}{ll} V(p) = \{f\} & \Longrightarrow \forall x(P(x) \Leftrightarrow x = f) \\ V(q) = W - \{g\} & \Longrightarrow \forall x(Q(x) \Leftrightarrow x \neq g) \end{array}$$

By substituting such a valuation in $\forall a(FO_a)(A)$ we obtain:

$$\begin{array}{l} \forall abcfg\exists de(Rcab \wedge (Rebd \wedge d = f \Rightarrow e \neq g) \Rightarrow (Rgfc \wedge \top \Rightarrow \bot)) \Longleftrightarrow \\ \forall cfg(\exists abRcab \wedge \forall de(d = f \wedge e = g \Rightarrow \neg Rebd) \Rightarrow \neg Rgfc) \Longleftrightarrow \\ \forall bcfg(\exists aRcab \wedge \neg Rgbf \Rightarrow \neg Rgfc) \end{array}$$

But since we know that, $\forall cRc0c$, it is always the case that, for $c = b$, $\exists aRcab$, so we end up with the first-order restriction:

$$\forall cfg(Rgfc \Rightarrow Rgcf)$$

That is, the restriction imposed on R is the commutativity of its second and third arguments. It remains to be shown that whenever we have the commutativity of R's second and third arguments, the formula A is valid; such a proof can be found in [Kur94]. It follows that $(p \rightarrow q) \rightarrow (q \leftarrow p)$ corresponds to the restriction of 2,3-commutativity over ternary frames. Note that it is well known that $(p \rightarrow q) \rightarrow (q \leftarrow p)$ is a theorem of substructural logics that allow for commutativity of premises in a sequent deduction [Doš93].

The really interesting part of the procedure above is to know whether the second-order formula generated is equivalent to a first-order one and what is the substitution that will lead to it. This is the basic task of our algorithm developed in [Fin00]; as there is no space for a full presentation of the method, we only briefly present it next.

2.1 The SLaKE-Tableaux Method

We compute a first order formula equivalent to a substructural sequent (or formula) by means of a construction of a tableau. This method is called SLaKE-tableau (Substructural Labelled KE).

Each formula in a SLaKE-tableau is signed with T or F and receives a label; the signed labeled formulas $T\ A : a$ and $F\ A : b$ are called ***opposites***. The original sequent $A_1, \ldots, A_n \vdash C$ is associated with an initial SLaKE-tableau:

$$\begin{array}{c} T\ A_1 : a_1 \\ \vdots \\ T\ A_n : a_n \\ F\ C : a \end{array}$$

and with a first-order formula:

$$\psi = \neg\exists aa_1 \ldots a_n[V_{a_1}(A_1) \wedge \ldots \wedge V_{a_n}(A_n) \wedge \neg V_a(C) \wedge Ra(a_1 \ldots a_{n-1})a_n \wedge \sharp_1]$$

where $V_a(A)$ is the valuation of the formula A at label a and is defined as follows:

- $V_a(A) =_{def} \top$ if A is not atomic
- $V_a(p) =_{def} (a \neq a_1) \wedge \ldots \wedge (a \neq a_n)$, where $p : a_1, \ldots, p : a_n$ occur in a branch above $p : a$ with opposite sign. If no opposite formula occurs above $p : a$, $V_a(p) =_{def} \top$.

Each of the tableau linear expansion rules is associated with an expansion of the correspondence formula of the form $\sharp_i := \psi(R, A_1, \ldots, A_n, \sharp_{i+1})$, where R is the ternary accessibility relation, $A_1, \ldots, A_n$ are the formulas generated in the expansion, and $\sharp$ is the "substitution place" for next expansion and can be read simply as *truth*. The tableau rules for SLaKE-tableaux are illustrated in Figure 1.

SLaKE Expansion	*Formula Expansion*
$\dfrac{T\ B \to A : a \quad T\ B : b}{T\ A : c \text{ (new } c)}$	$\sharp_i := \forall c(Rcab \Rightarrow (V_c(A) \wedge \sharp_{i+1}))$
$\dfrac{F\ B \to A : a}{T\ B : b \text{ (new } b) \quad F\ A : c \text{ (new } c)}$	$\sharp_i := \exists b \exists c(Rcab \wedge V_b(B) \wedge \neg V_c(A) \wedge \sharp_{i+1})$
$\dfrac{T\ A \leftarrow B : a \quad T\ B : b}{T\ A : c \text{ (new } c)}$	$\sharp_i := \forall c(Rcba \Rightarrow (V_c(A) \wedge \sharp_{i+1}))$
$\dfrac{F\ A \leftarrow B : a}{T\ B : b \text{ (new } b) \quad F\ A : c \text{ (new } c)}$	$\sharp_i := \exists b \exists c(Rcba \wedge V_b(B) \wedge \neg V_c(A) \wedge \sharp_{i+1})$
$\dfrac{T\ A \bullet B : a}{T\ A : b \text{ (new } b) \quad T\ B : c \text{ (new } c)}$	$\sharp_i := \exists b \exists c(Rabc \wedge V_b(A) \wedge V_c(B) \wedge \sharp_{i+1})$
$\dfrac{F\ A \bullet B : a \quad T\ A : b}{F\ B : c \text{ (new } c)}$	$\sharp_i := \forall c(Rabc \Rightarrow (\neg V_c(B) \wedge \sharp_{i+1}))$
$\dfrac{}{T\ A : x \quad F\ A : x}$	$\sharp_i := \forall x((V_x(A) \wedge \sharp^1_{i+1}) \vee (\neg V_x(A) \wedge \sharp^2_{i+1}))$

Figure 1: SLaKE rules

In each linear rule in Figure 1, the formulas above the horizontal line are the ***premises*** of the rule, and those below it are the ***conclusions*** of the rule. There are one-premised and two-premised rules, but each rule has exactly one premise that is a ***compound*** formula, which is called the ***main premise***; other premises are called ***auxiliary***. Two-premised rules are $\forall$-rules and one-premised rules are $\exists$-rules. If either of the conclusions of an $\exists$-rule is present on the current branch, it is not added again with a new label. $\forall$-rules always generate a new conclusion.

The last rule in Figure 1 is the Principle of Bivalence (PB), the only branching rule. It introduces two "substitution places" in the correspondence formula, $\sharp^1_{i+1}$

and $\sharp^2_{i+1}$, one for each new branch. A branch that can still be expanded is called *active*. Each active branch in a SLaKE tableau always has exactly one substitution place.

The importance of substitution places is that they guarantee that each formula introduced in the correspondence formula will "see the correct context", that is, it will be in the scope of the correct quantifiers.

A full presentation of the method is beyond the scope of this paper. Here we repeat Example 1 using the SLaKE-tableau method.

EXAMPLE 2 Consider the sequent $q \to p \vdash p \leftarrow q$. Its associated SLaKE tableau is:

1.	$T\, q \to p : a$		
2.	$F\, p \leftarrow q : a$		$\psi = \neg\exists a(\sharp_1)$
3.	$T\, q : b$	from 2	
4.	$F\, p : c$	from 2	$\sharp_1 := \exists b \exists c(Rcba \wedge \top \wedge \top \wedge \sharp_2)$
5.	$T\, p : d$	from 1, 3	$\sharp_2 := \forall d(Rdab \Rightarrow d \neq c \wedge \sharp_3)$

By putting together all substitution places we obtain the formula:

$$\psi = \neg\exists a \exists b \exists c(Rcba \wedge \forall d(Rdab \Rightarrow d \neq c))$$

which is equivalent to $\forall a \forall b \forall c(Rcba \Rightarrow Rcab)$, the commutativity of the second and third R-positions.

A tableau as above is ***deterministic***, that is, at all expansions of a branch, there is only a single expansion rule to be applied. In [Fin00] it has been shown that:

PROPOSITION 3 If the SLaKE tableau generated by a sequent is finite, saturated and deterministic, then the associated first-order formula ψ it computes is the sequent's correspondence formula.

We note that SLaKE-tableaux may be infinite, in which case no first-order formula is computed. I may also be non-deterministic, in which case we have to take the conjunction of the formulas associated to all possible SLaKE-tableaux.

2.2 Extending the Method

As the example above shows, the method is based on the semantics of the connectives. We can in this way extend the method to other connectives, such as *classical negation* ($\neg$) and *classical conjunction* ($\wedge$) given by their semantical definitions:

$$\begin{array}{l} \mathcal{M}, a \models \neg A \quad \text{iff } \mathcal{M}, a \not\models A \\ \mathcal{M}, a \models A \wedge B \text{ iff } \mathcal{M}, a \models A \textit{ and } \mathcal{M}, a \models B \end{array}$$

These semantical rules translate generate the following tableau rules:

$$\frac{T\, \neg A : a}{F\, A : a} \quad \sharp_i := \neg V_a(A) \wedge \sharp_{i+1} \qquad\qquad \frac{F\, \neg A : a}{T\, A : a} \quad \sharp_i := \neg V_a(A) \wedge \sharp_{i+1}$$

The computational results in [Fin00] do not immediately apply to such extensions, so we cannot affirm that it is a decidable process. However, the method can still be applied to particular examples with success.

But the point we are going to make here is that such a method (even if not fully automated for larger fragments) can be applied to the study of structural conditions for paraconsistency.

3 Consistent and Paraconsistent Restrictions on Ternary Frames

A ***consistency condition*** is a formula that one wants to see valid so that the system is considered consistent. As a consequence, a system will be ***paraconsistent*** with respect to a consistency condition if such a formula is invalidated.

We want to apply the techniques described above to associate a constraint over ternary frames with a consistency formula. The rejection of such constraint will therefore characterize paraconsistency over ternary models.

Usually, consistency formulas have to deal with negation. So we introduce *classical negation* ($\neg$) in our language with its usual semantics:

$$\mathcal{M}, a \models \neg A \textit{ iff } \mathcal{M}, a \not\models A$$

The obvious extension of the first-order translation is: $FO_a(\neg A) = \neg FO_a(A)$. We can thus explore the constraint associated with consistency conditions related to the principle of *non-contradiction.*

Consistency Condition 1: $\neg(p \otimes \neg p)$

We start by computing the first order translation of $\neg(p \otimes \neg p)$:

1.	$F\ \neg(p \otimes \neg p) : \mathbf{0}$	$\psi := \neg\sharp_1$
2.	$T\ (p \otimes \neg p) : \mathbf{0}$	$\sharp_1 := \neg V_{\mathbf{0}}(p \otimes \neg p) \wedge \sharp_2$
3.	$T\ p : b$	
4.	$T\ \neg p : c$	$\sharp_2 := \exists bc(R\mathbf{0}bc \wedge V_b(p) \wedge V_c(\neg p) \wedge \sharp_3)$
5.	$F p : c$	$\sharp_3 := b \neq c$

Putting everything together and doing some classical equivalences, we get the formula

$$\forall bc(R\mathbf{0}bc \rightarrow b = c)$$

That is, for the consistency condition to be valid on ternary frames, the the special world $\mathbf{0}$ is related only to pairs of identical worlds. A structural condition to paraconsistency in this case would be:

$$\exists bc(R\mathbf{0}bc \wedge b \neq c)$$

Hence for a paraconsistency that rejects the consistency condition above, it suffices that in every model there is a triple $\langle \mathbf{0}, b, c \rangle \in R$ with distinct last two arguments.

Consistency Condition 2: $\neg(p \wedge \neg p)$

Suppose now that we want to add boolean conjunction in our language so that we can study the constraint associated with the usual boolean consistency condition $\neg(p \wedge \neg p)$.

For that, first, we add the obvious semantic definition

$$\mathcal{M}, a \models A \wedge B \textit{ iff } \mathcal{M}, a \models A \textit{ and } \mathcal{M}, a \models B$$

together with its obvious first-order translation

$$FO_a(A \wedge B) = FO_a(A) \wedge FO_a(B)$$

and the tableau rules

$$\frac{T\ A \wedge B : a}{\begin{array}{l} T\ A : a \\ T\ B : a \end{array}} \quad \sharp_i := V_a(A) \wedge V_a(B) \wedge \sharp_{i+1}$$

$$\frac{\begin{array}{l} F\ A \wedge B : a \\ T\ A : a \end{array}}{F\ B : a} \quad \sharp_i := V_a(A) \wedge \neg V_a(B) \wedge \sharp_{i+1}$$

If we now apply our method to $\vdash \neg(p \wedge \neg p)$ we see that it is logically equivalent to $\top$; details omitted. This is not at all surprising, since we are dealing with both boolean negation and conjunction, which are enough to define all classical connectives, thus rejecting inconsistency.

Intuitionistic Negation

The main idea of intuitionistic negation (which we represent here as $\sim$) is to assert the negation of a formula in a world provided that this formula is not asserted at any other world "above" it. In our ternary models, if $Rabc$ then a is above b and c, which we write $a > b$ and $a > c$. Formally:

$$a > b \textit{ iff } \exists c(Rabc)$$

Such a definition is inspired on a similar one in [RM73][2]. We then have, for ternary frames, the usual intuitionistic definition of negation over Kripke models [Fit69]:

$$\mathcal{M}, a \models \sim A \textit{ iff } \forall b(b > a \Rightarrow \mathcal{M}, b \not\models A)$$

This definition generates a first-order translation:

$$FO_a(\sim A) = \forall b(b > a \Rightarrow \neg FO_b(A))$$

and SLaKE-tableau rules

$$\frac{T\ \sim A : a}{F\ A : b} \quad \sharp_i := \forall b(b > a \Rightarrow \neg V_b(A) \wedge \sharp_{i+1})$$

$$\frac{F\ \sim A : a}{T\ A : b} \quad \sharp_i := \exists b(b > a \wedge V_b(A) \wedge \sharp_{i+1})$$

We then choose as a consistency condition the formula $\sim(p \otimes \sim p)$. For space reasons we omit here the details, but when we develop the expansion we get that $\vdash \sim(p \otimes \sim p)$ corresponds to the first-order restriction:

$$\forall abc(Rabc \Rightarrow b > c)$$

[2]In fact, since we do not assume any properties of R, we could define two orders, the other one being $a >_2 c$ *iff* $\exists c(Rabc)$.

imposing the order $>$ on all R-related worlds. The paraconsistency condition here states that in every model there must exist an R-related triple $Rabc$ such that b is not above c.

Similarly, a consistency condition of the form $\sim (\sim p \otimes p)$ would generate a restriction of the form $\forall abc(Rabc \Rightarrow c > b)$, leading to a different imposition of $>$-ordering.

If both consistency conditions are required, a structural condition for paraconsistency should be that in every model there must exist an R-related triple $Rabc$ such that neither b nor c is above the other. This is expressed by the following structural condition:

$$\exists abc(Rabc \wedge \neg(b > c) \wedge \neg(c > b)).$$

Finally, we consider the consistency condition $\sim (p \wedge \sim p)$. The development of a SLaKE-tableau for $\vdash \sim (p \wedge \sim p))$ leads us to the first-order condition

$$\forall a(a > a)$$

Thus the intuitionistic consistency condition $\sim (p \otimes \sim p)$ imposes $>$-reflexivity, which is a condition normally expected in intuitionistic models. Those models support the semantic of $\wedge$ in exactly the terms defined here[3]; see e.g. [Fit69].

So a paraconsistent condition that rejects this intuitionistic view of consistency requires that every ternary model contains a $>$-irreflexive world:

$$\exists a \neg(a > a).$$

Consistency as Trivialization

Another possible way of defining a consistency condition, perhaps more in conformity with the original formulation of paraconsistency [dC74], is to state that an inconsistency trivializes implication, that is, from p and its negation we can derive any q. If we focus only on boolean conjunction, two new consistency conditions arise, namely:

1. $(p \wedge \neg p) \rightarrow q$;
2. $(p \wedge \sim p) \rightarrow q$.

By applying our method, we get their correspondent first-order restriction over ternary frames, respectively as:

1. $\top$;
2. $\forall a(a > a)$.

Item 1 implies that the consistency conditions for boolean negation based on non-contradiction and triviality lead exactly to the same restrictions over ternary frames, and hence to the same paraconsistent condition. Item 2 tells us that exactly the same fact occurs for intuitionistic negation, and the structural restriction of $>$-reflexivity is the same for both non-contradiction and triviality conditions.

[3]the transitivity of $>$ found in intuitionistic Kripke models is imposed by intuitionistic implication.

3.1 Summary and Analysis

Consistency Condition	*Structural Restriction*
$\neg(p \otimes \neg p)$	$\exists bc(R0bc \wedge b \neq c)$
$(p \otimes \neg p) \to q$	$\exists abc(Rabc \wedge b \neq c)$
$\sim (p \otimes \sim p)$	$\exists abc(Rabc \wedge \forall d \neg Rbcd)$
$(p \otimes \sim p) \to q$	$\exists abc(Rabc \wedge \forall d \neg Rbcd)$
$\sim (p \wedge \sim p)$	$\exists a \forall b \neg Raab$
$(p \wedge \sim p) \to q$	$\exists a \forall b \neg Raab$
$\neg(p \wedge \neg p)$	impossible to violate
$(p \wedge \neg p) \to q$	impossible to violate

Table 1: Structural conditions for paraconsistency

Table 1 summarizes the results obtained by our method. Each consistency condition is associated to the structural restriction that violates it, and is expressed in terms of the ternary R relation.

What calls the attention in this result is that the pairs:

$$\begin{array}{cc} \sim (p \otimes \sim p) & (p \otimes \sim p) \to q \\ \sim (p \wedge \sim p) & (p \wedge \sim p) \to q \\ \neg(p \wedge \neg p) & (p \wedge \neg p) \to q \end{array}$$

generate the same structural conditions for paraconsistency. That is, non-contradiction and the corresponding trivialization condition yield the same structural condition.

The other pair examined here is

- $\neg(p \otimes \neg p)$
- $(p \otimes \neg p) \to q$

where the latter leads to a structural restriction for paraconsistency that is implied by the the structural condition of the former.

But it is widely known that there are logics for which the non-contradiction and trivialization conditions are totally independent.

The conclusion is that such logics employ a kind of negation that is neither classical (in the sense of the semantic definition: $\mathcal{M}, a \models \neg A$ iff $\mathcal{M}, a \not\models A$) nor intuitionistic, also semantically defined. In fact, the semantics of negation may take extra parameters in these logics; for example, in [Res00] we find semantics for substructural negations that depend not only on the ternary relation R but also in a partial order $\sqsubseteq$ of ***information refinement*** where $Rabc$ does not necessarily imply $b \sqsubseteq a$. Other kinds of semantical definitions for negation can be found in [Dun94].

In the cases where intuitionistic or classical negation is employed with its fixed semantics, trivialization and non-contradiction always yield structural conditions that are either identical or strongly connected. As a last example of such connection, we will examine the structural conditions associated with ***relevant negation.***

4 Relevant Negation

There are a great range of relevant logics defined in the literature [AB75]. In several of the proposed systems, and in particular in system R, a kind of negation is used, which is represented as $\overline{A}$, meaning that it is inconsistent with the formula A.

To provide a semantics for such a negation over ternary frames, Routley and Meyer [RM73] postulated the existence of a unary function $^* : W \to W$ such that, for every $a, b, c \in W$:

1. $a^{**} = a$
2. $Rabc \Rightarrow (Ra^*bc^* \wedge Ra^*b^*c)$

With such a function, the System-R's relevant negation [AB75] is defined as:

$$\mathcal{M}, a \models \overline{A} \text{ iff } \mathcal{M}, a^* \not\models A$$

Note that in such a system, it is possible not to have neither A nor $\overline{A}$ holding at a possible world a.

With such semantics we apply our method to the following consistency conditions:

- $\overline{(A \otimes \overline{A})}$
- $(A \otimes \overline{A}) \to q$

By applying our method to it, we see that the first one imposes on the model the condition:

$$\forall bc(R\mathbf{0}^*bc \Rightarrow b = c^*)$$

whose negation leads to the paraconsistency condition:

$$\exists bc(R\mathbf{0}^*bc \wedge b \neq c^*)$$

On the other hand, by applying our method to the trivialization formula $(A \otimes \overline{A}) \to q$ we obtain the frame condition:

$$\forall abc(Rabc \Rightarrow b = c^*)$$

which is leads to the following structural restriction:

$$\exists abc(b \neq c^* \wedge Rabc)$$

Again, we see that the latter paraconsistency condition — associated with trivialization — is logically implied by the former one — associated with non-contradiction.

5 Conclusions

We have provided a method that allows us to find structural conditions on ternary Kripke frames to support paraconsistency. Our method is not biased towards any particular definition of paraconsistency. The examples developed here were based on possible definitions of consistency conditions to be refuted by a paraconsistent model.

Admittedly, the examples of consistency condition displayed here were quite simple. For the cases of consistency conditions based on the principle of non-contradiction and involving boolean conjunction and the use of boolean and intuitionistic negation, namely the formulas $\neg(p \wedge \neg q)$ and $\sim (p\wedge \sim p)$, the results obtained were the expected ones; the corresponding conditions based on the trivialization principle provided coincident conditions. This represents a validation of the method presented here.

More importantly, the examples presented show that the method, whether automated or not, is really quite flexible and may, in principle, be applicable to more daring definitions of paraconsistency than those presented here. There are several candidates for alternative negation, such as those in [Res00]:

- split negation;
- simple negation;
- De Morgan Negation;
- ortho-negation; and
- Strict De Morgan Negation;

These negations need a more refined semantics, for which the simple ternary semantics used in this paper is a limit case. We know that in such cases the formula computed by our SLaKE-tableau method is *implied* by the correspondence formula, but we do not know if the formula thus computed is *the* correspondence formula (nor do we know whether the method can decide in the generic case, as it can in the simple fragment of $\{\otimes, \rightarrow, \leftarrow\}$, whether the condition does have a first-order correspondence formula.

References

[AB75] A. R. Anderson and N. D. Belnap Jr. *Entailment: The Logic of Relevance and Necessity, volume 1.* Princeton University Press, 1975.

[BS84] R. Bull and K. Segerberg. Basic Modal Logic. In D. Gabbay and F. Guenthner, editors, *Handbook of Philosophical Logic*, volume II, pages 1–88. D. Reidel Publishing Company, 1984.

[CA81] W. A. Carnielli and L. P. Alcantara. Paraconsistent algebras. Technical Report IME-RT-I59-1981-v21-e1, Instituto de Matemática e Estatística, Universidade de Sã o Paulo, 1981.

[Car97] Bob Carpenter. *Type-Logical Semantics.* MIT Press, 1997.

[Car98] W. A. Carnielli. Possible-translation semantics for paraconsistent logics. In *Frontiers in Paraconsistency — Proceedings of the I World Congress on Paraconsistency*. King's College Publications, Ghent, 1998.

[Che80] B. F. Chellas. *Modal Logic — an Introduction*. Cambridge University Press, 1980.

[dC74] N. C. A. da Costa. On the theory of inconsistent formal systems. *Notre Dame Journal of Formal Logic*, 15(4):497–510, 1974.

[Doš93] K. Došen. A Historical Introduction to Substructural Logics. In P. Schroeder-Heister and K. Došen, editors, *Substructural Logics*, pages 1–31. Oxford University Press, 1993.

[Dun94] Dunn, J.M. "Star and Perp," *Philosophical Perspectives* 7, pp. 331-357, 1994.

[DM97] J. M. Dunn and R. K. Meyer. Combinators and Structurally Free Logic. *Logic Journal of the IGPL*, 5(4):505–538, July 1997.

[Fin00] M. Finger. Algorithmic correspondence theory for substructural categorial logic. To appear in *Proceedings of the 3rd Workshop on Advances in Modal Logic*, Leipzig, Germany; October 4-7, 2000.

[Fit69] M. Fitting. *Intuitionistic Logic Model Theory and Forcing*. North-Holland Publishing Co., 1969.

[Kur94] Natasha Kurtonina. *Frames and Labels — A Modal Analysis of Categorial Inference*. PhD thesis, Research Institute for Language and Speech (OTS), Utrecht, and Institute of Logic, Language and Information (ILLC), Amsterdam, 1994.

[Moo97] M. Moortgat. Categorial type logics. In J. Van Benthem and A. ter Meulen, editors, *Handbook of Logic and Language*, pages 93–178. Elsevier North-Holland/The MIT Press, 1997.

[Res00] G. Restall. *An Introduction to Substructural Logics*. Routledge, 2000.

[Roo91] Dirk Roorda. *Resource Logics: Proof-theoretical Investigations*. PhD thesis, Institute of Logic, Language and Information (ILLC), University of Amsterdam, 1991.

[RM73] Richard Routley and Robert K. Meyer. The semantics of entailment. In Hugues Leblanc, editor, *Truth, Syntax and Modality*, volume 68 of *Studies in Logic and the Foundations of Mathematics*, pages 199–243. North-Holland, 1973.

[vB84] J. van Benthem. Correspondence theory. In *Handbook of Philosophical Logic*, volume II, pages 167–248. D. Reidel Publishing Company, 1984.

Beyond Truth(-Preservation) *

R.E. JENNINGS Laboratory for Logic and Experimental Philosophy, Simon Fraser University, Burnaby, BC, Canada
jennings@sfu.ca

D. SARENAC Laboratory for Logic and Experimental Philosophy, Simon Fraser University, Burnaby, BC, Canada
sarenac@stanford.edu

Abstract
In this paper, we present some further results in the area of preservationist treatment of implication [[1], [2]]. In particular, we present a set of finite matrices with an emphasis on the matrix for the implicational connective. We begin with a brief discussion of the preservationist approach to logic, and then move on to present the system SX and examine its status with respect to paraconsistency.

1 PRELIMINARIES

At first blush the idea seems plausible enough. An argument is invalid if and only if the truth of its premisses does not force the truth of its conclusion. Therefore if the inference from $P \wedge \neg P$ to Q is invalid, there must be a model in which $P \wedge \neg P$ is true and Q false. If the truth-conditions of conjunction and negation do not permit such a model, then one or the other of the truth-conditions must be altered. It is not therefore that the inference from $P \wedge \neg P$ to Q is invalid, but that the inference from $P \wedge' \neg' P$ to Q is invalid, so that one or the other of conjunction or negation is not (or perhaps only not′) what we thought it was. It is natural to suppose that the only guide available to us for the devising of truth-conditions for these connectives is our conversational understanding of the English words *and* and *not*, and their correspondents in other natural languages. But if that is so, then it will surprise the classicalist that, of all the places where our understanding might have been misguided or insufficient, it should have been there that it has let us down rather than elsewhere. To put the matter another way, if earlier classical logicians had

*The authors wish to thank N.A. Friedrich who developed PARA, a computer application which greatly facilitated this research. The application is available for download from http://www.sfu.ca/llep/software.html. We also wish to thank anonymous commentators for numerous useful suggestions.

been told that controversy would eventually arise about truth-conditions of some connectives, it seems unlikely that they would have been able to predict conjunction or negation as the locus of debate. Even if our conversational understanding of their natural counterparts could let us down, one might have supposed that their standard truth-conditions could be agreed by convention as the most useful that could be devised. To be sure, we should not ever take for granted the theoretical adequacy of conversational understanding. Consider only our improverished understanding of, say, *truth* and *inference*. But one might have thought that we could rely on our ordinary understanding of conjunction and negation.

Now these last remarks would apply with even greater force in attempts to defuse contradictions in implicational, as distinct from illative, systems. For if our conversational understanding of *truth* and *inference* are an infirm ground for theory-construction, our conversational understanding of *implication* is a quagmire, and of the natural language conditional a quicksand for serious scientific purpose. Nevertheless the rational presumption of inadequacy could be set aside in theorizing about inference and implication, because there one *ought* to expect a plurality of theories corresponding to the plurality of their applications. That is to say, even if we ought not to expect a sufficient prior understanding of what inference or implication is, we ought to know what we want a particular inferential or implicational system to be used for. For different applications, different degrees of laxity in the matter of inferences from inconsistent premisses, and for that matter, inferences to tautologous conclusions, can be tolerated. Paraconsistent logic is a research programme; it is neither a quest nor a crusade. What we want to know is: what, in a given context, are plausible semantic constraints on implication and inference? What implicative and illative systems correspond to these constraints?

Even if we had a theoretically adequate understanding of *truth*, there would remain a doubt whether, for any serious scientific purpose, the notion of truth was quite what was wanted. In the physical sciences, the best we can hope for is that a sentence should remain compatible with ever finer-grained experimental outcomes. There is always the risk that the whole language in which the sentence occurs should be thrown over in favour of another having no translation for the sentences of the old. In such an eventuality, there seems little point in regarding the discarded sentences as false, rather than, from the newer perspective, as non-existent, merely as *hors de combat*, or invalided out. In the course of such theoretical refinement, we have no reason to suppose that there would be no transient phases in which the best theory we can muster harbours inconsistencies; however, no application of the theory can require us to regard *any* of the sentences that are party to the inconsistency true, let alone all of them, however many it might take. Perhaps the best sentences we can hope for in physical theory are those that persist and are experimentally confirmed throughout all such successive refinements.

None of this counts against *reading* the values in $\{1, 0\}$ as truth-values. Rather it speaks against the supposition that this *reading* is also a metaphysically informative or reliable *interpretation*. It is a reading, no more. If the allegedly surprising cases of contradictions arise in circumstances conceivable only in the realm of abstruse mathematical physics, why would we suppose that the inherited folk-theoretic notion of truth provides the right currency for accounting for their status? Why would we *insist* upon assessing the sentences of late twentieth-century physics in the language of early seventeenth-century religion? Truth provides little mathematical nutrition. The predicament can be summed up in the words of Samuel Johnson's quip.

Truth, Sir, is a cow, which will yield such people no more milk, and so they are gone to milk the bull.

The preservationist attitude is this: $V(\alpha)$ gives a measure of α. To say that the material $\rightarrow$ preserves the value 1 is to say that $V(p \rightarrow q) = 1$ iff $V(p) \leqslant V(q)$. The *ex falso quodlibet* of classical implication merely reflects the fact that $0 \leqslant 0$ and $0 \leqslant 1$. But there are many mathematically well-defined measures under the sun, and some that might discriminate between two wffs α and β even when $V(\alpha) = V(\beta) = 0$. The basic preservationist question is this: which classes of measures (or of families of measures) determine which classes of formal systems? But even this approach is highly particularistic. Classical semantics requires that $\rightarrow$ be a subrelation of $\leqslant$ on 2. But after all, $\leqslant$ is just one relation. The more general preservationist question asks which relation on which family of measures determines which formal system of implication or inference.

Classical propositional logic (henceforth PL) provides a good illustration of a virtuous preservationist system. PL is such system in at least two significant and not unrelated respects. First, the $\models$–relation of PL preserves truth from left to right (or what turns out to be the same thing, falsity from right to left). Any augmentation of a set of true PL sentences by means of the $\models$–relation results in a larger set of true sentences. In other words, $\models$ augments the set only in ways which preserve truth. Second, the implicational connective of PL (i.e. material conditional) preserves truth. A material conditional receives the designated value if and only if the truth of the antecedent is preserved to the consequent. Or, less esoterically, a material conditional is true iff the consequent has at least as much truth as the antecedent.

The simplicity of PL appeals to the preservationist aesthetic. But it evokes somewhat wistful sighs. If only the domains of application of logic were as simple, beautiful and yielding! Not only do we often face situations in which truth cannot be guaranteed, we are also occasionally expected to behave reasonably in the face of inconsistency of data. In such situations, unless we make radical assumptions about truth-conditions, the preservation of truth alone will be of little help. What can we do? Russell is alleged to have said When all else fails, lower your standards. We prescribe the opposite remedy. The Preservationist program seeks to overcome the difficulties that PL encounters by requiring inference and implication to preserve more than truth. In other words, we raise our standards by resorting to a stricter preservational regime.

In that vein, and corresponding to two preservationist virtues of PL, we could proceed in one of two ways:

1. require the $\models$–relation to be more restrictive, i.e. require it to preserve not just truth but some additional properties as well.[1]

2. require the implication connective to preserve an additional hierarchy of the properties of truth, i.e. the hierarchy of meta-valuational properties.

[1] A great deal of work has been done along these research lines. Some of the most prominent examples are found in [3], [4], [5], [6], and [7].

This research takes the latter approach. We consider the preservational merits of an implicational connective. For better understanding, we will first inquire into the notions of preservational and non-alethic profiles.

2 PRESERVATIONAL AND NON-ALETHIC PROFILES

We present the notions of *preservational* and *non-alethic profile* for an arbitrary finite matrix. Let E be the set of matrix elements, and let a, b, c range over E. Let D be a set of designated elements, and let $P_i : E \to 2$ be a property assignment function. $\mathbb{P}$ is the set of such P_i's. We first transform the elements of the matrix into elements which are represented as lists of 2-valued properties.

2.1 Transformation of the Element Set

Let E' be the element set of some matrix set. A transformation is any function $f : E' \to E$ that satisfies the following four conditions:

1. $P_1(f(a)) = 1$ iff $a \in D$
2. if $a \in E \neq b \in E$ then $\exists P_i, P_i(f(a)) \neq P_i(f(b))$
3. There are finitely many properties (P_i)
4. $|E'| = |E|$

The matrices remain unchanged in that all the matrix relations are carried over. The only difference is that the elements are now represented as lists of properties. In other words, the elements are now represented as a sequence of 0's and 1's. Each place in the sequence represents a binary property. 1 signifies that the property is present, and 0 that it is absent.
A preservational profile captures the set of designated places in a matrix by reference to properties of the matrix elements. The notion as we use it applies to the implicational connective, although it can be extended to other connectives.

DEFINITION 1 A *preservational profile* is a sentence of the form

$$P_1(a \to b) = 1 \text{ iff } \phi,$$

where ϕ specifies what property relations must obtain between the antecedent and the consequent if the implication connective is to receive a designated value.

DEFINITION 2 A preservational profile ***proper*** is a preservational profile in which ϕ is a conjunction of clauses of the type $P_i(a) \leq P_i(b)$ and $P_i(b) \leq P_i(a)$ only (for any number of P_i's).

The characterization of the class of implicational matrices which can be represented as having a preservational profile proper is an open problem. Many of the commonly explored implicational matrices can be represented this way, and all the systems mentioned in this paper have a preservational profile proper.

EXAMPLE 3 The material conditional of PL. One property is sufficient for transformation of the element set of $\vdash_L$. Call it, as we did, P_1. Then, as one would expect, the preservational profile of the implication is

$$P_1(a \to b) = 1 \text{ iff } P_1(a) \leq P_1(b)$$

In other words, the implication preserves the designated property (truth).

PROPOSITION 4 Heyting's Intuitionist system in preservationist terms. Heyting's system is based on three-valued matrices. $E = \{1, 2, 3\}$, $D = \{3\}$ [See [8]]. The matrix for the implicational connective:

→	3	2	1
3	3	2	1
2	3	3	1
1	3	3	3

Now consider the following binary representation of the elements in which the leftmost place (P_1) represents the presence or absence of truth, and the rightmost place (P_2) stands for some other metalinguistic property:

$$3 = 11, 2 = 01 \text{ and } 1 = 00$$

The → matrix in binary terms

→	11	01	00
11	11	01	00
01	11	11	00
00	11	11	11

The binary matrix reveals that the arrow preserves truth and this (nameless) additional property. The preservational profile is:

$$P_1(a \to b) = 1 \text{ iff } P_1(a) \leq P_1(b) \text{ and } P_2(a) \leq P_2(b).$$

Representing the matrices in this way can be seen as an uncovering of additional, non-classical properties preserved by the corresponding implicational connective. In the case of Heyting's implication, we discover that intuitionism amounts to preservation of the property we called P_2 in addition to truth. In general, the discovered properties can be studied independently, and, in the ideal case, can be matched with some independently known property. Once the property is identified, one further preservationist questions arise. One significant one is whether the $\models$ preserves the same properties as the implicational connective. In other words, one can ask whether the logic is preservationally symmetrical with regards to → and $\models$.

2.2 Non-Alethic Profile

The preservational profile captures aspects of the implicational matrix directly related to designation. It divides the matrix elements into designated and non-designated ones. It assigns P_1 in the implicational matrix, but leaves the question of assigning other properties undetermined. If we want to determine a unique implicational matrix, we will have to answer the question of the assignment of $P_2 \dots P_n$. As it turns out, the strategy used in the case of preservational profile straightforwardly generalizes.

DEFINITION 5 A ***non-alethic profile*** is a set of sentences (for each $i(2 \leq i \leq n)$) of the form $P_i(a \to b) = 1$ iff ϕ, where ϕ specifies what property relations must obtain between the antecedent and the consequent if the implication connective is to assign 1 to $P_i(a \to b)$.

EXAMPLE 6 The following non-alethic profile together with the preservational profile uniquely determines the Heyting implicational matrix presented above.

$$P_2(a \to b) = \begin{cases} 1 & \text{when either } P_1(a \to b) = 1 \text{ or } P_2(b) = 1, \\ 0 & \text{otherwise.} \end{cases}$$

It is evident from the example that the non-alethic profile is not unique.

DEFINITION 7 A first-degree logic is a restriction of a logic to theorems that contain at most one occurrence of the implicational connective. The implicational connective, if it occurs, is the main connective.

Let $S^1_\to$ and $S^2_\to$ be logical systems differing only in the matrix for the implicational connective.

PROPOSITION 8 Let $\to_1$ and $\to_2$ be the implicational connectives of $S^1_\to$ and $S^2_\to$ respectively. If $\to_1$ and $\to_2$ share their preservational profile, then they have the same first-degree logic.

Proof: We note that for any matrix entries a, b, $a \to_1 b$ iff $a \to_2 b$. □

Two implicational connectives that share a preservational profile, but that have distinct non-alethic profiles commonly have distinct higher-order logics. The reason for our not being categorical on this point is that the non-alethic profile could consist of vacuously satisfied restrictions.

To demonstrate that logics sharing a preservational profile can be distinct provided that their respective non-alethic profiles differ, we introduce the system SX.

3 THE LOGIC OF SX

3.1 Semantics

In [1], Jennings and Johnston present ***paradox-tolerant logic*** (PTL). The implicational connective there presented is the first explicitly preservationist implication.

The connective preserves truth and a property that the authors call *fixity*. The intuitive idea of fixity is that a truth-value is fixed iff the truth-value cannot be changed. In the case of atomic wffs, the truth-value (first assigned value) is fixed iff the second assigned value (the fixity value) is 1. For a non-atomic wff, its truth-value is fixed iff no change in the truth-values of its component subwffs can change its truth-value. Fixed truth-values of component subwffs of course cannot change. So if alterations in unfixed values will not change the truth-value of the wff as a whole, then the truth-value of the wff is fixed (has fixity value 1). An entry in the implication matrix receives a designated value if both properties are preserved. The system uses matrices for conjunction and disjunction which turn out to be isomorphic to the matrices of the above mentioned Heyting's system when an extra application of Jaśkowski's Γ-function is performed [see [9]]. The negation is internal. The only property it reverses is the designation property (the one we call P_1). The negation and disjunction matrices are as follows:

α	$\neg\alpha$
00	10
01	11
10	00
11	01

$\vee$	00	01	10	11
00	00	00	10	11
01	00	01	10	11
10	10	10	10	11
11	11	11	11	11

The other PL connectives are defined in the standard way[2]. In fact, the matrices are characteristic for PL. The main differences are in the implicational connective and the *falsum* constant, which are both independent of the standard PL connectives.

$\to$	00	01	10	11
00	10	11	10	11
01	01	11	01	11
10	00	00	10	11
11	01	01	01	11

$\bot$
01

The logic, whose main aim was paradox-tolerance, behaves admirably in the first degree fragment and even in the second degree fragment to some extent. Many suspicious implicational theorems fail. Thus, to name a few interesting ones,

$$\bot \to \alpha$$

$$\alpha \to (\neg\alpha \to \beta)$$

$$\alpha \to (\alpha \vee \beta)$$

$$(\alpha \wedge \beta) \to \alpha$$

[2]The negation is distinct from Heyting's negation and such that disjunction and conjunction are interdefinable.

all fail. (For a more thorough list see [1]). As we note elsewhere (see [2]), however, higher order counterparts of some of these theorems hold in PTL. For instance, all of

$$\bot \to (\bot \to \alpha)$$
$$\neg\alpha \to (\alpha \to (\neg\alpha \to \beta))$$
$$\alpha \to (\alpha \to (\alpha \to (\alpha \vee \beta)))$$

are theorems of PTL.

From the point of view of this investigation, this need for an increase in nesting is a centrally interesting feature of PTL. We generalize it in [2] to a sequence of logics which take a somewhat different approach to inconsistency. The logics distinguish inconsistencies not by whether they are true or false but by how complex they are. It must be admitted, however, that for the more conservative approach to paraconsistency, this feature is undesirable. You still get everything from an inconsistency, you have only to work progressively harder. The question arises: what smallest change to PTL will produce a satisfactory paraconsistent logic of implication? Is there an interesting paraconsistent logic with entirely classical conjunction and negation? Since PTL seemed satisfactory for first-degree implications, we can keep the preservational profile of PTL. The obvious candidate for the change was the non-alethic profile. The question becomes how to fill in the second place in the following matrix skeleton:

$\to$	00	01	10	11
00	1	1	1	1
01	0	1	0	1
10	0	0	1	1
11	0	0	0	1

The undesirable feature of the logic is in nesting. Nesting iterated antecedents produces implicational wffs in which the properties not preserved in the first degree are gradually lost in higher degrees until a more or less distant relative of the unattractive sentence reappears in the logic.

EXAMPLE 9 ***Ex falso quodlibet.*** As we have already mentioned, $\bot \to \alpha$ is not a theorem of PTL. Since the $\bot$ always has a fixity value of 1, the formula fails when the truth-value of α is not fixed.

$\bot$	α	$\bot \to \alpha$
01	00	01
01	01	11
01	10	01
01	11	11

But, since the resulting table always receives a fixed value, $\bot \to (\bot \to \alpha)$ is a theorem. In other words, what we gain through preservational strictness, we lose through nesting.

Now, consider what would happen if the conditional always kept the second value of the consequent. On every nesting the same property profile would have to be preserved if the conditional is to receive a designated value. The logic of implication would behave in every degree as it does in the first degree. The desired non-alethic profile is:

$$P_2(a \to b) = \begin{cases} 1 & \text{when } P_2(b) = 1, \\ 0 & \text{otherwise.} \end{cases}$$

The completed implicational matrix becomes:

$\to$	00	01	10	11
00	10	11	00	11
01	00	11	00	11
10	00	01	10	11
11	00	01	00	11

3.2 Axiomatization and Completeness

It is often said that an axiomatization is a good one to the extent that it simplifies the completeness proof. An axiomatization, on this view, is a translation manual, and an elegant translation manual is easy to use. In devising an axiomatization for SX we take a somewhat literalist approach. We translate semantics into syntax, connective by connective, translating properties into corresponding syntactic expressions. What enables the translation is the following simple fact noted by Jennings and Johnston in [1]. The sentence $\bot \to \alpha$ is true if and only if α has a 1 in the place of the second property. We abbreviate $\bot \to \alpha$ $\Delta\alpha$ and use it to translate matrices directly into axioms of the logic. In the case of the implicational connective, the preservational and non-alethic profiles will mechanically give us the needed axioms. Thus, the preservational profile translates into

$$(\alpha \to \beta) \equiv ((\alpha \supset \beta) \wedge (\Delta\alpha \supset \Delta\beta)) \tag{1}$$

As is already mentioned, $\alpha \supset \beta$ is simply $\neg\alpha \vee \beta$, and $\alpha \equiv \beta$ is $(\alpha \supset \beta) \wedge (\beta \supset \alpha)$. The non-alethic profile is translated into

$$\Delta(\alpha \to \beta) \equiv \Delta\beta. \tag{2}$$

As it turns out, these are the only two implicational axioms needed for completeness. The other axioms needed are some or other axioms sufficient for the completeness of PL and the delta translation manual for disjunction and negation:

$$\Delta\alpha \equiv \Delta\neg\alpha \tag{3}$$

$$\Delta(\alpha \vee \beta) \equiv (((\Delta\alpha \wedge \Delta\beta) \vee (\Delta\alpha \wedge \alpha)) \vee (\Delta\beta \wedge \beta)) \tag{4}$$

3.3 Completeness construction

The proof itself is a standard Henkin-style construction. The fundamental theorem splits into two parts. For Σ an SX-maximal consistent set, and P_i^Σ suitably defined, it must be proved not only that

$$\alpha \in \Sigma \text{ iff } P_1^\Sigma(\alpha) = 1,$$

but also that

$$\Delta\alpha \in \Sigma \text{ iff } P_2^\Sigma(\alpha) = 1.$$

The PL base enables us to carry over a number of PL meta-theorems and definitions. The notion of maximal consistent set is that of PL.

DEFINITION 10 A set Σ is maximal SX-consistent iff

1. $\Sigma \nvdash \bot$
2. For any formula α, if $\alpha \notin \Sigma$ then $\Sigma \cup \alpha \vdash_{SX} \bot$.

It should be noted that, in SX an inconsistency can arise with respect to the syntactic counterpart of property P_2, namely Δ sentences.

We use the standard Henkin construction to show that every SX–consistent set has a model.

DEFINITION 11 An SX–Henkin model is an ordered triple:

$$\langle \Sigma, P_1^\Sigma, P_2^\Sigma \rangle$$

where

1. Σ is a SX–maximal consistent set,
2. for every atomic sentence, $p_i, P_1^\Sigma(p_i) = 1$ iff $p_i \in \Sigma$ and,
3. for every atomic sentence, $p_i, P_2^\Sigma(p_i) = 1$ iff $\Delta p_i \in \Sigma$.

3.4 The Fundamental Theorem

We prove

$$\alpha \in \Sigma \text{ iff } P_1^\Sigma(\alpha) = 1$$

only for the case $\alpha = \beta \to \gamma$. The other cases carry over unchanged from PL.

Proof:

($\Leftrightarrow$) By 1, I.H. and the preservational profile. □

For the second property, P_2, we prove three cases:

$$\alpha = \Delta\neg\beta$$
$$\alpha = \Delta(\beta \vee \gamma)$$
$$\alpha = \Delta(\beta \to \gamma)$$

Given that the respective definitions of the connective are:

$$P_2(\neg\beta) = 1 \text{ iff } P_2(\beta) = 1,$$

$$P_2(\beta \vee \gamma) = 1 \text{ iff } P_2(\beta) = P_2(\gamma) = 1, \text{ or } P_2(\beta) = P_1(\beta) = 1, \text{ or } P_2(\gamma) = P_1(\gamma) = 1,$$

and the above mentioned non-alethic profile in the case of the arrow, and given axioms 2, 3, and 4, the actual proof is straightforward, and will not be rehearsed here.

4 IS SX PARACONSISTENT?

Throughout the intricate history of paraconsistency, numerous moves have been made to set some sort of minimum standards which every paraconsistent logic ought to meet. Most prominently, S. Jaśkowski has initiated a set of standards upon which N.C.A. da Costa has later improved (for quick details, see [10] and [11]). The standards proposed often yield interesting systems, and we agree that for particular domains of application some standards will have to be set. In fact, the properties of the domain to which the logic is to be applied will dictate the appropriate standards. The job of a researcher is to explicate and meet those standards. Any theoretical move that ventures beyond that, however, seems to us unnecessarily restrictive. Paraconsistency is a relatively fresh academic discipline, and such premature circumscriptions can only stultify or even promote its decline. If paraconsistency as a research project is to survive, we will need lots of room for interesting applications and for systems suited to such applications. The insistence upon some illiberal set of conditions to be met, the dogma that there is one true paraconsistent logic only works to frustrate that project.

SX is all about implication, and all its preservationist virtues lie in the negation-implication fragment of the logic. The consequence relation is classical and all the connectives outside the implication behave classically. No contradiction in SX ever receives a designated value, and $\neg(\alpha \wedge \neg\alpha)$ is a theorem of the logic. And yet, we believe that SX models the paraconsistent context very realistically. Inconsistencies are undesirable features of our theories. We would much rather avoid them if we could. The fact that $\neg(\alpha \wedge \neg\alpha)$ is a theorem of SX seems to capture our dislike for inconsistencies. But experience teaches us that inconsistencies are omnipresent, and it also teaches us that once we discover their presence we do not go insane and start, as it were, speaking in strings, inferring whatever crosses our minds. We wisely deal with the inconsistent specification trying to preserve as many good properties of its context as we can. The implication of SX captures our preservationist's prudent good sense. In the inconsistent situations we infer only enough to remedy the problem. And short of that ideal goal, we infer only in such a way as to not make things any worse. *Primum non nocere*, as Hippocrates speaks through the mouth of the eminent preservationist researcher P.K. Schotch.

References

[1] R.E. Jennings and D.K. Johnston. Paradox tolerant logic. *Logique Et Analyse*, 1983.

[2] Darko Sarenac and R.E. Jennings. The preservation of meta-valuational properties and the meta-valuational properties of implication. In *Logical Consequence: Rival Approaches and New Studies in Exact Philosophy, Logic, Mathematics and Science (Vol I)*. Hermes Science Publishers, 2000.

[3] B. Brown. Adjunction and aggregation. *Nous*, 33(2):273–283, 1999.

[4] B. Brown. Yes, Virginia, there really are paraconsistent logics. *Journal of Philosophical Logic*, 28:489–500, 1999.

[5] R.E. Jennings and P.K. Schotch. Preservation of coherence. *Studia Logica*, 1984.

[6] P.K. Schotch and R.E. Jennings. Inference and necessity. *Journal of Philosophical Logic*, 9:327–340, 1980.

[7] P.K. Schotch and R.E. Jennings. On detonating. In R. Routley, G. Priest, and N. Norman, editors, *Paraconsistent Logic, Essays on the Inconsistent*, pages 306–327. Philosophia Verlag, 1989.

[8] A. Heyting. Die formalen regeln der intuitionistischen logik. *Sitzungsberichte der Preussischen Akademie der Wissenschaften, Physikalisch–mathematische Klasse*, 1930.

[9] S. Jaśkowski. Propositional calculus for contradictory deductive systems. *Studia Logica*, 24:143–157, 1969.

[10] S. Jaśkowski. Recherches sur le système de la logique intuitioniste. In *Congrès International de Philosophie Scientifique, part iv*, pages 58–61. Paris, 1935.

[11] N. C. A. da Costa. On the theory of inconsistent formal systems. *Notre Dame Journal of Formal Logic*, 15(4):497–510, 1974.

Paraconsistency in Chang's Logic with Positive and Negative Truth Values *

RENATO A. LEWIN Facultad de Matemáticas
Pontificia Universidad Católica de Chile
Casilla 306 Correo 22, Santiago, CHILE
rlewin@mat.puc.cl

MARTA S. SAGASTUME Departamento de Matemáticas
Universidad Nacional de La Plata
La Plata, ARGENTINA
marta@cacho.mate.unlp.edu.ar

Abstract
In [1], C. C. Chang introduced a natural generalization of Łukasiewicz infinite valued propositional logic Ł. In this logic the truth values are extended from the interval [0,1] to the interval [-1,1]. We will call $Ł^*$ the logic whose designated values are those greater or equal than 0. (Chang calls this logic $p^*[0]$.)

In this semantics, for a truth assignment v the value of the negation is $v(\neg\varphi) = -v(\varphi)$. This implies that there are sentences for which $v(\varphi) = v(\neg\varphi) = 0$, that is, both sentences are tautologies. Moreover, the sentence $\varphi \to (\neg\varphi \to \psi)$ is not a tautology so $Ł^*$ is paraconsistent.

Two are the main results of this paper. First we axiomatize the system $Ł_0^*$, the logic whose only designated truth value is 0, that is, the paraconsistent sentences of $Ł^*$. Then, we prove that the categories $\mathcal{MV}$ and $\mathcal{MV}^*$, whose objects are MV–algebras and MV^*–algebras respectively, with their corresponding morphisms, are equivalent. These categories are associated with Łukasiewicz' infinite valued calculus and with Chang's logic $Ł^*$, respectively.

1 Introduction

It is well known that $\mathcal{I} = (\langle[0,1];\to,\neg\rangle, \{1\})$ where the operations on $[0,1]$ are given by

$$x \to y = \min(1, 1 - x + y),$$
$$\neg x = 1 - x.$$

*Funding for the first author has been provided by FONDECYT grant 199–0433 and FOMEC.

is a characteristic matrix for Lukasiewicz infinite valued propositional logic **L**. In [1], C. C. Chang introduced a natural generalization of **L** whose characteristic matrix is $\mathcal{I}^* = (\langle [-1,1]; \rightarrow, \neg, \rangle, [0,1])$ with operations defined by

$$\begin{aligned} x \rightarrow y &= \min(1, \max(-1, y - x), \\ \neg x &= -x. \end{aligned}$$

We will call this logic $\mathbf{L}^*$.

In a previous paper, [2], Chang had established that **L** had the variety of MV–algebras as what we now call an equivalent algebraic semantics in the sense of [3]. In fact this variety is generated by the algebra $\langle [0,1]; \oplus, \neg, 1 \rangle$, whose operations are

$$\begin{aligned} x \oplus y &= \min(1, x + y), \\ \neg x &= 1 - x. \end{aligned}$$

The reader may consult [4, 5] for full information on **L**, MV–algebras and other related topics.

In the same sense that the truth values [-1, 1] extend [0, 1], in [1], Chang introduces MV*–algebras and proves that they correspond to the Lindenbaum–Tarski algebras of $\mathbf{L}^*$, so they are an equivalent algebraic semantics for $\mathbf{L}^*$. This variety is generated by the algebra $\langle [-1,1]; \oplus, \neg, 1 \rangle$, where the operations are

$$\begin{aligned} x \oplus y &= \min(1, \max(-1, x + y)), \\ \neg x &= -x, \end{aligned}$$

that is, addition is ordinary real number addition truncated below at -1 and above at 1, and negation is the additive inverse.

Chang gives no intuitive interpretation for negative truth values, nevertheless, in the last decade positive and negative truth values have appeared in different contexts, both theoretical and applied.

For instance in [6, 7], comparative logics are introduced to model situations in which propositions are either true or false, but not necessarily in the same way, thus one can admit that one proposition might be "truer" than another, so truth values are many shades of truth and falsehood. It should be noted that the algebraic semantics for comparative logics is the variety of pre–groups. In a pre–group there are two distinguished elements $\mathbf{0}$ for the least positive (or true) truth value and $-\mathbf{0}$ for the largest negative (or false) truth value. Observe that there might exist intermediate values between $-\mathbf{0}$ and $\mathbf{0}$. If $-\mathbf{0} = \mathbf{0}$ then the pre–group is an Abelian lattice ordered group (an l–group.) Moreover, an l–group is a subpre–group of a pre–group P if and only if P is an l–group.

Positive and negative truth values have also appeared in the context of uncertain information processing and inconsistency tolerance in expert systems. In [8], the authors make a theoretical analysis of uncertainty processing in a broad class of compositional expert systems similar to MYCIN and PROSPECTOR. In these, the knowledge of a questionnaire, that is, an assignment of a weight to each question, can be extended to all propositions in such a way that each rule in the Rule Base of the system contributes to the weight of each proposition according to some fixed function. So given a Rule Base $\Theta = \{R_1, \ldots, R_n\}$ and a weight function (or questionnaire) w, the weight of a proposition p is

$$W_\Theta(p \mid w) = w_1 \oplus \cdots \oplus w_n,$$

obtained from the contributions $w_1, \ldots, w_n$ of each rule through a *combining function* $\oplus$.

Some natural conditions imposed on the set of weights are that they are a linearly ordered set G, with a largest element $\top$ (*true*,) a least element $\bot$ (*false*) and a distinguished element o (*no preference*,) that is also a neutral element for the combining function (*addition*) $\oplus$. The addition is closed on $G - \{\top, \bot\}$ (uncertainties cannot give certainty), and $G - \{\top, \bot\}$ has a structure of an *ordered Abelian group*. Therefore, G with this operation and order is what is called an *extended ordered Abelian group*.

In the case of PROSPECTOR, its ordered Abelian group of weights is $\mathbf{PP} = \langle (0,1), \oplus, \leq \rangle$, defined by the usual order and

$$x \oplus y = \frac{xy}{xy + (1-x)(1-y)}.$$

The ordered Abelian group of certainty factors of MYCIN is isomorphic to **PP**. A review of these results and many references on the subject appear in [9].

Recent work on uninorms should also be mentioned in the context of positive and negative truth values. Uninorms are a generalization of t–norms and t–conorms having a neutral element e. There is a natural association of the interval $(e, 1)$ with positive values and of $(0, e)$ with negative values. Uninorms are also related to the combining functions mentioned in the previous paragraphs, for instance, in [10] it is proved that the combining function for MYCIN is a uninorm. In fact, in [8], though they do not use the concept, it is shown that all MYCIN–like expert systems combining functions are (representable) uninorms. For information on uninorms see for example [11, 12, 13, 10, 14]

Even without these interpretations, a system with positive and negative truth values such as $\mathbf{L}^*$ has a special interest in itself as a non–classical logic. We observe that certain sentences, like for instance $p \to p$, take value 0 for any valuation $v : Variables \longrightarrow [-1, 1]$. Now since 0 is a designated value, $p \to p$ is a tautology of $\mathbf{L}^*$. But then, $\neg(p \to p)$ is also a tautology. Furthermore, we observe that $v(p \to (\neg p \to q)) = v(q)$, that is, if q takes a negative value, $p \to (\neg p \to q)$ is not a tautology. So this is a non–trivial inconsistent system, that is, $\mathbf{L}^*$ is paraconsistent.

The paper is organized as follows. After this Introduction, the first section introduces Chang's logic $\mathbf{L}^*$, its semantics and shows that what is now called its equivalent algebraic semantics, is the class of MV*–algebras introduced by Chang, all this is a rephrasing of the contents of [1]. In the next section we give an axiomatization for the subsystem of all those formulas that take value 0 for any valuation v, that is, the paraconsistent fragment of $\mathbf{L}^*$, and we prove its algebraizability in the sense of [3]. In the last section we prove that $\mathcal{MV}$ and $\mathcal{MV}^*$, the categories whose objects are MV–algebras and MV^*–algebras respectively, with their corresponding morphisms, are equivalent.

2 THE LOGIC Ł*

In this section we reproduce part of the contents of [1]. Some of them have be rephrased to better suit our purposes.

2.1 Language

The set $\mathcal{F}m$ of the formulas of this logic is recursively generated from a denumerable set Va of propositional variables by a binary operation $\rightarrow$, a unary operation $\neg$, and a constant $\mathbf{1}$. As usual, $x \leftrightarrow y$ stands for $x \rightarrow y$ and $y \rightarrow x$. We will also use the following abbreviations.

$$\begin{aligned} x^+ &:= (x \rightarrow \mathbf{1}) \rightarrow \mathbf{1}, \\ x^- &:= (x \rightarrow \neg\mathbf{1}) \rightarrow \neg\mathbf{1}, \\ x \vee y &:= ((x^+ \rightarrow y^+)^+ \rightarrow (\neg x)^-) \rightarrow ((y^- \rightarrow x^-)^- \rightarrow x^-). \end{aligned}$$

2.2 Axioms

P1 $(x \rightarrow y) \leftrightarrow (\neg y \rightarrow \neg x)$,

P2 $x \leftrightarrow ((y \rightarrow y) \rightarrow x)$,

P3 $\neg(x \rightarrow y) \leftrightarrow (y \rightarrow x)$,

P4 $x \rightarrow \mathbf{1}$,

P5 $\mathbf{1} \leftrightarrow ((\mathbf{1} \rightarrow x) \rightarrow \mathbf{1})$,

P6 $((x \rightarrow \mathbf{1}) \rightarrow ((y \rightarrow \mathbf{1}) \rightarrow z)) \rightarrow ((y \rightarrow \mathbf{1}) \rightarrow ((x \rightarrow \mathbf{1}) \rightarrow z))$,

P7 $(x \rightarrow y) \leftrightarrow ((y^+ \rightarrow x^-) \rightarrow (x^+ \rightarrow y^-))$,

P8 $(x \rightarrow (\neg x \rightarrow y))^+ \leftrightarrow (x^+ \rightarrow (\neg(x^+) \rightarrow y^+))$,

P9 $(x \rightarrow (y \vee z)) \leftrightarrow ((x \rightarrow y) \vee (x \rightarrow z))$,

P10 $(x \vee (y \vee z)) \leftrightarrow ((x \vee y) \vee z)$.

2.3 Rules

R1 $x\,,\, x \rightarrow y \vdash_{\text{Ł}^*} y$,

R2 $x \rightarrow y\,,\, u \rightarrow v \vdash_{\text{Ł}^*} (y \rightarrow u) \rightarrow (x \rightarrow v)$,

R3 $x \vdash_{\text{Ł}^*} x^-$.

2.4 Semantics

We extend recursively any valuation $v : Va \longrightarrow [-1,1]$ to all formulas by defining

$$\overline{v} : \mathcal{F}m \longrightarrow [-1,1],$$

as follows.

1. $\overline{v}(x) = v(x)$,
   for any propositional letter $x \in Va$,
2. $\overline{v}(x \to y) = \min(1, \max(-1, \overline{v}(y) - \overline{v}(x)))$,
3. $\overline{v}(-x) = -\overline{v}(x)$.

The following lemma is immediate.

Lemma 1 *For any valuation* v,

1. $\overline{v}(x^+) = \max(0, \overline{v}(x))$,

2. $\overline{v}(x^-) = -\max(0, \overline{v}(-x))$,

3. If $\overline{v}(\varphi \to \psi) = 0$, *then* $\overline{v}(\varphi) = \overline{v}(\psi)$.

Observe that the value of x^- is not the usual negative part of x but minus the negative part of x.

The soundness and (weak) completeness theorem is the main result in [1].

Theorem 2

$\vdash_{\mathcal{L}^*} x$ *if and only if* $\models_{\mathcal{I}^*} \varphi$ *if and only if* $\overline{v}(x) \geq 0$ *for any valuation* v.

An easy induction on the complexity of the proof will show that the strong version of the soundness theorem also holds. It is enough to check that the three rules go from non-negative values to non-negative values.

Theorem 3

$$\Gamma \vdash_{\mathcal{L}^*} \varphi \quad \Rightarrow \quad \Gamma \models_{\mathcal{I}^*} \varphi .$$

2.5 Algebraization

In [1], Chang introduces MV*–algebras as a generalization of MV–algebras. MV*–algebras can be built from a totally ordered group $\mathbf{G}$ and a positive element u of $\mathbf{G}$ in the same way as MV–algebras are built from ordered groups and a positive element. Let $G(u) = \{x \in G : -u \leq x \leq u\}$ and define the algebra $\mathbf{G}(u) = \langle G(u); \oplus, \neg, \mathbf{1}\rangle$ where

$$\begin{aligned} x \oplus y &= \min(u, \max(-u, x + y)), \\ \neg x &= -x, \\ \mathbf{1} &= u. \end{aligned}$$

Then $\mathbf{G}(u)$ is an MV*–algebra. In Theorem 2.21 Chang proves that every MV*–algebra is a subdirect product of algebras $\mathbf{G}(u)$.

The most interesting result for our paper is Theorem 3.13, where it is proven that the Lindenbaum–Tarski algebra of $\mathbf{L}^*$ is an MV*–algebra.

We rewrite this in the current terminology of algebraic logic, (see [3].)

Theorem 4 *The deductive system $Ł^*$ is algebraizable with one defining equation*

$$x \approx x^+$$

and one equivalence formula

$$x\Delta y := x \leftrightarrow y$$

(strictly speaking these are two formulas.) The equivalent algebraic semantics is the variety of MV^*-algebras.

3 The Logic $Ł_0^*$

In this section we introduce an axiomatization for all those sentences of the language of $Ł^*$ that take value 0 for any valuation.

3.1 Axioms

Ax1 $(x \to y) \leftrightarrow (\neg y \to \neg x)$,

Ax2 $x \leftrightarrow ((y \to y) \to x)$,

Ax3 $\neg(x \to y) \leftrightarrow (y \to x)$,

Ax4 $\mathbf{1} \leftrightarrow ((\mathbf{1} \to x) \to \mathbf{1})$,

Ax5 $((x \to \mathbf{1}) \to ((y \to \mathbf{1}) \to z)) \to ((y \to \mathbf{1}) \to ((x \to \mathbf{1}) \to z))$,

Ax6 $(x \to y) \leftrightarrow ((y^+ \to x^-) \to (x^+ \to y^-))$,

Ax7 $(x \to (\neg x \to y))^+ \leftrightarrow (x^+ \to (\neg(x^+) \to y^+))$,

Ax8 $(x \to (y \vee z)) \leftrightarrow ((x \to y) \vee (x \to z))$,

Ax9 $(x \vee (y \vee z)) \leftrightarrow ((x \vee y) \vee z)$.

3.2 Rules

MP $x\, ,\, x \to y \vdash_{Ł_0^*} y$.

TR $x \to y \vdash_{Ł_0^*} (y \to v) \to (x \to v)$.

K* $x \vdash_{Ł_0^*} y \to (x \to y)$.

3.3 Some Results

In this section we summarize several results that will be useful in the sequel. The main axioms, rules and previous results used in the proofs of the first two theorems are indicated.

Theorem 5 *Define* $\mathbf{0} := \mathbf{1} \to \mathbf{1}$. *The following hold in* $Ł_0^*$.

1. *If* $\vdash_{Ł_0^*} x \to y$ *and* $\vdash_{Ł_0^*} y \to z$, *then* $\vdash_{Ł_0^*} x \to z$.

2. $x \to y\,,\, u \to v \vdash_{L_0^*} (y \to u) \to (x \to v)$.

3. If $\vdash_{L_0^*} x \leftrightarrow y$*, then* $\vdash_{L_0^*} \neg x \leftrightarrow \neg y$.

4. If $\vdash_{L_0^*} x \leftrightarrow y$ *and* $\vdash_{L_0^*} u \leftrightarrow v$*, then* $\vdash_{L_0^*} (x \to u) \leftrightarrow (y \to v)$.

5. If $\vdash_{L_0^*} x \leftrightarrow y$ *and* $\vdash_{L_0^*} y \leftrightarrow z$*, then* $\vdash_{L_0^*} x \leftrightarrow z$.

6. Let ψ *be obtained from* φ *by replacing some instances of the subformula* σ *of* φ *by the formula* τ *. Assume also that* $\vdash_{L_0^*} \sigma \leftrightarrow \tau$ *. Then* $\vdash_{L_0^*} \varphi \leftrightarrow \psi$.

7. $\vdash_{L_0^*} x \to x$.

8. $\vdash_{L_0^*} (x \to x) \leftrightarrow (y \to y)$.

9. $\vdash_{L_0^*} \neg(x \to x) \leftrightarrow (x \to x)$.

10. $\vdash_{L_0^*} \neg\neg(x \to x) \leftrightarrow (x \to x)$.

11. $\vdash_{L_0^*} \neg\neg x \leftrightarrow x$.

12. $\vdash_{L_0^*} \neg x \leftrightarrow (x \to \mathbf{0})$.

13. $\vdash_{L_0^*} \neg(x \to y) \leftrightarrow (\neg x \to \neg y)$.

14. $x\,,\, y \vdash_{L_0^*} x + y$.

15. $x\,,\, y \vdash_{L_0^*} x \to y$.

16. $x \vdash_{L_0^*} \neg x$.

Proof: Item *1.* is obtained from (TR) by (MP). The second item is L^*'s Rule R2, of which the transitivity rule TR is a special case. We have chosen to replace R2 by TR since the latter is simpler. Assertions *3.* through *11.* were proven in [1] for L^*, without using axiom P4 or rule R3. Given that R2 can be replaced by TR, the same proof works here. Notice that *8.* states that for any x, $\vdash_{L_0^*} (x \to x) \leftrightarrow \mathbf{0}$.

12., *13.* and *14.* follow from Ax1, Ax2 and Ax3, while *15.* and *16.* follow from K^* and previous items of this theorem.

□

Theorem 6 *The following hold in* L_0^* .

1. $x \to y \vdash_{L_0^*} x^+ \to y^+$.

2. $x \vdash_{L_0^*} x^-$ *and* $x \vdash_{L_0^*} x^+$.

3. $\vdash_{L_0^*} x \leftrightarrow (x^+ + x^-)$.

4. $\vdash_{L_0^*} (\neg x)^+ \leftrightarrow \neg x^-$ *and* $\vdash_{L_0^*} (\neg x)^- \leftrightarrow \neg x^+$.

5. $\vdash_{L_0^*} x^{-+}$ *and* $\vdash_{L_0^*} x^{+-}$.

6. $\vdash_{L_0^*} x^+ \leftrightarrow x^{++}$ *and* $\vdash_{L_0^*} x^- \leftrightarrow x^{--}$.

7. $\vdash_{L_0^*} x \vee \mathbf{0} \leftrightarrow x^+$.

Proof: Item *1.* is a consequence of Thm. 5 (2), while *2.*, which is Chang's rule R2, follows from K^*. Item *3.* is a special case of Ax6. Item *4.* follows from several items in Thm. 5, while *5.* uses Ax4. Items *6.* and *7.* follow from all previous results. □

The next lemma contains some technical properties related to the order defined within the system. Item *4* is specially useful in the proof of the completeness theorem.

Lemma 7 *The following hold in* $Ł_0^*$.

1. $(x \to y)^- \vdash_{Ł_0^*} y \leftrightarrow (x \vee y)$.
2. $(x \to y)^-, (y \to z)^- \vdash_{Ł_0^*} (x \to z)^-$.
3. $(x \to y)^- \vdash_{Ł_0^*} ((z \to x) \to (z \to y))^-$.
4. $(x \to y)^- (u \to v)^- \vdash_{Ł_0^*} ((y \to u) \to (x \to v))^-$.

Proof: Using Ax8 and Thm. 6 (7), we prove that

$$\vdash_{Ł_0^*} (y \to (x \vee y)) \leftrightarrow (y \to x)^+ , \qquad (*)$$

and then by Thm. 6 (4), we obtain *1.*

Item *2* follows from *1*, Ax9 and (*). The proof of item *3* uses *1*, Ax8 and (*).

Next we observe that by Ax1 and Thm. 5 (13), $\vdash_{Ł_0^*} ((u \to x) \to (u \to y)) \leftrightarrow ((y \to u) \to (x \to u))$, which together with *2* and *3* yield *4*. □

3.4 Soundness and Completeness

Consider the matrix $\mathcal{I}_0^* = (\langle[-1,1]; \to, \neg 1\rangle, \{0\})$. We first observe that the system is sound.

Theorem 8 *If* $\Gamma \vdash_{Ł_0^*} \varphi$, *then* $\Gamma \models_{\mathcal{I}_0^*} \varphi$.

Proof: Simply check that for any valuation all axioms take value 0 and the rules go from value 0 to value 0. □

We will now prove a weak completeness theorem.

Theorem 9 *If* $\models_{\mathcal{I}_0^*} \varphi$, *then* $\vdash_{Ł_0^*} \varphi$.

Proof: Assume that φ takes value 0 for any valuation. Then $\neg\varphi$ also takes value 0 for any valuation.

By the Completeness Theorem 2, in $Ł^*$ there are proofs $\langle \sigma_1, \sigma_2, \ldots, \sigma_n \rangle$ of φ and $\langle \tau_1, \tau_2, \ldots, \tau_m \rangle$ of $\neg\varphi$.

We claim that for $1 \le i \le n$ and $1 \le j \le m$ $\vdash_{Ł_0^*} \sigma_i^-$ and $\vdash_{Ł_0^*} \tau_j^-$. This is a straightforward induction on the complexity of the proof.

Case 1. If σ_i is an axiom different from Axiom P4, then by Thm. 6 (2), $\vdash_{Ł_0^*} \sigma_i^-$.

If σ_i is Axiom P4, then $\sigma_i^- = (\tau \to \mathbf{1})^-$. But then

$$\begin{array}{lr}
\vdash_{L_0^*} \mathbf{1} \to ((\mathbf{1} \to \tau) \to \mathbf{1}), & \text{Ax4,} \\
\vdash_{L_0^*} \mathbf{1} \to (\neg(\tau \to \mathbf{1}) \to \neg\neg\mathbf{1}), & \text{Ax3, Thm. 5 (8),} \\
\vdash_{L_0^*} \mathbf{1} \to \neg((\tau \to \mathbf{1}) \to \neg\mathbf{1}), & \text{Thm. 5 (13),} \\
\vdash_{L_0^*} ((\tau \to \mathbf{1}) \to \neg\mathbf{1}) \to \neg\mathbf{1}, & \text{Ax1,} \\
\vdash_{L_0^*} (\tau \to \mathbf{1})^-, & \text{Def.}
\end{array}$$

So for any axiom σ_i, $\Gamma \vdash_{L_0^*} \sigma_i^-$

Case 2. If σ_i is obtained in $\mathbf{L}^*$ by Rule R1, that is, for some $j, k < i$, $\sigma_k = \sigma_j \to \sigma_i$, so

$$\begin{array}{lr}
\vdash_{L_0^*} (\sigma_j \to \sigma_i)^-, & \text{Induction Hyp.,} \\
\vdash_{L_0^*} \sigma_j^-, & \text{Induction Hyp.,} \\
\vdash_{L_0^*} (\mathbf{0} \to \sigma_j)^-, & \text{Ax2,} \\
\vdash_{L_0^*} ((\sigma_j \to \sigma_j) \to (\mathbf{0} \to \sigma_i))^-, & \text{Lem. 7 (4),} \\
\vdash_{L_0^*} (\mathbf{0} \to \sigma_i)^-, & \text{Ax2,} \\
\vdash_{L_0^*} \sigma_i^-, & \text{Ax2.}
\end{array}$$

Case 3. If σ_i is obtained by Rule R2, $\vdash_{L_0^*} \sigma_i^-$ is obtained by Lem. 7 (4).

Case 4. If σ_i is obtained by Rule R3, then $\sigma_i = \sigma_k^-$, for some $k < i$, and thus, $\sigma_i^- = \sigma_k^{--}$. Since by inductive hypothesis, $\vdash_{L_0^*} \sigma_k^-$, by Theorem 6 (6), $\vdash_{L_0^*} \sigma_i^-$. The proof for the τ_j's is similar. So we have proven that both $\vdash_{L_0^*} \varphi^-$ and $\vdash_{L_0^*} (\neg\varphi)^-$. We know that the latter is equivalent to, $\vdash_{L_0^*} \neg(\varphi^+)$, so

$$\begin{array}{lr}
\neg(\varphi^+), \varphi^- \vdash_{L_0^*} \neg(\varphi^+) \to \varphi^-, & \text{Thm. 5 (15),} \\
\neg(\varphi^+), \varphi^- \vdash_{L_0^*} \varphi, & \text{Thm. 6 (3), Def.}
\end{array}$$

This completes the proof of $\vdash_{L_0^*} \varphi$ and thus, of our theorem. □

4 Algebraizability

Theorem 10 *The deductive system L_0^* is algebraizable with one defining equation*

$$x \approx \neg x$$

and one equivalence formula

$$x\Delta y := x \leftrightarrow y.$$

Proof: The proof that $x\Delta y$ defines a congruence on the algebra of terms follows immediately from the theorems in Section 3.2.

$x \vdash_{L_0^*} x \leftrightarrow \neg x$ follows from Thm. 5 (15) and (16).

$$\begin{array}{lr}
x \leftrightarrow \neg x \vdash_{L_0^*} x^+ \leftrightarrow (\neg x)^+, & \text{Thm. 6 (1),} \\
x \leftrightarrow \neg x \vdash_{L_0^*} x^+ \leftrightarrow \neg x^-, & \text{Thm. 6 (4),}
\end{array}$$

$x \leftrightarrow \neg x \vdash_{\mathrm{L}_0^*} (\neg x^- \to x^+) \leftrightarrow (\neg x^- \to \neg x^-)$, Thm. 5 (2),
$x \leftrightarrow \neg x \vdash_{\mathrm{L}_0^*} (x^- + x^+) \leftrightarrow \mathbf{0}$, Thm. 5 (8), Def.
$x \leftrightarrow \neg x \vdash_{\mathrm{L}_0^*} x$, Thm. 6 (3).

This proves that $x \dashv\vdash_{\mathrm{L}_0^*} \delta(x)\Delta\varepsilon(x)$, so by [3], Thm. 4.7, $\mathbf{L}_0^*$ is algebraizable. □

5 Equivalence of the categories $\mathcal{MV}$ and $\mathcal{MV}^*$

In this section we give a precise relation between the categories $\mathcal{MV}$ and $\mathcal{MV}^*$ of MV–algebras and MV^*–algebras, namely they are categorically equivalent.

An MV–algebra (Chang, [2] Mangani, [15]) is a system $\mathbf{A} = \langle A; \oplus, \neg, \mathbf{0}\rangle$ satisfying the following equations.

MV1 $x \oplus (y \oplus z) = (x \oplus y) \oplus z$

MV2 $x \oplus y = y \oplus x$

MV3 $x \oplus \mathbf{0} = x$

MV4 $\neg\neg x = x$

MV5 $x \oplus \neg\mathbf{0} = \neg\mathbf{0}$

MV6 $\neg(\neg x \oplus y) \oplus y = \neg(\neg y \oplus x) \oplus x$

In every MV–algebra we can define the constant $\mathbf{1}$ and the binary operator $\to$ by the formulas:

$$\begin{aligned} \mathbf{1} &:= \neg\mathbf{0} \\ x \to y &:= \neg x \oplus y \end{aligned}$$

The follo wing operations define a structure of MV–algebra $\langle [0,1]; \oplus, \neg, 0\rangle$ over the unit interval.

$$\begin{aligned} x \oplus y &= \min(1, x+y), \\ \neg x &= 1 - x. \end{aligned}$$

Theorem 11 *(Algebraic completeness.) An equation is satisfied in every MV–algebra if and only if it holds in the MV–algebra* $[0,1]$.

This theorem was first proved by C. C. Chang in [2] and [16]. Other proofs have been given by R. Cignoli in [17], Panti in [18]. An elementary proof appears in R. Cignoli and D. Mundici [19].

Lemma 12 *Let $\mathbf{A} = \langle A; \oplus, \neg, 0\rangle$ be an MV–algebra. Then, the following statements hold.*

a) If $x, y \in A$, $x \wedge y = 0$, then $x = x \odot \neg y$.

b) If $x, y \in A$, then $x \vee y = (x \odot \neg y) \oplus y$, $x \wedge y = (x \oplus \neg y) \odot y$.

An MV^*–algebra (Chang, [1]) is a system $\mathbf{B} = \langle B; +, C, \mathbf{0}, \mathbf{1}\rangle$ satisfying the following equations. We define

$$\begin{aligned} -\mathbf{1} &:= C\mathbf{1}, \\ x^+ &:= \mathbf{1} + (-\mathbf{1} + x), \\ x^- &:= -\mathbf{1} + (\mathbf{1} + x), \\ x \vee y &:= [x^+ + (C(x^+) + y^+)^+] + [x^- + (C(x^-) + y^-)^+]. \end{aligned}$$

Bx1 $x + y = y + x$

Bx2 $(\mathbf{1} + x) + (y + (\mathbf{1} + z)) = ((\mathbf{1} + x) + y) + (\mathbf{1} + z)$

Bx3 $x + Cx = \mathbf{0}$

Bx4 $(x + \mathbf{1}) + \mathbf{1} = \mathbf{1}$

Bx5 $x + \mathbf{0} = x$

Bx6 $C(x + y) = Cx + Cy$

Bx7 $CCx = x$

Bx8 $x + y = (x^+ + y^+) + (x^- + y^-)$

Bx9 $(Cx + (x + y))^+ = C(x^+) + (x^+ + y^+)$

Bx10 $x \vee y = y \vee x$

Bx11 $x \vee (y \vee z) = (x \vee y) \vee z$

Bx12 $x + (y \vee z) = (x + y) \vee (x + z)$

Theorem 13 ([1], Thms. (2.15) and (2.16).) *Let $\mathbf{B} = \langle B; +, C, \mathbf{0}, \mathbf{1}\rangle$ be an MV^*–algebra. Define $B^+ = \{x^+ : x \in B\}$ and $B^- = \{x^- : x \in B\}$.*

(a) $\mathbf{B}^+ = \langle B^+; +, \widehat{\ }, \mathbf{0}\rangle$ is an MV–algebra, where $\widehat{\ }$ is defined by $\widehat{x} = \mathbf{1} + Cx$.

(b) $\mathbf{B}^- = \langle B^-; +, \widetilde{\ }, \mathbf{0}\rangle$ is an MV–algebra, where $\widetilde{\ }$ is defined by $\widetilde{x} = C(\mathbf{1} + x)$.

We note that $\widetilde{(\widetilde{x} + y)} + y = x \vee y$ and that $\widehat{(\widehat{x} + y)} + y = x \wedge y$.

Theorem 14 *Let $\mathbf{A} = \langle A; \oplus, \neg, \mathbf{0}\rangle$ be an MV–algebra and let*

$$A^* = \{(a, b) \in A \times A : a \wedge b = \mathbf{0}\}.$$

Define the following operations over A^*.

$$\begin{aligned} (a, b) + (c, d) &:= (((a \oplus c) \odot \neg(b \oplus d)), ((b \oplus d) \odot \neg(a \oplus c)), \\ C(a, b) &:= (b, a), \\ \mathbf{0} &:= (\mathbf{0}, \mathbf{0}), \\ \mathbf{1} &:= (\mathbf{1}, \mathbf{0}). \end{aligned}$$

Then $\mathbf{A}^* = \langle A^*; +, C, \mathbf{0}, \mathbf{1}\rangle$ is an MV^*–algebra.

Proof: Axiom Bx1 holds by commutativity of $\oplus$. Axiom Bx3 states that $(a,b)+(b,a)=(\mathbf{0},\mathbf{0})$ which is true because in every MV–algebra $x\odot\neg x=\mathbf{0}$. Next we observe that $(a,b)+(\mathbf{1},\mathbf{0})=(\neg b,\mathbf{0})$ and that $(\neg b,\mathbf{0})+(\mathbf{1},\mathbf{0})=(\mathbf{1},\mathbf{0})$, and thus, axiom Bx4 holds. From the identity $x\odot\neg x=\mathbf{0}$ we deduce $(a,b)+(b,a)=(\mathbf{0},\mathbf{0})$, so Bx5 holds. It is immediate that Bx6 and Bx7 also hold.

We now prove the axiom of restricted associativity Bx2. Let $x=(a,b)$, $y=(c,d)$ and $z=(e,f)$. Then, by direct calculations using Lem. 12, $\mathbf{1}+x=(\neg b,\mathbf{0})$ and $(\mathbf{1}+x)+y=((\neg b\oplus c)\odot\neg d, b\odot d)$, so

$$((\mathbf{1}+x)+y)+(\mathbf{1}+z)=(A,B)$$

where

$$A=(((\neg b\oplus c)\odot\neg d)\oplus\neg f)\odot(\neg b\oplus\neg d)$$
$$B=(b\odot d)\odot\neg(((\neg b\oplus c)\odot\neg d)\oplus\neg f)$$

Similarly $(\mathbf{1}+x)+(y+(\mathbf{1}+z))=(C,D)$, where

$$C=(((\neg f\oplus c)\odot\neg d)\oplus\neg b)\odot(\neg f\oplus\neg d)$$
$$D=(f\odot d)\odot\neg(((\neg f\oplus c)\odot\neg d)\oplus\neg b)$$

By the completeness theorem (Thm. 11,) it suffices to prove $A=C$, $B=D$ in $[0,1]$. In this case, $c\wedge d=0$ implies $c=0$ or $d=0$. If $d=0$, then $A=C=(\neg b\oplus c)\oplus\neg f$ and $B=D=0$. If $c=0$, we have $\neg A=((b\oplus d)\odot f)\oplus(b\odot d)$ and $\neg C=((f\oplus d)\odot b)\oplus(f\odot d)$. The equality follows from the analysis of the six cases that arise in the comparison of b, f and $\neg d$.

Following Chang we define

$$\begin{aligned}
\mathbf{-1} &:= C\mathbf{1}\\
x^{+} &:= \mathbf{1}+(\mathbf{-1}+x)\\
x^{-} &:= \mathbf{-1}+(\mathbf{1}+x)\\
x\vee y &:= (x^{+}+(C(x^{+})+y^{+})^{+})\ +\ (x^{-}+(C(x^{-})+y^{-})^{+})
\end{aligned}$$

Evaluating these terms in our case we get.

$$\begin{aligned}
\mathbf{-1} &= (\mathbf{0},\mathbf{1})\\
x^{+} &= (a,\mathbf{0})\\
x^{-} &= (\mathbf{0},b)\\
x\vee y &= (a\vee c, b\wedge d)
\end{aligned}$$

We prove the last equality for $x=(a,b)$, $y=(c,d)$.

$$\begin{aligned}
x^{+}+(C(x^{+})+y^{+})^{+} &= (a,\mathbf{0})+((\mathbf{0},a)+(c,\mathbf{0}))^{+}\\
&= (a,\mathbf{0})+(c\odot\neg a,\mathbf{0})\\
&= (a\oplus(c\odot\neg a),\mathbf{0})\\
&= (a\vee c,\mathbf{0})
\end{aligned}$$

Similarly $x^- + (C(x^-) + y^-)^+ = (\mathbf{0}, b \wedge d)$.

To conclude this proof we observe that if $t \wedge s = \mathbf{0}$, then $(t, \mathbf{0}) + (\mathbf{0}, s) = (t, s)$, so $(a \vee c, \mathbf{0}) + (\mathbf{0}, b \wedge d) = (a \vee c, b \wedge d)$. So Bx10 and Bx11, commutativity and associativity respectively, hold.

To prove Bx8, we observe that $x^+ + y^+ = (a \oplus c, \mathbf{0})$ and $x^- + y^- = (\mathbf{0}, b \oplus d)$.

The computation of the first member of Bx9 gives

$$(Cx + (x + y))^+ = ((b \oplus [(a \oplus c) \odot \neg(b \oplus d)]) \odot \neg(a \oplus [(b \oplus d) \odot \neg(a \oplus c)]), \mathbf{0})$$

The second member is $Cx^+ + (x^+ + y^+) = ((a \oplus c) \odot \neg a, \mathbf{0})$, that is $Cx^+ + (x^+ + y^+) = (\neg a \wedge c, \mathbf{0})$.

To prove the identity $((b \oplus [(a \oplus c) \odot \neg(b \oplus d)]) \odot \neg(a \oplus [(b \oplus d) \odot \neg(a \oplus c)]) = \neg a \wedge c$ in [0,1] we can proceed by cases. Recall that $a \wedge b = 0$ and $c \wedge d = 0$ imply that $a = 0$ or $b = 0$, and $c = 0$ or $d = 0$.

The computation of the l.h.s member of Bx12 gives:

$$(a, b) + (c \vee e, d \wedge f) = ((a \oplus (c \vee e)) \odot \neg(b \oplus (d \wedge f)), (b \oplus (d \wedge f)) \odot \neg(a \oplus (c \vee e))),$$

and that of the r.h.s. member is

$$((a \oplus c) \odot \neg(b \oplus d) \vee (a \oplus e) \odot \neg(b \oplus f), (b \oplus d) \odot \neg(a \oplus c) \vee (b \oplus f) \odot \neg(a \oplus e))$$

We can prove the equalities

$$(a \oplus (c \vee e)) \odot \neg(b \oplus (d \wedge f)) = (a \oplus c) \odot \neg(b \oplus d) \vee (a \oplus e) \odot \neg(b \oplus f)$$

and

$$(b \oplus (d \wedge f)) \odot \neg(a \oplus (c \vee e)) = (b \oplus d) \odot \neg(a \oplus c) \vee (b \oplus f) \odot \neg(a \oplus e)$$

in [0,1] by the analysis of the eight possible cases. □

Theorem 15 *Let $\mathcal{MV}$ and $\mathcal{MV}^*$ be the categories whose objects are MV–algebras and MV^*–algebras respectively, with their corresponding morphisms. Let $\mathcal{F}$ be the map defined for an MV–algebra $\mathbf{A}$ by $\mathcal{F}(\mathbf{A}) = \mathbf{A}^*$, where $\mathbf{A}^*$ is the algebra defined in Theorem 14 and for a MV–morphism $f : \mathbf{A} \longrightarrow \mathbf{B}$ by $\mathcal{F}(f) = (f \times f) \restriction_{\mathbf{A}^*}$. Then $\mathcal{F}$ is a functor that defines a categorical equivalence.*

Proof: It is easy to see that $\mathcal{F}$ is well defined in the sense that $\mathcal{F}(f)$ is an MV^*–morphism that takes A^* into B^*. Functoriality is also straightforward.

Let $\mathcal{G}$ be the map defined for an MV^*–algebra B by $\mathcal{G}(B) = B^+$ and for a MV^*–morphism $h : B \longrightarrow C$ by $\mathcal{G}(h) = h \restriction_{B^+}$. Then, $\mathcal{G}$ is well defined and is a functor. We now define the natural equivalences s and t as follows, $s : \mathcal{G} \circ \mathcal{F} \approx \mathbf{1}_{\mathcal{MV}}$, $t : \mathcal{F} \circ \mathcal{G} \approx \mathbf{1}_{\mathcal{MV}^*}$. In what follows we denote * and $^+$ the functors $\mathcal{F}$ and $\mathcal{G}$ respectively.

For an MV–algebra A, let $s_A : (A^*)^+ \longrightarrow A$ be defined by $s_A = pr_1$.

For an MV^*–algebra B, let $t_B : (B^+)^* \longrightarrow B$ be defined by $t_B((a, b)) = a + Cb$.

It is easy to see that for each A, s_A is bijective (note that $(a, b) \in (A^*)^+$ if and only if $b = \mathbf{0}$) and that it preserves addition and $\mathbf{0}$. Also $s_A(\widehat{(a,\mathbf{0})}) = s_A((\mathbf{1},\mathbf{0})+C(a,\mathbf{0})) = s_A((\neg a, \mathbf{0})) = \neg a = \neg s_A((a,\mathbf{0}))$. So, s_A is an MV–isomorphism.

To prove that for each B, t_B is injective, we first note that $a \wedge b = \mathbf{0}$ implies that $(a + Cb) \vee \mathbf{0} = a$, $(a + Cb) \wedge \mathbf{0} = Cb$. Surjectivity follows by Bx8, since $x = x^* + x^- = t_B((x^+, Cx^-))$. Also it is trivial to prove that t_B preserves $\mathbf{0}$, $\mathbf{1}$ and C. We now prove that addition is also preserved.

Indeed, $t_B((a, b) + (c, d)) = t_B(X, Y)$, where $X = \mathbf{1} + [(-\mathbf{1} + (a + c)) + C(b + d)]$, $CY = -\mathbf{1} + [(\mathbf{1} + C(b+d)) + (a+c)]$. By associativity (see [1], Thm. (2.5),) we have that $X = U^+$ and $CY = U^-$, were $U = (a + c) + C(b + d)$, so $t_B((a, b) + (c, d)) = (a + c) + C(b + d) = (a + Cb) + (c + Cd) = t_B((a, b)) + t_B((c, d))$, by Bx 8.

It is straightforward to prove that s and t are natural. □

We describe briefly some relationships between the categorical equivalence $\mathcal{F}$ defined by theorem 15 and known categorical equivalences Γ of Chang–Mundici and Σ of Di Nola–Lettieri (originally called $\mathcal{G}$ in [20]). The functors Γ and Σ are treated in Ch. 2 and 7 of the book [5].

Let $\mathcal{LGU}$ be the category of l–groups with strong order unit and l–group unital homomorphisms (see [5].) The functor Γ assigns to every lattice–ordered group $\mathbf{G}$ with unit u the MV–algebra given by the interval $[0, u]$ of $\mathbf{G}$ with truncated sum and negation $\neg x = u - x$ and to every unital l–group homomorphism its restriction to $[0, u]$. We can obtain a categorical equivalence between $\mathcal{LGU}$ and $\mathcal{MV}^*$ by composition of Γ and the functor $\mathcal{F}$.

An MV–algebra $\mathbf{A}$ is called *perfect* if for every element x, either x or $\neg x$ is of finite order. Let $\mathcal{P}$ be the full subcategory of $\mathcal{MV}$ whose objects are the perfect MV–algebras. The functor Σ establishes a categorical equivalence between the category $\mathcal{LG}$ of l–groups and $\mathcal{P}$. For any l–group $\mathbf{G}$, $\Sigma(\mathbf{G}) = \Gamma(\mathbb{Z} \otimes \mathbf{G}, (1, 0))$, where $\mathbb{Z} \otimes \mathbf{G}$ is the *lexicographic product* of $\mathbb{Z}$ and $\mathbf{G}$ (that is the product group with the lexicographic order) and $(1, 0)$ is the unit.

Let $\mathcal{P}^*$ be the full subcategory of $\mathcal{MV}^*$ whose objects are images of the objects of $\mathcal{P}$ by the functor $\mathcal{F}$. Otherwise, an algebra $\mathbf{B}$ is an object of $\mathcal{P}^*$ if and only if $\mathbf{B}^+$ is perfect. Therefore, the composition of Σ and $\mathcal{F}$ gives a categorical equivalence between $\mathcal{LG}$ and $\mathcal{P}^*$.

References

[1] C. C. Chang. A logic with positive and negative truth values. *Acta Philosophica Fennica*, 16:19–39, 1963.

[2] C. C. Chang. Algebraic analysis of many–valued logics. *Transactions of the A.M.S.*, 88:467–490, 1958.

[3] W. J. Blok and D. Pigozzi. *Algebraizable Logics.* Memoirs of the A.M.S., 77, Nr. 396, 1989.

[4] R. Cignoli, I. M. L. D'Ottaviano, and D. Mundici. *Algebras das Lógicas de Łukasiewicz.* UNICAMP, Centro de Lógica, Epistemologia e História da Ciência, 1994.

[5] R. Cignoli, I. M. L. D'Ottaviano, and D. Mundici. *Algebraic Foundations of Many–Valued Reasoning.* Trends in Logic, Studia Logica Library, Vol. 7, Kluwer Academic Publishers, 2000.

[6] E. Casari. Conjoining and disjoining on different levels,. *Logic and Scientific Method (M. L. dalla Chiara, K. Doetz, D. Mundici and J. van Benthem, eds.), Kluwer Academic Pub.*, pages 261–288, 1995.

[7] E. Casari. Comparative logics and abelian l–groups. *Logic Colloquium '88, (Ferro, Bonotto, Valentini and Zanardo eds.), Elsevier Science Pub.*, pages 161–190, 1989.

[8] P. Hájek, R. Havránek, and R. Jiroušek. *Uncertain Information Processing in Expert Systems.* CRC Press Inc., 1992.

[9] P. Hájek and J. Valdés. An analysis of MYCIN–like expert systems. *Mathware and Soft Computing*, 1:45–68, 1994.

[10] B. De Baets and J. Fodor. Van Melle's combining function in MYCIN is a representable uninorm: An alternative proof. *Fuzzy Sets and Systems*, 104:133–136, 1999.

[11] R. Yager and A. Rybalov. Uninorm aggregation operators. *Fuzzy Sets and Systems*, 80:111–120, 1996.

[12] J. Fodor, R. Yager, and A. Rybalov. Structure of uninorms. *Int. J. Uncertainty, Fuzziness Knowledge–Based Systems*, 5:411–427, 1997.

[13] B. De Baets and J. Fodor. Residual operators of uninorms. *Soft Computing*, 3:89–100, 1999.

[14] B. De Baets. Idempotent uninorms. *European J. of Operations Research*, 118:631–642, 1999.

[15] P. Mangani. Su certe algebre connesse con logiche a piú valori. *Bollettino Unione Matematica Italiana*, 8:68–78, 1973.

[16] C. C. Chang. A new proof of the completeness of the Łukasiewicz axioms. *Transactions of the A.M.S.*, 93:74–80, 1959.

[17] R. Cignoli. Free lattice–ordered abelian groups and varieties of MV–algebras. *Proceedings of the IX SLALM, Vol. I, Notas de Lógica Matemática, Univ. Nac. del Sur*, I:113–118, 1993.

[18] G. Panti. A geometric proof of the completeness of the calculus of Łukasiewicz. *J. Symbolic Logic*, 60:563–578, 1995.

[19] R. Cignoli and D. Mundici. An elementary proof of Chang's completeness theorem for the infinite–valued calculus of Łukasiewicz. *Studia Logica*, 58:79–97, 1997.

[20] A. Di Nola and A. Lettieri. Perfect MV-algebras are categorically equivalent to abelian l–groups. *Studia Logica*, 53:417–432, 1994.

Fault-tolerance and Rota-Metropolis cubic logic *

DANIELE MUNDICI Department of Computer Science, University of Milan, Via Comelico 39-41, 20135 Milan, Italy
mundici@mailserver.unimi.it

for Newton C. A. da Costa on occasion of his 70th birthday

Abstract
The cubic algebras of Rota and Metropolis determine a three-valued logic which is related to Łukasiewicz logic, and to the Ulam-Rényi game of Twenty Questions with one lie/error. The latter is a chapter of fault-tolerant search, and of error-correcting codes with feedback, first considered by Dobrushin and Berlekamp. We discuss these relationships, with particular reference to the issue of non-triviality vs. inconsistency-tolerance.

1 THE ORDERED SET OF FACES OF THE n-CUBE

For each $n = 1, 2, \ldots$, Rota and Metropolis [33] equipped the partially ordered set $\mathcal{F}_n$ of all nonempty faces of the n-cube $[0,1]^n$ with the following operation:

the smallest face $A \sqcup B$ containing two faces A and B,

together with the following two *partially defined* operations:

the intersection $A \sqcap B$ of any two intersecting faces A and B,

and

the antipodal $\triangle(B, A)$ of A in B whenever A is contained into B. (See below for more details)

DEFINITION 1 A *finite cubic algebra* is a partial structure $C = (C, \sqcup, \sqcap, \triangle)$ which for some integer $n \geq 1$ is isomorphic to $(\mathcal{F}_n, \sqcup, \sqcap, \triangle)$. This uniquely determined n is called [1] the *dimension* of C.

*Partially supported by MURST Project on Logic
[1] If necessary, we shall say that C is an n-cubic algebra.

First realization

As a possible realization of an n-cubic algebra, [2] Rota and Metropolis considered the set $\mathcal{P}_n$ of all pairs $A = (A_0, A_1)$ of disjoint subsets of $\{1, 2, \ldots, n\}$, with the understanding that A_0 (resp., A_1) is the set of coordinates where all points of A constantly have value 0 (resp., have value 1). The cubic operation $\sqcup$ is given by

$$(A_0, A_1) \sqcup (B_0, B_1) = (A_0 \cap B_0, A_1 \cap B_1). \tag{1}$$

The partial operation $\sqcap$ is defined whenever $A_0 \cap B_1 = \emptyset = A_1 \cap B_0$, by the stipulation

$$(A_0, A_1) \sqcap (B_0, B_1) = (A_0 \cup B_0, A_1 \cup B_1). \tag{2}$$

The partial operation $\triangle$ is defined whenever $A_0 \supseteq B_0$ and $A_1 \supseteq B_1$, by the stipulation

$$\triangle((B_0, B_1), (A_0, A_1)) = (B_0 \cup (A_1 \setminus B_1), B_1 \cup (A_0 \setminus B_0)). \tag{3}$$

Second realization

Let us map each pair $(A_0, A_1) \in \mathcal{P}_n$ into the function $f\colon \{1, \ldots, n\} \to \{0, 1/2, 1\}$ given by $A_0 = f^{-1}(0)$ and $A_1 = f^{-1}(1)$. One then obtains the set $\{0, 1/2, 1\}^{\{1,\ldots,n\}}$ of all such functions. To construct an isomorphic copy $\mathcal{Q}_n$ of $\mathcal{P}_n$ one defines the (partial) operations $\sqcup, \sqcap, \triangle$ by stipulating that, for all $f, g \in \{0, 1/2, 1\}^{\{1,\ldots,n\}}$ and each $j \in \{1, \ldots, n\}$,

$$(f \sqcup g)(j) = \begin{cases} 0 & \text{if } f(j) = g(j) = 0 \\ 1 & \text{if } f(j) = g(j) = 1 \\ 1/2 & \text{otherwise.} \end{cases} \tag{4}$$

Further, whenever f *intersects* g (in the sense that $|f - g| \leq 1/2$) we stipulate

$$(f \sqcap g)(j) = \begin{cases} 0 & \text{if either } f(j) \text{ or } g(j) \text{ equals } 0 \\ 1 & \text{if either } f(j) \text{ or } g(j) \text{ equals } 1 \\ 1/2 & \text{if } f(j) = g(j) = 1/2. \end{cases} \tag{5}$$

Finally, assuming $f \sqcup g = g$ (which is the same as assuming that whenever $g(j) \neq 1/2$ then $g(j) = f(j)$) we stipulate

$$(\triangle(g, f))(j) = \begin{cases} f(j) & \text{if } f(j) = g(j) \\ 1 - f(j) & \text{otherwise.} \end{cases} \tag{6}$$

A moment's reflection yields the following equivalent reformulation of (4):

$$(f \sqcup g)(j) = \begin{cases} g(j) & \text{if } f(j) = g(j) \\ 1/2 & \text{if } 1/2 \leq |f(j) - g(j)|. \end{cases} \tag{7}$$

Further, whenever f and g intersect, (5) is equivalent to

$$(f \sqcap g)(j) = \begin{cases} g(j) & \text{if } f(j) = 1/2 \\ f(j) & \text{if } f(j) \in \{0, 1\}. \end{cases} \tag{8}$$

Finally, under the assumption $f \sqcup g = g$, condition (6) is equivalent to

$$(\triangle(g, f))(j) = \begin{cases} f(j) & \text{if } f(j) = g(j) \\ 1 - f(j) & \text{if } g(j) = 1/2. \end{cases} \tag{9}$$

[2] See [7] for further results on infinite cubic algebras.

2 ULAM-RÉNYI GAMES

In this section, the nonempty faces of the n-cube shall be reobtained in a game-theoretic context.

Stanisław Ulam [35, p.281] raised the following question:

> Someone thinks of a number between one and one million (which is just less than 2^{20}). Another person is allowed to ask up to twenty questions, to each of which the first person is supposed to answer only yes or no. Obviously the number can be guessed by asking first: Is the number in the first half-million? and then again reduce the reservoir of numbers in the next question by one-half, and so on. Finally the number is obtained in less than $\log_2(1,000,000)$. Now suppose one were allowed to lie once or twice, then how many questions would one need to get the right answer? One clearly needs more than n questions for guessing one of the 2^n objects because one does not know when the lie was told. This problem is not solved in general.

The same problem was also considered by Alfrèd Rényi in his book "A Diary on Information Theory" [32, p.47]. Here is a relevant passage:

> Assume that the number of questions which can be asked to figure out the "something" being thought of is fixed and the one who answers is allowed to lie a certain number of times. The questioner, of course, doesn't know which answer is true and which is not. Moreover, the one answering is not required to lie as many times as is allowable.

Naturally enough, the game of Twenty Questions with lies is called the *Ulam-Rényi game*, and the problem of minimizing the number of questions is the Ulam-Rényi problem. Once lies are understood as errors arising from of distortion, the Ulam-Rényi game becomes a chapter of the theory of error-correcting codes with feedback, [19], [5], [21], and of fault-tolerant search [1]. After the appearance of the papers [29] and [12], respectively giving the solution for one and for two lies, the literature on the Ulam-Rényi problem and its variants has been rapidly increasing [4, 3, 2, 6, 17, 22, 27, 28, 31, 30, 34, 8, 9, 26]. These developments are surveyed, among others, in [10], [18] and [20].

Our main interest in the present paper is different: following [25] we shall focus attention on how information is manipulated in a Ulam-Rényi game with one lie.

Initially, the two players agree to fix a finite set, say, $S = \{1, \ldots, n\}$, called the *search space*. Then the first player chooses a number $x_{\text{secret}} \in S$, and the second player must find x_{secret}, by asking the smallest possible number of questions, to each of which the first player can only answer "yes" or "no", being allowed at most one lie in his answers. [3] By definition, a *question* is a subset of S: thus for instance, the question

$$\text{is } x_{\text{secret}} \text{ an even number ?}$$

[3] Searching games with $L \geq 2$ lies were also investigated, both in the context of adaptive error-correcting codes, and as a natural tool for the semantics of $(L+2)$-valued logic. For bibliographical information see [25], [26] and [11] and references therein.

is identified with the set of all even numbers in S. The *opposite* question $\overline{Q}$ is the complementary set $S \setminus Q$.

We can safely assume Pinocchio to be first player, and identify ourselves with the second player. All we know, i.e., our "state of knowledge" about x_{secret}, is determined by the "conjunction" of Pinocchio's positive or negative answers to our questions. These answers in general do not obey the rules of classical logic, for,

(* * *) *The conjunction of two opposite answers to the same repeated question need not lead us to the incompatible state of knowledge, where all numbers falsify more than one answer, and hence there are left no candidates for* x_{secret}.

To understand the notion of "resulting state of knowledge from the conjunction of two answers", we shall first focus attention on a single question-answer step.

If Pinocchio's answer to our question Q is positive, then all elements of Q are said to *satisfy* the answer, while elements of $S \setminus Q$ *falsify* the answer. Conversely, if Pinocchio's answer to Q is negative, then all elements of $S \setminus Q$ are said to *satisfy* the answer, while elements of Q *falsify* the answer.

In the traditional game of Twenty Questions, when every answer is sincere, a number $y \in S$ is a possible candidate for x_{secret} if, and only if, it satisfies (the boolean conjunction of) all answers; skipping all syntactical details, our current state of knowledge during the game is completely represented by a function $f: S \to \{0,1\}$, where $f^{-1}(1)$ is the set of elements of S satisfying all answers. The conjunction of two opposite answers to the same repeated question does lead to the incompatible state of knowledge where no candidates for x_{secret} are left.

By contrast, if up to one answer may be mendacious, then a number ceases to be a candidate for x_{secret} if, and only if, it falsifies two or more answers. Again skipping all syntactical details, generalizing the case of no lies, our *state of knowledge* is now given by a function $f: S \to \{0, 1/2, 1\}$, assigning to each $y \in S$ one of the *three truth-values*, as follows:

$$f(y) = \begin{cases} 1 & \text{iff } y \text{ satisfies all answers} \\ 0 & \text{iff } y \text{ falsifies two or more answers} \\ 1/2 & \text{iff } y \text{ falsifies exactly one answer.} \end{cases} \tag{10}$$

As a particular case of (10), let us consider the state of knowledge $f_{Q,\beta}$ resulting from Pinocchio's answer $\beta \in \{no, yes\}$ to our question $Q \subseteq S$. Evidently, the 0 truth-value is not in the range of f. Two cases are possible:

Case 1: $\beta = yes$. Then $f_{Q,\beta}$ is given by

$$f_{Q,\beta}^{-1}(1) = Q, \quad f_{Q,\beta}^{-1}(1/2) = S \setminus Q. \tag{11}$$

Case 2: $\beta = no$. Then $f_{Q,\beta}$ is given by

$$f_{Q,\beta}^{-1}(1) = S \setminus Q, \quad f_{Q,\beta}^{-1}(1/2) = Q. \tag{12}$$

As expected, a negative answer to Q has the same effect as a positive answer to the opposite question $\overline{Q}$.

The following easy proposition gives a simple characterization [4] of our current state of knowledge in a Ulam-Rényi game with one lie, in terms of the *Lukasiewicz conjunction* $\odot$ of (the states resulting from) single answers. As usual, for any two real numbers $x, y \in [0,1]$ we let

$$x \odot y = \max(0, x + y - 1). \tag{13}$$

PROPOSITION 2 [25] Suppose we have received answers $\beta_1, \ldots, \beta_t$ to our questions $Q_1, \ldots, Q_t$, where $\beta_i \in \{no, yes\}$. Let f be our resulting state of knowledge as given by (10). Let $f_i = f_{Q_i,\beta_i}$ be as in (11) and (12). Then

$$f = f_1 \odot \cdots \odot f_t.$$

Thus, starting from the *initial* state (i.e., the constant function 1 over $S = \{1, \ldots, n\}$) our evolving states of knowledge form a decreasing set of functions

$$1 \;\geq\; f_1 \;\geq\; f_1 \odot f_2 \;\geq\; f_1 \odot f_2 \odot f_3 \;\geq \cdots,$$

where the *natural pointwise order* $f \geq g$ is understood as "f is less informative than g."

The *Lukasiewicz negation*

$$\neg f = 1 - f \tag{14}$$

yields the least informative state of knowledge g which is incompatible with f, in the sense that $f \odot g = 0$. We say that 0 is the *incompatible* state.

In terms of the two operations $\neg$ and $\odot$ one can now express the natural order of states of knowledge, by writing $f \leq g$ iff $f \odot \neg g = 0$.

One can also define a *disjunction*

$$f \oplus g = \neg(\neg f \odot \neg g), \tag{15}$$

and *Lukasiewicz implication* $f \to g = \neg f \oplus g$. The latter relates to the natural order as follows: $f \leq g$ iff $f \to g = 1$. More generally, $f \to g$ is the least informative state k such that $f \odot k \leq g$. This relates implication to conjunction and order via a well known adjunction.

To clarify the above phenomenon $(***)$, suppose we ask the same question Q twice and we receive two opposite answers *yes* and *no*. By (11)-(12) the two resulting states separately resulting from these two answers are given by $f_{Q,yes}$ and $f_{Q,no}$. Recalling that $f_{Q,no} = f_{\overline{Q},yes}$, we can naturally regard $f_{Q,no}$ as (the effect of) Pinocchio's negation of $f_{Q,yes}$, in symbols,

$$f_{Q,no} \;=\; \sim f_{Q,yes}.$$

Similarly,

$$f_{Q,yes} \;=\; \sim f_{Q,no}.$$

Thus $\sim$ is obtained by pointwise application of the map $1/2 \mapsto 1$ and $1 \mapsto 1/2$. The partially defined operation $\sim$ can be *uniquely* extended to an involutive operation

[4] The third truth-value was defined to be 1/2 precisely for the purpose of having this natural extension of the boolean case.

(also denoted $\sim$, and called *Pinocchio's negation*) acting on all states of knowledge—by the additional stipulation that $\sim: 0 \mapsto 0$. While Lukasiewicz negation $\neg$ has the property that $f \odot \neg f = 0$ for all states of knowledge f, this is not the case of Pinocchio's negation: for instance, for any question Q, $f_{Q,yes} \odot \sim f_{Q,yes}$ is the *self-negated* [5] state of knowledge ϵ coinciding with the constant function 1/2 over $\{1,\ldots,n\}$. In this state of knowledge, all numbers of S still are possible candidates for x_{secret}. A direct pointwise inspection over $\mathbf{L}_3$, in the light of (13)-(15), shows the following

PROPOSITION 3 Pinocchio's negation is definable in terms of $\epsilon, \neg, \odot$ by

$$\sim f = (\epsilon \oplus \neg f) \odot (f \oplus f). \tag{16}$$

3 Cubic faces = States of knowledge

The set $\mathcal{F}_n \cong \mathcal{Q}_n = \{0, 1/2, 1\}^{\{1,\ldots,n\}}$ of all nonempty faces of the n-cube (second realization) coincides with the set $\mathcal{S}_n$ of states of knowledge in a Ulam-Rényi game over a search space of cardinality n. The dimension of the cube equals the cardinality of the search space S.

Under this identification we immediately have

PROPOSITION 4 The initial state coincides with the 0-dimensional face of the n-cube $[0,1]^n$ given by its vertex $(1,\ldots,1)$. Similarly, the incompatible state is the same as the vertex $(0,\ldots,0)$. The self-negated state ϵ is the largest face—the n-cube itself. The antipodal partial operation $\triangle(g,f)$ for g containing f, coincides with the Lukasiewicz negation $1 - f(j)$ whenever $g(j) = 1/2$, and otherwise coincides with $f(j)$ $(j = 1,\ldots,n)$.

Set-theoretic inclusion makes $\mathcal{Q}_n$ into an upper-lattice structure, which is also closed under taking the infimum of any two intersecting faces. On the other hand, $\mathcal{S}_n$ comes equipped with the natural pointwise order of $\{0, 1/2, 1\}$. Further, as a subset of the function space $[0,1]^S$ of all $[0,1]$-valued functions over S, $\mathcal{S}_n$ is equipped with the *sharpening order* $\leq'$ defined by De Luca-Termini in [15, p. 305], [16, p. 299]. The latter is given by $f \leq' g$ iff for all $j \in \{1,\ldots,n\}$ we either have $f(j) \leq g(j) \leq 1/2$ or $f(j) \geq g(j) \geq 1/2$. Thus, sharpest elements are the vertices of the cube (i.e., states of knowledge only taking the boolean truth-values 0 and 1); the coarsest element is the cube itself (i.e., the self-negated state of knowledge). As a matter of fact we have the following result, whose proof is immediate:

PROPOSITION 5 The natural set-theoretic order on the faces of the n-cube (second realization) coincides with the De Luca-Termini sharpening order on the set of states of knowledge in a Ulam-Rényi game over the n-element search space with one lie.

We conclude that all the Rota-Metropolis order-algebraic structure of the faces of the n-cube has an equivalent counterpart in the context of Ulam-Rényi games with one lie.

There is still one more aspect of cubic faces to be considered: as a matter of fact, Rota and Metropolis [33, p.694] gave the following logical interpretation of finite cubic algebras (first realization):

[5] with respect to Lukasiewicz negation

Each face $A = (A_0, A_1)$ of the n-cube is the result of sampling a population $S = \{1, \ldots, n\}$, with a view of testing the validity of a yes-no hypothesis. Here A_1 and A_0 are the subsets of S where the hypothesis does or does not hold, respectively. A third truth-value "not-yet-known" can be assigned to each element in $S \setminus (A_0 \cup A_1)$. Two results A and B of this sampling are said to be incompatible if the two faces A and B are disjoint.

In the rest of this paper we shall further explore this intriguing three-valued logical interpretation of $\mathcal{Q}_n$. Naturally enough, we shall work in a purely algebraic framework, so as to be able to develop adequate (equational) logical machinery for both cubic algebras and Ulam-Rényi games with one lie.

4 MV_3 ALGEBRAS WITH A SELF-NEGATED ELEMENT

We shall assume familiarity with the following well known class of algebras:

DEFINITION 6 [11, and references therein] An MV_3 *algebra* is an algebra $M = (M, 1, \neg, \odot)$, with a distinguished element 1, a unary operation $\neg$, and a binary operation $\odot$ satisfying the following equations:

$$\begin{aligned} x \odot (y \odot z) &= (x \odot y) \odot z \\ x \odot y &= y \odot x \\ x \odot 1 &= x \\ x \odot \neg 1 &= \neg 1 \\ \neg\neg x &= x \\ y \odot \neg(y \odot \neg x) &= x \odot \neg(x \odot \neg y) \\ (x \odot x) \odot x &= x \odot x. \end{aligned}$$

MV_3 algebras stand to three-valued Lukasiewicz logic as boolean algebras stand to classical logic. [6] They are deeply related to other mathematical structures. [7] Equipping the set $\{0, 1/2, 1\}$ with the operations $\neg x = 1 - x$ and $x \odot y = \max(0, x + y - 1)$ we obtain the MV_3 algebra Ł_3, in symbols,

$$\text{Ł}_3 = (\{0, 1/2, 1\}, 1, \neg, \odot).$$

This algebra has the same role for MV_3 algebras as the two-element algebra $\{0, 1\}$ has for boolean algebras: thus, an equation holds for all MV_3 algebras iff it holds for Ł_3 iff it can be obtained from the equations in Definition 6 by substituting equals for equals.

[6] While in this paper MV_3-algebras are defined in terms of Lukasiewicz conjunction $\odot$, in [24] and [11, 1.1 and 8.5] they are equivalently defined in terms of Lukasiewicz disjunction $\oplus$.

[7] In [24] the author showed that 3-subhomogeneous AF C*-algebras with Hausdorff structure space are in canonical one-one correspondence (via Grothendieck's functor K_0) with countable MV_3 algebras.

Abbreviations. As usual, we shall consider the $\neg$ operation more binding than all other operations. We shall also use the abbreviations

$$\begin{aligned}
0 &= \neg 1\\
x^2 &= x \odot x\\
x \ominus y &= x \odot \neg y\\
x \oplus y &= \neg(\neg x \odot \neg y)\\
x \to y &= \neg x \oplus y\\
|x - y| &= (x \ominus y) \oplus (y \ominus x)\\
2x &= x \oplus x\\
x \leftrightarrow y &= (x \to y) \odot (y \to x).
\end{aligned}$$

Cubic algebras and MV_3 algebras

Let M be a finite MV_3 algebra with a self-negated element $\epsilon = \neg\epsilon$. Let us enrich M with the following operation:

$$x \sqcup y = (y \odot (x \leftrightarrow y)^2) \oplus (\epsilon \odot (\epsilon \to |x - y|)^2), \tag{17}$$

together with the following two partially defined operations:

$$x \sqcap y = (y \odot (x \leftrightarrow \epsilon)^2) \oplus (x \odot (x \leftrightarrow 2x)^2), \tag{18}$$

for all $x, y \in M$ such that $|x - y| \to \epsilon = 1$, and

$$\triangle(z, y) = (y \odot (z \leftrightarrow y)^2) \oplus (\neg y \odot (z \leftrightarrow \epsilon)^2), \tag{19}$$

for all $z, y \in M$ such that $y \sqcup z = z$.

THEOREM 7 Let M be a finite MV_3 algebra with an element $\epsilon = \neg\epsilon$. Then ϵ is unique, and the structure $(M, \sqcup, \sqcap, \triangle)$ is an n-cubic algebra, where n is the number of homomorphisms of M into Ł_3. Further, all finite cubic algebras arise from this construction, and non-isomorphic finite MV_3 algebras with self-negated elements determine non-isomorphic finite cubic algebras.

Proof: We assume familiarity with MV algebras [11]. To prove the uniqueness of ϵ, by way of contradiction assume $\eta \in M$ to be another self-negated element, and let J be a prime ideal of M such that $\epsilon/J \neq \eta/J$. The existence of J is ensured by Chang's subdirect representation theorem [11, Theorem 1.3.3]. In the totally ordered quotient MV algebra M/J we can safely assume $\eta/J < \epsilon/J$. By [11, Lemma 1.1.4] we get the contradiction

$$\eta/J < \epsilon/J = \neg\epsilon/J < \neg\eta/J = \eta/J.$$

Thus ϵ is unique.

As a finite MV algebra, M is isomorphic to a finite product of finite MV chains [11, Proposition 3.6.5], in symbols,

$$M \cong C_1 \times \ldots \times C_n, \tag{20}$$

where n is the number of maximal ideals of M. For each $i = 1, \ldots, n$, the MV chain C_i is a homomorphic image of M, whence it satisfies the equation $x \odot x = x \odot x \odot x$, and there is a self-negated element $\epsilon_i \in C_i$. By [11, Corollary 8.2.4], C_i must coincide with the three-element chain $\mathbf{Ł}_3$, and M is the product of n copies of $\mathbf{Ł}_3$, whence we can write

$$M \cong \mathbf{Ł}_3^n. \tag{21}$$

It turns out that n equals the number of homomorphisms of M into $\mathbf{Ł}_3$. [8] As a matter of fact, the map taking homomorphisms into their kernels sends the set of such homomorphisms of M into the set of maximal ideals of M (see [11, 1.2]). The map is onto because every maximal ideal I of M is the kernel of the quotient homomorphism $M \to M/I \cong \mathbf{Ł}_3$ (see [11, 3.5]). The map is one-one because any two homomorphisms of M into $\mathbf{Ł}_3$ having the same kernel must coincide (see [11, Corollary 7.2.6]).

We shall now prove that the partial structure $(M, \sqcup, \sqcap, \triangle)$ is an n-cubic algebra. By (21) we can safely identify M with the MV algebra $\mathbf{Ł}_3^n$ of all functions $f\colon \{1, \ldots, n\} \to \{0, 1/2, 1\}$ with pointwise operations of $\mathbf{Ł}_3$. Using (17)-(19) we shall prove (7)-(9). We shall argue pointwise, for arbitrary $x, y, z \in \{0, 1/2, 1\}$, letting $\epsilon = 1/2$.

The familiar properties of the operations $\neg, \odot, \oplus, \to, \leftrightarrow, \ominus$ and of the MV algebraic distance function ([11, Proposition 1.2.5], where one finds the notation $d(x, y)$ instead of $|x - y|$) yield for all elements $x, y, z \in \mathbf{Ł}_3$

$$(x \leftrightarrow y)^2 = \begin{cases} 1 & \text{if } x = y \\ 0 & \text{otherwise,} \end{cases} \tag{22}$$

whence

$$y \odot (x \leftrightarrow y)^2 = \begin{cases} y & \text{if } x = y \\ 0 & \text{otherwise.} \end{cases} \tag{23}$$

Similarly,

$$(\epsilon \to |x - y|)^2 = \begin{cases} 1 & \text{if } 1/2 \le |x - y| \\ 0 & \text{otherwise,} \end{cases} \tag{24}$$

whence

$$\epsilon \odot (\epsilon \to |x - y|)^2 = \begin{cases} 1/2 & \text{if } 1/2 \le |x - y| \\ 0 & \text{otherwise.} \end{cases} \tag{25}$$

Thus from (17) we have obtained (7).

We now note that the assumption $|x - y| \to \epsilon = 1$ is the same as $|x - y| \le 1/2$. It follows that

$$y \odot (x \leftrightarrow \epsilon)^2 = \begin{cases} y & \text{if } x = 1/2 \\ 0 & \text{otherwise.} \end{cases} \tag{26}$$

Similarly,

[8] The existence of the self-negated element $\epsilon \in M$ ensures that any such homomorphism is automatically onto $\mathbf{Ł}_3$.

$$x \odot (x \leftrightarrow 2x)^2 = \begin{cases} x & \text{if } x = x \oplus x \text{ (i.e., } x \in \{0,1\}) \\ 0 & \text{otherwise.} \end{cases} \quad (27)$$

Thus from (18) we have obtained (8).

We finally note that the assumption $y \sqcup z = z$ amounts to asking that whenever $z \in \{0, 1\}$ then $z = y$. Therefore,

$$y \odot (z \leftrightarrow y)^2 = \begin{cases} y & \text{if } z = y \\ 0 & \text{otherwise,} \end{cases} \quad (28)$$

and

$$\neg y \odot (z \leftrightarrow \epsilon)^2 = \begin{cases} \neg y & \text{if } z = 1/2 \\ 0 & \text{otherwise.} \end{cases} \quad (29)$$

Thus from (19) we have obtained (9). The rest is clear. □

Ulam-Rényi games and MV_3 algebras

The tight relations between Ulam-Rényi games with one lie and MV_3 algebras with a self-negated element are shown by the following result:

THEOREM 8 Suppose $X_1, \ldots, X_n$ are variable symbols for arbitrary states of knowledge in Ulam-Rényi games with one lie. Let the two terms $\tau = \tau(X_1, \ldots, X_n)$ and $\rho = \rho(X_1, \ldots, X_n)$ be obtained from these variable symbols and the constant symbols $0, 1, \epsilon$ by repeated applications of the connectives $\neg$ and $\odot$. Then the following conditions are equivalent:

(i) The equation $\tau = \rho$ holds in every Ulam-Rényi game with one lie, where ϵ is interpreted as the constant function 1/2 over the search space;

(ii) The equation holds in $\mathbf{L}_3$, where ϵ is interpreted as 1/2;

(iii) The equation holds in every MV_3-algebra having a self-negated element interpreting ϵ.

Proof: This is a routine consequence of the completeness theorem for MV_3 algebras, [11, Corollary 8.2.4], and of the basic properties of self-negated elements. □

5 Conclusions

The logic arising from the partially ordered set of faces of the n-cube was proposed by Rota and Metropolis as an interesting alternative to boolean logic. This is so because, on the one hand boolean functions over $\{1, \ldots, n\}$ are in canonical one-one correspondence with the faces of the n-simplex, and, on the other hand, for all $n \geq 5$ the only two possible order structures arising from the faces of regular polyhedra in euclidean n-space are those obtained from the n-cube and the n-simplex.

Cubic logic is naturally three-valued: cubic faces coincide with states of knowledge in Ulam-Rényi games with one lie.

The antipodal of a face f with respect to the whole cube coincides with the Łukasiewicz negation of f. More generally, the cubic antipodal operation is expressible in terms of Łukasiewicz negation $\neg$ and conjunction $\odot$, once a self-negated element ϵ is available.

The natural set-theoretic order between faces coincides with the De Luca-Termini sharpening order: the latter, too, can be expressed in terms of $\neg, \odot, \epsilon$.

Algebras of three-valued Łukasiewicz logic (with a distinguished self-negated element) have a simple axiomatization, [9] and also have a well developed representation of free algebras and their ideals, allowing one to define the appropriate notion of consequence and Lindenbaum algebras of theories [11]. Ulam-Rényi games can be used to give a semantical interpretation of the Łukasiewicz connectives; the partially ordered set of cubic faces yields one more interpretation of the sharpening order.

Pinocchio's negation $\sim$ is definable in three-valued Łukasiewicz logic with a distinguished self-negated element. By contrast with Łukasiewicz negation, this negation happens to be "inconsistency-tolerant" with respect to Łukasiewicz conjunction $\odot$, in the sense that the state of knowledge $\phi\odot\sim\phi$ need not be incompatible. For instance, when as in $(***)$ above, Pinocchio gives opposite answers to the same repeated question, respectively resulting in the states of knowledge ϕ and $\sim\phi$, then our state of knowledge $\phi\odot\sim\phi$ still allows all elements of S to be candidates for Pinocchio's secret number. Bearing in mind that Pinocchio has certainly spent his lie, we can continue our search of x_{secret} using the classical methods [1]. In this case, inconsistency-tolerance is just a logical reformulation of fault-tolerance, and we also have an instance of the following [14, p.149, item 5.]:

> (Paraconsistent logic) must contribute to a correct appreciation of the concepts of negation and contradiction. Paraconsistent logic clarifies the fact that there are various kinds of negation, as well as different kinds of implication.

During our excursion in this paper, faces of the cube have been transformed into states of knowledge in a game of Twenty Questions with one lie. This game, on the one hand, is a basic chapter of communication with feedback, and, on the other hand, it provides a natural semantics for three-valued logic—just as the traditional game of Twenty Questions does for two-valued boolean logic. The reader will decide if this is one more instance of the following phenomenon [13, p.62, Conclusion, IV]:

> It is frequently the case that mathematical constructions which prima facie may look extravagant and singular, are extended, find applications, and eventually become commonplaces (for instance, non-euclidean geometries, many-valued logics, and n-dimensional geometries).

****

It goes without saying that this paper is motivated by, and dedicated to, my friend Newton: among others, I owe him the conviction that a logic which is capable of dealing with contradictory statements need not be trivial—quite the contrary.

[9] In [23] it is proved that the tautology problem for Łukasiewicz three-valued logic has the same complexity as its boolean counterpart.

REFERENCES

[1] M. Aigner, *Combinatorial Search*, Wiley–Teubner, New York–Stuttgart, 1988.

[2] M. Aigner, *Searching with lies*, J. Combin. Theory Ser. A, **74** (1995), pp. 43-56.

[3] A. Ambainis, S. A. Bloch, D. L. Schweizer, *Delayed Binary Search, or Playing Twenty Questions with a Procrastinator*, in Proc. of 10th AMC SIAM SODA (1999), pp. 844-845.

[4] J. Aslam, A. Dhagat, *Searching in the presence of linearly bounded errors*, In: Proc. of the 23rd ACM Symposium on the Theory of Computing (1991), pp. 486-493.

[5] E. R. Berlekamp, *Block coding for the binary symmetric channel with noiseless, delayless feedback*, In: Error-correcting Codes, H.B. Mann (Editor), Wiley, New York (1968), pp. 61-88.

[6] R.S. Borgstrom, S. Rao Kosaraju, *Comparison-based Search in the Presence of Errors*, In: Proc. of the 25th ACM Symposium on the Theory of Computing (1993), pp. 130-136.

[7] W.Y.C. Chen, J.S. Oliveira, *Implication algebras and the Metropolis-Rota axioms for cubic lattices*, Journal of Algebra, **171** (1995), pp. 383–396.

[8] F. Cicalese, D. Mundici, *Perfect 2-fault tolerant search with minimum adaptiveness*, Advances in Applied Mathemathics, **25** (2000), pp. 65-101.

[9] F. Cicalese, D. Mundici, *Optimal coding with one asymmetric error: below the Sphere Packing bound*, In: Proc. of COCOON-2000, Lecture Notes in Computer Science, **1858** (2000), 159-169.

[10] F. Cicalese, D. Mundici, U. Vaccaro, *Rota-Metropolis cubic logic and Ulam-Rényi games*, In: Algebraic Combinatorics and Computer Science: a tribute to Gian Carlo Rota, (H. Crapo, D. Senato, Eds.,) Springer-Verlag Italia, Milan, 2001, pp. 197-244.

[11] R. Cignoli, I.M.L. D'Ottaviano, D. Mundici. *Algebraic Foundations of many-valued Reasoning.* Trends in Logic, Studia Logica Library, vol.7, Kluwer Academic Publishers, Dordrecht, 2000.

[12] J. Czyzowicz, D. Mundici, A. Pelc, *Ulam's searching game with lies*, J. Combin. Theory Ser. A, **52** (1989), pp. 62-76.

[13] N.C.A. da Costa, *Sistemas Formais Inconsistentes*, Published by Editora UFPR, Curitiba, Paraná, Brazil, 1964, 1993.

[14] N.C.A. da Costa, *Ensaio sobre os Fundamentos da Lógica*, Published by Editora HUCITEC EDUSP, S.P., Brazil, 1980.

[15] A.De Luca, S. Termini, *A definition of nonprobabilistic entropy in the setting of fuzzy sets theory,* Information and Control, **20** (1972), pp. 301-312.

[16] A.De Luca, S. Termini, *Dispersion measures on fuzzy sets,* In: Approximate Reasoning in Expert Systems, (M.M.Gupta, et al., Eds.,) Elsevier Publishers B.V. (North-Holland), 1985, pp. 199-216.

[17] A. Dhagat, P. G'acs, P. Winkler, *On Playing "Twenty Questions" with a liar*, In: Proc. of the 3rd Annual ACM-SIAM Symposium on Discrete Algorithms (1992), pp. 16-22.

[18] D.Z. Du, F.K. Hwang, *Combinatorial Group Testing and its Applications*, World Scientific, Singapore, 1993.

[19] R.L.Dobrushin, *Information transmission in a channel with feedback*, Theory of Probability and Applications, **34** (1958) pp. 367-383. Reprinted in: D.Slepian (Ed.), Key papers in the development of information theory, IEEE Press, NY, 1974.

[20] R. Hill, *Searching with lies*, In: Surveys in Combinatorics, Rowlinson, P. (Editor), Cambridge University Press (1995), pp. 41–70.

[21] R. Hill, J. Karim, E. R. Berlekamp, *The solution of a problem of Ulam on searching with lies*, In: Proceedings of IEEE ISIT 1998 (Cambridge, USA), p. 244.

[22] A. Malinowski, *K-ary searching with a lie*, Ars Combinatoria, **37** (1994), pp. 301-308.

[23] D. Mundici, *Satisfiability in many-valued sentential logic is NP-complete*, Theoretical Computer Science, **52** (1987), pp. 145-153.

[24] D. Mundici, *The C^*-algebras of three-valued logic*, In: Proceedings Logic Colloquium 1988, Padova. (R. Ferro et al., Eds.), Amsterdam: North-Holland, 1989, pp. 61-77.

[25] D. Mundici, *The logic of Ulam's game with lies*, In: Knowledge, Belief and Strategic Interaction, (C.Bicchieri, M.L.Dalla Chiara, Eds.,) Cambridge Studies in Probability, Induction, and Decision Theory, 1992, pp. 275-284.

[26] D. Mundici, A. Trombetta, *Optimal comparison strategies in Ulam's searching game with two errors*, Theoretical Computer Science, **182** (1997), pp. 217-232.

[27] S. Muthukrishnan, *On optimal strategies for searching in presence of errors*, In: Proc. of the 5th ACM-SIAM SODA (1994), pp. 680-689.

[28] A. Negro and M. Sereno, *Ulam's searching game with three lies*, Advances in Applied Mathematics, **13** (1992), pp. 404-428.

[29] A. Pelc, *Solution of Ulam's problem on searching with a lie*, J. Combin. Theory, Ser. A, **44** (1987), pp. 129-142.

[30] A. Pelc, *Searching with known error probability*, Theoretical Computer Science, **63** (1989), pp. 185-202.

[31] A. Pelc, *Searching with permanently faulty tests*, Ars Combinatoria, **38** (1994), pp. 65-76.

[32] A. Rényi, *Napló az információelméletről*, Gondolat, Budapest, 1976. (English translation: *A Diary on Information Theory*, J.Wiley and Sons, New York, 1984).

[33] G-C. Rota, N. Metropolis, *Combinatorial structure of the faces of the n-cube*, SIAM J. Applied Math., **35** (1978), pp. 689-694.

[34] J. Spencer, *Ulam's searching game with a fixed number of lies*, Theoretical Computer Science, **95** (1992), pp. 307-321.

[35] S.M. Ulam, *Adventures of a Mathematician*, Scribner's, New York, 1976.

The Annotated Logics $OP_{\mathbf{BL}}$ *

GUILLERMO ORTIZ RICO Universidad del Valle - Colombia
Pontificia Universidad Católica de Chile

gortizri@mafalda.univalle.edu.co

Abstract
In classical logic semantics, inconsistent theories have no models and hence are meaningless from a model-theoretic point of view, therefore classical logic is not the appropriate formalism for reasoning about inconsistent databases. As a step towards the solution of this problem, V. S. Subrahmanian introduced annotated logics. Several aspects of formal logic were developed in the annotated logic $P\mathcal{T}$. It is proved that these systems are paraconsistent and that they could be a basis for a programming language for reasoning about databases that contain inconsistencies. These systems are not structural, thus one cannot find its algebraic counterpart. Some relationships between annotated logics and algebraizability were presented by R. Lewin et. al., they introduced a structural version $SP\mathcal{T}$ and SAL respectively of annotated logics and they proved that $SP\mathcal{T}$ is "equivalent" to the original $P\mathcal{T}$ systems. We herein present the systems $OP_{\mathcal{BL}}$ and $COP_{\mathbf{BL}}$. These are close to the $SP\mathcal{T}$ and SAL system, and are based on APC (annotated predicate calculus) and the original systems C_n of da Costa. But there are important differences in the concept of "well behaved formulas", and we will allow annotations of annotations. Also, some of the axioms are simplified. The systems $OP_{\mathbf{BL}}$ are algebraizable if, and only if, BL is finite. Therefore $COP_{\mathbf{BL}}$ is also algebraizable if BL is finite. The annotations in this paper belong to the interlaced bilattices.

Introduction

It is known that large databases can be inconsistent in many ways. Nevertheless, certain inconsistencies should not be allowed to significantly alter the intended meaning of such knowledge bases. In classical logic semantics inconsistent theories have no models and hence are meaningless from a model-theoretic point of

*Funding to travel to the WCP II was provided by FONDECYT grant 199–0433 and DIPUC, Pontificia Universidad Católica de Chile. We thank the anonymous referees of WCP2000, who helped us to improve the presentation of our paper.

view, classical logic is not the appropriate formalism for reasoning about inconsistent databases. As a step towards the solution of this problem, Subrahmanian [1] introduced annotated logics which were subsequently studied in [2], where it is proved that these systems are paraconsistent and that they could form a basis of a programming language for reasoning about databases that contain inconsistencies. Later, its foundational aspects from the model theoretic and proof theoric points of view were developed in [3], [4] and [5]. However, since there are several kinds of axioms (some for complex formulas, others for atomic formulas and still others for arbitrary formulas), these systems are not structural in the sense that their consequence relation is not closed under substitutions. One of the difficulties with non-structural systems is that one cannot find its algebraic counterpart.

We are interested in the algebraizability of annotated logics. Some relations between annotated logics and algebraizability (in the sense of Blok-Pigozzi [6]) were presented in [7] and [8], they introduced structural versions $SP_{\mathcal{T}}$ and SAL respectively of annotated logics and they proved that these are "equivalent" to the original $P_{\mathcal{T}}$ systems. The system $SP_{\mathcal{T}}$ is algebraizable if, and only if, τ is finite.

In the present article we present the systems $OP_{\mathcal{BL}}$ and $COP_{\mathbf{BL}}$. These are close to $SP_{\mathcal{T}}$ and SAL, but there are important differences in the concept of "well behaved formulas", and in the presentation of the annotations of annotations. Also, some of the axioms are simplified. The systems $OP_{\mathbf{BL}}$ are algebraizable if, and only if, BL is finite. Therefore $COP_{\mathbf{BL}}$ is algebraizable if BL is finite. The annotations in this paper belong to the interlaced bilattices.

We considered pertinent to stress that the significance in the conclusions of the annotated logics depends exclusively on the skill in the use of the annotations, which cross many items such as: degree of credibility, probability, quality of data, weather conditions, and any other type that allows us to draw conclusions (see [9], [10] and [11]).

1 The annotation algebra: bilattices

Although the annotations are originally defined for complete lattices, we assume the annotations belong to an algebra $\mathbf{A} = \langle A, \wedge, \vee, \sim, \otimes, \oplus \rangle$ of type $(2, 2, 1, 2, 2)$. In order to give a relatively easy handling of the annotations of annotations we have chosen interlaced bilattices, since these algebras have been widely studied in [12], [13], [14] and [15].

Our object of study is really the annotated logics and not bilattices themselves, nevertheless for the purpose of the self-containment of the paper we will fix here the basic developments of the bilattices, which appear in the above references.

The bilattices are algebras with two separate lattice structures. They have been used as the basis for denotational semantics for systems of inference that arise in artificial intelligence and knowledge-based logic programming (see [12] and [13]). In particular, they have been used to provide a general framework for an efficient procedural semantics of logic programming languages that can deal with incomplete as well as contradictory information. In [16] a much more efficient procedural semantics is introduced, which proves to be equivalent to general semantics for a special class of bilattices that includes all finite distributive bilattices.

DEFINITION 1 A **bilattice** is a structure $\mathbf{BL} = (B, \leq_1, \leq_2, \sim)$ such that B is a non empty set containing at least two elements; $(B, \leq_1), (B, \leq_2)$ are complete [1] lattices and $\sim$ is a unary operation [2] on B that has the following properties:

1. If $a \leq_1 b$, then $\sim a \geq_1 \sim b$,
2. If $a \leq_2 b$, then $\sim a \leq_2 \sim b$,
3. $\sim\sim a = a$.

According to custom, we shall use $\wedge$ and $\vee$ for the lattice operations which correspond to $\leq_1$, and $\otimes$, $\oplus$ for those that correspond to $\leq_2$. While $\wedge$ and $\vee$ can be associated with their usual intuitive meanings of "least" and "greatest", one may understand $\otimes$ and $\oplus$ as the "consensus" and the "gullibility" ("accept all") operator, respectively; $a \otimes b$ is the most that a and b can agree on, while $a \oplus b$ accepts the combined knowledge of a and b [3]. Thus, a bilattice can effectively be regarded as an algebra of type (2,2,1,2,2) $[\ B, \wedge, \vee, \sim, \otimes, \oplus\ ]$. *The annotation constants* are elements of B that can be thought of as confidence factors, or as degrees of belief, or as truth values. We will denote by f and t the least element and greatest element (respectively) of B w.r.t. $\leq_1$, while $\perp$ and $\top$ will denote the least element and greatest element of B w.r.t. $\leq_2$. Note that the negation keeps order $\leq_2$. This reflects the intuition that $\leq_2$ corresponds to differences in our knowledge about formulas and not to their degrees of truth. Hence, while one expects negation to invert the notion of truth, the role of negation w.r.t. $\leq_2$ is somewhat less transparent: we know no more and no less about $\neg p$ than we know about p.

DEFINITION 2 A bilattice is called **distributive** [12] if all the twelve possible distributive laws concerning $\wedge, \vee, \otimes$ and $\oplus$ hold. It is called **interlaced** [13] if each one of $\wedge, \vee, \otimes$ and $\oplus$ is monotonic with respect to both $\leq_1$ and $\leq_2$.

PROPOSITION 3 [13] Every distributive bilattice is interlaced.

PROPOSITION 4 Let $\mathbf{B} = (B, \leq_1, \leq_2, \sim)$ be a bilattice, and let $a, b \in B$. Then

1. $\sim (a \wedge b) = \sim a \vee \sim b \qquad \sim (a \vee b) = \sim a \wedge \sim b$
   $\sim (a \otimes b) = \sim a \otimes \sim b \qquad \sim (a \oplus b) = \sim a \oplus \sim b$,
2. $\sim f = t \qquad \sim t = f \qquad \sim \perp = \perp \qquad \sim \top = \top$.

PROPOSITION 5 Let $\mathbf{B} = (B, \leq_1, \leq_2, \sim)$ be an interlaced bilattice, and let $a, b \in B$. Then $\perp \wedge \top = f$; $\perp \vee \top = t$; $f \otimes t = \perp$; $f \oplus t = \top$.

[1] This is Ginsberg's original definition in [12]. Some authors have dropped this requirement of completeness.

[2] Fitting dropped the unary operation. In such cases, we say *bilattice without negation*.

[3] This is Fitting's idea.

EXAMPLES 6 **Five and Default**[4]

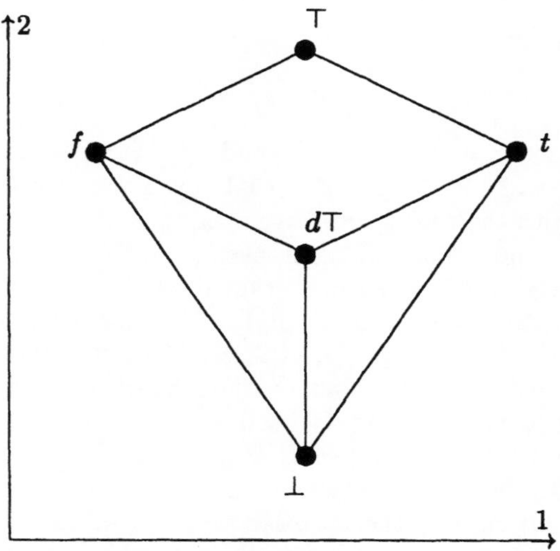

Figure 1: Bilattice **Five**.

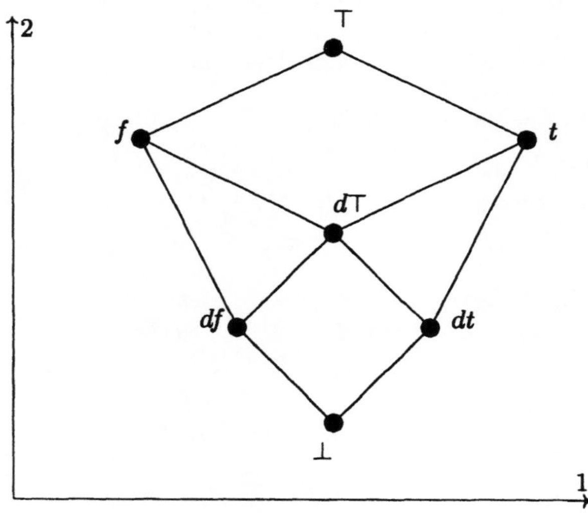

Figure 2: Bilattice **Default**.

[4] **Five** was considered in [12] and **Default** was considered in [14].

Now we present a bilattice construction which is due to Ginsberg.

DEFINITION 7 Suppose $(C, \leq)$ and $(D, \leq)$ are lattices. We give two orders on $C \times D$, $\leq_2$ and $\leq_1$, as follows;
$(c_1, d_1) \leq_2 (c_2, d_2)$ if $c_1 \leq c_2$ and $d_1 \leq d_2$
$(c_1, d_1) \leq_1 (c_2, d_2)$ if $c_1 \leq c_2$ and $d_2 \leq d_1$. Then we define
$BL(C, D) = (C \times D, \leq_2, \leq_1)$.

PROPOSITION 8 If $(C, \leq)$ and $(D, \leq)$ are complete lattices, then $BL(C, D)$ is an interlaced bilattice without negation.

PROPOSITION 9 If $(C, \leq)$ is a complete lattice, then $BL(C, C)$ is an interlaced bilattice.

A truth value $(x, y) \in BL(C, C)$ may intuitively be understood so that x represents the amount of belief for an assertion, and y is the degree of belief against it. Using this intuition in the original annotated logics, applications have been developed to the robotics (see [17]).

EXAMPLES 10 Next we will present some results about **Four** [5] and **Nine**, the proofs can be found in [14]. **Four**, drawn in Figure three, is the smallest non-degenerated bilattice. **Four** and **Nine**, which are drawn in Figure four, are distributive, but **Default** is not. In addition, if we denote the two element lattice $\{0, 1\}$ by **Two**, and the three element lattice $\{0, \frac{1}{2}, 1\}$ by **Three**, then **Four** is isomorphic to $BL(\mathbf{Two}, \mathbf{Two})$ and similarly **Nine** is isomorphic to $BL(\mathbf{Three}, \mathbf{Three})$.

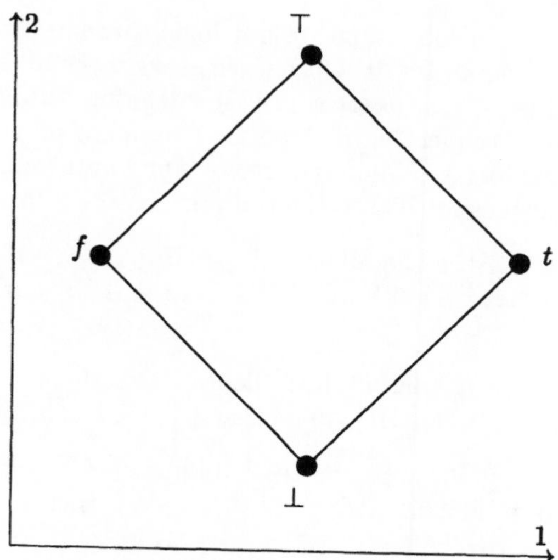

Figure 3: Bilattice **Four**.

[5]**Four** was introduced by Belnap in [18].

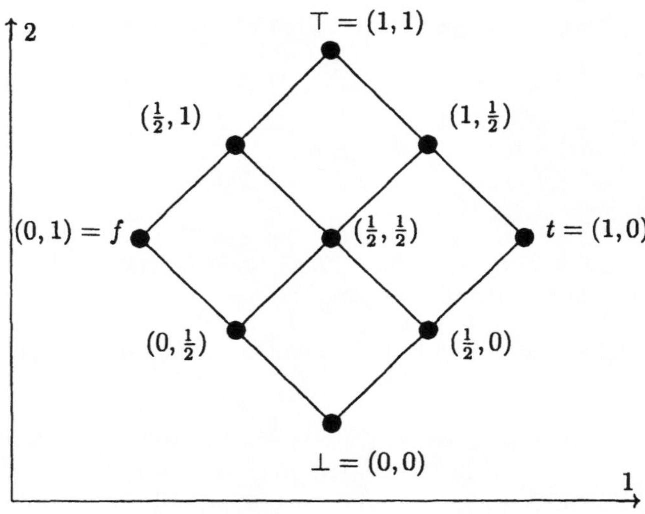

Figure 4: Bilattice **Nine**.

Ginsberg's **Default** was introduced in [12] as a tool for non-monotonic reasoning. The truth values that have a prefix "d" in their names are supposed to represent values of default assumptions (for instance *dt* is true by default). The negations of $\top, t, f, \top$ are identical to those of **Four**, and $\sim df = dt$, $\sim dt = df$, $\sim d\top = d\top$. Thus **Default** is not interlaced, for instance $f <_t df$, while $f \otimes d\top = d\top$, $df \otimes d\top = df$, but $d\top \not<_t df$.

1.1 Bifilters

Since later on we will develop studies of our logic through matrices, to search for an equivalent matrix semantics, it will be necessary to establish the subset of the designated truth values. This subset is used for defining validity of formulas and a consequence relation. Frequently, in algebraic treatment of the subject, the set of the designated valued forms a filter relative to some natural ordering of the truth values. Natural analogues for bilattices of filters, are the bifilters.

DEFINITION 11 A **bifilter** of a bilattice $\mathbf{B} = (B, \leq_1, \leq_2, \sim)$ is a nonempty subset $\mathbf{F} \subset B, \mathbf{F} \neq B$, such that $a \wedge b \in \mathbf{F}$ iff $a \in \mathbf{F}$ and $b \in \mathbf{F}$; and $a \otimes b \in \mathbf{F}$ iff $a \in \mathbf{F}$ and $b \in \mathbf{F}$.

EXAMPLES 12 **Four, Five** and **Default** contain exactly one bifilter: $\{\top, t\}$. **Nine** contains two bifilters: $\{\top, (1, \frac{1}{2}), t\}$, as well as $\{\top, (\frac{1}{2}, 1), (\frac{1}{2}, \frac{1}{2}), (1, \frac{1}{2}), (\frac{1}{2}, 0), t\}$.

PROPOSITION 13 [14](Basic properties of bifilters)
Let $\mathbf{F}$ be a bifilter of a bilattice $\mathbf{BL} = (B, \leq_1, \leq_2, \sim)$, then :

1. $\mathbf{F}$ is upward-closed w.r.t. both $\leq_1$ *and* $\leq_2$.
2. $t, \top \in \mathbf{F}$, while $f, \perp \notin \mathbf{F}$.

2 THE SYSTEMS $OP_{\mathbf{BL}}$ AND $COP_{\mathbf{BL}}$.

In this section, we present the systems $OP_{\mathbf{BL}}$ and $COP_{\mathbf{BL}}$, of which the annotations belong to the interlaced bilattices. In order to simplify, we make the following convention as of this moment, except for otherwise explicit mention, the expression "bilattice" will represent the expression "interlaced bilattice". These systems are based on APC (annotated predicate calculus) and the original systems C_n of da Costa ([19] and [20]). The first relations between annotated logics and algebraizability were presented in [7] and [8] using the systems $SP_{\mathcal{T}}$ and SAL, which are closed to the original $P_{\mathcal{T}}$ systems presented in [3].
The systems $OP_{\mathbf{BL}}$ and $COP_{\mathbf{BL}}$ are close to the $SP_{\mathcal{T}}$ and SAL system, but there are important differences since we simplified some axioms. We define the concept of "well behaved formulas" in da Costa's style, which simplifies the number of axioms to a great extent, as well as the handling of the logical system. In addition, we will allow annotations of annotations. It is important to emphasize that while we define the "well behaved formulas" within the $OP_{\mathbf{BL}}$ and $COP_{\mathbf{BL}}$ systems, $SP_{\mathcal{T}}$ and SAL are handled with additional symbol, which implies the inclusion of a great numbers of axioms. Furthermore, in the $SP_{\mathcal{T}}$ and SAL systems the annotations of annotations are considered updating, i.e., each last annotation cancels the last one, while we "nest" them through the annotation operators. Those are homomorphisms with respect to binary connectives but not with respect to negation, and when they are applied to a formula can be interpreted with an extra operation, namely the "$\oplus$" of the bilattice. This allows us to see that the annotation algebra will be chosen in a wider sense in forthcoming work and which is out of the scope of this paper.

Although the development of this type of formalisms does not enjoy practical applications in the form that they are presented here, we wish to emphasize that the annotations of annotations have been the object of study in other contexts, for example in "An Inductive Annotated Logic" by Newton da Costa and Decio Krause in [21] and "Fuzzy Operator and Fuzzy Resolution" by Weigert et. al. in [22]. In the latter they present the fuzzy resolution principle for this logic and show its completeness as an inference rule. In this logic the annotation operators are not homomorphisms with respect to binary connectives.

2.1 The language of $OP_{\mathbf{BL}}$

Let BL be a (fixed) bilattice. The language L of $OP_{\mathbf{BL}}$ is formed by the following primitive symbols:

1. A countable set P of propositional letters $p,q,r,\ldots$
2. Logical connectives $\wedge, \vee, \rightarrow, \neg,$
3. A unary annotation function g_b for each $b \in BL$
4. Auxiliary symbols: parentheses, comma.

The set F of formulas is defined recursively as follows:

1. If p is a propositional letter, then p is a formula.
2. If A is a formula, then $\neg$A is a formula.

3. If A and B are formulas, then $A \wedge B, A \vee B,$ and $A \to B$ are formulas.
4. If A is a formula and b is a constant of annotation, then $g_b A$ is a formula.
5. An expression is a formula if, and only if, it is obtained by finitely many applications of the above rules.

Intuitively, the annotated formula $g_b A$ may be interpreted as *the degree of evidence of A is at least b w.r.t.* $\leq_2$ and $\leq_1$. This can also represent a degree of belief or uncertainty associated to A by a reasoning agent.

We will distinguish two types of formulas: hyperliterals and complex formulas. For each $n \in \omega$, we define inductively $\neg^n$ by $\neg^0 A = A$ and $\neg^{n+1} A = \neg(\neg^n A)$. The formulas of the form $\neg^{k_n} g_{b_n} \neg^{k_{n-1}} g_{b_{n-1}} \ldots \neg^{k_1} g_{b_1} \neg^{k_0} p$, are called *hyperliterals,* where $k_i \in \omega$ and $b_i \in B$ for $0 \leq i \leq n$. A formula that is not a hyperliteral one will be called *a complex formula.*

Before we present the formal system, following the da Costa style we define the "well behaved formulas" and the abbreviation of equivalence.

DEFINITION 14 $A^0 = \neg(g_\top A \wedge g_\top \neg A)$

Intuitively, the symbol "small ball" intends to represent the classically behaved formulas. If a formula is classical (a good one), one cannot be certain neither of what it asserts nor of what its negation asserts.

DEFINITION 15 $A \leftrightarrow B$ if and only if $(A \to B) \wedge (B \to A)$.

2.2 Axioms of OP_{BL}

Axioms for $\to, \wedge, \vee$: positive intuitionistic logic.

[$\to_1$] $A \to (B \to A)$

[$\to_2$] $(A \to B) \to ((A \to (B \to C)) \to (A \to C))$

[$\wedge_1$] $A \wedge B \to A$

[$\wedge_2$] $A \wedge B \to B$

[$\wedge_3$] $A \to (B \to (A \wedge B))$

[$\vee_1$] $A \to A \vee B$

[$\vee_2$] $B \to A \vee B$

[$\vee_3$] $(A \to C) \to ((B \to C) \to (A \vee B \to C))$

Axioms for $\neg$

[$\neg_1$] $(A^0 \wedge A \wedge \neg A) \to B$

[$\neg_2$] $A^0 \to (A \vee \neg A)$

[$\neg_3$] $\neg\neg A \leftrightarrow A$

[$\neg_4$] $\neg(A^0) \to (\neg g_b A \leftrightarrow g_{\sim b}\neg A)$

[$\neg_5$] $A^0 \to (\neg g_b A \leftrightarrow g_b \neg A)$

[$\neg_6$] $(A \wedge B)^0 \wedge (A \vee B)^0 \wedge (A \to B)^0$

We want to remark that $\neg_1, \neg_2$ and $\neg_4$ are translations of axioms of P_T. Intuitively $\neg_1$ corresponds to reductio ad absurdum for classical formulas. $\neg_4$ corresponds to a technical negation in the bilattice, which is given by our particular choice for these logics: $g_b a = b \oplus a$ thus $\sim (b \oplus a) = \sim b \oplus \sim a$. But $\neg_5$ corresponds to classic behaviour.

Axioms for annotations

For any $b, c \in BL$:

[BL_1] If $b \leq_2 c$ and $b \leq_1 c$ then $g_b A \to g_c A$.

[BL_2] $g_b(A * B) \leftrightarrow g_b A * g_b B$, where $*$ represents $\wedge$ or $\vee$ or $\to$.

[BL_3] $g_b g_c A \leftrightarrow g_{b \oplus c} A$.

[BL_4] $\neg(A^0) \to ((A \leftrightarrow B) \to (g_c A \leftrightarrow g_c B))$.

[BL_5] $(g_b B \wedge g_c B) \to g_{b \vee c} B$.

Intuitively BL_2 asserts that g_b is a homomorphism with respect to the binary operations, but $\neg_4$ and $\neg_5$ assert that, in general, it is not a homomorphism for negation. BL_3 and BL_4 reflect the intuition that the operation g_b can be defined in BL as $g_b a = b \oplus a$.

The inference rule of $OP_{\mathbf{BL}}$:

[R_1] $\dfrac{A, A \to B}{B}$ for any A and B formulas of $OP_{\mathbf{BL}}$.

The notions of proof, $\vdash$ consequence relation in the system $OP_{\mathbf{BL}}$ and theorem are defined as usual.

PROPOSITION 16 Σ, A $\vdash$ B if, and only if, $\Sigma \vdash A \to B$.

Proof: This is an immediate consequence of $\to_1$ and $\to_2$ and the fact that R_1 is the only inference rule.

□

The proof of the following result is straightforward since the axioms for binary connectives correspond to positive intuitionistic logic.

PROPOSITION 17 Let φ $(x_1, x_2, \ldots, x_n)$ be a classical tautology. Then

1. $\{A_1^0, A_2^0, \ldots, A_n^0\} \vdash \varphi(A_1, A_2, \ldots, A_n)$.
2. If φ $(x_1, x_2, \ldots, x_n)$ does not contain negations, then $\vdash \varphi(A_1, A_2, \ldots, A_n)$.

In our system, certain axiom schemes of other annotated systems are theorems, for instance Peirce's law for "well behaved formulas" (M. Guillaume). The proofs of such are standard, and therefore we will omit them.

PROPOSITION 18 For any A, B and C formulas in F,

1. $\vdash B^0 \rightarrow ((A \rightarrow B) \rightarrow ((A \rightarrow \neg B) \rightarrow \neg A))$.
2. $\vdash A^0 \rightarrow ((A \rightarrow (\neg A \rightarrow B))$.
3. $\vdash A^0 \rightarrow (((A \rightarrow B) \rightarrow A) \rightarrow A)$.

We want to emphasize that in the system $OP_{\mathbf{BL}}$, Peirce's law for "well behaved formulas" is proved, therefore all the theorems of the propositional calculus classic positive are valid for "well behaved formulas".

PROPOSITION 19 (Rule of reductio ad absurdum)
If $\Gamma, A \vdash A^0 \wedge B^0 \wedge B \wedge \neg B$ then $\Gamma \vdash \neg A$

Proof: In trivial form $\Gamma, A \vdash B^0$, therefore by Proposition 16 $\Gamma \vdash A \rightarrow B^0$, $\Gamma \vdash A \rightarrow B$ and $\Gamma \vdash A \rightarrow \neg B$. Thus by Proposition 18 item one and Proposition 16 $\Gamma \vdash A \rightarrow \neg A$. On the other side $\Gamma \vdash \neg A \rightarrow \neg A$. Therefore using $A^0, \neg_2$ and $\vee_3$ we conclude that $\Gamma \vdash \neg A$. □

PROPOSITION 20 $A^0, B^0, \rightarrow \neg B \vdash B \rightarrow \neg A$

Proof: Since $A^0, B^0, A \rightarrow \neg B, B, A \vdash B^0, B, \neg B$ then by Proposition 16 $A^0, B^0, A \rightarrow \neg B, B \vdash \neg A$ and $A^0, B^0, A \rightarrow \neg B \vdash B \rightarrow \neg A$. □

THEOREM 21 $\vdash (A^0)^0$.

Proof: Since $A^0 = \neg((g_\top A \wedge g_\top \neg A)$, so $\neg A^0 = \neg\neg(g_\top A \wedge g_\top \neg A)$ and then $\neg A^0 \leftrightarrow (g_\top A \wedge g_\top \neg A)$. Thus $\neg(A^0)^0 \leftrightarrow (g_\top A^0 \wedge g_\top \neg A^0)$. But by $\neg_4$ if $\neg(A^0)^0$ then $g_\top \neg A^0 \leftrightarrow \neg g_\top A^0$ and by $\neg_6$ $(g_\top A^0)^0$. Thus $\neg(A^0)^0 \vdash g_\top A^0, \neg g_\top A^0, (g_\top A^0)^0$ and since by $\neg_6$ $(\neg(A^0)^0)^0$ by Rule of reductio ad absurdum we conclude $\vdash \neg\neg(A^0)^0$ so by $\neg_3$ $\vdash (A^0)^0$. □

COROLLARY 22 $A^0 \wedge \neg A^0 \vdash B$.

Proof: Is immediate consequence of Theorem 21 and $\neg_1$. □

The following results prove that the "well behaved formulas" are preserved by negations and annotations.

PROPOSITION 23 $\vdash A^0 \leftrightarrow (g_b A)^0$.

Proof: [$\Rightarrow$] By $\neg_3$ $\neg(g_b A)^0 \leftrightarrow g_\top(g_b A) \wedge g_\top(g_b(\neg A))$ so by $\neg_4$ and BL_3 we conclude $\neg(g_b A)^0 \leftrightarrow g_\top A \wedge \neg g_\top A$. Thus by $\neg_5$ $A^0 \vdash \neg(g_b A)^0 \rightarrow \neg A^0$. Besides by Theorem 21 $(A^0)^0$ and by $\neg_6$ $(\neg(g_b A)^0)^0$. Then by Proposition 20 $\vdash A^0 \rightarrow \neg\neg(g_b A)^0$ so by $\neg_3$ $\vdash A^0 \rightarrow (g_b A)^0$.

[$\Leftarrow$] By $\neg_3$ $\neg(g_b A)^0 \leftrightarrow g_\top(g_b A) \wedge g_\top(g_b(\neg A))$. Besides by $\neg_5$ $g_\top(g_b A) \leftrightarrow \neg(g_\top g_b A)$ and by BL_3 $g_\top(g_b A) \leftrightarrow \neg(g_\top A)$ so by $\neg_4$ $g_\top(g_b A) \leftrightarrow \neg(g_\top \neg A)$. Therefore $\neg(g_b A)^0 \leftrightarrow \neg A^0$. Thus $(g_b A)^0, \neg A^0 \vdash \neg(g_b A)^0$. Thus by Proposition 16 $(g_b A)^0 \vdash \neg A^0 \rightarrow \neg(g_b A)^0$. Besides by Theorem 21 $((g_b A)^0)^0$ and by $\neg_6$ $(\neg(g_b A)^0)^0$. Thus by Proposition 20 $\vdash (g_b A)^0 \rightarrow \neg\neg A^0$ so by $\neg_3$ $\vdash (g_b A)^0 \rightarrow A^0$.

□

PROPOSITION 24 $\vdash A^0 \leftrightarrow (\neg A)^0$.

Proof: [$\Rightarrow$] By $\neg_3$ $\neg(\neg A)^0 \leftrightarrow g_\top(\neg A) \wedge g_\top(\neg(\neg A))$. Since $g_\top(\neg(\neg A)) \leftrightarrow g_{\sim\top}(\neg(\neg A))$ then by $\neg_4$ we conclude $\neg(\neg A)^0 \leftrightarrow g_\top(\neg A) \wedge \neg g_\top(\neg A)$. So by $\neg_5$ and $\neg_3$ we conclude $\vdash \neg(\neg A^0) \rightarrow \neg A^0$ so by Proposition 20 and Theorem 21 $\vdash A^0 \rightarrow \neg\neg(\neg A)^0$ so by $\neg_3$ $\vdash A^0 \rightarrow (\neg A)^0$.

[$\Leftarrow$] By $\neg_3$ $\neg A^0 \leftrightarrow g_\top A \wedge g_\top(\neg A)$. Therefore BL_4 and $\neg_3$ $g_\top(g_b A) \leftrightarrow \neg(g_\top g_b A)$ and by BL_3 $\neg A^0 \leftrightarrow \neg(\neg A)^0$. So $(\neg A)^0, \neg A^0 \vdash (\neg A)^0, \neg(\neg A)^0$. Then by Theorem 21 and Proposition 16 $\vdash (\neg A)^0 \rightarrow \neg\neg A^0$ so by $\neg_3$ $\vdash (\neg A)^0 \rightarrow A^0$. □

LEMMA 25 All hyperliterals are equivalent to p, $\neg p$ $g_b p$, or $g_b \neg p$, for some propositional letter p.

Proof: Since the only operations involved in the formation of hyperliterals are those of negation and annotation, then after applying such operations we obtain formulas that are equivalent to one of the four cases. We prove our theorem by induction on the complexity of the formula. Let A be a hyperliteral. If $A = g_b B$ for some hyperliteral B, then when $B \leftrightarrow p$, or $B \leftrightarrow \neg p$ the result is immediate. In the case that $B \leftrightarrow g_b p$ or $B \leftrightarrow g_b \neg p$ by $\mathcal{BL}_3$ the result is obtained.

Now if $A = \neg B$ for some hyperliteral B, then when $B \leftrightarrow p$, the result is immediate. If $B \leftrightarrow \neg p$, from $\neg_3$ we get that A is equivalent to p. In the case that $B \leftrightarrow g_b p$ if $\neg(p)^0$ from $\neg_4$ we get that A is equivalent to $g_{\sim b} \neg p$, and in analogous form if $(p)^0$ from $\neg_5$ we get that A is equivalent to $g_b \neg p$. In the case that $B \leftrightarrow g_b \neg p$ if $\neg(\neg p)^0$ from $\neg_4$ and $\neg_3$ we get that A is equivalent to $g_{\sim b} p$, and in analogous form if $(\neg(p))^0$ from $\neg_5$ and $\neg_3$ we get that A is equivalent to $g_b p$. □

3 SEMANTICS FOR $OP_{\mathbf{BL}}$

Continuing with the proximity to P_τ we define a bivalent semantics for $OP_{\mathbf{BL}}$ on the set $\{0, 1\}$. As usual, 0 and 1 represent falsehood and truth in the metalanguage.

A ***interpretation*** I for the system $OP_{\mathbf{BL}}$ is a function $I : P \longrightarrow BL$. Each interpretation I defines a unique ***valuation*** $v_I : F \longrightarrow \{0, 1\}$, such that:

v_1 : $v_I\ (p) = 1$ if, and only if, $I(p) \geq_2 t$ and $I(p) \geq_1 \top$.

v_2 : $v_I\ (\neg p) = 1$ if, and only if, $I(p) \geq_2 f$ and $I(p) \leq_1 \top$.

v_3 : $v_I\ (g_b p) = 1$ if, and only if, $I(p) \oplus b \geq_2 t$ and $I(p) \oplus b \geq_1 \top$.

v_4 : $v_I\ (g_b \neg p) = 1$ if and only if $I(p) \oplus \sim b \geq_2 f$ and $I(p) \oplus \sim b \leq_1 \top$.

v_5 : $v_I\ (\neg g_b A) = v_I\ (g_{\sim b} \neg A)$, for any hyperliteral A.

v_6 : $v_I\ (\neg A) = 1 - v_I(\ A\ )$, for any complex formula A.

v_7 : $v_I\ (g_b g_c A) = v_I\ (g_{b \oplus c} A)$, for any $b, c \in BL$.

v_8 : $v_I\ (g_b(A * B)) = v_I\ ((g_b A) * (g_b B))$, where $*$ represents $\wedge$, $\vee$, or $\rightarrow$.

v_9 : $v_I\ (g_b \neg C) = v_I(\neg g_b C\ )$, where C is a complex formula.

v_{10} : $v_I(A * B) = v_I(A) * v_I(B)$, where $*$ on the left hand side represents $\wedge$, $\vee$, or $\rightarrow$, and $*$ on the right hand side represents the corresponding operation of the Boolean algebra in **Two**.

Intuitively v_1 and v_2 state that the designated (or true) values correspond to the smallest bifilter in the bilattice. In the case of the interlaced bilattice this corresponds to the interval $[t, \top]$, particularly in **Nine** this is $\{(1,0), (1, \frac{1}{2}), (1,1)\}$. With items v_4 and v_5 we would like to emphasize the difference between the negation of a complex formula and of a hyperliteral one. In the case of a complex formula, the negation is classical, whereas for the hyperliterals it obeys a particular choice for these logics already introduced before.

The notions of semantical consequence relation $\models$ and valid formula are defined as usual.

REMARK 26

- If I is an interpretation such that $I(p) = \top$, $v_I(p \wedge \neg p) = 1$. Thus the system is paraconsistent in the strong sense, that is, there are contradictions that are "true" under certain interpretations. Also, $(A \wedge \neg A) \rightarrow B$ does not hold in $OP_{\mathbf{BL}}$.
- The relevance of axiom $\neg_4$ is emphasized here, since if $c \in BL$ and I is an interpretation such that $I(p) = c$ it may happen that $v_I\ (g_b \neg p) = 1$ and $v_I(\neg g_b p) = 0$, for instance, if $c \geq_2 b$ and $c \geq_1 b$.

PROPOSITION 27 A is a complex formula if, and only if, $v_I(A^0) = 1$.

Proof: If A is a hyperliteral, by Lemma 25 there are only four cases.
Case 1: $A = p$. By v_3, the properties of $\oplus$ and $\top$ in the bilattice we obtain that $v_I(g_\top p) = 1$. In analogous form using v_4 in the place of v_3 we obtain that $v_I(g_\top \neg p) = 1$, thus $v_I(g_\top p \wedge g_\top \neg p) = 1$ and therefore $v_I(A^0) = 0$.
Case 2: $A = \neg p$. By $\neg_3$ and v_{10} is reduced to the Case 1.
Case 3: $A = g_b p$. By v_5, $\neg_3$ and v_{10} we get that $v_I(g_\top \neg g_b p) = v_I(g_\top p)$ and by Case 1 we know $v_I(g_\top p) = 1$, which implies that $v_I(A^0) = 0$.
Case 4: $A = g_b \neg p$. By v_3, v_{10}, BL_3, the properties of $\oplus$ and $\top$ in the

bilattice we obtain that $v_I(g_\top p) = 1$. In analogous form if we add v_8 we obtain that $v_I(g_\top \neg g_b \neg p) = 1$, thus $v_I(A^0) = 0$.
If A is complex formula, from v_6 we get that $v_I(A^0) = 1 - v_I(g_\top A \wedge g_\top \neg A)$. But $v_I(g_\top A \wedge g_\top \neg A) = 0$ since by v_6 $v_I(g_\top \neg A) = 1 - v_I\,(g_\top A)$. Thus $v_I(A^0) = 1$. □

PROPOSITION 28 If $\varphi\ (x_1, x_2, \ldots, x_n)$ is a classical tautology and $A_1, A_2, \ldots, A_n$ are complex formulas, then $\models \varphi(A_1, A_2, \ldots, A_n)$; besides if $\varphi\ (x_1, x_2, \ldots, x_n)$ does not contain negations, then for formulas $A_1, A_2, \ldots, A_n \models \varphi(A_1, A_2, \ldots, A_n)$.

PROPOSITION 29 If an *interpretation* I for the system $OP_{\mathcal{BC}}$ is such that $v_I(A) = 1$, then $v_I(g_b A) = 1$.

THEOREM 30 If $\Sigma \vdash$ A, then $\Sigma \models A$.

Proof: We do not need to verify the validity of axioms for $\wedge, \vee, \rightarrow$ and the rule R_1, since they are classical tautologies not involving negations and our valuations were defined accordingly. Moreover $\neg_1$, $\neg_6$ and $\neg_2$ are consequences of the Proposition 27. $\neg_3$ is straightforward. $\neg_4$ is deduced from Proposition 27 and the item v_4. In addition, $\neg_5$ is deduced from Proposition 27 and item v_5.
For BL_1, By Lemma 25 it is enough to check the four cases corresponding to hyperliterals;
Case 1: If $A = p$ then $v_I(g_b p) = 1$ if and only if, $I(p) \oplus b \geq_2 t$ and $I(p) \oplus b \geq_1 \top$ and then $v_I(g_c p) = 1$.
Cases 2, 3, 4 : $A = \neg p$, $A = g_c p$ and $A = g_c \neg p$ are reduced to the previous case.
If A is a complex formula, $v_I(g_b A) = 1$ implies $v_I(g_c A) = 1$ by induction and items v_7 and v_8. BL_2 is deduced from items v_8 and v_{10}, and BL_3 is proved by v_7.

For BL_4, let I be an interpretation such that $v_I\ (A) = a$ and $v_I\ (B) = b$. If $a = b = 1$ then by v_1 we get $a \geq_2 t$ and $a \geq_1 \top$, $b \geq_2 t$ and $b \geq_1 \top$. Then $a \oplus c \geq_2 t$ and $a \oplus c \geq_1 \top$, $b \oplus c \geq_2 t$ and $b \oplus c \geq_1 \top$. Thus, from v_3 we get $v_I\ (g_c A) = 1$.

For BL_5, let I be an interpretation such that $v_I\ (A \rightarrow g_b B) = 1$, $v_I\ (A \rightarrow g_c B) = 1$ and $v_I\ (A \rightarrow f_{b \vee c} B) = 0$. Then $v_I\ (A) = 1$ and $v_I\ (g_{b \vee c} B) = 0$. The case when B is a hyperliteral is reduced to case $B = p$. Therefore $I(p) \not\geq (b \oplus c)$, but $v_I(g_b p) = 1$ and $v_I(g_c p) = 1$, thus $I(p) \geq b$ and $I(p) \geq c$, and then $I(p) \geq (b \oplus c)$, which is a contradiction. For B complex, we are assuming that $v_I\ (g_b A) = 1$, thus by Proposition 29 $v_I(g_b B) = 1$, and therefore $v_I\ (g_{b \vee c} B) = 1$. □

The system is not complete since $\models_{OP_{\mathbf{BL}}}$ is not structural and $\vdash_{OP_{\mathbf{BL}}}$ is. In fact, it is easy to check that $\models g_\top p$, but if we let $\sigma(p) = C \wedge \neg C$ where C is a complex formula, then $\not\models g_\top \sigma(p)$. For the completeness theorem, we are going to modify the system $OP_{\mathbf{BL}}$ slightly and later we will consider a matrix semantics.

4 The Systems $COP_{\mathbf{BL}}$

In order to obtain a complete system we need to include a new rule of inference and two new axioms. The rule asserts that an accepted (or proved) formula has all levels of credibility. One of the axioms guarantees that all hyperliterals can be assigned a certain maximum degree of evidence. The other reflects the behavior of negations with respect to a given level of credibility. These are translations of axioms of system SAL [8]. In order to simplify the notation we make the following convention as of this moment, except for otherwise explicit mention, the symbol $\geq$ will represent symbols $\geq_2$ and $\geq_1$ jointly.

For any formula A and $b, c \in BL$, we define $[A]_b^c$ as follows.

$$[A]_b^c = \begin{cases} A \wedge \neg A & \text{if } c \geq b \text{ and } \sim c \geq b \\ A \wedge \neg(\neg A \wedge \neg A) & \text{if } c \geq b \text{ and } \sim c \not\geq b \\ \neg A \wedge \neg(A \wedge A) & \text{if } c \not\geq b \text{ and } \sim c \geq b \\ \neg(A \wedge A) \wedge \neg(\neg A \wedge \neg A) & \text{if } c \not\geq b \text{ and } \sim c \not\geq b \end{cases}$$

We also define $d_b(A) = g_b A \wedge \bigwedge_{c \not\geq b} \neg(g_c A \wedge g_c A)$. This term is true if and only if the degree of credibility of A is less than or equal to b. Observe that $g_c A \wedge g_c A$, being complex, behaves classically with respect to negations. Its use, instead of simply $g_c A$, is a technical device in order to obtain classical negations.

The systems $COP_{\mathbf{BL}}$ are based on the systems $OP_B L$, but we add a new rule and two new axioms.

4.1 New axioms for annotations

$[BL_6]$ $\neg(A^0) \to \bigvee_{b \in BL} d_b(A)$

$[BL_7]$ $d_b(A) \to \bigvee_{c \in BL} [A]_b^c$

Intuitively $\bigvee_{b \in BL} d_b(A)$ is true if and only if the maximum degree of credibility of A is b. The axiom BL_6 says that all hyperliterals have a maximum degree of credibility.

Likewise $\bigvee_{c \in BL} [A]_b^c$ gives the relation between a degree of credibility b and the degrees of credibility c and $\sim c$. This depends on the particular bilattice that is being used. Axiom BL_7 says that if A has maximum credibility b, then A and $\neg A$ behave as some given credibility c.

4.2 New inference rule of $COP_{\mathbf{BL}}$

$[R_2]$ $\dfrac{A}{f_b A}$ for any $b \in BL$.

REMARK 31 The system $COP_{\mathbf{BL}}$ is also sound with respect to the semantics defined in section 4.

5 Matrix Semantics for $COP_{\mathbf{BL}}$

In this section we will present a matrix semantic for the systems $COP_{\mathbf{BL}}$. In this way we have considered it pertinent to include some basic notations and definitions.

5.1 Matrix Semantics

By an L-**algebra** we mean a structure $\mathbf{A} = \langle A, w\rangle_{w\in L}$ where A is a non-empty set, called the universe of $\mathbf{A}$, and w is an operation on A of rank k for each connective w of rank k of $COP_{\mathcal{BL}}$. An L-**matrix** is a pair $\mathcal{A} = \langle \mathbf{A}, F\rangle$, where $\mathbf{A}$ is an L-algebra and F is an arbitrary subset of A; the elements of F are called *designated elements* of A.

A **valuation** on an L-matrix $\mathcal{A} = \langle \mathbf{A}, F\rangle$ is the unique homomorphism that extends a function (defined in the set of propositional letters) $u : P \longrightarrow A$ to $u^* : \mathbf{Form} \longrightarrow \mathbf{A}$.

The formula φ is a **consequence** of Γ in $\mathcal{A}$, in symbols $\Gamma \models_{\mathcal{A}} \varphi$ if for any valuation $u : P \longrightarrow A$, $u^*(\psi) \in F$, for any $\psi \in \Gamma$ implies $u^*(\varphi) \in F$.

For S a deductive system (a logic) we say that a matrix $\mathcal{A}$ is a **matrix model** of S if $\Gamma \vdash \varphi$ implies $\Gamma \models_{\mathcal{A}} \varphi$, in this case the subset F is called S-**filter**. Observe that if T is an S-theory, then $\langle \mathbf{Form_L}, T\rangle$ is a matrix model of S. These are called *Lindenbaum matrices* for S.

A class $\mathbf{M}$ of matrices is a **matrix semantics** for S if $\Gamma \vdash \varphi$ if and only if $\Gamma \models_{\mathcal{M}} \varphi$ for all $\mathcal{M} \in \mathbf{M}$. For instance, the class of all Lindenbaum matrices for S and the class of all matrix models of S are matrix semantics.

5.2 Soundness

We will now define a class $\mathbf{M}$ of matrices and prove that the matrices of $\mathbf{M}$ are matrix models for $COP_{\mathbf{BL}}$. Let BL be a bilattice, F a bifilter of BL and let 0 and 1 be objects not in BL. We define $\widehat{BL} = BL \cup \{0, 1\}$ and $\widehat{F} = F \cup \{1\}$. We extend the orders of BL to $\widehat{BL}$ as follows.

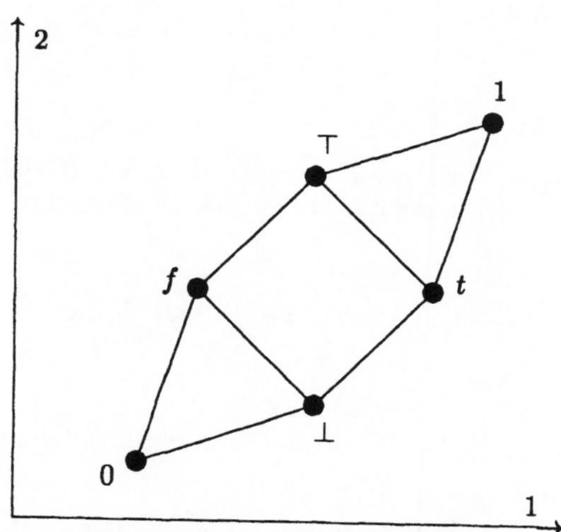

Figure 5: The orders of $\widehat{BL}$

In addition, we define the following operations on $\widehat{BL}$:

1. $a \vee b = \begin{cases} 1 & \text{if } a \in \widehat{F} \text{ or } b \in \widehat{F} \\ 0 & \text{otherwise.} \end{cases}$

2. $a \wedge b = \begin{cases} 1 & \text{if } a \in \widehat{F} \text{ and } b \in \widehat{F} \\ 0 & \text{otherwise.} \end{cases}$

3. $a \rightarrow b = \begin{cases} 1 & \text{if } a \notin \widehat{F} \text{ or } b \in \widehat{F} \\ 0 & \text{otherwise.} \end{cases}$

4. If $a \in BL$ then $\neg a = \sim a$, $\neg 1 = 0$ and $\neg 0 = 1$.

5. $g_b a = \begin{cases} b \oplus a & \text{if } a \neq 0 \text{ and } a \neq 1 \\ a & \text{if } a = 0 \text{ or } a = 1 \end{cases}$

6. $a^0 = \neg(g_\top a \wedge g_\top \neg a)$.

The class M is the set of the matrices $\mathbf{M} = \langle \widehat{BL}, \widehat{F} \rangle$.

PROPOSITION 32 If $a \in \widehat{BL}$ then $a^0 = \begin{cases} 0 & \text{if } a \in BL \\ 1 & \text{if } a \in \{0,1\} \end{cases}$

Proof: By definition $a^0 = \neg(g_\top a \wedge g_\top \neg a)$ thus $a^0 = 0$ iff $(g_\top a \wedge g_\top \neg a) = 1$ iff $\top \oplus a, \top \oplus \neg a \in F$ iff $a \in BL$, since if $a \in \{0,1\}$ then $(g_\top a \wedge g_\top \neg a) = 0$.
On the other hand $a^0 = 1$ iff $(g_\top a \wedge g_\top \neg a) = 0$ iff $g_\top a = 0$ or $g_\top \neg a = 0$ iff $a = 0$ or $\neg a = 1$ iff $a \in \{0,1\}$. □

PROPOSITION 33 Let $\mathcal{M} = \langle \widehat{BL}, \widehat{F} \rangle \in \mathbf{M}$ and $u : P \longrightarrow \widehat{BL}$ be such that $u(p) \in BL$ then for any hyperliteral B with propositional letter p, $u^*(B^0) = 0$ iff $u^*(B) \in BL$. Besides, if $u(p) \in \{0,1\}$ then $u^*(p^0) = 1$.

THEOREM 34 If a matrix $\mathcal{M} = \langle \widehat{BL}, \widehat{F} \rangle \in \mathbf{M}$ then $\mathcal{M}$ is a matrix model of $COP_{\mathbf{BL}}$.

Proof: $\rightarrow_1, \rightarrow_2, \wedge_1, \wedge_2, \wedge_3, \vee_1, \vee_2, \vee_3, \neg_3$, and R_1 are trivial, since by definition they have classical behavior and our operations are defined accordingly.

[$\neg_1$] $A^0 \wedge A \wedge \neg A \rightarrow B$
If $a^0 \in \widehat{F}$ then $a \in \{0,1\}$ so $a^0 \wedge a \wedge \neg a = 0$. From where the result follows immediately.

[$\neg_2$] $A^0 \rightarrow (A \vee \neg A)$
If $a^0 \in \widehat{F}$ then $a \in \{0,1\}$. From where the result follows immediately.

[$\neg_4$] $\neg(A^0) \rightarrow (\neg g_b A \leftrightarrow g_{\sim b} \neg A)$
If $\sim a^0 \in \widehat{F}$ then $a^0 \in BL$. Therefore $\neg g_b A \leftrightarrow \sim (b \oplus a)$. So $g_{\sim b} \neg A \leftrightarrow \sim b \oplus \sim a$. Thus, by Proposition 4 the result follows.

[$\neg_5$] $A^0 \rightarrow (\neg g_b A \leftrightarrow g_b \neg A)$
Case 1: If $a = 0$ then $g_b a = 0$. Thus $\neg g_b a = 1$ and $g_b \neg a = g_b 1$, so $g_b \neg a = 1$.
Case 2: If $a = 1$ then $g_b a = 1$. Thus $\neg g_b a = 0$ and $g_b \neg a = g_b 0$, so $g_b \neg a = 0$.

$[\neg_6]$ $(A \wedge B)^0 \wedge (A \vee B)^0 \wedge (A \to B)^0$

Is a consequence of the definition of the operation "small ball".

$[BL_1]$ $g_b A \to g_c A$ for any $b, c \in BL$ such that $b \leq_2 c$ and $b \leq_1 c$.

It is a consequence of basic properties of bifilters.

$[BL_2]$ $g_b(A * B) \leftrightarrow g_b A * g_b B$, where $*$ represents $\wedge$, *or* $\vee$ *or* $\to$.

It is immediate.

$[BL_3]$ $g_b g_c A \leftrightarrow g_{b \oplus c} A$ for any $b, c \in BL$.

It is a consequence of the associativity of $\oplus$ in the bilattice.

$[BL_4]$ $\neg(A^0) \to ((A \leftrightarrow B) \to (g_b A \leftrightarrow g_b B))$, for any $b \in BL$.

If $\neg a^0 \in \widehat{F}$ then $\neg a^0 = 1$. Thus $a^0 = 0 \leftrightarrow a \in BL$. Thus this property is reduced to $c \oplus a \in \widehat{F}$ iff $b \oplus a \in \widehat{F}$, this is satisfied since F is a bifilter.

$[BL_5]$ $(g_b B \wedge g_c B) \to f_{b \vee c} B$ for any $b, c \in BL$.

This is satisfied since F is a bifilter.

$[BL_6]$ $\neg(A^0) \to \bigvee_{b \in BL} d_b(A)$

Let S be the set $\{b : b \in F \text{ and } b \oplus a \in F\}$. Then $S \neq \phi$ since $\top \in S$. We define

$b_0 = infS$. So $(d_{b_0}(A))^{\mathcal{M}} = b \oplus a \wedge \bigwedge_{c \not\geq b_0} \neg(c \oplus a \wedge c \oplus a)$. Thus $d_{b_0}(A)^{\mathcal{M}} = 1$,

so $(\neg(A^0) \to \bigvee_{b \in BL} d_b(A))^{\mathcal{M}} = 1$.

$[BL_7]$ $d_b(A) \to \bigvee_{c \in BL} [A]_b^c$

If $d_b(a) \in \widehat{F}$ then $a \in BL$ and $a \notin F$. Since $g_b a = b \oplus a$ then $b \geq b_0$. On the other hand $c \not\geq b$ implies that $\neg(g_c a \wedge g_c a) = 1$. Then $b_0 \geq b$ so $b = b_0$. Besides, since $a \notin F$ then

$$([a]_{b_0}^a)^{\mathcal{M}} = \begin{cases} \neg a \wedge \neg(a \wedge a) & \text{if } \sim a \geq b_0 \\ \neg(a \wedge a) \wedge \neg(\neg a \wedge \neg a) & \text{if } \sim a \not\geq b_0 \end{cases}$$

Therefore $([a]_{b_0}^a)^{\mathcal{M}} = 1 \in \widehat{F}$. Thus we get $\bigvee_{c \in BL} [a]_b^c \in \widehat{F}$.

So $(d_b(A) \to \bigvee_{c \in BL} [A]_b^c)^{\mathcal{M}} = 1 \in \widehat{F}$.

We will now check R_2 : if $a \in \widehat{F}$. Let us consider two cases.

Case 1: $a = 1$. In this case $g_b a = 1$ by definition, and thus trivially $g_b a \in \widehat{F}$. So R_2 is sound.

Case 2: $a \in F$. Then $g_b a = b \oplus a \geq a$, so $g_b a \in F$ since F is a bifilter. So R_2 is sound.

□

5.3 Matrix semantics

We will now define a matrix **M** and prove that it is a matrix semantics for $COP_{\mathbf{BL}}$. Let BL be a (fixed) bilattice such that its bifilters are the set $\{F_1, F_2, ..., F_k\}$ and

let 0 and 1 be objects not in BL.
We define $\mathsf{F} = (\bigcup_{1 \le i \le k} \{i\} \times F_i) \cup \{1\}$ and $\mathsf{B} = (\bigcup_{1 \le i \le k} \{i\} \times BL) \cup \{0, 1\}$.

We extend the order of BL to B as follows.

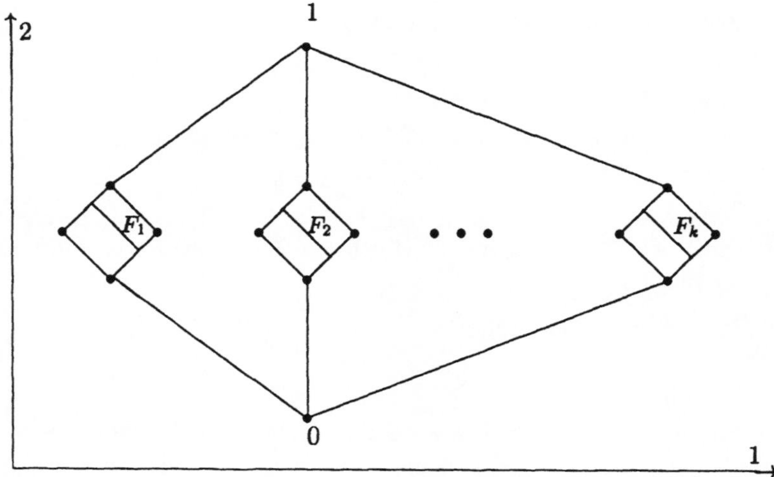

Figure 6: The orders for Matrix semantics.

In addition, we define the following operations on B :

1. $\mathsf{a} \vee \mathsf{b} = \begin{cases} 1 & \text{if } \mathsf{a} \in \mathsf{F} \text{ or } \mathsf{b} \in \mathsf{F} \\ 0 & \text{otherwise.} \end{cases}$

2. $\mathsf{a} \wedge \mathsf{b} = \begin{cases} 1 & \text{if } \mathsf{a} \in \mathsf{F} \text{ and } \mathsf{b} \in \mathsf{F} \\ 0 & \text{otherwise.} \end{cases}$

3. $\mathsf{a} \to \mathsf{b} = \begin{cases} 1 & \text{if } \mathsf{a} \notin \mathsf{F} \text{ or } \mathsf{b} \in \mathsf{F} \\ 0 & \text{otherwise.} \end{cases}$

4. If $\mathsf{a} = (i, a)$ where $a \in BL$ then $\neg\mathsf{a} = (i, \sim a)$, $\neg 1 = 0$ and $\neg 0 = 1$.

5. $g_b\mathsf{a} = \begin{cases} (i, b \oplus a) & \text{if } \mathsf{a} = (i, a) \\ \mathsf{a} & \text{if } \mathsf{a} = 0 \text{ or } \mathsf{a} = 1 \end{cases}$

6. $a^0 = \neg(g_\top a \wedge g_\top \neg a)$.

The matrix $\mathbf{M} = \langle \mathsf{B}, \mathsf{F} \rangle$.

THEOREM 35 The matrix $\mathbf{M}$ is a matrix model of $COP_{\mathbf{BL}}$.

Proof: The proof is similar to that of Theorem 34. □

The usual constructions tell us that any non-trivial set of formulas can be included in a maximal non-trivial consistent set of formulas, thus we can restrict ourselves to present the completeness for a maximal non-trivial consistent set of formulas.

THEOREM 36 Let Γ be a maximal non-trivial consistent set of formulas. Then there exists a valuation v: $P \longrightarrow \mathsf{B}$ such that for any formula A, $v^*(A) \in \mathsf{F}$ if and only if $A \in \Gamma$.

Proof: For p a propositional letter in P there are two possible cases: $p^0 \notin \Gamma$ or $p^0 \in \Gamma$. For case $p^0 \notin \Gamma$, let a_i be the least element of $\{\mathsf{b} : g_b(p) \in \Gamma\}$ and F_i the smallest bifilter of BL that contains a_i. In addition let b_i be the least element of $\{b : [p]_b^{\mathsf{a}_i} \in \Gamma\}$. We define

$$v(\mathrm{p}) = \begin{cases} 1 & \text{if } p^0 \in \Gamma \text{ and } p \in \Gamma \\ 0 & \text{if } p^0 \in \Gamma \text{ and } \neg p \in \Gamma \\ (i, b_i) & \text{if } p^0 \notin \Gamma \end{cases}$$

First of all, we prove the result for hyperliterals. By Lemma 25 it is enough to check the following four cases:

Case 1: $A = p$. Then $v^*(p) = v(p)$. Thus if $p^0 \in \Gamma$ then $v(p) = 1$, this is equivalent to $p \in \Gamma$. If $p^0 \notin \Gamma$, then $p \notin \Gamma$ by definition of b_i.

Case 2: $A = \neg p$. If $p^0 \in \Gamma$ then $v(p) \in \{0, 1\}$. So $v^*(\neg p) \in \mathsf{F}$ is equivalent to $v(\neg p) = 1$. This is equivalent to $v(p) = 0$, that is equivalent to, $\neg p \in \Gamma$. If $p^0 \notin \Gamma$, $\neg p^0 \in \Gamma$. Then $v^*(\neg p) \in \mathsf{F}$ is equivalent to $\sim b_i \in F_i$. So we must prove that this last assertion is equivalent to $\sim b_p \geq \mathsf{a}$. We see that $[A]_{b_i}^{a_i} = p \wedge \neg p$ or $\neg p \wedge \neg(p \wedge p)$. So $\vdash [A]_{b_i}^{a_i} \rightarrow \neg p$ by $\wedge_1$ and $\wedge_2$. Since $[A]_b^c \in \Gamma$, $\neg p \in \Gamma$, so $v^*(\neg p) \in \mathsf{F}$ implies that $\neg p \in \Gamma$. Now $v^*(\neg p) \notin \mathsf{F}$ is equivalent to $\sim b_i \ngeq \mathsf{a}_i$. Thus $[A]_{b_i}^{a_i} = p \wedge \neg(\neg p \wedge \neg p)$ or $[A]_{b_i}^{a_i} = \neg(p \wedge p) \wedge \neg(\neg p \wedge \neg p)$. Therefore $\vdash [A]_b^c \rightarrow \neg(\neg p \wedge \neg p)$ by $\wedge_1\, and \wedge_2$, and thus $\neg p \notin \Gamma$. Therefore $v^*(\neg p) \in \mathsf{F}$ is equivalent to $\neg p \in \Gamma$.

Case 3: $A = g_c p$. If $p^0 \in \Gamma$ then $v^*(A) \in \mathsf{F}$ iff $v^*(g_c p) \in \mathsf{F}$ iff $g_c\, v(p) \in \mathsf{F}$ iff $v(p) = 1$. This is equivalent to $p \in \Gamma$. So $v(p)$ cannot be equal to zero, since in such case $g_c\, v(p)$ would not belong to F. So it is enough to prove that if $p \in \Gamma$ then $g_c p \in \Gamma$, which is obtained by $R2$. If $p^0 \notin \Gamma$ then $v(p) = b_i$. So $v^*(g_c p) \in \mathsf{F}$ iff $g_c\, v(p) \in \mathsf{F}$ iff $c \oplus v(p) \in F_i$ iff $c \oplus b_i \in F_i$ iff $A \in \Gamma$.

Case 4: If $p^0 \in \Gamma$ is analogous to Case 3. For $p^0 \notin \Gamma$ then $v(p) = b_i$. From Case 2 we get that $v^*(\neg p) \in \mathsf{F}$ is equivalent to $\neg p \in \Gamma$. The rest of the proof is analogous to Case 3.

The rest of the result is obtained by induction.

□

COROLLARY 37 If Γ is a non-trivial set of formulas of $COP_{\mathbf{BL}}$ then $\Gamma \vdash \varphi$ iff $\Gamma \models_{\mathsf{M}} \varphi$.

Proof: As a consequence of Theorem 36 and fact that any non-trivial set of formulas is contained in a maximal non-trivial set of formulas.

□

6 THE ALGEBRAIZATION OF THE SYSTEMS $OP_{\mathbf{BL}}$

A deductive system S over a language L is **algebraizable** in the sense of Blok-Pigozzi if there exists a quasivariety K of L-algebras such that the S-consequence relation $\vdash$ and the equational consequence relation $\models_K$ over K are interpretable in one another in the following strong sense, that is to say, given a formula α there exists an equation $\widehat{\alpha}$ and given an equation $\sigma \approx \tau$ there exists a formula $\widetilde{\sigma \approx \tau}$ such that:

1. $\Gamma \vdash \alpha$ if, and only if, $\widehat{\Gamma} \models_K \widehat{\alpha}$,

2. $\Sigma \models_K \sigma \approx \tau$ if, and only if, $\widetilde{\Sigma} \models_K \widetilde{\sigma \approx \tau}$,

3. $\alpha \dashv\vdash \widetilde{\widehat{\alpha}}$,

4. $\sigma \approx \tau \;=\!\mid\models_K \widehat{\widetilde{\sigma \approx \tau}}$.

The quasivariety K is called the **equivalent algebraic semantics** of S.

REMARK 38

- Intuitively the items 1 and 4 say that any deduction in S can be carried out as an equational consequence within the quasivariety K, using all tools of model theory and universal algebra.
- Statements 2 and 3 say something similar in the other direction, that is, equational consequences in K can be carried out in the system S.

EXAMPLE 39 If S is classical logic and K is the variety of the Boolean algebras, $\widehat{\alpha} = \alpha \approx \top$ and $\widetilde{\sigma \approx \tau} = \sigma \leftrightarrow \tau$ satisfy the four conditions above mentioned.

Now we study formally the algebraization of the logics $OP_{\mathbf{BL}}$. Recall that algebraization here means algebraization in the sense of Blok-Pigozzi [6]. They proved that a system S is algebraizable if there are unary terms $\delta_i, \epsilon_i, i \in I$ and binary terms $\Delta_j, j \in J$, where I and J are finite, such that:

[1] $\vdash A\Delta A$

[2] $A\Delta B \vdash B\Delta A$

[3] $A\Delta B, B\Delta C \vdash A\Delta C$

[4]

[a] $A\Delta B \vdash \neg A\Delta\neg B$

[b] $A\Delta B \vdash g_b A\Delta g_b B$

[c] $A\Delta B, C\Delta D \vdash A * C\Delta B * D$, where $*$ represents $\wedge$, $\vee$, or $\rightarrow$

[5] $A \dashv\vdash \delta(A)\Delta\epsilon(A)$

where $\vdash A\Delta B$ means that $\vdash A\Delta_j B$, for any $j \in J$, and similarly for $\delta\, and\, \epsilon$. The set Δ is called ***a system of equivalence formulas for S*** and $\delta \approx \epsilon$ are called ***a set of defining equations for S***.

In addition, they proved that the equivalent quasivariety semantics associated with any fixed algebraizable deductive system is uniquely determined. The proof of this derives from the fact that Δ represents, within S, the relation of equality in the algebraic models of S.

THEOREM 40 Given a finite bilattice BL, $\delta(A) = A \wedge A \approx A \rightarrow A = \epsilon(A)$ is a set of defining equations for the logic $OP_{\mathbf{BL}}$ and

1. $\Delta_{\leftrightarrow}(A, B) = A \leftrightarrow B$
2. $\Delta_{\neg}(A, B) = \neg A \leftrightarrow \neg B$
3. $\Delta_b(A, B) = g_b A \leftrightarrow g_b B$

are a system of equivalence formulas for the logic $OP_{\mathbf{BL}}$, so the system $OP_{\mathbf{BL}}$ is algebraizable.

Proof: The proofs of [1], [2] and [3] are immediate. We will check [4];

Case [a] $A\Delta B \vdash \neg A \Delta \neg B$

1. $\neg A \leftrightarrow \neg B$ is deduced by $\Delta_{\neg}$.

2. $\neg\neg A \leftrightarrow \neg\neg B$, since $\neg\neg A \leftrightarrow A$, follows from $\Delta_{\leftrightarrow}$.

3. $g_b \neg A \leftrightarrow g_b \neg B$, for complex formulas is deduced from classical tautologies and for hyperliterals by the axiom BL_4.

Case [b] $A\Delta B \vdash g_b A \Delta g_b B$

1. $g_b A \leftrightarrow g_b B$ follows from Δ_b.

2. $\neg g_b A \leftrightarrow \neg g_b B$, for complex formulas is deduced from classical tautologies and for hyperliterals by axiom BL_4 together with $\Delta_{\leftrightarrow}$.

3. $g_c g_b A \leftrightarrow g_c g_b B$, for complex formulas is deduced from classical tautologies and the axiom BL_2. For hyperliterals, since by axiom BL_3 $g_c g_b A \leftrightarrow g_{c\oplus b} A$, from $\Delta_{\leftrightarrow}$ it is proved that $g_{c\oplus b} A \leftrightarrow g_{c\oplus b} B$, and so $g_c g_b A \leftrightarrow g_c g_b B$.

Case [c] is deduced from BL_2 and classical tautologies.

We will now check the item [5]; $\delta(A) = A \wedge A\, and\, \epsilon(A) = A \rightarrow A$. On the other hand $\rightarrow_1$ *and* $\rightarrow_2$ implies that $\vdash A \rightarrow A$, and thus $\vdash (A \wedge A) \rightarrow (A \rightarrow A)$, and trivially $A \vdash (A \wedge A) \rightarrow (A \rightarrow A)$. Also by $\wedge_3$ it is deduced $A \vdash (A \wedge A)$, then $A \vdash (A \rightarrow A) \rightarrow (A \wedge A)$, and then $A \vdash (A \wedge A) \leftrightarrow (A \rightarrow A)$. That is, $A \vdash \delta(A) \leftrightarrow \epsilon(A)$. Since $\delta(A)$ and $\epsilon(A)$ are complex formulas by classical tautologies we have $A \vdash \delta(A)\Delta\epsilon(A)$. On the other hand $\delta(A)\Delta\epsilon(A) \vdash (A \rightarrow A) \rightarrow (A \wedge A)$, but $(A \wedge A) \vdash A$ by $\wedge_1$, and so $\delta(A)\Delta\epsilon(A) \vdash A$. □

THEOREM 41 The annotated logic $OP_{\mathbf{BL}}$ is algebraizable if, and only if, BL is finite.

REMARK 42 Theorem 41 is also true about system $COP_{\mathbf{BL}}$, since it is stronger than the system $OP_{\mathbf{BL}}$.

The systems OP_{BL} and $COP_{\mathbf{BL}}$ are algebraizable, which partly tells us we are dealing with sound (good) logics, at least from the point of view of algebraic logic. This gives us some peace of mind!

REFERENCES

[1] V. S. Subrahmanian. On the Semantics of Quantitative Logic Programs. In *Proceedings of the 4th IEEE Symposium on Logic Programming*, pages 173–182. Computer Press, Washington D.C., 1987.

[2] H. A. Blair and V. S. Subrahmanian. Paraconsistent Logic Programming. *Theoretical Computer Science*, 68:135–154, 1989.

[3] N. C. A. da Costa, V. S. Subrahmanian, and C. Vago. The Paraconsistent Logics $P\tau$. *Zeitschrift für Math. Logic*, 37:139–148, 1991.

[4] N. C. A. da Costa, J. M. Abe, and V. S. Subrahmanian. Remarks on Annotated Logic. *Zeitschrift für Math. Logic*, 37:561–570, 1991.

[5] J. M. Abe. *Fundamentos da Lógica Anotada*. PhD thesis, São Paulo University, Brazil, 1992.

[6] W. J. Blok and D. Pigozzi. Algebraizable Logics. Memoirs of the A.M.S. **77**, Nr. 396. A.M.S., 1989.

[7] R. A. Lewin, I. F. Mikenberg, and M. G. Schwarze. On the Algebraizability of Annotated Logics. *Studia Logica*, 57:359–386, 1997.

[8] R. A. Lewin, I. F. Mikenberg, and M. G. Schwarze. Matrix Semantics for Annotated Logics. In X. Caicedo and C. Eds. Montenegro, editors, *Proceedings of the X SLALM, Bogotá 1995*, pages 279–293. Marcel Dekker, 1999.

[9] M. Kifer and E. L. Lozinskii. A Logic for Reasoning with Inconsistency. *Journal of Automated Reasoning*, 9(2):335–368, 1992.

[10] M. Kifer and V. S. Subrahmanian. Theory of Generalized Annotated Logic Programming and its Applications. *Zeitschrift für Math. Logic*, 12:335–367, 1992.

[11] M. Arenas, L. Bertossi, and M. Kifer. Applications of Annotated Predicate Calculus to Querying Inconsistent Databases. In *Lecture Notes in Artificial Intelligence 1861, Computational Logic CL 2000, Proceedings of the First International Conference London, UK, July 2000*, pages 926–942. Springer, 2000.

[12] M. Ginsberg. Bilattices and Modal Operators. In *Proc. 1990 Intl. Conf. On Theoretical Aspects of Reasoning About Knowledge*. Morgan Kaufmann, 1990.

[13] M. C. Fitting. Bilattices and the Semantics of Logic Programming. *Journal of Logic Programming*, 11:91–116, 1991.

[14] O. Arieli and A. Avron. Reasoning with Logical Bilattices. *Journal of Logic, Language and Information*, 5(1):25–63, 1996.

[15] O. Arieli and A. Avron. A Model-Theoretic Approach for Recovering Consistent Data from Inconsistent Knowledge Bases. *Journal of Automated Reasoning*, 22(1):263–309, 1999.

[16] B. Mobasher, D. Pigozzi, and G. Slutzki. Multi-valued Logic Programming Semantics An Algebraic Approach. *Theoretical Computer Science*, 171:77–109, 1997.

[17] N. C. A. da Costa, J. M. Abe, J. M. Silva Filho, J. I. Murolo, and C. F. Leite. *Lógica Paraconsistente Aplicada*. São Paulo, Editorial Atlas S.A., 1999.

[18] N. D. Belnap. A Useful Four-Valued Logic. In G. Epstein and J. M. Dunn, editors, *Modern Uses of Multiple-valued Logic*, pages 7–37. Reidel Publishing Company, 1977.

[19] N. C. A. da Costa. On Theory of Inconsistent Formal Systems. *Notre Dame Journal of Formal Logic*, 11:497–510, 1974.

[20] N. C. A. da Costa and E. H. Alves. A Semantical Analysis of the Calculi c_n. *Notre Dame Journal of Formal Logic*, 18:621–630, 1977.

[21] N. C. A. da Costa and D. Krause. An Inductive Annotated Logic. *This volume.*

[22] T. Weigert, J. P. Tsai, and X. Liu. Fuzzy Operator Logical and Fuzzy Resolution. *Journal of Logic Automated Reasoning*, 10:59–78, 1993.

Definability and Interpolation in Extensions of Johansson's Minimal Logic *

LARISA MAKSIMOVA Institute of Mathematics, Siberian Division of Russian Academy of Sciences, 630090, Novosibirsk, Russia
lmaksi@math.nsc.ru

Abstract
The exhaustive lists of positive logics in $E(J^+)$ with the interpolation property CIP or with the projective Beth property PBP are obtained. All positively axiomatizable extensions of Johansson's Minimal Logic with CIP or PBP are described.

We herein study analogs of Beth's theorem on implicit definability [1] as well as Craig's interpolation property CIP in the family $E(J)$ of extensions of Johansson's minimal logic J [2].

The logic J is paraconsistent unlike the intuitionistic logic Int. At the same time, J and Int have the same positive fragment J^+, and we can use known methods and results on superintuitionistic logics in our study of the family $E(J)$. Note that the paraconsistent systems $C_n, 1 < n < \omega$, and CC_ω introduced by N.C.A. da Costa [3, 4] and R.Sylvan [5] are extensions of J^+ but incomparable with J.

It follows from [6] that all extensions of J^+ or of J have the Beth property BP. In 1962 K.Schütte [7] proved CIP for the intuitionistic predicate logic. His proof implies CIP also for J and J^+. It is known that in the family $E(Int)$ of superintuitionistic logics there exist exactly sixteen logics with the projective Beth property PBP [8], among them eight logics have CIP [9]. In the present paper we find many examples of logics with CIP or PBP in $E(J)$ which do not contain Int.

Simultaneously we get exhaustive lists of positive logics in $E(J^+)$ with the interpolation property or with PBP. Also we find all positively axiomatizable logics with PBP in $E(J)$ and describe them.

1 Positive logics

In this section we examine positive logics containing the positive fragment J^+ of Johannson's minimal logic J. We take the propositional constant $\top$ ("true") and $\&, \vee, \to$ as primitive. Formulas of this language are called ***positive***. The language

*This research was supported by The Russian Foundation of Humanities, grant 00-03-00108

of J contains one more constant $\bot$ ("absurdity"); one can consider negation as the abbreviation $\neg A \rightleftharpoons A \to \bot$.

A logic L is usually identified with the set of its theorems. By a logic we mean any set of formulas (of some fixed language) closed under modus ponens and substitution rules. We denote by $L + Ax$ the logic obtained from L by adding extra axiom schemes Ax. A *proof* in $L + Ax$ is a sequence $B_1, \dots, B_n$ such that for each $i \leq n$ either B_i arises by some substitution from an axiom of L or of Ax, or B_i is an immediate consequence of some B_j and B_k, where $j, k < i$, by modus ponens. We write $\Gamma \vdash_L A$ if A is derivable from $L \cup \Gamma$ by modus ponens only.

A logic L is said to have *Craig's interpolation property (CIP)* if for each formula $(A \to B)$ in L there exists a formula C such that

(i) both $(A \to C)$ and $(C \to B)$ are in L and

(ii) every variable of C occurs in both A and B.

A logic L is said to have the *projective Beth property (PBP)*, if

$$A(P,Q,X), A(P,Q',Y) \vdash_L (X \leftrightarrow Y)$$

implies that there exists a formula $C(P)$ such that

$$A(P,Q,X) \vdash_L (X \leftrightarrow C(P)),$$

where P, Q, Q' are disjoint lists of variables not containing variables X and Y; the *Beth property BP* is a special case of PBP, when Q and Q' are empty.

It follows from [6] that all logics in the family $E(J^+)$ of positive logics extending J^+ possess the Beth property BP. In the same way as in [10], one can derive PBP from CIP in extensions of J^+.

First we reduce interpolation and projective Beth property in positive logics to superintuitionistic logics.

PROPOSITION 1 Let L be a positive logic in $E(J^+)$ and

$$L' = Int + (\neg p \vee \neg\neg p) + \{A' | A \in L\},$$

where A' is obtained from A by substituting $(x \vee \neg x)$ for each variable x of A. Then

(i) L has CIP iff L' has CIP,

(ii) L has PBP iff L' has PBP.

Proof: Follows from Theorem 3 and Lemma 7 below. □

It is known that $L'_1 = L'_2$ implies $L_1 = L_2$.

We described all superintuitionistic logics with CIP in [9]. Proposition 1(i) was stated in [11]; moreover, we proved in [11] that there exist exactly three non-trivial logics in $E(J^+)$ possessing CIP. All sixteen superintuitionistic logics with PBP were found in [8]. All these logics are finitely axiomatizable and have the finite model property. Using their description and Proposition 1, we prove

THEOREM 2 There exist exactly six non-trivial positive logics with PBP in $E(J^+)$. They are (1) J^+, and its extensions by axiom schemes

(2) $(p \to q) \vee (q \to p)$,

(3) $p \vee (p \to q)$,

(4) $r \vee (r \to (p \to q) \vee (q \to p))$,

(5) $r \vee (r \to p \vee (p \to q))$,

(6) $p \vee (p \to q) \vee (q \to r)$.

The logics J^+, $J^+ + (2)$, $J^+ + (3)$ possess CIP, and the others do not possess CIP.

The logic J^+ is complete with respect to Kripke models $\mathbf{M} = (W, \leq, \models)$, where W is a non-empty set, $\leq$ is a partial ordering of W, and $\models$ satisfies the conditions:

(K1) if $x \models p$ and $x \leq y$ then $y \models p$, for each variable p,
(K2) $x \models (A \to B)$ iff $\forall y((x \leq y$ and $y \models A) \Rightarrow y \models B)$,
(K3) $x \models (A\&B)$ iff ($x \models A$ and $x \models B$),
(K4) $x \models (A \vee B)$ iff ($x \models A$ or $x \models B$).

The logic $J^+ + (2)$ is characterized by linearly ordered models, and $J^+ + (3)$ by models with one-element set W. Further, $J^+ + (4)$ is determined by models satisfying the condition:

$(x < y \leq u$ and $x < y \leq v) \Rightarrow (u \leq v$ or $v \leq u)$;

$J^+ + (5)$ by models without three-element chains, and $J^+ + (6)$ is characterized by two-element chains.

We bring up only some ideas of our proof of Proposition 1 and of Theorem 2. To prove them, we essentially use the algebraic semantics. It is well known that there is a duality between the family $E(Int)$ and the family of varieties of Heyting algebras. With any superintuitionistic logic L one can associate a variety $V(L)$ of Heyting algebras. If L is a positive logic in $E(J^+)$ then its corresponding variety consists of relatively pseudo-complemented lattices [12]. As well as for superintuitionistic logics, we find algebraic equivalents of CIP and PBP.

Recall some definitions. Say that a variety V has *Amalgamation Property AP* if for any $\mathbf{B}, \mathbf{C} \in V$ with a common subalgebra $\mathbf{A}$ there exist $\mathbf{D} \in V$ and monomorphisms $\delta : \mathbf{B} \to \mathbf{D}$ and $\varepsilon : \mathbf{C} \to \mathbf{D}$ which coincide on $\mathbf{A}$; the variety V possesses *Strong Epimorphisms Surjectivity SES* if for any $\mathbf{A}, \mathbf{B} \in V$ such that $\mathbf{A}$ is a subalgebra of $\mathbf{B}$ and for any $b \in \mathbf{B} - \mathbf{A}$, there exist $\mathbf{D} \in V$ and homomorphisms $\delta, \varepsilon : \mathbf{B} \to \mathbf{D}$ which coincide on $\mathbf{A}$ and satisfy $\delta(b) \neq \varepsilon(b)$.

In the same way as in the case of superintuitionistic logics [9],[13], we prove

THEOREM 3 For every logic L in $E(J^+)$ or in $E(J)$:

(i) L has CIP iff $V(L)$ has AP;

(ii) L has PBP iff $V(L)$ has SES.

Proof: (i) In Theorem 1 of [9] we proved the statement for superintuitionistic logics. The same proof is valid for the logics under consideration.

(ii) Analogous to the proof of Theorem 3.1 in [13].

□

It is worth noting that some generalisation of Theorem 1 in [9] holds for all algebraizable logics [14], and our Theorem 3(i) follows from Corollary 5.28 of [14]. The property SES was introduced by the author, and the equivalence of SES and PBP for superintuitionistic, modal and related logics was proved in [15]. This equivalence also holds for algebraizable logics [16].

In order to prove Proposition 1, we recall [12] that each relatively pseudo-complemented lattice $\mathbf{A}$ can be extended to a Heyting algebra $\mathbf{A}'$ by adding a new bottom $\bot$. Then in $\mathbf{A}'$ we have $\neg x = \bot$ for $x \in \mathbf{A}$, and $\neg\bot = \top$.

A formula A is *valid* in $\mathbf{A}$ (in symbols, $\mathbf{A} \models A$) if $vA = \top$ for any valuation v in $\mathbf{A}$. We state the following

LEMMA 4 Let $\mathbf{A}$ be a relatively pseudo-complemented lattice. Then

(i) $\mathbf{A}' \models \neg A \vee \neg\neg A$,

(ii) for each positive formula $A(p_1, \ldots, p_n)$

$$\mathbf{A} \models A(p_1, \ldots, p_n) \text{ iff } \mathbf{A}' \models A(p_1 \vee \neg p_1, \ldots, p_n \vee \neg p_n).$$

As a corollary, we get

PROPOSITION 5 Let L be a positive logic in $E(J^+)$, L' defined in Proposition 1. Then for each positive formula $A = A(p_1, \ldots, p_n)$,

$$L \vdash A(p_1, \ldots, p_n) \text{ iff } L' \vdash A(p_1 \vee \neg p_1, \ldots, p_n \vee \neg p_n).$$

In addition, we prove

LEMMA 6 For $L \in E(J^+)$, the variety $V(L')$ of Heyting algebras is generated by the class $\{\mathbf{A}' | \mathbf{A} \in V(L)\}$. Moreover, if $V(L)$ is generated by a class K then $V(L')$ is generated by $K' = \{\mathbf{A}' | \mathbf{A} \in K\}$.

It follows that for each positive logic L in $E(J^+)$ the logic L' is characterized by Heyting algebras $\mathbf{A}'$, where $\mathbf{A} \in V(L)$.

At last, we need

LEMMA 7 Let L be a positive logic in $E(J^+)$. Then

(i) $V(L)$ has AP iff $V(L')$ has AP;

(ii) $V(L)$ has SES iff $V(L')$ has SES.

For the proof of this lemma we apply refined algebraic criteria for CIP and SES found in [9], Theorem 1, and in [13], Theorem 3.3, and special structure of Heyting algebras validating the formula $\neg A \vee \neg\neg A$.

Then Proposition 1 follows from Theorem 3 and Lemma 7.

2 Extensions of J

Now we turn to extensions of Johansson's minimal logic J. Their language contains the propositional constant $\bot$ of "absurdity" but J has the same axiom schemes as its positive fragment J^+. It is known that for any logic L in $E(J)$, either L contains $Int = J + (\bot \to p)$ or $L + \bot$ is non-trivial, i.e., differs from the set of all formulas.

First we formulate two preservation statements.

PROPOSITION 8 Let L be a positive logic in $E(J^+)$. If L has PBP (or CIP) then $J + L$ also has PBP (resp. CIP).

Proof: We note that any proof of $A \to B$ in $J + L$ can be transformed into a proof of a positive formula $A^* \to B^*$ in L by replacing $\bot$ with a new variable p. Then we get an interpolant of $A \to B$ in $J + L$ from an interpolant of $A^* \to B^*$ in L by substituting $\bot$ for p. The statement about PBP is proved by analogy. □

In particular, J and any of its extensions by one of the axiom schemes (2)-(6) of Theorem 2 have PBP.

PROPOSITION 9 If $L \in E(J)$ has CIP (or PBP) then the following logics also have CIP (resp. PBP):

$L + \bot$,

$L + (\bot \vee (\bot \to p))$,

$L + (\bot \vee ((\bot \to p)\&(\neg p \vee \neg\neg p)))$,

$L + (\bot \to p)$,

$L + ((\bot \to p)\&(\neg p \vee \neg\neg p))$.

Proof: We note that each of the mentioned additional axioms $A(p)$ is L-conservative [8], i.e., the formulas

$$A(p)\&A(q) \to A(p\&q), A(p)\&A(q) \to A(p \vee q), A(p)\&A(q) \to A(p \to q)$$

are provable in L as they are provable in J. We show that PBP and CIP are preserved by adding any L-conservative formulas as new axiom schemes.

Let $L \in E(J)$, $L_1 = L + A(p)$. One can easily see that for each L-conservative formula $A(p)$ and for any formula $D(p_1, \ldots, p_n)$

$$A(\bot), A(p_1), \ldots, A(p_n) \vdash_L A(D(p_1, \ldots, p_n)). \tag{1}$$

Further, assume

$$B(P, Q, X), B(P, Q', Y) \vdash_{L+A(p)} (X \leftrightarrow Y).$$

Due to (1) each instance of the axiom scheme $A(p)$ used in the derivation of $(X \leftrightarrow Y)$ from $B(P, Q, X), B(P, Q', Y)$ is derivable in L from the set $\{A(\bot), A(X), A(Y)\} \cup \{A(x) | x \in P \cup Q \cup Q'\}$, i.e.,

$$A(\bot), A(X), B(P, Q, X), \{A(x) | x \in P \cup Q \cup Q'\}, A(Y), B(P, Q', Y) \vdash_L (X \leftrightarrow Y).$$

If L has PBP then there exists $C(P)$ such that

$$A(\bot), A(X), B(P, Q, X), \{A(x) | x \in P \cup Q\} \vdash_L (X \leftrightarrow C(P)),$$

so

$$B(P, Q, X) \vdash_{L+A(p)} (X \leftrightarrow C(P)).$$

The proof for CIP is analogous. □

We note that any extension of J by positive axioms is paraconsistent. In addition, the axioms $\bot$, $(\bot \vee (\bot \to p))$, $(\bot \vee ((\bot \to p)\&(\neg p \vee \neg\neg p)))$ preserve paraconsistency. Using Propositions 8 and 9, we can find many examples of logics with CIP or PBP in $E(J)$ which do not contain Int and so are paraconsistent.

The logics with PBP in $E(J)$ differ from analogous extensions of Int in some aspects.

For instance, consider the *Hallden property*:

if $(A \vee B) \in L$ and A and B have no variable in common then $A \in L$ or $B \in L$.

We proved in [13] that the projective Beth property implies the Hallden property in superintuitionistic logics. On the contrary, the logic

$$J + (\bot \vee (\bot \to p))$$

does not possess the Hallden property but we see from Propositions 8 and 9 that this logic has PBP and even CIP.

Furthermore, it may happen that a positively axiomatizable logic in $E(Int)$ has CIP although its positive fragment does not possess CIP [9]. As an example, one can take

$$L = Int + (r \vee (r \to p \vee (p \to q))).$$

It was proved in [9] that L has CIP. Moreover, the positive formula

$$((p \to q) \to q)\&(q \to p) \to ((p \to r)\&((r \to p) \to p) \to r)$$

is in L but any of its interpolants must contain $\bot$. One can also find examples of logics with PBP in $E(Int)$ whose positive fragment does not possess PBP. On the other hand, we have

THEOREM 10 If Ax is a set of positive axiom schemes then the following are equivalent:

(i) $J + Ax$ has CIP (resp. PBP),

(ii) $J + Ax + \bot$ has CIP (resp. PBP),

(iii) the positive logic $J^+ + Ax$ has CIP (resp. PBP).

Proof: (iii) implies (i) by Proposition 8, and (i) implies (ii) by Proposition 9. We prove that (ii) implies (iii).

Assume

$$A(P,Q,X), A(P,Q',Y) \vdash_L (X \leftrightarrow Y),$$

where $L = J^+ + Ax$ is a positive logic in $E(J^+)$ and $A(P,Q,X)$ is a positive formula. Then we have

$$A(P,Q,X), A(P,Q',Y) \vdash_{L_1} (X \leftrightarrow Y),$$

where $L_1 = J + Ax + \bot$. If L_1 has PBP then there exists a formula $C(P)$ such that

$$A(P,Q,X) \vdash_{L_1} (X \leftrightarrow C(P)),$$

and

$$L_1 \vdash A(P,Q,X) \to (X \leftrightarrow C(P))$$

by the deduction theorem.

Consider a proof $B_1, \dots, B_n$ of $B_n = A(P,Q,X) \to (X \leftrightarrow C(P))$ in L_1. Let B^* arise from B by replacing $\bot$ with $\top$. Then all formulas $B_1^*, \dots, B_n^*$ are positive, and one can easily prove by induction that the sequence $B_1^*, \dots, B_n^*$ is a proof in $L = J^+ + Ax$. We note that B_n^* is of the form $A(P,Q,X) \to (X \leftrightarrow C^*(P))$, so

$$A(P,Q,X) \vdash_L (X \leftrightarrow C^*(P)).$$

The proof for CIP is analogous.

□

We see from Theorem 10 that for any positively axiomatizable and non-trivial logic in $E(J)$ with PBP its positive part must coincide with one of the positive logics listed in Theorem 2. As a consequence, one can prove the failure of CIP or PBP for many logics in $E(J) - E(Int)$. Moreover, we state

THEOREM 11 (i) There exist exactly six non-trivial positively axiomatizable extensions of J possessing PBP, namely, J, $J + (2)$, $J + (3)$, $J + (4)$, $J + (5)$ and $J + (6)$. The first three logics have CIP, and the others do not possess CIP.

(ii) There exist exactly six non-trivial logics with PBP in $E(J + \bot)$, and only three of them have CIP.

Proof: (i) Follows immediately from Theorems 2 and 10.

(ii) We note that the equivalence $\bot \leftrightarrow \top$ is provable in $J + \bot$. Therefore, by the replacement theorem, any formula A is equivalent in $J + \bot$ to a positive formula arising from A by replacing $\bot$ with $\top$. So any extension of $J + \bot$ is positively axiomatizable over $J + \bot$, and again we apply Theorems 2 and 10.

□

Recall [17] that the logic J is characterized by Kripke frames $\mathbf{F} = (W, \leq, Q)$, where W is a non-empty set, $\leq$ is a partial ordering of W, and Q is a hereditary subset of W, i.e., satisfies the condition

$$(\forall x, y)((x \in Q \text{ and } x \leq y) \Rightarrow y \in Q).$$

A Kripke model for J is a 4-tuple $\mathbf{M} = (W, \leq, Q, \models)$, where $(W, \leq, Q)$ is a frame and $\models$ satisfies the conditions (K1)-(K4) of Section 1 and, in addition,

(K5) $x \models \bot$ iff $x \in Q$.

The logic $J + (2)$ is characterized by linearly ordered frames, and $J + (3)$ by one-element frames. Further, $J + (4)$ is determined by frames satisfying the condition

$$(x < y \leq u \text{ and } x < y \leq v) \Rightarrow (u \leq v \text{ or } v \leq u);$$

$J + (5)$ by frames whose chains contain not more than two elements, and $J + (6)$ is characterized by any two-element chain. If we add $\bot$ as an axiom scheme, the condition $Q = X$ should be added (see [17]).

In addition to Theorem 11, we conclude from Proposition 9 and Theorems 2 and 10 that the following logics in $E(J) - E(Int)$ have CIP:

$$JE' = J + (\bot \vee (\bot \to p)),$$

$$JE'' = J + (\bot \vee ((\bot \to p) \& (\neg p \vee \neg\neg p))),$$

$$JE' + (p \to q) \vee (q \to p),$$

and the following logics have PBP and do not possess CIP:

$$JE' + (4), JE' + (5), JE' + (6), JE'' + (4), JE'' + (5).$$

One can prove that all above-mentioned logics are different.

Our proofs of Propositions 8-9 and of Theorem 10 are syntactical. In order to prove CIP or PBP for other extensions of J, one can apply semantic methods found in [9] and also algebraic methods of [9] and [8].

Semantical characterization for the logic J, and for a number of its extensions, was found in [17]. The algebraic semantics is presented in [12], [18]. Some algebraic representation for paraconsistent extensions of J is obtained in [19].

As we have seen, there are only finitely many logics possessing PBP or CIP in $E(Int)$ and in $E(J^+)$. So we formulate

PROBLEM 1. How many logics in $E(J)$ have CIP? PBP?

We proved that CIP and PBP are decidable on the class of finitely axiomatizable superintuitionistic calculi [9], [20]. This means that there exists an algorithm for determining, for any finite set Ax of axiom schemes, whether the logic $Int + Ax$ has CIP (or PBP) or not. Also we can prove decidability of CIP and PBP for positive calculi extending J^+. We leave open

PROBLEM 2. Is PBP or CIP decidable on the class of finitely axiomatizable logics extending J?

It would be interesting to solve the same problems for paraconsistent logics proposed by N.C.A. da Costa [3] and R.Sylvan [5].

References

[1] E.W. Beth. On Padoa's method in the theory of definitions. *Indagationes Math.*, 15(4):330–339, 1953.

[2] I. Johansson. Der Minimalkalkul, ein reduzierter intuitionistischer Formalismus. *Compositio Mathematica*, 4:119–136, 1937.

[3] N.C.A. da Costa. On the theory of inconsistent formal systems. *Notre Dame J. Formal Logic*, 15:497–510, 1974.

[4] N.C.A da Costa and E.H. Alves. A semantical analysis of the calculi C_n. *Notre Dame J. Formal Logic*, 18:621–680, 1977.

[5] R. Sylvan. Variations on da Costa C systems and dual-intuitionistic logics; I. Analyses of C_ω and CC_ω. *Studia Logica*, 59:47–65, 1990.

[6] G. Kreisel. Explicit definability in intuitionistic logic. *J.Symbolic Logic*, 25:389–390, 1960.

[7] K. Schütte. Der Interpolationssatz der intuitionistischen Pradikatenlogik. *Mathematische Annalen*, 148:192–200, 1962.

[8] L. Maksimova. Intuitionistic logic and implicit definability. *Annals of pure and applied logic*, 105:83–102, 2000.

[9] L.L. Maksimova. Craig's theorem in superintuitionistic logics and amalgamable varieties of pseudoboolean algebras. *Algebra and Logic*, 16(6):643–681, 1977.

[10] W. Craig. Three uses of Herbrand-Gentzen theorem in relating model theory and proof theory. *J. Symbolic Logic*, 22:269–286, 1957.

[11] L.L. Maksimova. Craig's interpolation theorem and amalgamable varieties. *Doklady AN SSSR*, 237(6):1281–1284, 1977.

[12] H. Rasiowa and R. Sikorski. *The Mathematics of Metamathematics*. Warszawa, PWN, 1963.

[13] L.L. Maksimova. Superintuitionistic logics and the projective Beth property. *Algebra and Logic*, 38(6):680–696, 1999.

[14] J. Chelakowski and D. Pigozzi. Amalgamation and interpolation in abstract algebraic logic. In X.Caicedo and C.H.Montenegro, editors, *Models, Algebras and Proofs, Selected papers of the X Latin American Symposium on Mathematical Logic held in Bogota*, pages 187–265. Marcel Dekker, Inc., New York, 1999.

[15] L. Maksimova. Explicit and implicit definability in modal and related logics. *Bulletin of Section of Logic*, 27(1/2):36–39, 1998.

[16] E. Hoogland. Algebraic characterisations of various Beth definability properties. *Studia Logica*, 65(1):91–112, 2000.

[17] K. Segerberg. Propositional logics related to Heyting's and Johansson's. *Theoria*, 34:26–61, 1968.

[18] W. Rautenberg. *Klassische und nicht-klassische Aussagenlogik*. Braunschweig, Wiesbaden: Vieweg, 1979.

[19] S.Odintsov. On j-algebras and j-frames. In *Intern. Maltsev conference on Math. Logic, Abstracts, Novosibirsk*, pages 101–102. 1999.

[20] L.L. Maksimova. Decidability of projective Beth's property in varieties of Heyting algebras. *Algebra and Logic (to appear)*.

Towards a Mathematics of Impossible Pictures

CHRIS MORTENSEN Dept of Philosophy, The University of Adelaide, North Tce, SA 5005, Australia
cmortens@arts.adelaide.edu.au

Abstract
In this paper are described recent developments in the mathematics of impossible pictures. Classical consistent theories by Cowan, Francis and Penrose are described. It is argued that only an inconsistent theory can capture the epistemic content of the experience. Approaches using inconsistent models which are heaps, greaps or cylinders are described.

1. Introduction
2. History of Impossible Pictures
3. Classical Mathematical Accounts: Cowan, Francis and Penrose.
4. Heap and Anti-Heap Models
5. Greap Models
6. Why Inconsistency?
7. The Routley and Cross Functors

1 Introduction

The theory of inconsistency, together with its parent discipline paraconsistent logic, is developing rapidly. One notable omission, however, is the theory of impossible pictures. So as to make no confusions, let it be stressed that this means real pictures of apparently impossible objects. We will see examples as we go along.

2 History of Impossible Pictures

Impossible pictures have been drawn for a long time. The earliest seems to have been on a Pompeii interior wall. There are also medieval alterpieces, and some parts of Piranesi's Carceri. Marcel Duchamp did a strange bed. But impossible pictures come of age with Oscar Reutersvaard. One day in 1934 in his high school Latin class in Stockholm, the 17 year old Oscar doodled like this:

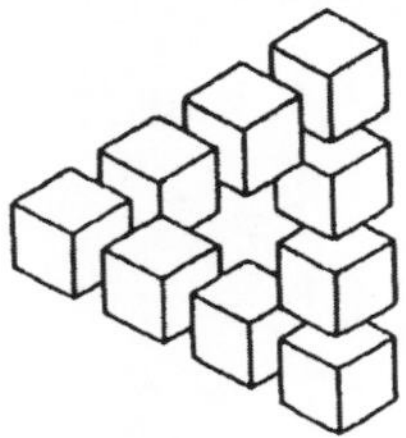

Thereby began a brilliant career in which he drew more than 4,000 pictures. He was honoured in the 1980s by the Swedish government in a number of stamps featuring his creations.

About 20 years after Reutersvaard began, the idea was re-discovered by the Penroses, and M. C. Escher. In [5] L. S. Penrose and his son Roger published a two page paper in the *British Journal of Psychology*, one page of which included both drawings and a photograph. Simultaneously, and then in communication with the Penroses, M. C. Escher drew masterpieces like *Belvedere*, *Waterfall*, and *Ascending and Descending*.

Since then, there have been many following in the construction of impossible pictures. One prominent person has been Bruno Ernst, who took a photograph of the impossible triangle.

There have also been very recently new pictures at Adelaide.

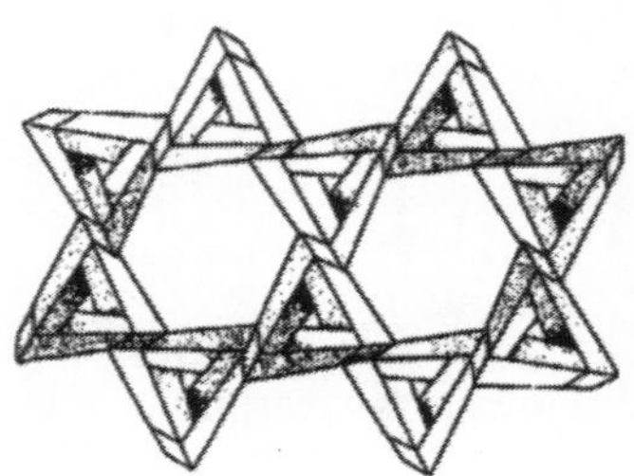

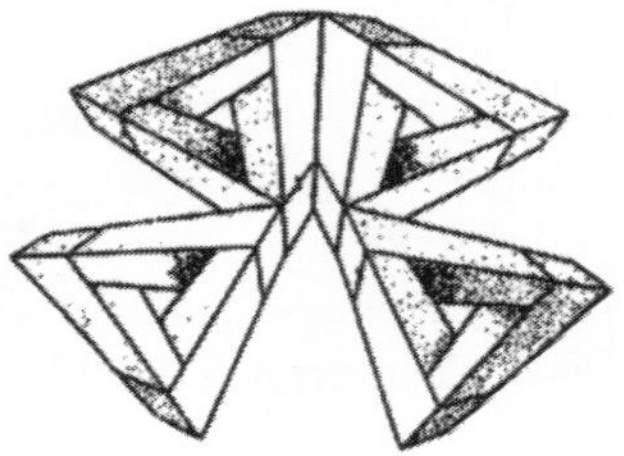

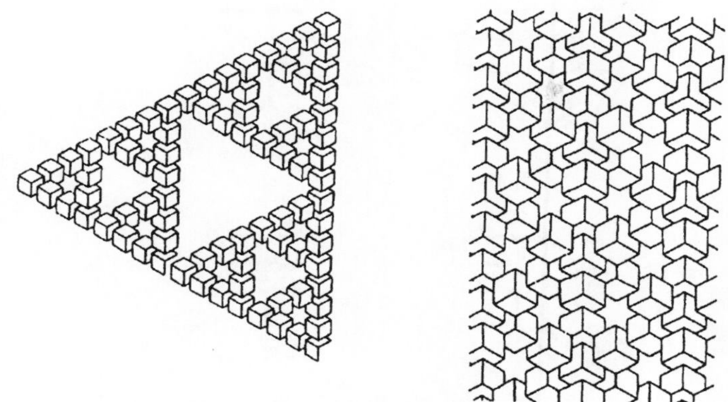

Mercier Wallpaper

3 Cowan, Francis and Penrose

There seem to be three attempts to describe these pictures in terms of classical mathematics: Thaddeus Cowan [1], George Francis [2] and Roger Penrose [6].

Thaddeus Cowan focusses on the case of four-sided figures with a hole in the middle, though the analysis applies to any n-sided figure, for $n > 2$, and thus the triangle.

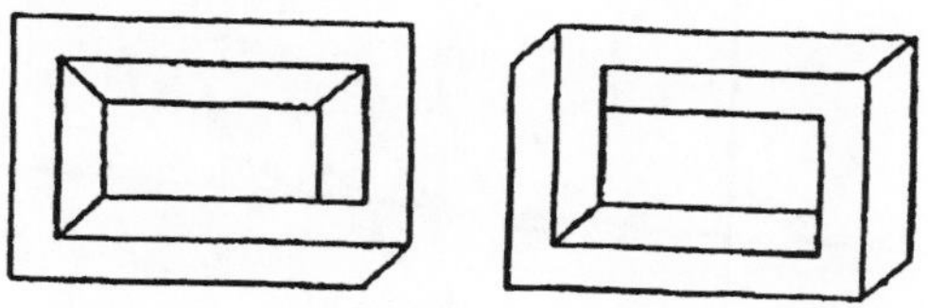

Different Inconsistencies in 4-sided figures

Cowan is able to classify 2-dimensional corner elements in terms of the 3-dimensional information that they represent, using the concepts front, back, inside and outside. This results in four kinds for each corner. With each figure having four corners, that is 256 different assembled figures. Using group theory, Cowan described necessary and sufficient conditions for such assembling to result in a picture of a possible object: one which always leaves these unchanged all the way around. Pictures of impossible objects are those which switch these in one or more places of assembly. The theory of the tricorn (impossible triangle) comes out as a special case, when two sides are identified, but the analysis applies to a much broader class of figures as well.

Corner elements

This is a very excellent and useful theory. However, it leaves something out. In effect it is telling us that if the conditions fail this is not a picture of a possible object. But that is a perfectly consistent state of affairs. It fails to capture our sense of what it is a picture of, its *content*: a object with impossible properties.

George Francis [2] made an advance when he asked the sensible question: what sort of consistent 3-D space could this picture represent? This immediately gives a positive heuristic to the problem. Francis pointed out that the triangle could exist in certain 3-D non-Euclidean spaces, specifically those which are $R^2 \times S^1$. It is difficult to represent this on the page, but the case of $R^1 \times S^1$ as embedded in R^3 is familiar as the cylinder:

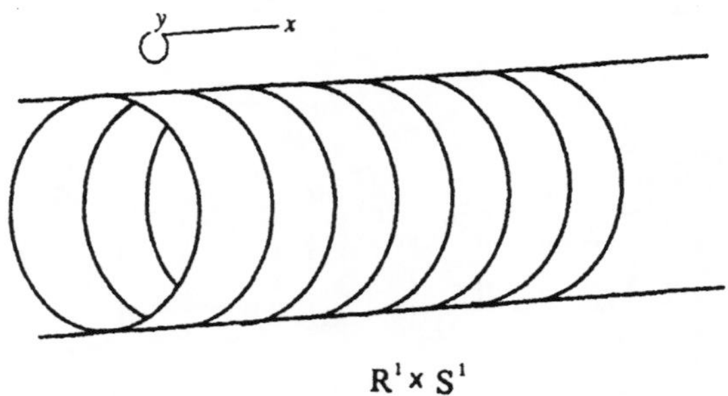

$R^1 \times S^1$

Francis is clearly right here. But there is another way to see Francis' point, and its limitations. The importance of Ernst's photograph is that it is a 3-D object. But it only looks this (impossible) way from a single camera angle. The real 3-D object simply does not have the connectivity it seems to have from that single angle. But what is it that it seems to be from that angle? Something with a hole, which the real physical object did not have. The experience has a content, one which leads us to say "I see it but it is impossible!" The mind is clearly projecting something onto its perceptions here. But what?

Roger Penrose [6] made a further significant advance when he described the situation using cohomology theory. The idea was similar, that a picture of an apparently impossible object can indeed represent a possible 3-D object, but one with different connectedness. For example, one can imagine that the triangle is a picture of three disassembled parts lying at different distances from the viewer, but lined up so that they look joined up. That is, they are lined up so that they project down onto the 2-D picture. The point is now, that in the consistent case it is possible to re-assemble the pieces into a connected whole, whereas in the inconsistent case it is not so possible. Penrose shows how to describe this in group theory.

This is clearly a useful advance. But again one feels that there is something left out. The mind doesn't think "Here are three disassembled parts", it doesn't perceive it as three disassembled parts, but as assembled. Nor does it simply "fail to dis-identify" separate points which are in a straight line with the eye. It is possible to see the photograph the way it is, but there remains the strong perception of something impossible. The mind completes the picture, it actively identifies points that it cannot distinguish. But to do that, its content must be represented by an inconsistent theory. We describe these below.

It is worth comparing with projective geometry. Perspective presents us with an obviously consistent perception: the railway lines ***look*** as if they meet. This is an objective phenomenon; if you take a photograph the lines will meet on the surface. Perspective does not look paradoxical. What we know is that the lines do not meet, but this is cognitive at a higher level. It is not part of the content of the percept, it is not "projected onto the percept" as it were. In this respect the content differs from that of the tricorn and other similar figures, which appear paradoxical. In each case, the content is a collection of propositions which form a rigorous theory. Why we should be interested in these theories, however, is because of the presence of the human perceptual apparatus.

4 Heaps and Anti-Heaps

In [4] the author described an inconsistent extension of Penrose's theory. The idea is to express the content of the apparently impossible perception by an inconsistent theory. The simplest example of this is to use the theory of heaps, developed by Meyer, van Bendegem, Priest and others. Heap theory began with heaps of natural numbers. A heap of natural numbers is represented by the counting sequence "One, two, three, heap". That is, all numbers beyond a certain maximum number H (for heap) are identified. There are consistent and inconsistent versions of such theories. The consistent theories add $H = H+1 = H+2 = \dots$ to the theory while removing $\neg H = H+1$, $\neg H = H+2$, ... etc. The inconsistent theories add $H = H+1\dots$ while retaining $\neg H = H+1\dots$ We consider the case for the inconsistent theories below.

There are also heaps of real numbers, where the idea is that all real numbers outside some (open) interval are identified with the respective bounds. Again, inconsistent versions retain the negated identities of classical real number theory. These can be represented by an interval of the real line with arrows outside the heap bounds to indicate identities outside the bounds. It is important to note, though, that these theories lose their character as real number theories because they must sacrifice some of their arithmetical functionality to avoid trivialisation (see Mortensen [3]). But this is no real problem. We are dealing with geometrical structures here rather than arithmetical theories. It would hardly be surprising if one could no longer do the arithmetic on different geometrical structures that one can do on the real line, since their algebraic properties may be radically different.

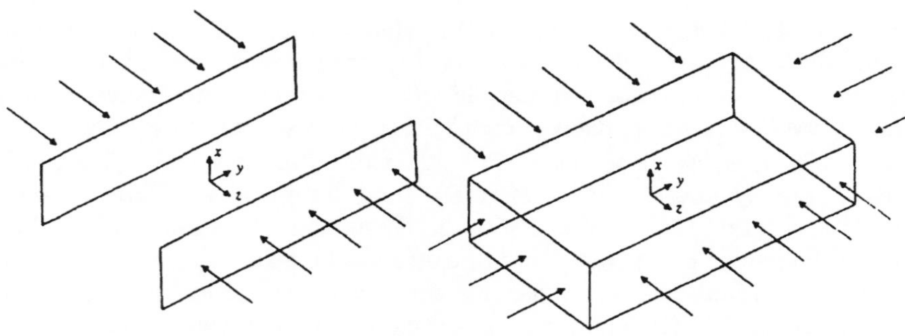

Heaps

In [4] it was proposed that the impossible triangle could be treated inconsistently by a "backdrop" universe, which is a heap. Assemble as much of Penrose's parts as one can, stick them out from a backdrop, and then draw in the remaining lines on the backdrop. Thus all points behind the backdrop were (inconsistently) identified with its surface. This certainly produces an inconsistent theory of a figure which will look like the triangle. But there is a more general theory using ***antiheaps***.

Antiheaps identify all those points which heaps keep distinct, and disidentify all those points which heaps identify. That is, antiheaps identify all points with a closed interval, and keep the classical disidentities outside.

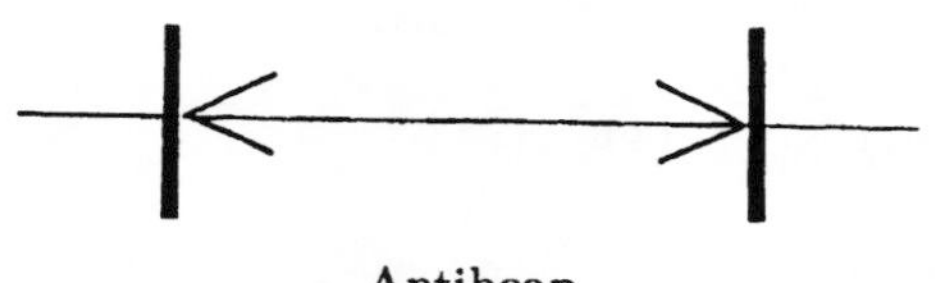

Antiheap

It is apparent that antiheaps provide a treatment for any impossible picture which can be obtained by photographing a 3-D model, such as Ernst's. Proceed as follows. Describe the actual 3-D model in polar co-ordinates (r, θ, ϕ) with the origin at the eye. Where a interval on a ray is identified by the eye, identify all points between the two bounds on the antiheap of the ray. Note that this is not a full projective geometry, because projective geometry aims to identify all points on a ray, which thus reduces the 3-D theory to a 2-D theory. Here we remain with a 3-D space in which only certain stretches are selectively identified and disidentified in the way that the eye-brain combination treats them. Then extend the diagram of the theory to a full theory using the model theory for any appropriate paraconsistent logic (for details of inconsistent model theory, see [3, Chap. 2])

5 Greap Models

Greaps are an intermediate structure between groups and heaps. They were suggested by a treatment such as Penrose's which proposes that a picture of a possible object is obtained just in case a certain structure forms a heap. We will not discuss these here, for details see [4].

6 Why Inconsistency?

It should be clear that no consistent theory can explain why it is that it seems to be an impossible object. It is proposed here that our sense that we see something impossible, the content of our perception, is represented by an inconsistent theory. The mind has a certain experience, but in addition has expectations which are incompatible, expectations of local Euclideanism, which are overlaid on our percept. These expectations are not removed by revealing the truth about the picture, since they are doubtless hard-wired in by evolution as outputs of the visual module. The mind puts both of these together and the result is represented by an inconsistent theory. Indeed, it is difficult to see what other explanation of the content of the impossible experience there could be.

It is also apparent that George Francis' account using $R^2 \times S^1$ space can be extended in a similar way. In this case, however, the inconsistent identifications required are given not by heaps or antiheaps, but by the mod function, where the modulus is the circumference of the S^1 circle. Construct the consistent space in which the problematic figure can comfortably live. Then extend the space by adding the disequations sufficient to make the space Euclidean. For example, in S^1, the circle, we have $0 = \text{circ}$, $0 + 1 = \text{circ} + 1, \ldots$ where circ is the circumference, but retain the Euclidean output of the visual module by keeping $\neg 0 = \text{circ}, \ldots$ etc.

Note that there is an important sense in which the percept of the inconsistent tricorn is ***locally consistent***, only becoming inconsistent when its properties are globally perceived as impossible (I owe this observation to a referee). It is certainly true that the inconsistency owes its presence to an overall ***gestalt*** of the picture as globally impossible. The present inconsistent extension of Francis reflects this, by not giving a preferred status to certain points or lines as the localised bearers of the inconsistency: all points on the surface of $R^2 \times S^1$ are equally the bearers of the inconsistent identifications which are the consequences of imposing the global R^3 condition.

Note that this approach is a thoroughly ***cognitive*** justification of the move to an inconsistent theory. There is no suggestion that there really exists an inconsistent object. On the other hand, this is an entirely appropriate application of paraconsistent methods, which falls under the ***epistemological*** justification of paraconsistency (see eg. [3, Chap. 1]).

7 The Routley and Cross Functors

There are two processes involved in forming all such inconsistent theories. (1) First there is a extension of the theory of Euclidean space *Th*1 by adding equations of the form $p_1 = p_2$, where p_1 and p_2 are points. Since we are insisting that this extension preserve the Euclidean character of its origin, we also have $\neg p_1 = p_2$, and thus an inconsistent theory, *Th*2. (2) Second, there is the strictly classical consistent theory *Th*3 which keeps the identifications $p_1 = p_2$, but drops the Euclidean overlay $\neg p_1 = p_2$. This is a general form of the "pasting" method of classical algebraic geometry, so it can be usefully thought of as the Paste functor on the category of theories. Its inverse functor is the Cut functor. The relation between these processes can be given a functorial description, as follows.

We can define two operations on any theory Th:

$$Th^* = \text{df}\{A : \neg A \text{ is not in } Th\} \quad \text{(The Routley star operation)}$$
$$Th^x = \text{df}\{A : \neg A \text{ is in } Th\}$$

In [3, Chap. 12], it is shown that if we begin with a simple inconsistent theory of closed set logic (where in a simple theory truth-values are just the null set, the whole space and any boundary), then the effect of the Routley operation is to snip out all those sentences A for which both A and $\neg A$ are in the original theory. It is convenient to re-name the Routley star as the Routley functor, since it is also a functor on the category of theories (with morphisms as set-inclusions). We further rename the x operation as the Cross functor.

Now notice that the Routley functor reverses the effect of the process described as (1) above. That is, applied to the inconsistent theory $Th2$, the result is $Th1$, which is to say that (1) is the inverse of the Routley functor. It is also shown that if we begin with a simple inconsistent theory of closed set logic, then the effect of the Cross functor is to excise all odd-numbered strings of negations of atomic sentences A and add all even numbered strings of negations, where the contradictory pair $(A, \neg A)$ are in the theory. So in particular A is retained and $\neg A$ removed. This is the inverse of the process described in (2) above.

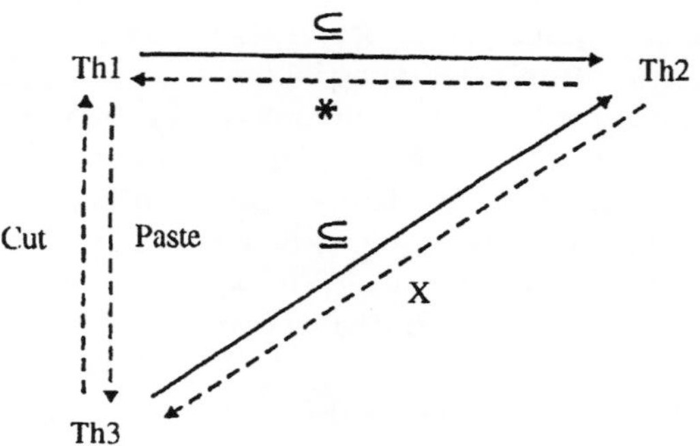

Thus we have:

THEOREM The Routley functor forms an adjunction with the Paste functor, with the Routley functor as the left adjoint and the Paste functor as the right adjoint. The Cross functor forms an adjunction with the Cut functor, with the Cross functor as the left adjoint and the Cut functor as the right adjoint. In each case, the unit of the natural transformation is a collection of inclusion morphisms which represent passing from a consistent theory to an inconsistent extension.

In conclusion, there is the obvious "dual" of impossible pictures, namely ambiguous pictures such as the duck-rabbit, Necker Cube etc. In this case, it is indeterminate what it is a picture of, but the ambiguous aspects are individually consistent. It is noted that a corresponding form of the above theorem holds for incomplete theories constructed on open set logic (intuitionism). Each of the two above adjunctions continue to hold, with similar natural inclusion relations holding, save that the inclusions represent passing from an incomplete theory to a complete extension. This will be developed further in later work.

The author wishes to express his thanks in this paper to very useful comments by the editors and several referees, as well as the audience of WCP2000 at beautiful Juquehy Beach, Brazil.

References

[1] Th. Cowan. The theory of braids and the analysis of impossible pictures. *Journal of Mathematical Psychology*, 11:190–212, 1974.

[2] G. A. Francis. *A Topological Picturebook*. Springer-Verlag, 1987.

[3] C. Mortensen. *Inconsistent Mathematics*. Kluwer, 1995.

[4] C. Mortensen. Peeking at the impossible. *Notre Dame Journal of Formal Logic*, 38:527–534, 1997.

[5] L. S. Penrose and R. Penrose. Impossible objects: a special type of illusion. *British Journal of Psychology*, 49, 1958.

[6] R. Penrose. On the cohomology of impossible pictures. *Structural Topology*, 17:11–16, 1991.

Ambiguity is not Enough

B. H. SLATER Department of Philosophy, University of Western Australia, Crawley, Australia
slaterbh@cyllene.uwa.edu.au

1 Introduction

In a recent paper on Paraconsistent Logic [1], Bryson Brown has presented a 'preservationist' approach to the subject, showing in particular that even Graham Priest's 'Logic of Paradox' can be re–interpreted in a preservationist manner. He does this by showing that the LP approach to paraconsistency is very closely related to treating sentence letters as ambiguous. Thus he presents LP as having three truth values, $T = \langle T \rangle$, $F = \langle F \rangle$, and $B = \langle T, F \rangle$, and then, given V is an LP valuation, he constructs another valuation V', such that, for every sentence letter, if $V(s) = T$ then $V'(s) = T$, if $V(s) = F$ then $V'(s) = F$, but if $V(s) = B$ then V' dispenses with s, and adds new letters s' and s'', where $V'(s') = T$, and $V'(s'') = F$. In this way he wants to 'consider the possibility of arriving at consistent images of sets of sentences by treating (some) sentence letters in them as ambiguous' [1, p 493]. I show in this paper that such forms of disambiguation will not handle all paraconsistent situations, in connection with some of the classical paradoxes of self–reference.

Brown's article starts with a discussion of a recent paper by myself, which mentions an approach to the subject which is immune to the paradoxes of self-reference. At the end of that paper I say [2, p 453]:

> Priest's ideas about the paradoxes...use a concept of truth which is surely not instantiated: the Tarskian concept of truth as a predicate of sentences. Since Montague we surely now know that syntactic treatments of modality must be replaced by operator formulations. And not only must that hold with the identity modality 'It is true that', as a consequence, but there are clearly no paradoxes with the operator notion, since it was proved many years ago, by Goodstein, to be consistent.

It is the Tarskian idea that truth is a feature of sentences which we principally have to get away from. The Tarskian T-schema, 'Ts ≡ p' is involved in the derivation of many self-referential paradoxes, see e.g. [3, p 31], but it ignores the possible ambiguity of 's', since otherwise there would be no appropriate substitution for 'p'. The operator approach allows for such ambiguity, and I first show below how that comes in. But allowing for ambiguity is not enough to solve all the paradoxes of self–reference, as we shall see through studying the tradition which followed Montague. The operator approach crucially allows also for ineffability, so I go on to indicate how and why that arises. I finish by modifying the operator approach in one important formal respect, before considering some further self–referential paradoxes in the light of what has been said.

2 How ambiguity comes in

In the 4th edition of Irving Copi's *Symbolic Logic* [4, p 301], in a passage which is in fact repeated in other editions of the book, there is a definition of heterologicality, and a supposed proof of the contradiction:

Het'het' ≡ ¬Het'het'.

But Copi has to explicitly assume '"het" is univocal' at one place in this 'proof', so his otherwise thorough analysis shows instead that the contradiction is not derivable as a matter of logic, but rests instead on that contingent premise, which *reductio ad absurdum*, it would seem, would simply require that we deny. Specifically Copi first takes the word 'heterological' 'to designate the property possessed by words which designate properties not exemplified by themselves' . Copi formalises this in an iota term manner as

Hs $=_{df}$ (∃F)(sDesF.(P)(sDesP ≡ P=F).¬Fs),

and his proof of the contradiction then has two parts. First, H'H' is shown to entail ¬H'H'. Then ¬H'H' is shown to entail H'H'. Now there is an unstated assumption in both of these formal proofs leading to the contradiction, namely: 'H'DesH. But there is also an openly expressed assumption in the next to the last line, which receives the gloss from Copi 'assuming 'H' is univocal'. Quite why Copi still repeatedly thought he had derived a contradiction which was inescapable is not too apparent, as a result, unless, by taking 'heterological' to 'designate the property...', he took its univocality to be guaranteed somehow by its definition. But he still needed a proof that there was one and only one such property. There is nothing which forces the designation of predicates to be 1–1, and in fact the denial of this is what can be concluded from the above proofs, by *reductio ad absurdum*. For the pair of assumptions cannot both be true, so if anything designates the above H, it demonstrably does not do so uniquely – on pain of contradiction. From Copi's 'proof' we thus see that what is necessary is just that

if 'het' just means one thing (Het) then Het'het' ≡ ¬Het'het',

and that might lead us to the plausible judgement that

'het' does not just mean one thing.

In other words, it may simply seem that a presumption of non-ambiguity has been suppressed, and that the remedy is for it to be negated.

A plausible demonstration that the Epimenides paradox does not arise, given a proper logical analysis, was provided in this style by Sayward [5], following the work on operators by Goodstein, who was notably preceded by Ramsey, and followed by Prior. And the point would generalise to all associated paradoxes, even Curry's. For, as before, the T-schema presumes univocality of the sentence said to be true, so the necessary fact is not that schema itself, but instead that

if s only means that p then Ts ≡ p.

But the now explicitly stated assumption shows the paradox does not arise when

s = '¬Ts',

without a further premise, that

s only means that ¬Ts,

and the contradiction which would otherwise be available may simply be taken to mean that this further premise must be denied. In fact, in Goodstein's initial case [6], if

A says that everything A says is false,

then he actually proves

Something A says is false,

and

Something A says is true.

So, in the case where A says just one thing, the point about ambiguity would seem to be established.

But even Prior, originally, was puzzled by this kind of result [7, p 261]. With the Paradox of the Preface, where there is a sentence in a book which says that everything said in the book is false, then Prior thought that [8, pp 84–87], by the above kind of proof, one could show by logic that there must be a second sentence in the book. The force of this idea might even lead to the belief which apparently Brown has, namely that, in the paradoxical case where there is just one sentence in the book this can be replaced with two. Certainly something said in the book must be true, and something said in the book must be false, but that does not require there to be two sentences in the book – if we allow sentences to be non-univocal. In 1986 I made this point, showing that all that is required by logic is that at least two different things are meant by any sole sentence in the book, as in the previous cases above. Prior himself [8, p 104] came to allow that sentences might have several meanings, so that if the operator form 'it is true that p' is written 'Vp' we can say

Ts ≡ (p)(s means that p ⊃ Vp).

And from this can be derived Sayward's conditionalised T-schema, for if

(∃!p) s means that p

then for that p

Ts ≡ p,

given that Vp is logically equivalent to p, i.e. that operator-truth is the null, identity modality.

3 Why ambiguity is not enough

But is ambiguity enough? If the non-univocality of the Liar, the Preface, Heterologicality, and therest were just ambiguity then one might get a very proper demand for what Ryle called 'namely–riders'. 'It is a bank' has two senses, which can be specified, namely 'It is a money bank' and 'It is a river bank'. It seems possible that Brown is expecting that such replacements will always be available, when he constructs the valuation V' so that, if V(s) = B then V' dispenses with s, and adds new letters s' and s", where V'(s') = T, and V'(s") = F. Maybe Brown's point is just that we can control the consequences of inconsistent premise sets, in certain cases, by appeal to a consequence relation that preserves consistency by treating some of the members of those sets as ambiguous. If so, then he is not strictly committed to the claim that disambiguating sentences can always be found. But if Brown wants to say that one can always find disambiguating sentences, then he needs to consider further the work which followed Montague on operators.

The developed tradition from Montague shows that the lack of namely-riders is essential in some cases, and specifically in the troubling cases of self-reference. One of the features of operator theories is that they do not incorporate 1-1 expression relations between sentences and propositions - thus they allow for ambiguity as above. But if only ambiguity was involved then in a disambiguated language, which restored the 1-1 relationship, paradoxes could still be generated. The further, crucial feature of operator theories is that they allow for ineffability, i.e. reference to meanings which cannot be linguistically expressed. The matter has a much broader aspect, as well. For the general demand that we should be able to discriminate meanings in parallel to explicit sayings is, of course, a common feature of compositional theories of meaning, as in, for instance, the classic, Tractarian Picture Theory of Meaning, and Fodor's Language of Thought. But meanings must be non-compositional, if there are to be no real paradoxes. Put briefly, the same sort of arguments as above can be generated with respect to intensions and meanings, if they have sentence-like structures, or are otherwise invariably expressible linguistically.

The first part of this was shown by Thomason, in his and Montague's arguments for the operator analysis of propositional attitudes. For Montague had shown, quite generally, in [9], that 'Indirect Discourse is not Quotational', as Thomason put it [10]. But Thomason not only supported Montague, he also showed that a comparable argument proved that indirect discourse was not about any structured intensional objects either. Thus it was not about some semantic analogues of sentences, which led Thomason to the conclusion [11] that representational theories of mind, such as Fodor's, had to be inconsistent, when fully worked out. As Asher and Kamp said [12, p 87]:

> To happy-go-lucky representationalists, Thomason (1980) is a stern warning of the obstacles that a precise elaboration of their proposals would encounter...[W]ith enough arithmetic at our disposal, we can associate a Gödel number with each such [representational] object, and we can mimic the relevant structural properties of, and relations between such objects by explicity defined arithmetical predicates of their Gödel numbers. This Gödelisation of representations can then be exploited to derive a contradiction in ways familiar from the work of Gödel, Tarski and Montague.

Asher and Kamp, however, found a comparable difficulty even in Montague's Intensional Logic. For intensional objects do not need a structure, they realised, for the same sort of result to hold: it suffices to have a mechanism for correlating the attitudinal objects with the sentences by which they are expressed [12, pp93– 4]. Then, if an arithmetisation of the notion of expression is formulated, by means of some relation which says that the formula with Gödel number n means that p, fixed points can again be generated, and standard difficulties with self–reference lead naturally to contradictions. But Asher and Kamp's proof still clearly presumes that the expression relation is 1–1 [12, p 94], and there is no guarantee that this is so. Indeed, as with Copi and Sayward, what Thomason, and Asher and Kamp have variously proved is merely that if the expression relation was 1–1 then there would be a contradiction. Hence meanings cannot always be linguistically expressed, and some meanings are semantically ineffable.

4 Propositional quantification

Asher and Kamp went on to state explicitly that only systems in which knowledge and belief are treated as sentential operators, and which therefore do not treat propositions as objects of reference and quantification, are protected from the paradoxes [12, p 94]. But one has to be clear just how the quantification involved with operator constructions completely works. What is the quantification over? Propositional quantification is sometimes taken to be substitutional, but one very forceful argument against such an account is that then there would not seem to be a stateable truth condition in the following kind of case:

$(\exists p)$(there is no linguistic expression in English for the thought that p).

Indeed, Michael Loux has recently made this difficulty his entire basis for rejecting Prior's account (c.f. [13, p 150]). For, of course, in such a case there expressly cannot be any namely–riders, if the statement is true. But it is not the case that the only complements in propositional attitude constructions are explicit used sentences, they can also be propositional referring phrases. Thus we can say, for example, 'Tom believed what Peter said', where 'what Peter said' just refers to a proposition without specifying it. And such a referential phrase can be provided in Loux' case, even if the proposition in question cannot be given directly. For we can talk about 'that thought there is no expression for in English'.

In fact a strict formulation of propositional quantification must allow for such referring phrases to propositions. Ramsey, Goodstein and Prior took it to be, somehow, over expressed sentences, as the above example illustrates, but apart from the difficulty of reading the quantifier in such a case, there are other constructions which clearly do not involve expressed sentences, but referring phrases instead. Thus we might re-formulate the above case as

$(\exists x)$(there is no linguistic expression in English for x),

which has the truth condition that that thought there is no linguistic expression for in English there is indeed no linguistic expression for in English – which follows the epsilon definition of the quantifiers at this level:

$(\exists x)\neg Lx \equiv \neg L\epsilon x \neg Lx.$

Certainly one type of propositional referring phrase has the form 'that p', with the demonstrative 'that' combining with the used sentence which is then displayed, to refer to the proposition that sentence expresses [14, p 585]. But that is a very special case, and in the generality of propositional referring phrases the proposition involved is not expressed. What proposition is referred to by some propositional referring phrase can in fact only be expressed if one also has an operator statement, like

what Tom believed was that p.

And indeed it is that fact which now gives us a clear view of the situation with the self-referential paradoxes.

For in the case of the ineffability caused by the expressive limits of a given language, referential phrases like that needed in Loux' example clearly cannot arise in such operator forms, which means, amongst other things, that the truth predicate cannot be eliminated. For if

that thought which cannot be expressed in English is true,

yet we cannot say anything of the form

that thought which cannot be expressed in English is that p,

then we cannot get

that p is true,

from which to obtain p, via the logical equivalence

it is true that $p \equiv p$.

But a similar thing happens in the case of the self-referential paradoxes. Phrases like 'what that sentence says', in connection with 'what this sentence says is not true' cannot occur in statements of the form

what that sentence says is that p,

i.e. what that sentence says cannot be identified. For if it could be identified we could eliminate the truth predicate in 'what this sentence says is not true' to derive a 'p' for which, contradictorily,

$p \equiv$ it is not true that p.

By *reductio ad absurdum* what the sentence says cannot be identified, which means it has no single meaning, and hence is either senseless or has, as before, several senses.

The demonstrative 'that' which identifies propositions is not commonly symbolised in standard logic. Susan Haack, however, introduced a symbol for 'the statement that...' [15, p 150], and '*' can be used for this purpose. Haack derives a paradox with this notion by first producing a sentence:

1. $(p)(c = {*}p \supset \neg p)$,

and then letting 'c' abbreviate 'the statement made by sentence numbered 1'. She goes on to say, as a result, that 'it can be established empirically' that

$c = {*}(p)(c = {*}p \supset \neg p)$,

deriving a contradiction analogous to one Tarski produced. But while any statement made by the sentence numbered 1 is certainly any statement made by '$(p)(c = {*}p \supset \neg p)$', that does not ensure that any identifiable statement is made by '$(p)(c = {*}p \supset \neg p)$'. There is no problem with defining self-referential sentences, for instance 'this sentence says just one thing, and it is false'. But with that sentence called 'c', and with 'Sxy' meaning 'x says y' (where 'y' is a propositional referring phrase), we merely have

$Sc{*}(\exists x)((y)(Scy \equiv y{=}x).\neg Tx)$,

and we can then prove that c must be ambiguous (c.f. [16, p 79]). Hence such a self-referential sentence cannot express a single proposition. A related point arises, with the same result, if one tries to let 'c' abbreviate a sentence in which it occurs, rather than, as before, a term referring to the statement made by that sentence. That might seem to directly produce a self-referential proposition. But if one tried to make 'c' abbreviate, for instance,

'$(p)((c \equiv p) \supset \neg p)$'

(c.f. [15, p 150]), then the sentence would include itself as a proper part, which is impossible.

As a consequence, there is no real problem with the Ramsey–Goodstein–Prior approach, merely with the idea that what sentences express needs to be, or indeed can always be made semantically explicit. Certainly Ramsey's formulation was problematic [15, p 130], but Goodstein's proof mentioned before can simply be put in the following variant form (Goodstein did not employ '*', and had 'x' where here there is 'Tx'):

> Given Say, where y=*(x)(Sax ⊃ ¬Tx). Assume, first (x)(Sax ⊃ Tx), then, since Say only if Ty, and Say, we must have Ty, and hence (x)(Sax ⊃ ¬Tx), but since also (x)(Sax ⊃ Tx) and (∃x)Sax, that is contradictory; hence ¬(x)(Sax ⊃ Tx), i.e. (∃x)(Sax.¬Tx). But assume, second, (x)(Sax ⊃ ¬Tx), then Ty, but also Say, hence (∃x)(Sax.Tx), and a contradiction; so ¬(x)(Sax ⊃ ¬Tx), i.e. (∃x)(Sax.Tx).

In the modified version of Goodstein's conclusions, therefore,

(∃x)(Sax.¬Tx), (∃x)(Sax.Tx),

one must first read the variables as nominal variables (unlike the 'p' above), but, second, one must not take the 'T' to be an operator, but a predicate, which means that there is no certainty that it can be eliminated. Certainly 'T*' is an operator, and T*p ≡ p, but 'Tx', in general, needs something of the form 'x=*p' before 'T' can be eliminated. The replacement of 'Tx' with 'x' was always criticised, in the Ramsey–Goodstein–Prior calculus, and accepting this criticism is, remarkably, just what gets us out of the self-referential paradoxes even more clearly, as above. But correcting it also means accepting that not everything then referred to can be expressed, and in particular, therefore, there need be no way to give an explicit disambiguation of what A says, if he just says, directly, one thing. Once the 1–1 nature of the expression relation is removed (for further argument to this effect see, e.g. [17, p 11]) we must accept that there may be things which can only be gestured towards, and not explicitly said.

5 Further cases

Does the above have any further relevance to other puzzles about self-reference which gripped the minds of philosophers and logicians early last century? I mentioned before the blindness which Copi seemed to suffer when thinking he was driven to an inescapable contradiction in connection with Heterologicality. A very similar blindness has also clearly affected many discussions on Truth – those which take Tarski's T-schema to be unexceptionable, for instance. But somewhat similar blinkers would seem to have been worn in other cases of self-reference.

As Kneale and Kneale say [14, pp 585–587], the above point about the difference between an expressed proposition 'p', and the demonstrative phrase which refers to it, 'that p', has a parallel in the case of concepts. The predicate 'is a P' expresses a concept, whereas the nominal phrase 'being a P' refers to it. This is not a distinction customarily symbolised in second–order logic, which is to that logic's great cost, since the paradox of predication arises because there is no way to say that what is expressed by some predicate is some concept. Thus, if we try to write

(∃F)(F=x.¬Fx) ≡ Gx,

then the double duty which terms like 'F' are doing – to express a concept, as in 'Fx', and to refer to a concept, as in '(∃F)' – makes it automatic that there is a concept the predicate on the left is expressing, namely λx(∃F)(F=x.¬Fx). Let us write 'x has F' in place of 'Fx', then there is nothing to guarantee that

(∃G)((∃F)(F=x.¬(x has F)) ≡ x has G).

Maybe 'λx(∃F)(F=x.¬(x has F))G' is still equivalent to '(∃F)(F=G.¬(G has F))', but there is nothing to show there is a concept H for which this is equivalent to 'G has H'. Indeed, if there was then there would be a contradiction; so there isn't.

This point clearly also arises in the case of Heterologicality. Indeed, Graham Priest, for instance, in his recent derivation of the seeming contradiction in this case, relies on the Satisfaction Schema [3, p 31]; but that again presumes univocality, and so must be treated in the same way as the T-Schema. With Berry's Paradox, also, Priest required the denotation relation to be 1–1 [18, p 162]. Again the conclusion must be that all that Priest has proved is that if the denotation relation is univocal then there is a contradiction. Hence, we may deduce, the denotation relation is not univocal – although we must also realise that that does not mean a more liberal semantic denotation relation could be defined.

In brief, if 'the least ordinal not definable in less than 19 words' defined, i.e. uniquely referred to some object, then that object both would and would not be definable in less than 19 words. Hence, directly against Priest's pre-supposition, some identifying descriptions cannot have a unique denotation. But that does not mean that 'the least ordinal not definable in less than 19 words' is simply ambiguous. To think simple ambiguity must be involved, in this case, is clearly to forget the function of demonstratives, and purely referential terms like Donnellan's 'the man drinking Martini', whose semantics does not determine, not just a unique referent, but any referent at all. Their referents are not semantically given, and we have to allow for areas of meaning beyond semantics: centrally pragmatics. It is in such pragmatic cases that we find the paradigms for where language comes to an end, and yet meaning is still expressed – for instance, simply by means of the gesture which accompanies 'this', or some other purely referential expression. Beyond semantics, it must be remembered, there is 'hand waving'.

References

[1] B. Brown. Yes Virginia, there really are Paraconsistent Logics. *Journal of Philosophical Logic*, 28:489–500, 1999.

[2] B. H. Slater. Paraconsistent Logics? *Journal of Philosophical Logic*, 24:451–454, 1995.

[3] G. G. Priest. The Structure of the Paradoxes of Self– Reference. *Mind*, 103:25–34, 1994.

[4] I. M. Copi. *Symbolic Logic*. Macmillan, New York, 4 edition, 1973.

[5] C. Sayward. Prior's Theory of Truth. *Analysis*, 47:83–87, 1987.

[6] R. L. Goodstein. On the Formalisation of Indirect Discourse. *Journal of Symbolic Logic*, 23:417–419, 1958.

[7] A. N. Prior. Epimenides the Cretan. *Journal of Symbolic Logic*, 23:261–266, 1958.

[8] A. N. Prior. *Objects of Thought*. O.U.P., Oxford, 1971.

[9] R. Montague. Syntactical Treatments of Modality. *Acta Philosophica Fennica*, 16:153–167, 1963.

[10] R. Thomason. Indirect Discourse is not Quotational. *The Monist*, 60:340–354, 1977.

[11] R. Thomason. A note on Syntactical Treatments of Modality. *Synthese*, 44:391–395, 1980.

[12] N. Asher and H. Kamp. Self–Reference, Attitudes and Paradox. In G. Chierchia, B. H. Partee, and R. Turner, editors, *Properties, Types and Meaning*, pages 85–158. Kluwer, Dordrecht, 1989.

[13] M. J. Loux. *Metaphysics; A Contemporary Introduction*. Routledge, London, 1998.

[14] W. Kneale and M. Kneale. *The Development of Logic*. Clarendon, Oxford, 1962.

[15] S. Haack. *Philosophy of Logics*. C.U.P., Cambridge, 1978.

[16] B. H. Slater. Prior's Analytic. *Analysis*, 46:76–81, 1986.

[17] R. Turner. *Truth and Modality for Knowledge Representation*. M.I.T. Press, Cambridge MA, 1991.

[18] G. G. Priest. The Logical Paradoxes and the Law of the Excluded Middle. *Philosophical Quarterly*, 33:160–165, 1983.

Are paraconsistent negations negations? *

JEAN-YVES BÉZIAU Stanford University, California, USA
jybeziau@hotmail.com

Dedicated to Prof. Newton C.A. da Costa for his 70th birthday

Abstract
To know if paraconsistent negations are negations is a fundamental issue: if they are not, paraconsistent logic does not properly exist. In a first part we present a philosophical discussion about the existence of paraconsistent logic and the surrounding confusion about the emergence of possible paraconsistent negations. In a second part we have a critical look at the main paraconsistent negations as they appear in the literature.

Contents

*We acknowledge financial support from the *Swiss National Science Foundation*

Does paraconsistent logic exist?

The principle of non-contradiction can be expressed in many different ways (not necessarily equivalent). One of them is: *a proposition and its negation cannot be true together*. Because the principle of non-contradiction is generally admitted, if someone says for example: "It is raining and it is not raining", this seems quite absurd.

A paraconsistent logic is a logic in which there is a negation, a *paraconsistent* negation, which does not obey the principle of non-contradiction. Such an entity may appear as a paradoxical funny object, like a plane that does not fly, or champagne without alcohol. Even worse: one may think that such an object is a contradictory thing, an impossible object, like a round square or a rocket which goes faster than the speed of light.

In fact, in a recent paper, B.H.Slater [1] claims that paraconsistent logic does not exist. His claim is based on the fact that, according to him, paraconsistent negations are not negations. The reason why is not really convincing.[1]

But even if until now no one has really disproved that paraconsistent negations are negations, no one has provided serious philosophical or matemathical reasonings and evidences to show that they are. So the question of the existence of paraconsistent logic is still an open problem.

In a first part we deal with a general philosophical discussion about the possible existence of paraconsistent logic. The second part will be a more technical discussion about the properties of "existing" paraconsistent negations.

1 Existence, irrationality and confusion

1.1 The relative existence of paraconsistent logic: paraconsistentology, paraconsistentists and paraconsistent logic

In some sense paraconsistent logic exists : many different systems of paraconsistent logics have been presented and studied over the years. A section of *Mathematical Reviews* has been created. Three world congresses have been organized (one in Belgium, one in Poland and the present one in Brazil). But one must make a clear distinction between the subject and the object. What exists in fact are people, let us call them, *paraconsistentists*, who study some supposed paraconsistent logics. Their science can be called *paraconsistentology*, or paraconsistency for short (although we will not use this term since in general people confuse paraconsistency with paraconsistent logic and the point is to show the difference). For the present time there are no clear evidences of the existence of paraconsistent logic, but paraconsistentists and paraconsistentology exist for sure, and even if one day someone proves that paraconsistent logic does not exist, this will not necessarily entail the non existence of paraconsistentists and paraconsistentology.

Let us explain this by a metaphor. Imagine that there is a planet in the universe called Babakos whose inhabitants are called Babakons, and let us call the specialists of Babakons, Babakonologists and their science Babakonology. Imagine furthermore that the existence of the inhabitants of Babakos is not certain. The

[1] See [2], [3] and [4] for a criticism of Slater's arguments.

planet is very far from the Earth and the observations about the planet are not sufficient to guarantee their existence, although there is some kind of evidence of their existence. Imagine that one day it is proven that there are no Babakons. So what about Babakonologists and Babakonology? Certainly the proof, by observations and/or theoretical means, of the nonexistence of Babakons can be considered as part of Babakonology. And this is certainly not the end of Babakonologists, since they probably have a lot to say about the nonexistence of Babakons and this knowledge can be useful for the study of the existence or nonexistence of inhabitants of other planets in the universe. Babakonology is part of the study of ET-life (extra-terrestrial life) and it has a value as such whether Babakons exist or not.

Let us emphasize that (before any proof of the existence or nonexistence of Babakons is given) the *belief* in Babakons is independent of Babakonology. Someone who believes in Babakons is not necessarily a Babakonologist and someone who doesn't believe in Babakons can be a very good Babakonologist. The belief in Babakons can be a good motivation for one to turn into Babakonology, but the disbelief can also be a strong impulse.

Someone may believe that Babakons do not exist and for that reason be against the development of Babakonology. For example one may, for religious reasons, think that humans are the only beings in the universe. This kind of behaviour is not good for Babakonology, nor for science in general. A no better position would be the situation of someone who believes that there are people living in the Moon. His belief may be based on a book from, let us say, Ancient Egypt and he will try by any means, fractal topology, quantum astronomy, bi-polar logic, to prove the existence of inhabitants in the Moon.

The present situation in mathematical logic does not prove or disprove the existence of paraconsistent negations and we must keep in mind two things: 1) We cannot infer the existence of God from the existence of theology; 2) We cannot infer the nonexistence of God from the existence of atheists. This means: we cannot infer the existence of paraconsistent logic from paraconsistentology, or from the existence of paraconsistentists; we cannot deny the existence of paraconsistent logic just because there are people who don't believe in it.

1.2 Comparison with the existence of other logics

The situation of paraconsistent logic is quite different from the situation of some other logics. Let us take the example of the fashionable *linear logic*. At the present time there are no serious doubts about the existence of such a logic, nobody has written a paper entitled "Linear logic?", trying to show that there are no linear logics. The reason is very simple and can be found in the very name "linear logic". It is a technical mathematical term without any philosophical connotations. This term is related to some mathematical background which was, according to Girard, the origin of the idea of linear logic. However there are no clear connections between the mathematical background of this term and the philosophical ambitions of linear logic.

Slater deduces that there are no paraconsistent logics from an alleged proof that a negation which is paraconsistent is not a negation. In linear logic all the connectives are different from the classical ones. Someone maybe can say that linear negation is not a negation and that therefore there are no linear logics. But the aim of linear logic is not to provide a new negation, it is more general: to provide logical operators which are adequate to deal not with eternal truths but perishable recyclable data. It is very difficult to know exactly to what extent linear logic is a satisfactory solution to the problem. There is certainly a huge gap between the vernacular examples presented by Girard to motivate his logic and the way linear logic works. At the end the question is: "Do the mathematical operators developed match some operations of any sort of reasoning?". If it is not the case, one can claim, not that linear operators don't exist, but that linear logic doesn't exist, simply because it is not a *logic*. What exists is a mathematical system, and many mathematical systems have nothing to do with logic. It would be the same situation as if Babakonologists discovered that a kind of monkeys were living on Babakos instead of something similar to humans.

In some sense linear logic is the result of a formal game which consists of modifying a mathematical tool, sequent calculus, developed to represent classical logic. This can make sense if later on an interpretation is provided. This is the main problem not only of linear logic but of the other *substructural logics* which are obtained by modifying the structural rules of sequent calculus.

A similar situation is that of *many-valued logic*, which is the result of generalizing the standard two-valued matrix of classical logic, considering matrices with more than two values. Here we have the same question as in the case of substructural logic: to know if the operators defined by many-valued matrices have a logical interpretation, that is to say the question to know if many-valued logic is really a *logic*. As it is known, Łukasiewicz developed many-valued logic in order to catch the notion of possibility, but it seems that with many-valued logic it is not possible to properly define this notion. Even if one really succeeds to give a meaning to operators of many-valued logic, there is still the question of whether many-valued logic is really *many-valued*. Suszko has shown that it was possible to provide a two-valued semantics for Łukasiewicz's three-valued logic and Suszko pointed out that we must not confuse logical values with algebraic values (on this topic see [5], [6], [7]).

In the case of *modal logic*, we have mathematical operators designed to represent the notions of possibility and necessity. In fact there is a whole class of modal logics and it is not clear at all which modal logic represents rightly these notions, if any. In this sense the question of the existence of *modal logic* is still an open one. Modal logic has been developed these last years in a pure mathematical way as the general study of unary operators. One asks for example which kind of unary operators can be represented by a Kripke structure. In this sense modal logic includes paraconsistent logic, since negations are unary operators. Nobody seems to be aware of this fact and this shows very well that most of the people working in modal logic do not really think about the interpretation of the "modal" operators.[2] It is not necessary wrong to call these operators "modalities", since a modality in fact is any variation of a given statement, including negation and affirmation. From this point of view, if the existence of paraconsistent logic was proved this will entail the existence of

[2]Of course many interpretations are in the air: knowledge, belief, information, etc. But one thing it to have a general intuitive idea and another one is to carry on a systematic investigation to see if the mathematical properties really fit with the interpretation. The difficulty here is that on the one hand we have something precise and the other hand something rather fuzzy, one has to check if a precise thing match with a fuzzy thing.

modal logic, at least a modal logic different from classical logic, since classical logic can itself be considered as a modal logic.

1.3 The supremacy of classical negation

Classical negation of mathematical logic (hereafter *Clanemalo*) is one representation of negation. The claim that it is the right representation of negation is very controversial. A less controversial claim is that classical negation is the right representation of negation as it appears in mathematical reasoning. An even less controversial claim is that it is the right representation of negation as it appears in *classical* mathematical reasoning.

Outside the sphere of mathematical reasoning, negation appears in many forms, some of them having very few connections with Clanemalo, so for this reason it seems totally absurd to say that Clanemalo is the right and only negation. Mathematical reasoning is certainly different from vernacular reasoning.

One may think however that mathematical reasoning is the only right reasoning, that vernacular reasoning is obscure and ambiguous and that Clanemalo is the only right negation, that vernacular negation is obscure and ambiguous. In this case Clanemalo has to be clearly taken as a *normative* definition of negation and not a *descriptive* one, since it does not describe properly vernacular negation.

One may want to give a descriptive definition, through mathematical logic, of vernacular negation. But this is not necessarily an obvious task. The classicist says: "Vernacular negation is ambiguous, Clanemalo works good, we must use Clanemalo". This leaves open the question of why, how and to which extent vernacular negation is ambiguous. The anti-classicist thinks that Clanemalo is a caricature, that it is not a good picture of *real* negation. He thinks that the classicist pejoratively says that vernacular negation is ambiguous only because it does not fit into the simplified schema of Clanemalo. He would say that vernacular negation is not ambiguous, but more *complex* that the oversimplified Clanemalo. Trying to give some other representations of negation the anti-classicist may shed som light on the nature of vernacular negation.

On the one hand the classicist tends to reject vernacular negation as ambiguous preferring a pure platonic idealization, on the other hand the anti-classicist tends to venerate the vernacular negation, with a kind of blind respect for concrete empirical data.

Maybe it is important to recall to someone fascinated by the "incredible complexity" of vernacular negation and who doesn't want to deal with classical negation, that one of the fundamental basis of science is the process of abstraction by simplification. It is interesting here to recall what Gentzen was saying about constructivist mathematics versus classical mathematics:

> We might consider still another example which, in its relation to physics, seems to provide even more striking analogies to the relationship between constructivist mathematics and actualist mathematics:
>
> I am thinking of the occasional attempt to construct a 'natural geometry', i.e. a geometry which is better suited to physical experience than the usual (Euclidean) geometry, for example. In this natural geometry, the theorem 'precisely one straight line passes through two distinct points' holds only if the points are not lying too close together. For if

> they are lying very close together, then *several* adjacent straight lines can obviously be drawn through the two points. The *draftsman* must take these considerations into account; in *pure geometry, however*, this is not done because here two points are *idealized*. The *extended* points of experience are replaced by the *ideal*, unextended, 'points' of theoretical mathematics which, in reality, have no existence. That this procedure is beneficial is borne out by its success: It results in a mathematical theory which is of a much simpler and considerably smoother form than that of natural geometry, which is continually concerned with unpleasant exceptions.
>
> The relationship between actualist mathematics and constructivist mathematics is quite analogous: Actualist mathematics idealizes, for example, the notion of 'existence' by saying: A number *exists* if its existence can be proved by means of a proof in which the logical deductions are applied to *completed infinite* totalities in the same form in which they are valid for finite totalities; entirely as if these infinite totalities were actually present quantities. In this way the concept of existence therefore inherits the advantages and the disadvantages of an ideal element: The *advantages* is, above all, that a considerable simplification and elegance of the theory is achieved - since intuitionist existence proofs are, as mentioned, mostly very complicated and plagued by unpleasant exceptions-, the *disadvantage*, however, is that this ideal concept of existence is no longer applicable to the same degree to physical reality as, for example, the constructive concept of existence. (...)
>
> The question now arises: what use are elegant bodies of knowledge and particularly *simple* theorems if they are not applicable to physical reality in their literal sense? Would it no be preferable in that case to adopt a procedure which is more *laborious* and which yields *more complicated* results, but which has the advantage of making these results immediately meaningful in reality? The answer lies in the *success* of the former procedure: Again consider the example of geometry. The great achievements of mathematics in the advancement of physical knowledge stem precisely from this method of *idealizing* what is physically given and thereby *simplifying* its investigation. ([8], pp.248-249).

It is important to keep Gentzen's remarks in mind at a time where a lot of intricate ugly "draftsman logics" are presented, which contrast so much with the beauty and simplicity of classical logic. Of course one can think that to venerate only classical logic and Clanemalo would be the same as thinking that natural numbers are very nice and that we don't need real numbers, ugly ambiguous imprecise things. But the people who think that Clanemalo is an absurd simplification of vernacular negation that must be banished certainly are not aware of the incredible jump that was made in Greece, more than two thousands years ago, when the people started to use the principle of non-contradiction (hereafter PNC). It is probably not wrong to say that the use of the principle of non-contradiction was the start of mathematics and science in general. Interesting enough the appearance of the PNC coincides with a rejection of empiricism.[3]

[3] On this topic see the remarkable book of Szabo [9]. What Szabó discusses is essentially the

One can have two opposite perspectives on paraconsistent negation:

1. Paraconsistent negation would be something less idealized than classical negation, and paraconsistent logic would be like a draftsman geometry.
2. Paraconsistent negation can be seen as an extension of the sphere of rationality, in the same sense that irrational numbers or transfinite numbers can be conceived as an extension of the sphere of rationality rather than a drawback toward a "draftsman mathematics".

It seems that the terminology "Transconsistent Logic" coined by G.Priest [10] is good to express this second perspective.[4] This second perspective, which can also be traced back to Vasiliev with his notion of "Imaginary Logic" or "Non-Aristotelean Logic" (on Vasiliev see for example [11], [12]), is very challenging but also is very controversial for several reasons. If we have a look at the birth of science in the Greek world, the PNC can itself be considered as the foundation of rationality, so it is not quite the same to consider the move from natural numbers toward irrational numbers and the move from classical negation toward paraconsistent negation. Of course the "crisis" of irrationals was really a crisis, but it was not a crisis of rationality, despite the expression "irrational".

People sometimes like to make an opposition between occidental rationality based on the PNC and oriental wisdom. This is the case of Kosko in a popular book about fuzzy logic (cf [13]). Anyway we must recall that within the occidental Greek tradition there were people like Heraclitus or Hegel who defended a kind of rationality not based on the PNC, even based in fact on something which appears as the contrary of the PNC. But if we look at the history of science, we see that until now this has led to nowhere. Maybe paraconsistentology can be the first stone in the construction of a new rationality, but we still don't know if any building based on paraconsistent logic will stay erect.

1.4 Irrationality and paraconsistency

Some people think that paraconsistent logic is dangerous, that to give away the PNC will lead to nonsense, chaos, confusion. They think that the PNC is the foundation of rationality, and that without it, there will be no more distinction between truth and falsity. Similar criticisms have been addressed to fuzzy logic.

Of course such kind of criticisms make sense in a world where we are surrounded by contradictory statements by politicians, advertisements and in fact at all levels of information. Someone who is supporting paraconsistent logic may appear as supporting the surrounding confusion.

Can we take seriously someone who says "I believe in God and I do not believe in God"? The classical rationalist will say no. But someone can say: "Well, not everything in life is black or white, it can be grey (cf. the famous "grey zone" of Kosko [13]). Someone can be beautiful and ugly, republican and Quacker, rich and communist". We must be very careful at this point because there are several important issues which are mixed.

use of the *reductio ad absurdum*, which is the strongest form of the PNC, in particular Clanemalo can be defined only with the *reductio*.

[4]The philosophical position of Priest himself is however not very clear, sometimes it seems that his perspective rather falls under 1.

If someone asks us "Do you believe in God ?", we can have no answer to this question, we can say "This question makes no sense to us because we don't know exactly what do you mean by God", or we can say "In some sense we believe in God, in some other sense we don't believe in God". Does this mean that we are rejecting the PNC? Not necessarily.

Given a property P, the PNC divides a class of objects into two parts, the objects having this property, and the objects not having this property. Let us say that the property is "to be odd", using the PNC, we have the class of odd numbers and of non-odd numbers. Now we can ask: "Is God odd?". The question makes no sense because odd is a property which applies to numbers only. Someone could say "God is both odd and even", and claims that the PNC is not valid. But in the best case, this has to be taken only as (bad) poetry and not a serious challenge to the PNC.

A number which is not odd is called even. Even means nothing more, nothing less that non-odd. It is clear that in natural language there are a lot of pairs of words that don't work like that, for example the pair blonde/intelligent: a woman can be blonde and intelligent without infringing the PNC. This at first seems a kind of terrible triviality. But it seems that this triviality is not so blatant for some of the people who want to reject the PNC.

However what is very interesting in paraconsistentology is the attempt to develop a negation which should be able to deal with pairs of concepts which work in a way very similar to contradictories but at the same time admit a common intersection.

On the other hand there are no good reasons to radically reject the PNC. It is clear that the PNC is useful in some sense, and that it is working quite well in many situations. Paraconsistent logic is not in fact necessarily based on a rejection of the PNC. If we define, as we did, paraconsistent logic as a logic in which there is a paraconsistent negation, then we may also have a classical negation, therefore in this case a paraconsistent logic is an extension of classical logic. A paraconsistent negation is an additional operator. Sometimes, like in the case of da Costa's logic $C1$, it is possible to define the classical negation with the paraconsistent negation. And what about the converse? We have pointed out that it is possible to define something which looks very much like a paraconsistent negation within first-order logic (cf [14]).

From this point of view it is clear that paraconsistent logic appears rather as an extension, than a rejection of classical rationality. If it exists ! Because we must not play with words, we still don't really know if there are any paraconsistent logic.[5]

1.5 Three ways to confusion

According to Slater (see [1]), the existence of paraconsistent logic is a result of a verbal confusion. Paraconsistent logic are dealing with subcontraries and not contradictories. Slater claims that paraconsistentists say that they are talking about negation, because they switch contradictories for subcontraries. It would be the same as to call women "men" and men "women". In this case one would be able to claim that women produce sperm, but of course the reality would not have

[5] About the topic of paraconsistency and irrationality one may consult the interesting book of G.G.Granger [15] dedicated entirely to irrationality and which inlcudes a chapter on paraconsistent logic.

changed. Let us call this kind of confusion ***switching confusion***. If someone claims that women produce sperm, it can be the result of an important discovery about the physionomy of women or just the result of a switching confusion. Two very different cases indeed.

We have shown in another paper (cf [4]) that paraconsistentists cannot be accused of such an easy trick,[6] that paraconsistent logic is not the result of a switching confusion, but it may be the result of other confusions.

Imagine that we extend the concept of inhabitants in order to include monkeys, dogs, or even rats and that small rats are discovered on Babakos, then we can say that there are inhabitants on Babakos. In the same way, if we extend the notion of negation to any unary operator, then we can say that paraconsistent negations exist. We can call this kind of confusion ***global confusion***. Certainly many paraconsistentists make implicitly this kind of global confusion when they start to speak about such or such operator they called paraconsistent negation just because it does not obey the PNC.

Finally we would like to talk about *Christopher Columbus* confusion. As it is known Columbus wanted to go to India, but he reached America instead and the inhabitants of America were called "Indians" because at first he thought that he was in India. The discover of America was a very important fact, but of course "Indians" are not Indians. This does not mean that they don't exist or that they are not interesting people, but they are different kind of people. One can say in some sense that Łukasiewicz made a kind of Columbus confusion. He wanted to reach modalities, but reached something else by many-valued logic. Maybe it is what is happening with paraconsistentists. They are looking for negations, but perhaps the operators they are discovering are something else, very interesting, but not negations. And perhaps, in the same way that nobody calls nowadays Łukasiewicz's logic $L3$ a modal logic, nobody will call in the future $C1$, LP or $P1$ paraconsistent logics. It is true that sometimes the power of words is really strong and that until now it is still quite common to call "Indians" people originally from North or South-America, although the terminology "American Indians" has been introduced. Anyway the important thing is that despite the name, few people believe that these Indians are from India.

Paraconsistentists may escape the most trivial confusion, the switching confusion, but it seems that in general they are not very careful about the use of the word "negation", they eat their cake before cooking it and made a lot of global confusions, and at the end they may even be into a big Colombus confusion.

[6] As Slater rightly recalled during our talk at the WCP2, he just generalized an idea which was first proposed by R.Routley and G.Priest in [16]. Routley and Priest were arguing that da Costa's negation was not a negation, that it was a subcontrary forming relation rather than a contradictory one. Slater showed that their argument could also be applied to Priest's negation and any paraconsistent negation. Later on Priest recognized that we should rather considered erroneous his original argument against da Costa's negation than to think that Slater's generalized argument is right.

2 A GUIDED TOUR IN THE LAND OF PARACONSISTENCY

The question to know if paraconsistent negations are really negations, if there really is any paraconsistent logic, must necessarily lead to a systematic study of the technical aspects of paraconsistent negations.

All paraconsistentists are united by a negative criterium: the rejection of the *ex contradictione sequitur quod libet* (EC for short). Mathematically speaking they say that if a negation $\neg$ is ***paraconsistent*** then

$$a, \neg a \nvdash b$$

Note that it would be absurd to say that if $a, \neg a \nvdash b$ then $\neg$ is paraconsistent. It is clear that the rejection of EC is a necessary condition but not a sufficient one. In order to be a paraconsistent ***negation***, $\neg$ must have some ***positive*** properties.

On the one hand it is not clear at all which properties are enough to define a negation. On the other hand given a set of properties for negation, one has to investigate if these properties are compatible. So let us say that people agree that a given set of properties SCN is enough to define a negation, then one has to check if this set together with the rejection of EC form a compatible set of assumptions. One may want to prove that there are no paraconsistent negations by considering a set NCN of necessary properties for negation and showing that it is not compatible with the rejection of EC.

The point is that it is not clear what should be sets of properties like SCN and NCN. The question is difficult for mainly two reasons: on the mathematical side, the propreties for negation can be of very different natures, on the philosophical side, it is not easy to have a coherent and intuitive interpretation of an operator having such or such property.

To study this problem one has first to describe and classify the properties of negation. This is what we will do in the next section. Then there are two methods: you can construct a logic, showing that some given properties are compatible, or you can get negative results showing that given properties are not compatible. What did happen in the field of paraconsistency until now is closer to the first method: people have built logics. But most of the time they claim that these logics are paraconsistent without investigating really which properties the underlying "negations" have and if these properties are sufficient to justify the name.[7]

In the next sections we will make critical reviews of the main paraconsistent negations which have been presented thus far.[8]

[7]During many years negation was not a notion in focus. Different negations were presented, like intuitionistic or minimal negations, but negation was not by itslef a subject of a systematic investigation. An exception maybe is the work of Curry. In [17] he presented a comparative study of four types of negation. This work is very interesting but Curry's treatement does not allow paraconsistent negations (in a previous paper [18], we have presented a generalization of Curry's work). A couple years ago, Gabbay started to investigate negation [19], trying to propose a definition and so on. Following his interesting initiative, a group of people have work on this direction (cf. [20], [21]). However until few years ago they didn't know nearly nothing about the extended literature on paraconsistent logic (it is obvious when we look at the lists of references of their works), and didn't include explicitly paraconsistent negation in their treatement.

[8]Not every paraconsistent negation "existing" under the sun will be discussed. Despite the present high speed of circulation via information highways, we are not necessary aware of everything, of some exotic paraconsistent negations elaborated in the florest of Transylvania, where the access to the internet is still limited.

2.1 Classification of the properties of negation

The classification of properties for negation depends on a general framework for the study of logics. Following our idea of *Universal Logic* (cf [22], [23], [24]), we just consider a *logic* as a mathematical structure of type

$$\langle L; \vdash \rangle$$

where L is any set and $\vdash$ is a relation between sets of objects of L and objects of L. A *negation* is a function defined on L having such or such property.[9]

We will use the abbreviation $a \dashv\vdash b$ for $a \vdash b$ and $b \vdash a$.

Pure laws[10]

1. Reductio ad absurdum[11]

if $\neg a \vdash b$ and $\neg a \vdash \neg b$, then $\vdash a$ [12]
if $a \vdash b$ and $a \vdash \neg b$, then $\vdash \neg a$
if $\neg a \vdash a$ then $\vdash a$
if $a \vdash \neg a$ then $\vdash \neg a$

2. Contraposition

if $\neg a \vdash \neg b$ then $b \vdash a$
if $a \vdash b$ then $\neg b \vdash \neg a$
if $a \vdash \neg b$ then $b \vdash \neg a$
if $\neg a \vdash b$ then $\neg b \vdash a$

3. Double negation

$\neg\neg a \vdash a$ [13]
$a \vdash \neg\neg a$

The two fundamental laws

1. Law of non-contradiction (LNC for short)
$\vdash \neg(a \wedge \neg a)$ [14]
2. Law of excluded middle (EM for short)
$\vdash a \vee \neg a$

---

[9] As the reader should have understood after our discussion in the first part of this paper, not every function defined on L can be called a "negation".

[10] By *law* we mean here statements about the relation $\vdash$. This relation is not considered as a proof-theoretical notion. It is important not to confuse a law with a ***rule of deduction***, mistake too much common in the literature nowadays. The properties of negations that we present here are not proof-theoretical properties, even less syntactic properties. They are properties of a function in a structure.

[11] The different forms of ***reductio ad absurdum*** are not equivalent, but there are good reasons to use the same generic name for these four versions; on this topic the reader can consult the book of J.-L.Gardies [25] entirely dedicated to the ***reductio***.

[12] This should be considered as an abbreviation of the following statement: for any set of formulas T, and any formulas a and b:

if $T, \neg a \vdash b$ and $T, \neg a \vdash \neg b$, then $T \vdash a$

This kind of abbreviation can be ambiguous since in substructural logics, including relevant logic, this abbreviation is not always equivalent to the thing abbreviated. Anyway we will use it and also use the same kind of abbreviations hereafter for the other laws.

[13] This law could be expressed equivalently, modulo some basic properties of $\vdash$, in the following way:

if $\vdash \neg\neg a$ then $\vdash a$.

For more details about this, see [18].

[14] This law of non-contradiction LNC should not be confused with the informal principle of non-contradiction PNC, one may think that the correct formulation of PNC is the ***ex contradictione*** EC ; for more discussion about this, see the section about full paraconsistent logic.

De Morgan Laws

1. De Morgan laws for conjunction

$\neg(a \wedge b) \dashv\vdash \neg a \vee \neg b$

$\neg(\neg a \wedge \neg b) \dashv\vdash a \vee b$

$\neg(\neg a \wedge b) \dashv\vdash a \vee \neg b$

$\neg(a \wedge \neg b) \dashv\vdash \neg a \vee b$

2. De Morgan laws for disjunction

$\neg(a \vee b) \dashv\vdash \neg a \wedge \neg b$

$\neg(\neg a \vee \neg b) \dashv\vdash a \wedge b$

$\neg(\neg a \vee b) \dashv\vdash a \wedge \neg b$

$\neg(a \vee \neg b) \dashv\vdash \neg a \wedge b$

3. De Morgan laws for implication

$a \to b \dashv\vdash \neg a \vee b$

$\neg a \to \neg b \dashv\vdash a \vee \neg b$

$\neg a \to b \dashv\vdash a \vee b$

$a \to \neg b \dashv\vdash \neg a \vee \neg b$

Self-Extensionality

The property of self-extensionality, expression coined by Wójcicki (see [26]; and [27] for comments about this terminology), corresponds to the validity of the replacement theorem and can be expressed in the following way:

if $a \dashv\vdash b$ then $T \vdash c$ iff $(T \vdash c)[b/a]$

where $(T \vdash c)[b/a]$ means that b replaces a in T and c.

Representability properties

Several general properties related to the ***representability*** of logics can be considered: a negation is ***truth-functional*** iff it can be expressed by a finite matrix, it is ***leibnizian*** iff it can be described by a possible world semantics, it is ***effective*** iff it can be defined with a recursive proof-system, etc...

Other properties

Now let us finish by stating properties not directly connected with negation but which are important for the discussion. A paraconsistent logic has generally the same basic language as classical logic, that is to say we have a negation, and three binary connectives: a conjunction, a disjunction, and an implication. Of course since the negation of a paraconsistent logic has not the same features as the negation of classical logic, the binary connectives cannot have exactly the same behaviour as the classical one, if we take into account the fact that the behaviour of a connective depends on the whole context. Anyway the binary connectives can be quite similar to the classical ones in the sense that a paraconsistent logic can be a conservative extenstion of ***positive classical logic***. On the other hand we can have a paraconsistent logic where this does not happen. There are mainly two basic cases:

1. *Non adjunctive paraconsistent logics* These are logics in which the conjunction fails to obey the following law of adjunction: $a, b \vdash a \wedge b$

2. *Non implicative paraconsistent logics* These are logics in which the implication fails to obey the following law of implicativity: if $\vdash a \to b$ then $a \vdash b$.

We have summarized above the main properties a negation can have.[15] Obviously the strongest properties are the various laws of *reductio ad absurdum* and contraposition. Unfortunately none of these, except the last two forms of *reductio*, are compatible with the rejection of EC. Or to be more exact with the rejection of EC and its weak form: $a, \neg a \vdash \neg b$. For a detailed account about this fact see [18].

So the only hope for the paraconsistentist is to gather other properties.

2.2 Non self-extensional paraconsistent logics

Most of the well-known non classical logics, like modal logics, intuitionistic logic, linear logic, etc., are self-extensional. Many people think that a logic must be self-extensional. However this is rather because it is a nice technical and practical property than for any precise philosophical reason. As we have argued elsewhere (cf [29]), there are no reasons *a priori* to reject a logic just because it is not self-extensional. Moreover it seems that any logic that wants to capture intensionality should be non self-extensional (cf [27]).

Nevertheless if a logic is not self-extensional, the counter examples of self extensionality must have an intuitive explanation. Unfortunately it seems that it is not the case with several paraconsistent logics which are not self-extensional.

In da Costa's logic $C1$, the formulas $a \wedge b$ and $b \wedge a$ are logically equivalent (i.e. $a \wedge b \dashv\vdash b \wedge a$) but not their negations and nobody has presented a philosophical idea to support this failure. In Priest's logic LP and in da Costa and D'Ottaviano's logic $J3$, the formulas $a \vee \neg a$ and $b \vee \neg b$ are logically equivalent but not their negations and here again no philosophical justification for this failure has been presented.[16]

Several results show that some properties of negation are incompatible with the idea of a self-extensional paraconsistent negation (see [35]). Maybe one can conclude from this that paraconsistent negations are not negations. However this will we a controversial conclusion as long as one gives a convincing reason why a negation should be self-extensional. On the other hand one may argue that it is not a problem since a paraconsistent negation should be an intensional operator. First, let us note that not any non self-extensional operator is intensional. Second, one should be able to provide an intuitive explanation of the failure of the replacement theorem, based on a discussion about intensionality or not.

2.3 Full paraconsistent logics

We say that a paraconsistent logic is full when we have:

$\vdash \neg(a \wedge \neg a)$.

There was a time when the people didn't make any distinction between this law of non-contradiction LNC and EC. In fact the question is still open to know if we can find an intuitive interpretation of an operator which obeys EC and not LNC or obeys LNC and not EC.

In the three-valued logic $L3$ of Łukasiewicz, LNC is not valid, but EC is. And this seems quite odd following the interpretation of his third value. If the value of a formula and its negation are undetermined, then the value of their conjunction

[15] Notice that we didn't mention non classical properties, that is to say properties which are not valid for classical negation, like for example: if $\neg a \vdash a$. For a discussion about this, see [28].

[16] about $C1$, see [30] and [31]; about LP, see [32] and [10]; about $J3$, see [33] and [34]

is undetermined and so is the negation of this conjunction. As undetermined is not a distinguished value, then LNC fails; but at the same time why should one be able to deduce anything from a formula and its negation when they are both undetermined?

There are many three-valued paraconsistent logics where the negation is defined exactly in the same way as in Lukasiewicz's logic, but the undetermined value is considered as distinguished, the effect of this interchanging is that LNC is valid but not EC. The problem is that the same oddity as in $L3$ appears in an inverted way.

The gap between LNC and EC is in fact not so huge. How can one jump from LNC to EC? This can be done in three easy steps: by the use of self-extensionality, involution[17] and adjunction (see [35]).

At the end it is not clear at all that the idea of a full paraconsistent logic is meaningful. In fact the initial idea of da Costa, the Pope of paraconsistent logic, was to reject both LNC and EC.

2.4 Paraconsistent classical logics

Paraconsistent classical logics (hereafter PCL) are logics which have the same theorems as classical logic, they differ only at the level of the consequence relation.

Note therefore that any PCL is full, so that a paraconsistent negation in a PCL cannot be at the same time involutive, self-extensional and adjunctive (due to results of [35]). Priest's logic LP is a PCL which is involutive but not self-extensional. Urbas's dual-intuitionistic logic LDJ is a PCL which is self-extensional but not involutive. Jaśkowski's discussive logic is a PCL which is self-extensional, involutive but not adjunctive.[18]

Moreover a PCL cannot be implicative, since in a PCL we have

$\vdash a \to (\neg a \to b)$.

If it is implicative we will get

$a, \neg a \vdash b$.

So Priest's logic LP, Urbas's LDJ and Jaśkowski's discussive logic are all non implicative.

What is the problem with non implicative logics? In a logic which is not implicative, there are formulas a and b such that

$\vdash a \to b$ and $a \nvdash b$

therefore under the assumption of the transitivity of $\vdash$ in such a logic the following version of the *modus ponens* cannot be valid

$a, a \to b \vdash b$.

Under the assumption of monotonicity the following version of the *modus ponens* cannot be valid

if $\vdash a$ and $\vdash a \to b$, then $\vdash b$.

The aim of this paper is not to discuss implication, but one can think that an implication without *modus ponens* is something as paradoxical as free money.[19]

---

[17] A negation is said to be involutive when both double negation laws hold.

[18] On Urbas's logic, see [36]; on Jaśkowski's logic, see [37] and [38]; a general paper on PCL is [39].

[19] There are a lot of discussions about implication, and many people, like relevantists, think that a connective which does not obey the *modus ponens* can still be called an implication. We have here a problem similar with paraconsistent negation. Relevantists are probably not making a switching confusion, but maybe a global confusion or a Columbus confusion (cf Section 1.5) Some

2.5 Paraconsistent atomical logics

Paraconsistent atomical logics (PAL for short) are logics where only atomic formulas have a paraconsistent behaviour. Molecular formulas have a standard behaviour. That means you can have

$a, \neg a \nvdash b$

only when a is an atomic formula.

The intuitive motivation can be the following: contradictions may appear at the level of facts, data, information (or whatever the atomic formulas are supposed to represent) and at this level they should not entail triviality, but reasoning is classical, so when you go to the logical level, i.e. non atomical, everything should work in the normal classical way.

This idea at first seems reasonable, but if you think about it two seconds you will see that PAL sound quite absurd: for example from an atomic formula a and its negation $\neg a$ you cannot deduce anything but you can do so from the molecular formula $a \wedge a$ and its negation $\neg(a \wedge a)$. One can seriously wonder how such duplication can draw a line between the field of paraconsistent reasoning and the field of classical reasoning. Moreover this same example shows that any PAL is not self-extensional.

So PAL seem really not a good solution to the paraconsistent problem, i.e. the problem of finding a negation which is paraconsistent and has a coherent intuitive interpretation. Several PAL have been presented. Sette's logic $P1$ is one of them [40]. It is interesting to remember how this logic was created. Sette wanted to solve the maximal problem set by da Costa and he presented $P1$ as a solution to this problem. $P1$ is certainly a solution to this problem, but the example of $P1$ just shows that solving this problem does not necessarily solve the paraconsistent problem.[20] If we are interested in paraconsistent logic, we must always keep the paraconsistent problem in mind which involves both technical and philosophical aspects. Blind technicality can be a fun game for those who practice it but most of the time leads only to formal nonsense.

Another PAL is the logic V, which has been presented by Puga and da Costa as a possible formalization of Vasiliev's logic (cf [41]). From the above considerations, we can conclude that either V is not a good formalization of Vasiliev's logic, either Vasiliev's logic is not a good solution to the paraconsistent problem.[21]

2.6 Truth-functional paraconsistent logics

In the subsection about full paraconsistent logics, we already said a word about three-valued logics. We noticed that some of them have the same negation as the one of Łukasiewicz's logic $L3$. The only difference is that the undetermined value in these paraconsistent logics is considered as distinguished. We have seen that this leads to a strange unsatisfactory feature.

people think that relevant implication is the right implication and that classical implication should not be called implication. But implication of classical logic represents perfectly the implication of mathematical reasoning, which is certainly an important aspect of human reasoning.

[20]The maximal problem is the problem of finding a paraconsistent logic which is maximal in the sense that it has no strict extension other than classical logic or the trivial logic; J.Marcos [34] has recently shown that there exists more than 8K solutions to this problem, and most probably some of these solutions are more interesting from the viewpoint of paraconsistency than Sette's $P1$.

[21]For more discussion about this question see [42]

What are the other possibilities? There are not a lot of them: the other ways to define another kind of paraconsistent negation in a three-valued logic leads, either to the validation of LNC (so we get the same problem as before), or to a paraconsistent atomical logic, as the reader can easily check.[22]

Maybe the only solution to get a reasonable paraconsistent negation it to work with four values. This has not yet been investigated systematically. The only four-valued paraconsistent logics which have been really studied are Nelson's logic (cf [43], [44]) and Belnap's logic. In Belnap's logic neither LNC, nor EC are valid, but the excluded middle is not valid either and if you add it to Belnap's logic you get classical logic.[23]

2.7 Paraconsistent morganian logics

We say that a paraconsistent logic is morganian if all De Morgan laws are valid as well as the two double negation laws. In most paraconsistent logics you have only some parts of the De Morgan laws. There are also paraconsistent logics in which all De Morgan laws for conjunction and disjunction are valid. This in particular the case of standard truth-functional paraconsistent logics like J3, LP or Belnap's logic (see also [49]).

Paraconsistent morganian logics cannot be self-extensional (on the assumption of adjunctivity). Let us prove this:

From the fact that we obviously have $\vdash a \rightarrow a$, we get $\vdash \neg a \vee a$, by application of the De Morgan law for implication. So we have $a \vee \neg a \dashv\vdash b \vee \neg b$. By self-extensionality, we get $\neg(a \vee \neg a) \dashv\vdash \neg(b \vee \neg b)$. Now applying De Morgans's law for disjunction and self-extensionality (or transitivity), we get: $\neg a \wedge a \dashv\vdash \neg b \wedge b$. From this it is easy to see that we have $\neg a \wedge a \vdash b$. And finally applying adjunction, we get $\neg a, a \vdash b$.

From this proof it is possible to see also that a logic cannot admit the excluded middle, be morganian, self-extensional and paraconsistent.

This fact corresponds to an obvious algebraic result: if you have a De Morgan lattice and you add the excluded middle, it will be the greatest element of the lattice, now by De Morgan law, this means that there is also a smallest element, which is of the form $a \wedge \neg a$ (see [50]).

In conclusion: the problem with morganian paraconsistent logics is that we have to choose between self-extensionality and the excluded middle.

---

[22]This is true unless one admits a non conservative matricial definition of conjunction (i.e. conjunction defined as *min*) or is ready to consider as negation, an operator such that $a \leftrightarrow \neg\neg\neg a$.

[23]The reader can find a good study of three-valued paraconsistent logics in [34], and [45]. [46] is a tentative to generalize truth-functional semantics which could be fruitful for the developement of paraconsistent logic. About Belnap's logic, see [47]. Belnap didn't know what was paraconsistent logic when he developed his logic. The fact that Belnap's logic is paraconsistent was already noticed by da Costa in his ***Mathematical Review*** (58 5021) of Belnap's paper. But of course it is not clear to which extent Belnap's negation is a paraconsistent ***negation***.

Historically it seems that the first person who had the idea to use logical matrices to develop paraconsistent logic was Asenjo (see [48]). He was followed later on by da Costa and D'Ottaviano with their system *J*3, Sette with his system *P*1 and finally by Priest with his system *LP*.

2.8 Paraconsistent leibnizian logics

By paraconsistent leibnizian logics we mean logics constructed with a possible world semantics.

There are two very simple ideas that we will discuss here. They are not of course the only possible but it seems that they are the two basic ones. Furthermore they are the only two which have been investigated in details, so in the present paper we will limit our discussion to them.[24]

We recall that possible world semantics is based on the idea to consider sets of possible worlds. In the basic semantics discussed here, no relation of accessibility is involved (the same as to consider a universal relation of accessibility). Possible worlds can be whatever your imagination can conceive (including Babakos) but in fact we can consider without loss of generality that they are only (bi)valuations. It is less poetical but simpler.

Jaśkowski's logic

The first idea is the following. We consider sets of ***classical*** valuations. Let us call any such a set a ***Jaśkowski frame***. Then in a Jaśkowski frame we say that a formula is true iff it is true in ***at least one*** valuation of the frame. With this we define a Jaśkowski's logic, by saying that a formula is valid (or is a theorem) iff it is true in any Jaśkowski frame, and we define the consequence relation accordingly. This logic is a full paraconsistent classical self-extensional, but it is non adjunctive.

It is based on a very intuitive idea which is the main idea of Jaśkowski's ***discussive*** logic: when you have a group of people discussing, you can say that something is true if at least one of them thinks it is true. It sounds a little bit chaotic, but it is very democratic (maybe too much). Apparently it should yield to something quite different from classical logic, but it yields to something surprisingly close since Jaśkowski's logic has the same theorems as classical logic. Moreover we have: $a \wedge \neg a \vdash b$ but not $a, \neg a \vdash b$.

How can this be understood? $a \wedge \neg a$ is always false in a Jaśkowski frame, because such a frame is a set of classical valuations. Intuitively: any member of the discussion group reasons in a classical way. Therefore if we have a contradiction of the form $a \wedge \neg a$, any individual of any group will deduce anything from it.

Nevertheless we can find a Jaśkowski frame where a formula a is true and its negation $\neg a$ is true. Intuitively: we can have a discussion group of two people, one who thinks that a is true, and the other one that a is false, therefore that $\neg a$ is true, and they can both agree that b is false. At the end we have a frame which validates a and $\neg a$ but not b.

Despite the very intuitive motivation of Jaśkowski's logic, one can wonder if it really works. If one forgets the intuitive idea and concentrate only on the basic mathematical features of its negation, what is the picture? We have on the one hand a negation which has too much properties, which is in fact quite similar to classical negation, and on the other hand a conjunction that has too few properties.

One can maybe improve the situation by generalizing the idea of Jaśkowski frame, by considering sets of non classical valuation (see [53]).

[24] Paraconsistent logics developed using semantics close to possible world semantics have been also presented in [3], [51] and [52].

Molière's logic

Let us examine now another kind of leibnizian paraconsistent logic. Let us call a Molière frame, a set of valuations which is defined in the following way. The conditions for binary connectives are the usual ones. A formula like $a \wedge b$ is true in a Molière frame iff it is true in any valuation of the frame and it is true in a given valuation iff both a and b are true in this given valuation. Now the condition for negation is as follows: in a given valuation of the frame $\neg a$ is false iff a is true in every valuation of the frame.

The idea is also quite intuitive: we are sure that $\neg a$ is false iff we are absolutely sure that a is true. In a case of doubt about a, let us say, to pursue Jaśkowski's metaphor, if there is someone in the discussion group who thinks that a is false, then $\neg a$ can be false too.[25]

This definition of negation is exactly dual to the definition of negation in the possible world semantics for intuitionistic logic (the difference is that we don't consider accessibility relations). As it is known there is a close connection between intuitionistic logic and the modal logic $S4$. So one may expect a connection between Molière's logic and a modal logic. In fact Molière's logic is nothing else than $S5$ itself. In a given valuation v of a Molière frame, $\neg a$ is true iff there exists a valuation w in which a is false. This means that the classical negation $\perp a$, of a is true in w. Therefore $\diamond \perp a$ is true in v. So the negation of Molière's logic is nothing else than the connective $\diamond \perp$ of $S5$ (where $\perp$ is classical negation).

Although Molière negation enjoys some nice properties (it is self-extensional, obeys several de Morgan laws, etc.) and have an intuitive interpretation, it has some drawbacks. For example it is a full paraconsistent logic. Anyway maybe Molière negation can be considered at the present time the best paraconsistent negation. What is funny is that people were trying to construct paraconsistent negations and there was one pretty ready just nearby. But at the end logicians are all funny *bourgeois gentilhommes*, aren't they?[26]

WAITING FOR NICE PARACONSISTENT NEGATIONS

What can we conclude combining our philosophical investigations and our little tour in the land of paraconsistency?

Certainly until now, no paraconsistent negations having "nice" features have been presented. By "nice", we mean having interesting mathematical properties together with a coherent intuitive interpretation. That does not mean that there are no such things, but at least they have not been discovered yet.

The present investigations do not permit one to be very optimistic about the chance to discover such things, since many classical techniques of mathematical logic, such as logical matrices, possible world semantics, sequent calculus, etc., have been applied - not in a real systematic way, it is true - without success.

But we can still hope. Maybe an entirely new technique must be developed to generate the challenging objects paraconsistent negations are.

[25] It is clear that Molière's logic could also be considered as a formalization of Jaśkowski's idea.

[26] On Molière's logic, see [54], [14]. We would like to thank Claudio Pizzi with whom we had the opportunity to discuss Molière's logic in his castle of Copacabana.

References

[1] B.H.Slater. Paraconsistent logics? *Journal of Philosophical Logic*, (24):451–454, 1995.

[2] B.Brown. Yes, Virginia, there really are paraconsistent logics. *Journal of Philosophical Logic*, (28):489–500, 1999.

[3] N. Rescher and R.Brandom. *The logic of inconsistency*. Rowman and Littlefield, Totowa, 1979.

[4] J.-Y. Béziau. Paraconsistent logic! *Preprint, National Laboratory for Scientific Computing*, 1995. Submitted.

[5] R.Suszko. The Fregean axiom and Polish mathematical logic in the 1920s. *Studia Logica*, (36):377–380, 1977.

[6] J.-Y. Béziau. What is many-valued logic? In *Proceedings of the 27th International Symposium on Multiple-Valued Logic*, pages 117–121, Los Alamitos, 1997. IEEE Computer Society.

[7] M.Tsuji. Many-valued logics and Suszko's Thesis revisited. *Studia Logica*, (60):299–309, 1998.

[8] G.Gentzen. The present state of research into the foundations of mathematics. In M.E.Szabo, editor, *The collected papers of Gerhard Gentzen*, 1938. Translation of the German original.

[9] A.Szabó. *The beginnings of Greek mathematics*. D.Reidel, Dordrecht, 1978.

[10] G.Priest. *In contradiction. A study of the transconsistent*. Martinus Nijhoff, Dordrecht, 1987.

[11] A.I. Arruda. N.A.Vasiliev : a forerunner of paraconsistent logic. *Philosophia Naturalis*, (21):472–491, 1984.

[12] V.A. Bazhanov. The imaginary geometry of N.I. Lobachevski and the imaginary logic of N.A. Vasiliev. *Modern Logic*, (4):148–156, 1994.

[13] B.Kosko. *Fuzzy thinking : the new science of fuzzy logic*. 1993.

[14] J.-Y. Béziau. S5 is a paraconsistent logic and so is first-order classical logic. Submitted.

[15] G.G.Granger. *L'irrationnel*. Odile Jacob, Paris, 1998.

[16] G. Priest and R. Routley. Systems of paraconsistent logic. In *[62]*, pages 151–186, 1989.

[17] H.B. Curry. *Lecons de logique algébrique*. Gauthier-Villars, Paris - Nauwelaerts, Louvain, 1952.

[18] J.-Y. Béziau. Théorie législative de la négation pure. *Logique et Analyse*, (147):209–225, 1994.

[19] D.M.Gabbay. What is a negation in a system? In *Logic Colloquium' 86*, pages 95–112, Amsterdam, 1988. Elsevier.

[20] D.M.Gabbay and H.Wansing (eds). *What is negation?* Kluwer, Dordrecht, 1999.

[21] H.Wansing (ed). *Negation - A notion in focus.* W.de Gruyter, Berlin, 1996.

[22] J.-Y. Béziau. Universal logic. In T.Childers and O.Majer, editors, *Logica'94 - Proceedings of the 8th International Symposium*, pages 73–93, Prague, 1994. Czech Academy of Sciences.

[23] J.-Y. Béziau. *Recherches sur la logique universelle.* PhD thesis, Department of Mathematics, University of Paris 7, Paris, 1995.

[24] J.-Y. Béziau. From paraconsistent logic to universal logic. *Sorites*, (12):1–33, 2000. http://www.ifs.csic.es/sorites.

[25] J.-L.Gardies. *Le raisonnement par l'absurde.* Presses Universitaires de France, Paris, 1991.

[26] R. Wójcicki. *Theory of logical calculi.* Reidel, Dordrecht, 1988.

[27] J.-Y. Béziau. The philosophical import of Polish logic. In M.Talasiewicz, editor, *Logic, Methodology and Philosophy of Science at Warsaw University*, 2001. to appear.

[28] J.-Y. Béziau. What is paraconsistent logic? In G.Priest D.Batens, C.Mortensen and J.P. van Bendegem, editors, *Frontiers in paraconsistent logic*, pages 95–111, Baldock, 2000. Research Studies Press.

[29] J.-Y. Béziau. Logic may be simple. *Logic and Logical Philosophy*, (5):129–147, 1997.

[30] N.C.A. da Costa. Calculs propositionnels pour les systèmes formels inconsistants. *Comptes Rendus de l'Académie des Sciences de Paris*, (257):3790–3793, 1963.

[31] J.-Y. Béziau. Nouveaux résultats et nouveau regard sur la logique paraconsistante $C1$. *Logique et Analyse*, (141):45–58, 1993.

[32] G. Priest. Logic of paradox. *Journal of Philosophical Logic*, (8):219–241, 1979.

[33] I.M.L. D'Ottaviano and N.C.A. da Costa. Sur un problème de Jaśkowski. *Comptes Rendus de l'Académie des Sciences de Paris*, (270):1349–1353, 1970.

[34] J. Marcos. 8K solutions and semi-solutions to a problem of da Costa. Forthcoming. Paper presented at the 2nd World Congress on Paraconsistency.

[35] J.-Y. Béziau. Idempotent full paraconsistent negations are not algebraizable. *Notre Dame Journal of Formal Logic*, (39):135–139, 1998.

[36] I.Urbas. Dual-intuitionistic logic. *Notre Dame Journal of Formal Logic*, (37):440–451, 1996.

[37] S. Jaśkwoski. Rachunek zdań dla systemów dedukcyjnych sprzecznych. *Studia Societatis Scientiarum Toruniensis*, (1), 1948.

[38] M.Urchs. Discursive logic. Towards a logic of rational discourse. *Studia Logica*, (54):231–249, 1995.

[39] R.Sylvan and I.Urbas. Paraconsistent classical logic. *Logique et Analyse*, (141):3–24, 1993.

[40] A.M.Sette. On the propositional calculus $P1$. *Mathematica Japonae*, (16):173–180, 1973.

[41] L.Puga and N.C.A. da Costa. On the imaginary logic of N.A.Vasiliev. *Zeitschrift für mathematische Logik und Grundlagen der Mathematik*, (34):205–211, 1988.

[42] A.S. Karpenko. Atomic and molecular paraconsistent logics. *Logical Studies*, (2), 1999. http://www.logic.ru.

[43] D. Nelson. Constructible falsity. *Journal of Symbolic Logic*, (14):16–26, 1949.

[44] H.Wansing. *The logic of information structures*. Springer, Berlin, 93.

[45] R.Tuziak. Finitely many-valued paraconsistent systems. *Logic and Logical Philosophy*, (5):121–127, 1997.

[46] W. Carnielli. Possible-translations semantics for paraconsistent logics. In G.Priest D.Batens, C.Mortensen and J.P. van Bendegem, editors, *Frontiers in paraconsistent logic*, Baldock, 2000. Research Studies Press.

[47] N.D. Belnap. A useful four-valued logic. In *Modern uses of multiple-valued logic*, pages 5–37, Dordrecht, 1977. D.Reidel.

[48] F. Asenjo. A calculus of antinomies. *Notre Dame Journal of Formal Logic*, (7):103–105, 1966.

[49] N.C.A. da Costa and J.-Y. Béziau. Overclassical logic. *Logique et Analyse*, (157):31–44, 1997.

[50] J.-Y. Béziau. De Morgan lattices, paraconsistency and the excluded middle. *Boletim da Sociedade Paranaense de Matemática.*, (18):169–172, 1998.

[51] A. Buchsbaum. A family of paraconsistent and paracomplete logics with recursive semantics. Submitted.

[52] A. Buchsbaum and T.Pequeno. A game characterization of paraconsistent negation. Submitted. Paper presented at 2nd World Congress on Paraconsistency.

[53] J.-Y. Béziau. The logic of confusion. In *Proceedings of the International Conference on Artificial Intelligence IC-AI 2001*, 2001. to appear.

[54] J.-Y. Béziau. The paraconsistent logic Z. Forthcoming. Paper presented at the Jaśkwoski Memorial Symposium, Torun, July 15-18, 1998.

[55] J.-Y. Béziau. Logiques construites suivant les méthodes de da costa. *Logique et Analyse*, (131):259–272, 1990.

[56] J.-Y. Béziau. Classical negation can be expressed by one of its halves. *Logic Journal of the Interest Group on Pure and Applied Logics*, (7):145–151, 1999.

[57] J.-Y. Béziau. The future of paraconsistent logic. *Logical Studies*, (2):1–20, 1999. http://www.logic.ru.

[58] W. Bychovski. Polar paraconsistent logic. *Proceedings of the Poldavian Academy of Sciences*, (37):123–156, 1987.

[59] N.C.A. da Costa. *Logiques classiques et non classiques.* Masson, Paris, 1997.

[60] J.M. Dunn. Two treatments of negation. In J.Tomberlin, editor, *Philosophical perspectives*, 1994.

[61] J. Łukasiewicz. O logice trójwartościowej. *Ruch Filozoficny*, (5):170–171, 1920.

[62] R.Routley G.Priest and J.Norman (eds). *Paraconsistent logic : Essays on the inconsistent.* Philosophia, Munich, 1989.

[63] H.Montgomery and R.Routley. Contingency and non-contingency bases for normal modal logics. *Logique et Analyse*, (9):318–328, 1966.

[64] I.Urbas. Paraconsistency. *Studies in Soviet Thought*, (39):343–354, 1990.

ACKNOWLEDGMENTS

I would like to thank Newton C.A. da Costa, João Marcos de Almeida and Arthur de Vallauris Buchsbaum for many discussions related to this paper. I am sure however that they don't agree with all the ideas expressed here.

I would like also to thank the organizers and the participants of the 12th ESSLLI (European Summer School in Logic Language and Information 2000) during which I presented an introductory course in paraconsistent logic. This was an opportunity for me to ameliorate several aspects of this paper.

Moreover, criticisms and commentaries by five referees allow me to greatly improve an earlier version of this paper. I am also grateful to Linda Eastwood who revised the English of my paper and turned it into something which looks like International English rather than French English.

Finally I would like to thank Patrick Suppes who invited me to work at Stanford University where this work was written.

Jean-Yves Béziau
Stanford University
Centre for the Study of Language and Information
Ventura Hall
Stanford, CA, 94305-4115
USA

On the role of adjunction in para(in)consistent logic *

MAX URCHS University of Konstanz, Germany
max.urchs@uni-konstanz.de

To Jerzy Kotas

Abstract
The non-adjunctive way in paraconsistent logic started out with Jaśkowski's construction of a propositional calculus for inconsistent deduction. Another contribution to this tradition is Rescher and Brandoms approach towards a logic of inconsistency. Also Lewis is suggestive of a non-adjunctive approach to accommodating contradiction.

Jaskowski's construction of discussive logic is probably the best-bashed project in the field. There is no room to correct all unfair or objectionable criticism concerning this approach. In my opinion, it is one of the most interesting and promising approaches in inconsistency-tolerant reasoning. Moreover, it has an excellent philosophical motivation. Numerous classical ideas by Leibniz or Kant can be restated within this setting. Discussive logic is evidently motivated by the weak approach to paraconsistency. According to a proposal by Peranowski, the philosophical background of Jaskowski's system and related calculi should be called "parainconsistency".

The reason for preferring discussive logic as the most promising approach to paraconsistency is not only because of its formal elegance and deep philosophical motivation. Besides, discussive logic represents an ideology that is, to my mind, the most appropriate one for paraconsistency. To put it informally: at the very core of paraconsistency lies not negation, but conjunction. The paradigm for paraconsistency, I take it, is scientific discourse. For any logical formalization, the observer's perspective is the only appropriate one. Hence, when creating a paraconsistent logic we have typically to assume the position of a keeper of the minutes of scientific discussion. Why should we then insinuate that a person who asserts a negated sentence has in mind some exotic kind of negation? Such an assumption could be justified only by referring to the overall consistency of the ongoing discussion. That motive, however, cannot be attributed to the person who made the claim. Writing down her claim in terms of non-classical negation thus means systematically to distort it. We therefore should rather concentrate on conjunction. With respect to inconsistency-tolerating calculi, this connective seems to be the most important one – at least for those approaches which are directed towards application.

*I acknowledge financial support from the DFG

1 The principle of contradiction

According to my experiences in Germany, paraconsistent logic still awaits its warm welcome by the philosophical community. In order to achieve that, one has to ensure two things:

1. to show, how the topic relates to fully established philosophical issues;
2. to convince all the suspicious colleagues that despite of its unfortunate name, paraconsistency has a solid and reasonable philosophical foundation.

People watch us playing around with contradictions and ask a most natural question: Shouldn't we be afraid of contradictions? My answer is: Yes! We certainly should! Logicians, as well as all other respectable men, abhor contradictions. It is not only a matter of Western cultural tradition – contradictions actually indicate a deviation from normality, from the usual standards of rationality. If a contradiction is manifest in some situation, whatever happens in that situation ceases to be rational. For an irrational situation, no sensible prediction can be made. Just anything is possible. Therefore it seems perfectly in order to assume the *ex contradictione quodlibet sequitur* principle. This principle leads to explosion of the system whenever one single contradiction occurs.

This being said, you may wonder why somebody preaching such messages was invited to this congress. So let me ask a second question: should we be afraid of inconsistencies? Of course, the answer is a resolute No! Inconsistencies are often interesting, even fascinating and deserve thorough logical investigation.

The concept of an inconsistency is by no means a sharp one. Besides logical contradictions there are refutations, collisions of thesis and antithesis, incoherencies, antinomies, contrasts, oppositions, self-defeating assertions, there is "contradictio-in-adiecto", paradoxical pronouncements, not to forget oxymorons and a number of other cases. Furthermore, there are dozens of adjectives intended to denote in a very subtle and refined way the slightest shades in meaning of various forms of inconsistencies. Actually, this reservoir of terminological material is still largely ignored in philosophy. In this paper, I do not aim at a classification either. Instead I just take the word "inconsistency" to be a generic term for all kinds of incompatibility and polarity. Contradictions are conjunctions of a thesis and its negation: *H et non–H*. Contradictions thus form a special kind of inconsistencies. For lack of a better term I will call them *et-non*-sentences. Thus on a formal level a conjunction *H et non–H* is depicted by $H \wedge \neg H$ and called, as usual, *counter-tautology* or (*formal*) *contradiction*.

Thesis and antithesis appearing together in some set of premises render that set inconsistent. $\{H, \neg H\}$ is thus a (formal) inconsistency. Obviously, if the rule of adjunction $H, F/H \wedge F$ is admissible, then inconsistency and contradiction are synonymous concepts. In general, however, they are not.

In order to make my position unmistakable, let us state the principle of *ex contradictione quodlibet*:

From *A et non–A* follows anything.

In *ex contradictione quodlibet*, the formula *A et non–A* expresses a contradiction in the above sense. The connective *et* is thus Boolean conjunction. From a philosophical position, it makes sense to differentiate between this principle and a similar, but different one. Let us call this second assumption the *ex falso quodlibet* principle. What is this principle about? Naturally, from a false claim made e.g. in discussion not everything follows. It is quite common to utter, willful or not, false sentences. In sciences, the whole reasoning process is often about to find out the truth. To that purpose one has first to take also into consideration the false sentence. That means, in a logical framework appropriate for scientific discourse the *ex falso quodlibet* is not a permissible rule.

To be more explicit, for any given A we may assume that *A* or *non-A* is false. *Ex falso quodlibet* says that from a false sentence anything follows. Hence,

From *A* follows anything or from *non–A* follows anything;

in other words

From *A and non–A* follows anything.

Here the *non*-connective is not to be confused with the *et* occurring in *ex contradictione quodlibet*. Rather the *non* represents some non-classical conjunction operator.

From that perspective, *ex falso quodlibet* seems methodologically unsound. It does not suit to what happens in rational discourse. Therefore it is adequate to dismiss the *ex falso quodlibet* principle in discussive logic. This, by the way, is how the paraconsistency of a calculus is usually defined: a system is paraconsistent if it lacks the *ex falso quodlibet* principle. I agree with this definition in letter, but not in spirit. It rest on the usual confusion, or, let me say more neutrally: on the identification, of both principles. The intention of the above definition of paraconsistency, I take it, is actually the exclusion of the principle of contradiction. Yet I think that we have good reason not to dismiss the *ex contradictione quodlibet principle*. We should accept it as the keystone of all Western rationality and dismiss instead the companion. To do this, of course, one has to block the rule of adjunction.

2 Discussive logic

The non-adjunctive way in paraconsistent logic sets out with Jaśkowski's construction of a propositional calculus for inconsistent deductive systems ([1], [2], English translation: [3]). Another contribution to this tradition is Rescher and Brandoms approach towards a logic of inconsistency ([4]). Also Lewis ([5]) is suggestive of a non-adjunctive approach to accommodating contradiction.

The central idea is this: What is really dangerous is not inconsistency as such, but contradictions. Thus one should hinder inconsistencies to condense into contradictions. Thesis and anti-thesis bouncing together bring about an explosion – quite as two subcritical radioactive masses detonate when they jointly cross the critical limit.

Jaśkowski's construction of discussive logic is probably the best bashed project in the field. There is no room to correct all unfair or objectionable criticism concerning this approach. In my opinion, it is one of the most interesting and promising approaches in inconsistency-tolerant reasoning. Moreover, it has an excellent philosophical motivation. Numerous classical ideas by Leibniz or Kant can be restated within this setting. Discussive logic is evidently motivated by the weak approach to paraconsistency. According to a proposal by Perzanowski, the philosophical background of Jaśkowski's system and related calculi should be called "para*in*consistency", since in fact we are dealing with apparent inconsistencies only and in that sense we are beyond inconsistency.

Jaśkowski was the first to observe that it makes sense to differentiate contradictory logical calculi[1] from trivial ones[2] (cf. [3]).

DEFINITION 1 A logical calculus L is *contradictory* in the formal language FOR [symb.: $L \notin Cons$] $\Longleftrightarrow_{df} \exists H \in FOR : H, \neg H \in L$.

If a contradictory system in Jaśkowski's sense, i.e. a calculus containing an inconsistency, is based on classical logic, then every formula is a thesis of the system. But, not so in general.

DEFINITION 2 A logical calculus L is *trivial* in the formal language $FOR \Longleftrightarrow_{df} FOR = L$.

Here is Jaśkowski's original comment on the issue:

> This deviates from the terminology accepted so far: in the methodology of the deductive sciences such systems have so far been called contradictory, but for the purpose of the analysis presented in this paper it is necessary to make a distinction between two different meanings of the term "a contradictory system", and to use it only in one sense, as specified above. The over-complete systems have no practical significance: no problem may be formulated in the language of an over-complete system, since every sentence is asserted in that system. Accordingly, the problem of the logic of contradictory systems is formulated here in the following manner: the task is to find a system of the sentential calculus which: 1) when applied to contradictory systems would not always entail their over-completeness, 2) would be rich enough to enable practical inference, 3) would have an intuitive justification. Obviously, these conditions do not univocally determine the solution, since they may be satisfied in varying degrees, the satisfaction of condition 3) being rather difficult to appraise objectively." ([3], 145)

The essential point in Jaśkowski's construction is a slightly more sophisticated attempt to conjunction – in fact we have two of them: an inner one and an external. Here is a short definition[3] of Jaśkowski's system D_2: Let FOR_d be the set of all formulas freely generated from a denumerable set of propositional variables by means of some Boolean complete set of propositional functors and two additional two-argument "discussive" connectives: the discussive conjunction $\wedge_d$ and the discussive implication $\rightarrow_d$. Next we explain a translation $t : FOR_d \longrightarrow FOR_m$ from FOR_d to propositional modal language FOR_m . t leaves unchanged propositional

[1] According to the terminology used here, these calculi are inconsistent.

[2] Jaśkowski calls any system in which every formula is a tautology *over-crowded*. Other names are *over-filled*, as well as *trans-complete*.

[3] It would be possible, and perhaps easier as well, to explain D_2 as a conservative extension of standard modal logic. Nevertheless, as we are forced to put the emphasis on subtleties of the formal language, I prefer to choose the "save" definition.

variables as well as the Boolean connectives. Let $\Diamond$ be the $S5$–possibility. Then we establish:

(1)	$t(H)$	$=_{df}$	H	, for $H \in AT$;
(2)	$t(\neg H)$	$=_{df}$	$\neg t(H)$;	
(3)	$t(H \wedge G)$	$=_{df}$	$t(H) \wedge t(G)$;	
(4)	$t(H \wedge_d G)$	$=_{df}$	$t(H) \wedge \Diamond t(G)$;	
(5)	$t(H \rightarrow_d G)$	$=_{df}$	$\Diamond t(H) \rightarrow t(G)$.	

DEFINITION 3 $D_2 =_{df} \{H \in FOR_d; \Diamond t(H) \in S5\}$.

The above classical construction can be generalized considerably. It is possible to use a large class of modal systems, among them even non-normal ones, to obtain interesting discussive calculi. Let S be any regular modal logic containing $S3$. According to Polish tradition it seems quite natural to understand discussive logic as a logical system (i.e. a consequence operation in a formal language) rather than as a set of formulas. For every modal calculus S containing $S3$ we are able to define a consequence operation Cn_S in the discussive language FOR_d and to give a direct semantical characterization for the system $\langle FOR_d, Cn_S\rangle$ thus obtained. With every such S we correlate a discussive system S_d, i.e. a consequence operation Cn_S in the discussive language FOR_d.

DEFINITION 4 Let $H \in FOR_d$. H is *d–valid* in a point of the model M [symbolically: $M \models_d H[x]$] $\Longleftrightarrow_{df} \exists y \in R(x) : M \models t(H)[y]$.

The notion of d–validity in a point of a model is extended to models, frames, and classes of frames, as usual. In natural way the acceptance property in S leads to an inference relation Cn_S.

This definition does not fully match the one presented in Priest and Routley's "Systems of paraconsistent logic" ([6], 158). As a result, some of the criticism included in that paper does not hurt Jaśkowski's original system, while some of the conjectures have no affirmative answer. Others perhaps do. I will return to this point later. In any case, the final verdict, what sounds rather like a death sentence, namely that

> "...the non-adjunctive approach to paraconsistency should be dismissed"[4]

deserves a revision.

3 JAŚKOWSKI SYSTEMS

DEFINITION 5 Let $\mathcal{K}_S$ be the class of all Kripke-frames for S and let $X \subseteq FOR_d$. $Cn_S(X) =_{df} \{H \in FOR_d : \forall F \in \mathcal{K}_S(F \models_d X \Longrightarrow F \models_d H)\}$.

Let us state the following lemma and theorems:

LEMMA 6 Cn_S is a consequence operation in FOR_d.

[4] ibid., 162

Proof: Standard. □

For all calculi S containing the modal system $S3$ the following deduction theorem holds:

THEOREM 7 $\forall S3 \subseteq S \;\; \forall X \subseteq FOR_d \;\; \forall H, G \in FOR_d$:
$(H \to_d G) \in Cn_S(X) \Longleftrightarrow G \in Cn_S(X \cup \{H\})$.

Proof: For any X, $Cn_S(X)$ is closed w.r.t. *modus ponens* and includes $Cn_{S3}(X)$. It is easy to check that the last set contains $H \to_d H$, $H \to_d (F \to_d H)$, as well as $(H \to_d (F \to_d G)) \to_d ((H \to_d F) \to_d (H \to_d G))$. The proof proceeds as usual. □

Usually we interpret the deduction theorem as a characterization of the implication within the considered system. But now it goes the other way round: the discussive implication possesses the expected properties of a discussive inference. By theorem 1 they are transmitted to the consequence relation. Moreover, we have the following theorem:

THEOREM 8 $Cn_{S5}(\emptyset) = D_2$.

Proof: Immediately from Definitions 3, 4, 5, validity in $S5$–frames and completeness. □

The above theorem shows that the consequence version of Jaśkowski's discussive calculus D_2 belongs to the class of logical systems $\{\langle FOR_d, Cn_S\rangle;\ S3 \subseteq S\}$. For that reason and because of the properties emerging from the Deduction Theorem we call the calculi $\langle FOR_d, Cn_S\rangle$ ***discussive Jaśkowski-systems.***

There is a close relationship between these systems and so called M–counterparts of modal calculi.

DEFINITION 9 $M - S =_{df} \{H \in FOR_m; \Diamond H \in S\}$.

Let s be a translation from FOR_m to FOR_d which leaves unchanged propositional variables, $\bot$, and Boolean connectives and fulfills the condition $s(\Diamond H) = \neg\bot \wedge_d s(H)$.

A result obtained by Kotas ([7]) can be generalized for quite a few of discussive Jaśkowski-systems. To put it more precisely, let C_{M-S} be the inference relation based on the M–S–acceptance property. (C_{M-S} is actually a consequence operation fulfilling $C_{M-S}(\emptyset) = M - S$.) Then the following theorem holds true for regular S containing $S3$.

THEOREM 10
Let S contain $S3$. $\forall H \in FOR_d \, \forall G \in FOR_m \, \forall X \subseteq FOR_d \, \forall Y \subseteq FOR_m$:

1° $H \in Cn_S(X) \Longrightarrow t(H) \in C_{M-S}(X)$;

2° $G \in C_{M-S}(X) \Longrightarrow s(G) \in Cn_S(X)$;

3° $\Box(G \to t \circ s(G)) \wedge \Box(t \circ s(G) \to G) \in C_{M-S}(\emptyset)$;

4° $\neg(\neg\bot \wedge_d (H \wedge \neg s \circ t(H)) \wedge \neg(\neg\bot \wedge_d (s \circ t(H) \wedge \neg H) \in Cn_S(\emptyset)$.

Proof:

ad 1° By induction on the structure of H.

ad 3° Induction on H, *ad absurdum*. Boolean cases are straightforward. Let $G := \Diamond F$. Furthermore, by $|F|_N$ we denote the set of all normal worlds in the universe of a frame F. As usual, $R(w)$ denotes the set of possible worlds accessible from a given world w. From $\Box(\Diamond H \rightarrow \Diamond t \circ s(H)) \notin Cn_{M-S}(\emptyset)$ we obtain:

$$\exists\varphi' \, \exists w' \in |F|_N \, \forall v \in R(w') : < F, \varphi' > \not\models \Box(\Diamond H \rightarrow \Diamond t \circ s(H))[v]$$

By normality of w', and assumption of induction, there is a v', accessible from w' such that $\Box(H \rightarrow t \circ s(H))$ holds and yet $\Box(\Diamond H \rightarrow t \circ s(H))$ does not. Since the last formula does not hold, there must be another world with $\Diamond H$ and without $\Diamond t \circ s(H)$. That leads to still another $v'' \in R^2(v')$ accepting both H and (because of transitivity) $H \rightarrow t \circ s(H)$, but refuting $t \circ s(H)$ – a contradiction! Since we have to respect contradictions in decent paraconsistent logic, this ends the first half of the third point. The second part is to be proved analogously.

ad 2° Again, it is enough to show by induction that

$F \models_d s(G) \iff \forall\varphi \, \forall w \in |F|_N \, \exists v \in R(w) : < F, \varphi > \models G[v]$.

In Boolean cases nothing happens.

$F \models_d s(\Diamond F) \iff$

$\iff \forall\varphi \, \forall w \in |F|_N \, \exists v \in R(w) : < F, \varphi > \models t \circ s(\Diamond F)[v]$

$\iff \forall\varphi \, \forall w \in |F|_N \, \exists v \in R(w) : < F, \varphi > \models t(\neg\bot \wedge_d s(F))[v]$

$\iff \forall\varphi \, \forall w \in |F|_N \, \exists v \in R(w) : < F, \varphi > \models \Diamond t \circ s(F)[v]$

$\iff \forall\varphi \, \forall w \in |F|_N \, \exists v \in R(w) : < F, \varphi > \models \Diamond F[v]$

The last step follows easily from both parts of 3°.

ad 4° Analogously to 3°.

□

The previous theorem shows that results concerning M–fragments can be transmitted to the according discussive systems. For technical simplicity, we will therefore investigate M–fragments instead.

What happens, if we change the modal logic underlying the construction of discussive calculi? It turns out that Jaśkowski's construction is stable with respect to changing the underlying modal logic. A first result was obtained by Tomasz Furmanowski ([8]:

THEOREM 11 (Furmanowski)
$\forall S4 \subseteq S \subseteq S5 : M - S = M - S5.$

As Jerzy Błaszczuk and Wiesław Dziobiak found ([9]), $S4$ is not the least system that obtains this property.

THEOREM 12 (Błaszczuk and Dziobiak)
$T^\star := T \oplus \{\Diamond\Diamond H \setminus \Diamond H\}$ is the least system which M–counterpart equals $M - S5$.

Jerzy Perzanowski had shown in [10] that no normal system below D ($D := K \oplus \Diamond(H \rightarrow H)$) gives rise to non-trivial discussive Jaśkowski-style systems. What, then, about non-normal based systems? Jerzy Kotas and Newton da Costa expressed the conjecture that such systems would be too weak to be interesting ([11]). Yet there is good reason to take a closer look on discussive calculi based on non-normal modal systems. To that purpose, let us slightly generalize the notion of M–counterpart:

$$\text{for } k \geq 1, \text{ let } M^k - S = \{H \in FOR_m; \Diamond H \in M^{k-1} - S\} \ (M^0 - S = S).$$

Søren Halldén proved that $S3$ is the intersection of $S4$ and $S7$ (in [12]). It follows that $M^2 - S3 = M - S4$ (since $S7$ contains $\Diamond\Diamond p$ and its second M–fragment therefore vanishes).[5] Moreover, $M^2 - S4 = M - S4$. $S4$ is the normal counterpart of $S3$, the normal system next to $S4$. In that sense, the second M–fragments of $S3$ and of its goedelian closure coincide. This property is typical for the systems under consideration. At the second level of "possibilitation" non-normal systems containing $S3$ and their normal companions are identical. Let S_N be the "normal companion" of S, i.e. its closure w.r.t. $\{H \setminus \Diamond H\}$.

THEOREM 13 $\forall S3 \subseteq S: \ M^2 - S = M - S_N$.

Proof: If S is a normal system, then from $S3 \subseteq S$ and $S = S_N$ we obtain:

$$M - S_N = M - S = M^2 - S.$$

Let S be non-normal. Then $S \not\subseteq S$ and $M^2 \subseteq M^2 - S_N = M - S_N$.

Let $H \notin M^2 - S$. Then there is a falsifying S–model M' for $\Diamond\Diamond H$. Let w' the world in M' that does not verify $\Diamond\Diamond H$. Because of the transitivity of R, all worlds accessible from w' do not accept $\Diamond H$. They are thus normal. Reflexivity makes w' to accept $\Diamond H$. Since $\Diamond H \in S_N$, this leads to a contradiction in a subsequent normal world. Hence, M' is inconsistent and thereby not a countermodel for $\Diamond\Diamond H$.

□

However, the assumption in the above theorem is essential: it does not work for $S2$: one easily checks that

$$\Box\Box\Box(p \rightarrow p) \in M^2 - T \setminus M^2 - S2.$$

THEOREM 14 $\forall S3 \subseteq S \subseteq S5: \ M^{k+2} - S = M - S5$, for all $k \in \omega$.

Proof: For any S fulfilling the assumption, we have $S4 \subseteq S_N \subseteq S5$. Therefore on the basis of lemma 1 and theorem 3 we obtain for all such S:

$$M - S5 = M - S4 \subseteq M - S_N = M^2 - S = M - S5.$$

Since $\Diamond\Diamond\Diamond p \rightarrow \Diamond\Diamond p$ is a $S3$–tautology, we may "push up" the the M–fragment to any rank above 2. □

Consequently the manifold of discussive systems based on regular systems between $S3$ and $S5$ collapses into D_2 for higher dimensions. What happens on the first level? Here is a partial answer:

[5] I owe this observation to Wiesław Dziobiak.

THEOREM 15 $\forall S \subseteq S3.5: M^{k+2} - S \neq M - S$, for all $k \in \omega$.

Proof: Theorem 6 holds that $M^{k+2} - S = M - S5$ for any natural k and S between $S3.5$ and $S5$. From

$\Box\Box(p \to p) \in M^2 - S3.5 \setminus M - S3.5$ it follows, that

$$M - S \subseteq M - S3.5 \neq M^2 - S3.5 = M - S5 = M^{k+2} - S.$$

□

It thus seems that Jaśkowski's construction of discussive logic on the basis of modal calculi is amazingly stable. A lot of interesting semantical questions can be stated. Anyway, we shall turn to syntactical considerations again.

4 Intentional conjunctions

Once being given discussive implication, discussive conjunction suggests itself: conjunction of A and non-F is negated if H, then F. So what we get is the following definition[6]:

$$H \wedge_d F =_{df} \neg(\Diamond H \to \neg F)$$

Therefore, discussive conjunction turns out to be just $\Diamond H \wedge F$. That is apparently anything but acceptable: symmetry is the least one should demand of conjunction.

So let us stay for one moment with the notion of conjunction. The general term for bringing components of sentences together into a structured complex shall be named *concatenation*. *Conjunction* is the name reserved for the classical truth-functional minimum-connective. Special forms of concatenation may arise, e.g., from modalizing conjunction. This can be performed in many fancy ways. Any resulting concatenation is called "the outcome of an adjunction process"; *adjunction* would be fine for short. Yet in the literature, Jaśkowski's construction is widely called the non-adjunctive approach. Clearly, the outcomes of some adjunction processes are permissible in discussive logic. It sounds odd, if non-adjunctive logic permits adjunction. The name „adjunction" does not fit. Let us therefore choose as a *terminus technicus* for intentional conjunction the term *fusion*.

The general form of fusion in Jaśkowski systems is thus

$$p\&q =_{df} \odot(\odot p \wedge \odot q)$$

where $\odot$ is a variable ranging over variables to be substituted by any sequence of $\Box$ and $\Diamond$, or $\emptyset$.

Basically, conjunction is symmetric, but fusion is not. In general, fusion has no recursive truth condition. That is a frequent reproach to Jaśkowski's discussive conjunction. Yet, keeping in mind that fusion, and discusive conjunction in particular, is a modalized connective, its lack of recursive truth condition seems only natural. In $S5$, for instance, there are 21 various fusions.[7] Not all of them are of philosophical importance. Yet, e.g.,

$$p\&_p q =_{df} \Box(\Diamond p \wedge \Diamond q)$$

[6] This is not Ja'skowski's original discussive conjunction from [1]. However, it is logically – and ideologically – very close.

[7] There are many modal calculi based on modalized implication. The same can be done with fusion (see e.g. [13]).

might well be interesting.

5 Non-adjunctive systems

How could we possibly interpret a connective such as Jaśkowski's discussive conjunction? One way is suggested by the discussive interpretation of Jaśkowski's approach. A well founded assumption of cognitive linguistics says that, usually, the more important an information is, the more to the front it is transmitted. Important things come first. Now, suppose we already have some H. Subsequently, F comes up somehow. Therefore we have $F \wedge_d H$. However, F is new information. So it might be a good idea to accept it under proviso only. That, perhaps, sufficiently motivates the slanted conjunction.

But it is easy enough to find a symmetric version of $\wedge_d$. Let us first recall that the discussive implication is not very intuitive, either. Its main motivation was a purely technical one: just find a connective that meets modus ponens. By chance, $\rightarrow_d$ was the first appropriate one that came along in Jaśkowski's search procedure.

In order to proceed more systematically, one might try an interpretation in terms of practical discourse, again. Discussive implication plays the part of hypothesis testing, i.e. to answer the question, whether or not we should accept some F. One might reason as follows: well, if we would assume F to be true (hence $\Diamond F$), then G turns out to occur (hence G). Therefore, $\Diamond F \rightarrow G$, in other words $F \rightarrow_d G$.

Then, however, one might think about a slightly stronger version of hypothesis testing: if one would assume F to be true (hence $\Diamond F$), then we would need to accept also G – or alternatively: we would accept G under all circumstances (hence $\Diamond\Box G$). That would amount to the following definition:

$$F \rightarrow_{d_1} G = \Diamond F \rightarrow \Diamond\Box G$$

The new connective is not worse than the original one, since – as well as $\rightarrow_d$ – it also meets modus ponens. Moreover, it is correlated with the following discussive conjunction:

$$F \wedge_{d_1} H =_{df} \neg(F \rightarrow_{d_1} \neg H) \equiv \Diamond F \wedge \Diamond H.$$

So we have a discussive conjunction that is symmetric. What about its further properties? First of all, adjunction holds for $\wedge_{d_1}$:

$$H, F / H \wedge_{d_1} F.$$

And, as before, $H \wedge_{d_1} \neg H \rightarrow_{d_1} F$ is not a theorem of D_2. Therefore, $\wedge_{d_1}$ does not sabotage the system: it does not turn inconsistencies into explosive contradictions.

On the other hand, since both classical negation and alternative are in D_2, classical conjunction can be constructed in the system. However, as before, adjunction does not hold for $\wedge$. Therefore irrelevant triviality, $H \wedge \neg H \rightarrow F$, is no threat.

According to what was said at the beginning about rational paraconsistent logic and irrational paralogic, discussive logic is at the safe side. How far away, let us nevertheless ask, are these systems from collapsing in view of inconsistencies? It turns out that any discussive calculus explodes, if there occurs a pair of inconsistent necessary sentences. The reason is that for necessary sentences adjunction holds:

$$\Box H, \Box F / \Box(H \wedge F)$$

Therefore the safety of discussive calculi apparently only hinges on the possibility (or rather: impossibility) to define necessary contradictions. That, however, is easily done. In case of $\wedge_{d_1}$ it suffice to take $\neg(\neg H \wedge_{d_1} \neg H)$ that equals $\neg(\Diamond\neg H \wedge \Diamond\neg H)$, hence $\Box H$. For $\wedge_d$ the following is appropriate $\neg(\neg H \wedge_d (F \vee \neg F))$. Both cases could be explicitly excluded by pointing, again, to the basic methodological setting: there is no need to introduce superfluous elements like repeated identical conjuncts or tautological elements. But this defense would hardly be an elegant one. However, maybe there is no real danger by the mere possibility to cope with necessary formulae. What is needed is a pair of inconsistent necessary formulae. But it is by no means obvious that there are such entities like contingent *absoluta*. Quite to the contrary, there seem to be good philosophical reason to exclude such a possibility. The upshot is as before: if there are inconsistent necessary sentences, then the situation is non-normal and henceforth anything is possible.

There is one closely related observation. In his original paper, Jaśkowski refuted $\rightarrow_{d_1}$ as inappropriate discussive implication, because a formula $H \rightarrow_{d_1} G$ together with its negation would yield triviality. (By the way, the same is true for $\wedge_{d_1}$.) Jaśkowski was interested in "rich calculi", i.e. rich in theorems. Therefore the above observation gave him sufficient reason to dismiss the defined symmetric connectives. Yet it seems to me that they are interesting after all. And they perhaps can be saved. Taking into account Jaśkowski's original interpretation of discussive logic in terms of scientific or any other rational discourse, there is an important difference between inconsistent elementary propositions, that represent just observations or opinions or the like on the one hand, and inconsistent complex formulae on the other hand.

Under the given interpretation, modalized formulae result from operation on elementary, non-modalized input formulae. The latter are formal counterparts of empirical observations. These being inconsistent does not force inconsistent modalized, i.e., methodologically processed sentences in the system. Strong inconsistencies $\{\Box H, \Box\neg H\}$ would indicate that there is something wrong within the system, i.e., with the form of reasoning established by the system. In this case the consequence operation itself would seem to be defective. Jaśkowski ignored this path and so did his followers. Yet it seems that the observation presented will raise interesting new questions concerning a kind of *relative stability* of discussive logics with respect to (strong) inconsistencies.

6 Two points about language

From the perspective of applied logic, two points about language should be mentioned. The first one is that formal representation of the discourse shall be as true and adequate as possible. That, of course, is true for any form of applied logic, but it is especially important in this case. We have very small a margin, indeed, to manipulate deliberately the original discourse in the process of formalization. We should all time be aware that we are dealing with an extremely subtle and sensitive topic of logical processing. The strong traditional reservations with respect to the notion of inconsistency, of contradiction, complementarity etc. are always capable of subconsciously changing the mode of formalization so as to omit the bothersome peculiarities. There seems to be a natural tendency in logical proceeding: first clean up the area, then go ahead with formalization.

The second point pertains to a feature of natural language. As it was mentioned earlier in this paper, in German, and quite so in English and and any other language I know, we have a plenitude of words expressing all possible nuances of inconsistencies. Adequate formalization should mirror that variety.

Therefore, the construction of paraconsistent calculi should be flexible enough to allow definitions both of many variants of non-classical negation and of intentional conjunction.

7 A LOOK AHEAD

Let me end with a more general reflection. Clearly, my sympathy in paraconsistent logic lies at the side of weak paraconsistency. Weak paraconsistency is the version that seems to be acceptable also for, as Bryson Brown once put it (cf. [14], 490), conservative thinkers who are merely sensitive to our epistemic limits, and prepared to acknowledge the need to reason, from time to time, with premises that cannot all be true, without having them explode in a puff of logic. Indeed that's what I feel is right. Concerning strong paraconsistency I share the position of Cusanus. For him, there was no reconciliation between supraoppositional thought on the one side and discursive reasoning of formal logic on the other side. Real contradictions are intriguing and maybe they are an even indispensable element of our human condition. But they are not the right objects for human reason to investigate.

From such a perspective, the cause for preferring discussive logic as a promising paradigm in paraconsistency is not only because of its formal elegance and deep philosophical motivation. Furthermore, discussive logic represents an ideology that is, at least for me, most appropriate for para(in)consistency. To put it informally: at the very core of para(in)consistency lies not negation, but intentional conjunction. The ultimate application for para(in)consistency, I take it, is scientific discourse. For any logical formalization, the observer's perspective is the only appropriate one. Hence, when creating a paraconsistent logic we have typically to assume the position of somebody who keeps the minutes of scientific discussion. Why should we then insinuate that a person who claims a negated sentence has in mind some exotic kind of negation? Such an assumption could be justified only by referring to the overall consistency of the ongoing discussion. That motive, however, cannot be attributed to the person who made the claim. Writing down her claim in terms of non-classical negation thus means to systematically distort it. We therefore should better concentrate on intentional conjunction. With respect to inconsistency tolerating calculi, this connective seems to be the most important one – at least for those approaches which are directed towards application.

Let me repeat the *leitmotiv* of this talk by paraphrasing Barwise and Perry: logicians thoughtfully abhor contradictions, but they still love inconsistencies. Non-adjunctive calculi in Jaśkowski's tradition are an appropriate formal setting for such a romance.

References

[1] S. Jaśkowski. Rachunek zdań dla systemów dedukcyjnych sprzecznych. *Studia Societatis Scientiarum Torunensis*, 1(5):57–77, 1948.

[2] S. Jaśkowski. O konjunkcji dyskusyjnej w rachunkach zdań dla systemów dedukcyjnych sprzecznych. *Studia Societatis Scientiarum Torunensis*, 1(8):171–172, 1949.

[3] S. Jaśkowski. Propositional calculus for contradictory deductive systems. *Studia Logica*, 24:143–160, 1969.

[4] N. Rescher and R. Brandom. *The Logic of Inconsistency. A Study in Non-Standard Possible-World Semantics and Ontology.* Blackwell, Oxford, 1980.

[5] D. Lewis. Logic for equivocators. *Nous*, 16:431–441, 1982.

[6] G. Priest and R. Routley. *Paraconsistent Logic. Essays on the Inconsistent.* Philosophia, Munich, 1989.

[7] J. Kotas. The axiomatization of S. Jaśkowski's discussive system. *Studia Logica*, 33:195–200, 1974.

[8] T. Furmanowski. Remarks on discussive propositional calculus. *Studia Logica*, 34:39–43, 1975.

[9] J. Błaszczuk and W. Dziobiak. An axiomatization of M^k-counterparts for some modal calculi. *Reports on Mathematical Logic*, 6:3–6, 1976.

[10] J. Perzanowski. On M-fragments and L-fragments of normal propositional logics. *Reports on Mathematical Logic*, 5:63–72, 1976.

[11] N.C.A. da Costa and J. Kotas. *On some modal logical systems defined in connexion with Jaśkowski's problem*, volume Non-Classical Logics, Model Theory and Computability, pages 57–73. North-Holland, Amsterdam, 1977.

[12] S. Halldén. Results concerning the decision problem of Lewis' calculi S3 and S6. *Journal of Symbolic Logic*, 14:230–236, 1949.

[13] P. Steinacker. *Nichtklassische Negationen und Alternativen.* PhD thesis, Habilitationsschrift, Leipzig University, 1987.

[14] B. Brown. Yes, Virginia, there really are paraconsistent logics. *JPL*, 28:489–500, 1999.

Between Consistency and Paraconsistency: Perspectives from Evidence Logic

DON FAUST Department of Mathematics and Computer Science, Northern Michigan University, Marquette, Michigan 49855 USA
dfaust@nmu.edu

Abstract
Many application areas involve conflict and make a clarion call to logicians to provide formal frameworks for the efficacious representation and processing of such conflict. If this conflict is so dire that it involves the assertion of both φ and $\neg\varphi$ for some φ then paraconsistent logics may be helpful since they entertain $\varphi \wedge \neg\varphi$ without consequent trivialization. However, intermediate levels of conflict certainly often arise, for instance, when one has both confirmatory and refutatory evidence for φ, and Evidence Logic (EL) is an attempt to exemplify a reasonable framework for the representation and processing of such intermediate levels of conflict. Additionally, analytical studies of EL, just because EL is intermediate between classical and paraconsistent logics, can provide some insight into foundational aspects of paraconsistency. In this paper we will focus particularly on two of the most crucial aspects of logics of conflict, the concept of negation and its semantics. Using the framework of EL, we will suggest a logical place for 'negative facts', provide an interpretation and extension of Aristotle's concept of privation, and attempt clarification of the important distinction in Priest's dialetheism between rejecting φ and accepting $\neg\varphi$.

Introduction

Classical Logic (CL) provides a knowledge representation framework wherein predication is both only absolute and only confirmatory. On the other hand, our knowledge of the Real World (R), as well as, of course, the approximative models we build to help us represent that knowledge, are neither. Most often, indeed, our knowledge of R is *gradational* rather than absolute, and is ***both confirmatorily and refutatorily evidential*** rather than simply confirmatory.

Thus it is natural to look beyond CL to the construction of other logics which attempt to address the above disparity. Evidence Logic (EL) is such a logic [3-9]. In EL, for each s-ary predicate symbol P, $\mathrm{P}_c(t_1,\ldots,t_s) : e$ asserts that there is confirmatory evidence at the level e regarding $\mathrm{P}(t_1,\ldots,t_s)$ while $\mathrm{P}_r(t_1,\ldots,t_s) : e$ asserts that there is refutatory evidence at the level e regarding $\mathrm{P}(t_1,\ldots,t_s)$. Thus in EL a clear distinction is readily available between $\neg\, \mathrm{P}_c(t_1,\ldots,t_s) : e$ (absence of evidence) and $P_r(t_1, \ldots, t_s) : e$ (evidence of absence). So in EL there is an enriched framework for treating aspects of the concept of negation which are masked by the absolute and only confirmatory predications of CL: in fact, in this sense EL is seen to be conceptually antecedent to CL.

In many of the domains with which Artificial Intelligence (AI) finds itself routinely faced, knowledge is indeed both evidential and gradational. Further, even when the knowledge at hand is absolute, inferences drawn may be gradational, by necessity because a real time processing environment precludes the luxury of complete argumentation to an absolute conclusion, or by choice because absolute conclusions are not required for the triggering of appropriate action. EL is a tool for use in numerous such AI knowledge representation problems, and indeed a resolution-based implementation of EL has been constructed and utilized experimentally. In fact, [8] explores hierarchies of conflict toleration which may prove helpful in the construction of EL-based tools in AI.

In another direction, the direction we take in this paper, can the enriched explication of negation built into EL provide tools which shed some light on the long-standing problems in philosophical logic regarding ***the nature of negative facts***? EL semantics is, technically, a straightforward extension of CL semantics wherein, for each s-ary predication, an interpretation provides two sets of evidentially annotated s-tuples - one witnessing confirmatory predications and another witnessing refutatory predications. However, conceptually this semantics provides an enriched logical environment, in comparison to that of CL, wherein problematic aspects of negation may receive further illumination and possibly, in some cases, even resolution. In this paper we look at how this possibility plays out in particular with regard to the problem of negative facts.

Simple honesty forces the admission that we clearly don't yet know the ultimate nature of R. On the other hand, there is an ubiquitously witnessed functional level of constructed objects and predicates (like 'snow' and 'is white') wherein, in a rough-and-ready way, we proceed daily to construct efficacious representations, used in a variety of productive ways - making bricks, building houses, sending 'information' through wires, and so on. At this level we operate daily with rough but sufficiently effective evidence which confirms or refutes, and indeed often both confirms and refutes, as well as with absences of such evidence. And it is with regard to such absences of evidence that, we will argue, negation deals.

This view is derived from the more general perspective of "explorationism" [6]. On the basis of this point of view, we will revisit Aristotle's privatives and investigate their interpretation as refutatory predications in EL, carefully distinguishing them from negations of confirmatory predications, thus gaining further insight into the concept of negation itself. This will lead us to exploring an EL-based resolution of Russell's extensive ruminations regarding negative facts. What is attained, we will argue, is a reasonable logical place for negative facts which is free of any ontological commitments and which is both precise and useful.

In Section 1 we introduce EL and provide detail sufficient for the uses we will make of it here. In Section 2 we discuss our great ignorance of R, along with the approximative nature of our models of R and the extent to which these models are both tentative and only partial. In Section 3 we bring to bear both the tools of EL from Section 1 and the perspective of Section 2 in order to provide a suggestion for a logical place for negative facts. Finally, in Section 4 we make some observations about the path we have taken and cite some of what remains in question: in Section 4.1 we make a few remarks about the logical place we have suggested for negative facts; in Section 4.2 we suggest a way to interpret Aristotle's notion of privation in EL as refutatory predication, and we generalize his "square of opposition" to a "column of squares of opposition" in EL where predication is gradational; finally, in Section 4.3 we construct a framework, motivated by EL, which may help in clarifying the important distinction in Priest's dialetheism between rejecting φ and accepting $\neg\varphi$.

1 Evidence Logic (EL)

EL is a logic which goes beyond CL in providing machinery for the representation of both confirmatory and refutatory predication, further annotated with levels of evidential support. Precisely, for each $n > 1$ we define the Evidence Logic EL_n as follows. Let

$$\mathrm{E}_n = \{i/(n-1) : i = 1, \ldots, n-1\}$$

be the Evidence Space of size n - 1. The evidence spaces E_n are used in EL to provide measures of 'evidence levels'. In general, in applications, n shall have to be chosen sufficiently large to handle the granularity of the evidence while simultaneously the implementation environment will dictate an upper bound for n. For each s-ary predicate symbol P, and for any terms $t_1, \ldots, t_s$ and any e in E_n, EL_n contains atomic formulas

$$\mathrm{P}_c(t_1, \ldots, t_s) : e \quad \text{and} \quad \mathrm{P}_r(t_1, \ldots, t_s) : e$$

where the former asserts that there is evidence at level e confirming $\mathrm{P}(t_1, \ldots, t_s)$ while the latter asserts evidence at level e refuting $\mathrm{P}(t_1, \ldots, t_s)$. The only additional logical axioms of EL_n, beyond any usual set of logical axioms, are straightforward axioms asserting that any predication at level e entails all similar predications at levels e' where $e' < e$. Since n is finite in our setting, there are only a finite number of such additional axioms for languages with only a finite number of predicate symbols.

The semantics for EL_n is similarly straightforward. An s-ary predicate symbol P is interpreted in a structure $\mathcal{A} =< A, \ldots >$ by a pair $\mathrm{P}^{\mathcal{A}} = < \mathrm{P}_c^{\mathcal{A}}, \mathrm{P}_r^{\mathcal{A}}$¿ where each of $\mathrm{P}_c^{\mathcal{A}}$ and $\mathrm{P}_r^{\mathcal{A}}$ is a partial mapping from A^s to E_n.

In [9] the reader can find the formal definition of EL_n, its motivation, numerous examples, and theorems which characterize the structure of EL_n.

Previous work on EL has been abstracted. In [3] a characterization is given of the Boolean sentence algebras of EL_μ: the recursive isomorphism types depend upon the language similarity type μ, whether μ is monadic (stipulating p proposition symbols, k constant symbols, and u unary predicate symbols), functional (additionally stipulating exactly one unary function symbol), or undecidable (stipulating either at least two unary function symbols or at least one function or predicate symbol which is at least binary). In [4] results are given which begin to investigate how axiomatizable extensions of EL provide frameworks for conflict toleration. In [7] three hierarchies of conflict toleration are explicated, including the characterization of the Boolean sentence algebras at each level in each hierarchy.

Further, [5] explores how EL provides an explication of the concept of negation which extends that of classical logic. And [6] argues that EL may be preferable over classical logic as a "base logic" for science, while [9] explicates EL and provides proofs of the results abstracted in [3].

2 Epistemology and Ontology

Our ignorance of the Real World (R) remains great, and most would agree that it may well be a very long time before our ignorance is overcome. At the same time, we constantly deal well with uncertainty, constructing rough-and-ready objects and predicates which serve in the myriad circumstances of daily work and yet involve us constantly in weighing often conflicting confirmatory and refutatory evidence in regard to our predications. This perspective is independent of any philosophical committments, for example in the directions of realism, anti-realism or Machean sensationalism. As Ernst Mach said [10, p.2],

> "Colors, sounds, temperatures, pressures, spaces, times, and so forth, are connected with one another in manifold ways; and with them are associated dispositions of mind, feelings, and volitions. Out of this fabric, that which is relatively more fixed and permanent stands prominently forth, engraves itself on the memory, and expresses itself in language. Relatively greater permanency is exhibited, first, by certain complexes of colors, sounds, pressures, and so forth, functionally connected in time and space, which therefore receive names, and are called bodies. Absolutely permanent such complexes are not."

There is indeed a palpable "scientific humility" expressed, in my view, time and again in Mach's writings, a humility which is instructive, which helps us to put in perspective the constructions of science and technology. His writings reflect, in my view, that he felt deeply the tentative and partial nature of our models of R, their approximative and 'rough' nature, in spite of their undeniable efficacy both in increasing our scientific understanding of R and in building technology-based stuff that works reasonably well in R.

And of course this perspective is not something from the time of Mach which no longer applies: certainly we have made some progress since then, and we might even say we are doing well indeed, but our ignorance remains great. Some might even say, and I would agree, that part of the progress of our century has been a further clarification of this ignorance.

In any case, our ignorance is indeed great and our models of R are both tentative and partial. And from this perspective CL provides a woefully inadequate framework, absolute and only confirmatory, while the models we build are obviously neither. Our constructed ontologies are 'rough and ready' approximations, and our epistemologies are rife with evidence which is both confirmatory and refutatory, and often conflicting since it is both, evidence which yields at most degrees of certainty well below the level of absolute certainty. This situation, if we face it squarely, makes a clarion call we should heed.

3 EL AND 'THE FACTS'

So, given the above perspective, let us propose EL as a reasonable replacement for CL. At the basis is the real world R, about which our ignorance is great. Above R lie the myriad constructed approximative models of R, each full of constructed objects and predicates all only 'rough and ready' in the spirit of Mach. It is the predications realized in any such model that constitute the 'facts', the so-to-speak 'positive facts' in that model. And it is the absence of a positive fact in a model which, in any case in which such absence obtains, witnesses a negative fact: that's what a negative fact is.

This is the logical place, in the perspective we have constructed here, of negative facts. To wit, the negative predication $\neg\, P_c : .7$ holds in a structure $\mathcal{A}$ if the negative fact 'the absence of $P_c : .7$' holds in $\mathcal{A}$.

4 OBSERVATIONS

4.1 Negative facts

I suppose one feeling one might have, after reading this proposal, is that "it all does so little", and that is certainly right. But to some extent it is the long-standing yet unwarranted perspective that we are really talking about R in some absolute sense that supports the belief that more can be done. The fact seems to be that our ignorance is still great indeed, and we are not at all yet talking about R in any absolute sense, in which case the above solution is all we can construct for now.

Also, it might be noted that the above perspective fits exactly with the current implementation of semantics in mathematical logic. In such structures a fact corresponds to a certain s-tuple being in a certain set of s-tuples, and a negative fact corresponds to the absence of a certain s-tuple from a certain set of s-tuples.

Indeed, one gets the feeling, in reading Aristotle in his struggles with privatives and how they relate to negations and in reading Russell in his constructions and refinements (and responses to Demos and many others) addressing the problem of negative facts, that all was fine whenever it was clear that only a theoretical construction, devoid of any absolute ontological or absolute epistemological committments with respect to R, was being attempted. It was only when something absolute about the nature of R was seeming to be asserted, trying to be wrung out of the analysis, that perplexity and confusion was generated.

It would seem, at least for the present and indeed possibly for a long time yet to come, we really can't say anything absolute about R. And if that's true, then we need to honestly face our ignorance, with a humility Mach so epitomized, and be satisfied with the approximative modeling we are doing and with the forthright interpretation of 'negative facts' entailed in such modeling.

4.2 Privatives

If we interpret Aristotle's privation as a refutatory predication, then an interpretation of his theory of negation and privation can be found in an axiomatizable extension of EL, in AL as defined and characterized in [8].

For example, let Gx be the predication "x is green". Consider first the case of interpretation in EL_2 (EL with Evidence Space $E_2 = \{1\}$, so we omit annotation of the evidence value since it is always 1). Interpreting "x is green" with G_cx and "x is non-green" with G_rx, Aristotle's opinion (Ch. XLVI of [2]) that "x is non-green implies it is not the case that x is green" yields an axiomatizable extension AL_2 of EL_2 wherein his square of opposition (Ch. 10 of [1])

Gx	*contrary*	non-Gx
↓		↓
¬ non-Gx	*subcontrary*	¬ Gx

becomes in EL_2 the square of opposition

G_cx	*contrary*	G_rx
↓		↓
$\neg\ G_rx$	*subcontrary*	$\neg\ G_cx$

This generalizes easily to EL_n (with Evidence Space $E_n = \{\varepsilon = \frac{1}{n-1}, \frac{2}{n-1}, \ldots, \frac{n-2}{n-1}, 1\}$), wherein Aristotle's opinion is generalized with a hierarchy of Aristotelian logics $AL_n(d)$ allowing no conflict of "degree" d or greater, for each $d = 1, \ldots, n$. In this generalization Aristotle's axiom that $G_rx \to \neg\ G_cx$ becomes, for conflict level d, the axiom $G_rx : d\varepsilon \to \neg\ G_cx : d\varepsilon$. The "square" of opposition likewise generalizes to the "column" of opposition with generalized Aristotelian squares of opposition at each of the $n - d$ levels beginning at level d and continuing up through the top level of the column, as illustrated in the diagram $AL_{n,<0>}(d)$ on the following page.

In fact, as will be argued in a forthcoming paper, it seems reasonable to remove from contemporary elementary logic all reference to privation: it is superfluous to Classical Logic. Alternatively, as indicated above, one could use EL in elementary logic in which case privation survives, and indeed plays a useful role, namely as refutatory predication.

4.3 Dialetheic negation

An excellent discussion of dialetheic negation occurs in [12], especially pp. 20-22, 36-38, and also in [11]. In a dialetheic framework so much is possible, including true contradictions, that a clear statement of something impossible for dialetheists (in particular, Graham Priest) may help in understanding their concept of negation. For Priest [12, p. 36] and also [11], the joint acceptance and rejection of φ is impossible. Elaborating, while rejecting φ implies accepting $\neg\varphi$ (for dialetheists who buy into the lack of gaps), accepting $\neg\varphi$ does not imply rejecting φ (certainly not: for if so, there would be no gluts).

Let P_a assert that 'P is accepted' and P_r assert that 'P is rejected'. Then the above dialetheic position is in fact that P_a and P_r are contrary in the sense that they cannot both be the case. Hence, we apparently have the following "dialetheic square of opposition":

Dialetheic Square 0

$$\begin{array}{ccc} P_a & \text{contrary} & P_r \\ \downarrow & & \downarrow \\ \neg P_r & \text{subcontrary} & \neg P_a \end{array}$$

where the arrows here denote material implication.

$AL_{n,<0>}(d)$

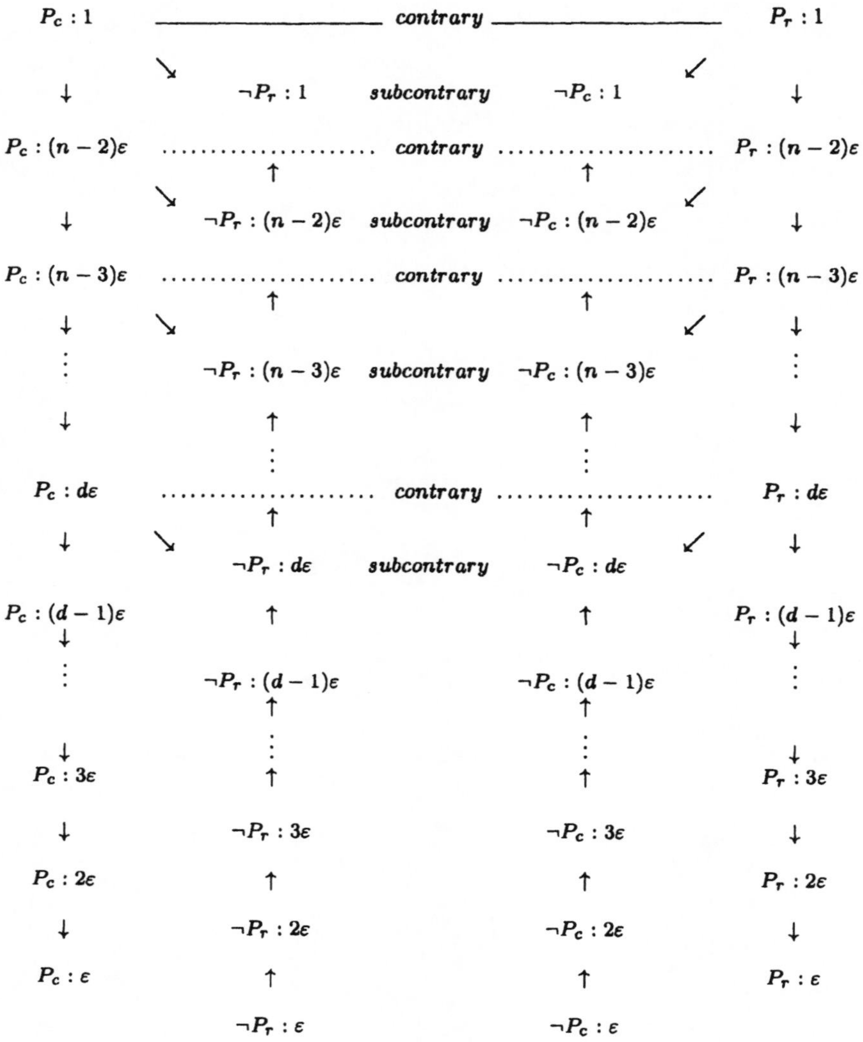

$P_c : 1$		———————	*contrary*	———————		$P_r : 1$
	↘				↙	
↓		$\neg P_r : 1$	*subcontrary*	$\neg P_c : 1$		↓
$P_c : (n-2)\varepsilon$		…………………	*contrary*	…………………		$P_r : (n-2)\varepsilon$
		↑		↑		
	↘				↙	
↓		$\neg P_r : (n-2)\varepsilon$	*subcontrary*	$\neg P_c : (n-2)\varepsilon$		↓
$P_c : (n-3)\varepsilon$		…………………	*contrary*	…………………		$P_r : (n-3)\varepsilon$
		↑		↑		
↓	↘				↙	↓
⋮		$\neg P_r : (n-3)\varepsilon$	*subcontrary*	$\neg P_c : (n-3)\varepsilon$		⋮
↓		↑		↑		↓
		⋮		⋮		
$P_c : d\varepsilon$		…………………	*contrary*	…………………		$P_r : d\varepsilon$
		↑		↑		
↓	↘				↙	↓
		$\neg P_r : d\varepsilon$	*subcontrary*	$\neg P_c : d\varepsilon$		
$P_c : (d-1)\varepsilon$		↑		↑		$P_r : (d-1)\varepsilon$
↓						↓
⋮		$\neg P_r : (d-1)\varepsilon$		$\neg P_c : (d-1)\varepsilon$		⋮
		↑		↑		
		⋮		⋮		
↓						↓
$P_c : 3\varepsilon$		↑		↑		$P_r : 3\varepsilon$
↓		$\neg P_r : 3\varepsilon$		$\neg P_c : 3\varepsilon$		↓
$P_c : 2\varepsilon$		↑		↑		$P_r : 2\varepsilon$
↓		$\neg P_r : 2\varepsilon$		$\neg P_c : 2\varepsilon$		↓
$P_c : \varepsilon$		↑		↑		$P_r : \varepsilon$
		$\neg P_r : \varepsilon$		$\neg P_c : \varepsilon$		

The dth Aristotelian Logic, with Evidence Space $E_n = \{\varepsilon, 2\varepsilon, \ldots, (n-2)\varepsilon, 1\}$ (where $\varepsilon = \frac{1}{n-1}$) and one proposition symbol P, with axiom $\neg\ (P_c : d\varepsilon \wedge P_r : d\varepsilon)$.

But observe: there's ambiguity lurking here. In fact, our use of the predications P_a and P_r, motivated by the confirmatory and refutatory predications of Evidence Logic, have helped to bring this ambiguity to the surface. What, indeed, is the predication $\neg P_a$? Is this $\neg(P_a)$, the absence of 'P is accepted'? Or is it $(\neg P)_a$, the acceptance of 'the absence of P'? Does it matter? And of course the analogous questions regarding $\neg P_r$ need to be raised as well.

Clearly Dialetheic Square 0 is ambiguous. First, following Priest's sentence (*) on p. 36 of [12] in a literal manner, and the arguments of both Priest and Smiley in regard to it, yields a substantial alteration,

Dialetheic Square 1

$$\begin{array}{ccc} P_a & \text{contrary} & P_r \\ \uparrow & & \downarrow \\ (\neg P)_r & \text{contrary} & (\neg P)_a \end{array}$$

wherein Priest goes along with the implications shown but understandably argues against their converses while Smiley argues for both the implications shown and their converses, and the contrarity is agreed to by both.

On the other hand, the alternative interpretation, while possibly a less faithful representation of the Priest-Smiley debate, yields the disambiguation

Dialetheic Square 2

$$\begin{array}{ccc} P_a & \text{contrary} & P_r \\ \downarrow & & \downarrow \\ \neg(P_r) & \text{subcontrary} & \neg(P_a) \end{array}$$

wherein both the implications and the subcontrarity are seemingly uncontroversial.

Although further analysis is clearly called for in our quest to better understand the foundational issues of negation raised in dialetheism, our analysis so far clearly shows that the dialetheic position, whether or not it is one we wish to advocate in some way, is a perfectly reasonable one. So although Smiley's critique of dialetheism [12] is penetrating, and indeed is helpful in clarifying substantial issues relating to negation and dialetheism, no problem of unreasonableness is here uncovered in dialetheism. Having a true contradiction, say $P \wedge \neg P$, is indeed not made impossible by a dialetheic position which maintains that it is impossible that P is both accepted and rejected, impossible that both P_a and P_r.

Of course, one might suggest that EL is susceptible to the same sort of ambiguity. But certainly that is not the case. For in EL the confirmatory and refutatory predications occur only at the atomic level, so in EL the questions above do not arise. In EL there is only the level of representation and enquiry in regard to $\neg(P_c)$ versus P_r, and indeed it is an important feature of EL that it clearly distinguishes the former, the absence of confirmatory evidence for P, from the latter, the presence of refutatory evidence for P.

Clearly there is room for further progress here. And such further clarification of the general concept of negation will surely help us better understand the deep and intricate ways negation is involved in foundational aspects of dialetheism.

References

[1] Aristotle. *On Interpretation*. The Loeb Classical Library, 1938.

[2] Aristotle. *Prior Analytics*. The Loeb Classical Library, 1938.

[3] D. Faust. The concept of evidence (abstract). *Journal of Symbolic Logic*, Vol. 59, 1994, pp. 347-348.

[4] D. Faust. Evidence and negation (abstract). *Bulletin of Symbolic Logic*, Vol. 3, 1997, p. 364.

[5] D. Faust. The concept of negation. *Logic and Logical Philosophy*, Vol. 5, 1997, pp. 35-48.

[6] D. Faust. Conflict without contradiction: noncontradiction as a Scientific *modus operandi*. *Proceedings of World Congress of Philosophy*, 1998, published at www.bu.edu/wcp.

[7] D. Faust. Evidence Logic (abstract). *Bulletin of Symbolic Logic*, Vol. 4, 1998, p. 86.

[8] D. Faust. Conflict without contradiction: paraconsistency and axiomatizable conflict toleration hierarchies in Evidence Logic. submitted.

[9] D. Faust. The concept of evidence. *International Journal of Intelligent Systems*, Vol. 15, 2000, pp. 477-493.

[10] E. Mach. *The Analysis of Sensations*. Dover Publications, 1959.

[11] G. Priest. What Not? A defense of the dialetheic theory of negation. In *Negation*, Kluwer, 2000.

[12] G. Priest and T. Smiley. Can contradictions be true? *Proc. of Aristotelian Society*, Suppl. Vol. 67, 1993, pp. 17-54.

Kinds of Inconsistency *†

GREGORY R. WHEELER Department of Philosophy, University of Rochester, Rochester, NY 14627, U.S.A.
wheeler@philosophy.rochester.edu

Abstract
Typically you find paraconsistency motivated by one of two types of arguments: either an ontological argument is given whereby it is claimed that inconsistent objects or events of some kind demand a logic able to reason about such items, or an epistemological argument is provided that denies that language or the world is infected with inconsistency but claims instead that the problem is all in a reasoning agent's head. Approaches to designing paraconsistent logics have thus far tended to heed this distinction. Stronger, dialetheic logics are called on to handle contradictory objects or events in a manner that does not trivialize inference, while they are out of favor with epistemological approaches which favor a weaker treatment that preserves as much of classical logic as possible. This latter approach is typically scene as one of introducing a set of techniques for how to manipulate an agent's inconsistent yet rational corpus to preserve a familiar form of logical consequence. This essay argues against the practice of linking epistemological concerns with weak paraconsistent logics. It introduces an epistemological motivation for adopting a strong paraconsistent logic by considering the results of measuring physical objects and how we might go about reasoning with these results. Measurement behaves in a way as to be a source of inconsistency that is neither best understood as a problem between agents nor the result of either a paradoxical property of language or the world.

1 Introduction

There are distinct kinds of inconsistency. My interest is to introduce a kind that occurs in languages whose domains include physical objects. In other words, I am interested in a kind of inconsistency that is fundamental to talk about domains like our very own –that is, domains consisting of things we observe, measure, and about which we reason.

The claim is not that there are many kinds of inconsistent theories. This claim can be easily substantiated by citing a pair of axiomatic systems or corpora of beliefs that, when represented in a classical first-order language, each entail a sentence

*This research was supported in part by grant SES 990-6128 from the National Science Foundation and generous support from the National Research Council of Brazil.

†Thanks to Eva Cadavid, Henry Kyburg, Graham Priest, Catherine McKeen, Gabriel Uzquiano and WCP2000 referees for their comments.

satisfying the schema $\ulcorner \phi \wedge \neg\phi \urcorner$. For instance, naïve set theory and the book of *Genesis* are at least *functionally* different kinds of theories: one and only one of the pair is a mathematical theory. But both are inconsistent. Whatever interest there is for this kind of classification, it will not be ours.

Instead, our interest concerns a source of inconsistency that is peculiar to physical theories. The resulting kind of inconsistency is existent in both folk and scientific theories, but its structure is better revealed in the empirical sciences. The source of the inconsistency is a probabilistic acceptance rule that is fundamental to experimental methods and measurement procedures common in scientific practice.[1] This paper raises the specter of a distinct 'kind' of inconsistency and presents a pair of probabilistically motivated examples to serve two ends. The first objective is to put to an end the practice of identifying certain philosophical motivations for paraconsistency with particular classes of paraconsistent logics. The tendency to view epistemological issues as solely treatable by 'weak' paraconsistent logics, for instance, is a mistake that impedes progress both in philosophy and logic. The second goal is of greater importance: an understanding of scientific inference. Paraconsistent logicians have often turned to the sciences for examples to apply paraconsistent systems. Below I introduce a structure within scientific inference that is both commonplace and ready made for paraconsistent logic.

2 Revising Theories, Revising Logic

Why concern ourselves with an inconsistent theory of *any* kind? After all, ours is an intellectual tradition that treats inconsistency as a pox rather than a provision. W. V. O. Quine [29, p. 81] refers to so-called logics that tolerate inconsistencies as a "change of subject" rather than logic at all. As for inconsistent theories, Karl Popper's well known polemics against the very idea rest on what he takes to be the fundamental requirement for theoryhood:

> The requirement of consistency plays a special role among the various requirements which a theoretical system, or an axiomatic system, must satisfy. It can be regarded as the first of the requirements to be satisfied by *every* theoretical system, be it empirical or non-empirical. [24, p. 91f.]

So, an *accepted* inconsistent theory is either by Quine's lights a non-logical program or, by Popper's, not something deserving of the name 'theory' at all. Pox indeed.

And the cure? Revision. A contradiction deduced from a set of axioms tells against that set: at least one of the axioms must be revised in order to restore consistency. How to go about this, though, is a matter of considerable controversy. Quine, for one, considers all statements fair game. Theories are partially-ordered sets of statements –a.k.a a 'web of belief'– whose ordering corresponds to each statement's proximity to observation. In most cases it will turn out cheaper and easier to revise the statements closer to the observational end of this scheme, but

[1] Some recent work in the philosophy of science and computer science on probabilistic acceptance rules and the distinction between an acceptance view and so-called subjective probability views are Deborah Mayo [22] and Choh Man Teng [33], the latter being technical work that builds upon Kyburg [15]. See also Kyburg [17] for a defense of acceptance and replies in the same volume.

nothing precludes tinkering with the axioms of a theory to restore consistency. Popper, on the other hand, accepts a sharp analytic/synthetic distinction, holds axioms of logic and mathematics fixed and certain, and demands severe testing of all 'corroborated' empirical statements. He does so while coaching us not to *accept* any non-mathematical or non-logical truth, no matter how rigorously tested. Despite considerable disagreement surrounding how to characterize revision, and even disagreement over how to regard theoretical structures in general, the discussion about revision is carried out with the assumption that inconsistency is intolerably problematic. The question, tradition tells us, is not *whether* to revise but *how.*[2]

Enter paraconsistent logic. We might view paraconsistent logic, and paraconsistency in general, as a proposal to make 'whether to revise' a reasonable question. Rather than view the problem of revising an inconsistent theory as simply one of setting a theory straight, paraconsistency suggests that we consider the costs involved in doing so. Determining just what those expenses are, however, has divided the field. On the one hand we find the reform-minded dialetheists. These logicians recommend adopting a new logic to accommodate *bona-fide* inconsistent objects. Their favored strategy is to argue for cases where revising an inconsistent theory would be an outright mistake. On the other hand we find neo-classicists who take logic and the world to be consistent, even if rather complicated. The motivation behind this movement is the recognition that fallible agents often reason with useful but imperfect knowledge bases. Given these circumstances, it may prove cognitively expensive to revise an inconsistent theory or the choices available at a particular time may take us farther away from the truth.

These two motivations for paraconsistency have settled into a peculiar form of partisanship.[3] Epistemic motivations are thought to be tied to conservative, neo-classical approaches to paraconsistency, while ontological concerns are thought to demand bolder, dialetheic approaches. This habit is unfortunate. One problem with conflating weak paraconsistent logics with epistemology, for instance, is an oversimplification of the epistemic issues at hand. Consider the probabilistic examples cited in the epistemic case for paraconsistency. These examples most often share the same underlying structure of Henry Kyburg's lottery paradox [15], [18]. An inconsistency can be generated by an n-member collection of accepted statements such that the chance of mistakenly rejecting each statement of the collection is no greater than ϵ, but that the chance of mistakenly rejecting the n-member conjunction when joined by adjunction is greater than ϵ. Most proposed solutions to the lottery paradox turn on some non-adjunctive strategy to contain the loss of probability mass that occurs by using the rule of adjunction. But, as we shall see, there is a problem which has

[2]The philosophical problem of how to revise an inconsistent empirical theory has its first explicit formulation in Pierre Duhem's *The Aim and Structure of Physical Theory*. For recent work on Duhem's problem, or the problem of how to revise a theory given disconfirming observational evidence that is contrary to a theory's predicted value for that observation, see Worrall [35], Mayo [22] and [23], Howson [11], and Wheeler [34]. For theory revision in computer science, see Alchourrón et. al. [1], the collection edited by Gärdenfors [10] and Kyburg and Teng [19].

[3]Exceptions hold, of course. Two to note are Brandon and Rescher's logic of inconsistency [5] and the recent debate concerning logical pluralism [30]. Brandon and Rescher's inconsistent logic is an ontologically motivated project which nevertheless retains classical logical consequence. They wind up handling inconsistent objects by introducing a semantics for inconsistent worlds instead. The second exception is to how reform-minded a dialethist must be. Logical pluralists argue that accepting a paraconsistent logic needn't entail the rejection of classical logic *per se*. See Priest [27] for a reply to Restall and Beall.

its roots in probabilistic acceptance rules yet is overlooked by this approach. The problem's structure suggests a stronger paraconsistent treatment, one which would otherwise be dismissed out of hand. I return to this discussion in the next section.

What is interesting about paraconsistency and the family of logics going by the same name is the possibility of using them to understand how what David Israel [12] has called "real, honest-to-goodness inference" works. Necessary to understanding such inference practices is an understanding of how working theories are revised with accepted but fallible evidence. For this latter problem, scientific inference presents an ideal test case. Assume that revisions to physical theories are due neither to capricious Quinean fancy nor merely to a trial-and-error Popperian hypothesis testing, but are instead evidence-driven events. If we understand scientific inference as an evidence driven enterprise, what we must first come to terms with is that the domain of such reasoning isn't the world *simpliciter*, but a model of the world. In the best case, this model is built in part from experimental observations and measurement practices that each come along with known error probabilities. Discussing paraconsistency in light of scientific inference raises a third question to add to the pair cited above. In addition to the questions of *how* to revise a theory and *whether* to revise a theory, we might add to the discussion the question of exactly *what* is to be revised. I turn to this question next.

3 What what?

What are the objects with which science reasons? Furthermore, what role, if any, should paraconsistent logic play in understanding scientific reasoning? Let's begin with the second question. A review of the paraconsistency literature turns up two examples from science that are often proposed as candidates for paraconsistent treatment. The first is the so-called dual nature of quantum physical states.[4] But this is a problem that has little to do with scientific inference, at least as far as paraconsistency is concerned, since Quantum mechanics is typically given a statistical interpretation, such as Born's. The idea here is to concede that Quantum theory doesn't predict what the measured value of X will be but, instead, provides a probability distribution over a set of possible measured values. The kind of inference involved then is statistical, not deductive. Hence, worries about paraconsistency are beside the point. The second example is the problem of reconciling the mathematical assumptions of quantum theory with those of relativity theory.[5]

[4]The textbooks tell us that the physical state of any isolated system behaves deterministically in accordance with Schrödinger's equation until a measurement of some physical magnitude (*e.g.*, position, energy, spin) is made. A pre-measured physical state, modeled by a state-vector Φ, is not a linear combination of vectors $\langle \phi_1, \phi_2, ..., \phi_n \rangle$ that represent particular values $\langle x_1, x_2, ..., x_n \rangle$: $x \in \mathbb{R}$ of the physical magnitude X in question, but rather measurement of some other observable not commensurate with X. Thus, $\Phi = a_1\phi_1 + a_2\phi_2 + ... + a_n\phi_n$, where each a_i is a complex scalar, many of which in fact are, typically, non-zero. The result then is that Φ does not have a definite value X. But on measurement of X, the state Φ changes, or "collapses", taking one of the measured values $x_1, x_2, ..., x_n$ of X. See Krips [14] for discussion.

[5]Relativity theory concerns macroscopic objects and treats collision as a primitive, or would if one assumes that the colliding particles are infinitely small. Quantum theory, in turn, concerns microscopic objects and cannot be said to treat collision as primitive since collision need not be isolated in space-time. This conflict may be traced to the mathematics underpinning of each theory: Relativity theory relies on a 4-dimensional Riemann space, whereas Quantum theory operates on

This problem, however, is primarily the problem of how best to unify two distinct yet incompatible *theories*. The problem of unification should be distinguished from the problem of revision. A serious answer to the unification problem is just to give up on the largely philosophical project of unifying science.[6] Notice that whatever the merits of this proposal, a similar reply to the problem of theory revision isn't an option. The revision problem has it that there is recalcitrant evidence within the domain described by a theory. Giving up on squaring theory with evidence is simply giving up doing science.

If we wish to understand scientific inference we'll need to focus on evidence. The temperatures of solutions, tensile strengths of compounds, and the masses of slabs function as evidential bedrock for evaluating scientific hypotheses and, in turn, scientific theories. Since the bulk of scientific reasoning concerns measured magnitudes of physical objects or physical events, it's here that we should look for a connection with paraconsistent logic.

Consider how the magnitude *length* is measured. The measurement of length is paradigmatic of all fundamental additive measurements in the physical sciences, such as mass, velocity, and the like. Imagine that a pole S is exactly 4 meters long. What, on the basis of measurement, does one conclude: that the length of S is 4 meters? No, not exactly. If careful, one concludes that the length of S is plus-or-minus d units of r, where r is the recorded value of the measurement. If measured carefully, the true value, 4, will very likely fall within the interval $r-d<r<r+d$, where $d \ll r$. The justification for this conclusion is the belief that it is simply incredible, given our knowledge of the methods and technology used to measure S, that the actual value which r is intended to represent (in this case, 4) does in fact fall outside the interval $r-d<r<r+d$. In standard practice we set the value of d only as high as our needs demand and as low as our instruments warrant to insure our incredulity. Once all this is in place we may *accept* that S is $r \pm d$ units long and then go about using or manipulating S, confident (although not certain) that its actual length is within specification.

Turning now to reasoning, consider a simple theory whose object language contains the sentence

(P) '$\texttt{Length\_of}(S)$ $\texttt{is}$ 3.95 ± 0.10 $\texttt{meters}$',

where S is the name of a flagpole, and r takes the value 3.95 and d, 0.10. If P is accepted, that is if P is assigned the value 1, we assume that there is a class of models which satisfy this sentence. Classically speaking we say that P is true in just the models making up this class, false otherwise. Standard declarative semantics has it that the actual length of S is an element we are firmly in command of; setting aside differences over how to represent S's 4-meter-longishness, S's *being* 4 meters long is generally taken to be the truth-maker for models of P.[7] But underpinning this abstraction is an assumption that, perhaps, we now are in a position to improve upon. The operative assumption behind explaining what counts as classical models of P is the assumption that physical properties and relations are readily accessible.

---

an infinite-dimensional Hilbert space.

[6]The philosopher Nancy Cartwright and physicist Richard Feynman are notable critics of the unification of science thesis.

[7]Or, S being r meters long is taken as a truth-maker for P when $r \in [3.85, 4.05]$.

Poles may be tall and not tall, but they aren't both 4 meters and not.[8] Likewise, the proposition expressed by (P) is either true or false. Yet the one thing conspicuously missing from any realistic method of ***discovering*** the truth value of statements like (P) is S's actual length. If S's length were directly accessible, we wouldn't need to measure S and science would be considerably easier than it in fact is. Instead, what stands between us and S's length is a fallible measurement procedure. So when considering arguments involving evidence statements like (P), rather than working with the propositions expressed by such statements we have instead just a collection of accepted statements –statements that give a reliable report about the physical state in question. So, while scientists reason ***about*** the length of a pole, very often they reason ***with*** accepted statements about the length of a pole. This collection of accepted statements is simply the brute data that can be shared or generated by researchers. Let's call this collection of statements a ***domain of discourse***.[9] In the best of circumstances we know quite a bit about how frequently such reports are mistaken and how such mistakes occur. Returning to poles, this knowledge is simply the know-how of measuring poles with a tape.[10]

Falling into error simply amounts to accepting a false report within the domain of discourse. A false report is just an accepted statement that either records a value for r (on the basis of a reliable measurement procedure) when in fact the value of r is outside this interval, or fails to record a value for r as falling within the acceptance interval that is in fact within that interval. Though erroneous, an accepted but false report is warranted. No report ***enters*** the domain of discourse by mistake. A reader troubled by the uncertainty introduced by measurement procedures might recommend that to decrease the chances of falling into error, simply increase the value for d. Increasing the value for d would increase the margin of error and thus, assuming that the recorded values for r are normally distributed, would reduce the number of false reports appearing in the domain of discourse. Though mathematically sound, this would be bad advice to follow. Remember that the final goal isn't error elimination but finding the length of poles. Our interest in the truth presses us to accept a minimal value for d whose associated frequency of error is known. The goal of finding the true length of poles is compromised by increasing the interval to stamp out error. Besides, error is eliminated completely only at the price of triviality.[11]

What is interesting about a domain of discourse –that is, a set of statements interpreted as accepted reports rather than expressing true propositions– is that it can be inconsistent. Consider just what good measurement practices deliver:

[8] In other words, the issue of vagueness is orthogonal to the topic at hand.

[9] I prefer 'domain of discourse' to 'data reports' or Deborah Mayo's 'models of data' only because these latter terms leave open the question of just where the statements stand with respect to data analysis –namely, pre- or post-analysis. I intend 'domain of discourse' to concern statements about pre-processed, raw observational data. 'Observation statement' might do as well so long as it is not taken to denote beliefs, a tempting move for those charmed by subjective interpretations of probability. At any rate, data analysis turns out to be one of the important tools needed to do revision –namely the one that does most of the clean-up work on some initial set of statements.

[10] There is an important point about measurement that needs to be brought out here. The choice of technology is irrelevant. A tape and the frustrations of using one are hopefully common enough to most readers to grasp the general point: measurements don't provide direct, certain access to physical properties.

[11] At extremes one could say that the length of r is a real number. But that tells you only what a magnitude is and not, say, where to cut the stock.

reliable reports about *bona fide* magnitudes. These reports are approximations at best, which is the reason they are interval-valued. However, these interval-valued reports are also fallible. This is the point behind accepted statements: a measurement may report of S that $r \in [x, y]$ when in fact $r \notin [x, y]$. It is important to remember that this situation describes what I prefer to call an error, not an inconsistency. The interesting case is when we happen to encode such errors by measurement, that is when we have an acceptable report of S that $r \in [x, y]$ and an acceptable report of S that $r \notin [x, y]$. In this case we have a pair of statements that ascribe inconsistent properties to S, namely that the value r is and is not in some interval x,y.

This last point is obscured by common, epistemological arguments for paraconsistency. An example should make this point clear. Consider a set of poles, each thought equal in length, yet nevertheless the cause of controversy when installed outside the UN. Though accepted as a set of poles of roughly equal length, seeing a nation's flag atop one of the new poles flying noticeably below those atop the others has the makings for a small scandal. Likewise, sawing boards and assembling bookcases presents a similar, perhaps even familiar, problem. The reader can well imagine a situation where someone accepts that each of the boards cut is a certain length but is flummoxed to learn that they don't all fit together to make a bookcase. If we're interested in representing such problems formally, the argument typically goes, then we are discussing a situation that is, even if temporarily, inconsistent.

That it is easy to see around these two problems has more to do with how the stories are told than about how the agents in each story found their troubles. The poles and planks aren't themselves in and out of specification, they are just out of specification. Furthermore, we know they are out of specification. In fact, the stories work the way they do only because we know so. What is important to consider is that we haven't always the benefit of knowing this last piece of information: when faced with an irregularity, it is a luxury to know immediately its source. In fact, science is simply the art of reasoning without this luxury.

Ignoring this point, it is tempting to think that all epistemological troubles rest with individual agents. Neo-classicists think so. The epistemological problem seems to them to be one epitomized by joining the craftsman's corpus (i.e., in particular, that the poles are equal in length) with the assembler's (i.e., they aren't). The structure of the problem, this view has it, lends itself to solution either by tossing out the obviously incorrect belief or, in tough cases, partitioning the joined sets into mutually exclusive but self-consistent sets (See Brown and Schotch [6], and Jennings and Schotch [13]).

But notice that this approach isn't as compelling when we aren't provided with the privileged perspective of knowing whom is to blame. As I've told these stories, we know that the poles and planks aren't all in spec and that their not being so is the key to solving the problems raised in each tale. But suppose now we develop the stories, adding to either one the presence of irregular mounting brackets, warped stock, optical illusions, paranoid diplomats–take your pick. Notice that changing a story in such a manner amounts to changing genre: rather than a morality tale, we have now a mystery. This is a critical point. In such cases we don't know what the cause of the problem is but only that there is one. The methods used to land us in such a fix –those which, at root, rely on a host of measurements reporting on relations and properties of the objects in play– are precisely those we must rely on to get us out.

The question then is how best to represent the problem of what to do in the face of wayward evidence. Following the neo-classical line of treating problems like this as essentially involving disagreement between reasoning agents passes over the state of affairs *any* agent confronts when reasoning about physical objects. All of us start with the same kind of evidence –a more-or-less reliable, inter-subjective representation of the properties and relations of the physical objects we're interested in manipulating. But this evidence can be –and as a practical matter *is*– inconsistent. This isn't to say that the world is inconsistent. Nor either is it to say that the problem is always one of mediating a dispute between agents, each of whom provides self-consistent accounts of some shared event. The problem, in short, is a dispute between us and the world: at our best we only get it almost right, nearly all of the time.

Once we recognize that domains of discourse about physical objects are likely inconsistent, then a proposal to partition the language into consistent cells misses the point. Of course it is preferable to work with a consistent set of statements. Hence, it is desirable to work toward removing errors from a domain of discourse. But what is important to recognize is that this work does not precede making that domain available for reasoning. In practice it may not even be possible to completely remove errors from the domain of discourse. And, even if it were, we still would need to deploy some kind of reasoning regime to clean up this sad set of sentences. So, if a paraconsistent logic is simply a logic whose consequence operation is nontrivial when it closes an inconsistent set of statements, then the set of sentences modeled true by the kind of statements described here presents itself as a prime candidate for a paraconsistent logic. Furthermore, since the source of inconsistency is not inter-agent but systemic –it is simply a side-effect of how we interact with our entire surroundings, not simply with each other– it is a mistake to imagine partitioning schemes to be the best approach to handling a collection of such sets of statements.

The idea of an inconsistent domain of discourse is likely to meet strong resistance. It might be thought, for instance, that the advice to measure twice and cut once holds at least the promise of eliminating error and, hence, a reduction of the problem under discussion to one already covered by neo-classicists. Notice that what this suggestion to measure twice/cut once amounts to is simply to run an experiment. With multiple measurement 'trials' we can expect to catch the very kind of errors under discussion. Unless one is a systematic incompetent, one could discover false reports by repeating the measurement procedure and tossing stray values out. The hope is, then, that we can dismiss this talk of accepted statements and holdout for honest-to-goodness ***propositions***. Unfortunately, while it is true that you can reduce the occurrence of error with this approach, you can't eliminate it.

To see why this is so, consider a standard experimental practice found in sciences as disparate as psychology, chemistry, and medicine. In each science, experiments are designed to test a null 'no-effect' hypothesis, h_0, by choosing a region of rejection within a well-defined sample space. If evidence lies in this region of rejection, then h_0 is rejected. The region is selected such that if the appropriate experimental assumptions of randomness, independence and their kin hold, then there is only a small chance, ϵ, that given the supposition that h_0 is true, evidence falling in the rejection region will be collected. Another way to put it is to say that if h_0

is true, the probability of mistakenly rejecting it is less than the specified value of ϵ, where ϵ is made as small as one likes. Put in practice, we sample, check that the experimental assumptions hold, and then, should the sample obtained fall in the rejection region, we reject the null h_0 which states that the controlled variable has no effect. Note that the rejection of h_0 isn't hedged, but full-out; for instance, we *reject* the hypothesis that cigarette smoking has no effect on cancer rates in mice and men. The *grounds* for rejecting h_0 rest on the statistical –and *ipso facto* uncertain– claim that there is only a small, preassigned chance that we shall do so mistakenly.

Notice that what we've spelled out are the very same rules that measurement follows.[12] The proposal to buy certainty at the cost of taking additional measurements fails because our best experimental methods are themselves fallible. There are three points to notice about this result. First, error cannot be eliminated but only, even under the best of circumstances, controlled. So, for those interested in understanding scientific inference, holding out for honest-to-goodness propositions is an exercise in wishful thinking.[13] Second, controlling error is expensive. We get the best (but not certain!) results when we are keenly aware of what kind of errors we're liable to commit and design experiments or conduct measurements in a manner that reduces those risks. Not knowing all the ways one can go wrong contributes, in part, to the difficulty of the problem of revision: we're constantly discovering new ways to err. What can make a case a tough one is figuring out whether one has stumbled upon a new way of bungling or is in the position of needing to reject part of a well confirmed theory. Finally, a comment on domains of discourse. What is compelling about accepted statements for recording the results of measurement and experiment alike is that they reflect the uncertain but likely results of a probabilistic acceptance rule that lies at the heart of these seemingly disparate activities. A domain of discourse so described is very reliable; it just isn't certain. Nearly all of what the reports state about the world is actually true. Taken together, these statements give a well confirmed yet likely inconsistent representation of the world. This structure is what science reasons with when turning to the world for physical evidence. So, any language capturing such a structure will likely contain double-talk that isn't a function of a particular agent's limited knowledge-base. The revision problem thus breaks down to at least two problems: the problem involving a noisy world and how to correct our immediate representations of it, and then the problem of squaring this imperfect representation with a given theory. Given the nature of the evidence used to revise a theory, a strong paraconsistent logic may present itself as part of a successful approach.

The moral to draw about revision is this. Tradition has it that the scientific enterprise comes full-stop when confronted with an inconsistency, which is clearly not the case. Some defenders of paraconsistency have suggested that there are cases where revising an inconsistent theory is mistaken because there are actual inconsistent objects–a controversial position, indeed. The main alternative to this view is that the best application of paraconsistency is to resolve disagreements

[12] Applied to measurement, we would propose some factor as responsible for getting an unreliable measurement. So in our example we might say that the null hypothesis is that having me measure S is *not* positively correlated with recording readings outside of the margin of error using a tape rule of kind K. Then we test to see whether or not my performance *is* outstanding and, if so, reject the null.

[13] We could chase certainty by starting a regimen of infinite trials, of course.

between agents. What the proposal in this essay amounts to is to recognize that the basic relations and properties of physical objects are the bedrock of scientific reasoning and that, *qua* objects of reasoning, they are best represented by accepted statements that are the result of a reliable measurement procedure. Taken together these statements make for a well-confirmed yet likely-inconsistent representation of the actual physical environment. Our world isn't inconsistent; just our evidence is. So while we strive to correct errors and mistakes when we can, we must face that this amounts to an ongoing project and is not one which we can pass off to inductive logicians while we wait for them to deliver *bona fide* properties and relations.

4 Conclusion

Hume put the matter this way: while we may know what a bad egg *is*, nothing *looks* so alike as two eggs. Traditionally, logicians think about domains containing bad eggs and good ones and reasoning about each, scoring arguments on truth preservation. But when we stop to worry about how to train this practice on our physical environment, Hume's comment becomes wise counsel. We haven't the luxury to crack every egg for inspection. This is what makes science so difficult.

What is desired is an account of how agents reason successfully about physical objects in worlds like ours with limitations very much like those we face. From a logician's point of view, we'd especially like to know the preferred way of doing this. Keeping this in mind, it may turn out that the ideal should not be constrained by representations in classical, first-order languages. I've suggested a reason here why it indeed shouldn't.[14]

Finally, a comment about honest-to-goodness inference. Another intriguing parallel between the study outlined here and 'real' inference concerns perception. Perception may be viewed as simply a measurement procedure. Within the field of computer vision, object recognition is treated as a stochastic process [32], and recent trends in computer speech recognition are toward modeling language understanding with probabilistic models [2],[21]. The point is that for perfectly epistemic reasons we may well have to work with formal representations that ascribe inconsistent properties to objects we wish to reason about. Moreover, the repair work on this domain may occur at any point during its useful life-span, not prior to its becoming available for reasoning.

On the view being developed, a well-behaved, classical, model-theoretic semantics is an afterthought. It's a tool brought to bear on a localized problem where the spread of error can be safely ignored. When dealing with empirical objects, one cannot ignore the threat of error. Indeed, the very best experimental methods catalogue the ways to err and the chances an agent stands of falling prey to each. Our logic should take this into account.

[14] Graham Priest has commented that even if one grants all that I've argued for, an issue remains whether there are in fact inconsistent objects. Hence, there remains the dispute which I labored at the outset to set aside. However, I understand the best dialetheic arguments only to establish the possibility of inconsistent objects. Establishing that there are actual inconsistent objects involves adopting a logic in which such a question could be posed. The dialetheist strategy, I mentioned, is to find objects to justify the logic. On the view proposed here, this has matters turned around. So, while the objects in this essay don't satisfy the dialethiests, they are a good reason to adopt the very kind of logic which is necessary to settle such a question.

REFERENCES

[1] Alchourrón, C. E.; Gärdenfors, P.; Makinson, D. 1985. "On the logic of theory change: Partial meet contraction and revision functions", *Journal of Symbolic Logic*, 50:510-530.

[2] Allen, J. 1995. *Natural Language Understanding, 2nd ed.* Redwood City, CA: Benjamin Cummings Publishing.

[3] Benferhat, S; Dubois, D.; Prade, H. 1997. "Some Syntactic Approaches to the Handling of Inconsistent Knowledge Bases: A Comparative Study", *Stud. Logic* 58:17-45.

[4] Boole, G. 1847. *The Mathematical Analysis of Logic.* Cambridge; Reprinted 1948. Oxford: Oxford Press.

[5] Brandon, R. and Rescher, N. 1979. *The Logic of Inconsistency: a study in non-standard possible-world semantics and ontology.* Totowa, NJ: Rowman and Littlefield.

[6] Brown, B. and Schotch, P. 1999. "Logic and Aggregation", *Journal of Philosophical Logic* 28:265-287.

[7] da Costa, N. C. A. 1982. "The Philosophical Import of Paraconsistent Logic", *Journal of Non-Classical Logic* 1:1-19.

[8] da Costa, N. C. A. 1997. "Overclassical Logic", *Logique et Analyse* 40(157):31-44.

[9] da Costa, N. C. A.; Bueno, O.; French, S. 1998. "The Logic of Pragmatic Truth", *Journal of Philosophical Logic* 27:603-620.

[10] Gärdenfors, P. [ed.] 1992. *Belief Revision.* Cambridge: Cambridge University Press.

[11] Howson, C. 1997. "A Logic of Induction", *Philosophy of Science* 64:268-290.

[12] Israel, D. 1980. "What's Wrong with Non-Monotonic Logic?", *Proceedings of the First Congress on Artificial Intelligence, Stanford University, August 18-21, 1980.* Palo Alto, CA: AAAI Press.

[13] Jennings, R. and Schotch, P. 1989. "On Detonating", appearing in Priest, G., and Routley, R. [ed.] *Paraconsistent Logic: Essays on the Inconsistent.* Hamden [Conn.]: Philosophia.

[14] Krips, H. 1999. "Measurement in Quantum Theory", *Stanford Encyclopedia of Philosophy.* <http://plato.stanford.edu/entries/qt-measurement/>.

[15] Kyburg, H. E., Jr. 1961. *Probability and the Logic of Rational Belief.* Middletown, CT: Wesleyan.

[16] Kyburg, H. E., Jr. 1984. *Theory and Measurement.* Cambridge: Cambridge University Press.

[17] Kyburg, H. E., Jr. 1994. "Believing on the Basis of Evidence", *Computational Intelligence* 10:3-22.

[18] Kyburg, H. E., Jr. 1997. "The Rule of Adjunction and Reasonable Inference", *Journal of Philosophy* 94:109-125.

[19] Kyburg, H. E., Jr., and Teng, C. 2000. *Uncertain Inference*. Forthcoming, Cambridge: Cambridge University Press.

[20] Laudan, L. 1997. "How about Bust? Factoring Explanatory Power Back into Theory Evaluation", *Philosophy of Science* 64:306-316.

[21] Manning, C. and Shütze, H. 1999. *Foundations of Statistical Natural Language Processing*. Cambridge: MIT Press.

[22] Mayo, D. 1996. *Error Statistics and the Growth of Experimental Knowledge*. Chicago: University of Chicago Press.

[23] Mayo, D. 1997. "Severe Tests, Arguing from Error, and Methodological Underdetermination", *Philosophical Studies* 86:243-66.

[24] Popper, K. 1959. *The Logic of Scientific Discovery*. New York: Routledge.

[25] Priest, G. 1979. "The Logic of Paradox", *Journal of Philosophical Logic* 8:219-241.

[26] Priest, G., and Routley, R. 1989. *Paraconsistent Logic: Essays on the Inconsistent*. Hamden [Conn.]: Philosophia.

[27] Priest, G. 1999. "Logic: One or Many?" Unpublished Manuscript.

[28] Quine, W. V. O. 1969. *Ontological Relativity and Other Essays*. New York: Columbia University Press.

[29] Quine, W. V. O. 1986. *Philosophy of Logic*. 2nd ed. Cambridge: Harvard Press.

[30] Restall, G. and Beall, J. C. 1999. "Logical Pluralism", Unpublished Manuscript.

[31] Routley, R., Meyer, R., Plumwood, V., and Brady, R. 1982. *Relevant logics and their rivals 1*. Atascadero, CA: Ridgeview.

[32] Sonka, M., Hlavac, V., and Boyle, R. 1999. *Image Processing, Analysis, and Machine Vision*. 2nd. ed. Pacific Grove, CA: Brooks/Cole Publishing.

[33] Teng, C. 1998. *Non-monotonic inference: characterization and combination*. Ph.D. thesis. Department of Computer Science, University of Rochester.

[34] Wheeler, G. 2000. "Error Statistics and Duhem's Problem", *Philosophy of Science* 67(3):410-420.

[35] Worrall, J. 1993. "Falsification, Rationality, and the Duhem Problem", appearing in *Philosophical Problems of the Internal and External Worlds: Essays on the Philosophy of Adolf Grünbaum*, edited by J. Earman, A. Janis, G. Massey, and N. Rescher. Pittsburg: University of Pittsburg Press.

Paraconsistent Logic
vs.
Meinongian Logic

JACEK PAŚNICZEK Maria Curie-Skłodowska University, Lublin/Poland
jpasnicz@bacon.umcs.lublin.pl

Abstract

The aim of this paper is to confront paraconsistent and Meinongian logics, where the latter is understood as a result of formalization of Meinong's theory of objects. Both logics share the liberal view of inconsistency rejecting the explosive character of contradiction. The two logics have, however, different philosophical backgrounds and are differently developed as formal systems. Developing general tolerance for various kinds of inconsistency (ontological, epistemological, methodological, etc.) is the primary goal of paraconsistent logics which deal with them mostly on the propositional level. Meinongian logics explain formal features of intentional objects, in particular, they account for contradictions encapsulated in these objects. In other words, these logics deal with inconsistency within the subject-predicate structure. Thus, it is argued in this paper that although paraconsistent and Meinongian logics need not reducible to each other, one can trace various mutual inspirations of the philosophical and logical character.

INTRODUCTION

Meinong is known particularly for his view that objects not only do not need to exist but they may even possess contradictory properties. It is then often suggested that in order to render Meinong's theory of objects consistent one should apply to it some kind of paraconsistent logic. However, such a formalization has never been carried out and so one may wonder whether such an undertaking is feasible at all. On the other hand, there have been proposed several formalizations of Meinong's ontological views which make no reference to paraconsistent logic of any kind.

The aim of this paper is to investigate theoretical connections between paraconsistent logic and Meinongian logic. By the term "Meinongian logic" I mean here the logic underlying Meinong's theory of objects.[1] It should however be stressed that there can be different interpretations of Meinong's views which give rise to different Meinongian logics. By "Meinongian logics" I mean also some recently developed formal theories which have no ambition to be adequate logical analyses or reconstructions of Meinong's ontological views but which are at least closely influenced by these views.[2] In this paper we will focus on Meinongian logic developed by Paśniczek as perhaps the simplest one, easy for presentation, and such that it accommodates basic principles of Meinong's ontology.

1 SOME PRINCIPLES OF MEINONG'S THEORY OF OBJECTS AND PROBLEMS WITH ITS FORMALISATION

Meinong's theory of objects is a highly sophisticated ontological theory. I am not going to present it here in details but I shall deal only with those points which are relevant to the topic of this paper. What is crucial in Meinong's ontology and what makes it distinctive among others theories is the concept of the ontological structure of objects. Objects may be constituted of any class of properties. In particular, they can be impossible (or: inconsistent; from now on I will rather use this latter, more technical term) and they can be incomplete. An object is inconsistent if it possesses contradictory properties - for example an object may possess the property of being round and the property of being square; we might then call that object *the round square*. According to a weaker definition an object a is inconsistent if for some property P, a is P and a is not P. An object is incomplete if for some property P it is neither P nor not P (incompleteness and inconsistency are here understood as concepts complementary to completeness and consistency respectively). Within Meinong's ontology the concept of completeness is in fact philosophically more important than that of consistency.[3] From a formal point of view however, these two concepts are independent of each other.[4]

Any formalization of the Meinongian ontology aims to render consistently and non-trivially impossible and incomplete objects. The only way to formulate inconsistency or incompleteness in the language of classical predicate logic is clearly the following: $\exists_P(Pa \wedge \neg Pa)$ and $\exists_P(\neg Pa \wedge \neg\neg Pa)$ (notice that the fact 'a is not P' is expressed by $\neg Pa$). This means that these two notions are rendered by equivalent inconsistent formulas. Of course, in any logic with classical negation the concepts of consistency and completeness - when expressed on the propositional level - coincide.

What is the source of Meinong's bizarre inconsistent and incomplete? Their being (or subsistence) is secured by the principle, called sometimes after Meinong *Annahmen Thesis*: for every class of properties there is an object which possesses those properties; we then say that the properties constitute the object. This principle can be formalised in a sufficiently rich language in the following way:[5]

(1) $\exists_a \forall_P(Pa \equiv A)$, where A is a formula not containing a.

A stronger version of *Annahmen Thesis* states that there is *exactly one* object constituted by a given class of properties. This uniqueness can be secured by:

(2) $\forall_P(Pa \equiv Pb) \supset a{=}b$

It should be stressed that 'is' in *Annahmen Thesis* expresses subsistence, not existence. This also means that an object need not exist in order to possess properties, in other words, predication is prior to existence. According to *Annahmen Thesis* there are incomplete and inconsistent objects.[6]

Needless to say these two principles do not exhaust the contents of Meinong's ontology. Although a more detailed presentation of this ontology goes beyond the scope of this paper , one may mention an intriguing principle which asserts two kinds of properties-internal and external ones-that can be predicated of objects. This dichotomy allows to avoid the objection raised by Russell against Meinong's theory of objects: ***the existent golden mountain*** possesses internally the property of existence while lacking externally this property, i.e. the existent golden mountain actually does not exist. Since the issue of inconsistent Meinongian objects is basically connected with internal predication, our further discussion of Meinongian logic is exclusively confined to this kind of predication.

2 CAN MEINONG'S THEORY OF OBJECTS BE ACCOMMODATED WITHIN CLASSICAL OR PARACONSISTENT LOGIC?

Meinongian objects can hardly be represented by any definite description within the classical first-order logic. In particular, inconsistent objects - when expressed by definite descriptions - possess no properties at all and, consequently, they are indistinguishable from each other. Neither can they be represented by Hilbert's operator or any other operator definable within the classical first-order logic.[7] However, the primary reason for rejecting the definite description operator and other operator representations of Meinongian objects is that those objects can be constituted of many properties and there is no ontological ground to assume that those properties are jointly equivalent to their conjunction. Moreover, there seems to be no way to cope by means of operators with objects constituted of infinitely many properties (and such objects play an important role in Meinong's ontology). Of course, the situation is different if we have a sufficiently rich second order logic. In that case, an object defined by $\exists_a \forall_P (Pa \equiv A)$ can be represented by the description $(\iota x)(\forall_P (Px \equiv A))$. One can check that the following equivalence is true: $P(\iota x)(\forall_P (Px \equiv A)) \equiv A$.

In the case of paraconsistent logic, the formula $A \wedge \neg A$ need not be inconsistent.[8] So we may express the fact that a Meinongian object a is impossible, i.e. that for some P, $Pa \wedge \neg Pa$, without falling into contradiction. On the ground of paraconsistent logic, from $Pa \wedge \neg Pa$ it does not follow that a possesses all properties. So far so good. However, this approach does not work for incomplete objects. As we pointed out above, in the classical language incompleteness is expressed by the same, or by a logically equivalent formula, as inconsistency, and this fact makes Meinong's distinction trivial. The concepts of consistency and completeness would become independent on the propositional level only when either the classical interdefinability of conjunction and disjunction or the law of double negation is suspended (depending on whether we define these two concepts as: $\forall_P \neg(Pa \wedge \neg Pa)$ and $\forall_P (Pa \vee \neg Pa)$ respectively or as: $\forall_P (\neg\neg Pa \vee \neg Pa)$ and $\forall_P (Pa \vee \neg Pa)$ respectively).

The fact that the formula $A \vee \neg A$ is a theorem in such pradigmatic systems of paraconsistent logic as that of da Costa's creates even worse problems for interpreting incomplete objects.[9] The consequence of this fact is that for any object a, $Pa \vee \neg Pa$ is a theorem. This in turn means that every object is complete.

Thus, the non-trivial character of Meinong's concepts of consistency and completeness can be maintained only at the cost of a profound reformulation of classical logic. Such reformulation must go further than do paraconsistent systems and may lead to unintuitive results.

There is an opinion that in developing many-valued logics Łukasiewicz was at least to some extent inspired by Meinong's views. In his logic the law of excluded middle and the law of non-contradiction do not hold and this is a welcome feature when incomplete and inconsistent objects are to be allowed in an ontological system. However, in Łukasiewicz's logic the two formulas: $A \vee \neg A$ and $\neg(A \wedge \neg A)$ are equivalent and this fact makes again the concepts of completeness and consistency trivial. It is also worth mentioning that according to commentators Meinong himself would presumably reject many-valued logics as a possible formal framework for his theory of objects.

Thus, we may then conclude that any attempt to render on the propositional level basic principles and concepts of Meinong's ontology is not very promising, if not hopeless. Even if we found a non-trivial formal strategy to secure inconsistent and incomplete objects, it would be tantamount to accepting a very weak non-classical logic with negation much different from the classical one (e.g. minimal Hilbert's logic). One should then rather expect that any Meinongian logic would express basic ontological notions on the predicative level and, generally, that any deduction on the basis of that logic would involve this level essentially.

3 MEINONGIAN LOGICS

In most of Meinongian logics the problem of inconsistent and incomplete objects is approached by introducing a distinction between ordinary propositional negation and predicate negation and by assuming that these two negations do not coincide. Consequently, a "negated" property P is to be associated with a property *non-P*. Let us express the predication "a possesses P" in an order of letters reverse to the order used in the language of classical logic: aP instead of Pa. We can then refer to predicate negation by the same symbol as we refer to the propositional negation. Thus, the formula $\neg aP$ expresses the fact that a does not possess P, while the formula $a\neg P$ expresses the fact that a possesses the property *non-P*. The concepts of completeness and consistency of objects may then be formally defined as follows:

Df1 a is consistent iff $\forall_P \neg(aP \wedge a\neg P)$
Df2 a is complete iff $\forall_P (aP \vee a\neg P)$

The concepts of consistency and completeness are dual in the following technical sense: if we replace 'a' by '$\neg a \neg$' then the consistency turns into completeness and *vice versa.*[10]

The approach to Meinongian objects based on predicate negation is certainly not uncontroversial. There is no clear evidence that predicate negation appearing in natural language is not reducible to the propositional negation. So the reasons for distinguishing these kinds of negation can only be theoretical, i.e. philosophical or logical. The straightforward consequence of introducing the negation of properties is that Meinongian logics can be two-valued. As a matter of fact most of those logics are based on classical logic.[11] In the simplest case we can assume that predicate negation is a primitive notion, not explicable by means of any other notions. The most severe Meinongian logic results then simply from classical logic by equipping the latter with predicate negation. Most often it is supposed that Meinongian logic is a second-order one, i.e. at least properties are to be quantified over in order to express consistency and completeness.[12]

The concept of predicate negation can be further explained by linking it to the concept of existence. The relation between P and *non-P* is then semantically characterized relatively to the domain D of existing (or possibly existing) objects which is a subdomain of the class of all Meinongian objects: the extension of *non-P* is the complement of the extension of P in D. Conspicuously, every existing object or every object that can exist - *an individual* - turns out to be consistent and complete although not every consistent and complete object is existing. It should however be noticed that the concept of negation of properties alone is not strong enough to distinguish the category of individuals from other Meinongian objects. In other words, given a consistent and complete class of properties, there can be more than one possible individual exemplifying all these properties. This quite obvious fact was overlooked by Meinong himself as well as by contemporary ontologists. Let us consider the object constituted - according to the *Annahmen* Thesis - by all properties which are negations of properties possessed by Clinton. That *non*-Clinton is obviously a consistent and complete object but it cannot be identified with any possible individual.

Although the negation of properties is an important concept for Meinongian theory of objects other kinds of complex properties are also needed to make the theory sufficiently general. Roughly speaking, we need complex properties to distinguish various categories of Meinongian objects and to define property entailment. Anyway, it seems to be quite a natural step. In particular we may introduce the conjunction of properties. Assuming that the extension of conjunction of some properties is the intersection of their extensions we notice that individuals must be closed under the conjunction of properties.

However, if complex properties are introduced the problem arises: how is simple predication logically related to predication involving complex properties? As we noticed earlier no entailment between the predicate and propositional negations is allowed. None can entail the other, for if it were the case then either inconsistent or incomplete objects would be expelled from the Meinongian ontology. The case of conjunction and other connectives that can build complex propositions and properties is however different. One can then ask whether an object possesses a conjunction of properties if and only if it possesses both conjuncts (or whether one of these predications entails the other); and in particular, whether $a(P\wedge\neg P)$ is equivalent to $aP\wedge\neg aP$.

It should be remarked that philosophers dealing with Meinongian theories of objects most often neglect the problem of complex properties and of entailment between those properties and simple ones. They even claim, for the sake of generality, that there is no such entailment. Undoubtedly, this 'generalization' makes the logical approach to Meinongian logic more trivial and controversial.[13] For if there were no such entailment then how can we explain for instance a possible fact that an object possesses a property P but not the property $P\wedge P$ or $\neg\neg P$ (and conversely)?

From the logical point of view, one may expect that the most straightforward and promising way of introducing complex properties consists in adopting usual definition axiom for properties:

(3) $\exists_P\forall_a(aP\equiv A)$, where A is a formula not containing P

Unfortunately, (3) is too strong - it is inconsistent with (2) (it's enough to notice that (2) and (3) implies respectively that on the one hand there are more objects than properties, on the other, there are more properties than objects). There is, however, a weaker way of introducing complex properties. Let us assume that they are given by a Boolean algebra: $\langle \boldsymbol{P},\cap,\cup,*,\mathbf{0},\mathbf{1}\rangle$, where $\boldsymbol{P}$ is a set of properties, $\cap$, $\cup$ are distributive lattice operations of join and meet respectively and * the operation of complement, 0 and 1 are the minimum and maximum elements respectively. This modelling is also treated as a semantics for a language in which we talk about complex properties by means of predicate variables and connectives (that language can be successively enriched). Let I be an interpretation function assigning to P, Q, etc. elements of $\boldsymbol{P}$ and assigning to complex properties their Boolean counterparts: $I(\neg P)=I(P)^*$, $I(P\wedge Q)=I(P)\cap I(Q)$, $I(P\vee Q)=I(P)\cup I(Q)$. According to the basic principles of Meinong's ontology, let I assign to object variables $a,b,\ldots$ subsets of the power set of $\boldsymbol{P}$, i.e. sets of properties. We say that these sets constitute respective objects. From now on let P,Q represent also simple or complex predicate expression. We then adopt the following truth condition:

(*) the proposition aP is true if and only if $I(P)\in I(a)$.

Perhaps the most natural requirement concerning complex predication is the following:

E if $I(P)=I(Q)$ then for every object a: $aP \equiv aQ$ holds,

and, of course, E is fulfilled in the proposed semantics. For example, an object a possesses a property P or a complex property $(\neg P\wedge Q)$ iff it possesses the property $\neg\neg P$ or $\neg(P\vee\neg Q)$ respectively. Obviously, this principle imposes the extensional view of properties, i.e. objects conform the Boolean structure of properties with all the bad consequences of this fact to which will turn back later in this paper.

Now suppose we assume additionally:

$\mathbf{D}^{\supset}$ $a(P\wedge Q) \supset aP\wedge aQ$

or/and:

$\mathbf{D}_{\supset}$ $aP\wedge aQ \supset a(P\wedge Q)$

We cannot accept both since we would have: $aP\wedge a\neg P \equiv a(P\wedge\neg P) \supset aQ$ which is unacceptable for obvious reasons. So we should choose between $\mathbf{D}^{\supset}$ and $\mathbf{D}_{\supset}$ and in both cases it leads to a non-classical understanding of conjunction, at least with respect to properties. Accepting $\mathbf{D}^{\supset}$ only (without accepting $\mathbf{D}_{\supset}$) can be associated with the so-called non-adjunctive approach.[14] Inconsistent objects may then be non-trivially interpreted only as possessing complementary properties separately: $aP\wedge a\neg P$. $a(P\wedge\neg P) \supset aQ$ remains valid according to this approach as being entailed by **E** and $\mathbf{D}^{\supset}$. The two principles are equivalent to:

C if $I(P)$ P $I(Q)$, where P is the Boolean order, then for every object a: $aP \supset aQ$ holds.

This principle expresses the closeness of objects under classical logic entailment and therefore it may be welcome for the theory of Meinongian objects.[15] For dealing with inconsistencies $\mathbf{D}_{\supset}$ looks apparently more promising than does $\mathbf{D}^{\supset}$ since neither $aP\wedge a\neg P$ nor $a(P\wedge\neg P)$ entails aQ for arbitrary Q. However, $\mathbf{D}_{\supset}$ is too weak in its deductive power and it does not guarantee the closeness of objects. On the other hand, the logic based on **E** and $\mathbf{D}^{\supset}$ turns out to be quite reasonable foundation for developing a deductively non-trivial logic which would be able to accommodate the essentials of Meinong's ontology. This logic can even be extended in such a way as to cover more content of the Boolean algebra of properties. Thus we may introduce quantifiers as categorematic expressions representing quantifier objects which can be considered as Meinongian entities. The universal quantifier will be interpreted as $\{\mathbf{1}\}$ and the existential quantifier as P-$\{\mathbf{0}\}$.[16] For a more developed version of Meinongian logic based on the principle **C** see the Appendix.

The content of the Meinongian ontology far exceeds the content of principles we have here sketched. And usually logicians aim to formalize other aspects of Meinong's theory of objects - those aspects are certainly philosophically more fascinating but they are irrelevant for the topic of this paper.[17] It should however be mentioned that formal Meinongian theories have to accommodate a wider class of complex properties than that here introduced[18] and for this and other reasons they are usually very sophisticated higher-order systems.

4 THE LOGIC OF NON-STANDARD POSSIBLE WORLDS

The concept of non-standard possible worlds was introduced by Nicholas Rescher and Robert Brandom almost twenty years ago as the generalization of the concept of "standard" possible worlds. A non-standard possible world (later: *N*-world) in contrast to a standard possible world, can be inconsistent and incomplete. An *N*-world is inconsistent if for some propositions *A*, both *A* and ¬*A* obtain in that world; an *N*-world is incomplete if for some proposition *A*, neither *A* nor ¬*A* obtains in that world.[19]

There is a close affinity between Meinongian logic sketched above and the logic of non-standard possible worlds which is considered by many as a paraconsistent logic. Roughly speaking, a non-standard possible world can be identified with a class of propositions obtaining in it and every class of propositions represents a non-standard possible world. Consider the following analogy:

Meinongian object	————	class of properties
non-standard possible world	——	class of propositions

We are now going to indicate a way which may lead from Meinongian logic to a logic which might be considered as paraconsistent. That is, we would like to indicate a path reverse to the usual ones leading from paraconsistent to Meinongian logics. Roughly, Meinongian objects can be identified with the classes of properties possessed by those objects. Analogously, worlds (or states of affairs) can be identified with sets of propositions obtaining in them. Let us assume that propositions are defined by means of the Boolean algebra $\mathcal{A}=\langle \mathbf{A},\wedge,\vee,\neg,\mathbf{0},\mathbf{1}\rangle$, where $\mathbf{A}$ is now a set of propositions and ∧, ∨, ¬ are interpreted as conjunction, disjunction, and complement operations respectively. *N*-worlds can be understood as represented by any set of propositions, i.e. they can be understood as subsets of $\mathbf{A}$ (excluding $\varnothing$ and $\{\varnothing\}$). If *a* is an atom of $\mathcal{A}$ then $\{a\}$ can be interpreted as a classical possible world and if this algebra is atomic and complete then all *N*-worlds are 'built out' of possible worlds.[20] Let α represent *N*-worlds and *A, B* - propositions, and let *I* be a function defining this semantic representation. The formula αA is then to be read as follows: in an *N*-world α it is the case that *A*, or: *A* obtains in an *N*-world α. Now we assume that:

(**) (a) αA is true if and only if $I(A)\in I(\alpha)$ or, (b) αA is true if and only if $I(B)\in I(\alpha)$ where $I(B)\leq I(A)$.

The logic of *N*-worlds may be conceived as an extension of the classical modal **S5**. Let us assume that □ and ◊ are now interpreted as the ***universal*** *N*-world: $I(\Box)=\{\mathbf{1}\}$ and as the ***existential*** *N*-world: $I(\Diamond)=\mathbf{A}-\{\mathbf{0}\}$. It is easy to show that this interpretation preserves the ordinary sense of modal operators.

The logic of *N*-worlds based on (**)(b), *N*-logic, is axiomatized by the following axioms and rules of inference (*N*-system).

N1 Truth functional tautologies
N2 $\Box(A\supset B)\supset(\alpha A\supset\alpha B)$
N3 $\Box A\supset A$
N4 $\neg\Box\neg\alpha A\supset\Box\alpha A$
N5 $\Diamond\neg A\supset\neg\Box A$
MP if $\vdash_N A\supset B$ and $\vdash_N A$ then $\vdash_N B$
NG if $\vdash_N A$ then $\vdash_N \alpha A$ and $\vdash_N \neg\alpha\neg A$

The formulas: αA and $\alpha\neg A$ may of course be simultaneously true or simultaneously false in N-semantics, i.e. a N-world associated with α may be inconsistent or incomplete, and these two formulas together do not entail αB for arbitrary B (the formula $\alpha A\wedge\alpha\neg A\supset\alpha B$ is not a law of N-logic). Also, according to NG, $\neg\alpha\neg(A\vee\neg A)$ is a thesis and consequently $\alpha(A\wedge\neg A)$ is inconsistent on the basis of N-logic. It means that $\alpha(A\wedge\neg A)$ cannot be true in N-semantics, i.e. contradictory propositions cannot obtain in any N-world.[21] Conspicuously, we may treat N-logic as a logic of generalized modal operators for their syntactic status is exactly the same as the status of $\Box$ and $\Diamond$, and their semantic interpretation could be just the sets of propositions. We can even introduce Jaśkowski's discussive implication (and other connectives) with relativisation to every modal operator: $A\supset_{\alpha}B =_{df} \alpha A\supset B$. Thus, generally, N-logic can be viewed as accommodating a non-adjunctive approach to inconsistency. The question is however, whether this approach is paraconsistent in spirit. At least some thinkers would certainly say a negative answer to this question.[22]

5 CONCLUDING REMARKS

Paraconsistent logics and Meinongian logics cannot be reduced to each other in a straightforward way and, as we have argued, there is no theoretical reason for attempting such a reduction. The two stem from different philosophical grounds. The only thing they share is tolerance for inconsistency. But in Meinongian logic there is no need to treat inconsistency on the propositional level since it is located within objects. On the other hand, paraconsistent logics are not ready to tolerate incompleteness since it would unnecessarily weaken their deductive power.

Yet, we may still ask whether these two kinds of logic can gain anything from each other. We demonstrated the path from Meinongian logic to N-logic. However, the status of the latter logic as a paraconsistent logic is a little bit controversial. We also showed that Meinongian logic, as it stands, must cope with inconsistent properties and predications. Imposing Boolean structure on properties - and this would be tantamount to treating complex properties in accordance with classical logic - is far from satisfactory. Let us notice that according to E for every P and Q, and every object a: $a(P\wedge\neg P)\equiv a(Q\wedge\neg Q)$ and if we additionally assume $D^{\supset}$ or C then we have: $a(P\wedge\neg P)\supset aQ$. Certainly this is not a welcome result even for Meinongians. So they may look for a logic different than classical which could provide the foundation for the ontology of properties. In particular, one may attempt to replace the Boolean algebra by a weaker structure, e.g. De Morgan's lattice.

APPENDIX: A SIMPLE MEINONGIAN LOGIC

M-language

The alphabet for M-language consists of the same symbols as the alphabet of classical logic, i.e.: (1) sentential connectives: $\neg$, $\supset$ (the other sentential connectives are introduced by means of the usual definitions); (2) the universal quantifier symbol: $\forall$; (3) the identity symbol: $=$; (4) individual variables: $x_1, x_2,...$; (5) constants: $a_1, a_2,...$; (6) predicate symbols: $P_1, P_2,...$; (7) brackets: (,).

Let us assume that metavariables s,t range over constants and quantifiers, i.e.

symbols listed in (2) and (5); $x, y, y_1, y_2, \ldots$ over variables. The grammar of M-language is defined as follows: (a) every expression of the form $Py_1 \ldots y_n$ and $x=y$ is a formula; (b) if A, B are formulas, then $\neg A$ and $(A \supset B)$ are formulas; (c) if A is a formula then xA is a *predicate*; (d) if Π is a predicate then $t\Pi$ is a formula.[23] Notice that the expression of the form xA is used instead of commonly adopted but more complicated λ-notation $[\lambda xA]$. Formulas $a\neg P$, $a(P \wedge Q)$ have in *M*-language the following forms: $ax\neg Px$, $ax(Px \wedge Qx)$ respectively.

M-system

M-system is defined by the following axiom-schemata and rules of inference:

M1 Classical truth-functional tautologies.
M2 $\forall x(A \supset B) \supset (axA \supset axB)$
M3 $A \supset \forall xA$, provided x is not free in A.
M4 $\forall xA \supset A(y \mid x)$
M5 $axA \supset ayA(y \mid x)$, provided y is not free in A.
M6 $x=x$
M7 $x=y \supset (A \supset A(y \| x))$
MP if $\vdash_M A \supset B$ and $\vdash_M A$ then $\vdash_M B$
MG if $\vdash_M A$ then $\vdash_M axA$ and $\vdash_M \neg ax\neg A$

As it is easy to notice, *M*-logic comprises the classical first-order logic - usually adopted axioms and rules of classical logic are included in the axiomatics of *M*-logic (in particular, the classical axiom: $\forall x(A \supset B) \supset (\forall xA \supset \forall xB)$ is an instance of M2).

M-semantics

Let us assume that the algebra of properties is extensional in the sense that for some non-empty set D, $P = \wp(D)$ ($\wp(D)$ is the power set of D). By a model of *M*-language, *M*-model, we mean a pair $m=[D,I]$ where D is a non-empty set called the domain of interpretation, I is a function defined on terms and predicates called the interpretation:

(a) $I(t) \subset \wp(D)$, where $\wp(D)$ is the power set of D, $I(t) \neq \varnothing$ and $I(t) \neq \{\varnothing\}$; in particular $I(\forall)=\{D\}$, $I(\exists)=\wp(D)-\{\varnothing\}$
(b) $I(P) \subset D^n$, for n-argument predicate symbol P

An assignment in D is a function V which assigns to every variable an element of D. Given V, by $V[d/x]$ we mean the function which is just like V, except possibly $V[d/x](x)=d$. In *M*-semantics truth conditions for atomic formulas, for negation and implication, are the same as in classical semantics. What is new in *M*-semantics is the truth condition (4) for predication.

(1) $\|Pv_1 \ldots v_n\|_V = 1$ iff $[V(y_1), \ldots, V(y_n)] \in I(P)$, $\|Pv_1 \ldots v_n\|_V = 0$ otherwise
(2) $\|x=y\|_V = 1$ iff $V(x)=V(y)$, $\|x=y\|_V = 0$ otherwise
(3) $\|\neg A\|_V = 1 - \|A\|_V$
(4) $\|A \supset B\|_V = \max[1-\|A\|_V, \|B\|_V]$
(5) $\|txA\|_V = 1$ iff $\forall_{X \in I(t)} X \sqsubset I_V(xA)$, where $I_V(xA)=\{d \in D: \|A\|_V = 1\}$

In particular, the formula $txPx$ is true in m iff there exists $X \in I(t)$ such that $X \sqsubset I(P)$. Notice also that the condition retains the meaning of the universal and existential quantifiers with

respect to a given interpretation. *A* formula *A* of *M*-language is *M*-valid iff it is true in every *M*-model with respect to any assignment (in short: $\models_M A$).
Completeness theorem for M-logic: For any *M*-formula *A*: $\vdash_M A$ iff $\models_M A$.[24]

NOTES

1. The term "Meinongian logic" is rarely used. It appears mainly in Jacquette's and Paśniczek's works; see [4], [7], and [8]. Other philosophers prefer to talk about formal Meinongian theories of objects.
2. See: Jacquette [4], Parsons [6], Paśniczek [8], Routley [13], Zalta [14], [15]. Also the theory developed by Castañeda [1] is in its spirit close to Meinong's theory. It should be noticed that Routley in his monumental [13] emphasises a paraconsistent character of Meinongian ontology. However, he does not propose any systematic reconstruction of Meinong's theory of objects.
3. Briefly speaking, in his ontology Meinong understands objects as possible objects of consciousness - intentional objects. Phenomenological description of consciousness reveals that every intentional object *qua* intentional is incomplete and some of them can be inconsistent.
4. Thus, four categories of Meinongian objects can be distinguished: complete and consistent, incomplete and consistent, complete and inconsistent, and incomplete and inconsistent.
5. Obviously, the supply of 'Meinongian objects' strongly hinges on the expressibility of an adopted language. In order to meet the content of *Annahmen Thesis* entirely, for every class of properties there must be a formula defining it. So, at least identity should be present in the language. Specifically, identity secures definability of every finite set of properties.
6. As it stands, this principle does not directly identify inconsistent and incomplete objects. That is, for any particular object *a* we cannot decide whether that object is consistent or complete, unless there is already in our ontology a distinction between inconsistent and complementary properties. If there are two objects *a* and *b* and *b* possesses all *a*'s properties and some more then all we can say is the following: if *a* is complete then *b* is inconsistent; if *a* is inconsistent then *b* is either; if *a* is complete then *b* is either. Thus, the concepts of contradictory and complementary properties are indispensable in order to make essential distinctions in the domain of Meinongian objects. In other words, completeness and consistency of objects are undetermined as long as we do not already know which properties are contradictory and which are complementary.
7. This general thesis can be proven by considering all nominal operators definable within classical logic. There were some - although not very successful - attempts to apply the definite description operator and Hilbert's operator to Meinongian theories of objects; see: Costa, Doria and Papavero [3], Jacquette [5].
8. Here, for simplicity, we prefer to talk about conjunction of formulas instead of sets of formulas. However, our considerations applies *mutatis mutandis* to a more general case.
9. There are neither 'ideological' nor technical reasons - which exist in intuitionistic logic - for avoiding completeness (expressed by the formula: $\alpha \vee \neg \alpha$) in paraconsistent logic.
10. When we formalize Meinong's ontology in an algebraic fashion, '$\neg a \neg$' corresponds to the Meinongian object which is the De Morgan's negation of *a*. In the theory of objects developed in Paśniczek [8] for every object *a* there exists a dual object '$\neg a \neg$' and this fact plays a highly important theoretical role in that system.
11. The only exception is Jacquette's logic which presupposes Łukasiewicz's three-valued logic.
12. However, first-order logics of a Meinongian kind may also exist. In those logics consistency and completeness are entailed by stronger extensional notions; see Paśniczek [8].
13. In order to be fair I should say that those philosophers quite often restrict their interest in formalizing Meinong's theory of objects to other ontological aspects of this theory.

14. See for instance: Priest and Routley [10].
15. As we mentioned earlier, Meinongian objects are primarily understood as intentional objects. According to phenomenological analyses, intentional objects are closed under some sort of entailment, although not necessarily under classical logic entailment.
16. It is easy to show that the formulas: $\forall(P\supset Q)\supset(aP\supset aQ)$ and $\forall P\supset aP$ are valid according to this interpretation. Also, all counterparts of the classical first-order laws expressible in this simple language hold: i.g.: $\neg\forall P\equiv\exists\neg P$. A more advanced extension of this logic to a logic comprising classical first-order logic has been developed in: Paśniczek [7] and [8].
17. We have in mind first of all the view that there is a double mode of predication or two classes of properties associated with Meinongian objects.
18. At least all properties represented by complex predicates built out of first-order formulas by means of lambda operator are to be assumed; see *Appendix*.
19. *N*-worlds should not be confused with the so-called "dead ends" and "non-normal worlds" used in the Kripkean semantics for S2 and S3.
20. It is possible to define on *N*-worlds the distributive lattice operation or even De Morgan's negation, see: Rescher and Brandom [12], Paśniczek [8], [9].
21. According to our earlier remarks, from the paraconsistent point of view incomplete *N*-worlds may seem redundant. We can relegate them all, except the universal *N*-world $\Box$, by adding to the *N*-system the following axiom:

$$\neg\Box(A\vee B)\supset(\alpha(A\vee B)\supset\alpha A\vee\alpha B)$$

See Paśniczek [9].
22. See: Priest and Routley [10], [11], da Costa and Marconi [2]. Priest and Routley claim that the non-adjunctive approach to inconsistency can hardly be recognized as paraconsistent logic and that "For all these sorts of reasons, the non-adjunctive modal approach to paraconsistency should be dismissed".
23. We can define the set of formulas in the simpler way replacing conditions (c) and (d) by the single condition: if A is a formula then ιxA is a formula.
24. For more informations on *M*-logic see Paśniczek [8].

REFERENCES

[1] Castañeda, Hector N. 'Thinking and the Structure of the World', Philosophia, 4, 1974.

[2] Costa, Newton C.A. da, and Marconi Diego, 'An Overview of Paraconsistent Logic in the 80s', *The Journal of Non-Classical Logic*, Vol. 6, No. 1, 1989.

[3] Costa, Newton C.A. da, Doria, F.A., and Papavero, N, 'Meinong's Theory of Objects and Hilbert's ε-Symbol', *Reports on Mathematical Logic* 25, 1991.

[4] Jacquette, Dale, *Meinongian Logic: The Semantics of Existence and Nonexistence*, Berlin, New York: Walter de Gruyter & Co, 1996.

[5] Jacquette, Dale, 'A Meinongian Theory of Definite Description', *Axiomates*, 5, 1994.

[6] Parsons, Terence, *Nonexistent Objects*, Yale University Press, New Haven & London, 1980.

[7] Paśniczek, Jacek, 'The Simplest Meinongian Logic', *Logique et Analyse*, 36e Annèe, 1993.

[8] Paśniczek, Jacek, *The Logic of Intentional Objects. A Meinongian Version of Classical Logic*. "Synthese Library vol. 269", Kluwer Dordrecht/Boston/London 1998.

[9] Paśniczek, Jacek, 'Beyond Consistent and Complete Possible Worlds', to appear in *Logique et Analyse*.

[10] Priest, Graham and Routley, Richard, 'Introduction: Paraconsistent Logics', *Studia Logica*, XLIII, 1/2 (special issue: G. Priest and R. Routley (eds.), *Paraconsistent Logics*), 1984.

[11] Priest, Graham and Routley, Richard, 'Systems of Paraconsistent Logics', in: G. Priest, R. Routley, Jean Norman (eds.), *Paraconsistent Logics. Essays on Inconsistency*, Philosophia Verlag, 1989.

[12] Rescher, Nicholas, and Brandom, Richard, *The Logic of Inconsistency. A Study in Non-Standard Possible- World Semantics and Ontology*, Basil Blackwell, Oxford, 1980.

[13] Routley, Richard, *Exploring Meinong's Jungle and Beyond*, Department Monograph #3, Philosophy Department, Research School of Social Sciences, Australian National University, Canberra 1980.

[14] Zalta, Edward, *Abstract Objects: An Introduction to Axiomatic Metaphysics*, D. Reidel, Dordrecht, 1983.

[15] Zalta, Edward, *Intensional Logic and the Metaphysics of Intentionality*, The MIT Press, Cambridge, Massachusetts, London, 1988.

CAN A PARACONSISTENT THEORIST BE A LOGICAL MONIST?*

OTÁVIO BUENO Department of Philosophy, California State University, Fresno, USA
otavio_bueno@csufresno.edu

To Newton da Costa, teacher and friend

Abstract

In recent years, the debate about logical pluralism has been fierce, with challenging arguments being provided by both sides. According to the logical pluralist, there are several different logics, that is, several answers to the question 'Is this argument valid?' (see, e.g., da Costa [1997], and Beall and Restall [2000] and [2001]). On the other hand, according to the logical monist, there is only one logic, that is, only one right answer to the question about the validity of arguments (see, e.g., Priest [2001]). In this paper, I examine the impact of this debate on our understanding of paraconsistent logic. After putting forward a defense of logical pluralism – along the lines articulated by da Costa [1997] – I argue that the best stance for the paraconsistent logician is provided by logical pluralism. After all, paraconsistent theorists cannot make sense of their own practice in a logical monist setting.

INTRODUCTION

There was a time, a long time ago, when the question 'What is the right logic?' was not an issue. Had it been raised, it would have been answered quite simply: the right logic is the only one that exists! Similarly to the situation in geometry before the emergence of non-Euclidean geometries, logic was identified with the logical system

* A number of the points put forward here emerged from discussions with Newton da Costa, who has always stressed the importance of adopting a pluralist view of logic, and who has articulated a comprehensive pluralist stance about the subject (see, in particular, da Costa [1997]). The paper is dedicated to him, for the incredible difference that meeting and working with him makes – a fact that all of his collaborators know so well. For illuminating discussions about the issues examined in this paper, my thanks go to him, as well as to JC Beall, Mark Colyvan, Steven French, Graham Priest and Scott Shalkowski. Thanks are also due to the helpful comments I received from three referees for this volume. Important changes were made as a result of their comments.

that existed at the time, and the choice between alternative logics didn't arise, simply because there was nothing there to choose from: there was only one logic.

The picture changed dramatically, of course, during the twentieth century, with the formulation of several logical systems: from extensions of classical logic (such as modal logic) to alternative and in some cases rival systems, such as paraconsistent logics, intuitionistic logics, and quantum logics (just to mention a few). Once there is a plurality of logics to consider, it becomes a substantial issue which of them (if any) is the right one.

In this paper, I will explore the implications of the plurality of logics to the debate over the nature of paraconsistency, examining in particular whether a paraconsistent logician can claim that there is only one right logic (that is, whether he or she can advocate *logical monism*). As I shall argue, there are several reasons why paraconsistency doesn't seem to be compatible with logical monism. I shall then indicate a version of logical pluralism that seems to be adequate to accommodate the nature of paraconsistent logic.

In talking about the *nature* of paraconsistent logic, I do not want to suggest that there is something peculiar (or peculiarly metaphysical) about this logic. I take the issue of the nature of a logic in the same way Duhem [1906/1954] took the issue of the nature of a physical theory. He characterized the *nature* of physical theories in terms of the role they played in scientific inquiry (the theories' *aim*), and the representational devices such theories provided to accommodate the phenomena (the theories' *structure*).[1] Similarly, in the case of logic (and of paraconsistent logic in particular), its *nature* is characterized by the role logic plays in drawing consequences (the logic's *aim*), and the representational devices a logic provides to achieve this aim (its *structure*). What I am calling here the *representational device* of a logic has to do with the mechanisms provided by a logic to achieve its aim. For example, in the case of paraconsistent logic, its *aim* has always been to accommodate inconsistencies in conceptual systems. The *relevant mechanisms* provided by this logic have to do with those semantic and proof-theoretic features of a paraconsistent logical system that allow one to avoid triviality even in the presence of "contradictions".

As I shall indicate below, paraconsistent logic is one logic among many. The idea is that, depending on our aims in a particular area of inquiry, different representational devices are required, and so different logics should be put forward. This is part and parcel of the pluralist view advocated here. Following da Costa [1997], this formulation of pluralism stresses that a logic ultimately depends on the domain of inquiry under consideration. Different domains – such as quantum mechanics and constructive mathematics – may require different logics, and that's

[1] See Chiappin [1989] for a systematic account of Duhem's view.

why there is no absolute answer to the question as to whether a given argument is valid or not.

1 LOGICAL PLURALISM AND LOGICAL MONISM

Two rival views about logic form the basis for the discussion that follows. According to *logical pluralism*, there are several logics that adequately characterize the notion of logical consequence. So for the logical pluralist, there is no absolute answer to the question "Does this sentence follow from those?" Any answer is always relative to a logic, and different logics provide adequate answers to such a question. On the other hand, according to *logical monism*, there is only one logic, and thus only one logic adequately instantiates the notion of consequence. In this way, for the logical monist the question about what follows from what has an absolute answer.

Given the way in which these two views have been formulated, it becomes clear that the notion of *adequacy* is crucial. But when do we say that a logic is *adequate*? What follows is not meant as a definition, but only as guiding idea to help us to avoid misunderstandings. Logic, as is now widely acknowledged, is basically the study of the relation of consequence. Given an argument, we typically have an informal idea as to whether it is valid or not. The adequacy of a logic depends on its capacity to justify as valid those inferences that are informally (or intuitively) taken to be so, and as invalid those that are not.

In this account, logic functions similarly to the way a scientific theory functions. Both logic and a scientific theory are put forward to accommodate certain types of phenomena: in the case of the former, the phenomena are the consequence relation in a given domain; in the case of the latter, observation reports obtained in a certain field of inquiry. Of course, new scientific theories often provide new descriptions and new ways of interpreting observation reports. For example, with the emergence of Newtonian mechanics, the scientific community no longer described a falling body in Aristotelian terms (that is, as a body moving toward its natural place); instead, the description of the body's movement was articulated in terms of gravity. Similarly, with the introduction of a new logic some inferences that have been initially taken as (intuitively) valid are no longer taken to be so. Thus, similarly to what happens in the case of a scientific theory, logic also plays a role in describing and reinterpreting the relevant phenomena.

But a scientific theory also helps us to shape and refine our intuitions about the physical world. Consider, for instance, the description of the movement of the sun relative to the earth in pre-Copernican physics, in contrast with how we describe this phenomenon now. Similarly, the adoption of new logics also helps us to shape and refine our judgments about logical consequence. For example, the emergence of free logic helped to identify and assess existential presuppositions found in standard first-order logic, similar to the way in which the emergence of first-order logic helped to identify and assess corresponding existential presuppositions in Aristotelian logic (for details, see Lambert (ed.) [1991], and references quoted therein).

Given these remarks, we can then say that a logic is *adequate* if the informal (or intuitive) consequence relation presupposed in a given domain has the same features as the consequence relation provided by that logic. If a logic is adequate, it formally represents the consequence relation adopted in a particular domain; that is, the logic generates the consequence relation among propositions that is informally (intuitively) found in the domain in question.

Of course, we are comparing here an intuitive notion of consequence with a formal notion. So there is no way of actually *proving* that a given logic is adequate (in the technical sense of 'proving'). The situation here is similar to the one found in the case of Church's thesis, which also establishes a relationship between intuitive and formal notions. And as is well known, although Church's thesis cannot be formally proved, an (informal) argument to support the thesis can be provided. Similarly, in the case of consequence, we also have to establish the relationship between an intuitive and a formal notion, and at best we can provide an (informal) argument to support the adequacy of a logic.

But the adequacy of a logic is not only a matter of the structural similarity between the informal consequence relation of a given domain and the formal counterpart provided by the logic. It is also a matter of the structural relationship between the objects and relations of the domain and the corresponding ordering of propositions yielded by the logic. In this sense, a logic should reflect the relations found among the objects of the domain in question. For example, quantum logic became such a fruitful tool to explore the quantum world because by studying the geometrical structure provided by this logic, we can determine relations between the propositions about quantum systems. In this sense, quantum logic is *heuristically* adequate for the quantum domain (see Putnam [1979]).

Similarly, one of the motivations for the introduction of paraconsistent logic was to avoid formulating *ad hoc* restrictions on the comprehension schema of set theory. In this way, the paraconsistent logician doesn't have to try to avoid the generation of the well-known set-theoretic paradoxes by somehow restricting that schema (and then, given the restriction, finding some way to add further set-theoretic axioms so that classical mathematics can be obtained). Instead of this, with the emergence of paraconsistent logic, a new way of investigating the foundations of set theory was devised. Paraconsistent logic allows one to study certain formulations of set theory in which the paradoxes are *not* avoided. But despite the inconsistency of the resulting theory, the latter is by no means trivial – or, at least, it is trivial if and only if classical (axiomatic) set theory is inconsistent (see da Costa [1986], da Costa

[1997], da Costa, Béziau and Bueno [1998], and Weir [1998]). In this way, paraconsistent logic is *heuristically* adequate for the investigation of inconsistent domains: it allows one to explore the inconsistent without triviality.

It might be argued that this is *not* a role that we should expect a *logic* to play. What has to be heuristically fruitful (or adequate) is a *theory* about the domain in question, not a logic. A logic is only concerned with the study of consequence relations in a language; heuristics has no part in this.

In reply, note that a theory typically presupposes the existence of a logic. In the so-called syntactic formulation, a theory is conceived as a set of sentences closed under *logical* consequence. Even if a theory is presented, following the semantic approach, as a family of models, a particular logic is assumed in the formulation of these models. (Models are not "free floating" entities, but their features depend on the theory in which they are expressed – typically set theory – which in turn presupposes a logic.) So to claim that heuristic fruitfulness is a property of a theory, and not of a logic, is to disregard a crucial feature of a theory: its dependence on logic. And depending on the logic one uses, different consequences are obtained, and so different theories are formulated. Thus, we can say that logic has a crucial role in the representation of the domain to which it is applied, in the sense that the structures that are generated by the logic in question reflect the relations among the objects in the domain (the examples of quantum logic and paraconsistent logic clearly illustrate this).

But if there are several consequence relations (that is, several answers to the question "What arguments are valid?"), the issue arises as to how we can choose between the resulting logics. Depending on the particular philosophical attitude one advocates toward logic, different answers are forthcoming. On the one hand, there are those that claim that a logic should be *true* to be acceptable. According to this *realist* proposal, truth provides the crucial criterion to decide which logic one should adopt. The main idea is that logic deals with truth-preserving methods, and that a logical system is true if it completely captures the consequence relation employed in a given domain.[2] Realists will use various (partial) criteria to assess whether a logic is true or not. Usually the outcome of such assessments is tentative, but it is enough for realists to provide an account of logic. On the other hand, there are those who do not advocate truth, but rather a weaker aim for logic. For example, extending to logic Field's [1980] anti-realist proposal for mathematics, an anti-realist can say that in order for a logic to be good it only has to be conservative over atomic statements (see Akiba [2000]). According to this proposal, no *new* consequence about atomic statements can be derived by adding a logic to a body of atomic statements. Given that in this account a logic doesn't have to be true to be good, this proposal provides a form of anti-realism about logic. And given that anti-realists do not take truth as an

[2] For two quite different versions of realism about logic, see Popper [1963] and Priest [1987].

aim for logic, the criteria for logic selection that they use are typically different from those employed by realists.[3]

As usually happens in debates about realisn., the realist is the one who advocates uniqueness in the selection of rival alternatives (whether they are rival scientific theories or rival logics). Moreover, it is typically part of realist proposals to suppose that there is only one complete description of the world (even if realists acknowledge that we may never be able to know when we have found such a description). As a result, realists require the development of criteria to select relevant alternatives (scientific theories or logics), so that the search for the true description of the world can be implemented. Anti-realists, on the other hand, usually don't require criteria that uniquely characterize their choices among alternatives. Tolerance with the existence of rival alternatives is typically greater among anti-realists. They use pluralism (of scientific theories or logics) as an argument against the realist's hope of obtaining one unique description of the world. The idea is that there are different theories (or different logics) that are equally adequate to describe a given domain, and that typically there isn't an *epistemic* reason to prefer one rather than the other. So anti-realists conclude that realists are not justified in their attempt to establish the existence of one unique description of the world.

It should be clear that when anti-realists put forward pluralism, they usually do not identify it with relativism. From the fact that we may not have epistemic criteria to decide among rival alternatives (scientific theories or logics), it doesn't follow that any scientific theory or logic is equally acceptable. It doesn't follow that "anything goes". A number of alternatives may be *in*adequate for a given domain. In the case of science, some scientific theories may be empirically inadequate; in the case of logic, some logics may not accommodate adequately the consequence relation used in a given domain. For example, classical logic is unable to deal with inconsistent domains, given that it identifies triviality and inconsistency. And the fact that not every alternative is acceptable for the pluralist is enough to avoid relativism.

So, depending on the philosophical attitude toward logic that one professes, different norms of evaluation of a logic are forthcoming. Realists and anti-realists provide different norms of evaluation, and so they typically have different attitudes toward logical monism and pluralism. I don't want to suggest that only realists are logical monists and that only anti-realists defend logical pluralism. This is certainly *not* the case in general. However, given that realists typically search for the unique correct description of the world, and that anti-realists usually resist this search, it

[3] In those cases in which realists and anti-realists agree about which criteria to adopt (for example, simplicity and unity), they do not assign the same significance to their satisfaction. Realists typically take such criteria to be *epistemic*; anti-realists take them as *pragmatic* at best (see van Fraassen [1980] and [1985]). I will return to this distinction below.

comes as no surprise that we often find logical monism defended by realists, whereas anti-realists so often put forward logical pluralism.

2 THREE FORMS OF LOGICAL PLURALISM

Logical pluralism comes in different *versions*. A highly original development and defense of logical pluralism was provided by da Costa, highlighting how a logic depends on the *domains* to which it is applied (see da Costa [1997]). On the other hand, Beall and Restall provided a systematic and thoughtful defense of logical pluralism in terms of the notion of *cases* (see Beall and Restall [2000] and [2001]).

The main difference between these two versions of logical pluralism is that da Costa highlights the role played by domains in the *selection* of a logic, whereas Beall and Restall emphasize the role played by cases in the *formulation* of a logic. The difference is not verbal, though. As conceived by Beall and Restall, a *case* indicates the particular metaphysical assumptions assumed by a given (formulation of) logic. In fact, according to the intuitive notion of consequence, α is a consequence of Γ if there is no *case* in which each premise in Γ is true, but in which α is not true. As Beall and Restall point out, *cases* range from (set-theoretic) models and possible worlds to situations (but are not restricted to these things). And depending on the nature of such *cases*, different answers to the question of which arguments are valid emerge. The crucial claim that makes Beall and Restall's position a version of pluralism is that, in their view, there are at least two different specifications of cases that satisfy the intuitive notion of consequence, and so there is more than one answer to the question about the validity of an argument.

Da Costa's domains, in turn, are employed primarily as a way of selecting the adequacy of a logic, indicating the domain to which it applies. In selecting which logic to apply we may consider particular features of the domain in question. For example, if our aim is to examine constructive features of mathematical reasoning, intuitionistic logic is far more adequate than classical logic; if our aim is to study the features of an inconsistent domain, paraconsistent logic fares systematically better than classical logic. Domains are then crucial in logic selection.

This is not the place, however, to compare these two versions of logical pluralism (I develop such a comparison in Bueno [2001]). And my argument in the present paper doesn't actually depend on the particular version one adopts. For simplicity, I will follow da Costa's version.

Independently of the version of pluralism one considers, there are three *forms* of logical pluralism.[4] (i) The most common type is *pluralism about pure logic* (see da

[4] Priest [2001] identifies and briefly examines the main features of these three forms of logical pluralism (his criticism focuses on the third version). As will emerge below, I don't think that Priest's criticism succeeds.

Costa [1997], and also Priest [2001]). We can call this type of pluralism "pure logical pluralism". Pluralism here amounts to the (undeniable) fact that there are several *pure* logics; that is, several purely formal structures that characterize logical consequence. The contrast between pure and applied logic can be clearly made in comparison with the similar contrast between pure and applied geometry. As da Costa [1997] points out, we can consider geometry from two different viewpoints: there is pure geometry (as the mathematical study of geometrical systems) and applied geometry (as the use of geometrical systems to describe physical space). Similarly, in the case of logic, there are pure logics (as the abstract study of consequence relations) and applied logics (as the use of such systems to describe different domains, from electronic circuits to grammatical structures). Pure logical pluralism is undeniable and uncontroversial. It is a mathematical fact that several different *pure* logics have been developed and are currently being studied.

(ii) The second level of logical pluralism focuses on *theoretical applications of logic*. (Priest [2001] calls this level of pluralism "theoretical pluralism".) According to this proposal, logics can be *applied* to several domains. So this form of pluralism focuses on the application of logic, rather than on its pure formulation. The domains of application range from grammatical structures through quantum systems to mathematical fields. According to the theoretical pluralist, typically there are several logics that can be *applied* to a given domain – although a given logic is usually more adequate to certain domains than to others. For example, both classical logic and paraconsistent logic can be applied to consistent domains, and typically they will generate the same results there. However, once we move to inconsistent domains, classical logic is no longer an adequate option (unless we want to allow the arbitrary rejection of some bits of information about such domains).[5] After all, by identifying inconsistency and triviality, classical logic precludes the possibility that, in inconsistent domains, we fail to validly infer at least one sentence of the language we use. On the other hand, paraconsistent logic allows inconsistency to be accommodated, since this logic doesn't *arbitrarily* require one to abandon some bits of information about the inconsistent domain, and so inconsistency need not be avoided at all costs. Of course, we still need to avoid triviality. But if we have a logic that distinguishes inconsistency from triviality, the avoidance of triviality provides no reason to avoid inconsistency.[6]

Finally, (iii) there is a version of logical pluralism that concentrates on the *canonical application of logic*, namely to reasoning in natural language (reasoning in

---

[5] Of course, depending on the amount of information we have about the inconsistent domain, we may have good reason to reject some bits of information about the domain. But when we are dealing with the construction of a new theory, for example, we typically have a very limited amount of information at our disposal, and so to exclude some bits of information is often precipitate and open to arbitrariness.

[6] And, of course, it is also *fruitful* to explore inconsistent domains. For example, the only way of studying the properties of the Russell set without triviality is by constructing appropriate paraconsistent set theories (see da Costa [1986], and da Costa, Béziau and Bueno [1998]).

the vernacular). For lack of a better term, this version of pluralism can be called "canonical pluralism". Canonical pluralism is a particular case of theoretical pluralism, when the application considered is the canonical one. According to the logical pluralist, even in the case of the canonical application, more than one logic are adequate to the domain under consideration. The plurality of logics goes "all the way down".

According to Priest [2001], only canonical pluralism provides an interesting case for logical pluralism. After all, pure logical pluralism and theoretical pluralism only claim that there are different logics – an undeniable, but ultimately uninteresting, mathematical fact. The only kind of pluralism that Priest takes to be challenging is *canonical* pluralism, and it is on this kind of pluralism that he concentrates his case. Priest takes it that theoretical pluralism and pure logical pluralism are not substantial enough.

But is this assessment correct?

3 Logical Pluralism Defended

I beg to differ. As opposed to Priest's suggestion, I don't think that pure logical pluralism is a trivial (i.e. uninteresting) proposal. On the contrary, the fact that pure logical pluralism is a recent phenomenon indicates that its emergence is anything but trivial. The idea that there are several distinct (pure) logics – the idea that one could study several different *bona fide* logical systems – is very recent; it was only in the twentieth century that the idea was seriously entertained. This indicates that it is a *substantial* issue whether one can explore different logical systems in the way suggested by the pure logical pluralist. That we can make such explorations is a fact; but not a trivial one.

As a *sociological* claim about logic, Priest's claim that "pure logical pluralism is trivial" is not contentious. It is a fact that currently there is a huge variety of pure logics around. But this is *not* the claim that the logical pluralist is concerned with. Logical pluralism is a *philosophical* claim about logic; it's not a sociological view. And from a philosophical point of view, it is far from trivial that a plurality of pure logics can legitimately be developed. Similarly to what has happened with geometry, it might be thought that it would be incoherent to develop alternative logical systems. It might be thought that such systems would never be consistent. Or it might be claimed that alternative pure logical systems are not even logic: such systems simply change the subject. (This is a reaction that we still find today, ever since Quine [1970] introduced the argument.) The fact that such systems *can* be developed, the fact that such systems can legitimately be considered *logics*, and the fact that there are substantial disagreements about this issue indicate that the plurality of pure logical systems is far from trivial. The question that the pure logical pluralist raises is clear: what should one make of this plurality? But the answer to the question is anything but trivial.

I also take it that there is a substantial issue as to whether *theoretical* pluralism is trivially true or not. Priest [2001] suggests that theoretical pluralism does not raise a serious problem for the logical monist. After all, he argues, the monist can always adopt the usual criteria of theoretical evaluation to choose between rival applied logics (such as simplicity, unity, no *ad hoc*ness, adequacy to the data etc.). But there are two difficulties with this suggestion: (1) some of these criteria (in particular, simplicity, unity and no *ad hoc*ness) only provide *pragmatic* reasons to accept a theory, and such reasons typically are *not* epistemic. A pragmatic reason indicates the *usefulness* of adopting a given theory (or applied logic). It is *easier for us* to work with theories and logics that are simple and unified. But why does this provide any reason for us to believe that such theories and logics are true? As opposed to pragmatic reasons, *epistemic* reasons are those that evaluate the relationship between the theory and the world, and so increase the chance that a given theory or logic is true. But simplicity, unity and no *ad hoc*ness on their own are not epistemic in this sense (see van Fraassen [1980] and [1985]). (2) Moreover, even if those criteria were epistemic, they typically do not uniquely select one theory in a given domain. After all, usually more than one theory satisfies the epistemic criteria. So we still end up having more than one theory underdetermined by the data. In other words, even with those criteria, theoretical pluralism, rather than monism, emerges.

The point of these remarks is to indicate that one typically cannot overcome a pluralist view about logic simply by suggesting that there are criteria of logic selection, anymore than one can overcome a pluralist view about science by mentioning that there are criteria of theory selection. In both cases, one has to show that the criteria *uniquely* determine the outcome. Otherwise, a pluralist view would still be justified. The fact that in both cases (in science and in applied logic) there is underdetermination of theories/logics by the data raises a serious problem for the (scientific and logical) monist. If there are more than one theory (or more than one logic) that is adequate for the domain in question, the monist view cannot be correct. (This is the point of (2) above.) And even if the criteria could uniquely select one theory (or one logic), given that the criteria are mostly pragmatic, they do not establish that the selected theory or logic is true – or, at least, that we have reason to believe that they are true. (This is the point of (1) above.)

In other words, given the underdetermination of theories and logics by the data, there will typically be more than one logic adequate to a given domain. Even if we were to use other criteria for theory/logic selection, typically there wouldn't be *epistemic* reasons for preferring a theory or a logic over another. And if the reasons in question are *pragmatic* at best, they fail to establish that the selected logic is *true*.

In other words, one should support logical and theoretical pluralism rather than their monist counterparts.

For these reasons I don't think that pure logical pluralism and theoretical pluralism are trivial. It is important to realize this point, since Priest often uses the strategy of turning the claims made by the logical pluralist into claims about pure logical pluralism or about theoretical pluralism, and then he just remarks that these claims are trivial. They are not. And as I tried to indicate above, one cannot claim that the logical pluralist's claims are trivial in this sense. What about *canonical* pluralism then? This is the only version of pluralism that Priest thinks is interesting, and I will discuss now the main argument against this form of pluralism that he provides (see Priest [2001]).

According to Priest, the main worry that the canonical pluralist faces is that his or her position is simply incoherent. Let us grant that logic depends on the domain under consideration as the pluralist claims. Why should we think that the different logics that are adequate to the domain in question are actually rival (with respect to the inferences under consideration)? Well, either these logics are rival or they are not. If they are not rival, there is no reason to support a pluralist view. After all, if there is no disagreement between the logics (in the sense that they actually generate the same set of consequences), the pluralist will have to acknowledge that there is no reason to distinguish them, no reason to choose one rather than the other. In this case, for all practical purposes, we have only one logic – in complete agreement with logical monism. On the other hand, if the logics under consideration are rival, then *only one of them* (if any) will be right about the representation of the inference in question. But in this case, once again, logical pluralism fails to be a viable option, since the conclusion supports logical monism! So in either case the logical pluralist doesn't provide an adequate response to the issue. To the extent that a response is discernible at all, it seems to conflict with the main features of logical pluralism.

According to Priest, this line of argument actually provides evidence for logical monism. And it certainly seems to raise a serious difficulty for the logical pluralist. But I think the difficulty is only apparent. Priest has introduced a false dilemma. For the logical pluralist, it is simply *not* the case that if two logics (applied to the same domain) are rival, then only one of them is right. Only a logical monist would grant that. The fact that there are logics that are inadequate for the domain in question *doesn't* entail that only one logic is adequate. More than one logic may be adequate – which is exactly what the logical pluralist claims! The logics may provide the *same* results with regard to inferences in the common domain (and that is why they are adequate with respect to that domain), although they may provide *different* results with regard to inferences beyond the domain (they are *different* logics, after all). The adequacy of a logic, as we saw above, depends on the aim and structure of the logic in question. Two different paraconsistent logics may provide the same results with respect to a given inconsistent domain, even though they may differ in other domains: one logic may be trivialized by a "contradiction" that doesn't trivialize the other (and both logics will agree with classical logic in a consistent domain).

Priest also criticizes da Costa [1997] for putting forward a pluralism that is domain dependent; a pluralism according to which depending on the domain we consider, different logics are adequate. For example, a constructivist logic is adequate for domains that involve constructive inferences; and a Schrödinger logic is adequate to the quantum domain.[7] According to Priest, this form of pluralism cannot make sense of the fact that sometimes we need to use a logic in *overlapping* domains. For example, we may need to study constructive inferences in the quantum domain.

I don't think this raises any difficulty for the domain-oriented pluralism of da Costa. After all, we can simply combine both logics (constructivist and Schrödinger logics) to accommodate the overlapping domain in question. And note that this is exactly what the logical pluralist should say. After all, when we have to consider the overlapping domain (in which *both* logics hold), strictly speaking, we have changed the domain. A different *aim* has to be achieved. Instead of only having to accommodate constructive inferences or having to accommodate inferences in quantum mechanics, the aim has changed to accommodate *constructive* inferences *in* quantum mechanics. And to achieve this aim a different *structure* is required: one that provides a constructive consequence relation defined for objects that may not have identity conditions. Once we realize this, it then becomes clear that more than one logic meets this requirement: several constructivist logics and several Schrödinger logics can be used to obtain a logic for the overlapping domain, with the result that more than one logic will be adequate to such a domain. For these reasons I think Priest failed to establish that logical monism should be the right answer to the question about what inferences are valid.

4 LOGICAL PLURALISM AND PARACONSISTENCY

Once we have seen that logical pluralism is a viable option, the question arises as to what attitude the *paraconsistent* logician should adopt about this issue. In other words, what is the status of paraconsistency in the logical pluralism debate?

It will come as no surprise now that I think the paraconsistent logician should be a logical pluralist. There are a number of reasons for this claim. First, there are different kinds of "contradictions", depending on the domain of knowledge we

[7] According to some interpretations of quantum theory, identity cannot be applied to quantum particles. To formally accommodate this situation, da Costa and Krause introduced Schrödinger logics (for details, see da Costa and Krause [1994] and [1997]).

consider, and there are different ways of dealing with them – different types of paraconsistent logics that are adequate to accommodate the inconsistencies in question. So in order to maintain a monist view, the paraconsistent logician would have to specify criteria of selection of alternative paraconsistent logics. But how can that be done?

Consider, for example, the first paraconsistent system that incorporated both the propositional and the predicate calculus (among other features): da Costa's C_n logics (for an overview, see da Costa [1974]; see also da Costa, Béziau and Bueno [1995]). This system provides a hierarchy of paraconsistent logics C_n ($1 \leq n \leq \omega$), that in a certain sense encompasses classical logic,[8] and is such that each logic in the hierarchy is strictly weaker than the previous one (and so a "contradiction" that trivializes C_i doesn't trivialize C_{i+1}). So there is not simply one paraconsistent logic but infinitely many. Which of them is the right one?

Of course, as a pure piece of formalism – as a *pure logic*, as da Costa [1997] would say – there is no issue of a logic being right or wrong (or being adequate or inadequate). At best there are constraints on the acceptability of the formalism, e.g. whether the formalism is sound, complete, decidable etc. But these constraints have nothing to do with the logic being right or wrong; rather they indicate whether the logic in question is *formally* acceptable. It goes without saying that there are substantial disagreements about the importance of these formal constraints. For example, in the debate between those that claim that second-order logic is not really logic and those that argue that it is, the first-order theorist takes the completeness of a logical system as a necessary condition for its acceptability. The second-order theorist disagrees of course (see Shapiro [1991]).

It is typically in the context of the *application* of a logic that the issue of its adequacy emerges.[9] As an *applied* logic, which of the infinitely many paraconsistent logics is the right one? The answer depends, of course, on the kind of inconsistency that we are dealing with. As noted above, there are some "contradictions" that trivialize even a paraconsistent system. For example, even in paraconsistent arithmetic, if it is established that 1 = 0, the resulting system is trivialized. Moreover, in da Costa's C_n logics, there are "contradictions" that trivialize certain logics C_i, although they may not trivialize the logic C_{i+1}. Despite these facts, there still are *several* paraconsistent logics adequate to accommodate inconsistent domains. And so although not every paraconsistent logic is equally adequate for a given inconsistent domain, more than one are.

As a result, it is difficult to see how paraconsistent logicians can make sense of their own practice if they were to adopt logical monism. Not only should the

[8] Classical logic is preserved in the sense that the principles of classical positive logic are valid in paraconsistent logic. Moreover, it is possible to introduce a negation in the paraconsistent system that behaves exactly like classical negation (see da Costa [1974]).

[9] Da Costa [1997] makes this point, and so does Priest [2001].

paraconsistent logician acknowledge the plurality of paraconsistent logics (both pure and applied), but also in *consistent* contexts, classical logic is contained in paraconsistent logic, and so in such contexts *both* logics are adequate. This is the case even of Priest's LP, which, as opposed to da Costa's C_n logics, is only one logic rather than a hierarchy of logics (for details, see Priest [1987]). In other words, if Priest claimed that the one true logic is LP, he still needs to acknowledge (as he does) that in consistent contexts LP and classical logic yield the same results. What this means is that such logics are equally adequate to deal with consistent domains. And so any commitment to one logic (in a consistent domain) will be extended to a commitment to the other logic (in such a domain). Paraconsistent logicians that advocate logical monism cannot deny this; if they do, they will end up being incoherent, given that they would be denying the very logic they advocate – after all, classical and paraconsistent logics *yield exactly the same results* in consistent domains. And *this* kind of incoherence is problematic – even for the paraconsistent logician!

The above argument is perfectly general, and it applies to whatever paraconsistent logic one considers: the C_n logics, LP, Jaskowski's logic (see D'Ottaviano and da Costa [1970], and da Costa, Bueno and French [1998]), LFI (see Carnielli, Marcos and de Amo [2001]), and so on. To the extent that in consistent domains paraconsistent logics agree with classical logic, a substantial form of pluralism will immediately emerge. And this is a pluralism that the paraconsistent logician cannot coherently deny. In other words, it is difficult to see how a paraconsistent logician can be a logical monist. Given that in a consistent context, classical logic and paraconsistent logic yield the same results, the paraconsistent logician needs to acknowledge the adequacy of classical logic in consistent domains.

Moreover, logical pluralism provides an adequate stance to make sense of non-classical logics in general. After all, there are close connections between classical logic and its non-classical counterparts. For example, in finite domains even the intuitionist logician recognizes that the excluded middle applies. And in domains involving macro-objects, the quantum logician recognizes that classical logic applies; not to mention that in order to provide a semantics for quantum logic, typically the quantum logician relies on classical logic. Once pure logical pluralism and theoretical pluralism are considered live and interesting options – as they should – the paraconsistent logician cannot coherently deny them.[10]

Let me elaborate on this. I've claimed that the quantum logician needs to be pluralist about logic, given that the semantics of quantum logic relies on classical logic. This is because the semantics for quantum logic is given in classical set

[10] It comes as no surprise that, in the process of defending logical monism, Priest *denies* that these two versions of pluralism (pure logical and theoretical pluralism) are relevant. But as I argued above, I don't think his assessment is right.

theory, and the latter presupposes classical logic. Of course, for the quantum logician to provide a semantics for quantum logic without relying on classical mathematics, he or she has to develop a quantum set theory. But even if we grant that there are excellent motivations for the development of quantum logic on the basis of quantum mechanical evidence – such as the failure of the distributive law in the outcome of certain spin measurements – there is no corresponding failure in set theory per se. Thus some independent evidence for the development of quantum set theory seems to be needed.[11]

The situation is quite different in the case of paraconsistent logic. *There are independent reasons to develop a paraconsistent set theory*, besides the need for providing a semantics for paraconsistent logic (that is not based on classical set theory). For example, with a paraconsistent set theory we can study genuinely inconsistent structures – such as the Russell set and the original version of the infinitesimal calculus – that cannot be properly studied otherwise. In other words, those structures cannot be studied *as inconsistent structures* without a paraconsistent set-theoretic framework (for details, see da Costa, Béziau and Bueno [1998]). The important point here is that, depending on the particular paraconsistent logic one adopts, *different* paraconsistent set theories emerge (see, again, da Costa, Béziau and Bueno [1998]). This provides a strong case for theoretical pluralism, since more than one paraconsistent logic can be used to articulate a paraconsistent set theory. Once again, this is a *significant* form of theoretical pluralism that a paraconsistent logician cannot coherently deny.

In this way, it becomes clear that logical pluralism provides an adequate framework for the paraconsistent logician to understand and articulate his or her own practice. It also provides a framework in which paraconsistency can flourish, given the integration it allows of the aim and structure of paraconsistency, and the search for a unified approach to study inconsistent domains.

5 CONCLUSION

The answer to the question posed in the title – "Can a paraconsistent theorist be a logical monist?" – should now be clear: *No*. Given the arguments above, it is difficult to see how a logical monist can *make sense* of certain systems of paraconsistent logic (and, of course, the whole point of advancing a philosophy of logic is to make sense of logic!). Da Costa's C_n logics, as we have seen, with its hierarchy of paraconsistent logics, immediately raises the issue of pluralism. In this

[11] I am *not* suggesting that quantum set theory should not be developed as a piece of pure mathematics. The question is whether we have motivation to develop such a set theory on the basis of *applied* mathematics – beyond the need for providing a semantics for quantum logic, of course.

system, there is not simply one paraconsistent logic but infinitely many. Which of them is the right logic? The logical monist has to provide an answer to this question, by spelling out criteria to decide what is the *right* paraconsistent logic. But, as we saw, there are substantial difficulties to achieve that. In the end, the best account of the plurality of logical systems is still provided by logical pluralism: there are several different logics each of them adequate to certain domains. Thus logical pluralism is still the best stance for the paraconsistent theorist – the best stance to make sense of paraconsistency.

REFERENCES

[1] K. Akiba. Logic and Truth. *Journal of Philosophical Research*, 25:101-123, 2000.

[2] JC Beall and G. Restall. Logical Pluralism. *Australasian Journal of Philosophy*, 78:475-493, 2000.

[3] JC Beall and G. Restall. Defending Logical Pluralism. In B. Brown and J. Woods, editors, *Logical Consequences*, forthcoming. Kluwer, 2001.

[3] O. Bueno. Logical Pluralism: A Pluralist View. Unpublished manuscript, Department of Philosophy, California State University, Fresno, 2001. Forthcoming.

[4] W.A. Carnielli, J. Marcos, and S. de Amo. Formal Inconsistency and Evolutionary Databases. *Logic and Logical Philosophy*, 2001, forthcoming.

[5] J.R.N. Chiappin. *Duhem's Theory of Science: An Interplay Between Philosophy and History of Science*. Ph.D. thesis, University of Pittsburgh, USA, 1989.

[6] N.C.A. da Costa. On the Theory of Inconsistent Formal Systems. *Notre Dame Journal of Formal Logic*, 15:497-510, 1974.

[7] N.C.A. da Costa. On paraconsistent set theory. *Logique et Analyse*, 115:361-371, 1986.

[8] N.C.A. da Costa. *Logiques classiques et non classiques*. Masson, 1997.

[9] N.C.A. da Costa, J.-Y. Béziau, and O. Bueno. Aspects of Paraconsistent Logic. *Bulletin of the Interest Group in Pure and Applied Logics*, 3:597-614, 1995.

[10] N.C.A. da Costa, J.-Y. Béziau, and O. Bueno. *Elementos de Teoria Paraconsistente de Conjuntos*. [*Elements of Paraconsistent Set Theory*.] Coleção CLE, 1998.

[11] N.C.A. da Costa, O. Bueno, and S. French. The Logic of Pragmatic Truth. *Journal of Philosophical Logic*, 27:603-620, 1998.

[12] N.C.A. da Costa and D. Krause. Schrödinger Logics. *Studia Logica*, 53:533-550, 1994.

[13] N.C.A. da Costa and D. Krause. An Intensional Schrödinger Logic. *Notre Dame Journal of Formal Logic*, 38:179-194, 1997.

[14] I.M.L. D'Ottaviano and N.C.A. da Costa. Sur un problème de Jaskowski. *Comptes Rendus de l'Académie des Sciences de Paris*, 270:1349-1353, 1970.

[15] P. Duhem. *The Aim and Structure of Physical Theory*. Princeton University Press, 1906/1954.

[16] H. Field. *Science without Numbers: A Defense of Nominalism*. Princeton University Press, 1980.

[17] K. Lambert, editor. *Philosophical Applications of Free Logic*. Oxford University Press, 1991.

[18] K.R. Popper. *Conjectures and Refutations*. Routledge and Kegan Paul, 1963.

[19] G. Priest. *In Contradiction*. Nijhoff, 1987.

[20] G. Priest. Logic: One or Many? In B. Brown and J. Woods, editors, *Logical Consequences*, forthcoming. Kluwer, 2001.

[21] H. Putnam. The Logic of Quantum Mechanics. In H. Putnam, *Mathematics, Matter and Method*, pages 174-197. Cambridge University Press, 1979.

[22] W.V.O. Quine. *Philosophy of Logic*. Prentice-Hall, 1970.

[23] S. Shapiro. *Foundations Without Foundationalism: A Case for Second-order Logic*. Clarendon Press, 1991.

[24] B.C. van Fraassen. *The Scientific Image*. Clarendon Press, 1980.

[25] B.C. van Fraassen. Empiricism in the Philosophy of Science. In P.M. Churchland and C.A. Hooker, editors, *Images of Science: Essays on Realism and Empiricism, with a Reply by Bas C. van Fraassen*, pages 245-308. The University of Chicago Press, 1985.

[26] A. Weir. Naive Set Theory is Innocent! *Mind*, 107:763-798, 1998.

Printed and bound by CPI Group (UK) Ltd, Croydon, CR0 4YY

29/10/2024

01780604-0001